Formula Weights

$AgBr$	187.78	$K_2Cr_2O_7$	294.19
$AgCl$	143.32	$K_3Fe(CN)_6$	329.26
Ag_2CrO_4	331.73	$K_4Fe(CN)_6$	368.36
AgI	234.77	$KHC_8H_4O_4$ (phthalate)	204.23
$AgNO_3$	169.87	$KH(IO_3)_2$	389.92
$AgSCN$	165.95	K_2HPO_4	174.18
Al_2O_3	101.96	KH_2PO_4	136.09
$Al_2(SO_4)_3$	342.15	$KHSO_4$	136.17
As_2O_3	197.84	KI	166.01
B_2O_3	69.62	KIO_3	214.00
$BaCO_3$	197.35	KIO_4	230.00
$BaCl_2 \cdot 2H_2O$	244.28	$KMnO_4$	158.04
$BaCrO_4$	253.33	KNO_3	101.11
$Ba(IO_3)_2$	487.15	KOH	56.11
$Ba(OH)_2$	171.35	$KSCN$	97.18
$BaSO_4$	233.40	K_2SO_4	174.27
$BaS_2O_3 \cdot H_2O$	267.48	$La(IO_3)_3$	663.62
Bi_2O_3	465.96	$Mg(C_9H_6ON)_2$	312.62
$(CH_2OH)_3CNH_2$ (THAM)	121.14	$MgCO_3$	84.32
CO_2	44.01	$MgNH_4PO_4$	137.32
$CaCO_3$	100.09	MgO	40.31
CaC_2O_4	128.10	$Mg_2P_2O_7$	222.57
CaF_2	78.08	$MgSO_4$	120.37
CaO	56.08	MnO_2	86.94
$CaSO_4$	136.14	Mn_2O_3	157.87
$Ce(HSO_4)_4$	528.40	Mn_3O_4	228.81
CeO_2	172.12	$Na_2B_4O_7 \cdot 10H_2O$	381.37
$Ce(SO_4)_2$	332.24	$NaBr$	102.90
$(NH_4)_2Ce(NO_3)_6$	548.23	$NaC_2H_3O_2$	82.03
$(NH_4)_4Ce(SO_4)_4 \cdot 2H_2O$	632.55	$Na_2C_2O_4$	134.00
Cr_2O_3	151.99	$NaCl$	58.44
CuO	79.54	$NaCN$	49.01
Cu_2O	143.08	Na_2CO_3	105.99
$CuSO_4$	159.60	$NaHCO_3$	84.01
$Fe(NH_4)_2(SO_4)_2 \cdot 6H_2O$	392.14	$Na_2H_2EDTA \cdot 2H_2O$	372.24
FeO	71.85	Na_2O_2	77.98
Fe_2O_3	159.69	$NaOH$	40.00
Fe_3O_4	231.54	$NaSCN$	81.07
$HC_2H_3O_2$ (acetic acid)	60.05	Na_2SO_4	142.04
$HC_7H_5O_2$ (benzoic acid)	122.12	$Na_2S_2O_3 \cdot 5H_2O$	248.18
$HClO_4$	100.46	NH_4Cl	53.49
$H_2C_2O_4 \cdot 2H_2O$	126.07	$(NH_4)_2C_2O_4 \cdot H_2O$	142.11
H_5IO_6	227.94	NH_4NO_3	80.04
HNO_3	63.01	$(NH_4)_2SO_4$	132.14
H_2O	18.015	$(NH_4)_2S_2O_8$	228.20
H_2O_2	34.01	$Ni(C_4H_7O_2N_2)_2$	288.93
H_3PO_4	98.00	$PbCrO_4$	323.18
H_2S	34.08	$Pb(NO_3)_2$	331.20
H_2SO_3	82.08	PbO	223.19
H_2SO_4	98.08	PbO_2	239.19
HgO	216.59	$PbSO_4$	303.25
Hg_2Cl_2	472.09	P_2O_5	141.94
$HgCl_2$	271.50	Sb_2S_3	339.69
KBr	119.01	SiO_2	60.08
$KBrO_3$	167.01	$SnCl_2$	189.60
KCl	74.56	SnO_2	150.69
$KClO_3$	122.55	SO_2	64.06
KCN	65.12	SO_3	80.06
K_2CrO_4	194.20	$Zn_2P_2O_7$	304.68

fundamentals of
analytical chemistry

fourth edition

douglas a. skoog   Stanford University

donald m. west   San Jose State University

SAUNDERS GOLDEN SUNBURST SERIES

SAUNDERS COLLEGE PUBLISHING

Philadelphia   New York   Chicago
San Francisco   Montreal   Toronto
London   Sydney   Tokyo   Mexico City
Rio de Janeiro   Madrid

Address orders to:
383 Madison Avenue
New York, NY 10017

Address editorial correspondence to:
West Washington Square
Philadelphia, PA 19105

This book was set in Caledonia by Progressive Typographers, Inc.
The editors were John Vondeling, Patrice L. Smith, Linda Kesselring.
The art & design director was Richrd L. Moore.
The text design was done by Arlene Putterman.
The cover design was done by Arlene Putterman.
The new artwork was drawn by ANCO/BOSTON.
The production manager was Tom O'Connor.
This book was printed by Fairfield Graphics.

FUNDAMENTALS OF ANALYTICAL CHEMISTRY   ISBN 0-03-058459-0
© 1982 by CBS College Publishing. Copyright 1963, 1969, 1976 by Holt Rinehart & Winston. All rights
reserved. Printed in the United States of America.
Library of Congress catalog card number 81-52763.

567   016   987

CBS COLLEGE PUBLISHING
Saunders College Publishing
Holt, Rinehart and Winston
The Dryden Press

preface

The aims of the fourth edition of *Fundamentals of Analytical Chemistry* are identical to those for the previous editions. A major goal is to provide the reader with a sound theoretical background in those chemical principles that are especially pertinent to chemical analysis. A second goal is to develop in the student an appreciation for the importance of the often difficult task of judging the accuracy and precision of experimental data and to provide the tools for sharpening these judgments. A third aim is to introduce the equipment as well as the experimental techniques of quantitative analysis and help the student develop skills in their use.

To these ends, the book is a mixture of theoretical and applied chemistry. Throughout, we have attempted to point out both the power and the limitations of theoretical calculations; that is, to show the reader that theory frequently serves as a useful guide to the solution of an analytical problem, but that verification in the laboratory is generally necessary in the end.

In preparing this revision, we have not only added new material and deleted other to keep the text current, but have also revised portions with the aim of making the presentation clearer and more understandable. To the latter end, the material on methods for solution of equilibrium problems in Chapters 4, 8, 9, 12, and 13 has been revised and expanded considerably. The review material on stoichiometric and equilibrium calculations, on the other hand, has been condensed into a single chapter (Chapter 2).

One change that the users of earlier editions will note is that we have chosen to abandon the unit of formal concentration, replacing it with the term *molar ana-*

lytical concentration. This change came about through the urging of about two-thirds of the users of the third edition.

We have expanded the material in Chapter 3 on statistical hypothesis testing and added a new section on regression analysis for the preparation of calibration curves. The sharp increase in the cost of silver nitrate has caused us to deemphasize its use as a volumetric reagent in Chapters 7 and 31, and to substitute material on mercurimetric titration of halides (Chapters 12 and 31). New OSHA regulations have also dictated revisions. Thus, we have substituted barium thiosulfate monohydrate for arsenious oxide as a primary standard for iodine, deleted the use of asbestos as a filtering medium, and substituted a coulometric procedure for ascorbic acid for one that involved the determination of arsenic(III).

The section on electrochemistry has been considerably revised. The theoretical material that was formerly dispersed throughout several chapters has been combined into a single introductory chapter (Chapter 13), thereby eliminating redundancies that existed in earlier editions. Furthermore, in order to gain space for new material on spectroscopy and chromatography, we have substantially condensed the chapter on polarography. The only additions to the electroanalytical chapters are brief sections dealing with Gran's plots and automatic titrators.

The reader will also note significant changes in the section dealing with spectroscopy. Most of the theoretical material concerned with emission, absorption, and fluorescence has been combined in a single location (Chapter 19); discussions of the applications and limitations to Beer's law are to be found there as well. Subsequent chapters treat instrumentation, molecular spectroscopy, and atomic spectroscopy, respectively. Sections devoted to ICP, fluorometry, and quantitative infrared methods have been added. The treatment of instrumental error has been expanded to include situations in which the magnitude of such error is not independent of transmittance.

Material on chromatography has been expanded and now is covered in three chapters. Chapter 27 consists of a general presentation of theory that is applicable to both liquid and gas chromatography. Chapter 28 is devoted to many of the rapidly expanding developments in high-performance liquid chromatography, and Chapter 29 is concerned with gas chromatography.

Chapters 30 and 31, dealing with laboratory techniques and methods, have seen much revision. The section on two-pan balances has been greatly shortened to make way for a brief discussion of electronic balances. The number of gravimetric procedures presented has been decreased from six to four. Directions for argentometric methods have been modified to permit the use of much smaller quantities of silver nitrate. The Volhard procedure has been deleted; in its place is a method for the determination of chloride by titration with mercury(II). Another new complexometric method involves the separation of nickel, cobalt, and zinc on an anion-exchange column, followed by titration with EDTA. Other new methods in Chapter 31 include a determination of ascorbic acid by bromate titration, a direct potentiometric determination of fluoride in water and in toothpaste, a coulometric titration of cyclohexene, a polarographic determination of copper and zinc in brass, a fluorometric determination of quinine in tonic water, a determination of lead in brass and leachable lead in pottery by atomic absorption spectroscopy, and finally, a separation and determination of C_6 ketones by gas chromatography.

Many colleagues have generously shared suggestions for improvements in this edition, among whom were some 161 users of the third edition who responded to a questionnaire regarding what should be added, deleted, and changed. It was impossible to take heed of all suggestions; not surprisingly, contradictions abound among them. We did, however, consider all and have made use of many individual suggestions. Moreover, we have responded where a consensus existed (as in the abandonment of formal concentrations). In any event, we offer our gratitude to all 161 who took the trouble to offer ideas.

As before, we are particularly grateful to Professor Alfred R. Armstrong of the College of William and Mary, who read the entire manuscript in detail and contributed many incisive criticisms. We also acknowledge with gratitude the comments of Professor M. D. Ryan of Marquette University, and of Professor John E. DeVries of California State University, Hayward, both of whom provided helpful suggestions on the entire manuscript. Finally, we offer special thanks to Professor J. D. Ingle of Oregon State University, for his careful reading and suggestions for the chapter on the evaluation of data, to Dr. Carol W. Mosher of Stanford Research Institute, for her comments on the chapter on liquid chromatography, and to Professor A. A. Gyberg of Augsburg College, for his suggestions as to a method of presentation of equilibrium calculations in Chapter 4.

D.A.S.
D.M.W.

contents overview

contents

31 Selected methods of analysis   731

Answers to problems   801

Appendixes   825

Index   841

introduction

1

Analytical chemistry deals with methods for the identification of one or more of the components in a sample of matter and the determination of the relative amounts of each. The identification process is called a *qualitative analysis* while the determination of amount is termed a *quantitative analysis*. This text is concerned largely with the latter.

A quantitative analysis provides numerical information concerning the quantity of some species (the *analyte*) in a measured amount of matter (the *sample*). The results of an analysis are expressed in such relative terms as parts of analyte per hundred (percent), per thousand, per million, or perhaps per billion of the sample. Also encountered are the weight (or volume) of analyte per unit volume of sample and the mole fraction.

Applications of chemical analyses are to be found everywhere in an industrialized society. For example, measurement of the parts per million of hydrocarbons, nitrogen oxides, and carbon monoxide in exhaust gases defines the effectiveness of automotive smog control devices. Determination of the concentration of ionized calcium in blood serum is important in the diagnosis of hyperparathyroidism in human patients. The nitrogen content of breakfast cereals and other foods can be directly related to their protein content and thus their nutritional qualities. Periodic quantitative analyses during the production of steel permit the manufacture of a product having a desired strength, hardness, ductility, and corrosion resistance. The continuous analysis for mercaptans in the household gas supply assures the presence of sufficient odorant to warn of dangerous leaks in the gas distribution system. The analysis of soils for nitrogen, phosphorus, potassium, and moisture throughout the growing season enables the farmer to tailor fertilization and irrigation schedules to meet plant needs efficiently and economically.

In addition to practical applications of the types just cited, quantitative analytical data are at the heart of research activity in chemistry, biochemistry, biology, geology, and the other sciences. Thus, for example, much of what is known of the mechanisms by which chemical reactions occur has been learned through kinetic studies employing quantitative measurements of the rates at which reactants are consumed or products formed. Recognition that the conduction of nerve signals in animals and the contraction or relaxation of muscles involve the transport of sodium and potassium ions across membranes was the result of quantitative measurements for these ions on each side of such membranes. Studies concerned with the mechanisms by which gases are transported in blood have required methods for continuously monitoring the concentration of oxygen, carbon dioxide, and other species within a living organism. An understanding of the behavior of semiconductor devices has required the development of methods for the quantitative determination of impurities in pure silicon and germanium in the range of 1×10^{-6} to $1 \times 10^{-10}\%$. The amounts of minor elemental constituents serve to identify the source of obsidian samples, an observation that has enabled archeologists to trace prehistoric trade routes for tools and weapons that were fashioned from these materials.

For many investigators in chemistry and biochemistry, as well as some of the biological sciences, the acquisition of quantitative information represents a significant fraction of their experimental efforts. Analytical procedures, then, are among the important tools employed by such scientists in pursuit of their research goals. Both an understanding of the basis of the quantitative analytical process and the competence and confidence to perform analyses are therefore prerequisites for research in these fields. The role of analytical chemistry in the education of chemists and biochemists is analogous to the role of calculus and matrix algebra for aspiring theoretical physicists and to the role of ancient languages in the education of scholars of the classics.

1A PERFORMANCE OF QUANTITATIVE ANALYSES

The results of a typical quantitative analysis are based upon two measurements (or sometimes two series of measurements), one of which is related to the quantity of sample taken and the second to the quantity of analyte in that sample. Examples of the quantities measured include mass, volume, light intensity, absorption of radiation, fluorescent intensity, and quantity of electricity. It is important to recognize, however, that these measurements are but a part of the typical quantitative analysis. Indeed, some of the preliminary steps are as important and often more difficult and time-consuming than the measurements themselves.

For the most part, the early chapters of this text are devoted to the final measurement steps; the other aspects of an analysis are not treated in detail until near the end of the book (Chapters 23 to 29). Thus, to lend perspective, it is useful at the outset to identify the several steps that make up the analytical process and to indicate their importance.

1A-1 SAMPLING

To produce meaningful information, an analysis must be performed on a sample whose composition faithfully reflects that of the bulk of material from which it was taken. Where the bulk is large and inhomogeneous, great effort is required to procure a representative sample. Consider, for example, a railroad car containing 25 tons of silver ore. Buyer and seller must come to agreement regarding the value of the shipment, based primarily upon its silver content. The ore itself is inherently heterogeneous, consisting of lumps of varying size and of varying silver content. The actual assay of this shipment will be performed upon a sample that weighs perhaps 1 g; its composition must be representative of the 25 tons (or approximately 22,700,000 g) of ore in the shipment. It is clear that the selection of a small sample for this analysis cannot be a simple one-step operation; in short, a systematic preliminary manipulation of the bulk of material will be required before it becomes possible to select 1 g and have any confidence that its composition is typical of the nearly 23,000,000 g from which it was taken.

Many sampling problems are not as formidable as the one just described. Nevertheless, the chemist needs the same assurance that the laboratory sample is representative of the whole before proceeding with the analysis.

1A-2 PREPARATION OF THE LABORATORY SAMPLE FOR ANALYSIS
Most solid materials must be ground to reduce particle size and then thoroughly mixed to ensure homogeneity. In addition, removal of adsorbed moisture is often required for solid samples. Adsorption or desorption of water causes the percentage composition of a substance to depend upon the humidity of its surroundings at the time of the analysis. To avoid the problems arising from such variations, it is common practice to base the analysis on a dry sample.

1A-3 MEASUREMENT OF THE SAMPLE
Quantitative analytical results are usually reported in relative terms; that is, in some way that expresses the quantity of the desired component present per unit weight or volume of sample. It is, therefore, necessary to know the weight or volume of the sample upon which the analysis is performed.

1A-4 SOLUTION OF THE SAMPLE
Most, but certainly not all, analyses are performed on solutions of the sample. Ideally, the solvent should dissolve the entire sample (not just the analyte) rapidly and under sufficiently mild conditions that loss of the analyte cannot occur. Unfortunately, such solvents do not exist for many of the materials that are of interest to the scientist—a silicate mineral, a high-molecular-weight polymer, or a specimen of animal tissue, to mention a few. Conversion of the analyte in such materials into a soluble form can be a formidable and time-consuming task.

1A-5 SEPARATION OF INTERFERING SUBSTANCES
Few, if any, chemical or physical properties of importance in analyses are unique to a single chemical species; instead, the reactions used and the properties measured are characteristic of a number of elements or compounds. This lack of truly specific reactions and properties adds greatly to the difficulties faced by the chemist when undertaking an analysis; it means that a scheme must be devised for isolating the species of interest from all others present in the original material that produce an effect upon the final measurement. Compounds or elements that prevent the direct measurement of the species being determined are called *interferences;* their separation prior to the final measurement constitutes an important step in most analyses. No hard and fast rules can be given for the elimination of interferences; this problem is often the most demanding aspect of the analysis.

1A-6 THE COMPLETION OF THE ANALYSIS

All preliminary steps in an analysis are undertaken to make the final measurement a true gauge of the quantity of the species being determined.

The chapters that follow contain descriptions of many types of final measurement, along with discussions of the chemical principles upon which such measurements are based.

1B CALIBRATION STANDARDS

All analytical methods are based upon the measurement of some physical quantity X that varies with the concentration C_A of the analyte. Ideally, the relationship between these two quantities is linear. That is,

$$C_A = kX$$

where k is a proportionality constant. With two exceptions, analytical methods require the empirical determination of k employing chemical standards for which C_A is known (the two exceptions are gravimetric and coulometric methods, which are discussed in Chapters 5 and 17, respectively). The process of determining k is an important step in most analyses and is termed *calibration*.

Smith and Parsons have listed desirable properties of chemical standards.[1] These include

1. *Stability*
   a. The compound should not be deliquescent, efflorescent, or hygroscopic.
   b. The compound should not undergo a chemical change upon drying.
   c. The compound should be easily weighable.
   d. Both the solution and the compound from which it is prepared should be stable in air and in solution.
2. *Purity*
   a. The compound should be available to a purity of 99.5% or better.
3. *Solubility*
   a. Since most analyses are performed on aqueous solutions, the compound should be readily soluble in water, common acids, or bases.
4. *Availability*
   a. The compound should be readily available from common chemical suppliers.
   b. The compound should not be unnecessarily costly.
5. *High Molecular Weight*
   a. The compound should have as high a molecular weight as possible to permit accurate weighing.
6. *Toxicity*
   a. The compound should have as low a toxicity as possible.

Few chemical compounds meet all of these criteria. The few that do are called *primary standards* and are discussed in detail in the several chapters on volumetric methods.

---

[1] B. W. Smith and M. L. Parsons, *J. Chem. Educ.*, **50**, 679 (1973).

A list of compounds suitable for preparation of solutions containing known concentrations of various common elements is provided in the end pages of this text.

1C CHOICE OF METHODS FOR AN ANALYSIS

The chemist or scientist who needs analytical data is frequently confronted with an awesome array of methods that could provide the desired information. Such considerations as speed, convenience, accuracy, availability of equipment, number of analyses, amount of sample that can be sacrificed, and concentration range of the analyte must all be considered; the success or failure of an analysis is often critically dependent upon the proper selection of method. Because no generally applicable rules exist, the choice of method is a matter of judgment. Such decisions are difficult; the ability to make them comes only with experience.

This text presents many of the common unit operations associated with chemical analyses and includes a variety of methods for the final measurement of analytes. Both theory and practical details are treated. Mastery of this material will permit the reader to perform many useful analyses and, in addition, will provide the background from which the judgment needed for the prudent choice among analytical methods can be developed.

a review of some

elementary concepts

2

Most quantitative analytical measurements are performed on solutions of the sample. The study of analytical chemistry, therefore, makes use of solution concepts with which students should already be familiar. The purpose of this chapter is to review the most important of these concepts.

2A THE CHEMICAL COMPOSITION OF SOLUTIONS

Both aqueous and organic solvents find widespread use in chemical analysis. Aqueous solvents, including solutions of the common inorganic acids and bases, are more widely used, however. Thus, our discussion will focus on the behavior of solutes in water; reactions in nonaqueous media will be considered in less detail.

2A-1 ELECTROLYTES

Electrolytes are solutes which ionize in a solvent to produce an electrically conducting medium. *Strong electrolytes* ionize completely whereas *weak electrolytes* are only partially ionized in the solvent. Table 2-1 summarizes the common strong and weak electrolytes in aqueous media.

2A-2 ACIDS AND BASES

For analytical chemists, the most useful acid-base concept is that proposed independently by Brønsted and Lowry in 1923. According to the Brønsted-Lowry view, *an acid is any substance that is capable of donating a proton; a base is any substance that can accept a proton.*

It is important to recognize that the proton-donating character of an acid manifests itself only in the presence of a proton acceptor or base; similarly, basic behavior requires the presence of an acid. Many solvents act as proton acceptors or donors and thus induce acidic or basic behavior of the solutes dissolved in them. For example, when nitrous acid is dissolved in water, the solvent acts as a proton acceptor and thus behaves as a base:

$$HNO_2 + H_2O \rightleftharpoons NO_2^- + H_3O^+ \qquad (2\text{-}1)$$
$$\text{acid}_1 \qquad \text{base}_2 \qquad \text{base}_1 \qquad \text{acid}_2$$

On the other hand, when ammonia is dissolved in water, the solvent provides a proton and is thus an acid:

$$NH_3 + H_2O \rightleftharpoons NH_4^+ + OH^- \qquad (2\text{-}2)$$
$$\text{base}_1 \qquad \text{acid}_2 \qquad \text{acid}_1 \qquad \text{base}_2$$

Water is the classic example of an *amphiprotic* solvent because it exhibits both acidic and basic properties, depending on the solute. Other useful amphiprotic solvents include methyl alcohol, ethyl alcohol, and glacial acetic acid. When nitrous acid or ammonia is dissolved in one of these, reactions similar to those shown by Equations 2-1 and 2-2 occur. With methanol, for example, we may write

$$HNO_2 + CH_3OH \rightleftharpoons NO_2^- + CH_3OH_2^+ \qquad (2\text{-}3)$$

and

$$NH_3 + CH_3OH \rightleftharpoons NH_4^+ + CH_3O^- \qquad (2\text{-}4)$$

Conjugate acids and bases. After an acid has donated a proton, the species that remains is capable of accepting a proton to re-form the original acid. In Equation 2-1, for example, it is seen that nitrite ion is the product of the acidic action of nitrous acid; nitrite ion, however, can behave as a base and accept a proton from a suitable donor. Thus, when sodium nitrite is dissolved in water, nitrous acid is formed by the reaction

Table 2-1   Classification of Electrolytes

Strong Electrolytes	Weak Electrolytes
1. The inorganic acids HNO_3, $HClO_4$, H_2SO_4,* HCl, HI, HBr, $HClO_3$, $HBrO_3$	1. Many inorganic acids such as H_2CO_3, H_3BO_3, H_3PO_4, H_2S, H_2SO_3
2. Alkali and alkaline-earth hydroxides	2. Most organic acids
3. Most salts	3. Ammonia and most organic bases
	4. Halides, cyanides, and thiocyanates of Hg, Zn, and Cd

* H_2SO_4 is completely dissociated into HSO_4^- and H_3O^+ ions and for this reason is classified as a strong electrolyte. However, it should be noted that the HSO_4^- ion is a weak electrolyte, being only partially dissociated.

$$NO_2^- + H_2O \rightleftharpoons HNO_2 + OH^-$$
$$\quad\text{base}_1 \quad\quad \text{acid}_2 \quad\quad\quad \text{acid}_1 \quad\quad \text{base}_2$$

Every Brønsted-Lowry acid, then, is paired with a corresponding base called its *conjugate base*, and every Brønsted-Lowry base is paired with a *conjugate acid*. In Equation 2-1, nitrite ion is seen to be the conjugate base of nitrous acid; the hydronium ion, H_3O^+, is the conjugate acid of the base water. Note also (Equation 2-2) that reaction between the base ammonia and the acid water results in the formation of the conjugate acid ammonion ion and the conjugate base hydroxide ion, respectively.

Autoprotolysis. Amphiprotic solvents undergo self-ionization or *autoprotolysis* to form a pair of ionic species. Autoprotolysis is an acid-base reaction, as illustrated by the following equations:

$$
\begin{array}{lllll}
\text{acid}_1 & + \text{base}_2 & \rightleftharpoons \text{base}_1 & + \text{acid}_2 \\
H_2O & + H_2O & \rightleftharpoons OH^- & + H_3O^+ \\
CH_3OH & + CH_3OH & \rightleftharpoons CH_3O^- & + CH_3OH_2^+ \\
HCOOH & + HCOOH & \rightleftharpoons HCOO^- & + HCOOH_2^+ \\
NH_3 & + NH_3 & \rightleftharpoons NH_2^- & + NH_4^+
\end{array}
$$

The positive ion formed by the autoprotolysis of water is called the *hydronium* ion, the proton being covalently bonded to the parent molecule by one of the unshared electron pairs of the oxygen. Higher hydrates such as $H_5O_2^+$ and $H_9O_4^+$ also exist, but they are significantly less stable than H_3O^+. Essentially no unhydrated hydrogen ions appear to exist in aqueous solutions.[1]

To emphasize the extraordinary stability of the singly hydrated proton, many chemists use the notation H_3O^+ when writing equations for reactions in which the proton is a participant. Others use H^+ to symbolize the proton, whatever its actual degree of hydration may be, as a matter of convenience. This notation has the advantage of simplifying the writing of equations that require the proton for balance. We shall use both notations, as convenient, in various sections of the text.

2A-3 STRENGTHS OF ACIDS OR BASES

Figure 2-1 shows the reactions of a few common acids with water. The first two entries are *strong acids* because their reaction with water is sufficiently complete to leave essentially no undissociated molecules, as such, in the solvent. The remaining acids are *weak acids*, which react incompletely to give solutions that contain significant quantities of both the parent acid and the conjugate base. Note that acids may be cationic, anionic, or electrically neutral.

The acids shown in Figure 2-1 become progressively weaker from the top to the bottom of the list. Thus, for the purpose of classification, perchloric and hydrochloric acid are completely dissociated; in contrast, only a few thousandths of a percent of the ammonium ions in an ammonium chloride solution are converted to ammonia molecules. It is also important to note that ammonium ion, the weakest acid, forms the strongest conjugate base of the group; that is, NH_3 has a much stronger affinity for protons than any base above it in Figure 2-1.

[1] See: P. A. Giguère, *J. Chem. Educ.*, **56**, 571 (1979).

The extent of reaction between a solute acid (or base) and a solvent is critically dependent upon the tendency of the latter to donate or accept protons. Thus, for example, perchloric, hydrochloric, and hydrobromic acids are all classified as strong acids in water. If anhydrous acetic acid, a poorer proton acceptor than water, is used *as the solvent*, none of the acids undergoes complete dissociation and remains a strong acid; instead, equilibria such as the following develop:

$$HClO_4 + HC_2H_3O_2 \rightleftharpoons ClO_4^- + H_2C_2H_3O_2^+$$

$$\text{acid}_1 \qquad \text{base}_2 \qquad \text{base}_1 \qquad \text{acid}_2$$

It is noteworthy that in this solvent, perchloric acid is a considerably stronger acid than either of the other two; the extent of its dissociation is 5000 times as great as that for hydrochloric acid, for example. Thus, glacial acetic acid serves as a *differentiating solvent* toward these acids in the sense that it reveals differences in the inherent acidities of the three compounds. On the other hand, water is termed a *leveling solvent* because the acidities of all acids that are strong are identical in this medium. Thus, no difference can be detected in the hydronium ion concentration of aqueous solutions containing hydrochloric and perchloric acids, provided their molar concentrations are the same.

2B UNITS OF WEIGHT AND CONCENTRATION

The mass of a substance is ordinarily determined in such metric units as the kilogram (kg), the gram (g), the milligram (mg), the microgram (μg), the nanogram (ng), or the picogram (pg).[2] For chemical calculations, however, it is necessary to employ units that express the weight relationship or *stoichiometry* among reacting species in terms of small whole numbers. The gram formula weight, the gram molecular weight, and the gram equivalent weight serve this purpose in analytical work. These terms are often shortened to formula weight, molecular weight, and equivalent weight.

2B-1 EMPIRICAL FORMULAS, CHEMICAL FORMULAS, AND THE MOLE

An *empirical formula* expresses the simplest whole-number combining ratio for the atoms in a substance. The *chemical formula*, on the other hand, specifies the number of atoms in a molecule. An empirical formula may be shared by more than one substance. To illustrate, CH_2O is the empirical formula for formaldehyde, CH_2O, glyceraldehyde, $C_3H_6O_3$, and glucose, $C_6H_{12}O_6$. The empirical formula can

Strongest acid

$$HClO_4 + H_2O \rightleftharpoons H_3O^+ + ClO_4^-$$
$$HCl + H_2O \rightleftharpoons H_3O^+ + Cl^-$$
$$H_3PO_4 + H_2O \rightleftharpoons H_3O^+ + H_2PO_4^-$$
$$Al(H_2O)_6^{3+} + H_2O \rightleftharpoons H_3O^+ + AlOH(H_2O)_5^{2+}$$
$$HC_2H_3O_2 + H_2O \rightleftharpoons H_3O^+ + C_2H_3O_2^-$$
$$H_2PO_4^- + H_2O \rightleftharpoons H_3O^+ + HPO_4^{2-}$$
$$NH_4^+ + H_2O \rightleftharpoons H_3O^+ + NH_3$$

Weakest base

Weakest acid

Strongest base

Figure 2-1. Relative strengths of some common weak acids and their conjugate bases.

[2] The relationship among these units is 10^{-3} kg = 1 g = 10^3 mg = 10^6 μg = 10^9 ng = 10^{12} pg.

be evaluated from the percentage composition of a substance; assignment of a chemical formula requires knowledge of the molecular weight.

The *molecular formula* may additionally provide structural information. Thus, C_2H_6O is both the empirical and the chemical formula for the chemically different ethanol, C_2H_5OH, and dimethyl ether, CH_3OCH_3.

The mole. The *mole* (mol) is a fundamental unit describing the amount of a chemical species.[3] It is always associated with a chemical formula and represents one Avogadro's number (6.02×10^{23}) of atoms, ions, molecules, or electrons. The weight of a mole of a substance is its *gram formula weight* or *formula weight* (fw), which is the summation of the atomic weights in grams for all of the atoms in the chemical formula for the species. For example, the gram formula weight of formaldehyde is given by

$$\text{fw } CH_2O = 12.0 \times 1 + 1.0 \times 2 + 16.0 \times 1 = 30.0 \text{ g/mol}$$

and for glucose, it is

$$\text{fw } C_6H_{12}O_6 = 12.0 \times 6 + 1.0 \times 12 + 16.0 \times 6 = 180.0 \text{ g/mol}$$

Thus, a mole of formaldehyde weighs 30.0 g while a mole of glucose is contained in 180.0 g. As further examples, a mole of sodium ion weighs 23.0 g and a mole of chloride ion weighs 35.5 g. These numbers are the formula weights of the two ions, respectively. The weight of a mole of NaCl is then the sum of the atomic weights of Na and Cl, or 58.5 g.

We will sometimes substitute the term *gram molecular weight* or molecular weight (mw) for formula weight to describe a mole of a real chemical compound. Thus, we may state that the molecular weight for sucrose is 180.0 g.

The millimole. Laboratory quantities are frequently more conveniently expressed in terms of *millimoles* (mmol) or *milliformula weights* (mfw), which are $\frac{1}{1000}$ of the mole or formula weight, respectively.

2B-2 METHODS FOR THE EXPRESSION OF CONCENTRATION OF SOLUTIONS
Chemists express the concentration of solutions in several ways. Definitions of some of the most important of these follow.

Molarity or molar concentration. The *molarity, M,* expresses the moles of a solute contained in one liter of solution; the molarity also expresses the *millimoles* in one milliliter. In using this concentration term, it is important to distinguish between the *analytical* concentration and the *equilibrium* or *species* concentration of a solute.

The molar analytical concentration of a solution gives the total moles of a solute present in one liter of the solution; that is, it specifies a recipe by which the so-

---

[3] In the International System (SI) of Units, proposed by the International Bureau of Weights and Measures, the only chemical unit for amount of substance is the *mole*. The mole is defined as the quantity of a material that contains as many elementary entities (these may be atoms, ions, electrons, ion-pairs, or molecules and must be explicitly defined) as there are atoms of carbon in exactly 0.012 kg of carbon-12 (that is, Avogadro's number). It seems probable that a shift to SI units will ultimately occur. It is equally important to have an understanding of the units upon which the present chemical literature is based, even though these may eventually disappear.

lution can be prepared. For example, the analytical concentration of H_2SO_4 in an aqueous 1.0 M solution is 1.0 mol H_2SO_4/L, which corresponds to 98 g/L (98 is the gram formula weight for H_2SO_4). On the other hand, the equilibrium or species concentration of H_2SO_4 is 0.0 M because the sulfuric acid is entirely dissociated into a mixture of H_3O^+, HSO_4^-, and SO_4^{2-} ions. The equilibrium or species concentrations of these three ions can be shown to be 1.01, 0.99, and 1.2×10^{-2} M, respectively.

Equilibrium molar concentrations are often symbolized by placing square brackets around the chemical formula for the species. Thus, for a 1.0 M solution of H_2SO_4 we may write.

$$[H_2SO_4] = 0.00 \qquad [H_3O^+] = 1.01$$
$$[HSO_4^-] = 0.99 \qquad [SO_4^{2-}] = 0.012$$

EXAMPLE 2-1. Calculate the analytical and equilibrium molar concentrations of the solute species in (a) an aqueous solution that contains 2.30 g of ethanol, C_2H_5OH (fw = 46.1), in 3.50 L and (b) an aqueous solution that contains 285 mg of trichloroacetic acid, Cl_3CCOOH (fw = 163), in 10.0 mL (the acid is 73% ionized in water).

a. The analytical concentration of ethanol is given by

$$M_{C_2H_5OH} = \frac{2.30 \text{ g } C_2H_5OH}{3.50 \text{ L}} \times \frac{\text{mol } C_2H_5OH}{46.1 \text{ g}} = 0.0143 \text{ mol/L}$$

The only solute species present in any significant quantity in aqueous solution is C_2H_5OH. Therefore, its species concentration and its analytical concentration are numerically identical; that is, its equilibrium molar concentration is also given by

$$M = [C_2H_5OH] = 0.0143$$

b. Employing HA as the symbol for Cl_3CCOOH, we write that the analytical or total concentration is

$$M_{HA} = \frac{285 \text{ mg HA}}{10.0 \text{ mL}} \times \frac{\text{mmol HA}}{163 \text{ mg}} = 0.175 \text{ mmol/mL}$$

Because all but 27% of the Cl_3CCOOH is dissociated as H_3O^+ and Cl_3CCOO^-, the species concentration of Cl_3CCOOH is given by

$$[HA] = \frac{(0.27 \times 285) \text{ mg HA}}{10.0 \text{ mL}} \times \frac{\text{mmol HA}}{163 \text{ mg}} = 0.047 \text{ mmol/mL}$$

The molarity of H_3O^+ as well as Cl_3CCOO^- will be equal to the analytical concentration of the acid minus the species concentration of undissociated acid; that is,

$$M_{H_3O+} = [H_3O^+] = [A^-] = 0.175 - 0.047 = 0.128 \text{ mmol/mL}$$

It should be noted that some chemists prefer to distinguish between species and analytical concentrations by restricting the use of molar concentration for the former and using formal concentration (F) for the latter.

EXAMPLE 2-2. Describe the preparation of 2.00 L of 0.100 M Na_2CO_3 from the solid.

$$\text{mol } Na_2CO_3 \text{ needed} = 0.100 \frac{\text{mol } Na_2CO_3}{L} \times 2.00 \text{ L} = 0.200$$

$$\text{wt } Na_2CO_3 = 0.200 \text{ mol} \times 106 \frac{\text{g } Na_2CO_3}{\text{mol}} = 21.2 \text{ g}$$

Therefore, dissolve 21.2 g of the Na_2CO_3 in water, and dilute to exactly 2.00 L.

EXAMPLE 2-3. Describe the preparation of 2.00 L of 0.100 M Na^+ from pure Na_2CO_3.

$$\text{mol } Na_2CO_3 \text{ needed} = 0.100 \frac{\text{mol } Na^+}{L} \times 2.00 \text{ L} \times \frac{1 \text{ mol } Na_2CO_3}{2 \text{ mol } Na^+} = 0.100$$

$$\text{wt } Na_2CO_3 = 0.100 \text{ mol } Na_2CO_3 \times 106 \frac{\text{g } Na_2CO_3}{\text{mol}} = 10.6 \text{ g}$$

Dissolve 10.6 g Na_2CO_3 in water, and dilute to 2.00 L.

Normality or normal concentration. This specialized method for expressing concentration is based upon the number of equivalents of solute that are contained in a liter of solution. Normality and equivalent weight are defined in Chapter 6.

Titer. Titer defines concentration in terms of the weight of some species with which a unit volume of the solution reacts. The applications of titer are considered in Chapter 6.

p-Functions. Frequently scientists express the concentration of a species in a dilute solution in terms of its *p-function* or *p-value: the p-function is defined as the negative logarithm (to the base 10) of the molar concentration of that species.* Thus, for the species X

$$pX = -\log [X]$$

As shown by the following examples, p-values offer the advantage of providing concentration information in terms of small, usually positive numbers that can be written in integer form.

EXAMPLE 2-4. Calculate the p-value for each ion in a solution that is 2.00×10^{-3} M in NaCl and 5.4×10^{-4} M in HCl.

$$pH = -\log [H_3O^+] = -\log (5.4 \times 10^{-4})$$
$$pH = -\log 5.4 - \log 10^{-4}$$
$$= -0.73 - (-4) = 3.27$$

To obtain pNa, we write

$$pNa = -\log (2.00 \times 10^{-3}) = -\log 2.00 - \log 10^{-3}$$
$$= -0.301 - (-3.00) = 2.699$$

The chloride ion concentration is given by the sum of the two concentrations; that is,

$$[Cl^-] = 2.00 \times 10^{-3} + 5.4 \times 10^{-4} = 2.00 \times 10^{-3} + 0.54 \times 10^{-3}$$
$$= 2.54 \times 10^{-3}$$
$$pCl = -\log 2.54 \times 10^{-3} = 2.595$$

EXAMPLE 2-5. Calculate the molar concentration of Ag^+ in a solution having a pAg of 6.372.

$$pAg = -\log [Ag^+] = 6.372$$
$$\log [Ag^+] = -6.372 = -7.000 + 0.628$$
$$[Ag^+] = \text{antilog} (-7.000) \times \text{antilog} (0.628)$$
$$= 10^{-7} \times 4.246 = 4.25 \times 10^{-7}$$

It is noteworthy that the p-value for a species becomes negative when its concentration is greater than unity. For example, in a 2.0 M solution of HCl

$$[H_3O^+] = 2.0$$
$$pH = -\log 2.0 = -0.30$$

Density and specific gravity. The *density* of a substance measures its mass per unit volume, whereas the *specific gravity* is the ratio of its mass to that of an equal volume of water at 4°C. In the metric system, density has units of kilograms per liter or grams per milliliter. Specific gravity, on the other hand, is dimensionless and thus not tied to any particular system of units; for this reason it is widely used in describing items of commerce. Because water at 4°C has a density of exactly 1.00 g/mL, and because we shall be employing the metric system throughout the text, density and specific gravity will be used interchangeably.

Parts per million; parts per billion. For very dilute solutions, it is convenient to express concentrations in terms of parts per million:

$$\text{ppm} = \frac{\text{weight of solute}}{\text{weight of solution}} \times 10^6$$

For even more dilute solutions, 1×10^9 rather than 1×10^6 is employed in the foregoing equation; the results are then given as parts per billion (ppb). The term parts per thousand (ppt) is also encountered.

If the solvent is water and the quantity of solute is so small that the density of the solution is still essentially 1.00 g/mL,

$$\text{ppm} = \frac{\text{mg solute}}{10^6 \text{ mg water}} = \frac{\text{mg solute}}{\text{L solution}}$$

EXAMPLE 2-6. What is the molarity of K^+ in a solution that contains 63.3 ppm of $K_3Fe(CN)_6$?

Because the solution is so dilute, it is safe to assume that the density of the solution is 1.00 g/mL. Therefore,

$$63.3 \text{ ppm } K_3Fe(CN)_6 = 63.3 \text{ mg } K_3Fe(CN)_6/L$$

$$[K^+] = \frac{63.3 \text{ mg } K_3Fe(CN)_6}{L} \times \frac{\text{mmol } K_3Fe(CN)_6}{329 \text{ mg}} \times \frac{3 \text{ mmol } K^+}{\text{mmol } K_3Fe(CN)_6} \times \frac{\text{mol}}{10^3 \text{ mmol}}$$

$$= 5.77 \times 10^{-4} \ M$$

Percentage concentration. Chemists frequently express concentrations in terms of percentage (parts per hundred). Unfortunately, this practice can be a source of ambiguity because of the many ways by which the percentage composition of a solution can be expressed. Common methods include:

$$\text{weight percent (w/w)} = \frac{\text{wt of solute}}{\text{wt of soln}} \times 100$$

$$\text{volume percent (v/v)} = \frac{\text{volume of solute}}{\text{volume of soln}} \times 100$$

$$\text{weight-volume percent (w/v)} = \frac{\text{wt of solute, g}}{\text{volume of soln, mL}} \times 100$$

It should be noted that the denominator in each of these expressions refers to the *solution* rather than to the solvent. Moreover, the first two expressions do not depend on the units employed (provided, of course, that there is consistency between numerator and denominator), whereas units must be defined for the third. Of the three expressions, only weight percentage has the virtue of being temperature-independent.

Weight percentage is frequently used to express the concentration of commercial aqueous reagents; thus, nitric acid is sold as a 70% solution, which means that the reagent contains 70 g of HNO_3 per 100 g of solution.

Weight-volume percentage is often employed to indicate the composition of dilute aqueous solutions of solid reagents; thus, 5% aqueous silver nitrate *usually* refers to a solution that is prepared by dissolving 5 g of silver nitrate in sufficient water to give 100 mL of solution.

To avoid uncertainty, it is necessary to specify explicitly the type of percentage composition that has been used. If this information is lacking, the user is forced to decide intuitively which of the several types is involved. The potential error resulting from a wrong choice is considerable. For example, commercial 50% (w/w) sodium hydroxide contains 763 g of the reagent per liter; thus it is 76.3% (w/v) sodium hydroxide.

EXAMPLE 2-7. Describe the preparation of 100 mL of 6.0 *M* HCl from the concentrated reagent. The label on the bottle states that the reagent is 37% HCl and has a specific gravity of 1.18.

Generally, the percentages employed in describing commercial reagents are weight-weight. Therefore,

$$\frac{\text{g HCl}}{\text{mL concd soln}} = \frac{1.18 \text{ g concd soln}}{\text{mL concd soln}} \times \frac{37 \text{ g HCl}}{100 \text{ g concd soln}} = 0.437$$

$$\text{g HCl required} = 100 \text{ mL} \times \frac{6.00 \text{ mmol HCl}}{\text{mL}} \times \frac{0.0365 \text{ g}}{\text{mmol HCl}} = 21.9$$

$$\text{vol} = 21.9 \text{ g HCl} \times \frac{\text{mL concd HCl}}{0.437 \text{ g}} = 50.1 \text{ mL concd HCl}$$

Dilute 50 mL of the concentrated reagent to a volume of about 100 mL.

Solution-diluent volume ratios. The composition of a dilute solution is sometimes specified in terms of the volume of a more concentrated reagent and the volume of solvent to be used in diluting it. The volume of the former is separated from that of the latter by a colon. Thus, a 1:4 HCl solution contains four volumes of water for each volume of concentrated hydrochloric acid taken. This method of notation is frequently ambiguous in that the concentration of the original reagent solution is not always obvious to the reader; the use of molar concentrations is preferable.

2C STOICHIOMETRIC RELATIONSHIPS

A balanced chemical equation is a statement of the combining ratios (in moles) that exist between reacting substances and their products. Thus, the equation

$$2NaI(aq) + Pb(NO_3)_2(aq) = PbI_2(s) + 2NaNO_3(aq)$$

indicates that 2 mol of sodium iodide in aqueous solution combine with 1 mol of lead nitrate to produce 1 mol of solid lead iodide and 2 mol of aqueous sodium nitrate.[4,5] Molar relationships such as the foregoing are called the *stoichiometry* of the reaction.

Experimental measurements are never obtained directly in terms of chemical mass (fw); instead they have units such as grams, milligrams, liters, and milliliters. Conversion of such data to a weight or volume of some other species or solution is a three-step process involving (1) transformation of the raw data in metric units into chemical mass units (mol or mmol), (2) multiplication by a factor to account for the stoichiometry, and (3) reconversion of the chemical mass data back into metric units. That is, the process involves:

$$\begin{matrix} \text{quantity} \\ \text{measured} \\ \text{(metric} \\ \text{units)} \end{matrix} \times \begin{matrix} \text{metric to chemi-} \\ \text{cal mass conver-} \\ \text{sion factor} \end{matrix} \times \begin{matrix} \text{stoichiometric} \\ \text{relationship} \end{matrix} \times \begin{matrix} \text{chemical mass} \\ \text{to metric con-} \\ \text{version factor} \end{matrix} = \begin{matrix} \text{desired} \\ \text{quantity} \\ \text{in metric} \\ \text{units} \end{matrix}$$

Table 2-2 lists methods for carrying out metric to chemical mass conversions; Table 2-3 provides the means for obtaining the analogous conversion in terms of concentration. These conversions—alone or in combination—will provide the solution to any stoichiometric problem; the ability to manipulate them must be cultivated. In setting up equations, it is helpful to supply units for all quantities that possess dimensions; the best proof that a correct relationship has been generated is agreement between the units that appear on the two sides of the equal sign.

EXAMPLE 2-8. Calculate the weight of $AgNO_3$ (fw = 170) required to convert 2.33 g of Na_2CO_3 (fw = 106) to Ag_2CO_3 (fw = 276).

$$2AgNO_3(aq) + Na_2CO_3(aq) \longrightarrow Ag_2CO_3(s) + 2NaNO_3(aq)$$
$$\text{mol } AgNO_3 \text{ required} = 2 \times \text{mol } Na_2CO_3$$

---

[4] Here, it is advantageous to depict the reaction in terms of chemical compounds. If we wish to focus on reacting species, the net ionic representation is preferable:

$$2I^-(aq) + Pb^{2+}(aq) = PbI_2(s)$$

[5] Chemists frequently include information about the physical state of substances in equations; thus, (g), (1), (s), and (aq) refer to gaseous, liquid, solid, and aqueous solution states, respectively.

Table 2-2 Expression of Mass in Chemical Units

Chemical Unit for Mass	Weight of Unit in Grams Given by	Method of Conversion From Metric Units to Chemical Units
Mole (mol)	fw	$mol = \dfrac{g}{fw}$
Millimole (mmol)	mfw	$mmol = \dfrac{g}{mfw}$
Equivalent (eq)	eqw	$eq = \dfrac{g}{eqw}$
Milliequivalent (meq)	meqw	$meq = \dfrac{g}{meqw}$

Step 1. Conversion of metric data to chemical mass units.

$$2.33 \text{ g Na}_2\text{CO}_3 \times \frac{1 \text{ mol Na}_2\text{CO}_3}{106 \text{ g Na}_2\text{CO}_3} = \frac{2.33}{106} \text{ mol Na}_2\text{CO}_3$$

Step 2. Multiply by stoichiometric relationship.

$$\frac{2.33}{106} \text{ mol Na}_2\text{CO}_3 \times \frac{2 \text{ mol AgNO}_3}{1 \text{ mol Na}_2\text{CO}_3} = \frac{2.33 \times 2}{106} \text{ mol AgNO}_3$$

Step 3. Reconversion of chemical mass data to metric units.

$$\frac{2.33 \times 2}{106} \text{ mol AgNO}_3 \times \frac{170 \text{ g AgNO}_3}{\text{mol AgNO}_3} = 7.47 \text{ g AgNO}_3$$

To summarize:

$$\text{wt AgNO}_3 = \overset{\text{Conversion to Chemical Mass}}{2.33 \text{ g} \times \frac{\text{mol Na}_2\text{CO}_3}{106 \text{ g}}} \times \overset{\text{Stoichiometry}}{\frac{2 \text{ mol AgNO}_3}{1 \text{ mol Na}_2\text{CO}_3}} \times \overset{\text{Conversion to Metric Units}}{\frac{170 \text{ g}}{\text{mol AgNO}_3}}$$

$$= 7.47 \text{ g AgNO}_3$$

EXAMPLE 2-9. How many milliliters of 0.0669 M AgNO$_3$ will be needed to convert 0.348 g of pure Na$_2$CO$_3$ to Ag$_2$CO$_3$?

Because the volume V in milliliters is desired, it will be more convenient to base our calculations on millimoles than on moles. Thus,

$$V = \overset{\text{Conversion to Chemical Mass}}{0.348 \text{ g} \times \frac{\text{mmol Na}_2\text{CO}_3}{0.106 \text{ g}}} \times \overset{\text{Stoichiometry}}{\frac{2 \text{ mmol AgNO}_3}{\text{mmol Na}_2\text{CO}_3}} \times \overset{\text{Conversion to Metric Units}}{\frac{1 \text{ mL AgNO}_3}{0.0669 \text{ mmol AgNO}_3}}$$

$$= 98.2 \text{ mL of } 0.0669 \text{ } M \text{ AgNO}_3$$

EXAMPLE 2-10. What weight of Ag$_2$CO$_3$ is formed after mixing 25.0 mL of 0.200 M AgNO$_3$ with 50.0 mL of 0.0800 M Na$_2$CO$_3$?

We must first determine which of the reactants exists in the stoichiometrically lesser amount because this species will limit the quantity of silver carbonate that can

Table 2-3 Expression of Concentration in Chemical Units

Chemical Term for Concentration	Definition of Term	Method of Calculation
Molarity, M	$\dfrac{\text{mol}}{\text{L}}$	$M = \dfrac{\text{g}}{\text{fw}} \times \dfrac{1}{\text{L}}$
	$\dfrac{\text{mmol}}{\text{mL}}$	$= \dfrac{\text{g}}{\text{mfw}} \times \dfrac{1}{\text{mL}}$
Normality, N	$\dfrac{\text{eq}}{\text{L}}$	$N = \dfrac{\text{g}}{\text{eqw}} \times \dfrac{1}{\text{L}}$
	$\dfrac{\text{meq}}{\text{mL}}$	$= \dfrac{\text{g}}{\text{meqw}} \times \dfrac{1}{\text{mL}}$

be produced. Again, step 1 involves a metric to chemical unit conversion:

$$\text{mmol AgNO}_3 = 25.0 \text{ mL} \times 0.200 \text{ mmol/mL} = 5.00$$
$$\text{mmol Na}_2\text{CO}_3 = 50.0 \text{ mL} \times 0.0800 \text{ mmol/mL} = 4.00$$
$$\text{mmol AgNO}_3 \text{ required} = 2 \times \text{no. mmol Na}_2\text{CO}_3$$

Thus, the reaction is limited by the number of millimoles of $AgNO_3$. From the stoichiometric relationship, we note that 2 mmol of $AgNO_3$ are required to produce 1 mmol Ag_2CO_3; therefore,

$$\text{mmol Ag}_2\text{CO}_3 = \frac{5.00}{2} = 2.50$$

$$\text{g Ag}_2\text{CO}_3 = 2.50 \text{ mmol} \times \frac{0.276 \text{ g Ag}_2\text{CO}_3}{\text{mmol}} = 0.690$$

2D A REVIEW OF SIMPLE EQUILIBRIUM CONSTANT CALCULATIONS

Most reactions that are useful for chemical analysis proceed rapidly to a state of *chemical equilibrium* in which reactants and products exist in constant and predictable ratios. A knowledge of these ratios, under various experimental conditions, often permits the chemist to decide whether or not a reaction is suitable for analytical purposes and to choose conditions which will minimize the error associated with an analysis.

Equilibrium constant expressions are algebraic equations that relate the concentrations of reactants and products in a chemical reaction to one another by means of a numerical quantity called an *equilibrium constant*. A chemist must know how to derive useful information from equilibrium constants; it is the purpose of this section to review the simple calculations that have been introduced in most general chemistry courses. Methods for treating more complex systems involving several equilibria will be dealt with in later chapters.

2D-1 THE EQUILIBRIUM STATE

For purposes of discussion, consider the equilibrium

$$2Fe^{3+} + 3I^- \rightleftharpoons 2Fe^{2+} + I_3^- \tag{2-5}$$

The rate of this reaction and the extent to which it proceeds to the right can be readily judged by observing the orange-red color imparted to the solution by the triiodide ion (at low concentrations, the other three participants in the reaction are essentially colorless). If, for example, 2 mmol of iron(III) are added to 100 mL of a solution containing 3 mmol of potassium iodide, color appears instantaneously; within a second or less, the color intensity becomes constant with time, showing that the triiodide concentration has become invariant.

A solution of identical color intensity (and hence triiodide concentration) can be produced by adding 2 mmol of iron(II) to 100 mL of a solution containing 1 mmol of triiodide ion. Here, an immediate decrease in color is observed as a result of the reaction

$$2Fe^{2+} + I_3^- \rightleftharpoons 2Fe^{3+} + 3I^- \tag{2-6}$$

Many other combinations of the four reactants could be employed to yield solutions indistinguishable from the two just described.

The foregoing examples illustrate that the concentration relationship at chemical equilibrium (that is, the *position of equilibrium*) is independent of the route by which the equilibrium state is achieved. On the other hand, it is readily shown that these relationships are altered by the application of stress to the system—for example, by changes in temperature, in pressure (if one of the reactants or products is a gas), or in the total concentration of a reactant or product. These effects can be predicted qualitatively from the *principle of Le Châtelier,* which states that the position of chemical equilibrium will always shift in a direction that tends to relieve the effect of an applied stress. Thus, an increase in temperature will alter the concentration relationships in the direction that tends to absorb heat; an increase in pressure favors those participants that occupy the smaller total volume. Of particular importance in an analysis is the effect of introducing an additional amount of a participating species to the reaction mixture; here, the resulting stress is relieved by a shift in equilibrium in a direction that partially consumes the added substance. Thus, for the equilibrium we have been considering, the addition of iron(III) would cause an increase in color as more triiodide ion and iron(II) are formed; the addition of iron(II) would have a reverse effect. An equilibrium shift brought about by changing the amount of one of the participants is called a *mass-action effect.*

If it were possible to examine a system at the molecular level, it would be found that interactions among the species present continue unabated even after equilibrium is achieved. The observed constant concentration relationship is thus the consequence of an equality in the rates of the forward and reverse reactions; that is, chemical equilibrium is a dynamic state.

2D-2 EQUILIBRIUM CONSTANT EXPRESSIONS

The influence of concentration (or pressure, if the species are gases) on the position of a chemical equilibrium is conveniently described in quantitative terms by means of an equilibrium constant expression. Such expressions are readily derived from thermodynamic theory; they are of great practical importance because they permit the chemist to predict the direction and the completeness of a chemical reaction. It is important to note, however, that equilibrium constant expressions yield no information concerning reaction rates.

The equilibrium constant expression for a reaction can take one of two forms—an exact form (called the *thermodynamic expression*) or an approximate form that is applicable to a limited set of conditions only. Approximate equilibrium constant expressions are more frequently encountered because they are more convenient to use; to be sure, calculations based upon them will be subject to a degree of uncertainty.

Implicit in an approximate equilibrium constant expression is the assumption that, at constant temperature and pressure, concentration is the sole factor affecting the influence of any ion (or molecule) upon the condition of equilibrium; that is, each species acts independently of its neighbors. For example, the influence of each iron(III) ion upon the equilibrium state in Equation 2-6 is assumed to be the same in very dilute solutions (where the ion is most likely surrounded by neutral water molecules) as it is in very concentrated solutions (where other charged species are in close proximity). A solution (or a gas) in which the ions or molecules act independently of one another is termed *ideal* or *perfect*. Truly ideal solutions and ideal behavior are seldom encountered in the laboratory.

Consider the generalized equation for a chemical equilibrium

$$m\mathrm{M} + n\mathrm{N} \rightleftharpoons p\mathrm{P} + q\mathrm{Q} \tag{2-7}$$

where the capital letters represent the formulas of participating chemical species and the italic letters are the small integers required to balance the equation. Thus, the equation states that m moles of M react with n moles of N to form p moles of P and q moles of Q. The approximate equilibrium constant expression for this reaction is

$$\frac{[\mathrm{P}]^p[\mathrm{Q}]^q}{[\mathrm{M}]^m[\mathrm{N}]^n} = K \tag{2-8}$$

where the letters in brackets represent the molar concentrations of dissolved solutes or partial pressures (in atmospheres) if the reacting substances are gases.

The letter K in Equation 2-8 is a temperature-dependent, numerical constant called the *equilibrium constant*. By convention, the concentrations of the products, *as the equation is written*, are always placed in the numerator and the concentrations of the reactants in the denominator. Note also that each concentration is raised to a power that is identical to the integer that accompanies the formula of that species in the balanced equation describing the equilibrium.[6]

The exact form of Equation 2-8 can be readily derived from thermodynamic concepts by assuming that the participants in a reaction exhibit ideal behavior.[7]

Equation 2-8 is useful for determining the approximate equilibrium composition of a dilute solution. More nearly exact results require the use of thermodynamic equilibrium constant expressions, which are considered in Chapter 4. The approximate form suffices to provide the chemist with adequate information for many purposes.

2D-3 THE ION PRODUCT CONSTANT FOR WATER

Aqueous solutions contain small amounts of hydronium and hydroxide ions as a consequence of the dissociation reaction

$$2\mathrm{H_2O} \rightleftharpoons \mathrm{H_3O^+} + \mathrm{OH^-}$$

---

[6] Derivation of equilibrium-constant expressions from theory reveals that the bracketed terms are in fact *concentration ratios* rather than absolute concentrations. The denominator in each of these ratios is the concentration of the species in its so-called *standard state*. The standard state is a 1 M solution for a solute. For pure elements or compounds, it is their state (solid, liquid, or gas) at 25°C and 1 atm pressure. Thus, if P in Equation 2-8 is a solute,

$$[\mathrm{P}] = \frac{\text{concn of P, } \cancel{\text{mol/L}}}{1.00 \ \cancel{\text{mol/L}}}$$

where the denominator is the concentration of solute in the standard state. An important consequence of the relative nature of [P] is that this quantity, as well as the equilibrium constant, is unitless. Strictly speaking then, [P] must be multiplied by 1.00 mol/L to obtain units of concentration. Ordinarily, this fine distinction is of no importance, and [P] is treated as an absolute concentration.

[7] See: L. K. Nash, *Elements of Classical and Statistical Thermodynamics*, Book 1, pp. 124–130. Reading, Mass.: Addison-Wesley, 1970.

An equilibrium constant for this reaction can be formulated as shown in Equation 2-8, namely,

$$\frac{[H_3O^+][OH^-]}{[H_2O]^2} = K \tag{2-9}$$

In dilute aqueous solutions, however, the concentration of water is large compared with the concentration of solutes and can be considered to be invariant. That is, the quantity of water in each liter of a dilute aqueous solution (approximately 55 mol) is enormous compared with the amount formed or lost through any equilibrium shift. Therefore, $[H_2O]$ in Equation 2-9 can be taken as constant, and we may write

$$[H_3O^+][OH^-] = K[H_2O]^2 = K_w \tag{2-10}$$

where the new constant K_w is given a special name, the *ion product constant for water*. Note that the inclusion of $[H_2O]$ in the equilibrium constant is equivalent to arbitrarily assigning $[H_2O]$ a value of $1.00 \ M$.

At 25°C, the ion product constant for water has a numerical value of 1.01×10^{-14} (for convenience, we will normally use the approximation that $K_w \cong 1.00 \times 10^{-14}$). Table 2-4 shows the dependence of this constant upon temperature.

The ion product constant for water permits the ready calculation of the hydronium and hydroxide ion concentrations of aqueous solutions.

Table 2-4   Variation of K_w With Temperature

Temperature °C	K_w
0	0.114×10^{-14}
25	$1.01 \ \times 10^{-14}$
50	$5.47 \ \times 10^{-14}$
100	$49 \ \ \times 10^{-14}$

EXAMPLE 2-11. Calculate the hydronium and hydroxide ion concentration of pure water at 25°C and at 100°C.

Because OH^- and H_3O^+ are formed from the dissociation of water only, their concentrations must be equal; that is,

$$[H_3O^+] = [OH^-]$$

Substitution into Equation 2-10 gives

$$[H_3O^+]^2 = [OH^-]^2 = K_w$$
$$[H_3O^+] = [OH^-] = \sqrt{K_w}$$

At 25°C,

$$[H_3O^+] = [OH^-] = \sqrt{1.00 \times 10^{-14}} = 1.00 \times 10^{-7}$$

At 100°C,

$$[H_3O^+] = [OH^-] = \sqrt{49 \times 10^{-14}} = 7.0 \times 10^{-7}$$

EXAMPLE 2-12. Calculate the hydronium and hydroxide ion concentrations in $0.200 \ M$ aqueous NaOH.

Sodium hydroxide is a strong electrolyte, and its contribution to the hydroxide ion concentration of this solution will be 0.200 mol/L. As in the previous example, hydroxide ions and hydronium ions are also formed *in equal amounts* from the dissociation of water. Therefore, we may write

$$[OH^-] = 0.200 + [H_3O^+]$$

where $[H_3O^+]$ accounts for the hydroxide ions contributed by the solvent. The concentration of OH^- from the water is vanishingly small when compared with 0.200; therefore,

$$[OH^-] \cong 0.200$$

We can then employ Equation 2-10 to calculate the hydronium ion concentration

$$[H_3O^+] = \frac{K_w}{[OH^-]} = \frac{1.00 \times 10^{-14}}{0.200} = 5.00 \times 10^{-14}$$

Note that the approximation

$$[OH^-] = 0.200 + 5.00 \times 10^{-14} = 0.200$$

will cause no significant error.

2D-4 EQUILIBRIUM INVOLVING SLIGHTLY SOLUBLE IONIC SOLIDS

Most sparingly soluble salts are essentially completely dissociated in saturated aqueous solution. For example, when an excess of silver carbonate is equilibrated with water, the dissociation process is adequately described by the equation

$$Ag_2CO_3(s) \rightleftharpoons 2Ag^+ + CO_3^{2-}$$

Application of Equation 2-8 leads to

$$\frac{[Ag^+]^2[CO_3^{2-}]}{[Ag_2CO_3(s)]} = K$$

The denominator represents the molar concentration of Ag_2CO_3 *in the solid.* The concentration of a compound in its solid state is, however, invariant. Therefore, the foregoing equation can be rewritten in the form

$$[Ag^+]^2[CO_3^{2-}] = K[Ag_2CO_3(s)] = K_{sp} \tag{2-11}$$

where the new constant is called the *solubility product constant* or the *solubility product.*

A table of solubility products for numerous inorganic salts is found in Appendix 3 of this text. In using these constants it is *essential* to keep in mind the expressions such as Equation 2-11 apply *only to saturated solutions that are in contact with an excess* of undissolved solutes.

The examples which follow demonstrate some typical uses of solubility-product expressions. Other applications will be considered in Chapters 4 and 7.

Solubility of a precipitate in pure water. The solubility constant expression permits the ready calculation of the solubility of an ionic precipitate in water.

> **EXAMPLE 2-13.** How many grams of $Ba(IO_3)_2$ (fw = 487) can be dissolved in 500 mL of water at 25°C?
>
> The solubility-product constant for $Ba(IO_3)_2$ is 1.57×10^{-9} (Appendix 3). Thus,
>
> $$Ba(IO_3)_2(s) \rightleftharpoons Ba^{2+} + 2IO_3^-$$

and

$$[Ba^{2+}][IO_3^-]^2 = 1.57 \times 10^{-9} = K_{sp}$$

It is seen from the equation describing the equilibrium that 1 mol of Ba^{2+} is formed for each mole of $Ba(IO_3)_2$ that dissolves. Thus,

$$\text{molar solubility of } Ba(IO_3)_2 = [Ba^{2+}]$$

We also see from the equation that the iodate concentration is twice that for barium ion. That is,

$$[IO_3^-] = 2[Ba^{2+}]$$

Substituting the latter into the equilibrium-constant expression gives

$$[Ba^{2+}](2[Ba^{2+}])^2 = 1.57 \times 10^{-9}$$

or

$$[Ba^{2+}] = \left(\frac{1.57 \times 10^{-9}}{4}\right)^{1/3} = 7.32 \times 10^{-4}$$

and

$$\text{solubility} = 7.32 \times 10^{-4} \ M$$

To obtain the solubility of $Ba(IO_3)_2$ in grams per 500 mL, we write

$$\text{solubility} = 7.32 \times 10^{-4} \ \frac{\text{mmol } Ba(IO_3)_2}{\text{mL}} \times 500 \text{ mL} \times 0.487 \ \frac{\text{g}}{\text{mmol } Ba(IO_3)_2}$$

$$= \frac{0.178 \text{ g}}{500 \text{ mL}}$$

where 0.487 is the milliformula weight of $Ba(IO_3)_2$.

Solubility of precipitates in the presence of a common ion.
The common-ion effect predicted from the Le Châtelier principle is demonstrated by the following examples.

EXAMPLE 2-14. Calculate the molar solubility of $Ba(IO_3)_2$ in a solution that is 0.0200 M in $Ba(NO_3)_2$.

In this example, the solubility is not directly related to $[Ba^{2+}]$ but can be described in terms of $[IO_3^-]$; that is,

$$\text{solubility of } Ba(IO_3)_2 = \tfrac{1}{2}[IO_3^-]$$

Here, barium ions arise from two sources, namely, $Ba(NO_3)_2$ and $Ba(IO_3)_2$. The contribution from the former is 0.0200 M while that from the latter is equal to the molar solubility, or $\tfrac{1}{2}[IO_3^-]$. Thus,

$$[Ba^{2+}] = 0.0200 + \tfrac{1}{2}[IO_3^-]$$

Substitution of these quantities into the solubility-product expression yields

$$(0.0200 + \tfrac{1}{2}[IO_3^-])[IO_3^-]^2 = 1.57 \times 10^{-9}$$

Since the exact solution for $[IO_3^-]$ will involve a cubic equation, it is worthwhile to seek an approximation that will simplify the algebra. The small numerical value for K_{sp} suggests that the solubility of $Ba(IO_3)_2$ is not large; therefore, it is reasonable to suppose that the barium ion concentration derived from the solubility of $Ba(IO_3)_2$ is small with respect to that from the $Ba(NO_3)_2$. That is, $\tfrac{1}{2}[IO_3^-] \ll 0.0200$

and

$$0.0200 + \tfrac{1}{2}[IO_3^-] \cong 0.0200$$

The original equation then simplifies to

$$0.0200[IO_3^-]^2 = 1.57 \times 10^{-9}$$
$$[IO_3^-] = 2.80 \times 10^{-4}$$

The assumption that

$$(0.0200 + \tfrac{1}{2} \times 2.80 \times 10^{-4}) \cong 0.0200$$

does not appear to cause serious error because the second term is only about 0.7% of 0.0200. Ordinarily we shall consider an assumption of this type to be satisfactory if the discrepancy is less than 5 to 10%. Therefore,

$$\text{solubility of } Ba(IO_3)_2 = \tfrac{1}{2}[IO_3^-] = \tfrac{1}{2} \times 2.80 \times 10^{-4} = 1.40 \times 10^{-4} \ M$$

If we compare this result with the solubility of barium iodate in pure water (Example 2-13), we see that the presence of a small concentration of the common ion has lowered the molar solubility of $Ba(IO_3)_2$ by a factor of about 5.

EXAMPLE 2-15. Calculate the solubility of $Ba(IO_3)_2$ in the solution that results when 200 mL of 0.0100 M $Ba(NO_3)_2$ are mixed with 100 mL of 0.100 M $NaIO_3$.

We must first establish whether either reactant will be present in excess at equilibrium. The amounts available are

$$\text{mmol } Ba^{2+} = 200 \text{ mL} \times 0.0100 \text{ mmol/mL} = 2.00$$
$$\text{mmol } IO_3^- = 100 \text{ mL} \times 0.100 \text{ mmol/mL} = 10.00$$

If formation of $Ba(IO_3)_2$ is complete,

$$\text{mmol excess } IO_3^- = 10.0 - 2(2.00) = 6.00$$

Thus,

$$[IO_3^-] = \frac{6.00 \text{ mmol}}{300 \text{ mL}} = 0.0200 \ M$$

As in Example 2-13,

$$\text{molar solubility of } Ba(IO_3)_2 = [Ba^{2+}]$$

Here, however,

$$[IO_3^-] = 0.0200 + 2[Ba^{2+}]$$

where $2[Ba^{2+}]$ represents the iodate contributed by the sparingly soluble $Ba(IO_3)_2$. We can obtain a provisional answer after making the assumption that

$$[IO_3^-] \cong 0.0200$$

Thus,

$$\text{solubility of } Ba(IO_3)_2 = [Ba^{2+}] = \frac{K_{sp}}{[IO_3^-]^2}$$

$$= \frac{1.57 \times 10^{-9}}{(0.0200)^2} = 3.93 \times 10^{-6} \text{ mol/L}$$

The approximation used in this calculation is seen to have been reasonable.

Note that the results from the last two examples demonstrate that the presence of excess iodate is more effective in decreasing the solubility of $Ba(IO_3)_2$ than is an equal excess of barium ions.

2D-5 DISSOCIATION EQUILIBRIA FOR WEAK ACIDS AND BASES

When a weak acid or base is dissolved in water, partial dissociation occurs. Thus, for nitrous acid we may write

$$HNO_2 + H_2O \rightleftharpoons H_3O^+ + NO_2^- \qquad \frac{[H_3O^+][NO_2^-]}{[HNO_2]} = K_a$$

where K_a is the *acid dissociation constant* for nitrous acid. In an analogous way, the *basic dissociation constant* for ammonia is given by

$$NH_3 + H_2O \rightleftharpoons NH_4^+ + OH^- \qquad \frac{[NH_4^+][OH^-]}{[NH_3]} = K_b$$

Note that a concentration term for water ($[H_2O]$) does not appear in either equation; as with the ion-product constant for water, the solvent concentration is assumed to be so large that its concentration is not significantly altered by the addition of the acid or base. Thus, the concentration of water is incorporated in the equilibrium constants K_a and K_b. Dissociation constants for weak acids and weak bases are found in Appendixes 4 and 5.

Relationship between dissociation constants for conjugate acid-base pairs.

Consider the dissociation-constant expression for ammonia and its conjugate acid, ammonium ion. Here, we may write

$$NH_3 + H_2O \rightleftharpoons NH_4^+ + OH^- \qquad \frac{[NH_4^+][OH^-]}{[NH_3]} = K_b$$

and

$$NH_4^+ + H_2O \rightleftharpoons NH_3 + H_3O^+ \qquad \frac{[NH_3][H_3O^+]}{[NH_4^+]} = K_a$$

Multiplication of the two equilibrium-constant expressions together gives

$$\frac{[\cancel{NH_3}][H_3O^+]}{[\cancel{NH_4^+}]} \times \frac{[\cancel{NH_4^+}][OH^-]}{[\cancel{NH_3}]} = [H_3O^+][OH^-] = K_aK_b$$

but

$$[H_3O^+][OH^-] = K_w$$

Therefore,

$$K_aK_b = K_w \tag{2-12}$$

This relationship is general for all conjugate acid-base pairs. Most tables of dissociation constants do not list both the acid- and the base-dissociation constants for conjugate pairs, since it is so easy to calculate one from the other by Equation 2-12.

EXAMPLE 2-16. What is K_b for the equilibrium

$$CN^- + H_2O \rightleftharpoons HCN + OH^-$$

Examination of Appendix 5 (dissociation constants for bases) reveals no entry for CN$^-$. Appendix 4, however, lists a K_a value of 2.1×10^{-9} for HCN. Thus,

$$\frac{[\text{HCN}][\text{OH}^-]}{[\text{CN}^-]} = K_b = \frac{K_w}{K_{\text{HCN}}}$$

$$K_b = \frac{1.00 \times 10^{-14}}{2.1 \times 10^{-9}} = 4.8 \times 10^{-6}$$

Applications of acid dissociation constants. Dissociation constants for many common weak acids are tabulated in Appendix 4. These constants find wide application for the calculation of the hydronium or hydroxide ion concentrations of solutions of such acids.

For example, the equilibrium established when the weak acid HA is dissolved in water may be written as

$$\text{HA} + \text{H}_2\text{O} \rightleftharpoons \text{H}_3\text{O}^+ + \text{A}^- \qquad \frac{[\text{H}_3\text{O}^+][\text{A}^-]}{[\text{HA}]} = K_a$$

In addition, hydronium ions result from the equilibrium

$$2\text{H}_2\text{O} \rightleftharpoons \text{H}_3\text{O}^+ + \text{OH}^-$$

Ordinarily, the hydronium ions produced from the first reaction will suppress the dissociation of water to such an extent that the concentration of hydronium and hydroxide ions from this source can be considered to be negligible. Under these circumstances, we see from the acid dissociation equilibrium that one A$^-$ ion is produced for each H$_3$O$^+$ ion; that is,

$$[\text{H}_3\text{O}^+] \cong [\text{A}^-] \tag{2-13}$$

Furthermore, the sum of the molar concentrations of the weak acid and its conjugate base must equal the analytical molar concentration of the acid because the solution contains no other species that contributes A$^-$. Thus,

$$M_{\text{HA}} = [\text{A}^-] + [\text{HA}] \tag{2-14}$$

Substituting Equation 2-13 into 2-14 and rearranging give

$$[\text{HA}] = M_{\text{HA}} - [\text{H}_3\text{O}^+] \tag{2-15}$$

When [A$^-$] and [HA] are replaced by Equations 2-13 and 2-15, the equilibrium expression becomes

$$\frac{[\text{H}_3\text{O}^+]^2}{M_{\text{HA}} - [\text{H}_3\text{O}^+]} = K_a \tag{2-16}$$

Equation 2-16 can be rearranged to the form

$$[\text{H}_3\text{O}^+]^2 + K_a[\text{H}_3\text{O}^+] - K_a M_{\text{HA}} = 0$$

The positive solution to this quadratic equation is

$$[\text{H}_3\text{O}^+] = \frac{-K_a + \sqrt{(K_a)^2 + 4K_a M_{\text{HA}}}}{2} \tag{2-17}$$

It is frequently possible to assume that [H$_3$O$^+$] is much smaller than M_{HA}; that is,

$[H_3O^+] \ll M_{HA}$. Equation 2-15 then becomes

$$[HA] \cong M_{HA}$$

Substituting this relationship and Equation 2-13 into the expression for K_a and rearranging yield

$$[H_3O^+] = \sqrt{K_a M_{HA}} \qquad (2\text{-}18)$$

The magnitude of the error introduced by the assumption that $[H_3O^+] \ll M_{HA}$ will increase as the molar concentration of acid becomes smaller and the acid dissociation constant becomes larger. This statement is supported by the data in Table 2-5. Note that the error introduced by the assumption is about 0.5% when the ratio M_{HA}/K_a is 10^4. The error increases to about 1.6% when the ratio is 10^3, to about 5% when it is 10^2, and to about 17% when it is 10. Figure 2-2 illustrates the effect graphically. It is noteworthy that the hydronium ion concentration from the approximate solution becomes equal to or greater than that for the acid itself when the ratio is 1 or smaller; clearly, the approximation leads to meaningless results under these circumstances.

In general, it is good practice to make the simplifying assumption and to obtain a trial value for $[H_3O^+]$ that may be compared with M_{HA} in Equation 2-15. If the trial value alters $[HA]$ by an amount smaller than the allowable error in the calculation, the solution may be considered satisfactory. Otherwise, the quadratic equation must be solved to give a better value for $[H_3O^+]$.

EXAMPLE 2-17. Calculate the hydronium ion concentration of an aqueous 0.120 M nitrous acid solution. The principal equilibrium in this solution is

$$HNO_2 + H_2O \rightleftharpoons H_3O^+ + NO_2^-$$

for which (Appendix 4)

Table 2-5 Errors Introduced by Assuming H_3O^+ Concentration Small Relative to M_{HA} in Equation 2-16

Value of K_a	Value of M_{HA}	Value for $[H_3O^+]$ from Equation 2-18	Value for $[H_3O^+]$ from Equation 2-17	Percent Error
1.00×10^{-2}	1.00×10^{-3}	3.16×10^{-3}	0.92×10^{-3}	244
	1.00×10^{-2}	1.00×10^{-2}	0.62×10^{-2}	61
	1.00×10^{-1}	3.16×10^{-2}	2.70×10^{-2}	17
1.00×10^{-4}	1.00×10^{-4}	1.00×10^{-4}	0.62×10^{-4}	61
	1.00×10^{-3}	3.16×10^{-4}	2.70×10^{-4}	17
	1.00×10^{-2}	1.00×10^{-3}	0.95×10^{-3}	5.3
	1.00×10^{-1}	3.16×10^{-3}	3.11×10^{-3}	1.6
1.00×10^{-6}	1.00×10^{-5}	3.16×10^{-6}	2.70×10^{-6}	17
	1.00×10^{-4}	1.00×10^{-5}	0.95×10^{-5}	5.3
	1.00×10^{-3}	3.16×10^{-5}	3.11×10^{-5}	1.6
	1.00×10^{-2}	1.00×10^{-4}	9.95×10^{-5}	0.5
	1.00×10^{-1}	3.16×10^{-4}	3.16×10^{-4}	0.0

$$\frac{[H_3O^+][NO_2^-]}{[HNO_2]} = K_a = 5.1 \times 10^{-4}$$

Thus,

$$[H_3O^+] = [NO_2^-]$$
$$[HNO_2] = 0.120 - [H_3O^+]$$

and

$$\frac{[H_3O^+]^2}{0.120 - [H_3O^+]} = 5.1 \times 10^{-4}$$

If we now assume $[H_3O^+] \ll 0.120$, we find

$$[H_3O^+] = \sqrt{0.120 \times 5.1 \times 10^{-4}} = 7.8 \times 10^{-3}$$

We must now examine the assumption that $(0.120 - 0.0078) \cong 0.120$; a difference of about 7% is involved. The relative error in $[H_3O^+]$ will be smaller than this figure, however, as we can see by calculating $\log M_{HA}/K_a = 2.4$; referring to Figure 2-2, we see that an error of about 3% results. If a more accurate figure were needed, solution of the quadratic equation would yield a value of 7.6×10^{-3} M.

EXAMPLE 2-18. Calculate the hydronium ion concentration of a solution which is 2.0×10^{-4} M in aniline hydrochloride $C_6H_5NH_3Cl$.

In aqueous solution, dissociation to Cl^- and $C_6H_5NH_3^+$ is complete. The weak acid $C_6H_5NH_3^+$ dissociates as follows:

$$C_6H_5NH_3^+ + H_2O \rightleftharpoons C_6H_5NH_2 + H_3O^+ \qquad \frac{[H_3O^+][C_6H_5NH_2]}{[C_6H_5NH_3^+]} = K_a$$

Inspection of Appendix 4 reveals no entry for $C_6H_5NH_3^+$, but Appendix 5 gives a basic constant for aniline, $C_6H_5NH_2$; that is,

$$C_6H_5NH_2 + H_2O \rightleftharpoons C_6H_5NH_3^+ + OH^- \qquad K_b = 3.94 \times 10^{-10}$$

Thus, Equation 2-12 is used to obtain a value for K_a:

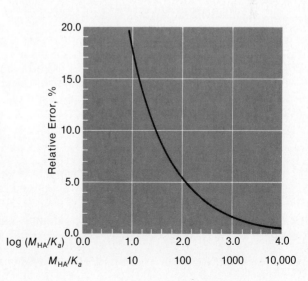

FIGURE 2-2 Relative error from the use of Equation 2-18.

$$K_a = \frac{1.00 \times 10^{-14}}{3.94 \times 10^{-10}} = 2.54 \times 10^{-5}$$

Proceeding as in the previous example,

$$[H_3O^+] = [C_6H_5NH_2]$$
$$[C_6H_5NH_3^+] = 2.0 \times 10^{-4} - [H_3O^+]$$

If we assume that $[H_3O^+] \ll 2.0 \times 10^{-4}$ and substitute the simplified value for $[C_6H_5NH_3^+]$ into the dissociation constant expression, we obtain

$$\frac{[H_3O^+]^2}{2.0 \times 10^{-4}} = 2.54 \times 10^{-5}$$

$$[H_3O^+] = 7.1 \times 10^{-5}$$

Comparison of 7.1×10^{-5} with 2.0×10^{-4} suggests that a significant error exists in this value for $[H_3O^+]$ (reference to Figure 2-2 indicates that this error is about 20%). Thus, unless only an approximate value for $[H_3O^+]$ is needed, it is necessary to use the more nearly exact expression

$$\frac{[H_3O^+]^2}{2.0 \times 10^{-4} - [H_3O^+]} = 2.54 \times 10^{-5}$$

which rearranges to

$$[H_3O^+]^2 + 2.54 \times 10^{-5}\,[H_3O^+] - 5.08 \times 10^{-9} = 0$$

and

$$[H_3O^+] = \frac{-2.54 \times 10^{-5} + \sqrt{(2.54 \times 10^{-5})^2 + 4 \times 5.08 \times 10^{-9}}}{2}$$

$$[H_3O^+] = 6.0 \times 10^{-5}$$

Application of dissociation constants for weak bases. The techniques discussed in previous sections are readily adapted to the calculation of the hydroxide ion concentration in solutions of weak bases.

Aqueous ammonia is basic by virtue of the reaction

$$NH_3 + H_2O \rightleftharpoons NH_4^+ + OH^-$$

Here, the predominant species has been clearly demonstrated to be NH_3. Nevertheless, such solutions are sometimes called ammonium hydroxide, the terminology being vestigial from the time when the substance NH_4OH rather than NH_3 was believed to be the undissociated form of the base. Application of the mass law to this equilibrium yields the expression

$$\frac{[NH_4^+][OH^-]}{[NH_3]} = K_b$$

The use of NH_3 in the denominator is preferable to the historical NH_4OH; in any event, the magnitude of K_b remains unchanged.

EXAMPLE 2-19. Calculate the hydronium ion concentration of a 0.075 M NH_3 solution. The predominant equilibrium in this solution is

$$NH_3 + H_2O \rightleftharpoons NH_4^+ + OH^-$$

From the table of basic dissociation constants (Appendix 5),

$$\frac{[NH_4^+][OH^-]}{[NH_3]} = 1.76 \times 10^{-5} = K_b$$

The equation for the equilibrium indicates that

$$[NH_4^+] = [OH^-]$$

and

$$[NH_4^+] + [NH_3] = M_{NH_3} = 0.075$$

If we substitute $[OH^-]$ for $[NH_4^+]$ in the second of these equations and rearrange, we find that

$$[NH_3] = 0.075 - [OH^-]$$

Substituting these quantities into the dissociation-constant expression yields

$$\frac{[OH^-]^2}{7.5 \times 10^{-2} - [OH^-]} = 1.76 \times 10^{-5}$$

Provided that $[OH^-] \ll 7.5 \times 10^{-2}$, this equation simplifies to

$$[OH^-]^2 \cong 7.5 \times 10^{-2} \times 1.76 \times 10^{-5}$$

and

$$[OH^-] = 1.15 \times 10^{-3}$$

Upon comparing the calculated value for $[OH^-]$ with 7.5×10^{-2}, we see that the error in $[OH^-]$ will be less than 2%. If needed, a better value for $[OH^-]$ could be obtained by solving the quadratic equation.

Finally, then

$$[H_3O^+] = \frac{K_w}{[OH^-]} = \frac{1.00 \times 10^{-14}}{1.15 \times 10^{-3}} = 8.7 \times 10^{-12}$$

EXAMPLE 2-20. Calculate the hydroxide ion concentration of a 0.010 M sodium hypochlorite solution.

The equilibrium between OCl^- and water is

$$OCl^- + H_2O \rightleftharpoons HOCl + OH^-$$

for which

$$\frac{[HOCl][OH^-]}{[OCl^-]} = K_b$$

Appendix 5 does not contain a value for K_b; an examination of Appendix 4, however, shows that the acid dissociation constant for HOCl is 3.0×10^{-8}. Therefore, employing Equation 2-12, we write

$$K_b = \frac{K_w}{K_a} = \frac{1.00 \times 10^{-14}}{3.0 \times 10^{-8}} = 3.3 \times 10^{-7}$$

Proceeding as in the previous example,

$$[OH^-] = [HOCl]$$
$$[OCl^-] + [HOCl] = 0.010$$

or

$$[OCl^-] = 0.010 - [OH^-] \cong 0.010$$

That is, we assume $[OH^-] \ll 0.010$. Substitution into the equilibrium constant expression gives

$$\frac{[OH^-]^2}{0.010} = 3.3 \times 10^{-7}$$

$$[OH^-] = 5.7 \times 10^{-5}$$

The error resulting from the approximation is clearly small.

2D-6 COMPLEX FORMATION EQUILIBRIA

An analytically important class of reactions involves the formation of soluble complex ions. Two examples are:

$$Fe^{3+} + SCN^- \rightleftharpoons Fe(SCN)^{2+} \qquad \frac{[Fe(SCN)^{2+}]}{[Fe^{3+}][SCN^-]} = K_f$$

$$Zn(OH)_2(s) + 2OH^- \rightleftharpoons Zn(OH)_4^{2-} \qquad \frac{[Zn(OH)_4^{2-}]}{[OH^-]^2} = K_f$$

where K_f is called the *formation constant* for the complex.[8] Note that the second constant applies only to a saturated solution that is in contact with the sparingly soluble zinc hydroxide. Note also that no concentration term for the zinc hydroxide appears in the formation constant expression because its concentration in the solid is invariant and is included in the constant K_f. In this regard, the treatment is analogous to that described for solubility product expressions.

Formation constants for numerous complex ions appear in Appendix 6. An application involving the use of formation constants is given in the following example.

> **EXAMPLE 2-21.** It has been found that the average person can see the red color imparted by $FeSCN^{2+}$ to an aqueous solution if the concentration of the complex is 6×10^{-6} M or greater. What minimum concentration of KSCN would be required to make it possible to detect 1 ppm of iron(III) in a natural water sample if the formation constant for the complex is 1.4×10^2?
>
> $$Fe^{3+} + SCN^- \rightleftharpoons FeSCN^{2+} \qquad K_f = 1.4 \times 10^2$$
>
> or
>
> $$\frac{[FeSCN^{2+}]}{[Fe^{3+}][SCN^-]} = 1.4 \times 10^2$$
>
> To convert the minimum concentration of iron(III) to a molar concentration, we recall (p. 13) that 1 ppm corresponds to 1 mg/L. Thus,
>
> $$M_{Fe^{3+}} = \frac{1\ mg}{L} \times \frac{1\ g}{1000\ mg} \times \frac{1\ mol}{55.8\ g}$$
>
> $$= 1.8 \times 10^{-5}\ mol/L$$

---

[8] Equilibria involving complex ions are sometimes described in terms of dissociation reactions and *instability constants;* the latter, which are found in the older literature, are the reciprocals of formation constants. For example,

$$Fe(SCN)^{2+} \rightleftharpoons Fe^{3+} + SCN^- \qquad \frac{[Fe^{3+}][SCN^-]}{[FeSCN^{2+}]} = K_{inst} = \frac{1}{K_f}$$

This minimum concentration will be distributed between two species. Fe^{3+} and $FeSCN^{2+}$. That is,

$$1.8 \times 10^{-5} = [Fe^{3+}] + [FeSCN^{2+}]$$

But $[FeSCN^{2+}]$ is 6×10^{-6} M if this species is to be seen. Thus,

$$[Fe^{3+}] = 1.8 \times 10^{-5} - 6 \times 10^{-6} = 1.2 \times 10^{-5}$$

We now substitute these concentrations into the formation constant expression in order to determine the SCN^- concentration that will be required; that is,

$$\frac{6 \times 10^{-6}}{1.2 \times 10^{-5}[SCN^-]} = 1.4 \times 10^2$$

$$[SCN^-] = 0.0036 = 0.004$$

Thus, if the test solution is made 0.004 M (or greater) in KSCN, a detectable amount of $FeSCN^{2+}$ will form, provided total iron(III) concentration is 1 ppm or greater.

2D-7 OXIDATION-REDUCTION EQUILIBRIA

Equilibrium constants for oxidation-reduction reactions can be formulated in the usual way. For example,

$$6Fe^{2+} + Cr_2O_7^{2-} + 14H_3O^+ \rightleftharpoons 6Fe^{3+} + 2Cr^{3+} + 21\ H_2O$$

$$\frac{[Fe^{3+}]^6[Cr^{3+}]^2}{[Fe^{2+}]^6[Cr_2O_7^{2-}][H_3O^+]^{14}} = K$$

As in earlier examples, no term for the concentration of water is needed.

Tables of equilibrium constants for oxidation-reduction reactions are not generally available because these constants are readily derived from more fundamental constants called *standard electrode potentials*. Derivations of this kind are treated in detail in Chapter 13.

2D-8 DISTRIBUTION EQUILIBRIA BETWEEN TWO IMMISCIBLE LIQUIDS

Another important type of equilibrium involves the distribution of a solute between two liquid phases. For example, if an aqueous solution of iodine is shaken with an immiscible organic solvent such as hexane or chloroform, a portion of the solute is extracted into the organic layer. Ultimately an equilibrium is established between the two phases which can be described by

$$I_2(aq) \rightleftharpoons I_2(org) \qquad \frac{[I_2]_{org}}{[I_2]_{aq}} = K$$

The equilibrium constant for this reaction, K, is often called a *distribution* or *partition coefficient*.

As we shall see in Chapters 26 and 27, distribution equilibria are of vital importance in understanding many separation processes.

2D-9 STEPWISE EQUILIBRIA

Many weak electrolytes associate or dissociate in a stepwise manner, and an equilibrium constant can be written for each step. For example, when ammonia is added to a solution containing silver ions, at least two equilibria involving the two species are established.

$$Ag^+ + NH_3 \rightleftharpoons AgNH_3^+ \qquad \frac{[AgNH_3^+]}{[Ag^+][NH_3]} = K_1$$

$$AgNH_3^+ + NH_3 \rightleftharpoons Ag(NH_3)_2^+ \qquad \frac{[Ag(NH_3)_2^+]}{[AgNH_3^+][NH_3]} = K_2$$

Here, K_1 and K_2 are stepwise formation constants for the two complexes. The two constants can be multiplied together to give an overall formation constant for the reaction

$$Ag^+ + 2NH_3 \rightleftharpoons Ag(NH_3)_2^+ \qquad \frac{[Ag(NH_3)_2^+]}{[Ag^+][NH_3]^2} = K_1K_2 = \beta_2$$

Overall formation constants of this type are often symbolized as β_n, where n corresponds to the number of moles of complexing species that combine with one mole of cation. Thus, for example,

$$Cd^{2+} + 3CN^- \rightleftharpoons Cd(CN)_3^- \qquad \frac{[Cd(CN)_3^-]}{[Cd^{2+}][CN^-]^3} = K_1K_2K_3 = \beta_3$$

$$Cd^{2+} + 4CN^- \rightleftharpoons Cd(CN)_4^{2-} \qquad \frac{[Cd(CN)_4^{2-}]}{[Cd^{2+}][CN^-]^4} = K_1K_2K_3K_4 = \beta_4$$

where K_1, K_2, K_3, and K_4 are the corresponding stepwise constants.

Many common acids and bases also undergo stepwise dissociation. For example, the following equilibria exist in an aqueous solution of phosphoric acid:

$$H_3PO_4 + H_2O \rightleftharpoons H_2PO_4^- + H_3O^+ \qquad \frac{[H_3O^+][H_2PO_4^-]}{[H_3PO_4]} = K_1 = 7.11 \times 10^{-3}$$

$$H_2PO_4^- + H_2O \rightleftharpoons HPO_4^{2-} + H_3O^+ \qquad \frac{[H_3O^+][HPO_4^{2-}]}{[H_2PO_4^-]} = K_2 = 6.34 \times 10^{-8}$$

$$HPO_4^{2-} + H_2O \rightleftharpoons PO_4^{3-} + H_3O^+ \qquad \frac{[H_3O^+][PO_4^{3-}]}{[HPO_4^{2-}]} = K_3 = 4.2 \times 10^{-13}$$

As illustrated by H_3PO_4, numerical values of K_n for an acid or a base become smaller with each successive dissociation step. Equilibria of this type are considered in Chapter 9.

PROBLEMS*

* 2-1. Indicate whether each of the following species is an acid or a base in water. Give the formula for its conjugate base or acid.
  (a) NH_4^+ (d) HSO_4^-
  (b) HCl   (e) NaCN
  (c) H_2SO_4 (f) CH_3NH_2

* 2-2. Indicate whether each of the following species is an acid or a base in water. Give the formula for its conjugate base or acid.
  (a) HCN   (d) $H_2PO_4^-$
  (b) H_3PO_4 (e) $C_6H_5NH_3^+$
  (c) Na_3PO_4 (f) HOCl

* Answers to problems or parts of problems marked with an asterisk are to be found at the end of the book.

2-3. Write autoprotolysis reactions for the following amphiprotic solvents. Indicate the acid and base formed by the reactions.
    * (a) H_2SO_4 * (c) ethanol, C_2H_5OH
       (b) ethylene diamine, $NH_2CH_2CH_2NH_2$ (d) H_2S

* 2-4. How many electrons are contained in 2.5 mol of electrons?

2-5. How many potassium ions are contained in 0.150 mmol of $K_4Fe(CN)_6$?

* 2-6. How many millimoles are contained in
    (a) 6.75 g of Al_2O_3?
    (b) 232 mg of Na_2SO_4?
    (c) 250 mL of 0.264 M $Na_2B_4O_7 \cdot 10 \ H_2O$?
    (d) 2.50 L of 2.00×10^{-4} M $K_2Cr_2O_7$?
    (e) 100 mL of an aqueous solution containing 3.72 ppm HCl?

2-7. How many millimoles are contained in
    (a) 18.3 mg of P_2O_5?
    (b) 20.3 g of dry ice (CO_2)?
    (c) 50.0 g of $MgNH_4PO_4$?
    (d) 63.4 mL of 9.86 M sulfuric acid?
    (e) 60.0 L of 0.0125 M $KMnO_4$?

* 2-8. How many milligrams are contained in
    (a) 0.160 mol of $CHCl_3$?
    (b) 120 mmol of acetic acid (CH_3COOH)?
    (c) 16.0 mol of HNO_2?
    (d) 5.5 mL of 0.50 M sucrose (fw = 342)?
    (e) 20.7 L of 3.0 M H_3PO_3?

2-9. How many grams are contained in
    (a) 2.64 mol of $CaSO_4$?
    (b) 22.1 mmol of I_2?
    (c) 450 mL of 0.250 M methanol (CH_3OH)?
    (d) 6.50 L of 0.0110 M $Ba(OH)_2$?
    (e) 6.10 mL of 2.00×10^{-4} M $NaNO_3$?

* 2-10. A solution was prepared by dissolving 273 mg of $Fe_2(SO_4)_3$ in dilute HCl and diluting to exactly 2.00 L. Calculate
    (a) the molar analytical concentration of $Fe_2(SO_4)_3$.
    (b) the molar equilibrium concentration of Fe^{3+}.
    (c) the molar equilibrium concentration of SO_4^{2-} (assume no HSO_4^- forms).
    (d) the weight-volume percentage of $Fe_2(SO_4)_3$.
    (e) the millimoles of Fe^{3+} in 25.0 mL of the solution.
    (f) the parts per million of Fe^{3+} in the solution.
    (g) the pFe(III) of the solution.
    (h) the pSO_4 of the solution.

2-11. A solution was prepared by dissolving 432 mg of $K_4Fe(CN)_6$ in water and diluting to 1500 mL. Calculate
    (a) the molar analytical concentration of $K_4Fe(CN)_6$.
    (b) the molar equilibrium concentration of K^+.
    (c) the weight-volume percentage of $K_4Fe(CN)_6$.
    (d) the molar concentration of $Fe(CN)_6^{4-}$.
    (e) the millimoles of K^+ in 100.0 mL of the solution.
    (f) the parts per million of K^+ in the solution.
    (g) the pK of the solution.
    (h) the $pFe(CN)_6$ of the solution.

* 2-12. Average sea water contains 1.27×10^3 ppm of Mg^{2+} and 400 ppm of Ca^{2+}.
    (a) Calculate the molar concentration of each of these ions if the density of sea water is 1.02 g/mL.

(b) Calculate the pMg and pCa of the solution.

2-13. Average human blood serum contains 360 mg Na^{2+} and 21 mg SO_4^{2-} per 100 mL.
(a) Calculate the molar concentration of each of these species.
(b) Calculate the pNa and pSO_4 of the solution.

2-14. Calculate the p-value of each of the indicated ions in the following solutions:
* (a) Na^+, SO_4^{2-}, and OH^- in a solution that was 0.115 M in Na_2SO_4 and 0.210 M in NaOH.
(b) Ba^{2+}, Mn^{2+}, and Cl^- in a solution that was 1.40×10^{-4} M in $BaCl_2$ and 1.60×10^{-3} M in $MnCl_2$.
* (c) H_3O^+, Cl^-, and Zn^{2+} in a solution that was 2.00 M in HCl and 0.350 M in $ZnCl_2$.
(d) Cu^{2+}, Zn^{2+}, and NO_3^- in a solution that was 0.150 M in $Cu(NO_3)_2$ and 0.360 M in $Zn(NO_3)_2$.
* (e) H_3O^+, Ba^{2+}, and ClO_4^- in a solution that was 6.04×10^{-4} M in $Ba(ClO_4)_2$ and 1.76×10^{-5} M in $HClO_4$.
(f) K^+, OH^-, and $Fe(CN)_6^{4-}$ in a solution that was 2.75×10^{-6} M in $K_4Fe(CN)_6$ and 4.64×10^{-5} M in KOH.

2-15. Calculate the molar hydronium ion concentration of a solution having a pH of
* (a) 11.97.
(b) 1.63.
* (c) 7.62.
(d) 10.16.
* (e) 4.44.
(f) 9.76.
* (g) -0.96.
(h) -0.271.

* 2-16. Calculate the molar analytical concentration of K_2SO_4 and the molar equilibrium concentration of K^+ in a 9.00% (w/w) K_2SO_4 solution that has a specific gravity of 1.07.

2-17. Calculate the molar analytical concentration of Na_2SO_4 and the molar equilibrium concentration of Na^+ in a 20.0% (w/w) Na_2SO_4 solution that has a specific gravity of 1.19.

* 2-18. Describe the preparation of
(a) 500 mL of 6.0% (w/v) aqueous ethylene glycol.
(b) 500 g of 6.0% (w/w) aqueous ethylene glycol.
(c) 500 mL of 6.0% (v/v) aqueous ethylene glycol.

2-19. Describe the preparation of
(a) 20.0 L of 24.0% (w/v) aqueous methanol.
(b) 20.0 kg of 24.0% (w/w) aqueous methanol.
(c) 20.0 L of 24.0% (v/v) aqueous methanol.

* 2-20. Describe the preparation of
(a) 6.00 L of a 0.0215 M solution of I_2 in CH_3OH.
(b) 600 mL of 0.0850 M K^+ from pure K_2SO_4.
(c) 600 mL of 0.200 M $K_2Cr_2O_7$ from a 1.25 M solution of the salt.
(d) 600 mL of 5.00% (w/v) $AgNO_3$ from a 1.25 M solution of $AgNO_3$.
(e) 6.00 L of a solution containing 5.00 ppm of chloride from pure $BaCl_2 \cdot 2H_2O$.

2-21. Describe the preparation of
(a) 750 mL of 0.200 M $(NH_4)_2C_2O_4$ from solid $(NH_4)_2C_2O_4 \cdot H_2O$.
(b) 1.67 L of 0.115 M K^+ from solid $K_3Fe(CN)_6$.
(c) 250 mL of 0.200 M $BaCl_2$ from a solution that was 0.330 M in the salt.
(d) 250 mL of a 0.200% solution (w/v) of $K_2Cr_2O_7$ from a 0.137 M solution of $K_2Cr_2O_7$.
(e) 2.00 L of a solution containing 2.5 ppm of NH_4^+ from a solution that was 2.00×10^{-3} M in $(NH_4)_2SO_4$.

* 2-22. Describe the preparation of
(a) 1.50 L of 0.165 M $K_2Cr_2O_7$ from the solid.
(b) 50.0 L of a solution that is 0.100 M in KCl from the solid.

(c) 500 mL of 1.00% (w/v) $Zn(NO_3)_2$ from a 0.175 M solution of $Zn(NO_3)_2$.

(d) 20.0 L of a solution that is 0.0113 M in Br^- from a 0.317 M solution of $AlBr_3$.

2-23. Describe the preparation of

(a) 2.00 L of a solution that is 0.150 M in $K_2Cr_2O_7$ from solid $K_2Cr_2O_7$.

(b) 500 mL of a solution that is 0.175 M in K^+ from solid $K_2Cr_2O_7$.

(c) 500 mL of a solution that is 0.0300 M in K^+ from a 0.700 M solution of $K_4Fe(CN)_6$.

(d) 500 mL of a solution that is 0.0400 M in $BaCl_2$ from a 1.04 M solution of $BaCl_2 \cdot 2H_2O$.

* 2-24. Concentrated HNO_3 has a specific gravity of 1.42 and is 69% (w/w) HNO_3.

(a) How many grams of HNO_3 are contained in 1.40 L of this reagent?

(b) Describe the preparation of 6.00 L of a solution having a concentration of about 0.15 M in HNO_3 from the concentrated reagent.

2-25. Concentrated HCl has a specific gravity of 1.185 and is 36.5% (w/w) in HCl.

(a) How many milliliters of HCl gas (measured at STP) are contained in 4.50 L of the reagent?

(b) Describe how 1.0 L of approximately 0.050 M HCl should be prepared from the concentrated reagent.

* 2-26. Describe the preparation of 500 mL of 6.0 M H_3PO_4 from the commercial reagent that is 85% (w/w) H_3PO_4 and has a specific gravity of 1.69.

2-27. Describe the preparation of 750 mL of 3.0 M H_2SO_4 from the concentrated reagent that is 95% (w/w) H_2SO_4 and has a specific gravity of 1.84.

2-28. In neutral solution, silver ion reacts with carbonate ions to form solid Ag_2CO_3 (fw = 276).

(a) How many grams of Na_2CO_3 are required to react completely with 1.75 g $AgNO_3$?

(b) How many grams of $AgNO_3$ are required to react with 200 mL of 0.150 M Na_2CO_3?

(c) How many grams of Ag_2CO_3 are formed when 5.00 g Na_2CO_3 are mixed with 2.45 g $AgNO_3$?

(d) How many grams of Ag_2CO_3 are formed when 5.00 g Na_2CO_3 are mixed with 30.0 g of $AgNO_3$?

(e) How many grams of Ag_2CO_3 are formed when 200 mL of 0.200 M Na_2CO_3 are mixed with 300 mL of 0.300 M $AgNO_3$?

* 2-29. Lanthanum ion reacts with iodate ion to form a slightly soluble precipitate having the formula $La(IO_3)_3$ and a gram formula weight of 664.

(a) How many grams of KIO_3 are required to react completely with 4.44 g of $La(NO_3)_3$ (fw = 325)?

(b) What weight of KIO_3 will react with 1.55 mol of $La(NO_3)_3$?

(c) How many grams of $La(IO_3)_3$ are formed when 2.00 g of $La(NO_3)_3$ are mixed with 3.00 g of KIO_3?

(d) How many grams of $La(NO_3)_3$ are consumed when 2.00 g of $La(NO_3)_3$ are mixed with 3.00 g of KIO_3?

(e) How many grams of $La(IO_3)_3$ are formed when 25.0 mL of 0.300 M $NaIO_3$ are mixed with 15.0 mL of 0.150 M $La(NO_3)_3$?

2-30. Silver ion reacts with arsenite ion to give the sparingly solute Ag_3AsO_3 (fw = 447).

(a) What weight of $AgNO_3$ will react completely with 2.00 g of $Mg_3(AsO_3)_2$ (fw = 319)?

(b) What weight of Ag_3AsO_3 can be formed from 0.500 mol of $Mg_3(AsO_3)_2$?

(c) What weight of Ag_3AsO_3 will form when 1.05 g of $AgNO_3$ are mixed with 3.33 g of $Mg_3(AsO_3)_2$?

(d) What weight of Ag_3AsO_3 will form when 1.50 L of 0.285 M $AgNO_3$ are mixed with 0.400 L of 0.165 M $Mg_3(AsO_3)_2$?

(e) What volume of 0.300 M $AgNO_3$ is needed to react completely with 4.12 g of $Mg_3(AsO_3)_2$?

2-31. Lead ion reacts with phosphate ion to form the sparingly soluble $Pb_3(PO_4)_2$ (fw = 812).
 (a) How many grams of $Pb(NO_3)_2$ are required to react completely with 0.216 g of Na_3PO_4 (fw = 164)?
 (b) How many grams of Na_3PO_4 are required to react with 2.05 L of 0.0125 M $Pb(NO_3)_2$?
 (c) How many grams of $Pb_3(PO_4)_2$ are formed upon mixing 1.75 g of $Pb(NO_3)_2$ with 13.2 g of Na_3PO_4?
 (d) How many grams of $Pb_3(PO_4)_2$ are formed when 0.512 g of Na_3PO_4 are added to 2.25 L of 0.00750 M $Pb(NO_3)_2$?
 (e) How many grams of $Pb_3(PO_4)_2$ are formed upon mixing 11.2 g of $Pb(NO_3)_2$ and 3.15 g Na_3PO_4?

* 2-32. How much of the substance in the second column is needed to react completely with the indicated amount of substance in the first column?
 (a) 12.0 mmol H_2SO_4 (a) g $Ba(OH)_2$
 (b) 6.50 mmol KOH   (b) mL of 0.111 M H_2SO_4
 (c) 0.700 g $Ba(OH)_2$ (c) mL of 0.250 M HCl
 (d) 18.6 mL of 0.300 M $BaCl_2$ (d) mL of 0.225 M $AgNO_3$
 (e) 5.00 mL HCl, sp gr = 1.12,   (e) mL of 0.110 M $Ba(OH)_2$
     % HCl (w/w) = 24.0

2-33. How much of the substance in the second column is needed to react completely with the indicated amount of substance in the first column?
 (a) 19.6 mmol HCl   (a) g $Ba(OH)_2$
 (b) 19.6 mmol $Ba(OH)_2$ (b) mL 0.500 M HCl
 (c) 0.671 g $Pb(NO_3)_2$ (c) g KCl
 (d) 10.2 mL $HClO_4$, sp gr = 1.60, (d) g $Zn(OH)_2$
     % $HClO_4$ (w/w) = 70.0
 (e) 63.0 mL of 0.500 M H_2SO_4 (e) mL of 0.429 M KOH

* 2-34. How many grams of $BaCl_2 \cdot 2H_2O$ are required to precipitate all of the SO_4^{2-} (as $BaSO_4$) in 500 mL of 0.122 M H_2SO_4?

2-35. How many grams of silver nitrate are required to precipitate all the chromium from 150 mL of a solution that is 0.0450 M in $K_2Cr_2O_7$?

$$K_2Cr_2O_7 + 4Ag^+ + H_2O \longrightarrow 2Ag_2CrO_4 + 2H^+ + 2K^+$$

* 2-36. When strontium oxalate is ignited at 500°C, the following reaction occurs.

$$SrC_2O_4(s) \longrightarrow SrCO_3(s) + CO$$

 (a) What weight of $SrCO_3$ is formed by igniting 2.14 g of SrC_2O_4?
 (b) What would be the weight loss upon ignition of 2.75 g of SrC_2O_4?

2-37. When $Ca(HCO_3)_2$ is ignited, the following reaction occurs:

$$Ca(HCO_3)_2(s) \longrightarrow CaCO_3(s) + H_2O + CO_2$$

 (a) How many grams of CO_2 are formed from ignition of 2.50 g of $Ca(HCO_3)_2$?
 (b) What loss in weight accompanies ignition of 1.77 g of $Ca(HCO_3)_2$?

2-38. Generate solubility-product expressions for
 * (a) $BaSO_4$.   (f) PbOHBr (products: Pb^{2+}, OH^-, Br^-).
   (b) Ag_2SO_3.   * (g) TlN_3 (products: Tl^+, N_3^-).
 * (c) BiI_3.   (h) $Pb_2Fe(CN)_6$.
   (d) $Ca_3(PO_4)_2$.   * (i) Hg_2Cl_2 (products: Hg_2^{2+}, Cl^-).
 * (e) $Zn(OH)_2$.   (j) $ZnHg(SCN)_4$ (products: Zn^{2+}, $Hg(SCN)_4^{2-}$).

* 2-39. Express the molar concentration of each cation in Problem 2-38 in water in terms of the equilibrium solubility, S, of the parent compound and also in terms of the solubility-product constant.

* 2-40. Express the molar concentration of each anion in Problem 2-38 in terms of the equilibrium solubility, S, of the parent compound.
* 2-41. Calculate the molar solubility of silver bromide in
    (a) water.
    (b) 0.0120 M AgNO$_3$.
    (c) 0.0250 M KBr.
  2-42. Calculate the molar solubility of lead iodide in
    (a) water.
    (b) 0.0300 M Pb(NO$_3$)$_2$.
    (c) 0.0200 M MgI$_2$.
* 2-43. Calculate the molar solubility of magnesium hydroxide in
    (a) water.
    (b) 0.0250 M Mg(NO$_3$)$_2$.
    (c) 0.0250 M KOH.
  2-44. Calculate the molar solubility of lanthanum iodate in
    (a) water.
    (b) 0.0500 M La(NO$_3$)$_3$.
    (c) 0.0500 M KIO$_3$.
  2-45. Write net-ionic equations and equilibrium-constant expressions for
    * (a) the basic dissociation of hypochlorite ion, OCl$^-$.
      (b) the acidic dissociation of the methylammonium ion, CH$_3$NH$_3^+$.
    * (c) the basic dissociation of the phenolate ion, C$_6$H$_5$O$^-$.
      (d) the basic dissociation of phosphate ion, PO$_4^{3-}$.
    * (e) the acidic dissociation of hydrogen carbonate ion, HCO$_3^-$.
      (f) the basic dissociation of hydrogen sulfide ion, HS$^-$.
  2-46. Write net-ionic equations and equilibrium-constant expressions for
    * (a) the acidic dissociation of hypochlorous acid, HOCl.
      (b) the basic dissociation of cyanide ion, CN$^-$.
    * (c) the acidic dissociation of hydrogen phosphate ion, HPO$_4^{2-}$.
      (d) the basic dissociation of hydrogen phosphate ion, HPO$_4^{2-}$.
    * (e) the basic dissociation of hydrogen carbonate ion, HCO$_3^-$.
      (f) the acidic dissociation of hydrogen sulfide ion, HS$^-$.
*2-47. Use the tables in the Appendix to calculate numerical values for the equilibrium constants in Problem 2-45.
*2-48. Use the tables in the Appendix to calculate numerical values for the equilibrium constants in Problem 2-46.
  2-49. Calculate the hydronium ion concentration of a solution that is 0.0800 M with respect to
    * (a) hydrochloric acid.   (d) hypochlorous acid, HOCl.
      (b) hydrazoic acid, HN$_3$.   * (e) iodic acid, HIO$_3$.
    * (c) hydrogen cyanide, HCN.   (f) hydroxylammonium ion, HONH$_3^+$.
  2-50. Calculate the hydroxide ion concentration of a solution that is 0.0500 M with respect to
    * (a) sodium hydroxide, NaOH.   (d) cyanide ion, CN$^-$.
      (b) barium hydroxide, Ba(OH)$_2$.   * (e) phenolate ion, C$_6$H$_5$O$^-$.
    * (c) ammonia, NH$_3$.   (f) trimethylamine, (CH$_3$)$_3$N.
  2-51. Write net-ionic equations and equilibrium-constant expressions for
    * (a) the formation of CuI$_2^-$ from CuI(s) and I$^-$.
      (b) the formation of Ni(CN)$_2$(s) from Ni^{2+} and Ni(CN)$_4^{2-}$.
    * (c) the formation of I$_3^-$ from I$_2$(s) and I$^-$.
  2-52. Write equilibrium-constant expressions for the following reactions:
    * (a) Cu^{2+} + 4NH$_3$ $\rightleftharpoons$ Cu(NH$_3$)$_4^{2+}$
      (b) 2CrO$_4^{2-}$ + 2H$^+$ $\rightleftharpoons$ Cr$_2$O$_7^{2-}$ + H$_2$O

       * (c) $Zn(s) + 2H^+ \rightleftharpoons Zn^{2+} + H_2(g)$

         (d) $2CH_3OH \rightleftharpoons CH_3OH_2^+ + CH_3O^-$

         (e) $MnO_4^- + 5Fe^{2+} + 8H^+ \rightleftharpoons Mn^{2+} + 5Fe^{3+} + 4H_2O$

2-53. Write net-ionic equations and equilibrium-constant expressions for

       * (a) dissociation of sulfurous acid, H_2SO_3.

         (b) dissociation of phosphoric acid, H_3PO_4.

       * (c) formation of HgI_4^{2-} from Hg^{2+} and I^-.

         (d) formation of CuI_2^- from Cu^+ and I^-.

       * (e) formation of $Ni(CN)_4^{2-}$ from Ni^{2+} and CN^-.

the evaluation of analytical data

3

Every physical measurement is subject to a degree of uncertainty that, at best, can only be decreased to an acceptable level. The determination of the magnitude of this uncertainty is often difficult and requires additional effort, ingenuity, and good judgment on the part of the scientist. Nevertheless, it is a task that cannot be neglected because an analysis of totally unknown reliability is worthless. On the other hand, a result that is not highly accurate may be of great use provided that the limits of probable error affecting it can be set with a reasonable degree of certainty. Unfortunately, there exists no simple, generally applicable means by which the quality of an experimental result can be assessed with absolute certainty; indeed, the work expended in evaluating the reliability of data is frequently comparable to the effort that went into obtaining them. This effort may involve a study of the literature to profit from the experience of others, the calibration of equipment, additional experiments specifically designed to provide clues to possible errors, and statistical analysis of the data. It should be recognized, however, that none of these measures is infallible. Ultimately, the scientist can only make a *judgment* as to the probable accuracy of a measurement; with experience, judgments of this kind tend to become harsher and less optimistic.

A direct relationship exists between the accuracy of an analytical measurement and the time and effort expended in its acquisition. A tenfold increase in reliability may require hours, days, or perhaps weeks of added labor. As a first step, then, the experienced scientist establishes how reliable the results of an analysis must be; this consideration will determine the amount of time and effort that must be invested in their acquisition. Careful thought at the outset of an investigation often provides major savings in time and effort. *It cannot be too strongly emphasized that a scientist cannot afford to waste time in the indiscriminate pursuit of the ultimate in accuracy when such is not needed.*

In this chapter, we consider the types of errors encountered in analyses, methods for their recognition, and techniques for estimating and reporting their magnitude.

3A DEFINITION OF TERMS

The chemist generally repeats the analysis of a given sample two to five times. The individual results for such a set of replicate measurements will seldom be exactly the same; it thus becomes necessary to select a central "best" value for the set. Intuitively, the added effort of replication can be justified in two ways. First, the central value of the set ought to be more reliable than any of the individual results; second, the variations among the results ought to provide some measure of reliability in the "best" value that is chosen.

Either of two quantities, the *mean* or the *median*, may serve as the central value for a set of measurements.

3A-1 THE MEAN AND MEDIAN

The *mean, arithmetic mean,* and *average* ($\bar{x}$) are synonymous terms for the numerical value obtained by dividing the sum of a set of replicate measurements by the number of individual results in the set.

The *median* of a set is that result about which all others are equally distributed, half being numerically greater and half numerically smaller. If the set consists of an odd number of measurements, selection of the median may be made directly; for a set containing an even number of measurements, the average of the central pair is taken.

EXAMPLE 3-1. Calculate the mean and median for 10.06, 10.20, 10.08, 10.10.

$$\text{mean} = \bar{x} = \frac{10.06 + 10.20 + 10.08 + 10.10}{4} = 10.11$$

Because the set contains an even number of measurements, the median is the average of the middle pair:

$$\text{median} = \frac{10.08 + 10.10}{2} = 10.09$$

Ideally, the mean and median should be numerically identical; more often than not, however, this condition is not realized, particularly when the number of measurements in the set is small.

3A-2 PRECISION

The term *precision* is used to describe the reproducibility of results. It can be defined as the agreement between the numerical values of two or more measurements that have been made *in an identical fashion*. Several methods exist for expressing the precision of data.

Absolute methods for expressing precision. The *deviation from the mean* ($x_i - \bar{x}$) is a common method for describing precision and is simply the numerical difference, *without regard to sign*, between an experimental value and the mean for the set of data that includes the value. To illustrate, suppose that a chloride analysis has yielded the following results:

Sample	Percent Chloride	Deviation From Mean $\lvert x_i - \bar{x} \rvert$	Deviation From Median
x_1	24.39	0.077	0.03
x_2	24.19	0.123	0.17
x_3	24.36	0.047	0.00
	3⟌72.94	3⟌0.247	3⟌0.20
$\bar{x} =$	24.313 = 24.31	av = 0.082 = 0.08	av = 0.067 = 0.07

$$w = x_{\max} - x_{\min} = 24.39 - 24.19 = 0.20$$

The mean value for the data is 24.31% Cl; the deviation of the second result from the mean is 0.12% Cl. The average deviation from the mean is 0.08% Cl.

Precision can also be reported in terms of *deviation from the median*. In the preceding example, deviations from 24.36 would be recorded, as shown in the last column of the table.

The *spread* or *range* (w) in a set of data is the numerical difference between the highest and lowest result; it is also a measure of precision. In the previous example, the spread is 0.20% chloride.

The most important measures of precision are the *standard deviation* and the *variance*. These terms will be defined in a later section of this chapter.

Relative methods for expressing precision. We have thus far calculated precision in absolute terms. It is often more informative, however, to indicate the precision relative to the mean (or the median) in terms of percentage or as parts per thousand (ppt). For example, for sample x_1 in the earlier example,

$$\text{relative deviation from mean} = \frac{0.077 \times 100}{24.31} = 0.32 = 0.3\%$$

Similarly, the average deviation of the set from the median can be expressed as

$$\text{relative average deviation from median} = \frac{0.067 \times 1000}{24.36} = 2.8 = 3 \text{ ppt}$$

3A-3 ACCURACY

The term *accuracy* denotes the nearness of a measurement to its accepted value and is expressed in terms of *error*. Note the fundamental difference between this term and precision. Accuracy involves a comparison with respect to a true or accepted value; precision compares a result with other measurements made in the same way.

The accuracy of a measurement is often described in terms of *absolute error*, E, which can be defined as the difference between the observed value x_i and the accepted value x_t.

$$E = x_i - x_t \tag{3-1}$$

The accepted value may itself be subject to considerable uncertainty; as a consequence, it is frequently difficult to arrive at a realistic estimate for the error of a measurement.

Returning to the previous example, suppose that the accepted value for the percentage of chloride in the sample is 24.36%. The absolute error of the mean is thus $24.31 - 24.36 = -0.05\%$ chloride; here, we ordinarily retain the sign of the error to indicate whether the result is high or low.

Often, a more useful quantity than the absolute error is the relative error, which is expressed as a percentage or in parts per thousand of the accepted value. Thus, for the chloride analysis we have been considering

$$\text{relative error} = -\frac{0.05 \times 100}{24.36} = -0.21 = -0.2\%$$

$$\text{relative error} = -\frac{0.05 \times 1000}{24.36} = -2.1 = -2 \text{ ppt}$$

3A-4 PRECISION AND ACCURACY OF EXPERIMENTAL DATA

The precision of a measurement is readily determined by performing replicate experiments under identical conditions. Unfortunately, an estimate of the accuracy is not equally available because this quantity requires sure knowledge of the very information that is being sought, namely, the true value. It is tempting to ascribe a direct relationship between precision and accuracy. The danger of this approach is illustrated in Figure 3-1, which summarizes the results of an analysis for nitrogen in two pure compounds by four analysts. The dots give the absolute errors of replicate measurements for each sample and each analyst. Note that analyst 1 obtained

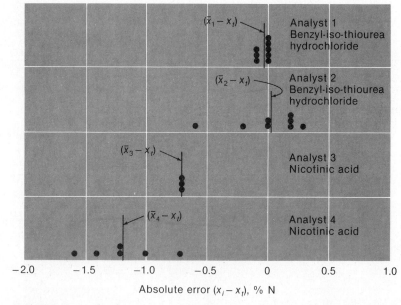

FIGURE 3-1 Absolute errors in nitrogen analyses by a micro Kjeldahl procedure. Each vertical line labeled $(\bar{x}_i - x_t)$ is the absolute deviation of the mean of the set from the true value. [Data taken from C. O. Willits and C. L. Ogg, *J. Assoc. Off. Anal. Chem.*, **32**, 561 (1949)].

relatively high precision and high accuracy. Analyst 2, on the other hand, had poor precision but good accuracy. The results from analyst 3 are of a kind that is by no means uncommon; the precision is excellent, but a significant error exists in the numerical average for the data. The scientist also encounters a situation similar to that recorded by analyst 4, in which both precision and accuracy are poor.

The behavior illustrated by Figure 3-1 can be rationalized by assuming that experimental measurements are afflicted by two general types of uncertainty and that the effect of one of these types is not revealed by the precision of the measurements.

3A-5 CLASSES OF ERRORS

The uncertainties that arise in a chemical analysis and are responsible for the behavior illustrated in Figure 3-1 may be classified into two broad categories: *determinate errors* and *indeterminate* or *random errors*. At the outset, it should be pointed out that it is frequently difficult or impossible to be certain of the category in which a given error belongs; indeed, assignment to one or the other may be largely a subjective judgment. Nevertheless, the classification is useful for discussing analytical errors.

Determinate errors.   Determinate errors (also called *systematic errors*) are those that have a definite value and an assignable cause; in principle (but not always in practice) the analyst can measure and account for these errors. A determinate error is often unidirectional; that is, it will cause all of a series of replicate analyses to be either high or low. Thus, the last two sets of data for nicotinic acid shown in Figure 3-1 appear to be affected by a negative determinate error, and the probable source of this error can be identified. The first step in this analysis requires the oxidation of the sample with concentrated sulfuric acid, a process that ordinarily converts the nitrogen present to ammonium sulfate. It has been found, however, that compounds containing a pyridine ring (such as nicotinic acid) are incompletely oxidized unless special precautions are taken; low results are the consequence. It is highly likely that the negative errors $[(\bar{x}_3 - x_t)$ and $(\bar{x}_4 - x_t)]$ shown in Figure 3-1 are determinate and attributable to this incomplete oxidation.

Indeterminate errors.   Indeterminate errors result from extending a system of measurements to its maximum. The sources of these errors cannot be positively identified, and the magnitudes of individual indeterminate errors are not measurable. The most important consequence of indeterminate errors is that they cause the data from replicate measurements to fluctuate in a *random* manner, in some instances causing high results and in others low.

The scatter of the individual errors about mean values $(x_n - \bar{x})$ in Figure 3-1 is a direct indication of indeterminate-type uncertainties. Larger indeterminate errors appear to be associated with the work of analysts 2 and 4 than with that of analysts 1 and 3.

3A-6 SOURCES OF ERRORS

It is not possible to list all conceivable causes of determinate and indeterminate errors. We can, however, recognize that limitations to both precision and accuracy can be traced to three general sources: (1) *instrument uncertainties* that are attributable to imperfections in measuring devices, (2) *method uncertainties* that arise from nonideal chemical or physical behavior of analytical systems, and (3) *personal uncertainties* that are caused by physical or psychological limitations of experimenters. Examples of these three sources appear in the sections that follow.

3B DETECTION AND CORRECTION OF DETERMINATE ERRORS

3B-1 TYPES OF DETERMINATE ERRORS

Instrument errors. All measuring devices are potential sources of determinate errors. For example, pipets, burets, and volumetric flasks frequently deliver or contain volumes slightly different from those indicated by their graduations. These differences may have such origins as use of the glassware at temperatures that differ significantly from the calibration temperature, distortions in the container walls due to heating while drying, errors in the original calibration, or contaminants on the inner surfaces of the containers. Many determinate errors of this type are readily eliminated by calibration.

Measuring devices powered by electricity are commonly subject to determinate errors. Examples include decreases in the voltage of battery-operated power supplies with use, increased resistance in circuits because of dirty electrical contacts, temperature effects on resistor and standard potential sources, and currents induced from 110-V power lines. Again, these errors are detectable and correctable; most are unidirectional.

Method errors. Determinate errors are often introduced from nonideal chemical or physical behavior of the reagents and reactions upon which an analysis is based. Such sources of nonideality include the slowness of some reactions, the incompleteness of others, the instability of some species, the nonspecificity of most reagents, and the possible occurrence of side reactions which interfere with the measurement process. For example, in a gravimetric analysis, the chemist is confronted with the problem of isolating the element to be determined as a solid of the greatest possible purity. If this solid is not washed sufficiently, the precipitate will be contaminated with foreign substances and have a spuriously high weight. On the other hand, the washing needed to remove these contaminants may cause weighable quantities to be lost owing to the finite solubility of the precipitate; here, a negative determinate error will result. In either event, the accuracy of the procedure is limited by a method error associated with the analysis.

A method error frequently encountered in volumetric analysis is due to the volume of reagent, in excess of theoretical, that is required to cause an indicator to undergo the color change that signals completion of the reaction. The ultimate accuracy of such an analysis is then limited by the very phenomenon that makes the determination possible.

Errors inherent in a method are frequently difficult to detect and are thus the most serious of the three types of determinate error.

Personal errors. Many measurements require personal judgments. Examples include estimating the position of a pointer between two scale divisions, the color of a solution at the end point in a volumetric analysis, the level of a liquid with respect to a graduation in a pipet, or the relative intensity of two light beams. Judgments of this type are often subject to systematic, unidirectional uncertainties. For example, one person may read a pointer consistently high, another may be slightly slow in activating a timer, and a third may be less sensitive to color changes and thus tend to employ excess reagent in a volumetric analysis. Color blindness or other physical handicaps often increase the probability of determinate personal errors.

A near-universal source of personal error is prejudice or bias. Most of us, no matter how honest, have a natural tendency to estimate scale readings in a direction that improves the precision in a set of results or causes the results to fall closer to a preconceived notion of the true value for the measurement. Number bias is another source of personal error that is widely encountered and varies considerably from person to person. The commonest bias encountered in estimating the position of a needle on a scale involves a preference for the digits 0 and 5. Also prevalent is a prejudice favoring small digits over large and even numbers over odd.

Scientists must actively fight prejudice; it does not suffice to assume that bias occurs only in others.

Gross mistakes. Errors of this type include arithmetic mistakes, transposition of numbers in recording data, reading a scale backward, reversing a sign, or using a wrong scale. Some gross errors will affect only a single result; others, such as using the wrong scale of an instrument, will affect an entire set of replicate measurements. Errors of this type are ordinarily the consequence of carelessness and can be eliminated by self-discipline. Many scientists always follow the practice of re-reading an instrument after the information has been recorded and then checking the new reading against the one that has been recorded.

3B-2 EFFECT OF DETERMINATE ERROR UPON THE RESULTS OF AN ANALYSIS

Determinate errors may be classified as being either *constant* or *proportional*. The magnitude of a constant error is independent of the size of the quantity measured. On the other hand, proportional errors increase or decrease in proportion to the size of the sample taken for analysis.

Constant errors. For any given analysis, a constant error will become more serious as the size of the quantity measured decreases. This problem is illustrated by the solubility losses that attend the washing of a precipitate.

> **EXAMPLE 3-2.** Suppose that 0.50 mg of precipitate is lost as a result of being washed with 200 mL of wash liquid. If 500 mg of precipitate are involved, the relative error due to solubility loss will be $-(0.50 \times 100/500) = -0.1\%$. Loss of the same quantity from 50 mg of precipitate will result in a relative error of -1.0%.

The amount of reagent required to bring about the color change in a volumetric analysis is another example of constant error. This volume, usually small, remains the same regardless of the total volume of reagent required for the titration. Again, the relative error will be more serious as the total volume decreases.

Clearly, one way of minimizing the effect of constant error is to use as large a sample as is consistent with the method at hand.

Proportional errors. Interfering contaminants in the sample, if not eliminated somehow, will cause a proportional error. For example, a method widely employed for the analysis of copper involves reaction of copper(II) ion with potassium iodide; the quantity of iodine produced in the reaction is then measured. Iron(III), if present, will also liberate iodine from potassium iodide. Unless steps are taken to prevent this interference, the analysis will yield erroneously high results for the percentage of copper because the iodine produced will be a measure of the sum of the copper and iron in the sample. The magnitude of this error is fixed by the *fraction* of iron contamination and will produce the same relative effect regardless of the size of sample taken for analysis. If the sample size is doubled, for example, the amount of iodine liberated by both the copper and the iron contaminant will also be doubled. Thus, the reported percentage of copper will be independent of sample size.

3B-3 DETECTION OF DETERMINATE INSTRUMENTAL AND PERSONAL ERRORS
Determinate instrumental errors are usually found and corrected by calibration procedures. Indeed, periodic recalibration of equipment is always desirable because the response of most instruments changes with time owing to wear, corrosion, or mistreatment.

Most personal errors can be minimized by care and self-discipline. Thus, most scientists develop the habit of systematically checking instrument readings, notebook entries, and calculations. Errors that result from a physical handicap can usually be avoided by a judicious choice of method, provided, of course, that the handicap is recognized.

3B-4 DETECTION OF DETERMINATE METHOD ERRORS
Determinate method errors are particularly difficult to detect. Identification and compensation for systematic errors of this type may take one or more of the courses described in the following paragraphs.

Analysis of standard samples. A method may be tested for determinate error by analysis of a synthetic sample whose overall composition is known and closely approximates that of the material for which the analysis is intended. Great care must go into the preparation of standard samples to ensure that the concentration of the constituent to be determined is known with a high degree of certainty. Unfortunately, the preparation of a sample whose composition truly resembles that of a complex natural substance is often difficult, if not impossible. Moreover, the problem is compounded by the requirement that the exact concentration of a particular constituent be known as a result of the method of preparation. These problems are frequently so imposing as to prevent the use of this approach.

Several hundred common substances that have been carefully analyzed for one or more constituents are available from the National Bureau of Standards.

These standard materials are valuable for the testing of analytical procedures for accuracy.[1]

Independent analysis. Parallel analysis of a sample by a method of established reliability that differs from the one under investigation is of particular value where samples of known purity are not available. In general, the independent method should not resemble the one under study to minimize the possibility that some common factor in the sample will have an equal effect on both methods.

Blank determinations. Constant errors affecting physical measurements can be frequently evaluated with a blank determination, in which all steps of the analysis are performed in the absence of a sample. The result is then applied as a correction to the actual measurement. Blank determinations are of particular value in exposing errors that are due to the introduction of interfering contaminants from reagents and vessels employed in an analysis. They also enable the analyst to correct titration data for the volume of reagent needed to cause an indicator to change color at the end point of a volumetric analysis.

Variation in sample size. As was demonstrated in the example on page 45, a constant error has a decreasing effect on a result as the size of the measurement increases. This fact can be helpful in detecting constant errors in a method. Here, the sample size is varied as widely as possible. In the presence of a constant error, the results will be found to increase or decrease systematically with sample size.

3C INDETERMINATE ERROR

As suggested by its name, indeterminate error arises from uncertainties in a measurement that are unknown and not controlled by the scientist. The effect of such uncertainties is to produce a *random* scatter of results for replicate measurements such as those for the four sets of data shown in Figure 3-1.

Table 3-1 illustrates the effect of indeterminate error upon the relatively simple process of calibrating a pipet. The procedure involves determining the weight of water (to the nearest milligram) delivered by the pipet. The temperature of the water must be measured to establish its density. The experimental weight can then be converted to the volume delivered by the pipet.

The data in Table 3-1 are typical of those that might be obtained by an experienced and competent worker who performs the weighings to the nearest milligram (which corresponds to 0.001 mL), with every effort being made to recognize and eliminate determinate errors. Even so, the average deviation from the mean of the 24 measurements is ±0.005 mL, and the spread is 0.023 mL. This dispersion among the data is the direct consequence of indeterminate error.

Variations among replicate results such as those in Table 3-1 can be rationalized by assuming that any measurement is affected by numerous small and individ-

---

[1] See the current edition of NBS Special Publication 260 entitled *Catalog of Standard Reference Materials*, U.S. Government Printing Office, Washington, D.C. 20402. For a description of the reference material programs of the NBS, see: J. P. Cali, *Anal. Chem.*, **48**, 802A (1976); W. W. Meinke, *Anal. Chem.*, **43**, 28A (1971); and G. A. Uriano, *ASTM Standardization News*, **7**, 8 (1979).

Table 3-1   Replicate Measurements From the Calibration of a 10-mL Pipet

Trial	Volume of Water Delivered, mL	Trial	Volume of Water Delivered, mL	Trial	Volume of Water Delivered, mL
1	9.990	9	9.988	17	9.977
2	9.993*	10	9.976	18	9.982
3	9.973	11	9.981	19	9.974
4	9.980	12	9.974	20	9.985
5	9.982	13	9.970**	21	9.987
6	9.988	14	9.989	22	9.982
7	9.985	15	9.981	23	9.979
8	9.970**	16	9.985	24	9.988

Mean volume = 9.9816 = 9.982 mL
Median volume = 9.982 mL
Average deviation from mean = 0.0052 mL = 0.005 mL
Spread = 9.993 − 9.970 = 0.023 mL
Standard deviation = 0.0065 mL = 0.006 mL

  * Maximum value
** Minimum value

ually undetectable instrument, method, and personal uncertainties caused by uncontrolled variables in the experiment. The cumulative effect of such uncertainties will be likewise variable. Ordinarily they tend to cancel one another and thus exert a minimal effect. Occasionally, however, they can act in concert to produce a relatively large positive or negative error. Sources for uncertainties in this calibration process might include such visual judgments as the liquid level with respect to the etch mark on the pipet, the mercury level in the thermometer, and the position of an indicator with respect to a scale in the balance (all personal uncertainties). Other sources include variation in the drainage time, the angle of the pipet as it drains (both method uncertainties), and temperature and, thus, volume change, resulting from the way the pipet is handled (an instrument uncertainty). Undoubtedly, numerous other uncertainties exist in addition to the ones cited. It is clear that many small and uncontrolled variables accompany even as simple an experiment as a pipet calibration. Although we are unable to detect the influence of any one of these uncertainties, their cumulative effect is an indeterminate error that accounts for the scatter of data about the mean.

3C-1 THE DISTRIBUTION OF DATA FROM REPLICATE MEASUREMENTS

In contrast to determinate errors, indeterminate errors cannot be eliminated from measurements. Furthermore, the scientist cannot ignore their existence on the basis of their small size. For example, it would probably be safe to assume that the average value of the 24 measurements in Table 3-1 is closer to the true volume delivered by the pipet than any of the individual data. Suppose, however, that only the first two measurements had been performed; the average of these two values, 9.992, differs by 0.010 mL from the mean of the 24 measurements. Note also that the average deviation of these two measurements *from their own mean* is only ±0.0015 mL. The result would be an unrealistic estimate of the indeterminate error associated with the process. Suppose further that the user of the pipet needed to deliver volumes known, let us say, to the nearest ± 0.002 mL. Here failure to recognize the true magnitude of the indeterminate error would create a totally false sense of security with respect to the performance of the pipet. If this pipet were employed for 1000 measurements, it can be shown that two to three of the transfers would, with high probability, differ from the mean of 9.982 mL by as much as 0.02 mL; more than 100 would differ by 0.01 mL or greater, despite every precaution on the part of the user.

To develop a qualitative grasp of the way small uncertainties affect the outcome of replicate measurements, let us first consider an imaginary situation in which just four uncertainties are the cause of indeterminate error. We shall specify that each of these uncertainties has an equal probability of occurring and can affect the final result in only one of two ways, namely, to cause it to be in error by plus or minus a fixed amount, U. Further, we shall stipulate that the magnitude of U is the same for each of the four uncertainties.

Table 3-2 shows all of the possible ways the four uncertainties can combine to give the indicated indeterminate errors. We note that there is only one way in which the maximum positive error of $4U$ can arise, compared with four combinations that lead to a positive error of $2U$, and six combinations that result in zero error. A similar relationship exists for negative indeterminate errors. This ratio of $6:4:1$ is a measure of the probability for an error of each size; if we made sufficient measurements, a frequency distribution of errors such as that shown in Figure 3-2a would be expected. Figure 3-2b shows the distribution for ten equal-sized uncertainties. Again, we see that the most frequent occurrence is zero error, while the maximum error of $10U$ would occur only occasionally (about once in 500 measurements).

Table 3-2   Possible Ways Four Equal-Sized Uncertainties U_1, U_2, U_3, and U_4 Can Combine

Combinations of Uncertainties	Magnitude of Indeterminate Error	Relative Frequency of Error
$+U_1 + U_2 + U_3 + U_4$	$+4U$	1
$-U_1 + U_2 + U_3 + U_4$ $+U_1 - U_2 + U_3 + U_4$ $+U_1 + U_2 - U_3 + U_4$ $+U_1 + U_2 + U_3 - U_4$	$+2U$	4
$-U_1 - U_2 + U_3 + U_4$ $+U_1 + U_2 - U_3 - U_4$ $+U_1 - U_2 + U_3 - U_4$ $-U_1 + U_2 - U_3 + U_4$ $-U_1 + U_2 + U_3 - U_4$ $+U_1 - U_2 - U_3 + U_4$	0	6
$+U_1 - U_2 - U_3 - U_4$ $-U_1 + U_2 - U_3 - U_4$ $-U_1 - U_2 + U_3 - U_4$ $-U_1 - U_2 - U_3 + U_4$	$-2U$	4
$-U_1 - U_2 - U_3 - U_4$	$-4U$	1

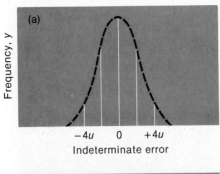

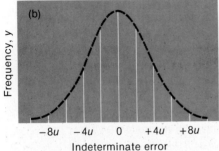

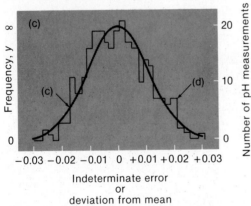

FIGURE 3-2 Theoretical distribution of indeterminate error arising from (a) 4 uncertainties, (b) 10 uncertainties, (c) a very large number of uncertainties. Curve (c) shows the normal error or Gaussian distribution. Curve (d) is an experimental distribution curve that might be obtained by plotting the deviations from the mean for about 250 replicate pH measurements against the number of times each deviation was observed.

If the foregoing arguments are extended to a very large number of uncertainties of smaller and smaller size, it can be demonstrated that the continuous distribution curve shown in Figure 3-2c will result. This bell-shaped curve is called a *Gaussian* or *normal error curve*.[2] Its properties include (1) a maximum frequency in occurrence of zero indeterminate error, (2) a symmetry about this maximum indicating that negative and positive errors occur with equal frequency, and (3) an exponential decrease in frequency as the magnitude of the error increases. Thus, a small indeterminate error will occur much more often than a very large one.

[2] In deriving the Gaussian curve, it is not necessary to assume, as we have, that the individual uncertainties have identical magnitudes.

Numerous *empirical* observations have shown that indeterminate errors in chemical analyses most commonly (but not always) distribute themselves in a manner which approaches a Gaussian distribution. For example, if the deviations from the mean of hundreds of repetitive pH measurements on a single sample were plotted against the frequency of occurrence of each deviation, a curve approximating that shown in Figure 3-2d would be expected to result.

The frequent experimental observation of Gaussian behavior lends credibility to the idea that the indeterminate error observed in analytical measurements can be traced to the accumulation of a large number of small, independent, and uncontrolled uncertainties. Equally important, the Gaussian distribution of most analytical data permits the use of statistical techniques to estimate the limits of indeterminate error from the precision of such data.

3C-2 CLASSICAL STATISTICS

Statistics permits a mathematical description of random processes such as the effect of indeterminate error on the results of a chemical analysis. It is important to realize, however, that the techniques of classical statistics apply exactly to an *infinite* number of observations only. When these techniques are applied to the two to five replicate analyses that the chemist can afford to make, conclusions as to the probable indeterminate error can be seriously incorrect and misleadingly optimistic. It is for this reason that modification of the classical techniques is necessary. Before considering these practical modifications, however, it is worthwhile to describe briefly some important relationships of classical statistics.

Properties of the normal error curve. The upper two curves of Figure 3-3 are normal error curves for two different analytical methods. The top-most curve represents data from the more precise of the two methods inasmuch as the results are distributed more closely about the central value.

As shown in Figure 3-3, normal error curves can be plotted in three different ways. In each, the ordinate is the frequency of occurrence y for each value of the abscissa. The observed values x of the measurement are plotted as abscissa a; here, the central value is the mean, which is symbolized by μ. Abscissa b consists of individual deviations from the mean, $x - \mu$; here, the most frequently occurring deviation has a value of zero. We shall consider the third type of plot, shown by abscissa c, presently.

It is important to emphasize that the curves under discussion are idealized because they represent the theoretical distribution of experimental results to be expected as the number of analyses involved approaches infinity. For a physically realizable set of results, a discontinuous distribution such as shown in Figure 3-2d would be observed. Classical statistics is based on curves such as those shown in Figure 3-3 rather than on curves such as Figure 3-2d.

The distribution data in Figure 3-3 can be described mathematically in terms of just three parameters, as shown by the expression

$$y = \frac{e^{-(x-\mu)^2/2\sigma^2}}{\sigma\sqrt{2\pi}} \tag{3-2}$$

In this equation, x represents values of individual measurements, and μ is the arithmetic mean for an infinite number of such measurements. The quantity $(x - \mu)$ is

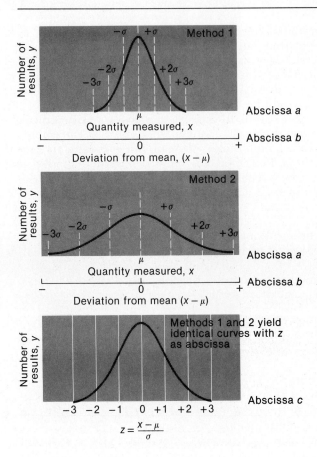

FIGURE 3-3 Normal error curves. The curves are for the measurement of the same quantity by two methods. Method 1 is more reliable; thus σ is smaller. Note the three types of abscissa. Abscissa a shows the measured quantity x with the maximum at μ. Abscissa b shows the deviation from the mean with the maximum at 0. Abscissa c shows z from Equation 3-2. Abscissa c reduces the two curves to a single one.

thus the deviation from the mean; y is the frequency of occurrence for each value of $(x - \mu)$. The symbol π has its usual meaning, and e is the base for Napierian logarithms, 2.718. . . . The parameter σ is called the *standard deviation* and is a constant that has a unique value for any set containing a large number of measurements. The breadth of the normal error curve is directly related to σ.

The exponent in Equation 3-2 can be simplified by introducing the variable

$$z = \frac{x - \mu}{\sigma} \tag{3-3}$$

Equation 3-2 can then be written as

$$y = \frac{e^{-z^2/2}}{\sigma\sqrt{2\pi}} \tag{3-4}$$

The quantity z gives the deviation from the mean in units of standard deviations. As demonstrated by abscissa c in Figure 3-3, the substitution of z produces a single curve for all values of σ.

The standard deviation.   Equation 3-2 indicates that a unique distribution curve exists for each value of the standard deviation. Regardless of the size of σ, however, it can be shown that 68.3% of the area beneath the curve lies within one standard deviation ($\pm1\sigma$) of the mean, μ. Thus 68.3% of the values lie within these boundaries. Approximately 95.5% of all values will be within $\pm2\sigma$; 99.7% will be within $\pm3\sigma$. Values of $(x - \mu)$ corresponding to $\pm1\sigma$, $\pm2\sigma$, and $\pm3\sigma$ are indicated by broken vertical lines in the upper curves of Figure 3-3. For the bottom curve, the units for z shown on the abscissa are $\pm\sigma$.

These properties of the normal error curve are useful because they permit statements to be made about the probable magnitude of the indeterminate error in a given measurement *provided the standard deviation of the method of measurement is known.* Thus, if σ were available, one could say that the chances are 68.3 out of 100 that the indeterminate error associated with any given single measurement is smaller than $\pm1\sigma$, that the chances are 95.5 out of 100 that the error is less than $\pm2\sigma$, and so forth. Clearly, the standard deviation for a method of measurement is a useful quantity for estimating and reporting the probable size of indeterminate errors.

For a very large set of data, the standard deviation is given by

$$\sigma = \sqrt{\frac{\sum_{i=1}^{N}(x_i - \mu)^2}{N}} \tag{3-5}$$

Here, the sum of the squares of the individual deviations from the mean $(x_i - \mu)$ is divided by the total number of measurements in the set, N. Extraction of the square root of this quotient gives σ.

Another precision term widely employed by statisticians is the *variance* which is equal to σ^2. Most experimental scientists prefer to employ σ rather than σ^2 because the units of the standard deviation are the same as those of the quantity measured. The variance, on the other hand, has the advantage of additivity. That is, if several independent causes of variation exist in a system, the total variance σ_T^2 is the sum of the individual variances. Thus,

$$\sigma_T^2 = \sigma_1^2 + \sigma_2^2 + \cdots \sigma_n^2$$

where σ_1^2, σ_2^2, and σ_n^2 are the individual variances.

3C-3 APPLICATIONS OF STATISTICS TO SMALL SETS OF DATA

It has been found that direct application of classical statistics to a small number of replicate measurements (2 to 20 results) often leads to false conclusions regarding the probable magnitude of the indeterminate error. Fortunately, modifications of the relationships have been developed to permit valid statements about the random error associated with as few as two or three values.

Equations 3-2 and 3-5 are not directly applicable to a small number of replicate measurements because μ, the mean value of an infinitely large number of measurements (and the true value in the absence of determinate error), is never known. In its stead, we are forced to employ $\bar{x}$, the mean of a small number of measurements. More often than not, $\bar{x}$ will differ somewhat from μ. This difference is, of course, the result of the indeterminate error whose probable magnitude we are trying to assess. It is important to note that *any error in $\bar{x}$ causes a corresponding error in σ* (Equation 3-5). Thus, with a small set of data, not only is the mean $\bar{x}$

likely to differ from μ but, *equally important, the estimate of the standard deviation also may be misleading.* In short, we have *two* uncertainties to cope with, the one residing in the mean and the other in the standard deviation.

The effect of a reduced number of measurements on the certainty of the mean and the standard deviation is illustrated in Table 3-3. Columns 3 and 4 of entry 1 show the mean and the standard deviation for the 24 pipet calibration data from Table 3-1; the standard deviation 0.0063 was obtained by means of Equation 3-5.

The second set of data (entry 2) was derived by dividing the data in Table 3-1 into two subsets, one consisting of measurements 1 through 12 and the other 13 through 24; the mean and the standard deviation for each subset are also tabulated. In entry 3, the data from Table 3-1 were divided in eight subsets, each containing three of the calibration data.

It is immediately evident that the means for the subsets containing three data fluctuate considerably but the mean of the means, as might be expected, is identical to the mean of the entire set. The spread of the means of the eight subsets is 0.008 mL. That is, if the data collected had consisted of only the three measurements labeled 10–12, the measured volume of the pipet would have been 9.977 mL; in contrast, if by accident the data had consisted of measurements 1–3, the volume would have been recorded as 9.985. Note, on the other hand, that the difference between the means of the two subsets of 12 (entry 2) is zero.

Column 4 illustrates two important points. First, considerable variation in the precision estimate develops as the number of measurements in a subset decreases. Thus, for the two subsets of 12, the standard deviations differ by 0.0014 mL while for the eight subsets the spread in standard deviation is

Table 3-3 Effect of Set Size on Standard Deviation
(Note that the data were all derived from Table 3-1)

Entry	Trial Number, Table 3-1	Mean, mL	Standard Deviation (σ) Calculated With Equation 3-5, mL	Standard Deviation (s) Calculated With Equation 3-6, mL
1	1–24	9.982	0.0063	0.0065
2	1–12	9.982	0.0070	0.0074
	13–24	9.982	0.0056	0.0058
	mean	9.982	0.0063	0.0066
	spread	0.000	0.0014	0.0016
3	1–3	9.985**	0.0088**	0.0108**
	4–6	9.983	0.0034	0.0042
	7–9	9.981	0.0079	0.0096
	10–12	9.977*	0.0029*	0.0036*
	13–15	9.980	0.0078	0.0095
	16–18	9.981	0.0033	0.0041
	19–21	9.982	0.0057	0.0070
	22–24	9.983	0.0037	0.0046
	mean	9.982	0.0054	0.0067
	spread	0.008	0.0059	0.0072

** Maximum value
* Minimum value

0.0059 mL with the largest being about 0.009 mL and the smallest only 0.003. Equally important however, the standard deviations from the small subsets have a *negative bias*. Thus, the mean of the standard deviations for the eight subsets is 0.0054 compared with 0.0063 for the entire set.

The negative bias in σ for small sets of data is a general phenomenon and is attributable to the fact that both a mean and a standard deviation must be extracted from the same small set. It can be shown that this bias can be largely eliminated by substituting the *number of degrees of freedom* $(N - 1)$ for N in Equation 3-5. That is, we define the standard deviation for a small number of measurements as

$$s = \sqrt{\frac{\sum_{i=1}^{N}(x_i - \bar{x})^2}{N - 1}} \tag{3-6}$$

Note that Equation 3-6 differs from Equation 3-5 in two regards. First, the denominator is now $(N - 1)$. Second, $\bar{x}$, the measured mean for the small set, replaces the true but unknown mean μ. To emphasize that the resulting standard deviation is but an approximation of the true value, it is given the symbol s rather than σ.

The rationale for the use of $(N - 1)$ in Equation 3-6 is as follows. When μ is unknown, we calculate two quantities, $\bar{x}$ and s, from our set of replicate data. The need to establish the mean $\bar{x}$ from the data removes one degree of freedom. That is, if their signs are retained, the individual deviations from $\bar{x}$ must total zero; once $(N - 1)$ deviations have been established, the final one is necessarily known. Thus, only $(N - 1)$ deviations provide independent measures of the precision for the set.

The fourth column of Table 3-3 illustrates that the negative bias in the standard deviation of small data sets is, on the average, eliminated by the use of Equation 3-6. Considerable variations in s are observed among the eight subsets in entry 3, but their average is essentially the same as the standard deviation for the entire set. Generally, the differences between s and σ become smaller as the number of data in a set increases; for 20 or more data, the difference is ordinarily negligible.

EXAMPLE 3-3. Calculate the standard deviation s for a subset consisting of the first five values in Table 3-1.

x_i	$\lvert x_i - \bar{x} \rvert$	$(x_i - \bar{x})^2$
9.990	6.4×10^{-3}	4.1×10^{-5}
9.993	9.4×10^{-3}	8.8×10^{-5}
9.973	10.6×10^{-3}	11.2×10^{-5}
9.980	3.6×10^{-3}	1.3×10^{-5}
9.982	1.6×10^{-3}	0.3×10^{-5}

$$\frac{5\lvert49.918}{9.9836} = \bar{x} \qquad \sum_{i=1}^{5}(x_i - \bar{x})^2 = 25.7 \times 10^{-5}$$

From Equation 3-6,

$$s = \sqrt{\frac{25.7 \times 10^{-5}}{5 - 1}} = 8.0 \times 10^{-3} = \pm0.008$$

Note that the data were not rounded until the end.

Estimation of s from w.

For a small number of replicate results (up to 15), it is also possible to estimate s from the spread w of the data by means of the relationship

Table 3-4 Factors for Calculating Standard Deviation *s* From Spread *w*
Employing *s* = *w*/*d*

Number of Samples, N	d	Number of Samples, N	d	Number of Samples, N	d
2	1.128	7	2.704	12	3.258
3	1.693	8	2.847	13	3.336
4	2.059	9	2.970	14	3.407
5	2.326	10	3.078	15	3.472
6	2.534	11	3.173		

$$s = \frac{w}{d} \tag{3-7}$$

where d is a statistical factor, which is dependent upon the number of measurements (see Table 3-4). Equation 3-7 is simpler to use but gives a somewhat less reliable estimate of s than does Equation 3-6.

It is interesting to note that the values of d in Table 3-4 are roughly equal to the square root of N. Thus, an equally rough approximation of s is given by

$$s = \frac{w}{\sqrt{N}} \tag{3-8}$$

where N is the number of measurements. It should be noted that Equation 3-8 is applicable when N is small. When N becomes large, the equation predicts that $s \to 0$.

3C-4 THE USES OF STATISTICS

Experimentalists employ statistical calculations to sharpen their judgment concerning the effects of indeterminate error.[3] Some of the applications include:

1. Definition of the interval around the mean of a set within which the true mean can be expected to be found with a certain probability.
2. Determination of the number of times a measurement should be replicated in order for the experimental mean to be included, with a certain probability, within a predetermined interval around the true mean.
3. Guidance concerning whether or not an outlying value in a set of replicate results should be retained or rejected in calculating a mean for the set.
4. Estimation of the probability that two samples analyzed by the same method are significantly different in composition; that is, whether or not a difference in experimental results is likely to be the consequence of indeterminate error or of a real composition difference or of a determinate error in analyzing one of the samples.

---

[3] For other applications of statistical calculations, see W. J. Dixon and F. J. Massey, Jr., *Introduction to Statistical Analysis*, 3d ed. New York: McGraw-Hill, 1969; H. A. Laitinen and W. E. Harris, *Chemical Analysis*, 2d ed., Chapter 26. New York: McGraw-Hill, 1975; E. B. Wilson, Jr., *An Introduction to Scientific Research*, Chapters 9 and 10. New York: McGraw-Hill, 1952.

5. Estimation of the probability that a difference in precision exists between two methods of analysis.

In the sections that follow, we shall examine these applications.

3C-5 CONFIDENCE INTERVALS

The true mean value (μ) of a measurement is a constant that must always remain unknown. With the aid of statistical theory, however, limits may be set about the experimentally determined mean ($\bar{x}$) within which we may expect to find the true mean with a given degree of probability; the limits obtained in this manner are called *confidence limits*. The interval defined by these limits is known as the *confidence interval*.

Some of the properties of the confidence interval are worthy of mention. For a given set of data, the size of the interval depends in part upon the odds for correctness desired. Clearly, for a prediction to be absolutely correct, we would have to choose an interval about the mean large enough to include all conceivable values that x_i might take. Such an interval, of course, has no predictive value. On the other hand, the interval does not need to be this large if we are willing to accept the probability of being correct 99 times in 100; it can be even smaller if 95% probability is acceptable. In short, as the probability for making a correct prediction is made less favorable, the interval included by the confidence limits becomes smaller.

The confidence interval, which is derived from the standard deviation s for the method of measurement, also depends in magnitude upon the certainty with which s is known. Often, the chemist will have reason to believe that experimental value for s is an excellent approximation of σ. In other situations, however, a considerable uncertainty in s may exist. Under these circumstances the confidence interval will necessarily be larger.

Methods for obtaining a good approximation of σ. Fluctuations in the calculated value for s decrease as the number of measurements N in Equation 3-6 increases (see Table 3-3); in fact, it is proper to assume that s and σ are, for all practical purposes, identical when N is greater than about 20. It thus becomes feasible for the chemist to obtain a good approximation of s when the method of measurement is not too time-consuming and when an adequate amount of sample is available. For example, if it were necessary to measure the pH of numerous solutions in the course of an investigation, it might prove worthwhile to evaluate s in a series of preliminary experiments. This particular measurement is simple, requiring only that a pair of rinsed and dried electrodes be immersed in the test solution; the potential across the electrodes serves as a measure of the pH. To determine s, 20 to 30 portions of a solution of fixed pH could be measured, following exactly all steps of the procedure. Normally, it would be safe to assume that the indeterminate error in this test would be the same as that in subsequent measurements and that the value of s calculated by means of Equation 3-6 would be a valid and accurate measure of the theoretical σ.

For analyses that are time-consuming, the foregoing procedure is not ordinarily practical. Here, however, precision data from a series of samples can often be pooled to provide an estimate of s which is superior to the value for any individual subset. Again, one must assume the same sources of indeterminate error among the samples. This assumption is usually valid provided the samples have similar compositions and each has been analyzed identically. To obtain a pooled estimate of s, deviations from the mean for each subset are squared; the squares for all of the subsets are then summed and divided by an appropriate number of degrees of freedom. The pooled s is obtained by extracting the square root of the quotient. One degree of freedom is lost for each subset. Thus, the number of degrees of freedom for the pooled s is equal to the total number of measurements minus the number of subsets. An example of this calculation follows.

EXAMPLE 3-4. The mercury in samples of seven fish taken from Lake Erie was determined by a method based upon absorption of radiation by elemental mercury. The results are given in the accompanying table. Calculate a standard deviation for the method, based upon the pooled precision data.

Sample Number	Number of Replications	Results, Hg Content, ppm	Mean, ppm Hg	Sum of Squares of Deviations from Mean
1	3	1.80, 1.58, 1.64	1.67	0.0258
2	4	0.96, 0.98, 1.02, 1.10	1.02	0.0116
3	2	3.13, 3.35	3.24	0.0242
4	6	2.06, 1.93, 2.12, 2.16, 1.89, 1.95	2.02	0.0611
5	4	0.57, 0.58, 0.64, 0.49	0.57	0.0114
6	5	2.35, 2.44, 2.70, 2.48, 2.44	2.48	0.0685
7	4	1.11, 1.15, 1.22, 1.04	1.13	0.0170
Number of measurements	28		Sum of squares =	0.2196

The values in columns 4 and 5 for sample 1 were calculated as follows:

(x_i)	$|(x_i - \bar{x}_1)|$	$(x_i - \bar{x}_1)^2$
1.80	0.127	0.0161
1.58	0.093	0.0086
1.64	0.033	0.0011
$3\overline{)5.02}$	Sum of squares =	0.0258

$\bar{x}_1 = 1.673 = 1.67$

The other data in column 5 were obtained similarly. Then

$$s = \sqrt{\frac{0.0258 + 0.0116 + 0.0242 + 0.0611 + 0.0114 + 0.0685 + 0.0170}{28 - 7}}$$

$$= 0.10 \text{ ppm Hg}$$

Because the number of degrees of freedom is greater than 20, this estimate of s can be considered to be a good approximation of σ.

Confidence interval when s is a good approximation of σ. As indicated earlier (p. 51), the breadth of the normal error curve is determined by σ. For any given value of σ the area under a part of the normal error curve *relative* to the total area can be related to the parameter z by means of Equation 3-4. This ratio of areas (usually expressed as a percent) is called the *confidence level*, and it measures the probability for the absolute deviation $(x - \mu)$ being less than $z\sigma$. Thus, the area under the

Table 3-5 Confidence Levels for Various Values of z

Confidence Level, %	z
50	0.67
68	1.00
80	1.29
90	1.64
95	1.96
96	2.00
99	2.58
99.7	3.00
99.9	3.29

curve encompassed by $z = \pm 1.96\sigma$ corresponds to 95% of the total area. Here, the confidence level is 95%, and we may state that 95 times out of 100 the calculated value of $(x - \mu)$ for a large number of measurements will be equal to or less than $\pm 1.96\sigma$. Table 3-5 lists confidence intervals for various values of z.

The confidence limit for a single measurement can be obtained by rearranging Equation 3-3 and remembering that z can be either plus or minus. Thus,

$$\text{confidence limit for } \mu = x \pm z\sigma \tag{3-9}$$

EXAMPLE 3-5. Calculate the 50% and the 95% confidence limits for the first entry (1.80 ppm Hg) in the example on page 58.

Here, we calculated that $s = 0.10$ ppm Hg and had sufficient data to assume $s \to \sigma$. From Table 3-5 we see that $z = \pm 0.67$ and ± 1.96 for the two confidence levels. Thus, from Equation 3-9

$$
\begin{aligned}
50\% \text{ confidence limit for } \mu &= 1.80 \pm 0.67 \times 0.10 \\
&= 1.80 \pm 0.07 \\
95\% \text{ confidence limit for } \mu &= 1.80 \pm 1.96 \times 0.10 \\
&= 1.80 \pm 0.20
\end{aligned}
$$

The chances are 50 in 100 that μ, the true mean (and in the absence of determinate error the true value), will be in the interval between 1.73 and 1.87 ppm Hg; there is a 95% chance that it will be in the interval between 1.60 and 2.00 ppm Hg.

Equation 3-9 applies to the result of a single measurement. It can be shown that the confidence interval is decreased by $\sqrt{N}$ for the average of N replicate measurements. Thus, a more general form of Equation 3-9 is

$$\text{confidence limit for } \mu = \bar{x} \pm \frac{z\sigma}{\sqrt{N}} \tag{3-10}$$

EXAMPLE 3-6. Calculate the 50% and the 95% confidence limits for the mean value (1.67 ppm Hg) for sample 1 in the example on page 58. Again, $s \to \sigma = 0.10$.

For the three measurements,

$$50\% \text{ confidence limit} = 1.67 \pm \frac{0.67 \times 0.10}{\sqrt{3}} = 1.67 \pm 0.04$$

$$95\% \text{ confidence limit} = 1.67 \pm \frac{1.96 \times 0.10}{\sqrt{3}} = 1.67 \pm 0.11$$

Thus, the chances are 50 in 100 that the true mean will lie in the interval of 1.63 to 1.71 ppm Hg and 95 in 100 that it will be between 1.56 and 1.78 ppm.

EXAMPLE 3-7. Calculate the number of replicate measurements needed to decrease the 95% confidence interval for the calibration of a 10-mL pipet to 0.005 mL, assuming that a procedure similar to the one for obtaining the data in Table 3-1 has been followed.

The standard deviation for the measurement is 0.0065 mL. Because s is based on 24 values, we may assume $s \rightarrow \sigma = 0.0065$.

The confidence interval is given by

$$\text{confidence interval} = \pm \frac{z\sigma}{\sqrt{N}}$$

$$0.005 \text{ mL} = \frac{1.96 \times 0.0065}{\sqrt{N}}$$

$$N = 6.5$$

Thus, by employing the mean of seven measurements, we would have a somewhat better than 95% chance of knowing the true mean volume delivered by the pipet to ± 0.005 mL.

A consideration of Equation 3-10 indicates that the confidence interval for an analysis can be halved by employing the mean of four measurements. Sixteen measurements would be required to narrow the limit by another factor of two. It is apparent that a point of diminishing return is rapidly reached in acquiring additional data. Thus, the chemist ordinarily takes advantage of the relatively large gain afforded by averaging two to four measurements but can seldom afford the time required for further increases in confidence.

Confidence limits when σ is unknown. Frequently, a chemist must make use of an unfamiliar method. Furthermore, limitations in time or amount of available sample may preclude an accurate estimation of σ. Here, a single set of replicate measurements must provide not only a mean value but also a precision estimate. As we have shown earlier (Table 3-3), s calculated from a limited set of data may be subject to considerable uncertainty; thus, the confidence limits must be broader under these circumstances.

To account for the potential variability of s, use is made of the parameter t, which is defined as

$$t = \frac{x - \mu}{s} \tag{3-11}$$

In contrast to z in Equation 3-3, t is dependent not only on the desired confidence level but also upon the number of degrees of freedom available in the calculation of s. Table 3-6 provides values for t for a few degrees of freedom; much more extensive tables are found in various mathematical handbooks. Note that the values for t become equal to those for z (Table 3-5) as the number of degrees of freedom becomes infinite.

The confidence limit can be derived from t by an equation analogous to Equation 3-10; that is,

Table 3-6   Values of t for Various Levels of Probability

Degrees of Freedom	Factor for Confidence Interval, %				
	80	90	95	99	99.9
1	3.08	6.31	12.7	63.7	637
2	1.89	2.92	4.30	9.92	31.6
3	1.64	2.35	3.18	5.84	12.9
4	1.53	2.13	2.78	4.60	8.60
5	1.48	2.02	2.57	4.03	6.86
6	1.44	1.94	2.45	3.71	5.96
7	1.42	1.90	2.36	3.50	5.40
8	1.40	1.86	2.31	3.36	5.04
9	1.38	1.83	2.26	3.25	4.78
10	1.37	1.81	2.23	3.17	4.59
11	1.36	1.80	2.20	3.11	4.44
12	1.36	1.78	2.18	3.06	4.32
13	1.35	1.77	2.16	3.01	4.22
14	1.34	1.76	2.14	2.98	4.14
∞	1.29	1.64	1.96	2.58	3.29

$$\text{confidence limit for } \mu = \bar{x} \pm \frac{ts}{\sqrt{N}} \tag{3-12}$$

EXAMPLE 3-8. A chemist obtained the following data for the alcohol content in a sample of blood: percent ethanol = 0.084, 0.089, and 0.079. Calculate the 95% confidence limit for the mean assuming (a) no additional knowledge about the precision of the method and (b) that on the basis of previous experiences $s \rightarrow \sigma = 0.006\%$ ethanol.

(a) $\bar{x} = \dfrac{(0.084 + 0.089 + 0.079)}{3} = 0.084$

$s = \sqrt{\dfrac{(0.000)^2 + (0.005)^2 + (0.005)^2}{3 - 1}} = 0.005$

Table 3-6 indicates that $t = \pm 4.30$ for two degrees of freedom and 95% confidence. Thus,

$$95\% \text{ confidence limit} = 0.084 \pm \frac{4.3 \times 0.005}{\sqrt{3}}$$

$$= 0.084 \pm 0.012$$

(b) Because a good value of σ is available,

$$95\% \text{ confidence limit} = 0.084 \pm \frac{z\sigma}{\sqrt{N}}$$

$$= 0.084 \pm \frac{1.96 \times 0.0060}{\sqrt{3}}$$

$$= 0.084 \pm 0.007$$

Note that a sure knowledge of σ decreased the confidence interval by almost half.

Table 3-7*   Critical Values for Rejection Quotient Q

Number of Observations	Q_{crit} (Reject if $Q_{exp} > Q_{crit}$)		
	90% Confidence	96% Confidence	99% Confidence
3	0.94	0.98	0.99
4	0.76	0.85	0.93
5	0.64	0.73	0.82
6	0.56	0.64	0.74
7	0.51	0.59	0.68
8	0.47	0.54	0.63
9	0.44	0.51	0.60
10	0.41	0.48	0.57

* Reproduced from W. J. Dixon, *Ann. Math. Stat.,* **22**, 68 (1951).

3C-6 REJECTION OF DATA

When a set of data contains an outlying result that appears to differ excessively from the average, the decision must be made to retain or to reject it. The choice of criterion for the rejection of a suspected result has its perils. If we set a stringent standard that makes the rejection of a questionable measurement difficult, we run the risk of retaining results that are spurious and have an inordinate effect on the average of the data. On the other hand, if we set lenient limits on precision and thereby make the rejection of a result easy, we are likely to discard measurements that rightfully belong in the set; we thus introduce a bias to the data. It is an unfortunate fact that no universal rule can be invoked to settle the question of retention or rejection.

Of the numerous statistical criteria suggested to aid in deciding whether to retain or reject a measurement, the Q test[4] is to be preferred. Here, the difference between the questionable result and its nearest neighbor is divided by the spread of the entire set. The resulting ratio, Q, is then compared with rejection values that are critical for a particular degree of confidence. Table 3-7 provides critical values of Q at the 90% confidence level.

> **EXAMPLE 3-9.** The analysis of a calcite sample yielded CaO percentages of 55.95, 56.00, 56.04, 56.08, and 56.23, respectively. The last value appears anomalous; should it be retained or rejected?
>
> The difference between 56.23 and 56.08 is 0.15%. The spread (56.23 − 55.95) is 0.28%. Thus,
>
> $$Q_{exp} = \frac{0.15}{0.28} = 0.54$$
>
> For five measurements, Q_{crit} is 0.64. Because 0.54 < 0.64, retention is indicated.

Notwithstanding its superiority over other criteria, the Q test must be used with good judgment as well. For example, there will be situations in which the dispersion associated with the bulk of a set will be fortuitously small and the indis-

[4] R. B. Dean and W. J. Dixon, *Anal. Chem.,* **23**, 636 (1951).

criminate application of the Q test will result in rejection of a value that actually should be retained; indeed, in a three-number set containing a pair of identical values, the experimental value for Q will inevitably exceed the critical value. On the other hand, it has been pointed out[4] that the magnitudes of rejection quotients for small sets are likely to cause the retention of erroneous data.

The blind application of statistical tests to the decision for retention or rejection of a suspect measurement in a small set of data is not likely to be much more fruitful than an arbitrary decision. Indeed, the application of good judgment based upon an estimate of the precision to be expected may be a more sound approach, particularly if this estimate is based upon broad experience with the analytical method being employed. In the end, however, the only entirely valid reason for rejecting an experimental result from a small set is the sure knowledge that a mistake has been made in its acquisition. If one lacks this knowledge, a *cautious approach to the rejection of data is desirable.*

In summary, there are a number of recommendations for the treatment of a small set of results that contains a suspect value.

1. Reexamine carefully all data relating to the suspected result to see if a gross error has affected its value. *A properly kept laboratory notebook containing careful notations of all observations is essential if this recommendation is to be helpful.*
2. If possible, estimate the precision that can be reasonably expected from the procedure to be sure that the outlying result actually is questionable.
3. Repeat the analysis if sufficient sample and time are available. Agreement of the newly acquired data with those that appear to be valid will lend weight to the notion that the outlying result should be rejected. Furthermore, the questionable result will have a smaller effect on the mean of the larger set of data if its retention is still indicated.
4. If more data cannot be secured, apply the Q test to the existing set to see if the doubtful result should be retained or rejected on statistical grounds.
5. If the Q test indicates retention, give consideration to reporting the median of the set rather than the mean. The median has the great virtue of allowing inclusion of all data in a set without undue influence from an outlying value. Moreover, it has been demonstrated that the median of a normally distributed set containing three measurements is more likely to provide a reliable estimate of the correct value than will the mean of the set after the outlying value has been arbitrarily discarded.[5]

3C-7 STATISTICAL AIDS TO HYPOTHESIS TESTING

Much of scientific and engineering endeavor is based upon hypothesis testing. Thus, in order to explain an observation, a hypothetical model is advanced which then serves as a basis for experimental testing to determine its validity. If the results from these experiments do not support it, the model is rejected and a new one is sought. If, on the other hand, there is agreement between the experimental results and what would be expected from the properties of the hypothetical model, then the hypothesis can serve as the basis for further experiments. When the hypothesis is

---

[5] National Bureau of Standards, *Technical News Bulletin* (July 1949); *J. Chem. Educ.*, **26**, 673 (1949).

supported by sufficient experimental data, it becomes recognized as a useful theory until such time as data are developed that refute it.

Seldom will experimental results agree exactly with those predicted by a theoretical model. As a consequence, scientists and engineers are frequently confronted with the necessity of making a judgment as to whether a numerical difference is a real one that calls for rejection of the hypothesis or whether it is a manifestation of indeterminate error that is inevitably associated with the measurements. Certain statistical tests are useful in sharpening these judgments.

In approaching a test of this kind, a *null hypothesis* is employed, which assumes that the numerical quantities being compared are, in fact, the same. The probability of the observed differences appearing as a result of indeterminate error is then computed from statistical theory. Usually, if the observed difference is as large or larger than the difference that would occur 5 times in 100 (the 5% probability level), the null hypothesis is considered questionable and the difference is judged to be significant. Other probability levels such as 1 in 100 or 10 in 100 may also be adopted, depending upon the certainty desired in making the judgment.

The kinds of testing that chemists use most often include the comparison of means, $\bar{x}_1$ and $\bar{x}_2$, from two sets of analyses, the mean from an analysis $\bar{x}_1$ and what is believed to be the true value μ, the standard deviations, s_1 and s_2 or σ_1 and σ_2 from two sets of measurements, and the standard deviation s of a small set of data with the standard deviation σ of a larger set of measurements. The sections that follow consider some of the methods for dealing with these comparisons.

Comparison of an experimental mean with a true value. A common way of testing for determinate errors is to employ the method for the analysis of a sample whose composition is accurately known (see p. 46). In all probability, the experimental mean $\bar{x}$ will differ from the true value μ; the judgment must then be made whether this difference is the consequence of indeterminate error in the measurement or of the presence of a determinate error in the method.

The statistical treatment for this type of problem involves comparing the difference $(\bar{x} - \mu)$ with the difference that would normally be expected as the result of indeterminate error. If the observed difference is less than that which is computed for a chosen probability level, the null hypothesis that $\bar{x}$ and μ are the same cannot be rejected; that is, no significant determinate error has been demonstrated. It is important to realize, however, that this statement does not say that no determinate error exists; its presence has just not been demonstrated. If $(\bar{x} - \mu)$ is significantly larger than the expected or critical value, it may be assumed that the difference is real and that a determinate error exists.

The critical value for the rejection of the null hypothesis can be obtained by rewriting Equation 3-12 in the form

$$\bar{x} - \mu = \pm \frac{ts}{\sqrt{N}} \tag{3-13}$$

where N is the number of replicate measurements employed in the test. If a good estimation of σ is available, Equation 3-13 can be modified by replacing t and s with z and σ, respectively.

EXAMPLE 3-10. A new procedure for the rapid analysis of sulfur in kerosenes was tested by the analysis of a sample which was known from its method of preparation to contain 0.123% S. The results obtained were: %S = 0.112, 0.118, 0.115, and 0.119. Do the data indicate the presence of a negative determinate error in the new method?

$$\bar{x} = \frac{0.112 + 0.118 + 0.115 + 0.119}{4} = 0.116$$

$$\bar{x} - \mu = 0.116 - 0.123 = -0.007$$

$$s = \sqrt{\frac{(0.004)^2 + (0.002)^2 + (0.001)^2 + (0.003)^2}{4 - 1}} = 0.0033$$

From Table 3-6, we find that at the 95% confidence level, t has a value of 3.18 for three degrees of freedom. Thus,

$$\frac{ts}{\sqrt{N}} = \frac{3.18 \times 0.0033}{\sqrt{4}} = \pm 0.0052$$

But

$$\bar{x} - \mu = -0.007$$

Five times out of 100, an experimental mean can be expected to deviate by ± 0.0052 or more. Thus, if we conclude that -0.007 is a significant difference and that a determinate error is present, we will, on the average, be right 95 times and wrong 5 times out of 100 judgments.

If we make a similar calculation employing the value for t at the 99% confidence level, $ts/\sqrt{N}$ assumes a value of 0.0096. Thus, if we insist upon being wrong no oftener than 1 time out of 100, we would have to say that no difference between the results has been *demonstrated*. Note that this statement is different from saying that no determinate error exists.

Comparison of two experimental means. A chemist will frequently employ analytical data in an effort to establish whether two materials are different or identical. Here, the judgment must be made as to whether a difference in analytical results is the consequence of indeterminate errors in the two measurements or if it represents a real difference. To illustrate, let us assume that the N_1 replicate analyses were made on material 1 and N_2 analyses on material 2. Applying Equation 3-12, we may write

$$\mu_1 = \bar{x}_1 \pm \frac{ts}{\sqrt{N_1}}$$

and

$$\mu_2 = \bar{x}_2 \pm \frac{ts}{\sqrt{N_2}}$$

where $\bar{x}_1$ and $\bar{x}_2$ are the two experimental means. In order to establish the existence or absence of a real difference between $\bar{x}_1$ and $\bar{x}_2$, we make the null hypothesis that μ_1 and μ_2 are identical. With this assumption, it can be shown that

$$\bar{x}_1 - \bar{x}_2 = \pm ts \sqrt{\frac{N_1 + N_2}{N_1 N_2}} \tag{3-14}$$

The numerical value for the term on the right is computed employing t for the particular confidence level desired. (The number of degrees of freedom for finding t in Table 3-6 will be $N_1 + N_2 - 2$.) If the experimental difference, $\bar{x}_1 - \bar{x}_2$, is smaller than the computed value, the null hypothesis is not rejected, and no significant difference between the means has been demonstrated. On the other hand, an experimental difference greater than the value computed from t indicates the existence of a significant difference.

If a good estimate of σ is available, Equation 3-14 can be modified by insertion of z and σ for t and s, respectively.

> **EXAMPLE 3-11.** The composition of a flake of paint found on the clothes of a hit-and-run victim was compared with that of paint from the car suspected of causing the accident. Do the following data for the spectroscopic analysis for titanium in the paints suggest a difference in composition between the two materials? From previous experience, the standard deviation for the analysis is known to be 0.35% Ti; that is, $s \rightarrow \sigma = 0.35\%$ Ti.
>
> paint from clothes % Ti = 4.0, 4.6
> paint from car   % Ti = 4.5, 5.3, 5.5., 5.0, 4.9
>
> $$\bar{x}_1 = \frac{4.6 + 4.0}{2} = 4.3$$
>
> $$\bar{x}_2 = \frac{4.5 + 5.3 + 5.5 + 5.0 + 4.9}{5} = 5.0$$
>
> $$\bar{x}_1 - \bar{x}_2 = 4.3 - 5.0 = -0.7\% \text{ Ti}$$
>
> Modifying Equation 3-14 to take into account our knowledge that $s \rightarrow \sigma$ and abstracting values of z from Table 3-5, we calculate for the 95 and 99% confidence levels
>
> $$\pm z\sigma \sqrt{\frac{N_1 + N_2}{N_1 N_2}} = \pm 1.96 \times 0.35 \sqrt{\frac{2 + 5}{2 \times 5}} = \pm 0.57$$
>
> $$= \pm 2.58 \times 0.35 \sqrt{\frac{2 + 5}{2 \times 5}} = \pm 0.76$$
>
> We see that only 5 out of 100 data should differ by 0.57% Ti or greater and only 1 out of 100 should differ by as much as 0.76% Ti. Thus, it seems reasonably probable (between 95 and 99% certain) that the observed difference of -0.7% does not arise from indeterminate error but in fact is caused, at least in part, by a real difference between the two paint samples. Hence, we would conclude the suspected vehicle was probably not involved in the accident.

> **EXAMPLE 3-12.** Two barrels of wine were analyzed for their alcohol content in order to determine whether they were from different sources. The average content of the first barrel was established, on the basis of six analyses, to be 12.61% ethanol. Four analyses of the second barrel gave a mean of 12.53% alcohol. The ten analyses yielded a pooled value of $s = 0.070\%$. Is a difference between the wines indicated by the data?
>
> Here, we employ Equation 3-14, using t for eight degrees of freedom $(10 - 2)$. At the 95% confidence level
>
> $$\pm ts \sqrt{\frac{N_1 + N_2}{N_1 N_2}} = 2.31 \times 0.070 \sqrt{\frac{6 + 4}{6 \times 4}} = 0.10\%$$

The observed difference is

$$\bar{x}_1 - \bar{x}_2 = 12.61 - 12.53 = 0.08\%$$

As often as 5 times in 100, indeterminate error will be responsible for a difference as great as 0.10%. At the 95% confidence level, then, no difference has been established.

In the last example, it was found that no significant difference existed at the 95% probability level. It should be noted that this statement is not equivalent to saying that $\bar{x}_1$ is equal to $\bar{x}_2$; nor do the tests prove that the wines came from the same source. Indeed, it is conceivable that one could have been a red and the other a white. Establishment, with a reasonable probability, that the two wines were derived from the same source would require extensive testing of other characteristics such as taste, color, odor, refractive index, acetic acid concentration, sugar content, and trace element content. If, for all of these tests and others, no significant differences were revealed, then it might be possible to judge the two as having a common genesis. In contrast, the finding of *one* significant difference among these would clearly show that the two were different. Thus, the establishment of a significant difference by a single test is much more revealing than the establishment of an absence of difference.

Equation 3-14 is useful for estimating the detection limit for a measurement. Here, the standard deviation from several blank determinations is calculated. The minimum detectable quantity Δx_{min} will be given by

$$\Delta x_{min} = \bar{x}_1 - \bar{x}_b > ts_b \sqrt{\frac{N_1 + N_b}{N_1 N_b}} \tag{3-15}$$

where the subscript b refers to the blank determination.

EXAMPLE 3-13. A method for the analysis of DDT gave the following results when applied to pesticide-free foliage samples: μg DDT = 0.2, -0.5, -0.2, 1.0, 0.8, -0.6, 0.4, 1.2. Calculate the DDT-detection limit (at the 99% confidence level) of the method for (a) a single analysis and (b) the mean of five analyses.

Here, we find

$$\bar{x}_b = (0.2 - 0.5 - 0.2 + 1.0 + 0.8 - 0.6 + 0.4 + 1.2)/8 = 0.3 \ \mu g$$

$$s_b = \sqrt{\frac{(0.1)^2 + (0.8)^2 + (0.5)^2 + (0.7)^2 + (0.5)^2 + (0.9)^2 + (0.1)^2 + (0.9)^2}{8 - 1}}$$

$$= 0.68 \ \mu g$$

(a) For a single analysis, $N_1 = 1$, and the number of degrees of freedom will be $(1 + 8 - 2)$ or 7. From Table 3-6, we find $t = 3.5$ and

$$\Delta x_{min} > 3.5 \times 0.68 \sqrt{\frac{1 + 8}{1 \times 8}} > 2.5 \ \mu g \ DDT$$

Thus, 99 times out of 100, a result greater than 2.5 μg DDT will indicate the presence of the pesticide on the plant.

(b) Here, N_1 is 5 and the number of degrees of freedom is 11. Therefore, $t = 3.11$ and

$$\Delta x_{min} = 3.11 \times 0.68 \sqrt{\frac{5 + 8}{5 \times 8}} = 1.2 \ \mu g \ DDT$$

Comparison of precision of measurements. The *F test* provides a simple method for comparing the precision of two sets of identical measurements. The sets do not necessarily have to be obtained from the same sample as long as the samples are sufficiently alike that the sources of indeterminate error can be assumed to be the same. The *F* test is also based upon the null hypothesis, which assumes that the precisions are identical. Here, the quantity *F*, which is defined as the ratio of the variances of the two measurements, is computed and compared with the maximum values of *F* which would be expected (at a certain probability level) if no difference in precision existed between the measurements. If the experimental *F* exceeds the critical value found in probability tables, a statistical basis exists for questioning the null hypothesis that the two standard deviations are alike. Table 3-8 provides critical values for *F* at the 5% probability level. These values of *F* will be exceeded only 5 times in 100 if the standard deviations of the two measurements are the same. Much more extensive tables of *F* values at various probability levels are found in most mathematics handbooks.

The *F* test may be used to provide insights into either of two questions, specifically, (1) whether method 1 is more precise than method 2 and (2) whether there is a difference in the precisions of the two methods. For the first of these applications, the variance of the supposedly more precise procedure is always placed in the denominator and that for the less precise in the numerator; for the second, the larger variance always appears in the numerator. This arbitrary placement of the larger of the two variances in the numerator in the second case has the effect of making the outcome of the test less certain; thus, the uncertainty level of the *F* values in Table 3-8 is doubled from 5 to 10%.

EXAMPLE 3-14. A standard method for the determination of carbon monoxide in gaseous mixtures is known from many hundreds of measurements to have a standard deviation of 0.21 ppm CO. A modification of the method has yielded an *s* of 0.15 ppm CO for a pooled set of data with 12 degrees of freedom. A second modification, also based on 12 degrees of freedom, has a standard deviation of 0.12 ppm CO. Is either of the modifications significantly more precise than the original?

Because an improvement is claimed, the variances of the modifications are placed in the denominator. For the first,

$$F_1 = \frac{s_{\text{std}}^2}{s_1^2} = \frac{(0.21)^2}{(0.15)^2} = 1.96$$

Table 3-8 Critical Values for *F* at the Five Percent Level

Degrees of Freedom (Denominator)	Degrees of Freedom (Numerator)							
	2	3	4	5	6	12	20	∞
2	19.00	19.16	19.25	19.30	19.33	19.41	19.45	19.50
3	9.55	9.28	9.12	9.01	8.94	8.74	8.66	8.53
4	6.94	6.59	6.39	6.26	6.16	5.91	5.80	5.63
5	5.79	5.41	5.19	5.05	4.95	4.68	4.56	4.36
6	5.14	4.76	4.53	4.39	4.28	4.00	3.87	3.67
12	3.89	3.49	3.26	3.11	3.00	2.69	2.54	2.30
20	3.49	3.10	2.87	2.71	2.60	2.28	2.12	1.84
∞	3.00	2.60	2.37	2.21	2.10	1.75	1.57	1.00

and for the second,

$$F_2 = \frac{(0.21)^2}{(0.12)^2} = 3.06$$

For the standard procedure, $s \to \sigma$ and the number of degrees of freedom for the numerator can be taken as infinite. The critical value of F is seen to be 2.30 (Table 3-8).

The value of F for the first modification is less than 2.30 and the null hypothesis, at the 95% probability level, has not been disproved. The second modification, however, does appear to have a significantly greater precision.

It is interesting to note that if we ask whether the precision of the second modification is significantly better than the first, the F test indicates that such a difference has not been demonstrated. That is,

$$F = (s_1)^2/(s_2)^2 = (0.15)^2/(0.12)^2 = 1.56$$

In this instance, the critical value for F is 2.69.

EXAMPLE 3-15. The skill of two analysts was compared by having each perform identical replicate nitrogen analyses on a pure organic compound. The standard deviations of the six data obtained by one analyst was 0.12% N, while s for the five determinations by the second was 0.06% N. Does this difference in precision suggest a difference in abilities between the two analysts?

Here, we obtain F by placing the larger variance in the numerator. Thus,

$$F = (0.12)^2/(0.06)^2 = 4.00$$

In Table 3-8, we find that the critical value of F for five degrees of freedom in the numerator and four in the denominator is 6.26. Thus, the null hypothesis is not rejected, and in this test, at least, the skills of the two workers appear to be the same. Note that because we had no basis for postulating that one set of data was more precise than the other, the confidence level is, in this example, only 90% rather than 95% as in the previous example.

3C-8 CALIBRATION CURVES

Most analytical methods require a calibration step in which standards containing known concentrations of the analyte are treated in the same way as the samples. The data are frequently plotted to give a calibration curve such as that shown in Figure 3-4. Typically such plots approximate a straight line; it is seldom, however, that all of the data will fall exactly on that line because of the indeterminate error in the measuring process. Thus, the investigator must try to derive a "best" straight line from the points. Statistics provides a mechanism for objectively obtaining such a line and also for specifying the uncertainties associated with its use for analyses. The techniques involved are called *regression analyses* by statisticians. Here, we will be treating only the most simple regression procedure, called the *method of least squares*.

Assumptions. In applying the method of least squares to the derivation of a calibration curve, two assumptions will be made. The first of these is that a linear relationship does indeed exist between analyte concentration and the measured variable. Second, it will be assumed that no significant error exists in composition of the standards—that is, the concentrations of the standards are known exactly. Thus, the deviations of points from the straight line shown in Figure 3-4 are en-

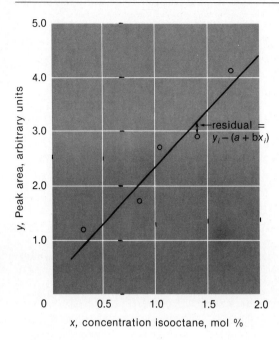

x, concentration isooctane, mol %

FIGURE 3-4 Calibration curve for the determination of isooctane in a hydrocarbon mixture.

tirely a consequence of the indeterminate error in y, the area of a chromatographic peak. Both of these assumptions are appropriate for most analytical methods.

Derivation of a least-squares line. The linear relationship between the measured or dependent variable y and the independent variable, the analyte concentration x, is assumed to be

$$y = a + bx$$

where a is the intercept or value of y when x is zero and b is the slope of the line. The method of least squares permits evaluation of a and b as well as the standard deviation for these quantities.

The method of least squares is based upon the derivation of that straight line for which the squares of the deviations for the individual points from that line (on the y axis) Q_i, are minimized. Here, Q_i is defined as

$$Q_i = [y_i - (a + bx_i)]^2$$

where the subscripts i refer to the various individual pairs of points. Note in Figure 3-4 that the deviations or *residuals* represent the *vertical* displacements of experimental data from the least-squares line.

Although the mathematical equations necessary for a least-squares analysis are readily derived, we shall, in the interest of saving space, simply present them and focus instead upon their use.[6]

[6] For detailed discussions of regression analysis and the least-squares method, see: *Statistical Methods in Research and Production*, 4th ed., O. L. Davies and P. L. Goldsmith, Eds., Chapter 7. New York: Hafner Publishing Co., 1972; W. J. Dixon and F. J. Massey, Jr., *Introduction to Statistical Analysis*, 3d ed. New York: McGraw-Hill, 1969.

For convenience, we shall define three quantities A^2, B^2, and AB as follows:

$$A^2 = \Sigma(x_i - \bar{x})^2 = \Sigma x_i^2 - (\Sigma x_i)^2/n \qquad (3\text{-}16)^7$$

$$B^2 = \Sigma(y_i - \bar{y})^2 = \Sigma y_i^2 - (\Sigma y_i)^2/n \qquad (3\text{-}17)^7$$

$$AB = \Sigma(x_i - \bar{x})(y_i - \bar{y}) = \Sigma x_i y_i - \Sigma x_i \Sigma y_i/n \qquad (3\text{-}18)$$

Here, x_i and y_i are individual pairs of data for x and y that are used to define the least-squares line. The quantity n is the number of pairs of data used in preparation of the calibration curve and $\bar{x}$ and $\bar{y}$ are the average values for the variables. That is,

$$\bar{x} = \Sigma x_i/n$$

$$\bar{y} = \Sigma y_i/n$$

Note that A^2 and B^2 are simply the sum of the squares of the deviations from the means for the individual values for x and y. In order to ease the computational process, the equivalent expressions shown on the right are ordinarily used.

We will be interested in deriving five quantities from A^2, B^2, and AB, namely the intercept a, the slope b, the standard deviation for b, s_b, the standard deviation for the residuals, s_r, and the standard deviation s_c for the results of an analysis based upon this calibration curve. These quantities are given by

$$b = AB/A^2 \qquad (3\text{-}19)$$

$$a = \bar{y} - b\bar{x} \qquad (3\text{-}20)$$

$$s_r = \sqrt{\frac{B^2 - b^2 A^2}{n - 2}} \qquad (3\text{-}21)$$

$$s_b = \sqrt{s_r^2/A^2} = s_r/A \qquad (3\text{-}22)$$

$$s_c = \frac{s_r}{b} \sqrt{\frac{1}{m} + \frac{1}{n} + \frac{(\bar{y}_c - \bar{y})^2}{b^2 A^2}} \qquad (3\text{-}23)$$

The last equation permits the calculation of the standard deviation from the mean $\bar{y}_c$ of a set of m replicate analyses when a calibration curve that contains n points is used; recall that $\bar{y}$ is the mean value of y for the n calibration data.

EXAMPLE 3-16. The first two columns of Table 3-9 contain the experimental data that are plotted in Figure 3-4. Carry out a least-squares analysis of the data.

Columns 3, 4, and 5 of the table contain computed values for y_i^2, x_i^2, and $x_i y_i$; their sums appear as the last entry of each column. Note that the number of figures carried in the computed values should be the *maximum allowed by the calculator; that is, rounding should not be performed until the end.*

We now use Equations 3-16, 3-17, and 3-18 to give[7]

$$A^2 = \Sigma x_i^2 - (\Sigma x_i)^2/n = 6.90201 - (5.365)^2/5 = 1.14537$$

$$B^2 = \Sigma y_i^2 - (\Sigma y_i)^2/n = 36.3775 - (12.51)^2/5 = 5.07748$$

$$AB = \Sigma x_i y_i - \Sigma x_i \Sigma y_i/n = 15.81992 - 5.365 \times 12.51/5 = 2.39669$$

[7] The student should be careful to distinguish between Σx_i^2 and $\Sigma(x_i)^2$ or Σy_i^2 and $\Sigma(y_i)^2$. The first is obtained by first squaring the value of x_i or y_i and then summing. For the second, the values of x_i and y_i are first summed; the sums are then squared.

Table 3-9   Calibration Data for a Chromatographic Method for the Determination of Isooctane in a Hydrocarbon Mixture

Mole Percent Isooctane, x_i	Peak Area, y_i	x_i^2	y_i^2	$x_i y_i$
0.352	1.09	0.12390	1.1881	0.38368
0.803	1.78	0.64481	3.1684	1.42934
1.08	2.60	1.16640	6.7600	2.80800
1.38	3.03	1.90140	9.1809	4.18140
1.75	4.01	3.06250	16.0801	7.01750
5.365	12.51	6.90201	36.3775	15.81992

Substitution of these quantities into Equations 3-19 through 3-22 yields

$$b = 2.39669/1.14537 = 2.0925 = 2.09$$

$$a = \frac{12.51}{5} - 2.0925 \times \frac{5.365}{5} = 0.2567 = 0.26$$

Thus, the equation for the least-squares line is

$$y = 0.26 + 2.09x$$

The standard deviation for the residuals is

$$s_r = \sqrt{\frac{B^2 - b^2 A^2}{n - 2}} = \sqrt{\frac{5.07748 - (2.0925)^2 \times 1.14537}{5 - 2}}$$
$$= 0.144 = 0.14$$

and the standard deviation of the slope is

$$s_b = \sqrt{s_r^2/A^2} = \sqrt{(0.144)^2/1.14537} = 0.13$$

The confidence limit for the slope can be derived using t from Table 3-6. Here, the number of degrees of freedom is two less than the number of points because one degree of freedom was lost in calculating a and one for b. The 90% confidence limit in this example is

$$90\% \text{ C.L.} = 2.09 \pm ts_b = 2.09 \pm 2.35 \times 0.13 = 2.09 \pm 0.31$$

EXAMPLE 3-17.   The calibration curve derived in Example 3-16 was used in conjunction with the chromatographic determination of isooctane in a hydrocarbon mixture. A peak area of 2.65 was obtained. Calculate the mole percent isooctane and the standard deviation for the result assuming the area was (a) the result of a single measurement and (b) the mean of four measurements.

In either case

$$x = \frac{y - 0.26}{2.09} = \frac{2.65 - 0.26}{2.09} = 1.14 \text{ mol } \%$$

(a) Substituting into Equation 3-23, we obtain

$$s_c = \frac{0.14}{2.09} \sqrt{\frac{1}{1} + \frac{1}{5} + \frac{(2.65 - 12.51/5)^2}{(2.09)^2 \times 1.145}}$$
$$= 0.074 \text{ mol } \%$$

(b) For the mean of four measurements

$$s_c = \frac{0.14}{2.09} \sqrt{\frac{1}{4} + \frac{1}{5} + \frac{(2.65 - 12.51/5)^2}{(2.09)^2 \times 1.145}}$$

$$= 0.046 \text{ mol } \%$$

3D PROPAGATION OF ERRORS IN COMPUTATION

The scientist must frequently estimate the error in a result that has been computed from two or more data, each of which has an error associated with it. The way in which the individual errors accumulate depends upon the arithmetic relationship between the term containing the error and the quantity being computed. In addition, the effect of determinate errors on a computed result differs from the effect of indeterminate errors.

3D-1 ACCUMULATION OF DETERMINATE ERRORS

The way in which determinate errors accumulate in a sum or difference varies from that for a product or quotient.

Errors in a sum or difference.   Let us consider the relationship

$$y = a + b - c$$

where a, b, and c are the values for three measurable quantities. If Δa, Δb, and Δc are the absolute determinate errors associated with the measurement of these quantities, the actual measurements are $(a + \Delta a)$, $(b + \Delta b)$, and $(c + \Delta c)$. The resulting error in y, then, is Δy and

$$y + \Delta y = (a + \Delta a) + (b + \Delta b) - (c + \Delta c)$$

The error in the computed result can be obtained by subtracting the first equation from the second. That is,

$$\Delta y = \Delta a + \Delta b - \Delta c \tag{3-24}$$

It is seen that *for addition or subtraction, the absolute error for the sum of difference is determined by the absolute error of the numbers forming the sum or difference.*

    EXAMPLE 3-18.   Calculate the error in the result of the following calculation:

$$
\begin{array}{ll}
+0.50 & (+0.02) \\
+4.10 & (-0.03) \\
\underline{-1.97} & \underline{(-0.05)} \\
2.63 &
\end{array}
$$

where the numbers in parentheses are the absolute determinate errors. The absolute error of the summation is

$$\Delta y = 0.02 + (-0.03) - (-0.05) = +0.04$$

Errors in a product or quotient.   Let us first consider the product

$$y = a \times b$$

We shall again assume determinate errors of Δa and Δb which result in the error Δy.

Thus,

$$y + \Delta y = (a + \Delta a)(b + \Delta b)$$
$$= ab + a\Delta b + b\Delta a + \Delta a \Delta b$$

Subtraction of the first equation from the last gives

$$\Delta y = b\Delta a + a\Delta b + \Delta a \Delta b$$

We now divide this equation by the first, which yields

$$\frac{\Delta y}{y} = \frac{\Delta a}{a} + \frac{\Delta b}{b} + \frac{\Delta a \Delta b}{ab}$$

The third term on the right side of the foregoing equation will generally be much smaller than the other two inasmuch as the numerator is the product of two small terms and the denominator the product of two much larger ones. Therefore, $\Delta a \Delta b / ab \ll (\Delta a/a + \Delta b/b)$, and

$$\frac{\Delta y}{y} = \frac{\Delta a}{a} + \frac{\Delta b}{b}$$

Note that the three terms correspond to *relative* determinate errors rather than absolute as was the case for a sum or a difference.

An analogous relationship can be derived for the error in a quotient. Thus,

$$y = \frac{a}{b}$$

Then

$$y + \Delta y = \frac{(a + \Delta a)}{(b + \Delta b)}$$

It is convenient to rewrite these two equations as

$$yb = a$$
$$yb + b\Delta y + y\Delta b + \Delta y \Delta b = a + \Delta a$$

The two equations can be combined to give

$$b\Delta y + y\Delta b + \Delta y \Delta b = \Delta a$$

Dividing by $yb = a$ and rearranging yields

$$\frac{\Delta y}{y} = \frac{\Delta a}{a} - \frac{\Delta b}{b} - \frac{\Delta y \Delta b}{yb}$$

Here again, we may ordinarily assume that $\Delta y \Delta b / yb \ll (\Delta a/a - \Delta b/b)$, and

$$\frac{\Delta y}{y} = \frac{\Delta a}{a} - \frac{\Delta b}{b}$$

For the more general case,

$$y = \frac{ab}{c}$$

it can be shown by the same type of argument that

$$\frac{\Delta y}{y} = \frac{\Delta a}{a} + \frac{\Delta b}{b} - \frac{\Delta c}{c}$$

(3-25)

Thus, for multiplication or division, the relative error of the product or quotient is determined by the relative errors of the numbers forming the computed result.

EXAMPLE 3-19. Compute the error in the result of the following calculation, where the numbers in parentheses are absolute determinate errors:

$$y = \frac{4.10(-0.02) \times 0.0050(+0.0001)}{1.97(-0.04)} = 0.01041$$

Here, we must base the calculation on *relative* errors. Thus,

$$\frac{\Delta y}{y} = \frac{-0.02}{4.10} + \frac{0.0001}{0.0050} - \frac{-0.04}{1.97}$$

$$= -0.0049 + 0.020 + 0.020 = 0.035$$

To obtain the absolute error Δy in y, we write

$$\Delta y = 0.035 \times y = 0.035 \times 0.01041 = 0.0004$$

and

$$y = 0.0104(+0.0004)$$

3D-2 ACCUMULATION OF INDETERMINATE ERRORS

As we have noted earlier, the absolute or the relative standard deviation serves as the most convenient measure of indeterminate errors of an experimental result. In contrast to a determinate error, however, no sign can be attached to a standard deviation, there being an equal probability of its being positive and negative. This fact leads to a range of possible standard deviations for a computed result. For example, consider the summation

$$
\begin{array}{r}
+0.50 \ (\pm 0.02) \\
+4.10 \ (\pm 0.03) \\
-1.97 \ (\pm 0.05) \\
\hline
2.63
\end{array}
$$

where the numbers in parentheses are now the absolute indeterminate errors expressed as standard deviations. The uncertainty associated with the sum could be as large as ± 0.10 if the signs of the three individual standard deviations happened by chance to be all positive or all negative. On the other hand, under fortuitous circumstances the three uncertainties could combine to give an accumulated error of zero. Neither of these is as probable as a combination leading to an uncertainty intermediate between these extremes. It can be shown from statistical theory that the most probable uncertainty in the case of sums or differences can be found by taking the square root of the sum of individual *absolute variances*. Thus, in the present example

$$s_y^2 = s_a^2 + s_b^2 + s_c^2$$

where s_y is the standard deviation of the sum and s_a, s_b, and s_c are the standard deviations of the three terms in the sum. We see then that

$$s_y = \sqrt{(\pm 0.02)^2 + (\pm 0.03)^2 + (\pm 0.05)^2}$$
$$= \pm 0.06$$

and the sum could be reported as $2.63(\pm 0.06)$.

3D-3 PROPAGATION OF ERROR IN MULTIPLICATION AND DIVISION

The following example demonstrates how the standard deviation of a product or quotient is found.

EXAMPLE 3-20.

$$\frac{4.10(\pm 0.02) \times 0.0050(\pm 0.0001)}{1.97(\pm 0.04)} = 0.0104(\pm \,?)$$

Note that the standard deviations of two of the numbers in this calculation are larger than the result itself. It is evident, then, that we cannot obtain the desired standard deviation by direct combination of the uncertainties as in an addition or a subtraction. Indeed, as for determinate errors, the *relative* error of the product or quotient is determined by the *relative* errors of the numbers forming the computed result. Thus, in this example we must first compute relative standard deviations.

$$(s_a)_r = \frac{(\pm 0.02)}{4.10} = \pm 0.0049$$

$$(s_b)_r = \frac{(\pm 0.0001)}{0.0050} = \pm 0.020$$

$$(s_c)_r = \frac{(\pm 0.04)}{1.97} = \pm 0.020$$

Again, the *relative* variance of the result $(s_y)_r^2$ is equal to the sum of the individual *relative* variances.

$$(s_y)_r^2 = (s_a)_r^2 + (s_b)_r^2 + (s_c)_r^2$$
$$s_r = \sqrt{(\pm 0.0049)^2 + (\pm 0.020)^2 + (\pm 0.020)^2} = \pm 0.029$$

In order to complete the calculation, we must find the *absolute* standard deviation of the result. Thus,

$$s_y = 0.0104 \times (\pm 0.029) = \pm 0.0003$$

and we can indicate the uncertainty of the answer as $0.0104(\pm 0.0003)$.

The following example demonstrates the calculation of the standard deviation of the result for a more complex calculation.

EXAMPLE 3-21. Calculate the standard deviation of the result of the following computation:

$$\frac{[14.3(\pm 0.2) - 11.6(\pm 0.2)] \times 0.050(\pm 0.001)}{[820(\pm 10) + 1030(\pm 5)] \times 42.3(\pm 0.4)} = 1.725(\pm \,?) \times 10^{-6}$$

First, we must calculate the standard deviation of the sum and the difference. For the difference in the numerator,

$$s_a = \sqrt{(\pm 0.2)^2 + (\pm 0.2)^2} = \pm 0.28$$

and for the sum in the denominator,

$$s_b = \sqrt{(\pm 10)^2 + (\pm 5)^2} = \pm 11$$

We may then rewrite the equation as

$$\frac{2.7(\pm 0.28) \times 0.050(\pm 0.001)}{1850(\pm 11) \times 42.3(\pm 0.4)} = 1.725 \times 10^{-6}$$

The equation now contains only products and quotients; we thus compute the relative standard deviations of the individual quantities.

$$(s_a)_r = \frac{\pm 0.28}{2.7} = \pm 0.104$$

$$(s_b)_r = \frac{\pm 0.001}{0.050} = \pm 0.020$$

$$(s_c)_r = \frac{\pm 11}{1850} = \pm 0.0060$$

$$(s_d)_r = \frac{\pm 0.4}{42.3} = \pm 0.0095$$

and

$$(s_y)_r = \sqrt{(\pm 0.104)^2 + (\pm 0.020)^2 + (\pm 0.0060)^2 + (\pm 0.0095)^2}$$
$$= 0.106$$

The absolute standard deviation of the result is

$$s_y = (1.725 \times 10^{-6} \times (\pm 0.106) = \pm 0.18 \times 10^{-6}$$

and our answer is thus written as

$$1.7(\pm 0.2) \times 10^{-6}$$

Note the amplification of error that results from the subtraction process in the numerator of the equation.

3D-4 PROPAGATION OF ERRORS IN EXPONENTIAL CALCULATIONS

To show how errors are propagated when the power or the root of an experimental result a is to be calculated, we write

$$y = a^x$$

where x is the power or the root and contains no uncertainty. In order to obtain the uncertainty Δy that results from the determinate error Δa in a, we take the derivative of the foregoing expression. Thus,

$$dy = xa^{(x-1)}da$$

Dividing by the original expression gives

$$\frac{dy}{y} = \frac{xa^{(x-1)}da}{a^x}$$

But

$$\frac{a^{(x-1)}}{a^x} = \frac{1}{a}$$

Therefore,

$$\frac{dy}{y} = x\frac{da}{a}$$

or for finite increments

$$\frac{\Delta y}{y} = x \frac{\Delta a}{a} \tag{3-26}$$

Here, Δy is the absolute error in y that results from the determinate error Δa in a. Clearly, the *relative error* $\Delta y/y$ of the computed result is simply the *relative error* of the experimental number $\Delta a/a$ multiplied by the exponent x. For example, the relative error in the square of a number is twice that for the number itself, whereas the relative error of the cube root of a number is simply one third that of the number.

It is also important to note that the propagation of an *indeterminate* error in raising a number of a power is treated differently from a multiplication because the possibility of errors canceling one another does not exist. Recall that the relative standard deviation in the product $a \times b$ will lie between the sum and the difference of the standard deviations of the two numbers, and a probable value can be calculated by taking the square root of the sum of the squares of the uncertainties. This technique is not applicable to the product of a single measurement $a \times a$; here, the signs are necessarily identical since the numbers are identical. Thus, the relative uncertainty in a^2 must be twice that of a. Equation 3-26, therefore, applies to indeterminate errors as well; here, Δy and Δa are replaced by s_y and s_a. That is,

$$(s_y)_r = x(s_a)_r$$

EXAMPLE 3-22.   The standard deviation in measuring the diameter d of a sphere is ± 0.02 cm. What is the standard deviation in its calculated volume V if $d = 2.15$ cm?

$$V = \frac{4}{3} \pi \left(\frac{d}{2}\right)^3 = \frac{4}{3} \pi \left(\frac{2.15}{2}\right)^3 = 5.20 \text{ cm}^3$$

Here, we may write

$$\frac{s_V}{V} = 3 \times \frac{s_d}{d}$$

$$= 3 \times \frac{0.02}{2.15} = 0.028$$

The absolute standard deviation in V is then

$$s_V = 5.20 \times 0.028 = 0.15$$

Thus,

$$V = 5.2(\pm 0.2) \text{ cm}^3$$

EXAMPLE 3-23.   The solubility product K_{sp} for the silver salt AgX is $4.0(\pm 0.4) \times 10^{-8}$. What is the uncertainty associated with the calculated solubility of AgX in water? Here, (see p. 21),

$$\text{solubility} = (4.0 \times 10^{-8})^{1/2} = 2.0 \times 10^{-4}$$

$$(s_a)_r = \frac{0.4 \times 10^{-8}}{4.0 \times 10^{-8}}$$

$$(s_y)_r = \frac{1}{2} \times \frac{0.4}{4.0} = 0.05$$

$$s_y = 2.0 \times 10^{-4} \times 0.05 = 0.1 \times 10^{-4}$$

and

$$\text{solubility} = 2.0(\pm 0.1) \times 10^{-4}$$

3D-5 PROPAGATION OF ERROR IN LOGARITHM AND ANTILOGARITHM CALCULATIONS

To show how errors are propagated when logarithms or antilogarithms are computed we take the derivative of the expression

$$y = \log a = 0.434 \ln a$$

where ln symbolizes the natural logarithm. Thus,

$$dy = 0.434 \frac{da}{a}$$

Conversion to finite increments gives

$$\Delta y = 0.434 \frac{\Delta a}{a} \tag{3-27}$$

Note that the *absolute* uncertainty in y is determined by the *relative* uncertainty in a and conversely. In employing Equation 3-27, the relative standard deviation can, as in exponentiation, be substituted for $\Delta a/a$ and the absolute standard deviation in y for Δy. That is,

$$s_y = 0.434(s_a)_r \tag{3-28}$$

EXAMPLE 3-24. Calculate the absolute standard deviations of the results of the following computations. The absolute standard deviation for each quantity is given in parentheses.

   (a) $y = \log [2.00(\pm 0.02) \times 10^{-3}] = -2.6990 \pm ?$
   (b) $a = \text{antilog} [1.200(\pm 0.003)] = 15.849 \pm ?$
   (c) $a = \text{antilog} [45.4(\pm 0.3)] = 2.5119 \times 10^{45} \pm ?$

   (a) Referring to Equation 3-28, we see that we must multiply the *relative* standard deviation by 0.434. That is,

$$s_y = \pm 0.434 \times \frac{0.02 \times 10^{-3}}{2.00 \times 10^{-3}} = \pm 0.004$$

Thus,

$$\log [2.00(\pm 0.002) \times 10^{-3}] = -2.699(\pm 0.004)$$

   (b) Rearranging Equation 3-28

$$(s_a)_r = \frac{s_a}{a} = \frac{s_y}{0.434} = \frac{\pm 0.003}{0.434} = \pm 0.0069$$

$$s_a = \pm 0.0069 \times a = \pm 0.0069 \times 15.849 = 0.11$$

Thus,

$$\text{antilog} [1.200(\pm 0.002)] = 15.8 \pm 0.1$$

   (c) $\dfrac{s_a}{a} = \dfrac{\pm 0.3}{0.434} = \pm 0.69$

$$s_a = 0.69 \times a = 0.69 \times 2.5119 \times 10^{45}$$
$$= 1.7 \times 10^{45}$$

Thus,

$$\text{antilog}\,[45.4(\pm 0.3)] = 2.5(\pm 1.7) \times 10^{45}$$

Note that a large absolute error is associated with an antilogarithm of a number with few digits beyond the decimal point. The large uncertainty here arises from the fact that the numbers to the left of the decimal (the characteristic) serve only to locate the decimal point. In the last example, the large error in the antilogarithm results from the relatively large uncertainty in the *mantissa* of the number (that is, 0.4 ± 0.3).

3E THE SIGNIFICANT FIGURE CONVENTION

In reporting a measurement, the experimenter should include not only what is considered to be its best value, be it a mean or a median, but also an estimate of its uncertainty. The latter is preferably reported as the standard deviation of the result; the deviation from the mean, the deviation from the median, or the spread may sometimes be encountered because these precision indicators are easier to calculate. Common practice also dictates that an experimental result should be rounded off so that it contains only the digits known with certainty plus the first uncertain one. This practice is called the *significant figure convention*.

For example, the average of the experimental quantities 61.60, 61.46, 61.55, and 61.61 is 61.555. The standard deviation of the sum is ± 0.069. Clearly, the number in the second decimal place is subject to uncertainty. Such being the case, all numbers in succeeding decimal places are without meaning, and we are forced to round the average value accordingly. The question of taking 61.55 or 61.56 must be considered, 61.555 being equally spaced between them. A good guide to follow when rounding a 5 is always to round to the nearest even number; in this way, any tendency to round in a set direction is eliminated, since there is an equal likelihood that the nearest even number will be the higher or the lower in any given situation. Thus, we could report the foregoing results as 61.56 ± 0.07. On the other hand, if we had reason to doubt that ± 0.07 was a valid estimate of the precision, we might choose to present the result as 61.6 ± 0.1.

Often the significant figure convention is used in lieu of a specific estimate of the precision of a result. Thus, by simply reporting 61.6 in this example, we would be saying in effect that we believe the first 6 and the 1 are certain digits but the value of the second 6 is in doubt. The disadvantage of this technique is obvious; all the reader can discern is the range of the uncertainty—here, greater than ± 0.05 and smaller than ± 0.5.

In employing the significant figure convention, it is important to appreciate that the zero not only functions as a number but also serves to locate decimal points in very small and very large numbers. A case in point is Avogadro's number. The first three digits, 6, 0, and 2, are known with certainty; the next is uncertain but is probably 3. Since the digits that follow the 6023 are not known, we substitute 20 zeros after the digit 3 to place the decimal point. Here, the zeros indicate the order of magnitude of the number only and have no other meaning. It is clear that a distinction must be made between those figures that have physical significance (that is, *significant figures*) and those that are either unknown or meaningless owing to the inadequacies of measurement.

Zeros bounded by digits only on the left may or may not be significant. Thus, the mass of a 20-mg weight that carries no correction (to a tenth of a milligram) is known to three significant figures, 20.0 mg; when this is expressed as 0.0200 g, the number of significant figures does not change. If, on the other hand, we wish to express the volume of a 2-L beaker as 2000 mL, the latter number will contain only one significant figure. The zeros simply indicate the order of magnitude. It can, of course, happen that the beaker in question has been found by experiment to contain 2.0 L; here, the zero following the decimal point is significant and implies that the volume is known to at least ± 0.5 L and might be known to ± 0.05 L. If this volume were to be expressed in milliliters, the zero following the 2 would still be significant, but the other two zeros would not. The use of exponential notation eliminates the difficulty. We could thus indicate the volume as 2.0×10^3 mL.

A certain amount of care is required in determining the number of significant figures to carry in the result of an arithmetic combination of two or more numbers. For addition and subtraction, the number of significant figures can be seen by visual inspection. For example,

$$3.4 + 0.02 + 1.31 = 4.7$$

Clearly, the second decimal place cannot be significant because an uncertainty in the first decimal place is introduced by the 3.4.

When data are being multiplied or divided, it is frequently assumed that the number of significant figures of the result is equal to that of the component quantity that contains the least number of significant figures. For example,

$$\frac{24 \times 0.452}{100.0} = 0.108 = 0.11$$

Here, the 24 has two significant figures and the result has therefore been rounded to agree. Unfortunately, the rule does not always apply well. Thus, the uncertainty in 24 in this calculation could be as small as 0.5 or as large as 5. The uncertainty in the quotient for these two limits is

Assumed Absolute Uncertainty in 24	Relative Uncertainty	Absolute Uncertainty in 0.108	Round to
>0.5	0.5/24 = 0.02	0.108 × 0.02 = 0.002	0.108
<5	5/24 = 0.2	0.108 × 0.2 = 0.02	0.11

Calculations such as this demonstrate the dilemma faced by the scientist who has only the significant figure convention to use as a guide to indeterminate errors.

Particular care is needed in rounding logarithms and antilogarithms to a proper number of significant figures. Consider again the example on page 79; proper rounding of $\log 2.00 \times 10^{-3}$ will yield

$$\log 2.00 \times 10^{-3} = -2.699$$

Here, the mathematical manipulation appears to give a gain in the number of significant figures. In fact, however, the initial 2 in the result only indicates the location of the decimal point in the original number and hence is not significant; information about 2.00 is contained in the three digits in 0.699. Thus, there is agreement in number of significant figures between this part of the result and the original.

As was shown by Example 3-24c on page 79, an apparent reduction by one in the number of significant figures appears to accompany computation of an antilog. The reason for this apparent anomaly lies again in the function of the characteristic of a logarithm.

Generally, it is a good practice in performing calculations to carry one extra digit beyond the last significant figure, saving the rounding to the final result. In this way rounding errors are avoided. The effect of intermediate rounding of data is demonstrated in the example that follows.

EXAMPLE 3-25.   All of the data in the following equations are known to four significant figures. Determine the value for y (a) without intermediate rounding and (b) by rounding the sum and differences before completing the calculations.

$$y = \frac{(47.33 \times 0.1000 - 6.42 \times 0.1200)(110.5 - 10.04)}{(4.912 + 0.2324)}$$

(a) $y = \dfrac{3.9626 \times 100.46}{5.1444} = 77.382$

  $= 77.38$

(b) $y = \dfrac{3.963 \times 100.5}{5.144} = 77.426$

  $= 77.43$

PROBLEMS*

3-1. Calculate the absolute error and the percent relative error in the following:

	Experimental Value	Accepted Value
*A	4.167	4.082
B	84.69	85.00
*C	33.86	38.57
D	9.87	10.04
*E	49.88	50.04
F	21.10	20.96

3-2. Consider the following sets of data:

*A	B	*C	D	*E
61.45	3.37	12.06	24.66	9.961
61.51	3.36	12.24	24.77	10.034
61.12	3.34	12.19	24.72	10.012
61.40	3.36	12.29	24.85	9.977

Calculate the (a) mean, (b) median, (c) range, (d) average deviation from the mean in parts per thousand, (e) absolute average deviation from the mean, (f) percent relative range, and (g) absolute standard deviation.

* Answer to problems or parts of problems marked with an asterisk are to be found at the end of the book.

3-3. Which of the indicated parts of data in Problem 3-2 has the larger relative range, the larger absolute deviation, the larger relative deviation (in parts per thousand), and the larger standard deviation?

     * (a) Set A and Set E   (d) Set B and Set E
     (b) Set D and Set B   * (e) Set C and Set D
     * (c) Set C and Set B   (f) Set A and Set C

3-4. Accepted values for the data in Problem 3-2 are as follows:

     * Set A 61.45   Set D 24.60
     Set B 3.33   * Set E 10.02
     * Set C 12.26

     Calculate (a) the absolute error and (b) the percent relative error of the mean for each set.

* 3-5. Use Equation 3-7 (p. 56) to estimate values of s for the sets in Problem 3-2. Compare these values with those calculated with Equation 3-6.

3-6. A particular method for the analysis of copper yields results that are low by 0.5 mg. What will be the percent relative error due to this source if the weight of copper in a sample is

     * (a) 25 mg?   (b) 100 mg?   * (c) 250 mg?   (d) 500 mg?

3-7. The method described in Problem 3-6 is to be used to analyze an ore that contains about 4.8% copper. What minimum sample weight should be taken if the relative error due to a 0.5-mg loss is to be smaller than

     * (a) 0.1%?   (b) 0.5%?   * (c) 0.8%?   (d) 1.2%

* 3-8. A constant solubility loss of 1.8 mg Se is associated with a gravimetric analysis for the element. A sample containing approximately 18% Se was analyzed by this method. Calculate the relative error (in parts per thousand) in the result if the sample taken for the analysis weighed 0.400 g.

* 3-9. Following are data from a continuing study of calcium ion in the blood plasma of several individuals:

Subject	Mean Calcium Content, mg/100 mL	No. of Observations	Deviation of Individual Results From Mean Values
1	3.16	5	0.14, 0.09. 0.06, 0.00, 0.11
2	4.08	4	0.07, 0.12, 0.10, 0.01
3	3.75	5	0.13, 0.05, 0.08, 0.14, 0.07
4	3.49	3	0.10, 0.13, 0.07
5	3.32	6	0.07, 0.10, 0.11, 0.03, 0.14, 0.05

     (a) Calculate s for each set of values.
     (b) Pool the data and calculate s for the analysis.

3-10. Calculate a pooled estimate of s for a method developed to determine trace amounts of iron(III).

Sample	ppb Iron(III)
1	100, 117, 107, 104
2	264, 257, 276
3	148, 131, 140, 133
4	37.5, 40.3, 36.1, 38.9

* 3-11. A method for determining the particulate lead content of air samples is based upon drawing a measured quantity of air through a filter and performing the analysis on circles cut from the filter. Calculate the individual values for s as well as a pooled value for the accompanying data.

Sample	μg Pb/m^3 Air
1	1.5, 1.2, 1.3
2	2.0, 2.3, 2.3, 2.2
3	1.8, 1.7, 1.4, 1.6
4	1.6, 1.3, 1.2, 1.5, 1.6

* 3-12. Based on extensive past experience, it is known that the standard deviation for an analytical method for gold in sea water is 0.025 ppb. Calculate the 99% confidence interval for an analysis using this method, based on
   (a) a single measurement.
   (b) three measurements.
   (c) five measurements.

3-13. Extensive use of a particular method for the analysis of chloride has established a standard deviation (σ) of 0.041. Triplicate measurements with this method yielded an average value of 21.46% Cl. For this determination, calculate
   (a) the 80% confidence interval.
   (b) the 90% confidence interval.
   (c) the 99% confidence interval.

3-14. An established method of analysis for chlorinated hydrocarbons in air samples has a standard deviation of 0.030 ppm.
   (a) Calculate the 95% confidence limit for a group of four measurements obtained by this method.
   (b) How many measurements should be made if the 95% confidence limit is to be ± 0.017?

* 3-15. The standard deviation in a method for the analysis of carbon monoxide in automotive exhaust gases has been found, on the basis of extensive past experience, to be 0.80 ppm.
   (a) Estimate the 90% confidence limit for a triplicate analysis.
   (b) How many measurements would be needed for the confidence limit for the set to be 0.50 ppm?

* 3-16. The method described in Problem 3-15 was significantly modified and found to have a standard deviation of 0.70, based upon four measurements. Establish the 90% confidence limit for the modified method.

3-17. A method for the analysis of potassium in blood serum yielded the accompanying data:

Trial	Concentration, mg K$^+$/100 mL
1	15.3
2	15.6
3	15.4
4	16.3

Calculate the 90% confidence interval for the set, assuming that these are the only data available.

* 3-18. A new method for the analysis of mercury was tested against an ore sample that was known to assay 12.63% Hg.

Trial	% Hg
1	12.76
2	12.57
3	12.72
4	12.79
5	12.76

(a) Calculate the standard deviation, s, for these data.

(b) Calculate the 95% confidence interval for the analysis.

(c) Is the assay mean within the bounds of (1) the 95% confidence interval and (2) the 80% confidence interval?

3-19. Using (1) the pooled standard deviation and (2) the individual standard deviation, calculate the 90% confidence interval for each of the means in

* (a) Problem 3-9.

  (b) Problem 3-10.

  (c) Problem 3-11.

* 3-20. Analysis of a group of samples yielded the following data:

       29.03, 29.08, 28.97, 29.24

Apply the Q test to see if the outlying result should be retained or rejected at the 90% confidence level.

3-21. Apply the Q test to the data in Problem 3-17 to ascertain whether justification exists (90% confidence) for neglecting the outlying result.

3-22. Each of the following sets of data has what appears to be an outlying result. Apply the Q test (90% confidence) to determine whether this value should be retained or rejected.

*A	B	*C	D	*E	F
75.97	14.64	31.42	31.42	9.22	9.22
76.36	14.41	31.40	31.40	9.06	9.06
76.04	14.46	31.04	31.04	9.20	9.20
76.13	14.44		31.44		9.24

* 3-23. A new method for the determination of vitamin C was tested against tablets known to contain 500 mg of ascorbic acid. The accompanying results were obtained:

      502, 500, 505, 501, 504   mean = 502.4 $s = 2.1$

Establish whether a positive error is indicated at the 95% confidence level.

* 3-24. A new spectrometric method was tested against a brass specimen of established composition. The standard deviation for each determination is given in parentheses beneath the percentage values.

	Ni	P	Pb	Sb	Sn	Zn
Accepted	0.0520	0.010	0.106	0.018	0.055	32.20
New method	0.0524	0.012	0.103	0.017	0.053	32.48
s	(0.007)	(0.010)	(0.012)	(0.020)	(0.011)	(0.10)

If the data for the new method represent the average of four measurements, is the existence of determinate error demonstrated in the analysis of

(a) nickel at the 90% confidence level?

     (b) phosphorus at the 95% confidence level?
     (c) lead at the 99% confidence level?
     (d) antimony at the 99% confidence level?
     (e) tin at the 95% confidence level?
     (f) zinc at the 90% confidence level?

3-25. Suppose each experimental value in Problem 3-24 is the mean of 20 replicate measurements. Is the existence of determinate error demonstrated at the indicated confidence levels?

* 3-26. Determine whether a significant difference has been demonstrated between experimental means $\bar{x}_1$ and $\bar{x}_2$ at the 95% confidence level. A standard deviation based on extensive experience is indicated by σ; a pooled estimate is symbolized by s. The quantities N_1 and N_2 are the number of replicate measurements used in obtaining $\bar{x}_1$ and $\bar{x}_2$.

	$\bar{x}_1$	$\bar{x}_2$	N_1	N_2	Standard Deviation
(a)	21.3	21.5	2	2	$s = 0.15$
(b)	21.3	21.5	5	6	$s = 0.15$
(c)	21.3	21.5	5	6	$\sigma = 0.15$
(d)	0.663	0.654	4	3	$s = 0.004$
(e)	0.663	0.654	4	3	$\sigma = 0.004$
(f)	0.663	0.654	2	3	$s = 0.004$

3-27. Establish whether a significant difference has been demonstrated between the experimental means $\bar{x}_1$ and $\bar{x}_2$ at the 90% confidence level. A standard deviation based on extensive past experience is indicated by σ; a pooled estimate is symbolized by s. The quantities N_1 and N_2 are the number of replicate measurements used in obtaining $\bar{x}_1$ and $\bar{x}_2$.

	$\bar{x}_1$	$\bar{x}_2$	N_1	N_2	Standard Deviation
(a)	31.68	31.74	2	4	$s = 0.04$
(b)	31.68	31.74	3	6	$s = 0.04$
(c)	31.68	31.74	4	8	$s = 0.04$
(d)	31.68	31.74	2	4	$\sigma = 0.04$
(e)	31.68	31.74	3	6	$\sigma = 0.04$
(f)	31.68	31.74	4	8	$\sigma = 0.04$

3-28. An attempt to fix responsibility for an oil spill is to be based upon comparison of the sulfur content in the spillage with specimens taken from the suspect vessel. Analysis of five samples from each source by an established method for which $s \rightarrow \sigma = 0.05$ has yielded values of 0.12% S for the collected oil and 0.16% S for oil from this ship. Is there reason to believe that the two specimens had different origins at the 95% confidence level?

* 3-29. A method for the analysis of codeine in prescription drugs yielded the following results when applied to a codeine-free blank: 0.2, 0.4, 0.3, 0.0, 0.2, 0.1 mg codeine. Calculate the detection limit (in terms of mg codeine) at the 99% confidence level, based upon the means of (a) two analyses, (b) four analyses, (c) six analyses.

3-30. Calculate the minimum detection limits at the 99% confidence level for the mean of N_1 measurements, given the accompanying data:

Individual Blank Determinations	N_1
*(a) 0.4, 0.1, 0.6, 0.3, 0.2	8
(b) 1.3, 1.7, 0.9, 1.5	6
*(c) 0.8, 1.1, 0.6, 1.4, 1.2, 1.0	5
(d) 0.67, 0.82, 0.75, 0.77, 0.69	7

3-31. An established method for the determination of manganese in alloy samples has a standard deviation (σ) of 0.11. Determine whether a modification to the method appears to yield improved precision if the standard deviation (s) is 0.07, based upon a pooled set with
* (a) 6 degrees of freedom.
   (b) 12 degrees of freedom.
* (c) 20 degrees of freedom.

3-32. A single alloy specimen was used to compare the results of two testing laboratories. The standard deviation and degrees of freedom (D.F.) in pooled sets for four analyses are given in the accompanying tabulation.

	Element	Laboratory A Standard Deviation, s	D.F.	Laboratory B Standard Deviation, s	D.F.
*(a)	Fe	0.10	6	0.12	12
(b)	Ni	0.07	12	0.04	20
*(c)	Cr	0.05	20	0.07	6
(d)	Mn	0.20	20	0.10	6

Use the F test to determine whether the results from one laboratory appear to be statistically more precise than those from the other.

* 3-33. How many significant figures are there in
   (a) 91.4?
   (b) 9.014?
   (c) 4137.98?
   (d) 7.00×10^{-4}?
   (e) 0.0063?
   (f) 1.0063?
   (g) 2.3×10^5?
   (h) 5612.66?

3-34. How many significant figures are there in
   (a) 27.6?
   (b) 0.00667?
   (c) 8.004?
   (d) 5.46×10^{-5}?
   (e) 3.094?
   (f) 4.0×10^{12}?
   (g) 77205?
   (h) 13.6×10^{-5}?

* 3-35. Estimate the absolute and relative standard deviations in the results of the accompanying calculations (the numbers in parentheses are standard deviations associated with the individual data). Round the result to the appropriate number of significant figures.
   (a) $16.9286(\pm 0.0001) + 16.8797(\pm 0.0001) = 33.8083$
   (b) $16.9286(\pm 0.0001) - 16.8797(\pm 0.0001) = 0.0489$
   (c) $0.0354(\pm 0.003) + 7.147(\pm 0.002) - 2.861(\pm 0.0003) = 4.3214$
   (d) $1047(\pm 3) + 6244(\pm 1) - 3976(\pm 2) = 3315$
   (e) $23.6(\pm 0.2) + 0.184(\pm 0.006) = 23.784$
   (f) $3.69(\pm 0.04) \times 10^{-3} + 7.87(\pm 0.06) \times 10^{-4} = 4.477 \times 10^{-3}$

3-36. Estimate the absolute and relative standard deviations in the results of the accompanying calculations (the numbers in parentheses are standard deviations associated with the individual data). Round the result to the appropriate number of significant figures.

(a) $41.49(\pm 0.03) + 12.37(\pm 0.04) = 53.86$
(b) $8.575(\pm 0.003) - 8.4128(\pm 0.0007) = 0.1662$
(c) $19.47(\pm 0.02) + 21.66(\pm 0.04) - 4.67(\pm 0.03) = 36.46$
(d) $1.493(\pm 0.004) + 12.41(\pm 0.01) - 8.947(\pm 0.002) = 4.956$
(e) $313.7(\pm 0.1) + 97.0(\pm 0.1) - 14.36(\pm 0.08) = 396.34$
(f) $2.04(\pm 0.07) \times 10^{-2} - 8.61(\pm 0.03) \times 10^{-3} = 1.179 \times 10^{-2}$

* 3-37. Estimate the absolute and relative standard deviations in the results of the accompanying calculations (the numbers in parentheses are standard deviations associated with the individual data). Round the result to the appropriate number of significant figures.
(a) $64.4(\pm 0.2) \times 0.381(\pm 0.007) = 24.5364$
(b) $18.18(\pm 0.03) \times 4.764(\pm 0.009) = 86.60952$
(c) $26.94(\pm 0.08) \div 0.0496(\pm 0.0004) = 543.14516. \ . \ . \ . \ .$
(d) $0.9194(\pm 0.0008) \div 46.18(\pm 0.03) = 0.01990905. \ . \ . \ . \ .$
(e) $\sqrt{4.593(\pm 0.006)} = 2.143129. \ . \ . \ . \ .$
(f) $[8.47(\pm 0.02) \times 10^{-3}]^4 = 5.146757. \ . \ . \ . \ \times 10^{-9}$

3-38. Estimate the absolute and relative standard deviations in the results of the accompanying calculations (the numbers in parentheses are standard deviations associated with the individual data). Round the result to the appropriate number of significant figures.

* (a) $\dfrac{29.67(\pm 0.03) - 8.51(\pm 0.01)}{6.36(\pm 0.02) + 4.83(\pm 0.02)} = 1.89097. \ . \ . \ . \ .$

(b) $\dfrac{7.614(\pm 0.008) - 6.923(\pm 0.005)}{14.2468(\pm 0.0002) - 13.6719(\pm 0.0001)} = 1.2019. \ . \ . \ . \ .$

* (c) $\dfrac{[3.44(\pm 0.01) \times 10^{-5}]^2}{0.100(\pm 0.004) + 0.250(\pm 0.001)} = 3.38103. \ . \ . \ . \ \times 10^{-9}$

(d) $\dfrac{[44.41(\pm 0.02) - 3.12(\pm 0.01)] \times 0.2048(\pm 0.0006)}{12.6349(\pm 0.0001) - 12.2775(\pm 0.0001)} = 23.6603. \ . \ . \ . \ .$

* (e) $\dfrac{8.40(\pm 0.01) \times [1.697(\pm 0.002)]^2}{186.1(\pm 0.3) \times [0.5843(\pm 0.0007)]^3} = 0.65161. \ . \ . \ . \ .$

(f) $\sqrt{\dfrac{765(\pm 1) \times 3.564(\pm 0.004)}{192.5(\pm 0.2)}} = 3.76343. \ . \ . \ . \ .$

* 3-39. Estimate the absolute standard deviation in the result derived from the following operations (the numbers in parentheses are absolute standard deviations for the numbers they follow). Round the result to the appropriate number of significant figures.
(a) $y = \log 878(\pm 3) = 2.94349. \ . \ . \ . \ .$
(b) $y = \log 0.4957(\pm 0.0004) = -0.30478. \ . \ . \ . \ .$
(c) $y = \log 1.64(\pm 0.03) \times 10^{-5} = -4.7852. \ . \ . \ . \ .$
(d) $a = \text{antilogarithm } 3.64(\pm 0.01) = 4365.16. \ . \ . \ . \ .$
(e) $a = \text{antilogarithm } -7.191(\pm 0.002) = 6.44169. \ . \ . \ . \ . \ \times 10^{-8}$
(f) $a = \text{antilogarithm } 0.30103(\pm 0.00005) = 2.00000. \ . \ . \ . \ .$

3-40. Estimate the absolute standard deviation in the result derived from the following operations (the numbers in parentheses are absolute standard deviations for the numbers they follow). Round the result to the appropriate number of significant figures.
(a) $y = \log 3.222(\pm 0.004) = 0.50813. \ . \ . \ . \ .$
(b) $y = \log 285.2(\pm 0.8) = 2.4551. \ . \ . \ . \ .$
(c) $y = \log 70.79(\pm 0.03) = 1.84997. \ . \ . \ . \ .$

(d) a = antilogarithm $16.84(\pm 0.01)$ = $6.9183. \ldots \times 10^{16}$
(e) a = antilogarithm $-5.93(\pm 0.04)$ = $1.174898. \ldots \ldots \times 10^{-6}$
(f) a = antilogarithm $-0.3251(\pm 0.0006)$ = $0.47304. \ldots \ldots$

3-41. Sulfate in natural water can be determined by measuring the turbidity that results when an excess of $BaCl_2$ is introduced into a measured quantity of the sample. A turbidimeter, the instrument used for this analysis was calibrated with a series of standard Na_2SO_4 solutions whose concentrations were known exactly. The following data were obtained:

mg SO_4^{2-}/L	Turbidimeter Reading	mg SO_4^{2-}/L	Turbidimeter Reading
0.00	0.06	15.0	3.98
5.00	1.48	20.0	4.61
10.00	2.28		

Assume that a linear relationship exists between the instrument reading and concentration, and
(a) plot the data and draw a straight line through them by eye.
* (b) derive a least-squares equation for the relationship between the variables.
(c) compare the straight line from the relationship derived in (b) with that in (a).
* (d) calculate the standard deviation for the slope and for the residuals of the least-squares line.
* (e) calculate the concentration of sulfate in a sample yielding a turbidimeter reading of 3.67. Calculate the absolute and relative standard deviations of the result.
* (f) repeat the calculations in (e) assuming that the 3.67 was a mean of 6 turbidimeter readings.

* 3-42. The following data were obtained in calibrating a calcium ion electrode for the determination of pCa. A linear relationship between the potential E and pCa is known to exist.

pCa	E, mV	pCa	E, mV
5.00	-53.8	2.00	+31.9
4.00	-27.7	1.00	+65.1
3.00	+2.7		

(a) Plot the data and draw a line through the points by eye.
(b) Derive a least-squares expression for the best straight line through the points. Plot this line.
(c) Calculate the standard deviation for the slope of the least-squares line.
(d) Calculate the standard deviation for the residuals from the least-squares line.
(e) Calculate the pCa of a serum solution in which the electrode potential was found to be 20.3 mV.
(f) Calculate the absolute and relative standard deviations for pCa if the millivolt reading in (e) was the mean of two replicate measurements. Repeat the calculation based upon the mean of eight measurements.
(g) Calculate the *calcium ion concentration* for the sample described in (e).
(h) Calculate the absolute and relative standard deviations in the calcium ion concentration if the measurement was performed as described in (f).

3-43. The following data were obtained in calibrating a colorimeter for the determination of trace amounts of iron in aqueous solution.

Fe Concentration, ppm	Colorimeter Reading	Fe Concentration, ppm	Colorimeter Reading
0.510	8.2	3.060	36.4
1.020	15.1	3.570	45.0
1.530	18.0	4.080	49.6
2.040	23.7	4.590	56.1
2.550	32.1	5.100	61.2

(a) Plot the data and draw a straight line through the points by eye.

(b) Derive a least-squares expression for the best straight line through the points. Plot this line.

(c) Calculate the standard deviation of the slope and the residuals.

(d) Calculate the parts per million iron in solutions that yielded readings of 9.9 and 43.7.

(e) Calculate the absolute and relative standard deviations of the concentrations found in (d) assuming the colorimetric data were for a single measurement. Repeat the calculations based upon the mean of five measurements.

3-44. The following are polarographic diffusion currents for standard solutions of methyl vinyl ketone (MVK).

Concn MVK, mmol/L	Current, μA	Concn MVK, mmol/L	Current, μA
0.500	3.76	3.50	20.42
1.50	9.16	4.50	25.33
2.50	15.03	5.50	31.97

(a) Derive a least-squares expression assuming the variables bear a linear relationship to one another.

(b) Plot the least-squares line as well as the experimental points.

(c) Calculate the uncertainty of the points around the line as well as the uncertainty of the slope.

(d) Two samples containing MVK yielded currents of 6.3 and 27.5 μA. Calculate the concentration of MVK in each solution.

(e) Assume that the results in (d) represent a single measurement as well as the mean of four measurements. Calculate the respective absolute and relative standard deviations.

the solubility of precipitates

4

Reactions that yield products of limited solubility find application in three important analytical processes: (1) the separation of an analyte as a precipitate from soluble substances that would otherwise interfere with its ultimate measurement; (2) gravimetric analysis, in which a precipitate is formed whose weight is chemically related to the amount of analyte; and (3) volumetric analysis, based on determining the volume of a standard reagent required to precipitate the analyte essentially completely. The success of each of these applications requires that the solid produced have a relatively low solubility, be reasonably pure, and have a suitable particle size. In this chapter, we consider the variables that influence the first of these three physical properties.

Examples illustrating the uses of solubility product constants to calculate the solubility of an ionic precipitate in water and in the presence of a common ion have been discussed in Chapter 2. The student should be thoroughly familiar with these principles before undertaking study of this chapter. Here, we will be concerned with the way such variables as pH, concentration of complexing agents, and concentration of electrolytes affect the solubility of precipitates.

4A EFFECT OF COMPETING EQUILIBRIA ON THE SOLUBILITY OF PRECIPITATES

The solubility of a precipitate increases in the presence of ionic or molecular species that react with the ions derived from the precipitate. This effect is illustrated by the examples shown in Table 4-1.

In the first example, the solubility of barium sulfate is enhanced by the presence of a strong acid because sulfate ion, the conjugate base of the weak acid HSO_4^-, tends to react with hydronium ions. From the Le Châtelier principle, it is evident that the added acid causes an increase in the hydrogen sulfate ion concentration. The consequent decrease in sulfate ion concentration is partially offset, however,

Table 4-1 Examples of Increases in Solubility Brought About by Competing Equilibria

Precipitate	Species Causing Solubility Increases	Equilibria
$BaSO_4$	H_3O^+	$BaSO_4(s) \rightleftharpoons Ba^{2+} + SO_4^{2-}$ $+$ H_3O^+ $\updownarrow$ $HSO_4^- + H_2O$
$AgBr$	NH_3	$AgBr(s) \rightleftharpoons Ag^+ + Br^-$ $+$ $2NH_3$ $\updownarrow$ $Ag(NH_3)_2^+$

by a shift of the first equilibrium to the right; the net result is an increase in solubility.

The second example shows that the solubility of silver bromide becomes greater in the presence of ammonia, which combines with silver ions to produce a silver ammine complex. Here, ammonia molecules tend to decrease the silver ion concentration; a shift of the solubility equilibrium to the right occurs, resulting in an increase in solubility.

As a more general case, consider the sparingly soluble AB, which dissolves to give A and B, where A and B are ions of unspecified charge:

$$AB(s) \rightleftharpoons \begin{array}{ccc} A & + & B \\ + & & + \\ C & & D \\ \updownarrow & & \updownarrow \\ AC & & BD \end{array}$$

If A and B react to form the species AC and BD, introduction of either C or D into the solution will cause a shift in the solubility equilibrium in the direction that increases the solubility of AB.

Determination of the solubility of AB in a system such as this requires knowledge of the molar concentrations of the added C and D, as well as equilibrium constants for all three equilibria. Generally, several algebraic expressions are needed to describe all of the concentration relationships in such a solution completely, and the solubility calculation requires solving several simultaneous equations. Unless a suitably programmed computer is available, the solution of these algebraic equations is frequently more formidable than the task of setting them up.

One point that should be constantly borne in mind when treating multiple equilibria is that *the validity and form of a particular equilibrium constant expression are in no way affected by the existence of additional competing equilibria in the solution.* Thus, in the present example, the solubility product expression for AB describes the relationship between the equilibrium concentrations of A and B regardless of whether or not C and D are present in the solution. That is, at a given temperature, the product [A][B] is a constant, provided only that some solid AB is present. To be sure, the *amount* of AB that dissolves is greater in the presence of C and D; the increase, however, is not because the ion product [A][B] has changed but rather because some of the precipitate has been converted to AC or BD.

A systematic approach by which any problem involving several equilibria can be attacked is presented in the following paragraphs. The approach will then be illustrated in the pages that follow.

4A-1 SYSTEMATIC METHOD FOR SOLVING PROBLEMS INVOLVING SEVERAL EQUILIBRIA

1. Write chemical equations for all the reactions that appear to have any bearing on the problem.
2. State in terms of equilibrium concentrations what is being sought.
3. Write equilibrium constant expressions for all of the equilibria shown in step 1; find numerical values for the constants from appropriate tables.
4. Write mass-balance equations for the system. These are algebraic expressions relating the equilibrium concentrations of the various species to one another and to the analytical concentrations of the substances present in the solution; they are derived by taking into account the way the solution was prepared (see Example 4–1 that follows).
5. Write a charge-balance equation. In any solution, the concentrations of the cations and anions must be such that the solution is electrically neutral. The charge-balance equation expresses this relationship. The method for deriving charge-balance equations is shown in Example 4-2 that follows.
6. Count the number of unknown quantities in the equations developed in steps 3, 4, and 5, and compare this number with the number of independent equations; the problem can be solved exactly by suitable algebraic manipulation if the number of equations is equal to the number of unknown concentrations. Should there be fewer equations than unknowns, attempt to derive additional independent equations; if this cannot be done, it must be concluded that an exact solution to the problem is not possible. To be sure, an approximate solution may still be realized.
7. Make suitable approximations to simplify the algebra or to decrease the number of unknowns so that the problem can be solved.
8. Solve the algebraic equations for the equilibrium concentrations that are necessary to give the answer as defined in step 2.
9. With the equilibrium concentrations obtained in step 8, check the approximations made in step 7 to be sure of their validity.

Step 6 in this scheme is particularly significant because it indicates whether an exact solution for the problem is theoretically feasible. If the number of independent equations is as great as the number of unknowns, the problem *becomes purely algebraic*, involving a solution to several simultaneous equations. On the other hand, if the number of equations is fewer than the number of unknowns, a search for other equations or approximations that will reduce the number of unknowns is essential. The student should never waste time in seeking a solution to a complex equilibrium problem without first establishing that sufficient data are available.

EXAMPLE 4-1. Write mass-balance expressions for the system formed when a 1.00×10^{-2} M NH_3 solution is saturated with AgBr.

Two of the equilibria in this solution are shown in Table 4-1. A third is

$$NH_3 + H_2O \rightleftharpoons NH_4^+ + OH^-$$

Because the only source of Br^-, Ag^+, and $Ag(NH_3)_2^+$ is the silver bromide in which silver and bromide ions exist in a $1:1$ ratio, it follows that one mass-balance expression is

$$[Ag^+] + [Ag(NH_3)_2^+] = [Br^-]$$

A second mass-balance expression is based upon the method of preparation. That is, the only source of ammonia-containing species is the 1.00×10^{-2} M NH_3. Therefore,

$$M_{NH_3} = 1.00 \times 10^{-2} = [NH_3] + [NH_4^+] + 2[Ag(NH_3)_2^+]$$

EXAMPLE 4-2. Write charge-balance expressions for (a) an aqueous solution of NaCl, (b) an aqueous solution of $MgCl_2$, and (c) an aqueous solution containing $Al_2(SO_4)_3$ and $MgCl_2$.

(a) All solutions have a net charge of zero although they may contain both positive and negative species. In the NaCl solution, this neutrality is a direct consequence of the following charge-balance relationship:

$$[Na^+] + [H_3O^+] = [Cl^-] + [OH^-]$$

Thus, the solution is neutral by virtue of the fact that the sum of the concentrations of the positively charged species is equal to the sum of the concentrations of the negative species.

(b) In this instance, we must write

$$2[Mg^{2+}] + [H_3O^+] = [Cl^-] + [OH^-]$$

Here, it is necessary to multiply the magnesium ion concentration by 2 in order to account for the two units of charge contributed by this ion; that is, charge balance is preserved because the chloride ion concentration is *twice* the magnesium ion concentration ($[Cl^-] = 2[Mg^{2+}]$). The concentration of a triply charged species, if present, would have to be multiplied by 3 for the same reason.

(c) For a solution containing $Al_2(SO_4)_3$, $MgCl_2$, and water, the charge-balance equation would be

$$3[Al^{3+}] + 2[Mg^{2+}] + [H_3O^+] = 2[SO_4^{2-}] + [HSO_4^-] + [Cl^-] + [OH^-]$$

4A-2 THE EFFECT OF pH ON SOLUBILITY

The solubilities of many precipitates are affected by the hydronium ion concentration of the solvent. Precipitates that exhibit this behavior contain an anion with basic properties, a cation with acidic properties, or both. An example of a precipitate containing an anion with basic properties is barium sulfate (Table 4-1). Its solubility increases with increases in acidity. Precipitates with acidic cations, on the other hand, become less soluble as the acid concentration becomes greater. For example, when water is saturated with bismuth iodide, the following equilibria are established:

$$BiI_3(s) \rightleftharpoons Bi^{3+} + 3I^-$$

$$Bi^{3+} + 2H_2O \rightleftharpoons BiOH^{2+} + H_3O^+$$

Addition of acid to this system decreases the concentration of $BiOH^{2+}$; the resulting increases in Bi^{3+} shift the first equilibrium to the left and cause the formation of additional precipitate.

Solubility product calculations when the hydronium ion concentration is fixed and known. Analytical precipitations are frequently performed in solutions in which the hydronium ion concentration is fixed at some predetermined and known concentration. Calculation of solubility losses under these circumstances is a relatively straightforward process using the systematic approach, as illustrated by the following example.

EXAMPLE 4-3. Calculate the molar solubility of calcium oxalate in a solution that has a constant hydronium ion concentration of $1.00 \times 10^{-4} \ M$.

Step 1. Chemical equations.

$$CaC_2O_4(s) \rightleftharpoons Ca^{2+} + C_2O_4^{2-} \tag{4-1}$$

Since they are conjugate bases of weak acids, both oxalate and hydrogen oxalate ions will be involved in equilibria with hydroxide ions.

$$C_2O_4^{2-} + H_2O \rightleftharpoons HC_2O_4^- + OH^- \tag{4-2}$$

$$HC_2O_4^- + H_2O \rightleftharpoons H_2C_2O_4 + OH^- \tag{4-3}$$

Thus, if this solution is to remain constant in pH, it must contain other species capable of replacing the hydronium ions used up by the hydroxide ions formed in establishing equilibria 4-2 and 4-3.

Step 2. Definition of the unknown. What is sought? We wish to know the solubility of CaC_2O_4 in moles per liter. Because CaC_2O_4 is ionic, its molar solubility S will be equal to the molar concentration of calcium ion; it will also be equal to the sum of the equilibrium concentrations of the oxalate species. That is,

$$S = [Ca^{2+}]$$

$$= [C_2O_4^{2-}] + [HC_2O_4^-] + [H_2C_2O_4]$$

Thus, if we can calculate either the calcium ion concentration or the sum of the concentrations of the oxalate species, we shall have obtained a solution to the problem.

Step 3. Equilibrium constant expressions.

$$[Ca^{2+}][C_2O_4^{2-}] = K_{sp} = 2.3 \times 10^{-9} \tag{4-4}$$

Application of Equation 2-12 (p. 24) gives

$$\frac{[HC_2O_4^-][OH^-]}{[C_2O_4^{2-}]} = \frac{K_w}{K_2} = \frac{1.00 \times 10^{-14}}{5.42 \times 10^{-5}} = 1.85 \times 10^{-10} \tag{4-5}$$

$$\frac{[H_2C_2O_4][OH^-]}{[HC_2O_4^-]} = \frac{K_w}{K_1} = \frac{1.00 \times 10^{-14}}{5.36 \times 10^{-2}} = 1.87 \times 10^{-13} \tag{4-6}$$

Step 4. Mass-balance expressions. Because the only source of Ca^{2+} and the various oxalate species is the dissolved CaC_2O_4, it follows that

$$[Ca^{2+}] = [C_2O_4^{2-}] + [HC_2O_4^-] + [H_2C_2O_4] \tag{4-7}$$

Furthermore, it is given that at equilibrium,

$$[H_3O^+] = 1.0 \times 10^{-4}$$

and therefore,

$$[OH^-] = 1.00 \times 10^{-10} \tag{4-8}$$

Step 5. Charge-balance expressions. A useful charge-balance equation cannot be written for this system because an amount of some unknown acid HX has been added to maintain $[H_3O^+]$ at 1.0×10^{-4}; an equation based on the electrical neutrality of the solution would require inclusion of the concentration of the anion $[X^-]$ associated with the unknown acid. As it turns out, an equation containing this additional unknown term is not needed.

Step 6. Comparison of equations and unknowns. We have four unknowns: $[Ca^{2+}]$, $[C_2O_4^{2-}]$, $[HC_2O_4^-]$, and $[H_2C_2O_4]$. We also have four independent algebraic relationships: Equations 4-4, 4-5, 4-6, and 4-7. Therefore, an exact solution is possible, and the problem has now become one of algebra.

Step 7. Approximations. An exact solution to this problem is readily obtained, and we will dispense with approximations.

Step 8. Solution of the equations. A convenient way to solve for the four unknowns is to make suitable substitutions into Equation 4-7 and thereby establish a relationship between $[Ca^{2+}]$ and $[C_2O_4^{2-}]$. We must first derive expressions for $[HC_2O_4^-]$ and $[H_2C_2O_4]$ in terms of $[C_2O_4^{2-}]$. Substitution of 1.00×10^{-10} for $[OH^-]$ in Equation 4-5 yields

$$\frac{[HC_2O_4^-](1.00 \times 10^{-10})}{[C_2O_4^{2-}]} = 1.85 \times 10^{-10}$$

Thus,

$$[HC_2O_4^-] = 1.85\,[C_2O_4^{2-}]$$

Upon substituting this relationship and the hydroxide ion concentration into Equation 4-6, we obtain

$$\frac{[H_2C_2O_4]\,(1.00 \times 10^{-10})}{1.85\,[C_2O_4^{2-}]} = 1.87 \times 10^{-13}$$

Thus,

$$[H_2C_2O_4] = \frac{1.87 \times 10^{-13} \times 1.85\,[C_2O_4^{2-}]}{1.00 \times 10^{-10}} = 0.0035\,[C_2O_4^{2-}]$$

These values for $[H_2C_2O_4]$ and $[HC_2O_4^-]$ are substituted into Equation 4-7 to give

$$[Ca^{2+}] = [C_2O_4^{2-}] + 1.85\,[C_2O_4^{2-}] + 0.0035\,[C_2O_4^{2-}]$$

$$= 2.85\,[C_2O_4^{2-}]$$

Substitution for $[C_2O_4^{2-}]$ in Equation 4-4 gives

$$[Ca^{2+}]\frac{[Ca^{2+}]}{2.85} = 2.3 \times 10^{-9}$$

$$[Ca^{2+}] = \sqrt{6.56 \times 10^{-9}} = 8.1 \times 10^{-5}$$

Thus, from step 2 we conclude that

$$S = 8.1 \times 10^{-5}\ M$$

The foregoing procedure can be applied to any precipitate containing an anion with basic properties provided the hydroxide or hydronium ion concentration is known and fixed. The steps involve:

1. Writing the solubility product expression and the basic dissociation expressions for the various anions resulting from the dissolution of the substance.

2. Deriving a mass-balance expression.
3. Starting with the least protonated anion, substitute the hydroxide ion concentration into the dissociation constant expressions in order to obtain the ratios for the concentration of the protonated to the deprotonated species.
4. Substituting the equations developed in step 3 into the mass-balance expression to give a relationship between the cation and the completely deprotonated anion concentration.
5. Combining the relationship developed in step 4 with the solubility product expression to determine the cation concentration.
6. Calculating the molar solubility from the cation concentration (for a $1:1$ salt, the two will be identical).

Solubility calculations where the hydronium ion concentration is variable. Solutes containing basic anions (such as barium carbonate) or acidic cations influence the hydronium ion concentration of their aqueous solutions. Thus, if no auxiliary reagent is employed to maintain a constant pH, as in the previous example, the hydronium ion concentration becomes dependent upon the extent to which such solutes dissolve. For example, an aqueous solution of saturated barium carbonate becomes basic as a consequence of the reactions

$$BaCO_3(s) \rightleftharpoons Ba^{2+} + CO_3^{2-}$$

$$CO_3^{2-} + H_2O \rightleftharpoons HCO_3^- + OH^-$$

$$HCO_3^- + H_2O \rightleftharpoons H_2CO_3 + OH^-$$

In contrast to the example just considered, the hydroxide ion concentration now becomes an unknown, and an additional algebraic equation must therefore be developed.

In many instances, the reaction of a precipitate with water cannot be neglected without introducing an error in the calculation. As shown by the data in Table 4-2, the magnitude of the error depends upon the solubility of the precipitate

Table 4-2 Calculated Solubility of MA from Various Assumed Values of K_{sp} and K_b

Solubility Product Assumed for MA	Dissociation Constant Assumed for HA, K_{HA}	Basic Constant for A⁻, $K_b = K_w/K_{HA}$	Calculated Solubility of MA, M	Calculated Solubility of MA Neglecting Reaction of A⁻ With Water, M
1.0×10^{-10}	1.0×10^{-6}	1.0×10^{-8}	1.02×10^{-5}	1.0×10^{-5}
	1.0×10^{-8}	1.0×10^{-6}	$1.2 \ \times 10^{-5}$	1.0×10^{-5}
	1.0×10^{-10}	1.0×10^{-4}	$2.4 \ \times 10^{-5}$	1.0×10^{-5}
	1.0×10^{-12}	1.0×10^{-2}	$10 \ \ \times 10^{-5}$	1.0×10^{-5}
1.0×10^{-20}	1.0×10^{-6}	1.0×10^{-8}	1.05×10^{-10}	1.0×10^{-10}
	1.0×10^{-8}	1.0×10^{-6}	$3.3 \ \times 10^{-10}$	1.0×10^{-10}
	1.0×10^{-10}	1.0×10^{-4}	$32 \ \ \times 10^{-10}$	1.0×10^{-10}
	1.0×10^{-12}	1.0×10^{-2}	$290 \ \times 10^{-10}$	1.0×10^{-10}

as well as the basic dissociation constant of the anion. The solubilities of the hypothetical precipitate MA, shown in column 4, were obtained by taking into account the reaction of A^- with water. Column 5 gives the calculated results when the basic properties of A^- are neglected; here, the solubility is simply the square root of the solubility product. Two solubility products, 1.0×10^{-10} and 1.0×10^{-20}, have been assumed for these calculations as well as several values for the basic dissociation constant of A^-. It is apparent that neglect of the reaction of the anions with water leads to a negative error, which becomes more pronounced both as the solubility of the precipitate decreases (smaller K_{sp}) and as the conjugate base becomes stronger. Note that the error becomes insignificant for anions derived from acids having dissociation constants greater than about 10^{-6}.

It is not difficult to write the algebraic relationships needed to calculate the solubility of such a precipitate; solution of the equations, however, is tedious. Fortunately, it is ordinarily possible to invoke one of two simplifying assumptions to decrease the algebraic labor.

1. The first simplification is applicable to moderately soluble compounds containing an anion that reacts extensively with water. Here, it is assumed that sufficient hydroxide ions are formed to permit neglect of the hydronium ion concentration in calculations. Another way of stating this assumption is to say that the hydroxide ion concentration of the solution is determined exclusively by the reaction of the anion with water and that the contribution of hydroxide ions from the dissociation of water itself is negligible by comparison.

2. The second assumption is applicable to precipitates of very low solubility, particularly those containing an anion that does not react extensively with water. In such a system, it can be assumed that the dissolving of the precipitate does not significantly change the hydronium or hydroxide ion concentrations and that at room temperature these concentrations remain essentially 10^{-7} mol/L. The solubility calculation then follows the course summarized on pages 96 and 97.

An example of each of these types of calculations follows.

EXAMPLE 4-4.   Calculate the solubility of $BaCO_3$ in water.
Step 1. Chemical equations

$$BaCO_3(s) \rightleftharpoons Ba^{2+} + CO_3^{2-} \tag{4-9}$$

$$CO_3^{2-} + H_2O \rightleftharpoons HCO_3^- + OH^- \tag{4-10}$$

$$HCO_3^- + H_2O \rightleftharpoons H_2CO_3 + OH^- \tag{4-11}$$

$$2H_2O \rightleftharpoons H_3O^+ + OH^- \tag{4-12}$$

Step 2. Definition of the unknown.

$$S = [Ba^{2+}]$$
$$= [CO_3^{2-}] + [HCO_3^-] + [H_2CO_3]$$

Step 3. Equilibrium constant expressions.

$$[Ba^{2+}][CO_3^{2-}] = K_{sp} = 5.1 \times 10^{-9} \tag{4-13}$$

$$\frac{[HCO_3^-][OH^-]}{[CO_3^{2-}]} = \frac{K_w}{K_2} = \frac{1.00 \times 10^{-14}}{4.7 \times 10^{-11}} = 2.13 \times 10^{-4} \qquad (4\text{-}14)$$

$$\frac{[H_2CO_3][OH^-]}{[HCO_3^-]} = \frac{K_w}{K_1} = \frac{1.00 \times 10^{-14}}{4.45 \times 10^{-7}} = 2.25 \times 10^{-8} \qquad (4\text{-}15)$$

and

$$[H_3O^+][OH^-] = 1.00 \times 10^{-14} \qquad (4\text{-}16)$$

Step 4. Mass-balance expression.

$$[Ba^{2+}] = [CO_3^{2-}] + [HCO_3^-] + [H_2CO_3] \qquad (4\text{-}17)$$

Step 5. Charge-balance expression.

$$2[Ba^{2+}] + [H_3O^+] = 2[CO_3^{2-}] + [HCO_3^-] + [OH^-] \qquad (4\text{-}18)$$

Step 6. Number of equations and unknowns. We have developed six equations (Equations 4-13 through 4-18), which should be sufficient to solve for the six unknowns, namely, $[Ba^{2+}]$, $[CO_3^{2-}]$, $[HCO_3^-]$, $[H_2CO_3]$, $[OH^-]$, and $[H_3O^+]$.

Step 7. Approximations. Here, we examine Equations 4-17 and 4-18 with the goal of eliminating one or more terms on the basis that their magnitudes are so small that their absence will not create a significant error. Generally, one cannot be certain whether or not it is appropriate to neglect a term. It is important to understand, however, that an invalid assumption here is not serious, for the error ultimately becomes obvious. It is usually more efficient, therefore, to make a simplifying assumption than to carry out a long and tedious calculation at the outset. If the assumption leads to an intolerable error, the longer calculation can then be performed.

Beginners generally find this step in an equilibrium calculation the most difficult with which to become comfortable. Thus, we will try to be as detailed as possible in showing the reasoning process that leads to the two assumptions we will make here.

First, we cannot eliminate $[Ba^{2+}]$ since this is the quantity we seek. Furthermore, we will surely need to use the solubility product expression ultimately. Thus, it is of no use to eliminate $[CO_3^{2-}]$. Our candidate then must be among $[HCO_3^-]$, $[H_2CO_3]$, $[H_3O^+]$, and $[OH^-]$.

We have already noted that the solution is basic. Therefore, $[H_3O^+]$ must be smaller than 10^{-7} M. An examination of Equation 4-13, however, suggests that $[Ba^{2+}]$ must be considerably larger than this figure. Thus, if no reaction occurred between CO_3^{2-} and H_2O, $[Ba^{2+}]$ would be simply the square root of K_{sp} or 7×10^{-5} M. The fact that reaction does occur means that $[CO_3^{2-}]$ is smaller than this figure and therefore $[Ba^{2+}] > 7 \times 10^{-5}$ M. It is clear then that we may assume that $[H_3O^+] \ll 2[Ba^{2+}]$ in Equation 4-18. We can, therefore, provisionally neglect $[H_3O^+]$ and delete Equation 4-16 from consideration.

If the formation of HCO_3^- and H_2CO_3 has a significant effect on the solubility, we will certainly need to use Equation 4-14 and/or 4-15. Thus, we cannot eliminate $[OH^-]$ and our candidates for rejection are reduced to $[H_2CO_3]$ and $[HCO_3^-]$. Equation 4-14 provides a basis for a decision as to whether or not either of these species can be neglected. We have already noted that $[OH^-] > 10^{-7}$. If it were just 10^{-7}, we find from Equation 4-15 that

$$\frac{[H_2CO_3]}{[HCO_3^-]} = \frac{2.25 \times 10^{-8}}{10^{-7}} = 0.225$$

If $[OH^-]$ is 10^{-6}, the ratio is about 0.02; if $[OH^-]$ is 10^{-5}, the ratio is 0.002. These calculations suggest that we may possibly be able to assume that $[H_2CO_3]$ is enough smaller than the sum of the other two terms in Equation 4-17 that it can be neglected. At this point, we cannot be sure that this assumption is valid, but it is surely worth trying.

As a result of the first assumption, Equation 4-18 becomes

$$2[Ba^{2+}] = 2[CO_3^{2-}] + [HCO_3^-] + [OH^-] \tag{4-19}$$

Further, insofar as

$$[HCO_3^-] \gg [H_2CO_3]$$

the mass-balance expression simplifies to

$$[Ba^{2+}] = [CO_3^{2-}] + [HCO_3^-] \tag{4-20}$$

Furthermore, Equations 4-15 and 4-16 are no longer needed. Thus, we have reduced the number of equations and unknowns to four.

Step 8. Solution of the equations. If we multiply Equation 4-20 by 2 and subtract the product from Equation 4-19, we obtain upon rearrangement

$$[OH^-] = [HCO_3^-] \tag{4-21}$$

Substitution of $[HCO_3^-]$ for $[OH^-]$ in Equation 4-14 gives

$$\frac{[HCO_3^-]^2}{[CO_3^{2-}]} = \frac{K_w}{K_2}$$

$$[HCO_3^-] = \sqrt{\frac{K_w}{K_2}[CO_3^{2-}]}$$

This expression permits elimination of $[HCO_3^-]$ from Equation 4-20:

$$[Ba^{2+}] = [CO_3^{2-}] + \sqrt{\frac{K_w}{K_2}[CO_3^{2-}]} \tag{4-22}$$

From Equation 4-13 we have

$$[CO_3^{2-}] = \frac{K_{sp}}{[Ba^{2+}]}$$

Substituting for $[CO_3^{2-}]$ in Equation 4-22 yields

$$[Ba^{2+}] = \frac{K_{sp}}{[Ba^{2+}]} + \sqrt{\frac{K_w K_{sp}}{K_2[Ba^{2+}]}}$$

It is convenient to multiply through by $[Ba^{2+}]$; upon rearranging terms,

$$[Ba^{2+}]^2 - \sqrt{\frac{K_w}{K_2}K_{sp}[Ba^{2+}]} - K_{sp} = 0$$

Finally, after numerical values have been supplied for the constants, we obtain

$$[Ba^{2+}]^2 - 1.04 \times 10^{-6}[Ba^{2+}]^{1/2} - 5.1 \times 10^{-9} = 0$$

This equation is readily solved by systematic approximations. If, for example, we let $[Ba^{2+}] = 0$, the left side of the equation has a value of -5.1×10^{-9}. On the other hand, when $[Ba^{2+}] = 1 \times 10^{-3}$, the equation yields a value of about 1×10^{-6}. That is,

$$(1 \times 10^{-3})^2 - (1.04 \times 10^{-6})(1 \times 10^{-3})^{1/2} - (5.1 \times 10^{-9}) = 9.6 \times 10^{-7}$$

Substituting $[Ba^{2+}] = 1 \times 10^{-4}$ gives

$$(1 \times 10^{-4})^2 - (1.04 \times 10^{-6})(1 \times 10^{-4})^{1/2} - (5.1 \times 10^{-9}) = -5.5 \times 10^{-9}$$

We see then that $[Ba^{2+}]$ must lie between 1×10^{-4} and 10^{-3}. The trial value $[Ba^{2+}] = 5 \times 10^{-4}$ yields 2.2×10^{-7}. The positive sign here indicates that the assumed value is too high. Further approximations reveal that

$$[Ba^{2+}] = 1.30 \times 10^{-4}$$

$$S = 1.3 \times 10^{-4} \, M$$

Step 9. Check of approximations. To check the two assumptions that were made, we must calculate the concentrations of most of the other ions in the solution. We can evaluate $[CO_3^{2-}]$ from Equation 4-13:

$$[CO_3^{2-}] = \frac{5.1 \times 10^{-9}}{1.30 \times 10^{-4}} = 3.9 \times 10^{-5}$$

From Equation 4-20

$$[HCO_3^-] = 1.30 \times 10^{-4} - 0.39 \times 10^{-4} = 9.1 \times 10^{-5}$$

From Equation 4-21

$$[OH^-] = [HCO_3^-] = 9.1 \times 10^{-5}$$

From Equation 4-15

$$\frac{[H_2CO_3][9.1 \times 10^{-5}]}{[9.1 \times 10^{-5}]} = 2.25 \times 10^{-8}$$

$$[H_2CO_3] = 2.2 \times 10^{-8}$$

Finally, from Equation 4-16

$$[H_3O^+] = \frac{1.00 \times 10^{-14}}{9.1 \times 10^{-5}} = 1.1 \times 10^{-10}$$

We see that the two assumptions should not lead to large errors; $[HCO_3^-]$ is about 4000 times greater than $[H_2CO_3]$, and $[H_3O^+]$ is clearly much smaller than any of the species in Equation 4-18.

Finally, failure to take account of the basic reaction of CO_3^{2-} would have yielded a solubility of 7.1×10^{-5}, which is only about one half the value yielded by the more rigorous method.

An example of the second type of calculation referred to on page 98 follows. Here, it is assumed that the hydronium and hydroxide ion concentrations are $10^{-7} \, M$ after the solid has dissolved.

EXAMPLE 4-5. Calculate the solubility of silver sulfide in pure water.
Step 1. Chemical equations.

$$Ag_2S(s) \rightleftharpoons 2Ag^+ + S^{2-} \tag{4-23}$$

$$S^{2-} + H_2O \rightleftharpoons HS^- + OH^- \tag{4-24}$$

$$HS^- + H_2O \rightleftharpoons H_2S + OH^- \tag{4-25}$$

$$2H_2O \rightleftharpoons H_3O^+ + OH^- \tag{4-26}$$

Step 2. Definition of unknown.

$$S = \tfrac{1}{2}[Ag^+] = [S^{2-}] + [HS^-] + [H_2S]$$

Step 3. Equilibrium-constant expressions.

$$[Ag^+]^2[S^{2-}] = 6 \times 10^{-50} \tag{4-27}$$

$$\frac{[HS^-][OH^-]}{[S^{2-}]} = \frac{K_w}{K_2} = \frac{1.0 \times 10^{-14}}{1.2 \times 10^{-15}} = 8.3 \tag{4-28}$$

$$\frac{[H_2S][OH^-]}{[HS^-]} = \frac{K_w}{K_1} = \frac{1.0 \times 10^{-14}}{5.7 \times 10^{-8}} = 1.8 \times 10^{-7} \tag{4-29}$$

$$[H_3O^+][OH^-] = 1.00 \times 10^{-14}$$

Step 4. Mass-balance expression.

$$\tfrac{1}{2}[Ag^+] = [S^{2-}] + [HS^-] + [H_2S] \tag{4-30}$$

Step 5. Charge-balance expression.

$$[Ag^+] + [H_3O^+] = 2[S^{2-}] + [HS^-] + [OH^-] \tag{4-31}$$

Step 6. Comparison of equations and unknowns. We see that the number of unknowns and equations is six. Thus, an exact solution is feasible.

Step 7. Approximations. The solubility product for Ag_2S is very small; therefore, it is probable that little alteration in the hydroxide ion concentration of the solution will occur as the precipitate dissolves. As a consequence, we will assume tentatively that

$$[OH^-] \cong [H_3O^+] = 1.0 \times 10^{-7}$$

This assumption will be correct, provided that in Equation 4-31

$$[Ag^+] \ll [H_3O^+] \quad \text{and} \quad (2[S^{2-}] + [HS^-]) \ll [OH^-]$$

We now proceed exactly as we did in Example 4-3 (p. 95). Substitution of 1.0×10^{-7} for $[OH^-]$ in Equations 4-28 and 4-29 gives

$$\frac{[HS^-]}{[S^{2-}]} = \frac{8.3}{1.0 \times 10^{-7}} = 8.3 \times 10^7$$

$$[H_2S] = \frac{1.8 \times 10^{-7}[HS^-]}{1.0 \times 10^{-7}} = 1.8 \times 8.3 \times 10^7[S^{2-}]$$

When these relationships are substituted into Equation 4-30, we obtain

$$\tfrac{1}{2}[Ag^+] = [S^{2-}] + 8.3 \times 10^7 [S^{2-}] + 14.9 \times 10^7 [S^{2-}]$$

$$[S^{2-}] = 2.3 \times 10^{-9} [Ag^+]$$

Substituting this relationship into the solubility product expression gives

$$2.3 \times 10^{-9} [Ag^+]^3 = 6 \times 10^{-50}$$

$$[Ag^+] = 3.0 \times 10^{-14}$$

$$S = \tfrac{1}{2}[Ag^+] = 1.5 \times 10^{-14} = 2 \times 10^{-14} \; M$$

Step 8. Check of approximations. The assumption that $[Ag^+]$ is much smaller than $[H_3O^+]$ is clearly valid. We can readily calculate a value for $(2[S^{2-}] + [HS^-])$ and confirm that this sum is likewise much smaller than $[OH^-]$. Therefore, we conclude that the assumptions made were reasonable and that the approximate solution obtained is satisfactory.

4A-3 SOLUBILITY OF METAL HYDROXIDES IN WATER

In determining the solubility of metal hydroxides, two equilibria may have to be considered. For example, with the divalent metal ion M^{2+}, these are

$$M(OH)_2(s) \rightleftharpoons M^{2+} + 2OH^-$$

$$2H_2O \rightleftharpoons H_3O^+ + OH^-$$

Three algebraic equations are readily derived for this system, namely,

$$[M^{2+}][OH^-]^2 = K_{sp} \tag{4-32}$$

$$[H_3O^+][OH^-] = K_w \tag{4-33}$$

and from charge-balance considerations

$$2[M^{2+}] + [H_3O^+] = [OH^-] \tag{4-34}$$

Manipulation of these three equations will permit a rigorous derivation of the solubility. In most instances, simplifications analogous to the ones described in the previous sections can be made, that is, (1) for moderately soluble precipitates, the hydronium ion will be sufficiently small so that it can be neglected or (2) for precipitates of low solubility, the hydroxide and hydronium ion concentrations are identical.

In the first case,

$$2[M^{2+}] = [OH^-]$$

Substitution of this expression into Equation 4-32 and rearrangement gives

$$[M^{2+}] = \left(\frac{K_{sp}}{4}\right)^{1/3} = S \tag{4-35}$$

In the second case, $2[M^{2+}]$ is much smaller than $[H_3O^+]$; Equation 4-34 then becomes

$$[H_3O^+] \cong [OH^-] = 1.00 \times 10^{-7}$$

Again, substitution into Equation 4-32 and rearrangement yields

$$[M^{2+}] = \frac{K_{sp}}{[OH^-]^2} = \frac{K_{sp}}{1.00 \times 10^{-14}} = S \tag{4-36}$$

EXAMPLE 4-6. Calculate the solubility of $Fe(OH)_3$ in water.

As a hypothesis, let us assume that the charge-balance expression simplifies to

$$3[Fe^{3+}] + [H_3O^+] \cong 3[Fe^{3+}] = [OH^-]$$

Substitution for $[OH^-]$ into the solubility product expression gives

$$[Fe^{3+}](3[Fe^{3+}])^3 = 4 \times 10^{-38}$$

$$[Fe^{3+}] = \left(\frac{4 \times 10^{-38}}{27}\right)^{1/4} = 2 \times 10^{-10}$$

and

$$S = 2 \times 10^{-10} \ M$$

We have assumed, however, that

$$[OH^-] \cong 3[Fe^{3+}] = 3 \times 2 \times 10^{-10} = 6 \times 10^{-10}$$

which means that

$$[H_3O^+] = \frac{1.00 \times 10^{-14}}{6 \times 10^{-10}} = 1.7 \times 10^{-5}$$

Clearly, $[H_3O^+]$ is not much smaller than $3[Fe^{3+}]$; indeed, the reverse appears to be the case. That is,

Table 4-3 Relative Errors Associated with Approximate Calculations for the Solubility of a Precipitate $M(OH)_2$

Assumed K_{sp}	Solubility Calculated Without Approximations	Solubility Calculated With Equation 4-35	Percent Error With Equation 4-35	Solubility Calculated With Equation 4-36	Percent Error With Equation 4-36
1.00×10^{-18}	6.3×10^{-7}	6.3×10^{-7}	0	1.00×10^{-4}	16,000
1.00×10^{-20}	1.24×10^{-7}	1.36×10^{-7}	9.7	1.00×10^{-6}	710
1.00×10^{-22}	8.4×10^{-9}	2.92×10^{-8}	250	1.00×10^{-8}	19
1.00×10^{-24}	1.00×10^{-10}	6.3×10^{-9}	6200	1.00×10^{-10}	0.00
1.00×10^{-26}	1.00×10^{-12}	1.36×10^{-9}	140,000	1.00×10^{-12}	0.00

$$3[Fe^{3+}] \ll [H_3O^+]$$

and the charge-balance equation reduces to

$$[H_3O^+] = [OH^-] = 1.00 \times 10^{-7}$$

Substitution for $[OH^-]$ in the solubility product expression yields

$$[Fe^{3+}] = \frac{4 \times 10^{-38}}{(1.00 \times 10^{-7})^3} = 4 \times 10^{-17}$$

$$S = 4 \times 10^{-17} \, M$$

The assumption that $3[Fe^{3+}] \ll [H_3O^+]$ is clearly valid. Note the very large error in the first calculation where the faulty assumption was made.

From the foregoing example, it is apparent that solubility calculations for metal hydroxides are analogous to those for compounds containing a weak conjugate base in the sense that they frequently can be made simpler by the proper choice of one of two assumptions. It is to be expected that a range of solubility products exists for which neither assumption is valid and for which Equations 4-32, 4-33, and 4-34 must be solved for all three variables. Table 4-3 shows the extent of this range for precipitates of the type $M(OH)_2$.

4A-4 EFFECT OF THE PRESENCE OF UNDISSOCIATED SOLUTE ON SOLUBILITY

Ordinarily, the assumption is made in solubility product calculations that the dissolved solute exists completely in the form of ions and that undissociated solute molecules are absent. In some instances, this assumption is not valid, and the neglect of the undissociated species may lead to serious errors in solubility calculations. For example, it has been found experimentally that in a saturated solution of calcium sulfate significant amounts of undissociated molecules are present as well as calcium and sulfate ions. Here, two equilibria are required to adequately describe the system. That is,

$$CaSO_4(s) \rightleftharpoons CaSO_4(aq)$$

$$CaSO_4(aq) \rightleftharpoons Ca^{2+} + SO_4^{2-}$$

The equilibrium constant for the first reaction takes the form

$$\frac{[CaSO_4(aq)]}{[CaSO_4(s)]} = K$$

where the numerator is the concentration of the undissociated species *in the solution* and the denominator is the concentration of calcium sulfate *in the solid phase*. The latter term is a constant, however (see p. 21), and the equation can, therefore, be written

$$[CaSO_4(aq)] = K[CaSO_4(s)] = K_s \tag{4-37}$$

It is evident from this equation that at a given temperature, the concentration of the undissociated calcium sulfate is constant and *independent* of the sulfate and calcium ion concentrations.

The equilibrium constant for the dissociation reaction is

$$\frac{[Ca^{2+}][SO_4^{2-}]}{[CaSO_4(aq)]} = K_d \tag{4-38}$$

The product of these two constants is equal to the solubility product. That is,

$$[Ca^{2+}][SO_4^{2-}] = K_d K_s = K_{sp}$$

As shown by the following example, both reactions contribute to the solubility of $CaSO_4$.

EXAMPLE 4-7. The dissociation constant K_d for $CaSO_4$ in aqueous solution has a value of 5.2×10^{-3}; its solubility product is 2.6×10^{-5}. Calculate the solubility of calcium sulfate in water and the percent of the dissolved solute that is present as the undissociated compound.

Here, the solubility S is given by

$$S = [CaSO_4(aq)] + [Ca^{2+}]$$

For the dissociation equilibrium, we write

$$\frac{[Ca^{2+}][SO_4^{2-}]}{[CaSO_4(aq)]} = 5.2 \times 10^{-3}$$

But the numerator of this equation is equal to K_{sp}. Thus,

$$\frac{K_{sp}}{[CaSO_4(aq)]} = 5.2 \times 10^{-3}$$

$$[CaSO_4(aq)] = \frac{2.6 \times 10^{-5}}{5.2 \times 10^{-3}} = 5.0 \times 10^{-3}$$

In a neutral solution, little HSO_4^- is present. Thus,

$$[Ca^{2+}] = [SO_4^{2-}]$$

or

$$[Ca^{2+}] = \sqrt{2.6 \times 10^{-5}} = 5.1 \times 10^{-3}$$

Therefore,

$$S = 5.0 \times 10^{-3} + 5.1 \times 10^{-3} = 1.01 \times 10^{-2} \ M$$

and the percent undissociated solute is

$$\% \, CaSO_4(aq) = \frac{5.0 \times 10^{-3}}{1.01 \times 10^{-2}} \times 100 = 50$$

Data on dissociation constants for salts are sparse and the assumption must usually be made that the concentration of the undissociated species is negligible. Fortunately, in contrast to the foregoing example, most sparingly soluble inorganic salts are essentially completely dissociated in aqueous solution. For example, in saturated solutions of the various silver halides, less than 3% of the solute is present in the molecular form. No evidence exists for the presence of undissociated $Ba(IO_3)_2$ in saturated solutions of that compound; less than 0.5% is present as $BaIO_3^+$. Often, then, serious errors are not incurred by assuming complete dissociation of the dissolved solute.

4A-5 COMPLEX ION FORMATION AND SOLUBILITY

The solubility of a precipitate may be greatly altered in the presence of some species that forms a soluble complex with the anion or cation of the precipitate. For example, the precipitation of aluminum with base is never complete in the presence of fluoride ion, even though aluminum hydroxide has an extremely low solubility; the fluoride complexes of aluminum(III) are sufficiently stable to prevent quantitative removal of the cation from solution. The equilibria involved can be represented by

$$Al(OH)_3(s) \rightleftharpoons Al^{3+} + 3OH^-$$
$$+$$
$$6F^-$$
$$\updownarrow$$
$$AlF_6^{3-}$$

Fluoride ions thus compete successfully with hydroxide ions for aluminum(III); as the fluoride concentration is increased, more and more of the precipitate is dissolved and converted to fluoroaluminate ions.

Quantitative treatment of the effect of complex formation on the solubility of precipitates. The solubility of a precipitate in the presence of a complexing reagent can be calculated, provided equilibrium constants for the complex formation reactions are known. The techniques used are similar to those discussed in the preceding sections.

EXAMPLE 4-8. Find the solubility of AgBr in a solution that is 0.10 M in NH_3.

Step 1. Chemical equations.

$$AgBr(s) \rightleftharpoons Ag^+ + Br^-$$
$$Ag^+ + NH_3 \rightleftharpoons AgNH_3^+$$
$$AgNH_3^+ + NH_3 \rightleftharpoons Ag(NH_3)_2^+$$
$$NH_3 + H_2O \rightleftharpoons NH_4^+ + OH^-$$

Step 2. Definition of the unknown.

$$S = [Br^-]$$
$$= [Ag^+] + [AgNH_3^+] + [Ag(NH_3)_2^+]$$

Step 3. Equilibrium constant expressions.

$$[Ag^+][Br^-] = K_{sp} = 5.2 \times 10^{-13} \tag{4-39}$$

$$\frac{[AgNH_3^+]}{[Ag^+][NH_3]} = K_1 = 2.0 \times 10^3 \tag{4-40}$$

$$\frac{[Ag(NH_3)_2^+]}{[AgNH_3^+][NH_3]} = K_2 = 6.3 \times 10^3 \tag{4-41}$$

$$\frac{[NH_4^+][OH^-]}{[NH_3]} = K_b = 1.76 \times 10^{-5} \tag{4-42}$$

Step 4. Mass-balance expression.

$$[Br^-] = [Ag^+] + [AgNH_3^+] + [Ag(NH_3)_2^+] \tag{4-43}$$

Since the NH_3 concentration was initially 0.10, we may also write

$$0.10 = [NH_3] + [AgNH_3^+] + 2[Ag(NH_3)_2^+] + [NH_4^+] \tag{4-44}$$

Furthermore, the reaction of NH_3 with water produces one OH^- for each NH_4^+. Thus,

$$[OH^-] \cong [NH_4^+] \tag{4-45}$$

Step 5. Charge-balance expression.

$$[NH_4^+] + [Ag^+] + [AgNH_3^+] + [Ag(NH_3)_2^+] = [Br^-] + [OH^-] \tag{4-46}[1]$$

Step 6. Comparison of equations and unknowns. Close examination of these eight equations reveals that there are only seven independent expressions, since Equation 4-46 is the sum of Equations 4-45 and 4-43. There are only seven unknowns, however, so an exact solution is possible.

Step 7. Approximations.

(a) $[NH_4^+]$ is much smaller than the other terms in Equation 4-44. This assumption seems reasonable in light of the rather small numerical value of the dissociation constant for NH_3 (Equation 4-42).

(b) $[Ag(NH_3)_2^+] \gg [AgNH_3^+]$ and $[Ag^+]$. The large values of the constants in Equations 4-40 and 4-41 suggest that this assumption is reasonable except for very dilute solutions of NH_3.

Application of these approximations to Equations 4-43 and 4-44 leads to the simplified expressions

$$[Br^-] \cong [Ag(NH_3)_2^+] \tag{4-47}$$

$$[NH_3] \cong 0.10 - 2[Ag(NH_3)_2^+] \tag{4-48}$$

Step 8. Solution of equations. Upon substituting Equation 4-47 into Equation 4-48, we obtain

$$[NH_3] = 0.10 - 2[Br^-] \tag{4-49}$$

Let us now multiply Equations 4-40 and 4-41 together to give

$$\frac{[Ag(NH_3)_2^+]}{[Ag^+][NH_3]^2} = K_1 K_2 = 1.26 \times 10^7 \tag{4-50}$$

Substitution of Equation 4-39 yields after rearrangement

---

[1] We have neglected the $[H_3O^+]$, since its concentration will certainly be negligible in a 0.10 M solution of NH_3.

$$\frac{[Br^-][Ag(NH_3)_2^+]}{[NH_3]^2} = 1.26 \times 10^7 \times 5.2 \times 10^{-13} = 6.55 \times 10^{-6}$$

When Equations 4-47 and 4-49 are introduced into this equation, we obtain

$$\frac{[Br^-]^2}{(0.1 - 2[Br^-])^2} = 6.55 \times 10^{-6}$$

Rearrangement yields the quadratic expression

$$[Br^-]^2 + 2.62 \times 10^{-6}\,[Br^-] - 6.55 \times 10^{-8} = 0$$

The solution for the quadratic equation is

$$[Br^-] = 2.6 \times 10^{-4}$$

$$S = 2.6 \times 10^{-4}\,M$$

A check of the assumptions will show that they were valid.

Complex formation involving a common ion of the precipitate. Many precipitates tend to react with one of their constituent ions to form soluble complexes. Well-known examples include the amphoteric hydroxides such as those of aluminum and zinc. Upon treatment with base, these ions first form sparingly soluble precipitates which, however, redissolve in the presence of excess hydroxide ions, giving the complex aluminate and zincate ions. For aluminum, the equilibria can be represented as

$$Al^{3+} + 3OH^- \rightleftharpoons Al(OH)_3(s)$$

$$Al(OH)_3(s) + OH^- \rightleftharpoons Al(OH)_4^-$$

Another common example involves silver chloride. Its solubility first decreases with the introduction of excess chloride, as predicted by the Le Châtelier principle. Subsequently, however, complex formation causes a reversal in this trend. The chloride ion concentration that corresponds to minimum solubility is readily calculated through use of the several equilibrium constants involved in the system.

EXAMPLE 4-9. Derive an expression for the solubility of AgCl as a function of the molar concentration of KCl and find the minimum in the solubility-concentration relationship. The equilibria that influence the system are

$$AgCl(s) \rightleftharpoons AgCl(aq) \tag{4-51}$$

$$AgCl(aq) \rightleftharpoons Ag^+ + Cl^- \tag{4-52}$$

$$AgCl(s) + Cl^- \rightleftharpoons AgCl_2^- \tag{4-53}$$

$$AgCl_2^- + Cl^- \rightleftharpoons AgCl_3^{2-} \tag{4-54}$$

The molar solubility of AgCl will be given by the sum of the concentration of all of the soluble, silver-containing species. That is, the molar solubility S is given by

$$S = [AgCl(aq)] + [Ag^+] + [AgCl_2^-] + [AgCl_3^{2-}] \tag{4-55}$$

Equilibrium constants found in the literature include

$$\frac{[Ag^+][Cl^-]}{[AgCl(aq)]} = K_d = 3.9 \times 10^{-4} \tag{4-56}$$

$$[Ag^+][Cl^-] = K_{sp} = 1.82 \times 10^{-10} \tag{4-57}$$

$$\frac{[AgCl_2^-]}{[Cl^-]} = K_1 = 2.0 \times 10^{-5} \tag{4-58}$$

$$\frac{[AgCl_3^{2-}]}{[AgCl_2^-][Cl^-]} = K_2 = 1 \tag{4-59}$$

For convenience, we will multiply the last two equations together to give

$$\frac{[AgCl_3^{2-}]}{[Cl^-]^2} = K_1 K_2 = 2.0 \times 10^{-5} \tag{4-60}$$

As noted earlier (p. 106), a few percent of silver chloride remains undissociated in solution as AgCl(aq). In this calculation, we will take its presence into account and demonstrate that under certain conditions it is the major constituent in the solution. To calculate [AgCl(aq)], we divide Equation 4-57 by Equation 4-56 and rearrange to give

$$[AgCl(aq)] = \frac{K_{sp}}{K_d} = \frac{1.82 \times 10^{-10}}{3.9 \times 10^{-4}} = 4.7 \times 10^{-7}$$

Note that the concentration of this species is constant and independent of chloride concentration.

Substitution of this expression as well as Equations 4-57, 4-58, and 4-60 into Equation 4-55 permits expression of S in terms of the chloride ion concentration and the several constants. That is,

$$S = \frac{K_{sp}}{K_d} + \frac{K_{sp}}{[Cl^-]} + K_1[Cl^-] + K_1 K_2[Cl^-]^2 \tag{4-61}$$

From charge-balance considerations we can write

$$[K^+] + [Ag^+] + [\cancel{H_3O^+}] = [Cl^-] + [AgCl_2^-] + 2[AgCl_3^{2-}] + [\cancel{OH^-}]$$

If the solution is neutral, $[H_3O^+]$ and $[OH^-]$ will cancel; furthermore, $[K^+]$ will be the analytical concentration of KCl, M_{KCl}. Thus,

$$M_{KCl} = [Cl^-] + [AgCl_2^-] + 2[AgCl_3^{2-}] - [Ag^+]$$

Let us assume initially that the minimum in the solubility will occur at a concentration of $[Cl^-]$ that is sufficiently large that

$$[Cl^-] \gg [AgCl_2^-] + 2[AgCl_3^{2-}] - [Ag^+]$$

Under these circumstances,

$$M_{KCl} \cong [Cl^-]$$

We must ultimately check the validity of this assumption. In the meanwhile, substitution into Equation 4-61 gives

$$S = \frac{K_{sp}}{K_d} + \frac{K_{sp}}{M_{KCl}} + K_1 M_{KCl} + K_1 K_2 M_{KCl}^2 \tag{4-62}$$

To find the minimum, we set the derivative of S with respect to M_{KCl} equal to zero. Thus,

$$\frac{dS}{dM_{KCl}} = 0 = -\frac{K_{sp}}{M_{KCl}^2} + K_1 + 2K_1 K_2 M_{KCl}$$

which rearranges to

$$2K_1 K_2 M_{KCl}^3 + K_1 M_{KCl}^2 - K_{sp} = 0$$

Substituting numerical values gives

$$4.0 \times 10^{-5} M_{KCl}^3 + 2.0 \times 10^{-5} M_{KCl}^2 - 1.82 \times 10^{-10} = 0$$

Following the procedure shown on page 100, this equation can be solved by systematic approximations to give

$$M_{KCl} = 0.0030 = [Cl^-]$$

In order to check the assumption, we will calculate the concentration of the various species. Substitutions into Equations 4-57, 4-58, and 4-60 yield

$$[Ag^+] = 1.82 \times 10^{-10}/0.0030 = 6.1 \times 10^{-8}$$

$$[AgCl_2^-] = 2.0 \times 10^{-5} \times 0.0030 = 6.0 \times 10^{-8}$$

$$[AgCl_3^{2-}] = 2.0 \times 10^{-5} \times (0.0030)^2 = 1.8 \times 10^{-10}$$

We note that the assumption that M_{KCl} is much larger than the concentrations of the ions of the precipitate is reasonable. The minimum solubility is obtained by substitution of these concentrations and [AgCl(aq)] into Equation 4-55:

$$S = 4.7 \times 10^{-7} + 6.1 \times 10^{-8} + 6.00 \times 10^{-8} + 1.8 \times 10^{-10}$$

$$= 5.9 \times 10^{-7} M$$

The solid curve in Figure 4-1 illustrates the effect of chloride ion concentration on the solubility of silver chloride; data for the curve were obtained by substi-

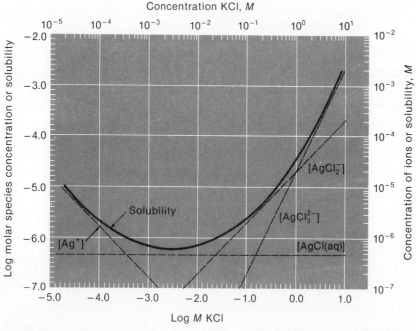

FIGURE 4-1 The effect of the concentration of chloride ion on the solubility of AgCl. The solid curve shows the total concentration of dissolved AgCl. The broken lines show the concentrations of the various silver-containing species.

tuting various chloride concentrations into Equation 4-62. Note that at high concentrations of the common ion, the solubility becomes greater than that in pure water. The broken lines represent the changes in equilibrium concentrations of the various silver-containing species as a function of the molar potassium chloride concentration. Note that at the solubility minimum, the undissociated silver chloride, AgCl(aq), is the major silver species in the solution, representing about 80% of the total dissolved silver. Note also that its concentration is invariant.

Unfortunately, reliable equilibrium data regarding undissociated species such as AgCl(aq) and complex species such as $AgCl_2^-$ are not abundant; consequently, solubility calculations are often, of necessity, based on solubility product equilibria alone. The example just considered and Example 4-7 show that under some circumstances such neglect of other equilibria can lead to serious error.

4B SEPARATION OF IONS BY CONTROL OF THE CONCENTRATION OF THE PRECIPITATING REAGENT

When two competing ions react with a reagent to form precipitates of different solubilities, the less soluble species will be produced at a lower reagent concentration. If the solubilities are sufficiently different, quantitative removal of the first ion from solution may be achieved without precipitation of the second. Such separations require careful control of the precipitating reagent concentration at some suitable, predetermined level. Many important analytical separations, notably those involving sulfide ion, hydroxide ion, and organic reagents, are based on this method.

4B-1 CALCULATION OF THE FEASIBILITY OF SEPARATIONS

An important application of solubility product calculations involves determining the feasibility and the optimum conditions for separations based on the control of reagent concentration. The following example illustrates such an application.

> **EXAMPLE 4-10.** Is the difference between the solubilities of their hydroxides sufficient to permit the quantitative separation of Fe^{3+} and Mg^{2+} in a solution that is 0.10 M in each cation? If the separation is possible, what range of OH^- concentrations is permissible? Solubility product constants for the two hydroxides are
>
> $$[Fe^{3+}][OH^-]^3 = 4 \times 10^{-38}$$
>
> $$[Mg^{2+}][OH^-]^2 = 1.8 \times 10^{-11}$$
>
> The K_{sp} for $Fe(OH)_3$ is so much smaller than that for $Mg(OH)_2$ as to suggest that the former will precipitate at a lower OH^- concentration.[2]
>
> We can answer the questions posed in this problem by (1) calculating the OH^- concentration required to achieve the quantitative precipitation of Fe^{3+} from this solution and (2) determining the OH^- concentration at which $Mg(OH)_2$ will just begin to precipitate. If (1) is smaller than (2), a separation is feasible, and the range of permissible OH^- concentrations will be defined by the two values.
>
> To determine (1), we must first decide what constitues a quantitative removal of Fe^{3+} from the solution. Under no conditions can every iron(III) ion be precipita-

---

[2] The reader should be aware, however, that it is only the enormous numerical difference between the two constants that permits this judgment. In computing the K_{sp} for $Mg(OH)_2$, the hydroxide ion concentration is squared while for $Fe(OH)_3$, it is cubed; strictly then, the two constants are not comparable.

ted; we must therefore *arbitrarily* set some limit below which, for all practical purposes, the further presence of this ion can be neglected. When its concentration has been decreased to 10^{-5} M, only 1 part in 10,000 of the original quantity of Fe^{3+} will remain in the solution; for most purposes, removal of all but this fraction of a species is more than adequate for a quantitative separation.

We can readily calculate the OH^- concentration in equilibrium with 1.0×10^{-5} M Fe^{3+} by substituting directly into the solubility product expression:

$$(1.0 \times 10^{-5})[OH^-]^3 = 4 \times 10^{-38}$$

$$[OH^-] = 2 \times 10^{-11}$$

Thus, if we maintain the OH^- concentration at about 2×10^{-11} mol/L, the Fe^{3+} concentration will be lowered to 1×10^{-5} mol/L. Note that quantitative precipitation of $Fe(OH)_3$ is achieved in a distinctly acidic solution.

We must now consider question (2), that is, what is the maximum OH^- concentration that can exist in solution without causing formation of $Mg(OH)_2$. Precipitation cannot occur until the Mg^{2+} concentration multiplied by the square of the OH^- concentration exceeds the solubility product, 1.8×10^{-11}. By substituting 0.1 (the molar Mg^{2+} concentration of the solution) into the solubility product expression, we can calculate the *maximum* OH^- concentration that can be tolerated:

$$0.10[OH^-]^2 = 1.8 \times 10^{-11}$$

$$[OH^-] = 1.3 \times 10^{-5}$$

When the OH^- concentration exceeds this level, the solution will be supersaturated with respect to $Mg(OH)_2$, and precipitation can begin.

From these calculations, we conclude that quantitative separation of $Fe(OH)_3$ can be expected if the OH^- concentration is greater than 2×10^{-11} mol/L and that $Mg(OH)_2$ will not precipitate until a concentration of 1.3×10^{-5} mol/L is reached. Therefore, it should, in principle, be possible to separate Fe^{3+} from Mg^{2+} by maintaining the OH^- concentration between these levels. In practice, the concentration of OH^- is kept as low as practical—perhaps about 10^{-10} M.

4B-2 SULFIDE SEPARATIONS

A number of important methods for the separation of metallic ions involve controlling the concentration of the precipitating anion by regulating the hydronium ion concentration of the solution. Such methods are particularly attractive because of the relative ease with which the hydronium ion concentration can be maintained at some predetermined level by the use of a suitable buffer.[3] Perhaps the best known of these methods makes use of hydrogen sulfide as the precipitating reagent. Hydrogen sulfide is a weak acid, dissociating as follows:

$$H_2S + H_2O \rightleftharpoons H_3O^+ + HS^- \qquad \frac{[H_3O^+]^2[HS^-]}{[H_2S]} = K_1 = 5.7 \times 10^{-8}$$

$$HS^- + H_2O \rightleftharpoons H_3O^+ + S^{2-} \qquad \frac{[H_3O^+][S^{2-}]}{[H_2S]} = K_2 = 1.2 \times 10^{-15}$$

These equations may be combined to give an expression for the overall dissociation of hydrogen sulfide into sulfide ion:

[3] The preparation and properties of buffer solutions are considered in Section 8C-3. An important property of a buffer is that it maintains the hydronium ion concentration at an approximately fixed and predetermined level.

$$H_2S + 2H_2O \rightleftharpoons 2H_3O^+ + S^{2-} \qquad \frac{[H_3O^+]^2[S^{2-}]}{[H_2S]} = K_1K_2 = 6.8 \times 10^{-23}$$

The constant for this overall reaction is simply the product of K_1 and K_2.

In sulfide separations, the solutions are continuously kept saturated with hydrogen sulfide; thus, the molar concentration of the reagent is essentially constant throughout the precipitation. Because it is such a weak acid, the actual molar concentration of hydrogen sulfide will correspond closely to its solubility in water, which is about 0.1 M. It is thus permissible to assume that throughout any sulfide precipitation

$$[H_2S] \cong 0.10 \text{ mol/L}$$

Substituting this value into the overall dissociation constant expression, we obtain

$$\frac{[H_3O^+]^2[S^{2-}]}{0.10} = 6.8 \times 10^{-23}$$

$$[S^{2-}] = \frac{6.8 \times 10^{-24}}{[H_3O^+]^2} \tag{4-63}$$

Thus, the molar concentration of the sulfide ion varies inversely as the square of the hydronium ion concentration of the solution. This relationship is useful for calculating the optimum conditions for the separation of cations by sulfide precipitation.

EXAMPLE 4-11. Find the conditions under which Cd^{2+} and Tl^+ can, in theory, be separated quantitatively with H_2S from a solution that is 0.1 M in each cation.

The constants for the two solubility equilibria are:

$$CdS(s) \rightleftharpoons Cd^{2+} + S^{2-} \qquad [Cd^{2+}][S^{2-}] = 2 \times 10^{-28}$$

$$Tl_2S(s) \rightleftharpoons 2Tl^+ + S^{2-} \qquad [Tl^+]^2[S^{2-}] = 1 \times 10^{-22}$$

CdS will precipitate at a lower S^{2-} concentration than the Tl_2S. By assuming again that lowering the Cd^{2+} concentration to $10^{-5} M$ or less constitutes quantitative removal and substituting this value into the solubility product expression, we can evaluate the required sulfide ion concentration.

$$10^{-5}[S^{2-}] = 2 \times 10^{-28}$$

$$[S^{2-}] = 2 \times 10^{-23}$$

This value should then be compared with the S^{2-} concentration needed to initiate precipitation of Tl_2S from a 0.1 M solution:

$$(0.1)^2[S^{2-}] = 1 \times 10^{-22}$$

$$[S^{2-}] = 1 \times 10^{-20}$$

Thus, to achieve a separation the S^{2-} concentration must be kept between 2×10^{-23} and $1 \times 10^{-20} M$. Now, we must compute the H_3O^+ concentrations necessary to hold the S^{2-} concentration within these confines. Substituting the two limiting values for $[S^{2-}]$ into Equation 4-63, we obtain

$$[H_3O^+]^2 = \frac{6.8 \times 10^{-24}}{2 \times 10^{-23}} = 3.4 \times 10^{-1}$$

$$[H_3O^+] = 0.58 \cong 0.6$$

and

$$[H_3O^+]^2 = \frac{6.8 \times 10^{-24}}{1 \times 10^{-20}} = 6.8 \times 10^{-4}$$

$$[H_3O^+] = 0.026 \cong 0.03$$

By maintaining the H_3O^+ concentration between 0.03 and 0.6 M, it should, in principle, be possible to separate CdS without precipitation of Tl_2S. From a practical standpoint, however, it is questionable whether conditions could be controlled closely enough to give a clean separation.

4C EFFECT OF ELECTROLYTE CONCENTRATION ON SOLUBILITY

It is found experimentally that precipitates tend to be more soluble in an electrolyte solution than in water, provided, of course, that the electrolyte contains no ions in common with the precipitate. The data plotted in Figure 4-2 demonstrate the magnitude of this effect for three precipitates. A twofold increase in the solubility of barium sulfate is observed when the potassium nitrate concentration is increased from 0 to 0.02 M. The same change in electrolyte concentration increases the solubility of barium iodate by a factor of only 1.25 and of silver chloride by 1.20.

The effect of an electrolyte upon solubility stems from the electrostatic attraction between the foreign ions and the ions of opposite charge in the precipitate. Such interactions shift the position of the equilibrium. It is important to realize that this effect is not peculiar to solubility equilibria but is observed with all other types as well. For example, the data in Table 4-4 show that the degree of dissociation of acetic acid increases significantly in the presence of sodium chloride. These experimental dissociation constants were obtained by measuring the equilibrium *concentrations* of hydronium and acetate ions in solutions containing the indicated salt concentrations. Again, the obvious shift in equilibrium can be attributed to the attraction between the ions of the electrolyte and the charged hydronium and acetate ions.

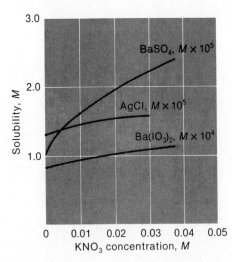

FIGURE 4-2 Effect of electrolyte concentration on the solubility of some salts.

Table 4-4 Dissociation Constants for Acetic Acid in Solutions of Sodium Chloride at 25°C*

Concentration of NaCl, M	Apparent K_a'
0.00	1.75×10^{-5}
0.02	2.29×10^{-5}
0.11	2.85×10^{-5}
0.51	3.31×10^{-5}
1.01	3.16×10^{-5}

* From H. S. Harned and C. F. Hickey, *J. Amer. Chem. Soc.* **59**, 1289 (1937). With permission of the American Chemical Society.

From data such as these, one must conclude that the equilibrium law, as we have presented it thus far, is a *limiting law* in the sense that it applies exactly only to very dilute solutions in which the electrolyte concentration is insignificant (that is, to ideal solutions only; see p. 18). We must now consider a more rigorous form of the law which can be applied to nonideal solutions.

4C-1 SOME EMPIRICAL OBSERVATIONS

Extensive studies concerned with the influence of electrolyte concentration upon chemical equilibrium have led to a number of important generalizations. One is that the magnitude of the effect is highly dependent upon the charges of the species involved in the equilibrium. Where all are neutral particles, little variation in the equilibrium constant is observed. On the other hand, the effects become greater as the charges on the reactants or products increase. Thus, for example, of the two equilibria

$$AgCl(s) \rightleftharpoons Ag^+ + Cl^-$$

$$BaSO_4(s) \rightleftharpoons Ba^{2+} + SO_4^{2-}$$

the second is shifted farther to the right in the presence of moderate amounts of potassium nitrate than is the first (see Figure 4-2).

A second important generality is that over a considerable electrolyte concentration range, the effects are essentially independent of the *kind* of electrolyte and dependent only upon a concentration parameter of the solution called the *ionic strength*. This quantity is defined by the equation

$$\text{ionic strength} = \mu = \tfrac{1}{2}(m_1 Z_1^2 + m_2 Z_2^2 + m_3 Z_3^2 + \ldots) \tag{4-64}$$

where $m_1, m_2, m_3, \ldots$, represent the molar concentrations of the various ions in the solution, and $Z_1, Z_2, Z_3, \ldots$, are their respective charges.

EXAMPLE 4-12. Calculate the ionic strength of (1) a 0.1 M solution of KNO_3 and (2) a 0.1 M solution of Na_2SO_4.
(1) For the KNO_3 solution, m_{K+} and $m_{NO_3^-}$ are 0.1 and
   $\mu = \tfrac{1}{2}(0.1 \times 1^2 + 0.1 \times 1^2) = 0.1$
(2) For the Na_2SO_4 solution, $m_{Na+} = 0.2$ and $m_{SO_4^{2-}} = 0.1$. Therefore,
   $\mu = \tfrac{1}{2}(0.2 \times 1^2 + 0.1 \times 2^2) = 0.3$

EXAMPLE 4-13. What is the ionic strength of a solution that is both 0.05 M in KNO_3 and 0.1 M in Na_2SO_4?

$$\mu = \tfrac{1}{2}(0.05 \times 1^2 + 0.05 \times 1^2 + 0.2 \times 1^2 + 0.1 \times 2^2)$$
$$= 0.35$$

It is apparent from these examples that the ionic strength of a strong electrolyte solution consisting solely of singly charged ions is identical with the total molar salt concentration. On the other hand, the ionic strength is greater than the molar concentration if the solution contains ions with multiple charges.

For solutions with ionic strengths of 0.1 or less, it is found that the electrolyte effect is independent of the *kind* of ions and dependent *only upon the ionic strength*. Thus, the degree of dissociation of acetic acid is the same in the presence of sodium chloride, potassium nitrate, or barium iodide, provided the concentrations of these species are such that the ionic strength is fixed. It should be noted that this independence with respect to electrolyte species disappears at high ionic strengths.

4C-2 ACTIVITY AND ACTIVITY COEFFICIENTS

In order to describe the effect of ionic strength on equilibria in quantitative terms, chemists use a concentration parameter called the *activity*, which is defined as follows:

$$a_A = [A]f_A \tag{4-65}$$

where a_A is the activity of the species A, $[A]$ is its molar concentration, and f_A is a dimensionless quantity called the *activity coefficient*. The activity coefficient (and thus the activity) of A varies with ionic strength such that employment of a_A instead of $[A]$ in an equilibrium constant expression frees the numerical value of the constant from dependence on the ionic strength. To illustrate, for the dissociation of acetic acid we write

$$K_a = \frac{a_{H_3O^+} \cdot a_{OAc^-}}{a_{HOAc}} = \frac{[H_3O^+][OAc^-]}{[HOAc]} \times \frac{f_{H_3O^+} \cdot f_{OAc^-}}{f_{HOAc}}$$

where $f_{H_3O^+}$, f_{OAc^-}, and f_{HOAc} vary with ionic strength to keep K_a numerically constant over a wide range of ionic strengths (in contrast to the *apparent* K_a' shown in Table 4-4).

Properties of activity coefficients. Activity coefficients have the following properties:

1. The activity coefficient of a species can be thought of as a measure of the effectiveness with which that species influences an equilibrium in which it is a participant. In very dilute solutions, where the ionic strength is minimal, this effectiveness becomes constant, and the activity coefficient acquires a value of unity. Under such circumstances, the activity and molar concentration become identical. As the ionic strength increases, however, an ion loses some of its effectiveness, and its activity coefficient decreases. We may summarize this behavior in terms of Equation 4-65. At

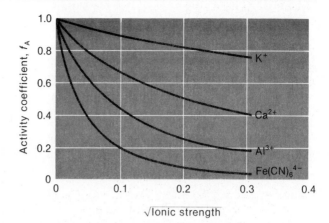

FIGURE 4-3 Effect of ionic strength on activity coefficients.

moderate ionic strengths $f_A < 1$; as the solution approaches infinite dilution, however, $f_A \rightarrow 1$ and thus $a_A \rightarrow [A]$.

At high ionic strengths, the activity coefficients for some species increase and may even become greater than one. Interpretation of the behavior of solutions in this region is difficult; we shall confine our discussion to regions of low or moderate ionic strengths (that is, where $\mu < 0.1$).

The variation of typical activity coefficients as a function of ionic strength is shown in Figure 4-3.

2. In solutions that are not too concentrated, the activity coefficient for a given species is independent of the specific nature of the electrolyte and dependent only upon the ionic strength.

3. For a given ionic strength, the activity coefficient of an ion departs farther from unity as the charge carried by the species increases. This effect is shown in Figure 4-3. The activity coefficient of an uncharged molecule is approximately unity, regardless of ionic strength.

4. For ions of the same charge, activity coefficients are approximately the same at any given ionic strength. The small variations that do exist can be correlated with the effective diameter of the hydrated ions.

5. The activity coefficient of a given ion describes its effective behavior in all equilibria in which it participates. For example, at a given ionic strength, a single activity coefficient for cyanide ion describes the influence of that species upon any of the following equilibria:

$$HCN + H_2O \rightleftharpoons H_3O^+ + CN^-$$

$$AgCN(s) \rightleftharpoons Ag^+ + CN^-$$

$$Ni(CN)_4^{2-} \rightleftharpoons Ni^{2+} + 4CN^-$$

Evaluation of activity coefficients. In 1923, Debye and Hückel derived the following theoretical expression, which permits the calculation of activity coefficients of ions:[4]

[4] P. Debye and E. Hückel, *Physik. Z.*, **24**, 185 (1923).

$$-\log f_{\mathrm{A}} = \frac{0.5085\, Z_{\mathrm{A}}^2 \sqrt{\mu}}{1 + 0.3281\, a_{\mathrm{A}}\sqrt{\mu}} \tag{4-66}$$

where

f_{A} = activity coefficient of the species A

Z_{A} = charge on the species A

μ = ionic strength of the solution

a_{A} = the effective diameter of the hydrated ion A in ångström units (Å)
$\qquad$ (1 Å = 10^{-8} cm)

The constants 0.5085 and 0.3281 are applicable to solutions at 25°C; other values must be employed at different temperatures.

$\qquad$ Unfortunately, considerable uncertainty exists regarding the magnitude of a_{A} in Equation 4-66. Its value appears to be approximately 3 Å for most singly charged ions; for these species, then, the denominator of the Debye-Hückel equation simplifies to approximately $(1 + \sqrt{\mu})$. For ions with higher charge, a_{A} may become as large as 10 Å. It should be noted that the second term of the denominator becomes small with respect to the first when the ionic strength is less than 0.01; under these circumstances, uncertainties in a_{A} are of little significance in calculating activity coefficients.

$\qquad$ Kielland[5] has calculated values of a_{A} for numerous ions from a variety of experimental data. His "best values" for effective diameters are given in Table 4-5. Also presented are activity coefficients calculated from Equation 4-66 using these values for the size parameter.

$\qquad$ Experimental verification of individual activity coefficients such as those shown in Table 4-5 is unfortunately impossible; all experimental methods give only a mean activity coefficient for the positively and negatively charged ions in a solution.[6] It should be pointed out, however, that mean coefficients calculated from the data in Table 4-5 agree satisfactorily with the experimental values.

$\qquad$ The Debye-Hückel relationship and the data in Table 4-5 give satisfactory activity coefficients for ionic strengths up to about 0.1. Beyond this value, however,

---

[5] J. Kielland, *J. Amer. Chem. Soc.*, **59**, 1675 (1937).

[6] The mean activity of the electrolyte $A_m B_n$ is defined as follows:

$f_{\pm}$ = mean activity coefficient = $(f_{\mathrm{A}}^m \cdot f_{\mathrm{B}}^n)^{1/(m+n)}$

The mean activity coefficient can be measured in any of several ways, but it is impossible experimentally to resolve this term into the individual activity coefficients for f_{A} and f_{B}. For example, if $A_m B_n$ is a precipitate, we can write

$K_{\mathrm{sp}} = [\mathrm{A}]^m [\mathrm{B}]^n \cdot f_{\mathrm{A}}^m \cdot f_{\mathrm{B}}^n = [\mathrm{A}]^m [\mathrm{B}]^n \cdot f_{\pm}^{(m+n)}$

By measuring the solubility of $A_m B_n$ in a solution in which the electrolyte concentration approaches zero (that is, where f_{A} and $f_{\mathrm{B}} \to 1$), we could obtain K_{sp}. A second solubility measurement at some ionic strength μ_1 would give values for [A] and [B]. These data would then permit the calculation of $f_{\mathrm{A}}^m \cdot f_{\mathrm{B}}^n = f_{\pm}^{(m+n)}$ for ionic strength μ_1. It is important to understand that there are insufficient experimental data to permit the calculation of the *individual* quantities f_{A} and f_{B} and that there appears to be no additional experimental information that would permit evaluation of these quantities. This situation is general; the *experimental* determination of individual activity coefficients appears to be impossible.

Table 4-5   Activity Coefficient for Ions at 25°C

Ion	a_A Effective Diameter, Å	Activity Coefficient at Indicated Ionic Strengths				
		0.001	0.005	0.01	0.05	0.1
H_3O^+	9	0.967	0.933	0.914	0.86	0.83
Li^+, $C_6H_5COO^-$	6	0.965	0.929	0.907	0.84	0.80
Na^+, IO_3^-, HSO_3^-, HCO_3^-, $H_2PO_4^-$, $H_2AsO_4^-$, OAc^-	4–4.5	0.964	0.928	0.902	0.82	0.78
OH^-, F^-, SCN^-, HS^-, ClO_3^-, ClO_4^-, BrO_3^-, IO_4^-, MnO_4^-	3.5	0.964	0.926	0.900	0.81	0.76
K^+, Cl^-, Br^-, I^-, CN^-, NO_2^-, NO_3^-, $HCOO^-$	3	0.964	0.925	0.899	0.80	0.76
Rb^+, Cs^+, Tl^+, Ag^+, NH_4^+	2.5	0.964	0.924	0.898	0.80	0.75
Mg^{2+}, Be^{2+}	8	0.872	0.755	0.69	0.52	0.45
Ca^{2+}, Cu^{2+}, Zn^{2+}, Sn^{2+}, Mn^{2+}, Fe^{2+}, Ni^{2+}, Co^{2+}, Phthalate^{2-}	6	0.870	0.749	0.675	0.48	0.40
Sr^{2+}, Ba^{2+}, Cd^{2+}, Hg^{2+}, S^{2-}	5	0.868	0.744	0.67	0.46	0.38
Pb^{2+}, CO_3^{2-}, SO_3^{2-}, $C_2O_4^{2-}$	4.5	0.868	0.742	0.665	0.46	0.37
Hg_2^{2+}, SO_4^{2-}, $S_2O_3^{2-}$, CrO_4^{2-}, HPO_4^{2-}	4.0	0.867	0.740	0.660	0.44	0.36
Al^{3+}, Fe^{3+}, Cr^{3+}, La^{3+}, Ce^{3+}	9	0.738	0.54	0.44	0.24	0.18
PO_4^{3-}, $Fe(CN)_6^{3-}$	4	0.725	0.50	0.40	0.16	0.095
Th^{4+}, Zr^{4+}, Ce^{4+}, Sn^{4+}	11	0.588	0.35	0.255	0.10	0.065
$Fe(CN)_6^{4-}$	5	0.57	0.31	0.20	0.048	0.021

the equation fails, and mean activity coefficients must be determined experimentally.

Solubility calculations employing activity coefficients.   The use of activities rather than molar concentrations in equilibrium constant calculations yields more accurate information. Unless otherwise specified, values for K_{sp} found in tables are generally constants based upon activities (activity-based constants are sometimes called *thermodynamic constants*). Thus, for the precipitate A_mB_n we may write

$$K_{sp} = a_A^m \cdot a_B^n = [A]^m[B]^n \cdot f_A^m \cdot f_B^n$$

or

$$[A]^m[B]^n = \frac{K_{sp}}{f_A^m \cdot f_B^n}$$
$$= K_{sp}'$$

where the bracketed terms are *molar concentrations* of A and B. Division of the thermodynamic constant, K_{sp}, by the product of the activity coefficients for A and B (or the mean activity coefficient) yields a *concentration constant* K_{sp}' that is applicable to a solution of a particular ionic strength. This constant can then be employed in the equilibrium calculations as discussed earlier. The following example will demonstrate the procedure.

EXAMPLE 4-14. Use activities to calculate the solubility of $Ba(IO_3)_2$ in a 0.033 M solution of $Mg(IO_3)_2$. The thermodynamic solubility product for $Ba(IO_3)_2$ has a value of 1.57×10^{-9} (Appendix 3).

At the outset, we may write

$$[Ba^{2+}][IO_3^-]^2 = \frac{1.57 \times 10^{-9}}{f_{Ba^{2+}} + f_{IO_3^-}^2} = K_{sp}'$$

We must next calculate activity coefficients for Ba^{2+} and IO_3^- ions from the ionic strength of the solution. Thus,

$$\mu = \tfrac{1}{2}[m_{Mg^{2+}} \times (2)^2 + m_{IO_3^-} \times (1)^2]$$
$$= \tfrac{1}{2}(0.033 \times 4 + 0.066 \times 1) = 0.099 \cong 0.1$$

In calculating μ, we have assumed that the Ba^{2+} and IO_3^- ions from the precipitate do not significantly affect the ionic strength of the solution. This simplification seems justified, considering the low solubility of barium iodate. In situations where it is not possible to make the assumption, the concentrations of the two ions can be approximated by an ordinary calculation, assuming activities and concentrations to be identical. These concentrations can then be introduced to give a better value for μ.

Turning now to Table 4-5, we find that at an ionic strength of 0.1

$$f_{Ba^{2+}} = 0.38 \qquad f_{IO_3^-} = 0.78$$

If the calculated ionic strength did not match that of one of the columns in the table, $f_{Ba^{2+}}$ and $f_{IO_3^-}$ could be calculated from Equation 4-66.

We may thus write

$$\frac{1.57 \times 10^{-9}}{(0.38)(0.78)^2} = 6.8 \times 10^{-9} = K_{sp}'$$

$$[Ba^{2+}][IO_3^-]^2 = 6.8 \times 10^{-9}$$

Proceeding now as for an ordinary solubility calculation,

$$S = [Ba^{2+}]$$

$$[IO_3^-] \cong 0.066$$

$$S(0.066)^2 = 6.8 \times 10^{-9}$$

$$S = 1.6 \times 10^{-6} \, M$$

It is of interest to note that the calculated solubility, neglecting the effects of ionic strength, is $3.6 \times 10^{-7} \, M$.

Omission of activity coefficients in equilibrium calculations.

We shall ordinarily neglect activity coefficients and simply use molar concentrations in applications of the equilibrium law. This recourse simplifies the calculations and greatly decreases the amount of data needed. For most purposes, the errors introduced by the assumption of unity for the activity coefficient will not be large enough to lead to false conclusions. It should be apparent from the preceding example, however, that disregard of activity coefficients may introduce a significant numerical error in calculations of this kind; as in the example, relative errors of 100% or more are not uncommon.

The student should be alert to the conditions under which the approximation of concentration for activity is likely to lead to the largest errors. Significant discrepancies will occur when the ionic strength is large (0.01 or larger) or when the ions involved have multiple charges (see Table 4-5). With dilute solutions (ionic strength < 0.01) of nonelectrolytes or of singly charged ions, the use of concentrations in a mass-law calculation often provides reasonably accurate results.

It is also important to note that the decrease in solubility resulting from the presence of an ion common to the precipitate is in part counteracted by the larger electrolyte concentration associated with presence of the salt containing the common ion. This effect is illustrated by the sample calculation just completed.

4D ADDITIONAL VARIABLES THAT AFFECT THE SOLUBILITY OF PRECIPITATES

The solubility of most precipitates is influenced by temperature and the presence of organic solvents. Heat is absorbed as most solids dissolve. Therefore, the solubility of precipitates generally increases with rising temperatures; correspondingly, solubility product constants for most sparingly soluble compounds become larger at high temperatures.

The solubility of most inorganic substances is markedly less in mixtures of water and organic solvents than in pure water. The data for calcium sulfate in Table 4-6 are typical of this effect.

4E RATE OF PRECIPITATE FORMATION

It is important to stress that no conclusions can be drawn about the rate of a reaction from the magnitude of its equilibrium constant. Many reactions with highly favorable equilibrium constants approach equilibrium at an imperceptible rate.

Precipitation reactions are often slow, with several minutes or even several hours being required for the attainment of equilibrium. Occasionally, the chemist can take advantage of a slow rate to accomplish separations that would not be feasible if equilibrium were approached rapidly. For example, calcium can be separated from magnesium by precipitation as the oxalate, despite the fact that the latter ion also forms an oxalate of only somewhat greater solubility. The separation is possible because equilibrium for magnesium oxalate formation is approached at a much slower rate than that for calcium oxalate formation; if the calcium oxalate is filtered shortly after precipitation, a solid that is essentially free of contamination by magnesium is obtained. On the other hand, a precipitate that remains in contact with the liquid will become contaminated with magnesium oxalate.

Table 4-6   Solubility of Calcium Sulfate in Aqueous Ethyl Alcohol Solution*

Concentration of Ethyl Alcohol (w/w percent)	Solubility of $CaSO_4$, g $CaSO_4$/100 g solvent
0	0.208
6.2	0.100
13.6	0.044
23.8	0.014
33.0	0.0052
41.0	0.0029

* From T. Yamamoto, *Bull. Inst. Phys. Chem. Res.* (Tokyo) **9**, 352 (1930); W. C. Linke, *Seidell Solubilities of Inorganic and Metal-Organic Compounds*, 4th ed., vol. I, p. 685. Washington, D.C.: American Chemical Society, 1958. With permission.

PROBLEMS*

* 4-1. Evaluate the solubility product constant for the following solutes from their equilibrium solubilities (in parentheses):
    (a) $RaSO_4$ (2.11 mg/L)
    (b) $Ba(BrO_3)_2$ (9.28×10^{-3} mol/L)
    (c) $Ag_3Fe(CN)_6$ (6.6×10^{-4} g/L)
    (d) $Sr(IO_3)_2$ (0.19 g/100 mL)

4-2. Evaluate the solubility product constant for the following solutes from their equilibrium solubilities (in parentheses):
    (a) $TlSCN$ (3.93 mg/mL)
    (b) $Cr(OH)_2$ (1.36×10^{-6} mol/L)
    (c) BiI_3 (7.76×10^{-3} g/L)
    (d) Hg_2SO_4 (42.8 mg/100 mL)

* 4-3. Calculate the molar solubility of silver iodide in
    (a) water.
    (b) 0.020 M $AgNO_3$.
    (c) 0.040 M KI.

4-4. Calculate the molar solubility of lead hydroxide in
    (a) water.
    (b) 0.020 $M Pb(NO_3)_2$.
    (c) 0.020 M $Ba(OH)_2$.

* 4-5. Calculate the molar sulfate concentration in a solution that is 0.200 M in
    (a) Ag^+ and saturated with Ag_2SO_4.   (Ksp $= 1.6 \times 10^{-5}$)
    (b) Ba^{2+} and saturated with $BaSO_4$.
    (c) Sm^{3+} and saturated with $Sm_2(SO_4)_3$. (K_{sp} $Sm_2(SO_4)_3 = 2.0 \times 10^{-10}$)

4-6. Calculate the molar iodate ion concentration in a solution that is 0.030 M in
    (a) Tl^+ and saturated with $TlIO_3$.   (K_{sp} $TlIO_3 = 3.1 \times 10^{-6}$)
    (b) UO_2^{2+} and saturated with $UO_2(IO_3)_2$. (K_{sp} $UO_2(IO_3)_2 = 1.9 \times 10^{-8}$)
    (c) Ce^{3+} and saturated with $Ce(IO_3)_3$.   (K_{sp} $Ce(IO_3)_3 = 3.2 \times 10^{-10}$)
    (d) Ce^{4+} and saturated with $Ce(IO_3)_4$.   (K_{sp} $Ce(IO_3)_4 = 4.7 \times 10^{-17}$)

* 4-7. How many milligrams of $Ba(IO_3)_2$ will be dissolved in 150 mL of a saturated solution that is also
    (a) 1.00×10^{-2} M in KIO_3?
    (b) 1.00×10^{-2} M in $BaCl_2$?

4-8. What weight (mg) of PbI_2 will be dissolved in 600 mL of a saturated solution that is also
    (a) 2.00×10^{-1} M in KI?
    (b) 2.00×10^{-1} M in MgI_2?

4-9. Establish whether a precipitate will form when the indicated volumes of solutions A and B are mixed.

Solution A	Solution B
(a) 40.0 mL 0.0080 M $HClO_4$	30.0 mL 0.030 M CsCl (K_{sp} $CsClO_4 = 4.0 \times 10^{-3}$)
* (b) 40.0 mL 0.0080 M KCl	20.0 mL 0.030 M $TlNO_3$ (K_{sp} $TlCl = 1.7 \times 10^{-4}$)
* (c) 25.0 mL 0.060 M $Pb(NO_3)_2$	50.0 mL 0.030 M KCl
(d) 25.0 mL 0.060 M $Pb(NO_3)_2$	50.0 mL 0.030 M $CaCl_2$

4-10. Establish whether a precipitate will form when 50.0 mL of 0.015 M HIO_3 are mixed with 10.0 mL of

* Answers to problems or parts of problems marked with an asterisk are to be found at the end of the book.

(a) 0.010 M KCl. (K_{sp} KIO$_3$ = 5 × 10^{-2})
(b) 0.010 M CuCl$_2$. (K_{sp} Cu(IO$_3$)$_2$ = 7.4 × 10^{-8})
(c) 0.010 M CaCl$_2$. (K_{sp} Ca(IO$_3$)$_2$ = 7.1 × 10^{-7})
(d) 0.010 M CeCl$_3$. (K_{sp} Ce(IO$_3$)$_3$ = 3.2 × 10^{-10})

* 4-11. Calculate the cation concentration needed to initiate precipitation of
  (a) BaSO$_4$ from a solution that is 0.036 M in SO$_4^{2-}$.
  (b) Ag$_2$SO$_4$ from a solution that is 0.036 M in SO$_4^{2-}$. (K_{sp} Ag$_2$SO$_4$ = 1.6 × 10^{-5})
  (c) BiI$_3$ from a solution that is 0.010 M in I$^-$. (K_{sp} BiI$_3$ = 8.1 × 10^{-19})

* 4-12. Calculate the anion concentration needed to lower the concentration of the cation in Problem 4-11 to 1.0 × 10^{-6} M.

4-13. Calculate the anion concentration needed to initiate precipitation of
  (a) AgSCN from a solution that is 0.050 M in Ag$^+$.
  (b) PbI$_2$ from a solution that is 0.020 M in Pb^{2+}.
  (c) Al(OH)$_3$ from a solution that is 0.025 M in Al^{3+}.

4-14. Calculate the anion concentration needed to lower the concentration of the cation in Problem 4-13 to 1.0 × 10^{-6} M.

4-15. Calculate the minimum value for a solubility product constant if the contribution of hydroxide from water (1.0 × 10^{-7} M) is to be less than 5% of the total hydroxide ion concentration in a saturated solution of a sparingly soluble hydroxide with the formula
  * (a) MOH.   (c) M(OH)$_3$.
    (b) M(OH)$_2$.   * (d) M(OH)$_4$.

* 4-16. Calculate the molar concentration of the cation in a saturated aqueous solution of
  (a) Cr(OH)$_2$. (K_{sp} = 1.0 × 10^{-17})
  (b) Cr(OH)$_3$. (K_{sp} = 2.4 × 10^{-23})
  (c) La(OH)$_3$. (K_{sp} = 2.0 × 10^{-19})
  (d) Th(OH)$_4$. (K_{sp} = 4 × 10^{-43})

* 4-17. Calculate the molar concentration of the cation in Problem 4-16 in a saturated aqueous solution that is also 0.040 M in KOH.

4-18. Calculate the molar concentration of the cation in a saturated aqueous solution of
  (a) Mn(OH)$_2$. (K_{sp} = 1.9 × 10^{-13})
  (b) Nd(OH)$_3$. (K_{sp} = 3.2 × 10^{-22})
  (c) In(OH)$_3$. (K_{sp} = 6 × 10^{-34})
  (d) Zr(OH)$_4$. (K_{sp} = 1 × 10^{-52})

4-19. Calculate the molar concentration of the cation in Problem 4-18 in a saturated aqueous solution that is also 0.020 M in Ba(OH)$_2$.

* 4-20.

Substance	K_{sp}
BiOOH	4 × 10^{-10}
Fe(OH)$_2$	8 × 10^{-16}
Ce(OH)$_3$	2 × 10^{-20}
Hf(OH)$_4$	4 × 10^{-26}

Use the accompanying solubility-product data to identify the hydroxide with
  (a) the largest molar solubility in water.
  (b) the smallest molar solubility in a solution that is 0.040 M in hydroxide ion.
  (c) the smallest molar solubility in a solution that is 0.040 M in the solute cation.
  (d) the largest cation concentration in a saturated aqueous solution.
  (e) the smallest hydroxide ion concentration in a saturated aqueous solution.

4-21.

Substance	K_{sp}
AgIO$_3$	3.0 × 10^{-8}
Pb(IO$_3$)$_2$	3.2 × 10^{-13}
In(IO$_3$)$_3$	3.3 × 10^{-12}
Ce(IO$_3$)$_4$	4.7 × 10^{-17}

Use the accompanying solubility-product data to identify the iodate with
(a) the smallest molar solubility in water.
(b) the largest molar solubility in a solution that is $0.025\ M$ in iodate ion.
(c) the largest molar solubility in a solution that is $0.025\ M$ in the solute cation.
(d) the smallest cation concentration in a saturated aqueous solution.
(e) the largest iodate ion concentration in a saturated aqueous solution.

* 4-22. Calculate the magnesium ion concentration in the solution that results when 40.0 mL of $0.0625\ M$ $MgCl_2$ are mixed with
(a) sufficient water to give 100 mL of solution.
(b) 60.0 mL of $0.0450\ M$ KOH.
(c) 60.0 mL of $0.0450\ M$ $Ba(OH)_2$.

4-23. Calculate the iodate ion concentration of the solution that results when 25.0 mL of $0.0800\ M$ $Mg(IO_3)_2$ are
(a) diluted to 100 mL with water.
(b) mixed with 75.0 mL of $0.0200\ M$ $AgNO_3$.
(c) mixed with 75.0 mL of $0.0200\ M$ $Ba(NO_3)_2$.
(d) mixed with 75.0 mL of $0.0200\ M$ $La(NO_3)_3$.

4-24. Write a material-balance and a charge-balance equation for a solution that is
    (a) $0.020\ M$ in HCl.
* (b) $0.100\ M$ in HNO_2.
    (c) $0.080\ M$ in H_2SO_3.
* (d) $0.050\ M$ in H_3PO_4.
    (e) $0.020\ M$ in HCl and $0.015\ M$ in $MgCl_2$.
    (f) $0.100\ M$ in HNO_2 and $0.040\ M$ in $NaNO_2$.
* (g) $0.080\ M$ in H_2SO_3 and $0.030\ M$ in $NaHSO_3$.
    (h) $0.080\ M$ in $NaHSO_3$ and $0.040\ M$ in Na_2SO_3.
* (i) $0.040\ M$ in NaH_2PO_4 and $0.020\ M$ in Na_2HPO_4.

4-25. Write a material-balance equation and a charge-balance equation for the following systems. Use the symbol S to indicate the molar solubility of a sparingly soluble species, if such is needed.
    (a) a saturated solution of silver iodide
* (b) a $0.040\ M$ solution of MgI_2 that has been saturated with AgI
    (c) a saturated solution of silver chromate
    (d) a $0.10\ M$ solution of sodium cyanide, to which $AgNO_3$ has been added (product: $Ag(CN)_2^-$) in an amount that is not sufficient to react with all of the available cyanide

* 4-26. A $0.100\ M$ solution of acetic acid is saturated with the sparingly soluble silver acetate; the silver ion concentration of the resulting solution is sought. Write material- and charge-balance equations that describe the system.

4-27. A saturated solution of lead oxalate that contains an excess of oxalate ion is produced by combining 75.0 mL of $0.0400\ M$ $Na_2C_2O_4$ with 25.0 mL of $0.0800\ M$ $Pb(NO_3)_2$. Write material- and charge-balance equations for the resulting system; use the symbol S to indicate the molar solubility of PbC_2O_4, if needed.

* 4-28. Calculate the molar solubility of lead tartrate ($K_{sp} = 5.0 \times 10^{-9}$) in a solution that has a fixed hydronium ion concentration of
    (a) $1.00 \times 10^{-1}\ M$.   (c) $1.00 \times 10^{-4}\ M$.
    (b) $1.00 \times 10^{-3}\ M$.   (d) $1.00 \times 10^{-5}\ M$.

* 4-29. Calculate the molar solubility of silver sulfite ($K_{sp} = 1.5 \times 10^{-14}$) in a saturated aqueous solution (a) neglecting the basic dissociation of sulfite ion and (b) taking account of the dissociation of the anion. (c) Calculate the percent relative error associated with neglect of dissociation processes in this problem.

4-30. Calculate the molar solubility of silver sulfite (see Problem 4-29) in a solution with a fixed hydronium ion concentration of

(a) $1.00 \times 10^{-2} M$.   (c) $1.00 \times 10^{-6} M$.

(b) $1.00 \times 10^{-4} M$.   (d) $1.00 \times 10^{-8} M$.

* 4-31. Calculate the molar solubility of silver cyanide in a saturated aqueous solution.

4-32. Calculate the molar solubility of silver cyanide in a solution with a fixed hydronium ion concentration of

(a) $1.00 \times 10^{-11} M$.   (c) $1.00 \times 10^{-7} M$.

(b) $1.00 \times 10^{-9} M$.   (d) $1.00 \times 10^{-5} M$.

4-33. Calculate the molar solubility of a saturated aqueous solution of

(a) CuS.   (c) Tl_2S.

* (b) MnS.   * (d) Cu_2S. $(K_{sp} = 3 \times 10^{-48})$

4-34. Calculate the molar solubility of a saturated aqueous solution of

(a) $Hg_2(C_2H_3O_2)_2$. $(K_{sp} = 3 \times 10^{-11})$ (c) MgC_2O_4.

$$Hg_2(C_2H_3O_2)_2(s) \rightleftharpoons Hg_2^{2+} + 2C_2H_3O_2^-$$

(b) Ag_2CO_3.   * (d) $FeCO_3$. $(K_{sp} = 3.5 \times 10^{-11})$

* 4-35. Equilibrium constants for reactions involving Ag^+ and SCN^- are

$$Ag^+ + 2SCN^- \rightleftharpoons Ag(SCN)_2^- \qquad \beta_2 = 2.5 \times 10^8$$
$$Ag(SCN)_2^- + SCN^- \rightleftharpoons Ag(SCN)_3^{2-} \qquad K_3 = 17$$
$$Ag(SCN)_3^{2-} + SCN^- \rightleftharpoons Ag(SCN)_4^{3-} \qquad K_4 = 1.9$$
$$AgSCN(s) \rightleftharpoons Ag^+ + SCN^- \qquad K_{sp} = 1.1 \times 10^{-12}$$

Calculate the solubility of AgSCN in a solution having an equilibrium molar concentration of SCN^- of

(a) 4.0.   (d) 5×10^{-2}.

(b) 2.0.   (e) 5×10^{-4}.

(c) 0.500.

4-36. The equilibrium constant for formation of $CuCl_2^-$ is given by

$$Cu^+ + 2Cl^- \rightleftharpoons CuCl_2^- \qquad K = \frac{[CuCl_2^-]}{[Cu^+][Cl^-]^2} = 8.7 \times 10^4$$

What is the solubility of CuCl in solutions having the following molar NaCl concentrations:

(a) 1.0?   (d) 1.0×10^{-3}?

(b) 1.0×10^{-1}?   (e) 1.0×10^{-4}?

(c) 1.0×10^{-2}?

* 4-37. The equilibrium constant for the formation of $Al(OH)_4^-$ is given by

$$Al(OH)_3(s) + OH^- \rightleftharpoons Al(OH)_4^- \qquad K = 10$$

How many milliliters of 1.0 M NaOH would be required to completely redissolve 1.00 g of $Al(OH)_3$ suspended in 100.0 mL of water?

4-38. What concentration of OH^- must be maintained in order to dissolve 0.200 g of $Pb(OH)_2$ in 200 mL of solution?

$$Pb(OH)_2(s) + OH^- \rightleftharpoons Pb(OH)_3^- \qquad K = 5.0 \times 10^{-2}$$

* 4-39. Formation constants for the reaction Ag^+ with $S_2O_3^{2-}$ are

$$Ag^+ + S_2O_3^{2-} \rightleftharpoons AgS_2O_3^- \qquad K_1 = 6.6 \times 10^8$$
$$AgS_2O_3^- + S_2O_3^{2-} \rightleftharpoons Ag(S_2O_3)_2^{3-} \qquad K_2 = 4.4 \times 10^3$$

Calculate the solubility of AgI in 0.200 M $Na_2S_2O_3$ (assume that $S_2O_3^{2-}$ does not combine with H_3O^+).

* 4-40. Sodium hydroxide is slowly added to a solution that is 0.050 M in $ZnCl_2$ and 0.035 M in $CrCl_3$ (for $Cr(OH)_3$, $K_{sp} = 6 \times 10^{-31}$).

(a) Which hydroxide will precipitate first?

(b) What will be the concentration of the cation that forms the less soluble hydroxide at the onset of precipitation by the more soluble hydroxide?

4-41. A solution that is 0.040 M in sulfate and 0.026 M in iodate is treated with a 0.050 M solution of $BaCl_2$.

(a) Which barium salt will precipitate first?

(b) What will be the concentration of the anion that forms the less soluble barium salt at the onset of precipitation by the more soluble one?

4-42. Determine whether the following separations are theoretically feasible. Indicate which species will precipitate first; use 1.0×10^{-6} M as the criterion for quantitative removal.

Species to be separated	Precipitant
* (a) 0.600 M Ag^+, 0.100 M Pb^{2+}	I^-
(b) 0.080 M Bi^{3+}, 0.075 M Ag^+	I^- (K_{sp} for BiI_3 = 8.1×10^{-19})
(c) 0.050 M Mg^{2+}, 0.040 M Cu^{2+}	OH^-
* (d) 0.012 M BiO^+, 0.093 M Ce^{3+}	OH^- (see Problem 4-20 for K_{sp} data)
(e) 0.036 M Ag^+, 0.012 M In^{3+}	IO_3^- (see Problem 4-21 for K_{sp} data)
* (f) 0.042 M Pb^{2+}, 0.064 M Ce^{4+}	IO_3^- (see Problem 4-21 for K_{sp} data)

4-43. Determine which of the following separations is feasible by controlling the hydronium ion concentration of a saturated H_2S solution. Assume that the initial concentration of each ion is 0.100 M and that lowering of a concentration to 1.0×10^{-6} M constitutes quantitative removal. If a separation is possible, specify the range of H_3O^+ concentrations that could be employed.

* (a) Fe^{2+} and Cd^{2+}

(b) Cu^{2+} and Zn^{2+}

* (c) La^{3+} and Mn^{2+} (K_{sp} for La_2S_3 = 2.0×10^{-13})

(d) Ce^{3+} and Fe^{2+} (K_{sp} for Ce_2S_3 = 6.0×10^{-11})

* (e) Cd^{2+} and Zn^{2+}

(f) Cd^{2+} and Tl^+

4-44. Calculate the ionic strength of a solution that is

* (a) 1.67×10^{-3} M in Na_2SO_4.

(b) 3.33×10^{-3} M in $MgCl_2$.

* (c) 1.33×10^{-3} M in $MgCl_2$ and 6.01×10^{-3} M in $NaCl$.

(d) 7.14×10^{-3} M in $Fe(NH_4)_2(SO_4)_2$.

* (e) 5.00×10^{-3} M in $K_4(Fe(CN)_6$ and 5.00×10^{-2} M in KCl.

(f) 9.00×10^{-3} M in $Mg(NO_3)_2$ and 2.30×10^{-2} M in $NaNO_3$.

4-45. Calculate the concentration constant K'_{sp} for

* (a) $Mg(OH)_2$ in the solution described in Problem 4-44(a).

(b) $MgNH_4PO_4$ in the solution described in Problem 4-44(b).

* (c) $Ba(IO_3)_2$ in the solution described in Problem 4-44(c).

(d) CaC_2O_4 in the solution described in Problem 4-44(d).

* (e) $Ag_4Fe(CN)_6$ in the solution described in Problem 4-44(e). (K_{sp} = 2.5×10^{-22})

(f) $Cu(IO_3)_2$ in the solution described in Problem 4-44(f). (K_{sp} = 7.4×10^{-8})

4-46. Use Equation 4-66 (p. 118) to evaluate activity coefficients for the following species:

Ion	Effective diameter, a_A, Å	Ionic strength, μ
* (a) Li^+	6	0.075
(b) Ba^{2+}	5	0.012
* (c) Fe^{3+}	9	0.064
(d) SCN^-	3.5	0.039
* (e) $S_2O_3^{2-}$	4.0	0.026
(f) $Fe(CN)_6^{4-}$	5	0.041

* 4-47. Calculate the ionic strength of the solution that results when 50.0 mL of 0.0200 M $AgNO_3$ are mixed with an equal volume of 0.0120 M KIO_3.

* 4-48. Calculate the concentration constant, K'_{sp} for $AgIO_3$ in the solution produced in Problem 4-47.

  4-49. Calculate the ionic strength of the solution that results when 1.50 mmol of $Mg(OH)_2$ are mixed with 100.0 mL of 0.0350 M HCl.

  4-50. A 0.232-g specimen of $Mg(OH)_2$ is treated with 100 mL of 0.0250 M H_2SO_4.

  (a) Calculate the ionic strength of the resulting solution.

  (b) Calculate the concentration constant, K'_{sp}, for $Mg(OH)_2$ in this solution.

  (c) Compare values for the hydroxide ion concentration computed with K_{sp} and with K'_{sp}. Calculate the percent relative error associated with neglect of activities in this system.

* 4-51. Derive an exact equation for calculating the solubility of the precipitate $M(OH)_3$ in pure water.

gravimetric analysis

5

A gravimetric analysis is based upon the measurement of the weight of a substance of known composition that is chemically related to the analyte. Two types of gravimetric methods exist. In ordinary *precipitation methods,* the species to be determined is caused to react chemically with a reagent to yield a product of limited solubility; after filtration and other suitable treatment, the solid residue of known chemical composition is weighed. Less frequently the difference in weight of a sparingly soluble reagent before and after reaction with the analyte forms the basis for some analyses. In *volatilization methods,* the substance to be determined is separated as a gas from the remainder of the sample; here, the analysis is based upon the weight of the volatilized substance or upon the weight of the nonvolatile residue. We shall be concerned principally with precipitation methods because these are more frequently encountered than methods involving volatilization.

5A CALCULATION OF RESULTS FROM GRAVIMETRIC DATA

A gravimetric analysis requires two experimental measurements; specifically, the weight of sample taken and the weight of a product of known composition derived from the sample. Ordinarily, these data are converted to a percentage of analyte by a simple mathematical manipulation.

If A is the analyte, we may write

$$\% \text{ A} = \frac{\text{weight of A}}{\text{weight of sample}} \times 100 \tag{5-1}$$

The weight of A is seldom measured directly. Instead, the species that is actually isolated and weighed either contains A or can be chemically related to A. In either case, a *gravimetric factor* is needed to convert the weight of the precipitate to the corresponding weight of A. The properties of this factor are conveniently demonstrated with examples.

EXAMPLE 5-1. How many grams of Cl (fw = 35.45) are contained in a precipitate of AgCl (fw = 143.3) that weighs 0.204 g?

This problem, and others like it, consists of (1) conversion from units of metric mass to those of chemical mass, (2) an accounting of the combining ratio or stoichiometry between the substance whose mass is sought and the substance whose mass is known, and finally (3) conversion back to metric units (see Section 2C). Here, the operations are

Conversion to Chemical Mass	Stoichi- ometry	Conversion to Metric Unit

$$0.204 \text{ g AgCl} \times \frac{1 \text{ mol AgCl}}{143.3 \text{ g AgCl}} \times \frac{1 \text{ mol Cl}}{1 \text{ mol AgCl}} \times \frac{35.45 \text{ g Cl}}{\text{mol Cl}}$$

$$= 0.0505 \text{ g Cl}$$

or simply

$$0.204 \text{ g AgCl} \times \frac{35.45 \text{ g Cl}}{143.3 \text{ g AgCl}} = 0.0505 \text{ g Cl}$$

EXAMPLE 5-2. To what weight of $AlCl_3$ (fw = 133.3) would 0.204 g of AgCl correspond? Proceeding as before, we write

Conversion to Chemical Mass	Stoichi- ometry	Conversion to Metric Unit

$$0.204 \text{ g AgCl} \times \frac{1 \text{ mol AgCl}}{143.3 \text{ g AgCl}} \times \frac{1 \text{ mol AlCl}_3}{3 \text{ mol AgCl}} \times \frac{133.3 \text{ g AlCl}_3}{\text{mol AlCl}_3}$$

$$= 0.0633 \text{ g AlCl}_3$$

$$0.204 \text{ g AgCl} \times \frac{133.3 \text{ g AlCl}_3}{3 \times 143.3 \text{ g AgCl}} = 0.0633 \text{ g AlCl}_3$$

Note how these two calculations resemble each other. In both, the weight of one substance is given by the product involving the known weight of some other substance and a ratio that contains their respective gram formula weights. This ratio is the gravimetric factor. In the second example, it was necessary to multiply the gram formula weight of silver chloride by 3 in order to balance the number of chlorides that appear in the numerator and denominator of the gravimetric factor.

EXAMPLE 5-3. What weight of Fe_2O_3 (fw = 159.7) can be obtained from 1.63 g of Fe_3O_4 (fw = 231.5)? What is the gravimetric factor for this conversion?

Here, it is necessary to assume that all Fe in the Fe_3O_4 is transformed into Fe_2O_3 and ample oxygen is available to accomplish this change. That is,

$$2Fe_3O_4 + \tfrac{1}{2}O_2 = 3Fe_2O_3$$

$$\text{g Fe}_3\text{O}_4 \times \frac{\text{mol Fe}_3\text{O}_4}{231.5 \text{ g}} \times \frac{3 \text{ mol Fe}_2\text{O}_3}{2 \text{ mol Fe}_3\text{O}_4} \times \frac{159.7 \text{ g}}{\text{mol Fe}_2\text{O}_3} = \text{g Fe}_2\text{O}_3$$

or

$$1.63 \text{ g} \times \frac{3 \times 159.7}{2 \times 231.5} = 1.687 = 1.69 \text{ g Fe}_2\text{O}_3$$

In this example,

$$\text{gravimetric factor} = \frac{3 \times \text{fw Fe}_2\text{O}_3}{2 \times \text{fw Fe}_3\text{O}_4} = 1.035$$

A general definition of a gravimetric factor is

$$\text{gravimetric factor} = \frac{a}{b} \times \frac{\text{fw of the substance sought}}{\text{fw of the substance weighed}}$$

where a and b are small integers that take such values as are necessary to make the number of formula weights in the numerator and denominator *chemically equivalent.*

Equation 5-1 can now be converted to the more useful form

$$\% \text{ A} = \frac{\text{wt ppt} \times \dfrac{a \times \text{fw A}}{b \times \text{fw ppt}} \times 100}{\text{wt sample}}$$

Additional examples of gravimetric factors are given in Table 5-1. Most chemical handbooks contain tabulations of these factors and their logarithms.

In all of the gravimetric factors considered thus far, chemical equivalence between numerator and denominator has been established by simply balancing the number of atoms of an element (other than oxygen) that is common to both. Occasionally this approach will be inadequate. Consider, for example, an indirect analysis for the iron in a sample of iron(III) sulfate that involves precipitation and weighing of barium sulfate. Here, the gravimetric factor will contain no element common to numerator and denominator, and we must seek further for the means of establishing the chemical equivalence between these quantities. We note that

$$2 \text{ mol Fe} \equiv 1 \text{ mol Fe}_2(\text{SO}_4)_3 \equiv 3 \text{ mol SO}_4 \equiv 3 \text{ mol BaSO}_4$$

The gravimetric factor for calculating the percent Fe will be

$$\text{gravimetric factor} = \frac{2 \times \text{fw Fe}}{3 \times \text{fw BaSO}_4}$$

Table 5-1 Typical Gravimetric Factors

Species Sought	Species Weighed	Gravimetric Factor
In	In_2O_3	$\dfrac{2 \times \text{fw In}}{\text{fw In}_2\text{O}_3}$
HgO	$\text{Hg}_5(\text{IO}_6)_2$	$\dfrac{5 \times \text{fw HgO}}{\text{fw Hg}_5(\text{IO}_6)_2}$
I	$\text{Hg}_5(\text{IO}_6)_2$	$\dfrac{2 \times \text{fw I}}{\text{fw Hg}_5(\text{IO}_6)_2}$
K_3PO_4	K_2PtCl_6	$\dfrac{2 \times \text{fw K}_3\text{PO}_4}{3 \times \text{fw K}_2\text{PtCl}_6}$

Thus, even though the species in the gravimetric factor are not directly related by a common element, we can establish their equivalence through knowledge of the stoichiometry between them.

The examples that follow illustrate the use of the gravimetric factor in the calculation of the results from analyses.

EXAMPLE 5-4. A 0.703-g sample of a commercial detergent was ignited at a red heat to destroy the organic matter. The residue was then taken up in hot HCl which converted the P to H_3PO_4. The phosphate was precipitated as $MgNH_4PO_4 \cdot 6H_2O$ by addition of Mg^{2+} followed by aqueous NH_3. After being filtered and washed, the precipitate was converted to $Mg_2P_2O_7$ (fw = 222.6) by ignition at 1000°C. This residue weighed 0.432 g. Calculate the percent P (fw = 30.97) in the sample.

$$\% \text{ P} = \frac{0.432 \times \dfrac{2 \times \text{fw P}}{\text{fw } Mg_2P_2O_7} \times 100}{0.703}$$

$$= \frac{0.432 \times 0.2783 \times 100}{0.703} = 17.1$$

EXAMPLE 5-5. At elevated temperatures, $NaHCO_3$ is converted quantitatively to Na_2CO_3:

$$2NaHCO_3(s) = Na_2CO_3(s) + CO_2(g) + H_2O(g)$$

Ignition of a 0.7184 g sample of impure $NaHCO_3$ yielded a residue weighing 0.4724 g. Calculate the percentage purity of the sample assuming any impurities are nonvolatile.

The difference in weight before and after ignition represents the amount of CO_2 and H_2O evolved from the $NaHCO_3$ in the sample. Thus, since

$$2 \text{ mol } NaHCO_3 \equiv 1 \text{ mol } CO_2 + 1 \text{ mol } H_2O$$

$$\% \text{ } NaHCO_3 = \frac{(0.7184 - 0.4724) \times \dfrac{2 \times \text{fw } NaHCO_3}{\text{fw } CO_2 + \text{fw } H_2O} \times 100}{0.7184}$$

$$= \frac{0.2460 \times \dfrac{2 \times 84.01}{44.01 + 18.02}}{0.7184} \times 100 = 92.75$$

Note that the denominator of the gravimetric factor in this example is the sum of the gram formula weights of the two volatile products and that it is the combined weight of these products that forms the basis for this analysis.

EXAMPLE 5-6. A 0.2795-g sample of an insecticide containing only lindane ($C_6H_6Cl_6$; fw = 290.8) and DDT ($C_{14}H_9Cl_5$; fw = 354.5) was burned in a stream of oxygen in a quartz tube. The products (CO_2, H_2O, and HCl) were passed through a solution of $NaHCO_3$. After acidification, the chloride in this solution yielded 0.7161 g of AgCl. Calculate the percent lindane and DDT in the sample.

Here, there are two unknowns, and we must, therefore, develop two independent equations that can be solved simultaneously.

One useful equation is

$$\text{wt } C_6H_6Cl_6 + \text{wt } C_{14}H_9Cl_5 = 0.2795 \text{ g}$$

A second equation is

$$\text{wt AgCl from } C_6H_6Cl_6 + \text{wt AgCl from } C_{14}H_9Cl_5 = 0.7161 \text{ g}$$

After inserting the appropriate gravimetric factors, the second equation becomes

$$\text{wt } C_6H_6Cl_6 \times \frac{6 \times \text{fw AgCl}}{\text{fw } C_6H_6Cl_6} + \text{wt } C_{14}H_9Cl_5 \times \frac{5 \times \text{fw AgCl}}{\text{fw } C_{14}H_9Cl_5} = 0.7161 \text{ g}$$

or

$$\text{wt } C_6H_6Cl_6 \times 2.957 + \text{wt } C_{14}H_9Cl_5 \times 2.021 = 0.7161$$

Substituting wt $C_{14}H_9Cl_5$ from the first equation gives

$$2.957 \text{ wt } C_6H_6Cl_6 + 2.021(0.2795 - \text{wt } C_6H_6Cl_6) = 0.7161$$

Thus,

$$\text{wt } C_6H_6Cl_6 = 0.1616 \text{ g}$$

and

$$\% \text{ } C_6H_6Cl_6 = \frac{0.1616}{0.2795} \times 100 = 57.82$$

$$\% \text{ } C_{14}H_9Cl_5 = 100 - 57.82 = 42.18$$

5B PROPERTIES OF PRECIPITATES AND PRECIPITATING REAGENTS

An ideal precipitating reagent for a gravimetric analysis would react specifically with the analyte to produce a solid that (1) has a sufficiently low solubility so that losses from that source are negligible, (2) is readily filtered and washed free of contaminants, and (3) is unreactive and of known composition after drying or, if necessary, ignition. Few precipitates or reagents possess all these desirable properties; thus, the chemist frequently finds it necessary to perform analyses using a product that is far from ideal.

The variables that influence the solubility of precipitates were discussed in Chapter 4; we must now consider what can be done to achieve a pure and easily filtered solid.

5B-1 FILTERABILITY AND PURITY OF PRECIPITATES

The ease with which a precipitate is isolated and purified depends on the particle size of the solid phase. The relationship between particle size and ease of filtration is straightforward. Coarse precipitates are readily retained by porous media and are thus rapidly filtered. Finely divided precipitates require dense filters; low filtration rates result. The effect of particle size upon the purity of a precipitate is more complex. More often than not, large particles contain less contaminants.

In considering the purity of precipitates, we shall use the term *coprecipitation*, which describes those processes by which *normally soluble* components of a solution are carried down during the formation of a precipitate. The student should clearly understand that contamination of a precipitate by a second substance whose solubility product has been exceeded *does not constitute coprecipitation.*

Factors that determine the particle size of precipitates. Particle size depends not only upon the chemical composition of a precipitate but also upon the conditions that exist at the time of its formation. Enormous variations are observed. At one extreme are *colloidal suspensions*, whose individual particles are so small as to be invisible to the naked eye (10^{-6} to 10^{-4} mm in diameter). These particles show no tendency to settle from solution nor are they retained upon common filtering media. At the other extreme are particles with dimensions on the order of several tenths of a millimeter. The temporary dispersion of such particles in the liquid phase is called a *crystalline suspension*. The particles of a crystalline suspension tend to settle spontaneously and are readily filtered.

No sharp discontinuities in physical properties occur as the dimensions of the particles in the solid phase increase from colloidal to those typical of crystals. Indeed, some precipitates possess characteristics that are between these defined extremes. The majority, however, are easily recognizable as predominately colloidal or predominately crystalline. Thus, although imperfect, this classification can be usefully applied to most solid phases.

The phenomenon of precipitation has long attracted the attention of chemists, but the mechanism of the process still is not fully understood. It is certain, however, that the particle size of a precipitate is influenced in part by such experimental variables as solubility of the precipitate in the medium in which it is being formed, temperature, reactant concentrations, and the rate at which reactants are mixed. The effect of these variables can be accounted for, at least qualitatively, by assuming that the particle size is related to a single property of the system called its *relative supersaturation*,[1] where

$$\text{relative supersaturation} = \frac{Q - S}{S} \tag{5-2}$$

In this equation, Q is the concentration of the solute at any instant and S is its equilibrium solubility.

During the formation of a sparingly soluble precipitate, each addition of precipitating reagent presumably causes the solution to be momentarily supersaturated (that is, $Q > S$). Under most circumstances, this unstable condition is relieved, usually after a brief period, by precipitate formation. Experimental evidence suggests, however, that the particle size of the resulting precipitate varies inversely with the average degree of relative supersaturation that exists after each addition of reagent. Thus, when $(Q - S)/S$ is large, the precipitate tends to be colloidal; when this parameter is low on the average, a crystalline solid results.

Mechanism of precipitate formation. The effect of relative supersaturation on particle size can be rationalized by postulating two competing precipitation mechanisms, *nucleation* and *particle growth*. The particle size of a freshly formed precipitate is governed by the extent to which one of these processes predominates over the other.

Nucleation is a process whereby some minimum number of ions or molecules (perhaps as few as four or five) unite to form a stable second phase. Further precipitation can occur either by the generation of additional nuclei or by the deposition of

[1] The name of P. P. von Weimarn is associated with the concept of relative supersaturation and its effect upon particle size. An account of von Weimarn's work is to be found in *Chem. Rev.*, **2**, 217 (1925). The von Weimarn viewpoint adequately suggests the general conditions that will lead to a satisfactory particle size for a precipitate. Other theories are superior in accounting for details of the precipitation process. See, for example, A. E. Nielsen, *The Kinetics of Precipitation*. New York: Macmillan, 1964.

additional ions on the nuclei that have already been produced. If nucleation predominates, a precipitate containing a large number of small particles results; if growth predominates, a smaller number of larger particles will be produced.

The rate of nucleation is believed to increase exponentially with relative supersaturation, whereas the rate of particle growth bears an approximately linear relationship to this parameter. Thus, when supersaturation is high, the nucleation rate far exceeds particle growth and is the predominant precipitation mechanism. At low relative supersaturations, on the other hand, the rate of particle growth may be the greater of the two. Under these circumstances deposition of solid on the particles already present may occur to the exclusion of further nucleation.

Experimental control of particle size.   Experimental variables that minimize supersaturation and thus lead to crystalline precipitates include elevated temperatures (to increase S), dilute solutions (to minimize Q), and slow addition of the precipitating agent with good stirring (also to lower the average value of Q).

The particle size of precipitates with solubilities that depend upon the acidity of the environment can often be enhanced by increasing S during precipitation. For example, large, easily filtered crystals of calcium oxalate can be obtained by forming the bulk of the precipitate in a somewhat acidic environment in which the salt is moderately soluble. The precipitation is then completed by slowly adding aqueous ammonia until the acidity is sufficiently low for quantitative removal of the calcium oxalate; the additional precipitate produced during this step forms on the solid.

A crystalline solid is much easier to manipulate than a colloidal suspension. Particle growth is thus preferable to further nucleation during the formation of a precipitate. If, however, the solubility S of a precipitate is very small, it is essentially impossible to avoid a momentarily large relative supersaturation as solutions are mixed; as a consequence, colloidal suspensions often cannot be avoided. For example, under conditions feasible for an analysis, the hydrous oxides of iron(III), aluminum, and chromium(III) and the sulfides of the most heavy metal ions can be formed only as colloids because of their very low solubilities. The same is true for the halide precipitates of silver ion.[2]

5B-2 COLLOIDAL PRECIPITATES

Individual colloidal particles are so small that they are not retained on ordinary filtering media; furthermore, Brownian motion prevents their settling from the solution under the influence of gravity. Fortunately, however, the individual particles of most colloids can be caused to coagulate or agglomerate to give a filterable, noncrystalline mass that rapidly settles from solution.

Coagulation of colloids.   Three experimental measures hasten the coagulation process: heating, stirring, and adding an electrolyte to the medium. To understand the effectiveness of these measures, we first need to account for the stability of a colloidal suspension.

The individual particles in a typical colloidal suspension bear either a posi-

---

[2] Silver chloride illustrates that the relative supersaturation concept is imperfect. This compound ordinarily forms as a colloid, yet its molar solubility is not significantly different from other compounds such as $BaSO_4$ which generally form as crystals.

tive or a negative charge owing to *adsorption* of cations or anions on their surfaces. The existence of this charge is readily demonstrated experimentally by observing the migration of the particles under the influence of an electric field.

Adsorption of ions upon an ionic solid has as its origin the normal bonding forces that are responsible for crystal growth. Thus, a silver ion at the surface of a silver chloride particle has a partially unsatisfied bonding capacity by virtue of its surface location. Negative ions are attracted to this site by the same forces that hold chloride ions in the silver chloride lattice. Chloride ions on the surface exert an analogous attraction for cations in the solvent.

The nature and magnitude of the charge on the particles of a colloidal suspension depend in a complex way on a number of variables. For colloidal suspensions of interest in analysis, however, the species adsorbed, and thus the charge on the particles, can be readily predicted from the empirical observation that lattice ions are generally more strongly adsorbed than any others. Thus, a silver chloride particle will be positively charged in a solution containing an excess of silver ions, owing to the preferential adsorption of those ions. It will have a negative charge in the presence of excess chloride ions for the same reason. The silver chloride particles formed in a gravimetric chloride analysis initially carry a negative charge but become positive as an excess of the precipitating agent is added.

The extent of adsorption increases rapidly with increases in the concentration of the absorbed ion. Ultimately, however, the surface of each particle becomes saturated; under these circumstances, further increases in concentration have little or no effect.

Figure 5-1 depicts schematically a colloidal silver chloride particle in a solution containing an excess of silver ions. Attached directly to the solid surface are silver ions in the *primary adsorption layer.* Surrounding the charged particle is a *region of solution* called the *counter-ion layer* within which there exists a sufficient excess of negative ions to just balance the charge of the adsorbed positive ions on the particle surface. The counter-ion layer forms as the result of electrostatic forces.

Considered together, the primarily adsorbed ions on the surface and their counter ions in the solution form an *electrical double layer* that exerts a repulsive force toward similarly constituted particles. This force is often sufficient to offset the normal cohesive forces that exist between small particles having the same chemical composition. Coagulation of a colloid, therefore, requires a reduction in the repulsive force of the double layer so that cohesion can occur.

The effect of the electrical double layer on stabilization of a colloid is readily seen in the precipitation of chloride ion with silver ion. With the initial addition of silver nitrate, silver chloride is formed in an environment that has a high chloride ion concentration. The negative charge on the silver chloride particles is therefore high; the volume of the positive counter-ion layer surrounding each particle must also be relatively large to contain enough positive ions (hydronium or sodium ions, for example) to neutralize the negative charge of the primary layer. Coagulation does not occur under these circumstances. As more silver ions are added, the charge per particle diminishes because the chloride ion concentration is decreased, and the number of particles is increased; the repulsive effect of the double layer thus becomes smaller. As chemical equivalence is approached, a sudden appearance of the coagulated colloid is observed. Here, the number of adsorbed chloride ions per particle becomes small, and the double layer shrinks to a point

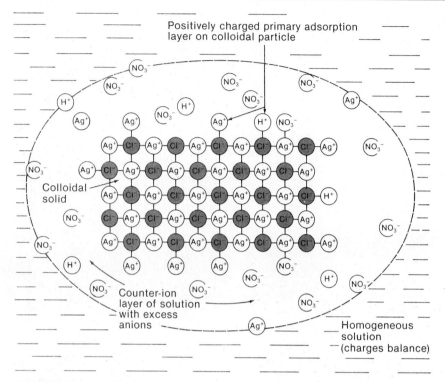

FIGURE 5-1   A colloidal AgCl particle suspended in a solution of AgNO$_3$.

where individual particles approach one another closely enough to permit agglomeration. The agglomeration process can be reversed by adding a large excess of silver ions; here, of course, the charge of the double layer is reversed because the counter-ion layer is now negative.

Coagulation is often brought about by a short period of heating, especially if accompanied by stirring. The decrease in adsorption at elevated temperatures causes a corresponding lowering of net charge on each particle; in addition, the particles acquire kinetic energies sufficient to overcome the barrier to close approach.

An even more effective method of coagulation is to increase the electrolyte concentration of the solution by adding a suitable ionic compound. Under these circumstances, the volume of solution that contains enough ions of opposite charge to neutralize the charge on the particle is lessened. Thus, the introduction of an electrolyte has the effect of shrinking the counter-ion layer, with the result that the surface charge on the particles is more completely neutralized. With their effective charge decreased, the particles can approach one another more closely and coagulation can take place.

Coprecipitation in coagulated colloids.   Adsorption is the principal type of coprecipitation that affects coagulated colloids; other types are encountered with crystalline solids.

A coagulated colloid consists of irregularly arranged particles that form a loosely packed, porous mass. Within this mass, large internal surface areas remain in contact with the solvent phase (see Figure 5-2). These surfaces will retain most of the primarily adsorbed ions that were on the uncoagulated particles. Even though the counter-ion layer surrounding the original colloidal particle is part of the solution, sufficient counter ions to impart electrical neutrality must accompany the particle (in the film of liquid surrounding the particle) through the processes of coagulation and filtration. *The net effect of surface adsorption is, therefore, the carrying down of an otherwise soluble compound as a surface contaminant.*

Peptization of colloids. Peptization refers to the process whereby a coagulated colloid reverts to its original dispersed state. Peptization frequently occurs when pure water is used to wash such a precipitate. Washing is not particularly effective in dislodging primarily adsorbed contaminants; it does tend, however, to remove the electrolyte responsible for coagulation from the internal liquid in contact with the solid. As the electrolyte is removed, the counter-ion layers increase again in volume. The repulsive forces responsible for the original colloidal state are thus reestablished, and the particles detach themselves from the coagulated mass. The washings may become cloudy as the freshly dispersed particles pass through the filter.

The chemist is thus faced with a dilemma in handling coagulated colloids. Although washing is needed to minimize contamination, there is also the risk of losses from peptization. This problem is commonly resolved by washing the agglomerated colloid with a solution containing a volatile electrolyte that can subsequently be removed from the solid by heating. For example, silver chloride precipitates are ordinarily washed with dilute nitric acid, giving a product that is contaminated with the acid; no harm results, however, because the nitric acid is volatilized when the precipitate is dried at 110°C.

Practical treatment of colloidal precipitates. Colloids are ordinarily precipitated from hot, stirred solutions to which sufficient electrolyte has been added to assure

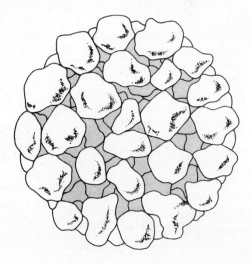

FIGURE 5-2 Coagulated colloidal particles.

coagulation. The filterability of a coagulated colloid is frequently improved by allowing it to stand for an hour or more in contact with the hot solution from which it was formed. During this process, which is known as *digestion,* weakly bound water appears to be lost from the precipitate; the result is a denser mass that is easier to filter.

A dilute solution of a volatile electrolyte is used to wash the filtered precipitate. Washing does not appreciably affect primarily adsorbed ions because the attraction between these species and the solid is too strong. Some exchange, however, may occur between the existing counter ions and one of the ions in the wash solution. Under any circumstance, it must be expected that the precipitate will still be contaminated to some degree, even after extensive washing. The error introduced into the analysis from this source can range from 1 or 2 ppt (as in the coprecipitation of silver nitrate on silver chloride) to an intolerable level (as in the coprecipitation of heavy metal hydroxides upon the hydrous oxides of trivalent iron or aluminum).

A drastic, but effective, way to minimize the effects of adsorption is *reprecipitation.* Here, the filtered solid is redissolved and again precipitated. The first precipitate normally carries down only a fraction of the contaminant present in the original solvent. Thus, the solution containing the redissolved precipitate will have a significantly lower contaminant concentration than the original. When precipitation is again carried out, less adsorption is to be expected. Reprecipitation adds substantially to the time required for an analysis; nevertheless, it is often necessary for such precipitates as the hydrous oxides of iron(III) and aluminium, which possess extraordinary tendencies to adsorb the hydroxides of heavy metal cations such as zinc, cadmium, and manganese.

5B-3 CRYSTALLINE PRECIPITATES
In general, crystalline precipitates are more easily manipulated than coagulated colloids. The size of individual crystalline particles can be varied to a degree. As a consequence, the physical properties and purity of the solid are determined by experimental variables over which the chemist has a measure of control.

Methods of improving particle size and filterability. The particle size of crystalline solids can be improved by keeping the relative supersaturation low during the period in which the precipitate is formed. Equation 5-2 suggests that minimizing Q, maximizing S, or both will accomplish this purpose.

The use of dilute solutions and the slow addition of precipitating agent with thorough mixing tend to minimize the momentary local supersaturation in the solution; moreover, S can usually be increased by precipitation from hot solution. Significant improvement in particle size can be obtained with these simple measures.

Purity of crystalline precipitates. The specific surface area[3] of crystalline precipitates is relatively small; consequently, coprecipitation by direct adsorption is

[3] The specific surface is defined as the exposed surface area by a unit weight of solid; it ordinarily is expressed in terms of square centimeters per gram.

negligible. However, other forms of coprecipitation, which involve the incorporation of contaminants within the interior of crystals, may cause serious errors.

Two types of coprecipitation, *inclusion* and *occlusion,* are associated with crystalline precipitates. The two differ in the manner in which the contaminant is distributed throughout the interior of the solid. Inclusion involves the random distribution of foreign ions or molecules throughout the crystal. Occlusion, on the other hand, involves a nonhomogeneous distribution of ions or molecules of the contaminant within imperfections in the crystal lattice.

Inclusions can be either isomorphic or nonisomorphic. Isomorphic inclusion (also known as mixed crystal formation) occurs when the contaminant possesses dimensions and composition that permit its incorporation within the crystal structure with little or no strain of the lattice. A nonisomorphic inclusion appears to involve a solid solution of the contaminant in the precipitate. Both types tend to cause a homogeneous distribution of contaminant throughout the precipitate.

Occlusion occurs when whole droplets of the solution containing the impurities are trapped and surrounded by a rapidly growing crystal. Because the contaminants are located within the crystal, washing does little to decrease their amount. A lower precipitation rate may significantly lessen the extent of occlusion by providing time for the impurities to escape before they are trapped in the growing crystal. Digestion of the precipitate for as long as several hours is even more effective in eliminating occluded contaminants.

Digestion of crystalline precipitates. The heating of crystalline precipitates (without stirring) for some time after formation frequently yields a purer, more filterable product. The improvement in purity during this digestion process undoubtedly results from the solution and recrystallization that occur continuously and at an enhanced rate at elevated temperatures. During these processes, many pockets of imperfection become exposed to the solution; the contaminant is thus able to escape from the solid, and more perfect crystals result.

Solution and recrystallization during digestion are probably responsible for the improvement in filterability as well. Bridging between adjacent particles yields larger crystalline aggregates that are more easily filtered. This view seems to be confirmed by the fact that little improvement in filtering characteristics occurs if the mixture is stirred during digestion.

5B-4 DIRECTION OF COPRECIPITATION ERRORS

Coprecipitated impurities may cause either negative or positive errors in an analysis. If the contaminant is not a compound of the ion being determined, positive errors will always result. Thus, a positive error will be observed when colloidal silver chloride adsorbs silver nitrate during a chloride analysis. On the other hand, when the contaminant contains the ion being determined, either positive or negative errors may be observed. In the determination of barium ions by precipitation as barium sulfate, for example, occlusion of other barium salts occurs. If the occluded contaminant is barium nitrate, a positive error will be observed because this compound has a larger formula weight than the barium sulfate that would have formed had no coprecipitation occurred. If barium chloride were the contaminant, however, a negative error would arise because its formula weight is less than that of the sulfate salt.

5B-5 PRECIPITATION FROM HOMOGENEOUS SOLUTION

In precipitation from homogeneous solution, the precipitating agent is chemically generated in the solution. Local reagent excesses do not occur because the precipitating agent appears slowly and homogeneously throughout the entire solution; the relative supersaturation is thus kept low. In general, homogeneously formed precipitates, both colloidal and crystalline, are better suited for analysis than solids formed by direct addition of a reagent.

Urea is often employed for the homogeneous generation of hydroxide ion. The reaction can be expressed by the equation

$$(H_2N)_2CO + 3H_2O = CO_2 + 2NH_4^+ + 2OH^-$$

This reaction proceeds slowly at temperatures just below boiling; typically 1 to 2 hours are needed to produce sufficient reagent for complete precipitation. The method is particularly valuable for the precipitation of hydrous oxides or basic salts. For example, the hydrous oxides of iron(III) and aluminum are bulky, gelatinous masses that are heavily contaminated and difficult to filter when they are formed by the direct addition of base. In contrast, these same products are dense, readily filtered, and have considerably higher purity when they are produced by the homogeneous generation of hydroxide ion, particularly in the presence of succinate ion.

Homogeneous precipitation of crystalline precipitates also results in marked increases in crystal size as well as improvement in purity.

Representative analyses based upon precipitation by homogeneously generated reagents are given in Table 5-2.

5B-6 DRYING AND IGNITION OF PRECIPITATES

After filtration, a gravimetric precipitate is heated until its weight becomes constant. Heating serves the purpose of removing the solvent and volatile electrolytes carried down with the precipitate; in addition, this treatment may induce chemical decomposition to give a product of known composition.

The temperature required to produce a suitable product varies, depending upon the precipitate. Figure 5-3 shows weight loss as a function of temperature for several common analytical precipitates. These data were obtained with an automatic thermobalance,[4] an instrument that measures the weight of a substance continuously as its temperature is increased at a constant rate in a furnace. The heating of three of the precipitates—silver chloride, barium sulfate, and aluminum oxide—serves simply to remove water and perhaps volatile electrolytes carried down during the precipitation process. Note the wide range in temperature required to produce an anhydrous precipitate of constant weight. Thus, moisture is completely removed from silver chloride at a temperature above 110 to 120°C; dehydration of aluminum oxide, on the other hand, is not complete until a temperature greater than 1000°C is achieved. It is of interest to note that aluminum oxide, formed homogeneously with urea, can be completely dehydrated at a temperature of about 650°C.

The thermal curve for calcium oxalate is considerably more complex than

---

[4] For descriptions of thermobalances, see: W. W. Wendlandt, *Thermal Methods of Analysis*, 2d ed. New York: Wiley, 1974.

Table 5-2 Methods for the Homogeneous Generation of Precipitants

Precipitant	Reagent	Generation Reaction	Elements Precipitated
OH^-	Urea	$(NH_2)_2CO + 3H_2O = CO_2 + 2NH_4^+ + 2OH^-$	Al, Ga, Th, Bi, Fe, Sn
PO_4^{3-}	Trimethyl phosphate	$(CH_3O)_3PO + 3H_2O = 3CH_3OH + H_3PO_4$	Zr, Hf
$C_2O_4^{2-}$	Ethyl oxalate	$(C_2H_5)_2C_2O_4 + 2H_2O = 2C_2H_5OH + H_2C_2O_4$	Mg, Zn, Ca
SO_4^{2-}	Dimethyl sulfate	$(CH_3O)_2SO_2 + 4H_2O = 2CH_3OH + SO_4^{2-} + 2H_3O^+$	Ba, Ca, Sr, Pb
CO_3^{2-}	Trichloroacetic acid	$Cl_3CCOOH + 2OH^- = CHCl_3 + CO_3^{2-} + H_2O$	La, Ba, Ra
S^{2-}	Thioacetamide	$CH_3\overset{S}{\overset{\|}{C}}NH_2 + H_2O = CH_3\overset{O}{\overset{\|}{C}}NH_2 + H_2S$	Sb, Mo, Cu, Cd
8-Hydroxyquinoline	8-Acetoxyquinoline	$+ H_2O = CH_3COH +$	Al, U, Mg, Zn

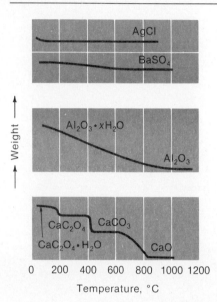

FIGURE 5-3   Effect of temperature on precipitate weights.

the others shown in Figure 5-3. At temperatures below about 135°C, unbound water is eliminated to give the monohydrate $CaC_2O_4 \cdot H_2O$; a temperature of about 225°C converts this compound to the anhydrous oxalate. The abrupt change in weight at about 450°C signals the decomposition of calcium oxalate to calcium carbonate and carbon monoxide. The final step in the curve depicts the conversion of the carbonate to calcium oxide and carbon dioxide. It is evident that the weighing form employed for a calcium oxalate precipitate will depend upon the ignition conditions.

5C A CRITIQUE OF THE GRAVIMETRIC METHOD

Some chemists are inclined to discount the present-day value of gravimetric methods on the grounds that they are inefficient and obsolete. We, on the other hand, believe that the gravimetric approach to an analytical problem, like all others, has strengths and weaknesses and that ample situations exist where it represents the best possible choice for the resolution of an analytical problem.

5C-1 TIME REQUIRED FOR A GRAVIMETRIC ANALYSIS

In all analytical methods, a physical quantity X is ultimately measured which varies with the amount or the concentration C_A of the analyte A in the original sample. Ideally, but by no means always, the functional relationship between C_A and X is linear; that is,

$$C_A = kX$$

Most analytical procedures require the determination of k by a standardization process or by an empirical calibration in which X is measured for one or more standards of known C_A. Two notable exceptions to this generality are the gravimetric method and the coulometric method (Chapter 17). With these procedures, k can be calculated directly from known, well-established physical constants; the need for calibration or standardization is thus avoided.

For a gravimetric method, k is simply the gravimetric factor, which can be calculated from data in a table of atomic weights. As we shall show, this direct knowledge of k sometimes makes a gravimetric procedure the most efficient one for solving an analytical problem.

In considering the time required to perform an analysis, it is necessary to differentiate between elapsed time and operator time. The first refers to the clock hours or minutes between the start of the analysis and the report of the result. Operator time, on the other hand, represents the actual time the chemist or technician must spend in performing the various manipulations required to complete the analysis and calculate the result. A characteristic of the gravimetric method is that the difference between elapsed time and operator time is large compared with other methods, because the most time-consuming steps in the analysis do not require the constant attention of the analyst. For example, drying of crucibles, evaporation of solutions, digestion of precipitates, and ignition of products frequently take several hours to complete but require, at the most, a few minutes of the analyst's time, thus leaving time for the performance of other tasks.

On the basis of operator time, the gravimetric approach for a particular analysis may prove to be the most efficient, particularly where only one or two samples are to be analyzed, because no time is needed for calibration or standardization (that is, for the evaluation of k). On the other hand, as the number of samples to be analyzed increases, the time required for calibration in a nongravimetric procedure becomes smaller *on a per sample basis* and often becomes negligible when eight or ten samples are involved (assuming, of course, that a single calibration will suffice for all of the samples). With a large number of samples, then, the time required for precipitating, filtering, washing, and weighing may become larger than the equivalent operations in a nongravimetric procedure. It is frequently but not always true that a gravimetric approach becomes less advantageous when a large number of samples is involved.

5C-2 SENSITIVITY OF THE GRAVIMETRIC METHOD

The sensitivity and accuracy of many methods are frequently limited by the device used for the analytical measurement. Such limitations seldom, if ever, affect a gravimetric analysis. Thus, with a suitable type of balance it is perfectly feasible to obtain the weight of a few micrograms of material to within a few parts per thousand of its true value; for larger masses, the weighing uncertainty can be decreased to a few parts per million.

The sensitivity of a gravimetric analysis is more likely to be limited by the difficulties encountered in attempting to separate a small amount of precipitate from a relatively large volume of solution that contains high concentrations of other constituents from the sample. In some situations, solubility losses may be troublesome; in others, the precipitation rate may be so slow as to make the process impractical. Coagulation of colloidal precipitates often becomes difficult or impossible when the amounts of material are low. Finally, mechanical losses of a significant fraction of the solid become inevitable during filtration of a minute amount of solid suspended in a large volume of solution.

Because of these problems, the chemist is wise to discard the idea of employing a gravimetric analysis for a constituent if its concentration is likely to be below 0.1%; the method is best applied where the concentration is greater than 1%.

5C-3 ACCURACY OF GRAVIMETRIC METHODS
It is difficult to generalize about the accuracy of gravimetric methods because each is subject in varying degree to uncertainties from such sources as solubility, coprecipitation, and variations in the chemical composition of the final product. Each of these factors is dependent upon the composition of the sample. For example, the iron content of a sample containing no other heavy metal ions is readily determined gravimetrically to within a few parts per thousand. On the other hand, the analysis will be subject to errors of several percent if such divalent cations as zinc, nickel, or copper are also present; elimination of coprecipitation by these ions requires the expenditure of substantially greater effort. Solubility losses are also affected by sample composition. With multicomponent samples, the likelihood of complex formation between the analyte and a constituent of the sample is enhanced; in addition, purification of the precipitate may require a larger volume of wash water which, in turn, may cause a concomitant increase in loss by solubility.

The accuracy of gravimetric analysis for simple samples containing more than 1% of the analyte is seldom surpassed by other methods; here, errors may often be decreased to 1 or 2 parts in 1000. With increasing sample complexity, larger errors are inevitable unless a great deal of time is expended in circumventing them. Here, the accuracy of the gravimetric method may be no better than, and sometimes poorer than, other analytical methods.

5C-4 SPECIFICITY OF GRAVIMETRIC METHODS
Gravimetric reagents are seldom specific; instead, virtually all are *selective* in the sense that they tend to form precipitates with groups of ions. Each ion within any group will then interfere with the analysis of any other ion in the group unless a preliminary separation is performed. In general, gravimetric procedures are less specific than some of the other methods that will be considered in later chapters.

5D APPLICATIONS OF THE GRAVIMETRIC METHOD
Gravimetric methods have been developed for most, if not all, inorganic anions and cations as well as for neutral species such as water, sulfur dioxide, carbon dioxide, and iodine. A variety of organic substances can also be readily determined gravimetrically. Examples include lactose in milk products, salicylates in drug preparations, phenolphthalein in laxatives, nicotine in pesticides, cholesterol in cereals, and benzaldehyde in almond extracts. Indeed, gravimetric methods are among the most widely applicable of all analytical procedures.

5D-1 INORGANIC PRECIPITATING AGENTS
Table 5-3 lists common inorganic precipitating agents. These reagents typically cause formation of a slightly soluble salt or a hydrous oxide. The weighing form is either the salt itself or else an oxide. The lack of specificity of most inorganic reagents is clear from the many entries in the table.

Detailed procedures for the gravimetric determination with inorganic reagents are given in Methods 1-2 and 1-3 in Chapter 31.

Table 5-3 Some Inorganic Precipitating Agents *

Precipitating Agent	Element Precipitated **
NH_3(aq)	**Be** (BeO), **Al** (Al_2O_3), **Sc** (Sc_2O_3), Cr (Cr_2O_3),* **Fe** (Fe_2O_3), Ga (Ga_2O_3), Zr (ZrO_2), **In** (In_2O_3), Sn (SnO_2), U (U_3O_8)
H_2S	Cu (CuO),* **Zn** (ZnO, or $ZnSO_4$), **Ge** (GeO_2), As ($\underline{As_2O_3}$, or As_2O_5), Mo (MoO_3), Sn (SnO_2),* Sb ($\underline{Sb_2O_3}$, or Sb_2O_5), Bi (Bi_2S_3)
$(NH_4)_2S$	Hg ($\underline{HgS}$), Co (Co_3O_4)
$(NH_4)_2HPO_4$	**Mg** ($Mg_2P_2O_7$), Al ($AlPO_4$), Mn ($Mn_2P_2O_7$), Zn ($Zn_2P_2O_7$), Zr ($Zr_2P_2O_7$), Cd ($Cd_2P_2O_7$), Bi ($BiPO_4$)
H_2SO_4	Li, Mn, **Sr, Cd, Pb, Ba** (all as sulfates)
H_2PtCl_6	K (K_2PtCl_6, or Pt), Rb ($\underline{Rb_2PtCl_6}$), Cs ($\underline{Cs_2PtCl_6}$)
$H_2C_2O_4$	Ca (CaO), Sr (SrO), **Th** (ThO_2)
$(NH_4)_2MoO_4$	Cd ($CdMoO_4$),* Pb ($\underline{PbMoO_4}$)
HCl	**Ag** (AgCl), Hg (Hg_2Cl_2), Na (as NaCl from butyl alcohol), Si (SiO_2)
$AgNO_3$	**Cl** (AgCl), Br ($\underline{AgBr}$), I($\underline{AgI}$)
$(NH_4)_2CO_3$	**Bi** (Bi_2O_3)
NH_4SCN	Cu ($Cu_2(SCN)_2$)
$NaHCO_3$	Ru, Os, Ir (precipitated as hydrous oxides; reduced with H_2 to metallic state)
HNO_3	Sn (SnO_2)
H_5IO_6	Hg ($Hg_5(IO_6)_2$)
NaCl, $Pb(NO_3)_2$	F (PbClF)
$BaCl_2$	SO_4^{2-} ($BaSO_4$)
$MgCl_2$, NH_4Cl	PO_4^{3-} ($Mg_2P_2O_7$)

* From W. F. Hillebrand, G. E. F. Lundell, H.A. Bright, and J. I. Hoffman, *Applied Inorganic Analysis.* New York: Wiley, 1953.

** Boldface type indicates that gravimetric analysis is the preferred method for the element or ion. The weighed form is indicated in parentheses. An asterisk indicates that the gravimetric method is seldom used. An underscored entry indicates the most reliable gravimetric method.

5D-2 REDUCING REAGENTS
Table 5-4 lists several reagents that convert the analyte to its elemental form for weighing.

5D-3 ORGANIC PRECIPITATING AGENTS
A number of organic reagents have been developed for the gravimetric analysis of inorganic species. Some of these reagents are more selective in their reactions than many of the inorganic reagents listed in Table 5-3.

Two types of organic reagents are encountered. One forms slightly soluble nonionic complexes called *coordination compounds;* the other forms products in which the bonding between the inorganic species and the reagent is largely ionic.

Organic reagents which yield sparingly soluble coordination compounds typically contain at least two functional groups, each of which is capable of bonding with the cation by donation of a pair of electrons. The functional groups are located in the molecule in such a way that a five- or six-membered ring results from the reaction. Coordination compounds which form complexes of this type are called *chelating agents;* their products with a cation are termed *chelates.*

Table 5-4   Some Reducing Reagents Employed in Gravimetric Methods

Reducing Agent	Analyte
SO_2	Se, Au
$SO_2 + H_2NOH$	Te
H_2NOH	Se
$H_2C_2O_4$	Au
H_2	Re, Ir
HCOOH	Pt
$NaNO_2$	Au
$TiCl_2$	Rh
$SnCl_2$	Hg
Electrolytic reduction	Co, Ni, Cu, Zn, Ag, In, Sn, Sb, Cd, Re, Bi

Neutral coordination compounds are relatively nonpolar; as a consequence, their solubilities are low in water but high in organic liquids. Chelates usually possess low densities and are often intensely colored. Because they are not wetted by water, coordination compounds are readily freed of moisture at low temperatures. At the same time, however, their hydrophobic nature endows these precipitates with the annoying tendency to creep up the sides of the filtering medium during the washing operation; physical loss of solid may result unless care is taken. Three coordinating reagents that exhibit varying degrees of specificity are described in the section that follows.

8-Hydroxyquinoline. Approximately two dozen cations form sparingly soluble coordination compounds with 8-hydroxyquinoline, which is also known as *oxine.*

Typical of these is the product with magnesium

The solubilities of metal oxinates vary widely from cation to cation and, moreover, are pH dependent because proton formation always accompanies the chelation reaction. Therefore, a considerable degree of selectivity can be imparted to the action of 8-hydroxyquinoline through pH control.

α-Nitroso-β-naphthol. This compound was one of the first selective organic reagents discovered (1885); its structure is

The reagent reacts with cobalt(II) to give a neutral cobalt(III) chelate having the structure CoA_3, where A^- is the conjugate base of the reagent. Note that formation of the product involves both oxidation and precipitation of the cobalt by the reagent; the precipitate is contaminated by reduction products of the reagent as a consequence. Therefore, it is common practice to ignite the chelate in oxygen to produce Co_3O_4; alternatively, the ignition is performed in a hydrogen atmosphere to produce the element as the weighing form.

The most important application of α-nitroso-β-naphthol has been for the determination of cobalt in the presence of nickel. Other ions that precipitate with the reagent include bismuth(III), chromium(III), mercury(II), tin(IV), titanium(III), tungsten(VI), uranium(VI), and vanadium(V).

Dimethylglyoxime. An organic precipitating agent of unparalleled specificity is dimethylglyoxime.

Its coordination compound with palladium is the only one that is sparingly soluble in acid solution. Similarly, only the nickel compound precipitates from a weakly alkaline environment. Nickel dimethylglyoxime is bright red and has the structure

This precipitate is so bulky that only small amounts of nickel can be handled conveniently; it also has an exasperating tendency to creep as it is filtered and washed. The solid is readily dried at 110°C and has the composition indicated by its formula. A procedure for the determination of nickel in a steel employing dimethylglyoxime is found in Method 1-4 in Chapter 31.

Sodium tetraphenylboron. Sodium tetraphenylboron, $(C_6H_5)_4B^-Na^+$, is an important example of organic precipitating reagents that form salt-like precipitates. In cold mineral acid solutions, it is a near-specific precipitating agent for potassium and ammonium ions. The precipitates are stoichiometric, corresponding to the potassium or the ammonium salt, as the case may be; they are amenable to vacuum filtration and can be brought to constant weight at 105 to 120°C. Only mercury(II), rubidium, and cesium interfere and must be removed by prior treatment.

5D-4 GRAVIMETRIC ORGANIC FUNCTIONAL GROUP ANALYSIS

Several reagents have been shown to react selectively with certain organic functional groups and thus can be used for the determination of most compounds containing these groups. A list of gravimetric functional group reagents is given in Table 5-5. Many of the reactions shown can also be used for volumetric and spectrophotometric determinations. For the occasional analysis, the gravimetric procedure will often be the method of choice since no calibration or standardization is required.

5D-5 GRAVIMETRIC METHODS FOR SPECIFIC ORGANIC COMPOUNDS

A few typical examples of methods developed for specific organic compounds are described in the paragraphs that follow.

Determination of salicylic acid. Salicylic acid can be determined by dissolving the compound in a sodium carbonate solution, adding an iodine solution, and digesting. The reaction is

$$8I^- + 2HCO_3^- + 6H_2O$$

The yellow tetraiodophenylenequinone is filtered, dried, and weighed. Acetylsalicylic acid (aspirin) can be determined in the same way after hydrolysis to give salicylic acid.

Determination of nicotine. Nicotine in pesticides or tobacco products can be determined, after steam distillation from the sample, by precipitation with silicotungstic acid. After filtering and washing, the residue is ignited to give $SiO_2 \cdot 12WO_3$, which is weighed.

Determination of phenolphthalein. A standard method for the determination of phenolphthalein in laxative preparations involves separation of the compound by alcohol extraction followed by evaporation to dryness. The phenolphthalein is then dissolved in dilute alkali and precipitated as the tetraiodo compound by addition of iodine solution. The product can be dried to constant weight at 110°C.

Table 5-5 Gravimetric Methods for Organic Functional Groups

Functional Group	Basis for Method	Reaction and Product Weighed*
Carbonyl	Weight of precipitate with 2,4-dinitrophenylhydrazine	$RCHO + H_2NNHC_6H_3(NO_2)_2 \rightarrow$ $R\text{—}CH{=}NN\underline{HC_6H_3(NO_2)_2}(s) + H_2O$ (RCOR' reacts similarly)
Aromatic carbonyl	Weight of CO_2 formed at 230°C in quinoline. CO_2 distilled, absorbed, and weighed	$ArCHO \xrightarrow[CuCO_3]{230°C} Ar + \underline{CO_2}(g)$
Methoxyl and ethoxyl	Weight of AgI formed after distillation and decomposition of CH_3I or C_2H_5I	$\left.\begin{array}{l} ROCH_3 + HI \rightarrow ROH + CH_3I \\ RCOOCH_3 + HI \rightarrow RCOOH + CH_3I \\ ROC_2H_5 + HI \rightarrow ROH + C_2H_5I \end{array}\right\}$ $CH_3I + Ag^+ + H_2O \rightarrow$ $\underline{AgI}(s) + CH_3OH$
Aromatic nitro	Weight loss of Sn	$RNO_2 + \tfrac{3}{2}\underline{Sn}(s) + 6H^+ \rightarrow RNH_2 + \tfrac{3}{2}Sn^{4+} + 2H_2O$
Azo	Weight loss of Cu	$RN = NR' + 2\underline{Cu}(s) + 4H^+ \rightarrow RNH_2 + R'NH_2 + 2Cu^{2+}$
Phosphate	Weight of Ba salt	$ROP(OH)_2 + Ba^{2+} \rightarrow RO\overset{O}{P}O_2\underline{Ba}(s) + 2H^+$
Sulfamic acid	Weight of $BaSO_4$ after oxidation with HNO_2	$RNHSO_3H + HNO_2 + Ba^{2+} \rightarrow ROH + \underline{BaSO_4}(s) + N_2 + 2H^+$
Sulfinic acid	Weight of Fe_2O_3 after ignition of Fe^{3+} sulfinate	$3ROSOH + Fe^{3+} \rightarrow (ROSO)_3Fe(s) + 3H^+$ $(ROSO)_3Fe \xrightarrow{O_2} CO_2 + H_2O + SO_2 + \underline{Fe_2O_3}(s)$

* The substance weighed is underlined.

5D-6 VOLATILIZATION PROCEDURES

The two most common gravimetric methods based on volatilization are for water and for carbon dioxide.

Water is quantitatively eliminated from many inorganic samples by ignition. In the direct determination, it is collected on any of several solid desiccants, and its mass is determined from the gain in weight of the desiccant. An apparatus for this procedure is shown in Section 24C-1.

Less satisfactory is the indirect method based upon the loss in weight suffered by the sample as the result of ignition. Here, it must be assumed that water is the only component that has been volatilized. This assumption is frequently unjustified; ignition of many samples results in decomposition and a change in weight, irrespective of the presence of water. Nevertheless, the indirect method has found wide use for the determination of water in items of commerce. For example a semiautomated instrument for the determination of moisture in grains can now be purchased. It consists of a balance platform upon which a 10-g sample is heated with an infrared lamp. The percent residue is read directly.

Carbonates are ordinarily decomposed by acids to give carbon dioxide, which is readily evolved from solution by heat. As in the direct analysis for water, the weight of carbon dioxide is established from the increase in the weight of a solid absorbent. Ascarite II,[5] which consists of sodium hydroxide on a nonfibrous silicate, serves to retain the carbon dioxide by the reaction

$$2NaOH + CO_2 \longrightarrow Na_2CO_3 + H_2O$$

The absorption tube must also contain a desiccant to prevent loss of the evolved water.

Sulfides and sulfites can also be determined by volatilization. Here, hydrogen sulfide or sulfur dioxide is evolved from the sample after treatment with acid and collected in a suitable absorbent.

Finally, the classical method for the determination of carbon and hydrogen in organic compounds is a gravimetric procedure in which the combustion products (H_2O and CO_2) are collected selectively on weighed absorbents. The increase in weight serves as the analytical parameter.

PROBLEMS*

5-1. Use chemical symbols to express gravimetric factors for the following:

	Sought	Weighed		Sought	Weighed
* (a)	$MgCl_2$	$AgCl$	(f)	$Na_2B_4O_7 \cdot 10H_2O$	B_2O_3
(b)	Cr	Ag_2CrO_4	* (g)	Fe_6S_{17}	Fe_2O_3
* (c)	Ag	Ag_2CrO_4	(h)	Fe_6S_{17}	$CdSO_4$
(d)	Cr_2O_3	Ag_2CrO_4	* (i)	K_2SO_4	$(C_6H_5)_4BK$
* (e)	Pb_3O_4	PbO_2	(j)	$(PbCO_3)_2 \cdot Pb(OH)_2$	PbO

* 5-2. Use chemical symbols to generate the gravimetric factor for an analysis in which the results are to be expressed in terms of Cu_2HgI_4 and the precipitate that is weighed is

[5] ® Arthur H. Thomas Co., Philadelphia, PA.

* Answers to problems or parts of problems marked with an asterisk are to be found at the end of the book.

(a) Hg.   (d) Cu.
(b) Hg_2S.   (e) AgI.
(c) $Cu_2(SCN)_2$.   (f) $In(IO_3)_3$.

5-3. Use chemical formulas to generate the gravimetric factors needed to express the results of an analysis in terms of percent $CoI_2 \cdot 6NH_3$ if the precipitate that is weighed is

(a) $Pb(IO_3)_2$.   (d) Co_2O_3.
(b) $K_3Co(NO_2)_6 \cdot H_2O$.   (e) $(C_6H_5)_4BNH_4$.
(c) AgI.   (f) Hg_2I_2.

* 5-4. Which of the weighing forms in Problem 5-2 will yield the largest weight of precipitate from a given weight of Cu_2HgI_4?

5-5. Which of the weighing forms in Problem 5-3 will yield the smallest weight of precipitate from a given weight of $CoI_2 \cdot 6NH_3$?

5-6. Use chemical symbols to express the gravimetric factor needed to calculate the percentage of $CaC_2O_4 \cdot 2H_2O$, based upon

* (a) the weight of the anhydrous salt after the hydrate water has been driven off by heating at 250°C (see Figure 5-3).
(b) the loss of weight that results from heating $CaC_2O_4 \cdot 2H_2O$ at 250°C.
* (c) the weight of $CaCO_3$ produced as the result of heating at 600°C.
(d) the loss in weight that occurs when $CaC_2O_4 \cdot 2H_2O$ is heated at 600°C.
* (e) the weight of CaO produced as the result of ignition at 1000°C.
(f) the loss in weight that occurs when $CaC_2O_4 \cdot 2H_2O$ is ignited at 1000°C.

* 5-7. The manufacturer of a household cleaner advertises that the product contains 0.9% phosphorous. Express the content as percent Na_3PO_4.

5-8. A plant-food preparation is advertised to contain 10% P and 10% K. Convert these figures to percent $(NH_4)_3PO_4$ and percent K_2O, respectively.

* 5-9. How many grams of potassium iodate are needed to produce 1.67 g of lead iodate?

5-10. What weight of Ag_2CrO_4 can be produced from a 0.500-g sample that is 64% (w/w) $AgNO_3$?

* 5-11. What weight of $Ag_4Fe(CN)_6$ can be produced from a 1.68-g sample that is 82.5% $K_4Fe(CN)_6$?

5-12. Calculate the weight of silver chloride produced when 0.364 g of AgI is heated in a stream of chlorine. Reaction:

$$2AgI(s) + Cl_2(g) \longrightarrow 2AgCl(s) + I_2(g)$$

* 5-13. Calculate the weight of calcium oxide that is produced when 3164 g of CaC_2O_4 are strongly ignited. Reaction:

$$CaC_2O_4(s) \longrightarrow CaO(s) + CO(g) + CO_2(g)$$

5-14. Treatment of a 0.7165-g sample with an excess of $BaCl_2$ yielded a precipitate of $BaSO_4$ that weighed 0.4329 g. Express the results of this analysis in terms of percent

(a) $MnSO_4 \cdot 7H_2O$.   (d) $RbFe(SO_4)_2 \cdot 12H_2O$.
(b) $Fe(NH_4)_2(SO_4)_2 \cdot 6H_2O$.   (e) $Th(SO_4)_2$.
(c) $Ga_2(SO_4)_3 \cdot 18H_2O$.   (f) $(NH_4)_2SO_4 \cdot Ce_2(SO_4)_3 \cdot 8H_2O$.

* 5-15. A 0.5881-g sample yielded 0.641 g of AgCl upon treatment with an excess of $AgNO_3$. Express the results of this analysis in terms of percent

(a) KCl.   (d) NH_4ClO_4.
(b) $BaCl_2 \cdot 2H_2O$.   (e) $ZnCl_2 \cdot 2NH_4Cl$.
(c) Au_2Cl_6.   (f) $FeCl_3$.

5-16. After appropriate preliminary treatment, the thallium in a 10.20-g pesticide sample was precipitated as TlI and was weighed as such. Calculate the percentage of Tl_2SO_4 in the sample if 0.1964 g of TlI was recovered.

* 5-17. The aluminum in a 0.764-g sample was precipitated as $Al_2O_3 \cdot xH_2O$ with an excess

of ammonia and subsequently ignited to Al_2O_3. Calculate the percentage of Al_2O_3 in the sample if 0.127 g of Al_2O_3 was recovered.

5-18. Analysis of the components of a 1.062-g sample of ship-nail brass yielded 0.0377 g of SnO_2, 0.1321 g of $PbSO_4$, 1.301 g of $Cu_2(SCN)_2$, and 0.6187 g of $Zn_2P_2O_7$. Calculate the percentage of each element in this alloy.

* 5-19. The mercury in a 0.8164-g mineral sample was brought into solution, converted to the +2 state, and precipitated with an excess of paraperiodic acid. Reaction:

$$5Hg^{2+} + 2H_5IO_6 \longrightarrow Hg_5(IO_6)_2(s) + 10H^+$$

After being filtered, washed, and dried, the precipitate was found to weigh 0.6729 g. Calculate the percentage of HgO in the mineral.

5-20. A 1.236-g sample of an herbicide was decomposed with metallic sodium in alcohol. The chloride ion liberated by this treatment was precipitated as AgCl and found to weigh 0.1840 g. Calculate the percentage of 2,4-dichlorophenoxyacetic acid $(C_8H_6O_3Cl_2$, fw = 221) in the sample.

* 5-21. The sulfur in five saccharine $(C_7H_5NO_3S$, fw = 183) tablets that weighed a total of 0.2140 g was oxidized to sulfate and then precipitated as $BaSO_4$. Calculate the average weight of saccharine in these tablets if 0.2070 g of $BaSO_4$ was recovered.

5-22. The analysis in Problem 5-21 contains one unnecessary step; identify it.

* 5-23. Chlorisondiamine chloride is an antihypertensive agent with the formula $C_{14}H_{18}Cl_6N_2$ (fw = 427). A 2.89-g sample of a medication containing this drug was heated in a closed tube to destroy the organic matter and to free the chloride. When reaction was complete, water was added and the carbonaceous residue was removed by filtration. Addition of excess $AgNO_3$ to the clear filtrate yielded 0.187 g of AgCl. Calculate the percentage of chlorisondiamine chloride in the sample, assuming that the drug is the only source of chloride.

5-24. The label on a bottle of the tranquilizer Thioridazine, $C_{21}H_{26}N_2S_2$ (fw = 371) was so damaged that the contents of each pill could no longer be discerned. A 12-tablet sample was decomposed to convert the sulfur to sulfate. A gravimetric analysis subsequently yielded 0.301 g of $BaSO_4$. Calculate the average number of milligrams of Thioridazine in each tablet.

* 5-25. The nitrogen in a 584.0-mg sample of ammonium sulfamate $(H_2NSO_3NH_4$, fw = 114) was converted to ammonium ion, which in turn was precipitated with chloroplatinic acid.

$$H_2PtCl_6 + 2NH_4^+ \longrightarrow (NH_4)_2PtCl_6(s) + 2H^+$$

The solid was filtered and washed, following which it was decomposed by ignition.

$$(NH_4)_2PtCl_6(s) \longrightarrow Pt(s) + Cl_2(g) + 2NH_3(g) + 2HCl(g)$$

Calculate the percentage of ammonium sulfamate in the sample if 327.4 mg of platinum were recovered.

* 5-26. Calculate the molar concentration of a hydrochloric acid solution if a 50.0-mL aliquot, treated with an excess of silver nitrate, yielded 0.4644 g of AgCl.

5-27. Calculate the molar concentration of a silver nitrate solution 25.0 mL of which yielded 0.1240 g of Ag_2CrO_4 when treated with an excess of K_2CrO_4.

* 5-28. Ignition in a stream of oxygen resulted in conversion of the carbon in a 1.367-g steel sample to CO_2, which was subsequently collected in a tube containing an absorbent for the gas. This tube and its contents weighed 14.3951 g at the outset, and 14.4278 g at the completion of the analysis. What was the percentage of carbon in the steel?

5-29. A 2.634-g sample containing the mineral eriochaleite, $CuCl_2 \cdot 2H_2O$, was dehydrated quantitatively by heating in a stream of dry air. The water vapor was collected in a tube containing a desiccant. The tube originally weighed 21.9457 g; it weighed

22.4549 g on completition of the analysis. Calculate the percentage of $CuCl_2 \cdot 2H_2O$, assuming that the mineral was the only source of water in the sample.

* 5-30. A 3.849-g sample of navy beans assayed 17.8% protein on an as-received basis. Calculate this percentage on an oven-dry basis if the sample contained 2.87% moisture.

5-31. Pesticide residues on a 4.79-g sample of oven-dried leaves were found to be 6.8 ppm. Calculate an as-received value for the data, given that the undried sample contained 9.64% moisture.

* 5-32. A 4.714-g ore sample assayed 32.84% Fe_2O_3, as received, and 34.09% after having been dried to constant weight. Estimate the water content of the sample as it was received.

5-33. W. W. White and P. J. Murphy [*Anal. Chem.*, **51**, 1864 (1979)] report that the anionic complex between silver(I) and thiosulfate is quantitatively precipitated with hexaaminecobalt(III) trichloride; the reaction is

$$Co(NH_3)_6^{3+} + Ag(S_2O_3)_2^{3-} \longrightarrow [Co(NH_3)_6][Ag(S_2O_3)_2](s)$$

The product (fw = 493.2) is dried to constant weight at 95°C. A 25.00-mL portion of a photographic fixer solution yielded 0.4161 g of precipitate when analyzed by this method. Calculate the grams of silver contained in each liter of this solution assuming an excess of $S_2O_3^{2-}$ was present in the solution.

* 5-34. The nitrobenzene (fw = 123) in a 0.739-g sample was determined by dilution with an HCl/methanol mixture, followed by the introduction of 0.465 g of pure tin. The mixture was refluxed for an hour, during which time the nitrobenzene was reduced to aniline:

$$2C_6H_5NO_2 + 3Sn(s) + 12H^+ \longrightarrow 2C_6H_5NH_2 + 4H_2O + 3Sn^{4+}$$

When reaction was complete, the unused tin was isolated by filtration, dried, and found to weigh 0.128 g. Calculate the percentage of nitrobenzene in the sample.

5-35. Chloride ion is quantitatively converted to chlorine by the action of manganese dioxide in an acidic solution:

$$MnO_2(s) + 2Cl^- + 4H^+ \longrightarrow Mn^{2+} + Cl_2(g) + 2H_2O$$

Use the accompanying data to calculate the percentage of $ZnCl_2 \cdot 2NH_4Cl$ in an impure sample.

wt of sample taken	1.280 g
wt of MnO_2 introduced	0.6270 g
wt of residual MnO_2	0.4183 g

* 5-36. A 3.000-g sample containing KCl, $(NH_4)_2SO_4$, and inert materials was dissolved in sufficient water to give exactly 500 mL. A 50.0-mL aliquot of this solution (treated with an excess of sodium tetraphenylboron) yielded 0.732 g of precipitate that consisted of $(C_6H_5)BK$ (fw = 358) and $(C_6H_5)_4BNH_4$ (fw = 337). A second 50.0-mL portion was made alkaline and heated to drive off the ammonia

$$NH_4^+ + OH^- \longrightarrow NH_3(g) + H_2O$$

following which treatment with an excess of the sodium tetraphenylboron gave 0.247 g of $(C_6H_5)_4BK$. Calculate the respective percentages of KCl and $(NH_4)_2SO_4$ in the sample.

5-37. A 4.08-g sample that contained ammonium sulfate, potassium nitrate, and inert materials was dissolved in sufficient water to give 500 mL of solution. A 50.0-mL aliquot of this solution yielded 0.267 g of $(C_6H_5)_4BNH_4$ (fw = 337) when treated with an excess of sodium tetraphenylboron. A second 50.0-mL aliquot was heated with Devarda's alloy (50% Cu, 45% Al, 5% Zn) in alkaline solution, which resulted in reduction of the nitrate to ammonia:

$$NO_3^- + 6H_2O + 8\,e \longrightarrow NH_3 + 9OH^-$$

Because this reaction occurs in alkaline solution, all of the ammonium ion in the aliquot was converted to ammonia, which was then distilled and collected in dilute acid. Treatment of the distillate with sodium tetraphenylboron yielded 0.720 g of $(C_6H_5)_4BNH_4$ (fw = 337). Calculate the respective percentages of $(NH_4)_2SO_4$ and KNO_3 in the sample.

* 5-38. A 0.8644-g mixture of calcium oxalate, calcium carbonate, and inert materials was heated to constant weight at 500°C (see Figure 5-3). Calculate the percentage of CaC_2O_4 in the sample if the residue weighed 0.7562 g.

  5-39. The solid produced in Problem 5-38 was subsequently ignited at 1000°C to a constant weight of 0.5702 g. Calculate the percentage of calcium carbonate in the original sample.

* 5-40. A 2.5000-g sample containing ammonium chloride, sodium iodate, and inert materials was dissolved in sufficient water to give exactly 250 mL of solution. Treatment of one 50.0-mL aliquot with an excess of silver nitrate yielded a mixture of AgCl and $AgIO_3$ weighing 0.6314 g; treatment of a second 50.0-mL aliquot with an excess of barium chloride resulted in formation of 0.4530 g of $Ba(IO_3)_2$. Calculate the percentages of NH_4Cl and $NaIO_3$ in the sample.

* 5-41. The addition of dimethylglyoxime to a solution containing nickel(II) results in formation of a precipitate (p. 147). This product is so bulky that amounts greater than 175 mg are difficult to manipulate. A German silver alloy is known to assay between 15 and 25% Ni. Calculate the sample size that should not be exceeded in the analysis of the alloy for its nickel content.

  5-42. A sample that is approximately 77% $BaCl_2\cdot2H_2O$ is to be analyzed for its chloride content.

    (a) What sample weight should be taken if the AgCl produced is to weigh 0.500 g?

    (b) What volume of 0.050 M $AgNO_3$ will be needed to provide a 5% excess over that needed to precipitate the chloride from a 0.600-g sample?

    (c) What weight of AgCl will be produced in (b)?

* 5-43. Analysis for the chloride content of crude cobalt(II) chloride is required. Experience has shown that the product will range between 76% and 94% $CoCl_2\cdot6H_2O$. The accompanying questions have arisen in the course of developing a method for a routine analysis based upon a gravimetric chloride determination.

    (a) What sample weight should be taken to assure that the AgCl precipitated shall weigh at least 0.400 g?

    (b) If the sample weight calculated in (a) is used, what will be the maximum precipitate weight to be expected?

    (c) What volume of 0.150 M $AgNO_3$ should be used with 0.500-g samples in order to assure, at a minimum, a 10% excess of this reagent?

    (d) What weight of silver nitrate will be needed to produce 3.00 L of 0.150 M reagent?

    (e) It is desired that 50.0 mL of a silver nitrate solution constitute a 5% excess over the maximum needed to precipitate the chloride from 0.500-g samples. What should be the molar concentration of this solution?

    (f) To simplify calculations, what sample weight should be taken in order to have the percentage of $CoCl_2\cdot6H_2O$ exceed the weight of AgCl produced by a factor of 100?

* 5-44. The composition of an alloy containing only magnesium and aluminum was determined by dissolving a 0.3764-g sample in acid, precipitating $Mg(OH)_2$ and $Al_2O_3\cdot xH_2O$ with base, and igniting the precipitate to MgO and Al_2O_3, respectively. Calculate the percentages of Mg and Al in the sample if the ignited residue weighed 0.6850 g.

  5-45. Several alloys that contained only silver and copper were analyzed by dissolving weighed quantities in nitric acid, adding an excess of iodate, and bringing the filtered mixture of iodates to constant weight. Use the accompanying data to calculate the percentage composition of the alloys.

	wt of sample, g	wt of precipitate, g
(a)	0.2119	0.7860
(b)	0.1874	0.5460
(c)	0.2465	1.0291
(d)	0.2008	0.6824
(e)	0.1981	0.5809

* 5-46. What weight of thioacetamide will be needed to provide a 5% excess over that needed for the homogeneous precipitation of silver from 50 mL of a solution that is 0.080 M in $AgNO_3$ (see Table 5-2 for reaction)?

5-47. The zinc in a 0.1220-g sample that assays 40% Zn is to be precipitated homogeneously with diethyl oxalate (see Table 5-2). Calculate the weight needed to provide a 10% excess of this reagent.

* 5-48. A loss corresponding to 0.7 mg SO_4^{2-} occurs during the filtering and washing of $BaSO_4$ according to a particular set of instructions. This method is to be used for a specimen that assays about 32% sulfate. Calculate the minimum sample weight that should be taken to insure that the relative error due to solubility losses shall not exceed 0.25%.

5-49. A gravimetric analysis for iodate is to be based upon the formation of $Pb(IO_3)_2$ ($K_{sp} = 3.2 \times 10^{-13}$).
   (a) Calculate the weight of $Pb(IO_3)_2$ (in mg) that will remain unprecipitated in 350 mL of supernatant liquid if the excess lead ion concentration is 0.008 M.
   (b) Calculate the maximum weight loss to be expected if the $Pb(IO_3)_2$ is washed with 180 mL of pure water.

* 5-50. A brass ingot has the following composition:

   Sn 1.43%; Pb 3.57%; Cu 85.6%; Zn 9.40%

   During the analysis of a 1.200-g sample, incomplete washing caused 1.08 mg of copper to be retained by the $H_2SnO_3 \cdot x H_2O$. This copper was converted to CuO during ignition of the $H_2SnO_3 \cdot x H_2O$ to SnO_2.
   (a) Calculate the percentage of copper yielded by the analysis and its relative error in parts per thousand.
   (b) Calculate the percentage of tin yielded by the analysis and its error in parts per thousand.

5-51. The benzaldehyde (fw = 106) in a 100-mL liqueur sample was precipitated with 2,4-dinitrophenylhydrazine (fw = 200). Reaction:

$$C_6H_5CHO + H_2NNHC_6H_5(NO_2)_2 \longrightarrow C_6H_5C{=}NNHC_6H_5(NO_2)_2(s) + H_2O$$

The solid was filtered and dried to constant weight. Calculate the weight (in mg) of benzaldehyde contained in the sample if 118 mg of product (fw = 287) were recovered. The gravimetric factor should be based upon the number of moles of benzaldehyde needed to produce a mole of product.

* 5-52. The methyl salicylate in a 1.234-g sample of linament was isolated from other components by extraction. Treatment of the extract with base converted the ester to salicylate ion ($C_7H_5O_3^-$) and methanol, both of which returned to the aqueous phase. Iodine was added; the tetraiodophenylenequinone (p. 148) produced was filtered, dried, and found to weigh 0.3487 g. Calculate the percentage of methyl salicylate (fw = 152) in the linament; base the gravimetric factor on the number of moles of methyl salicylate that are needed to produce 1 mol of tetraiodophenylenequinone (fw = 664).

an introduction to titrimetric

methods of analysis

6

A quantitative determination based upon the combining capacity of an analyte with a reagent is called a *titrimetric* analysis. The most common way of performing a determination of this kind is to find the volume of a standard reagent required to completely consume the analyte. A titrimetric analysis performed in this way can also be called a *volumetric* procedure. This chapter and the nine that follow deal with volumetric methods of analysis. Part of Chapter 17 is concerned with another type of titrimetric method in which the measurement of the combining capacity of the analyte for the electrons in a direct current serves as the basis for the determination.

Volumetric methods are much more widely used for routine analyses than gravimetric because they are usually more rapid, convenient, and often as accurate.

6A DEFINITION OF SOME TERMS

A volumetric method employs one or more *standard solutions,* which are reagents whose concentrations are known exactly. Standard solutions are used to perform *titrations* in which the quantity of analyte in a solution is determined from the volume of standard solution that it consumes. Ordinarily, a titration is performed by carefully adding the standard solution until reaction with the analyte is judged to be complete; the volume of standard reagent is then measured. Occasionally, it is convenient or necessary to add an excess of the reagent and then determine the excess by *back-titration* with a second standard reagent.

The accuracy of a volumetric method is limited by the accuracy with which the concentration of the standard solution is known; for this reason, much care is taken in the preparation of such solutions. The concentration of a standard solution is established in one of two ways:

1. Directly, by dissolving a carefully weighed quantity of the pure reagent and diluting to an exactly known volume.
2. Indirectly, by titrating a weighed quantity of a pure compound with the reagent solution.

In both methods, a highly purified substance—called a *primary standard*—is required as the reference material. The process whereby the concentration of a standard solution is determined by titration against a primary standard is called a *standardization.*

The goal of every titration is the addition of standard solution in an amount that is chemically equivalent to the substance with which it reacts. This condition is achieved at the *equivalence point.* For example, the equivalence point in the titration of sodium chloride with silver nitrate is attained when exactly one mole of silver ion has been introduced for each mole of chloride ion in the sample. The equivalence point in the titration of sulfuric acid with sodium hydroxide occurs when two moles of the latter have been introduced for each mole of the former.

The equivalence point in a titration is a theoretical concept; in actual fact, its location can be estimated only by observing physical changes associated with equivalence. These changes manifest themselves at the *end point* of the titration. It is to be hoped that any volume difference between the end point and equivalence point will be small. Differences do exist, however, owing to inadequacies in the physical changes and our ability to observe them; a *titration error* is the result.

A common method of end point detection in volumetric analysis involves the use of a supplementary chemical compound that exhibits a change in color as a result of concentration changes occurring near the equivalence point. Such a substance is called an *indicator.*

6B REACTIONS AND REAGENTS USED IN TITRIMETRIC ANALYSIS

It is convenient to classify volumetric methods according to four reaction types, specifically, precipitation, complex formation, neutralization (acid-base), and oxidation-reduction. Each reaction type is unique in such matters as nature of equilibria involved; the indicators, reagents, and primary standards available; and the definition of equivalent weight.

6B-1 PRIMARY STANDARDS

The accuracy of a volumetric analysis is critically dependent upon the primary standard used to establish, directly or indirectly, the concentration of the standard solution. Important requirements for a substance to serve as a good primary standard include the following:

1. Highest purity. Established methods should be available for confirming its purity.
2. Stability. It should not be attacked by constituents of the atmosphere.

3. Absence of hydrate water. If the substance is hygroscopic or effluorescent, drying and weighing are difficult.
4. Availability at reasonable cost.
5. Reasonably high equivalent weight to minimize the relative error associated with the weighing operation.

Few substances meet or even approach these requirements. As a result, the number of primary standard substances available to the chemist is limited.

Occasionally, it is necessary to use less pure substances in lieu of a primary standard. The assay (that is, the percent purity) of such a *secondary* standard must be established by careful analysis.[1]

6B-2 STANDARD SOLUTIONS

An ideal standard solution for titrimetric analysis would have the following properties:

1. Its concentration should remain constant for months or years after preparation to eliminate the need for restandardization.
2. Its reaction with the analyte should be rapid in order to minimize the waiting period after each addition of reagent.
3. Its reaction with the analyte should be reasonably complete. As we shall presently show, satisfactory end points generally require this condition.
4. It should be possible to describe the reaction of the reagent with the analyte by a balanced chemical equation; otherwise, the weight of the analyte cannot be calculated directly from the volumetric data. This requirement implies the absence of side reactions between the reagent and the analyte or with other constituents of the solution.
5. A method must exist for detecting the equivalence point between the reagent and the analyte; that is, a satisfactory end point is required.

Few volumetric reagents currently in use meet all of these requirements perfectly.

6C CALCULATIONS ASSOCIATED WITH TITRIMETRIC METHODS

In preparation for the discussion that follows, the student may find it helpful to review the material in Section 2B on chemical units of weight and concentration.

Chemists frequently make use of the *equivalent weight* (or the *milliequivalent weight*) as the basis for volumetric calculations; *normality* is the corresponding unit of concentration. The manner in which these units are defined depends upon the type of reaction that serves as the basis for the analysis, that is, whether the titration reaction involves neutralization, oxidation-reduction, precipitation, or complex formation. Furthermore, the chemical behavior of the substance

---

[1] Usage of the terms "primary standard" and "secondary standard" is a source of disagreement among chemists. Some reserve primary standard for the relatively few compounds that possess the properties listed at the start of this section (see, for example, H. A. Laitinen and W. B. Harris, *Chemical Analysis*, 2d ed., p. 101. New York: McGraw-Hill, 1975) and use secondary to denote a species that has a reliably established composition. Others prefer to consider all substances of known assay as primary standards and refer to solutions of known concentration as secondary standards. We prefer the former set of definitions.

must be carefully specified if its equivalent weight is to be defined unambiguously. If the substance can react in more than one way, it will likewise have more than one equivalent or milliequivalent weight. *Thus, the definition of equivalent weight or milliequivalent weight for a substance is always based on its behavior in a specific chemical reaction. Evaluation of the equivalent weight is impossible if the reaction involved is not specifically stated. Likewise, the concentration of a solution cannot be expressed in terms of normality without this information.*

6C-1 EQUIVALENT OR MILLIEQUIVALENT WEIGHTS IN NEUTRALIZATION REACTIONS

The gram equivalent weight (eqw) or the equivalent weight of a substance participating in a neutralization reaction is that weight which either contributes or reacts with one mole of hydrogen ion *in that reaction*. The milliequivalent weight (meqw) is the equivalent weight divided by 1000.

The relationship between the equivalent weight and the formula weight is straightforward for strong acids or bases and for other acids or bases that contain a single reactive hydrogen or hydroxide ion. For example, the equivalent weight for potassium hydroxide and hydrochloric acid must be equal to their formula weights, respectively, because each has but a single reactive hydrogen ion or hydroxide ion. Similarly, we know that only one hydrogen in acetic acid, $HC_2H_3O_2$, is acidic; therefore, the formula weight and equivalent weight for this acid must also be identical. Barium hydroxide, $Ba(OH)_2$, is a strong base containing two hydroxide ions that are indistinguishable. This base will necessarily react with two hydrogen ions in any acid-base reaction; thus, its equivalent weight will be one-half its formula weight. For sulfuric acid, dissociation of the second hydrogen ion in water is not complete; the hydrogen sulfate ion, however, is a sufficiently strong acid so that both hydrogens participate in all aqueous neutralization reactions. The equivalent weight of H_2SO_4 as an acid is, therefore, always one-half its formula weight in aqueous solution.

The situation becomes more complex for acids that contain two or more hydrogen ions with differing tendencies to dissociate. For example, certain indicators undergo a color change when only the first of the three protons in phosphoric acid has been neutralized; that is,

$$H_3PO_4 + OH^- \longrightarrow H_2PO_4^- + H_2O$$

Certain other indicators change color after two hydrogen ions have reacted:

$$H_3PO_4 + 2OH^- \longrightarrow HPO_4^{2-} + 2H_2O$$

For a titration involving the first reaction, the equivalent weight of phosphoric acid is equal to its formula weight; for the second, it is one-half its formula weight. (It is not practical to titrate the third proton; thus, an equivalent weight that is one-third of the formula weight for H_3PO_4 is not encountered in the context of a neutralization titration.) *Without knowing which of these reactions is involved, an unambiguous definition of the equivalent weight for phosphoric acid is impossible.*

6C-2 EQUIVALENT WEIGHT IN OXIDATION-REDUCTION REACTIONS

The equivalent weight of a participant in an oxidation-reduction reaction is that weight which directly or indirectly produces or consumes one mole of electrons.

The numerical value for the equivalent weight is conveniently established by dividing the formula weight of the substance of interest by the change in oxidation number associated with its reaction. As an example, consider the oxidation of oxalate ion by permanganate:

$$5C_2O_4^{2-} + 2MnO_4^- + 16H^+ \longrightarrow 10CO_2 + 2Mn^{2+} + 8H_2O$$

The change in oxidation number for manganese in this reaction is 5 because the element passes from the $+7$ to the $+2$ state; the equivalent weight for MnO_4^- or Mn^{2+} is therefore one-fifth of the corresponding formula weights. Each carbon atom in the oxalate ion is oxidized from the $+3$ to the $+4$ state. The equivalent weight of sodium oxalate is one-half its formula weight because each formula weight produces two moles of electrons. If the reverse reaction were considered, the equivalent weight of carbon dioxide would equal its formula weight because it consumes but a single mole of electrons. These and other examples of equivalent weights are given in Table 6-1. Note that the equivalent weight for a substance is evaluated from the change in oxidation state that occurs *in the titration* and *not* from its oxidation state in the compound in question. Thus, for example, the oxidation state of manganese in Mn_2O_3 is $+3$. Before the reaction just considered could be used to determine the Mn_2O_3 content of a sample, the manganese would first have to be converted completely to permanganate ion by some reagent. Each manganese would then be reduced from the $+7$ to the $+2$ state in the titration step as before. The equivalent weight is thus the formula weight of Mn_2O_3 divided by ten.

As in neutralization reactions, the equivalent weight for a given oxidizing or reducing agent is not invariant. Potassium permanganate, for example, reacts with reducing agents in four different ways, depending upon the conditions existing in the solution. The half-reactions are

Table 6-1 Equivalent Weights of Some Manganese and Carbon Species Based on the Reaction

$$5C_2O_4^{2-} + 2MnO_4^- + 16H^+ \rightarrow 10CO_2 + 2Mn^{2+} + 8H_2O$$

Substance	Equivalent Weight
Mn	$\dfrac{\text{fw Mn}}{5}$
$KMnO_4$	$\dfrac{\text{fw } KMnO_4}{5}$
$Ca(MnO_4)_2 \cdot 4H_2O$	$\dfrac{\text{fw } Ca(MnO_4)_2 \cdot 4H_2O}{2 \times 5}$
Mn_2O_3	$\dfrac{\text{fw } Mn_2O_3}{2 \times 5}$
CO_2	$\dfrac{\text{fw } CO_2}{1}$
$C_2O_4^{2-}$	$\dfrac{\text{fw } C_2O_4^{2-}}{2}$

$$MnO_4^- + e \longrightarrow MnO_4^{2-}$$

$$MnO_4^- + 3e + 2H_2O \longrightarrow MnO_2(s) + 4OH^-$$

$$MnO_4^- + 4e + 3H_2P_2O_7^{2-} + 8H^+ \longrightarrow Mn(H_2P_2O_7)_3^{3-} + 4H_2O$$

$$MnO_4^- + 5e + 8H^+ \longrightarrow Mn^{2+} + 4H_2O$$

The changes in oxidation number of manganese are 1, 3, 4, and 5. The equivalent weight of potassium permanganate would be equal to the gram formula weight for the first reaction and one-third, one-fourth, and one-fifth of the gram formula weight, respectively, for the others.

6C-3 EQUIVALENT WEIGHTS IN PRECIPITATION AND COMPLEX FORMATION REACTIONS

It is awkward to devise a definition for equivalent weight that is entirely free of ambiguity for compounds involved in precipitation or complex formation reactions. As a result, many chemists prefer to avoid the use of the concept for reactions of this type and use formula weights exclusively instead. We are in sympathy with this practice and will follow it in this text. However, it is likely that the student will encounter situations where equivalent or milliequivalent weights are specified for substances involved in precipitation or complex formation; it is thus important to know how these quantities are defined.

The equivalent weight of a participant in a precipitation or a complex formation reaction is that weight which reacts with or provides one gram formula weight of the *reacting* cation if it is univalent, one-half of a gram formula weight if it is divalent, one-third of a gram formula weight if it is trivalent, and so on. Our earlier definitions of equivalent weight were based either on one mole of hydrogen ions or on one mole of electrons. Here, one mole of a univalent cation or the equivalent thereof is used. *The cation referred to in this definition is always the cation directly involved in the reaction* and not necessarily the cation contained in the compound whose equivalent weight is being defined.

6C-4 EQUIVALENT WEIGHTS OF COMPOUNDS THAT DO NOT PARTICIPATE DIRECTLY IN A VOLUMETRIC REACTION

There is frequent need to define the equivalent weight for an analyte that is related only indirectly to the actual reactants of the titration. To illustrate, lead can be determined by an indirect method in which the cation is first precipitated as the chromate from an acetic acid solution. The precipitate is filtered, washed free of excess precipitant, and redissolved in dilute hydrochloric acid to give a solution of lead and dichromate ions. The latter may then be determined by an oxidation-reduction titration with a standard iron(II) solution. The reactions are

$$Pb^{2+} + CrO_4^{2-} \xrightarrow[\text{HOAc}]{\text{dil}} PbCrO_4(s) \qquad \text{(precipitate filtered and washed)}$$

$$2PbCrO_4(s) + 2H^+ \xrightarrow[\text{HCl}]{\text{dil}} 2Pb^{2+} + Cr_2O_7^{2-} + H_2O \qquad \text{(precipitate redissolved)}$$

$$Cr_2O_7^{2-} + 6Fe^{2+} + 14H^+ \longrightarrow 2Cr^{3+} + 6Fe^{3+} + 7H_2O \qquad \text{(titration)}$$

For the purpose of calculation, we must ascribe an equivalent weight to lead. *Because the titration is an oxidation-reduction process, the equivalent weight of lead will have to be based on a change of oxidation number.* Clearly, the lead exhibits no such change. It is, however, associated in a 1:1 ratio with chromium, and this element changes from the $+6$ to the $+3$ state in the titration. Therefore, we can say that a change in oxidation state of 3 is *associated* with each lead, and its equivalent weight in this sequence of reactions is one-third its formula weight.

It is often helpful to make an inventory of the chemical relationships existing between the substance whose equivalent weight is sought and the participants in the titration. In this example, the equations reveal that

$$2Pb^{2+} \equiv 2CrO_4^{2-} \equiv Cr_2O_7^{2-} \equiv 6Fe^{2+} \equiv 6e$$

Thus, the quantity of each substance associated with the transfer of one mole of electrons is

$$\frac{2 \text{ fw } Pb^{2+}}{6} \equiv \frac{2 \text{ fw } CrO_4^{2-}}{6} \equiv \frac{1 \text{ fw } Cr_2O_7^{2-}}{6} \equiv \frac{6 \text{ fw } Fe^{2+}}{6} \equiv \frac{6 \text{ mol } e}{6}$$

Let us consider an additional example. The nitrogen in the organic compound $C_9H_9N_3$ can be determined by quantitative conversion to ammonia, followed by titration with a standard solution of acid. The reactions are:

$$C_9H_9N_3 + \text{reagent} \longrightarrow 3NH_3 + \text{products}$$

$$NH_3 + H^+ \longrightarrow NH_4^+ \qquad\qquad\qquad \text{(titration)}$$

Here, titration involves a neutralization process; therefore, the equivalent weight for the analyte must be based on the consumption or production of hydrogen ions. Since each molecule of the analyte yields three ammonia molecules, it can be considered responsible for the consumption of three hydrogen ions. The equivalent weight, then, is the formula weight of $C_9H_9N_3$ divided by three. By the same reasoning, each nitrogen is converted to one ammonia molecule that reacts with one hydrogen ion; the equivalent weight of nitrogen, N, is thus equal to its formula weight.

6C-5 CONCENTRATION UNITS USED IN VOLUMETRIC CALCULATIONS

Some of the ways by which chemists express the concentrations of solutions were discussed in Section 2B-2; these included molar concentration and various types of percentage composition. We now need two additional terms, *normality* and *titer*, which are commonly used to define the concentration of solutions used in volumetric analysis.

Normality. *The normality, N, of a solution expresses the number of milliequivalents of solute contained in one milliliter of solution* or the number of equivalents contained in one liter. Thus, a 0.20 N hydrochloric acid solution contains 0.20 milliequivalent (meq) of this solute in each milliliter of solution or 0.20 equivalent (eq) per liter.

Titer. *The titer of a solution is the weight of a substance that is chemically equivalent to one milliliter of that solution.* Thus, a silver nitrate solution having a titer of 1.00 mg of chloride would contain just enough silver nitrate in each milliliter to react completely with that weight of chloride ion. The titer might also be expressed in terms of milligrams or grams of potassium chloride, barium chloride, sodium iodide, or any other compound that reacts with silver nitrate. The concentration of a reagent to be used for the routine analysis of many samples is advantageously expressed in terms of its titer.

EXAMPLE 6-1. What is the barium chloride titer of a 0.125 M $AgNO_3$ solution?

$$\text{titer} = \frac{0.125 \text{ mmol } AgNO_3}{\text{mL } AgNO_3} \times \frac{1 \text{ mmol } BaCl_2}{2 \text{ mmol } AgNO_3} \times \frac{208.2 \text{ mg } BaCl_2}{\text{mmol } BaCl_2}$$

$$= \frac{13.0 \text{ mg } BaCl_2}{\text{mL } AgNO_3}$$

EXAMPLE 6-2. The label on a bottle of dilute silver nitrate states that the solution has a titer of 8.78 mg NaCl. What is the molar concentration of this solution?

$$M_{AgNO_3} = \frac{8.78 \text{ mg NaCl}}{\text{mL } AgNO_3} \times \frac{1 \text{ mmol NaCl}}{58.44 \text{ mg NaCl}} \times \frac{1 \text{ mmol } AgNO_3}{\text{mmol NaCl}} = \frac{0.150 \text{ mmol}}{\text{mL}}$$

6C-6 SOME IMPORTANT WEIGHT-VOLUME RELATIONSHIPS

The raw data from a volumetric analysis are ordinarily expressed in units of milliliters, grams, and normality. Volumetric calculations involve conversion of such information into units of milliequivalents followed by reconversion into the metric weight of the desired chemical species. Two relationships, based on the foregoing definitions, are used for these transformations. The first involves converting the weight of a compound from units of grams to those of milliequivalents. This transformation is accomplished by dividing the weight of the substance by its milliequivalent weight; that is,

$$\text{meq A} = \frac{\text{wt A (g)}}{\text{meqw A (g/meq)}}$$

EXAMPLE 6-3. Calculate the number of milliequivalents of $H_2C_2O_4 \cdot 2H_2O$ (fw = 126.1) in 0.500 g of the pure compound, assuming that two of the hydrogens are to be titrated with standard base.

$$\text{meq } H_2C_2O_4 \cdot 2H_2O = \frac{\text{wt } H_2C_2O_4 \cdot 2H_2O \text{ (g)}}{\text{meqw } H_2C_2O_4 \cdot 2H_2O \text{ (g/meq)}}$$

Because each formula weight of the acid provides $2H_3O^+$, we may write

$$\text{meq } H_2C_2O_4 \cdot 2H_2O = 0.500 \text{ g } H_2C_2O_4 \cdot 2H_2O \times \frac{2 \text{ meq } H_2C_2O_4 \cdot 2H_2O}{0.1261 \text{ g } H_2C_2O_4 \cdot 2H_2O} = 7.93$$

The second relationship permits calculation of the number of milliequivalents of solute contained in a given volume V, provided the normality of the solution is known. Since by definition the normality is equal to the number of milliequivalents in each milliliter, it follows that

$$\text{meq A} = V_A(\text{mL}) \times N_A(\text{meq/mL})$$

EXAMPLE 6-4. The number of milliequivalents involved in a titration that required 27.3 mL of 0.200 N $KMnO_4$ is given by

$$\text{meq KMnO}_4 = 27.3 \text{ mL} \times 0.200 \text{ meq/L}$$

$$= 5.46$$

Further applications of these relationships are illustrated in the following examples.

EXAMPLE 6-5. What weight of primary standard $K_2Cr_2O_7$ (fw = 294.2) is needed to prepare exactly 2 L of 0.1200 N reagent by the direct method? Titrations with dichromate involve the half-reaction

$$\text{Cr}_2\text{O}_7^{2-} + 14\text{H}^+ + 6e \rightleftharpoons 2\text{Cr}^{3+} + 7\text{H}_2\text{O}$$

The number of milliequivalents of $K_2Cr_2O_7$ required is first calculated:

$$\text{meq K}_2\text{Cr}_2\text{O}_7 = \text{mL}_{\text{K}_2\text{Cr}_2\text{O}_7} \times N_{\text{K}_2\text{Cr}_2\text{O}_7}$$

$$= 2000 \text{ mL} \times 0.1200 \text{ meq/mL}$$

$$= 240.0$$

Conversion of a weight in milliequivalents to a weight in grams involves multiplication by the milliequivalent weight.

$$\text{wt K}_2\text{Cr}_2\text{O}_7 = 240.0 \text{ meq} \times 294.2 \frac{\text{g}}{\text{mol}} \times \frac{1 \text{ mol}}{6 \text{ eq}} \times \frac{1 \text{ eq}}{1000 \text{ meq}}$$

$$= 11.768 = 11.77 \text{ g}$$

EXAMPLE 6-6. What volume of 0.100 N HCl can be produced by diluting 150 mL of 1.24 N acid?

The number of milliequivalents of HCl must be the same in the two solutions. Therefore,

$$\text{meq HCl in dil soln} = \text{meq HCl in concd soln}$$

$$\text{mL dil soln} \times 0.100 \text{ meq/mL} = 150 \text{ mL} \times 1.24 \text{ meq/mL}$$

$$\text{mL dil soln} = \frac{150 \times 1.24}{0.100} = 1860$$

$$= 1.86 \times 10^3$$

Thus, a 0.100 N HCl solution would be obtained by diluting 150 mL of 1.24 N acid to 1.86×10^3 mL.

A fundamental relationship between quantities of reacting substances. By definition, one equivalent weight of an acid contributes one mole of hydrogen ions to a reaction. Also, one equivalent weight of a base consumes one mole of these ions. It then follows that at the equivalence point in a neutralization titration, the number of equivalents (or milliequivalents) of acid and of base will always be numerically equal. Similarly, at the equivalence point in an oxidation-reduction titration, the number of milliequivalents of oxidizing and reducing agent must also be equal. An identical relationship holds for precipitation and complex formation titrations. To generalize, we may state that *at the equivalence point in any titration, the number of milliequivalents of standard is exactly equal to the number of milliequivalents of the substance with which it has reacted.* The reason equivalent weights are defined as they are is to make this statement possible. Nearly all volumetric calculations are based on this relationship.

6C-7 CALCULATION OF CONCENTRATION OF STANDARD SOLUTIONS

The normality of a standard solution is computed either from the data from the titration of a primary standard (indirect method) or from the data related to its actual preparation (direct method).

EXAMPLE 6-7. A $Ba(OH)_2$ solution was standardized by titration against $0.1280\ N$ HCl, 31.76 mL of the base being required to neutralize 46.25 mL of the acid. Calculate the normality of the $Ba(OH)_2$ solution.

Provided that the end point in the titration corresponds to the equivalence point,

$$meq\ Ba(OH)_2 = meq\ HCl$$

$$mL_{Ba(OH)_2} \times N_{Ba(OH)_2} = mL_{HCl} \times N_{HCl}$$

$$31.76\ mL \times N_{Ba(OH)_2} = 46.25\ mL \times 0.1280\ meq/mL$$

$$N_{Ba(OH)_2} = \frac{46.25 \times 0.1280}{31.76} = 0.1864\ meq/mL$$

EXAMPLE 6-8. Calculate the normality of an iodine solution if 37.34 mL were required to titrate a 0.2040-g sample of primary standard As_2O_3 (fw = 197.8). The reaction is

$$I_2 + H_2AsO_3^- + H_2O \longrightarrow 2I^- + H_2AsO_4^- + 2H^+$$

At the equivalence point,

$$meq\ I_2 = meq\ As_2O_3$$

The number of milliequivalents of I_2 can be computed from the volume and normality; the number of milliequivalents of As_2O_3 can be calculated from the weight taken for the titration. Thus, we may write

$$mL_{I_2} \times N_{I_2} = \frac{wt\ As_2O_3}{meqw\ As_2O_3}$$

In its reaction with iodine, arsenic loses two electrons and is thus oxidized from the $+3$ to the $+5$ state. Therefore, a total change of 4 mol of electrons is associated with each mole of As_2O_3, making its equivalent weight one-fourth its formula weight. Thus,

$$37.34\ mL \times N_{I_2} = 0.2040\ g\ As_2O_3 \times \frac{4\ meq\ As_2O_3}{0.1978\ g\ As_2O_3}$$

$$N_{I_2} = \frac{0.2040}{37.34 \times 0.04945} = 0.1105$$

6C-8 CALCULATION OF RESULTS FROM TITRATION DATA

EXAMPLE 6-9. A 0.804-g sample of iron ore was dissolved in acid. The iron was then reduced to the $+2$ state and titrated with 47.2 mL of a $0.112\ N$ $KMnO_4$ solution. Calculate the results of the analysis in terms of percent Fe (fw = 55.85) as well as percent Fe_3O_4 (fw = 231.5).

The titration involves oxidation of Fe^{2+} to Fe^{3+}:

$$5Fe^{2+} + MnO_4^- + 8H^+ \longrightarrow 5Fe^{3+} + Mn^{2+} + 4H_2O$$

At the equivalence point,

$$\text{meq Fe} = \text{meq KMnO}_4$$

$$= 47.2 \text{ mL} \times 0.112 \text{ meq/mL}$$

Because Fe^{2+} loses one electron in this reaction, its milliequivalent weight is identical to its milliformula weight, and

$$\text{g Fe} = 47.2 \text{ mL} \times 0.112 \frac{\text{meq}}{\text{mL}} \times \frac{55.85 \text{ g}}{1000 \text{ meq}}$$

Therefore,

$$\% \text{ Fe} = \frac{47.2 \times 0.112 \times 0.05585}{0.804} \times 100 = 36.7$$

The percentage of Fe_3O_4 can be obtained in essentially the same way; thus, at the equivalence point,

$$\text{meq Fe}_3\text{O}_4 = \text{meq KMnO}_4$$

and by the same arguments,

$$\% \text{ Fe}_3\text{O}_4 = \frac{47.2 \times 0.112 \times 231.5/3000}{0.804} \times 100 = 50.7$$

Note that the oxidation state of Fe in the Fe_3O_4 was of no concern; only the change from $+2$ to $+3$ in the titration was important.

EXAMPLE 6-10. A 0.475-g sample containing $(NH_4)_2SO_4$ was dissolved in water and made alkaline with KOH. The liberated NH_3 was distilled into exactly 50.0 mL of 0.100 N HCl. The excess HCl was back-titrated with 11.1 mL of 0.121 N NaOH. Calculate the percent NH_3 (fw = 17.03) as well as the percent $(NH_4)_2SO_4$ (fw = 132.1) in the sample.

At the equivalence point, the number of milliequivalents of acid and base are equal. In this titration, however, two bases, NaOH and NH_3, are involved; thus,

$$\text{meq HCl} = \text{meq NH}_3 + \text{meq NaOH}$$

After rearranging,

$$\text{meq NH}_3 = \text{meq HCl} - \text{meq NaOH}$$

$$= (50.0 \times 0.100 - 11.1 \times 0.121)$$

Thus,

$$\% \text{ NH}_3 = \frac{(50.0 \times 0.100 - 11.1 \times 0.121) \times 17.03/1000}{0.475} \times 100$$

$$= 13.1$$

The number of milliequivalents of $(NH_4)_2SO_4$ is the same as the number of milliequivalents of NH_3 by definition. Therefore,

$$\% \text{ (NH}_4)_2\text{SO}_4 = \frac{(50.0 \times 0.100 - 11.1 \times 0.121) \times 132.1/2000}{0.475} \times 100$$

$$= 50.8$$

Here, the milliequivalent weight of $(NH_4)_2SO_4$ is one-half the milliformula weight because

$$\text{mol (NH}_4)_2\text{SO}_4 \equiv 2 \times \text{mol NH}_3 \equiv 2 \times \text{mol H}^+$$

EXAMPLE 6-11. The organic matter in a 3.77-g sample of a mercuric ointment was decomposed with HNO_3. After dilution, the Hg^{2+} was titrated with a 0.114 M solution of NH_4SCN; exactly 21.3 mL of the reagent were required. Calculate the percent Hg (fw = 200.6) and the percent $Hg(NO_3)_2$ (fw = 324.6) in the ointment.

This titration involves the formation of a stable neutral complex, $Hg(SCN)_2$; that is,

$$Hg^{2+} + 2SCN^- \longrightarrow Hg(SCN)_2(aq)$$

Because this is a complex formation titration, we shall not use equivalent weights but moles instead. At the equivalence point,

mmol Hg^{2+} = mmol $NH_4SCN/2$

$$= 21.3 \text{ mL } NH_4SCN \times 0.114 \frac{\text{mmol } NH_4SCN}{\text{mL } NH_4SCN} \times \frac{1 \text{ mmol Hg}}{2 \text{ mmol } NH_4SCN}$$

$$\text{wt Hg} = \left(21.3 \times 0.114 \times \frac{1}{2}\right) \text{ mmol Hg} \times \frac{0.2006 \text{ g Hg}}{\text{mmol Hg}}$$

$$\% \text{ Hg} = \frac{21.3 \times 0.114 \times 0.2006/2}{3.77} \times 100 = 6.46$$

The percentage of $Hg(NO_3)_2$ is calculated identically:

mmol $Hg(NO_3)_2$ = mmol $NH_4SCN/2$

and

$$\% \text{ Hg}(NO_3)_2 = \frac{21.3 \times 0.114 \times 0.3246/2}{3.77} \times 100 = 10.5$$

6D END POINTS IN TITRIMETRIC METHODS

End points are based upon physical changes that occur in a solution in the course of a titration. A variety of properties have been applied to end point detection including color; electrical potential, current, and conductivity; turbidity; and light scattering. The most widely used of these are change in color due to the reagent, the analyte, or an indicator and changes in potential of an electrode that responds to the analyte or reagent concentration.

Two general methods exist for deriving end points from the observation or measurement of physical properties. In one, the measurements are largely confined to a small region surrounding the equivalence point—typically ±1 or 2 mL. In the second, observations are made on both sides but well away from the point of chemical equivalence; indeed, measurements near this point are avoided. The first approach offers the advantage of speed and convenience; the latter is often more sensitive and does not require as favorable equilibrium constants for the reaction between the analyte and reagent.

6D-1 END POINTS BASED UPON OBSERVATIONS NEAR THE EQUIVALENCE POINT

End points of this type are the result of marked changes in reactant or product concentrations in the equivalence point region. To illustrate, the second column of Table 6-2 contains data that show the variations in hydronium ion concentration that occur when 50.00 mL of 0.1000 M HCl are titrated with standard 0.1000 M NaOH. To emphasize the *relative* changes in concentration that take place, the volumes selected for tabulation correspond to the increments needed to cause a ten-

Table 6-2   Concentration Changes During a Titration*

Vol 0.1000 M NaOH, mL	Concentration H_3O^+, mol/L	pH	pOH	Vol NaOH Required to Cause a Tenfold Change in H_3O^+ or a Unit Change in pH, mL
0.0	1.0×10^{-1}	1.00	13.00	
40.91	1.0×10^{-2}	2.00	12.00	40.91
49.01	1.0×10^{-3}	3.00	11.00	8.1
49.90	1.0×10^{-4}	4.00	10.00	0.89
49.990	1.0×10^{-5}	5.00	9.00	0.09
49.9990	1.0×10^{-6}	6.00	8.00	0.009
50.0000	1.0×10^{-7}	7.00	7.00	0.001
50.0010	1.0×10^{-8}	8.00	6.00	0.001
50.010	1.0×10^{-9}	9.00	5.00	0.009
50.10	1.0×10^{-10}	10.00	4.00	0.09
51.01	1.0×10^{-11}	11.00	3.00	0.91
61.1	1.0×10^{-12}	12.00	2.00	10.09

* Although these data are useful for illustration purposes, it should be understood that their realization in the laboratory would not be possible because measurements of volumes and normalities to six significant figures are ordinarily impossible.

fold increase or decrease in concentrations of the reacting species. Thus, we see that 40.91 mL of base are needed to reduce the hydronium ion concentration by an order of magnitude from 0.100 M to 0.0100 M. An additional 8.1 mL are required to lower the concentration to 0.00100 M; 0.89 mL will cause yet another tenfold decrease and so on. Analogous increases in the hydroxide ion concentration occur simultaneously.

It is apparent from the data in Table 6-2 that a maximum in the rate of change of relative concentrations of the reacting species is found at the equivalence point. Figure 6-1 consists of plots of the concentration data in Table 6-2. Note that the ordinate scale for these plots must be large to encompass the enormous concentration changes experienced by the reactants. As a result, changes occurring in the region of interest—that is, the equivalence point—are obscured.

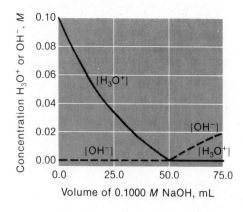

FIGURE 6-1   Changes in reactant concentrations during the titration of 50.00 mL of 0.1000 M HCl with 0.1000 M NaOH.

In order to show more clearly the concentration variations occurring in the equivalence point region, it is useful to substitute p-values (see Section 2B-2) for one of the reactants in place of molar concentrations. Thus, pH or pOH would be used for the titration illustrated by Table 6-2; columns 3 and 4 list p-values after the various additions of reagent.

When pH values are plotted against the volume of reagent added (Figure 6-2), a curve is obtained that gives a much clearer picture of the changes that occur in the solution near the equivalence point; it is evident from this figure that a marked increase in pH takes place here. Also shown in Figure 6-2 is a curve in which pOH rather than pH is plotted as the ordinate for the same titration; here again, a major change in the p-function marks the region of equivalence.

Curves such as those in Figure 6-2 are called *titration curves*. Plots of analogous data for titrations involving oxidation-reduction, precipitation, or complex formation reactions have the same general characteristics. Typical examples are derived and discussed in later chapters.

Visual end points. End points based upon visual observations of color change or of the appearance or disappearance of turbidity offer the considerable advantage of speed and simplicity of equipment. The applicability of visual end points is critically dependent, however, upon the magnitude of the relative concentration changes that occur in the equivalence point region. Typically, the sensitivity of the human eye is such that an increase or decrease in p-value of one to two units is required before the accompanying physical change can be detected. For an end point to be usable, a change of this magnitude must occur over a volume range that corresponds to the allowable titration error. Ideally, the change will take place symmetrically around the true equivalence point for the titration. For the titrations illustrated in Table 6-2 and Figure 6-2, a change of four pH units occurs over a volume range corresponding to ± 0.01 mL of the equivalence point pH of 7.00. Several indicators are available that exhibit a readily distinguishable color change within this pH range; the titration can thus be performed with a minimal titration error.

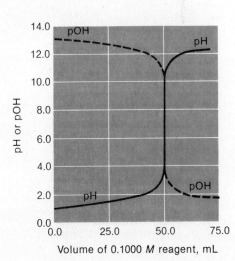

FIGURE 6-2 Titration curves for 50.00 mL of 0.1000 M HCl with 0.1000 M NaOH.

In contrast, consider curve A in Figure 6-3 for the titration of iodate ion with silver nitrate in which the product is solid silver iodate. Here, the solutions are more dilute than those in Figure 6-2 by a factor of 10; furthermore, the equilibrium constant for the formation of silver iodate is only 3×10^7 compared with 10^{14} for the formation of the product (water) in the earlier case. That is,

$$H_3O^+ + OH^- \rightleftharpoons H_2O \qquad \frac{1}{[H_3O^+][OH^-]} = \frac{1}{K_w} = 1.0 \times 10^{14}$$

$$Ag^+ + IO_3^- \rightleftharpoons AgIO_3(s) \qquad \frac{1}{[Ag^+][IO_3^-]} = \frac{1}{K_{sp}} = 3.0 \times 10^7$$

The consequence of the greater dilution and less favorable formation constant reduces significantly the rate of change in pAg at equivalence compared with the rate of pH change seen in Figure 6-2. Thus, approximately 6 mL of reagent are needed for 1-unit change in pAg; a 2-unit change requires approximately 16 mL; visual detection of the equivalence point for this reaction is impossible.

Potentiometric end points.   Titration curves such as those in Figures 6-2 and 6-3A are experimentally derivable because the potentials of electrodes that are responsive to titrant or analyte ions are directly proportional to p-values. For example, the potential for a glass electrode is linearly related to pH; similarly, the potential of a silver electrode is proportional to pAg. Such electrodes are widely used to generate experimentally titration curves from which end points can be ascertained by inspection (Chapter 16). End points of this type are somewhat more sensitive than visual ones; they do not, however, permit the accurate determination of species exhibiting a titration curve such as that involving silver and iodate ions (Figure 6-3A).

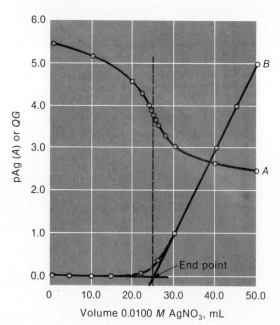

FIGURE 6-3 Titration curves for 25.0 mL of 0.0100 M NaIO$_3$ with 0.0100 M AgNO$_3$. A, End point from observations of pAg in the equivalence point region. B, End point from observations well away from equivalence. Here, a parameter G that is proportional to [Ag$^+$] is plotted (after correction for the volume change); thus, G = k [Ag$^+$].

6D-2 END POINTS BASED UPON OBSERVATIONS MADE WELL AWAY FROM EQUIVALENCE

End points of this type are nonvisual and are based upon the measurement of some physical quantity that is proportional to the concentration of the reactant, the reagent, or the reaction product (in some instances, the measuring device may respond to more than one of these species).

Consider the titration of V_X mL of an analyte X with a reagent R

$$X + R \rightleftharpoons XR \tag{6-1}$$

After the addition of V_R mL, the number of millimoles of X present will be equal to the number originally present minus the number of millimoles of R added. That is,

$$\text{mmol X} = \text{original mmol X} - \text{mmol R}$$

or

$$\text{mmol X} = V_X M'_X - V_R M'_R$$

where M'_X is the original molar concentration of the analyte and M'_R is the corresponding concentration of the reagent. As a consequence of the reaction and dilution of the reagent, the concentration of X will be reduced to M_X where

$$M_X = \frac{\text{mmol X remaining}}{\text{new volume}} = \frac{V_X M'_X - V_R M'_R}{V_X + V_R}$$

If sufficient X remains in the solution, it will force the reaction in Equation 6-1 far to the right, thus making the equilibrium concentration of X essentially the same as the analytical concentration, M_X. That is,

$$[X] = M_X = \frac{V_X M'_X - V_R M'_R}{V_X + V_R} \tag{6-2}$$

For this type of end point to be successful, initial measurements must be taken far enough removed from the equivalence point so that the common ion effect largely represses the dissociation of the product. Here, the foregoing approximation that the analytical and species concentrations are equal is valid; furthermore in this region, we may write

$$[R] \cong M_R = 0$$

where M_R is the analytical concentration of the reagent in the titration mixture.

Additional measurements are also performed well beyond the equivalence point where the analytical reaction is again forced far to the right; here, the species R is in excess and the concentration of X is vanishingly small. That is,

$$[R] \cong M_R = \frac{V_R M'_R - V_X M'_X}{V_X + V_R} \tag{6-3}$$

and

$$[X] \cong M_X = 0$$

If we employ G to symbolize the magnitude of some physical quantity such as conductivity, current, or light absorption and if G is proportional to $[X]$ or $[R]$ (or both), we may write Equation 6-2 in the form

$$G = k[X] = k\,\frac{V_X M_X' - V_R M_R'}{V_X + V_R}$$

where k is a proportionality constant. Upon rearranging, we obtain

$$G\left(\frac{V_X + V_R}{V_X}\right) = -\frac{kM_R'}{V_X}\,V_R + kM_X'$$

and

$$QG = K_1 V_R + K_2 \tag{6-4}$$

where K_1 and K_2 are constants and Q is a ratio of volumes that converts G to the value it would have if dilution by the titrant had not occurred.

If the measuring device is sensitive to the reagent as well ($G = k'M_R$), analogous equations can be derived for a region beyond equivalence. That is,

$$QG = K_3 V_R + K_4 \tag{6-5}$$

It is evident from Equations 6-4 and 6-5 that a plot of QG against the reagent volume V_R is linear as long as the assumption is valid that the analytical and species concentrations of the reactant are identical; that is, as long as the reaction is forced far to the right. Figure 6-3B illustrates this type of plot for the titration of iodate ion with silver nitrate. Here, it was assumed that iodate ion does not produce a measurable signal—that is, that K_1 and K_2 in Equation 6-4 are zero. A value for K_3 was arbitrarily chosen so that a common ordinate scale could be used for the two plots. Note that the points well away from the equivalence point form straight lines that can be extrapolated to give an end point. The dashed curvature in the equivalence point region shows the departure from linearity that would be observed owing to the incompleteness of the reaction. In this region, the species silver ion concentration is greater than the analytical silver nitrate concentration because the common-ion effect is not wholly effective in repressing the solubility equilibrium. Thus, the detector begins to show a response to silver ions before the equivalence point is reached.

If iodate rather than silver ion had provided the analytical signal, the titration curve would show high initial values that decreased linearly to zero at the end point. If both ions were responsive to the detector, a V-shaped curve would result, which would make the end point even more pronounced.

End points based on measurements well away from equivalence have the advantage that the reaction between analyte and reagent need not be nearly as complete as in the case of end points requiring measurements in the equivalence point region. This type of end point is considered in the chapters on potentiometry (16), voltammetry (18), and molecular spectroscopy (21).

PROBLEMS*
  * 6-1. Classify the accompanying reactions according to type, and express the equivalent weight of each listed substance as a fraction or a multiple of its formula weight.
    (a) $Mg(OH)_2(s) + 2H_3O^+ \rightarrow Mg^{2+} + 4H_2O$
         $HClO_4$, MgO, H_2SO_4, $Mg_3(PO_4)_2$

---

* Answers to problems or parts of problems marked with an asterisk are to be found at the end of the book.

(b) $Fe^{2+} + V(OH)_4^+ + 2H^+ \rightarrow Fe^{3+} + VO^{2+} + 3H_2O$
V_2O_5, Fe, $Tl_4V_2O_7$, Fe_3O_4

(c) $CO_3^{2-} + 2H_3O^+ \rightarrow CO_2(g) + 3H_2O$
Na_2CO_3, CO_2, $Al_2(CO_3)_3$, $K(HIO_3)_2$

(d) $Cr_2O_7^{2-} + 3U^{4+} + 2H^+ \rightarrow 3UO_2^{2+} + 2Cr^{3+} + H_2O$
Cr, Cr_3O_4, U_3O_8, $K_2Cr_2O_7$

6-2. Classify the accompanying reactions according to type, and express the weight of each listed substance as a fraction or a multiple of its formula weight.

(a) $MnO_4^- + 5Ti^{3+} + H_2O \rightarrow Mn^{2+} + 5TiO^{2+} + 2H^+$
TiO_2, $BaTiO_3$, $Mg(MnO_4)_2 \cdot 6H_2O$, Mn_3O_4

(b) $HONH_3^+ + OH^- \rightarrow HONH_2 + H_2O$
KOH, $(HONH_3)_2SO_4$, $Ba(OH)_2$, $HONH_2$

(c) $B_4O_7^{2-} + 2H_3O^+ + 3H_2O \rightarrow 4H_3BO_3$
$Na_2B_4O_7 \cdot 10H_2O$, B, B_2O_3, $NaBO_2 \cdot 4H_2O$

(d) $BrO_3^- + 3H_2AsO_3^- \rightarrow Br^- + 3H_2AsO_4^-$
$KBrO_3$, As, As_2O_3, Br_2

* 6-3. Nitrogen in organic compounds is commonly converted to ammonium ion by digestion in boiling sulfuric acid. The digested mixture is cooled and an excess of base is introduced; the liberated ammonia is distilled off and subsequently titrated with standard acid. Use chemical symbols to express the equivalent weight of the following compounds as fractions or multiples of their formula weights when determined by this method.

(a) methylamine, CH_3NH_2
(b) vitamin K_6, $C_{11}H_{12}N_2$
(c) anthranilic acid, $C_7H_7NO_2$
(d) WY-3654, $C_{19}H_{22}N_8O_2$
(e) methenamine, $C_6H_{12}N_4$
(f) urea, H_2NCONH_2
(g) glutamic acid, $C_5H_9NO_4$
(h) neohetramine, $C_{16}H_{21}N_4O$
(i) viomycin, $C_{25}H_{43}N_{13}O_{10}$
(j) diphenylguanidine, $(C_6H_5NH_2)_2CNH$

6-4. Organic compounds containing sulfur can be determined by combustion in a stream of oxygen to form CO_2, H_2O, and SO_2. The combustion products are then passed through a dilute solution of H_2O_2 which retains the SO_2 as a result of the reaction

$$H_2O_2 + SO_2 \longrightarrow H_2SO_4$$

The sulfuric acid is titrated with standard base. Use chemical formulas to express the equivalent weight of the following compounds as fractions or multiples of their formula weights when determined by this method.

(a) captan, $C_9H_4O_2NSCl_3$
(b) sulfanilamide, $C_6H_8N_2SO_2$
(c) oil of garlic, $C_6H_{10}S_2$
(d) sulfonal, $C_7H_{16}O_4S_2$
(e) direct blue 2B, $C_{32}H_{20}O_{14}N_6S_4Na_4$

* 6-5. Calculate the number of milliequivalents in

(a) 0.6498 g of HgO as a primary standard for acids:
$$HgO + 2H_3O^+ + 4Br^- \longrightarrow HgBr_4^{2-} + 3H_2O$$

(b) 0.7644 g of $AgNO_3$ as a reducing reagent for borohydride in the reaction:
$$8Ag^+ + BH_4^- + 8OH^- \longrightarrow 8Ag(s) + H_2BO_3^- + 5H_2O$$

(c) 0.4602 g of $NaH_2PO_4 \cdot 2H_2O$ in the reaction:
$$H_2PO_4^- + OH^- \longrightarrow HPO_4^{2-}$$

6-6. Calculate the number of milliequivalents in

(a) 0.4332 g of $Na_2B_4O_7 \cdot 10H_2O$ as a primary standard for acids:
$$B_4O_7^{2-} + 2H_3O^+ + 3H_2O \longrightarrow 4H_3BO_3$$

(b) 0.1771 g of KCN as a reductant in the reaction:
$$CN^- + 2MnO_4^- + 2OH^- \longrightarrow CNO^- + 2MnO_4^{2-} + H_2O$$

(c) 0.3563 g of $Na_2HPO_4 \cdot 2H_2O$ in the reaction:
$$HPO_4^{2-} + 2H_3O^+ \longrightarrow H_3PO_4 + 2H_2O$$

* 6-7. An aqueous solution contains 9.21 g of $K_4Fe(CN)_6$ in a total of 500 mL. Express the concentration of this solution in terms of its

(a) molar analytical concentration.

(b) molar potassium ion concentration.

(c) normality with respect to the reaction:
$$MnO_4^- + 5Fe(CN)_6^{4-} + 8H^+ \longrightarrow Mn^{2+} + 5Fe(CN)_6^{3-} + 4H_2O$$

(d) titer (as mg Zn/mL) with respect to the reaction:
$$3Zn^{2+} + 2Fe(CN)_6^{4-} + 2K^+ \longrightarrow K_2Zn_3[Fe(CN)_6]_2(s)$$

6-8. Potassium hydrogen iodate, $KH(IO_3)_2$, is a versatile primary standard. It has an acidic hydrogen; in addition, it is a relatively strong oxidizing agent. A solution is prepared by dissolving 8.77 g of $KH(IO_3)_2$ in sufficient water to give 500 mL of solution. Express the concentration of this solution in terms of

(a) the molar analytical concentration of $KH(IO_3)_2$.

(b) the molar concentration of iodate ion.

(c) its normality as an acid.

(d) its normality as an oxidizing agent in the reaction:
$$IO_3^- + H_2NNH_2 + 2H^+ + 2Cl^- \longrightarrow ICl_2^- + N_2 + 3H_2O$$

(e) its titer (in mg H_2NNH_2/mL) with respect to the reaction in (d).

* 6-9. How many grams of solute are contained in 25.0 mL of

(a) 0.0840 M $HClO_4$.

(b) 0.0465 M $Ba(OH)_2$.

(c) 0.0465 N $Ba(OH)_2$.

(d) 0.0573 M $AgNO_3$.

(e) 0.0573 M KCN.

(f) 0.0246 N $KMnO_4$. $MnO_4^- + 5Fe^{2+} + 8H^+ \longrightarrow Mn^{2+} + 5Fe^{3+} + 4H_2O$

6-10. How many grams of solute are contained in 40.0 mL of

(a) 0.0614 M H_2SO_4.

(b) 0.0614 N H_2SO_4.

(c) 0.0378 N Na_3PO_4. $PO_4^{3-} + 2H_3O^+ \longrightarrow H_2PO_4^- + 2H_2O$

(d) 0.0832 M KSCN.

(e) 0.0505 M KI.

(f) 0.0505 N KI. $5I^- + IO_3^- + 6H^+ \longrightarrow 3I_2 + 3H_2O$

* 6-11. A hydrogen iodide solution has an analytical concentration of 0.0400 M; calculate its normality as

(a) an acid.

(b) a reducing agent for iodate in strong HCl (product: ICl_2^-).

(c) a reducing agent for oxygen (product: I_2).

* 6-12. Calculate the titer of the HI solution in Problem 6-11 in terms of

(a) mg Na_2CO_3/mL [see 6-11(a)].

(b) mg $KH(IO_3)_2$/mL [see 6-11(b)].

(c) mg O_2/mL [see 6-11(c)].

6-13. A potassium permanganate solution has an analytical concentration of 0.0250 M; calculate its normality as an oxidizing reagent for

(a) H_3AsO_3 in acid solution (products: Mn^{2+}, H_3AsO_4).

(b) $NaHSO_3$ in a slightly alkaline solution [products: $MnO_2(s)$, SO_4^{2-}].

(c) Mn^{2+} in the presence of pyrophosphate ion [product: $Mn(P_2O_7)_3^{3-}$].

(d) CH_3OH in strong base (products: MnO_4^{2-}, CO_2, H_2O).

6-14. Calculate the titer of the permanganate solution in Problem 6-13 in terms of

(a) mg As_2O_3/mL [see 6-13(a)].   (c) mg Mn/mL [see 6-13(c)].

(b) mg SO_2/mL [see 6-13(b)].   (d) mg CH_3OH/mL [see 6-13(d)].

* 6-15. How many millimoles of solute are contained in 35.0 mL of 0.0200 N

(a) $Ba(OH)_2$?

(b) Na_2CO_3? $CO_3^{2-} + 2H_3O^+ \longrightarrow CO_2 + 3H_2O$

(c) $K_2Cr_2O_7$? $Cr_2O_7^{2-} + 6Fe^{2+} + 14H^+ \longrightarrow 2Cr^{3+} + 6Fe^{3+} + 7H_2O$

(d) $KBrO_3$? $BrO_3^- + 6Cu^+ + 6H^+ \longrightarrow Br^- + 6Cu^{2+} + 3H_2O$

6-16. How many millimoles of solute are contained in 24.0 mL of 0.0500 N
    (a) H_2SO_4?
    (b) H_3PO_4? $H_3PO_4 + 2OH^- \longrightarrow HPO_4^{2-} + 2H_2O$
    (c) $KMnO_4$? $2MnO_4^- + 5H_2C_2O_4 + 6H^+ \longrightarrow 2Mn^{2+} + 10CO_2 + 8H_2O$
    (d) I_2? $I_2 + 2S_2O_3^{2-} \longrightarrow 2I^- + S_4O_6^{2-}$
    (e) KI? $2HNO_2 + 2I^- + 2H^+ \longrightarrow 2NO + 2H_2O + I_2$

 * 6-17. Describe the preparation of 500 mL of 0.300 N $HClO_4$ from
    (a) a solution in which the analytical concentration of $HClO_4$ is 7.34 M.
    (b) the concentrated reagent (sp gr 1.60, 70% $HClO_4$).
    (c) a 14.3% (w/w) aqueous solution.

6-18. Describe the preparation of 750 mL of a 0.0500 N solution of HCl from
    (a) a stock solution in which the analytical concentration of HCl is 6.00 M.
    (b) the concentrated reagent (sp gr 1.18, 37% HCl).
    (c) constant-boiling HCl which is 20.27% HCl (w/w).

6-19. What weight of KI is needed to produce 1000 mL of a solution that is 0.0400 N in terms of
    * (a) a reaction in which the iodide is oxidized to the elemental state?
    (b) a reaction in which the iodide is oxidized to IO_4^-?

6-20. What weight of solute is contained in 48.6 mL of 0.0500 N
    (a) $NaHCO_3$?   (d) $Na_2S_2O_3$ (product: $S_4O_6^{2-}$)?
   * (b) $Ba(OH)_2$?   * (e) $Na_2S_2O_3$ (product: SO_4^{2-})?
    (c) I_2 (product: ICl_2^-)?   (f) Na_2SO_3 (product: SO_4^{2-})?

 * 6-21. A solution with a titer of 1.00 mg KSCN/mL is sought. What weight should be taken to prepare 1500 mL of this solution if the reagent is
    (a) $Hg(NO_3)_2$ [product: $Hg(SCN)_2$]?   (c) KIO_3 (products: CN^-, SO_4^{2-}, I^-)?
    (b) $AgNO_3$ [product: $AgSCN(s)$]?   (d) $Hg_2(OAc)_2$ [product: $Hg_2(SCN)_2(s)$]?

6-22. Describe the preparation of 250 mL of a solution with a titer of 4.00 mg HNO_2/mL if the reagent is
    (a) KOH.   (c) $KMnO_4$ (products: NO_3^-, Mn^{2+})?
    (b) KI (products: NO, I_2).   (d) $Ce(HSO_4)_4$ (products: NO_3^-, Ce^{3+})?

 * 6-23. Calculate the molarity of a potassium hydrogen iodate solution that has a titer of
    (a) 3.21 mg $Ba(OH)_2$/mL.
    (b) 5.09 mg $AgNO_3$/mL.
    (c) 6.78 mg I_2/mL. $IO_3^- + 2I_2 + 6H^+ + 10Cl^- \longrightarrow 5ICl_2^- + 3H_2O$
    (d) 4.00 mg KI/mL. $IO_3^- + 2I^- + 6H^+ + 6Cl^- \longrightarrow 3ICl_2^- + 3H_2O$

6-24. Calculate the molarity of a hydrogen iodide solution that has a titer of
    (a) 3.37 mg $Mg(OH)_2$/mL.
    (b) 4.62 mg HgO/mL. $Hg^{2+} + 4I^- \longrightarrow HgI_4^{2-}$
    (c) 6.75 mg $AgNO_3$/mL.

 * 6-25. What volume of 0.1120 N NaOH will be needed to react with
    (a) 25.00 mL of 0.0840 N $HClO_4$?
    (b) 25.00 mL of 0.0840 N H_2SO_4?
    (c) 0.9783 g of $KH(IO_3)_2$?
    (d) 0.3126 g of $H_2C_2O_4 \cdot H_2O$ (product: $C_2O_4^{2-}$)?
    (e) 0.4448 g of $NaH_2PO_4 \cdot H_2O$ (product: HPO_4^{2-})?

6-26. What volume of 0.1208 N HI will be needed to react with
    (a) 26.80 mL of 0.0750 N $Ba(OH)_2$?
    (b) 0.1200 g of KIO_3 (product: I_2)?
    (c) 0.1200 g of KIO_3 (product: ICl_2^-)?

 * 6-27. Calculate the normality of the solution in Column A if the volume indicated is needed to titrate the amount of primary standard (or standardized solution) shown in Column B.

A	B
(a) 41.59 mL H_2SO_4	0.2372 g Na_2CO_3 (products: CO_2, H_2O)
(b) 38.26 mL KOH	0.7915 g $KHC_8H_4O_4$ (product: $K_2C_8H_4O_4$)
(c) 23.93 mL NH_3	25.00 mL of the H_2SO_4 solution in (a)
(d) 29.08 mL $HClO_4$	50.00 mL of the KOH solution in (b)

6-28. Calculate the normality of the solution in Column A if the volume indicated is needed to titrate the amount of primary standard (or standardized solution) shown in Column B.

A	B
(a) 37.24 mL $KMnO_4$	183.9 mg $Na_2C_2O_4$ (products: CO_2, Mn^{2+})
(b) 29.70 mL $Ba(OH)_2$	0.1375 g C_6H_5COOH (product: $C_6H_5COO^-$)
(c) 40.60 mL $FeSO_4$	50.00 mL of the $KMnO_4$ solution in (a)
(d) 19.63 mL HBr	25.00 mL of the $Ba(OH)_2$ solution in (b)

* 6-29. Calculate the normality of a sulfuric acid solution, 50.0 mL of which yielded 0.494 g of $BaSO_4$.

6-30. What is the normality of a hydrochloric acid solution if 25.00 mL yield 0.5042 g of AgCl?

* 6-31. Calculate the normality of a sodium thiosulfate solution if 36.72 mL were needed to titrate the iodine derived from 117.7 mg of primary standard $KH(IO_3)_2$. Reactions:

$$IO_3^- + 5I^- + 6H^+ \longrightarrow 3I_2 + 3H_2O$$

$$I_2 + 2S_2O_3^{2-} \longrightarrow S_4O_6^{2-} + 2I^-$$

6-32. A 0.4046-g sample of mercury(II) oxide was dissolved in a solution containing an excess of KBr. Reaction:

$$HgO(s) + 4Br^- + H_2O \longrightarrow HgBr_4^{2-} + 2OH^-$$

Calculate the normality of a perchloric acid solution if 37.22 mL were needed to titrate the liberated base.

* 6-33. The iron in a 0.7975-g sample of ore was brought into solution, reduced to the $+2$ state, and titrated with 43.69 mL of 0.1008 N $K_2Cr_2O_7$. Reaction:

$$Cr_2O_7^{2-} + 6Fe^{2+} + 14H^+ \longrightarrow 2Cr^{3+} + 6Fe^{3+} + 7H_2O$$

Express the results of this analysis in terms of percent Fe_2O_3.

6-34. The phosphate in a 0.6850-g sample of plant food was converted to $H_2PO_4^-$ and subsequently titrated with 32.46 mL of 0.0894 M $AgNO_3$. Reaction:

$$3Ag^+ + H_2PO_4^- \longrightarrow Ag_3PO_4(s) + 2H^+$$

Express the results of the analysis as percent P_2O_5.

* 6-35. A 0.2178-g sample of impure magnesium hydroxide was dissolved in 50.00 mL of 0.1204 N HCl. Back-titration of the excess acid required 3.76 mL of 0.0948 N NaOH. Calculate the percentage of $Mg(OH)_2$ in the sample.

6-36. A 0.8146-g sample of impure potassium chlorate was treated with 50.00 mL of 0.1017 N $Fe(NH_4)_2(SO_4)_2$. Reaction:

$$ClO_3^- + 6Fe^{2+} + 6H^+ \longrightarrow Cl^- + 3H_2O + 6Fe^{3+}$$

When reaction was complete, the excess iron(II) was back-titrated with 8.35 mL of 0.1026 N $K_2Cr_2O_7$. Calculate the percentage of $KClO_3$ in the sample.

* 6-37. The cyanide in a 0.6913-g sample was oxidized to cyanate with potassium permanganate in a strongly alkaline medium:

$$2MnO_4^- + CN^- + 2OH^- \longrightarrow 2MnO_4^{2-} + CNO^- + H_2O$$

Calculate the percentage of KCN in the sample if the titration required 46.27 mL of a permanganate solution with a titer of 6.44 mg KCN/mL.

6-38. The thiourea in a 1.361-g sample was extracted into a dilute H_2SO_4 solution and titrated with 42.70 mL of a mercury(II) solution with a titer of 2.62 mg thiourea/mL. Reaction:

$$4(NH_2)_2CS + Hg^{2+} \longrightarrow [(NH_2)_2CS]_4Hg^{2+}$$

Calculate the percentage of $(NH_2)_2CS$ in the sample.

* 6-39. A 0.612-g sample containing $Ca(ClO_3)_2 \cdot 2H_2O$ was analyzed by reduction of the ClO_3^- to Cl^- which was precipitated by the addition of 25.0 mL of 0.200 M $AgNO_3$. The excess $AgNO_3$ was titrated with 3.10 mL of a 0.186 M KSCN solution (Ag^+ + $SCN^- \rightarrow AgSCN$). Calculate the percent $Ca(ClO_3)_2 \cdot 2H_2O$ in the sample.

6-40. A 0.500-g sample of steel was dissolved in acid and the chromium present oxidized to dichromate ($Cr_2O_7^{2-}$) by ammonium persulfate. To the resulting solution was added exactly 1.242 g of Mohr's salt, $FeSO_4(NH_4)_2SO_4 \cdot 6H_2O$, the iron(II) ion of which reduced the dichromate ion to the chromium(III) state. The excess iron(II) ion was titrated with 14.1 mL of 0.0463 N $KMnO_4$. Calculate the percent Cr in the steel.

* 6-41. A 50.0-mL aliquot of a solution containing uranium in the +6 state was passed through a reductor which reduced it to a mixture of the +3 and +4 states. Bubbling air through the solution converted all of the +3 to the +4 state which was then oxidized quantitatively back to the +6 form with 36.9 mL of 0.0624 N $K_2Cr_2O_7$.

$$3UO^{2+} + Cr_2O_7^{2-} + 8H^+ \longrightarrow 3UO_2^{2+} + 2Cr^{3+} + 4H_2O$$

What weight of uranium was contained in a liter of the sample solution?

6-42. A 2.00-g sample of chromite ($FeO \cdot Cr_2O_3$) was fused with sodium peroxide. The resulting mass was dissolved and the excess peroxide destroyed by boiling. After acidification, 50.0 mL of 0.160 N Fe^{2+} were added which reduced the $Cr_2O_7^{2-}$ to Cr^{3+}. A back titration of 3.14 mL of 0.0500 N $K_2Cr_2O_7$ was required to oxidize the excess Fe^{2+}. Calculate (a) the percent Cr in the sample and (b) the percent chromite in the sample.

* 6-43. The routine analysis for H_2SO_4 in an electroplating rinse is to be undertaken. A NaOH solution is to be prepared such that the volume used in titration is numerically ten times as great as the percent H_2SO_4 in a 20.0-g sample. What should be the normality of the NaOH solution?

6-44. The sulfur in an organic compound was determined by burning a 0.471-g sample in a stream of oxygen and collecting the resulting SO_2 in a neutral solution of H_2O_2, which converted the SO_2 to H_2SO_4.

$$SO_2 + H_2O_2 = H_2SO_4$$

The sulfuric acid was titrated with 28.2 mL of 0.108 N KOH. Calculate the percent sulfur in the sample.

* 6-45. An organic mixture was known to contain the compound, $C_6H_4Cl_2$. This compound was analyzed by treating a 1.17-g sample with metallic sodium which converted the chlorine quantitatively to NaCl. After destruction of the excess sodium metal, the chloride was titrated with 30.1 mL of a 0.0884 M solution of $Hg(NO_3)_2$.

$$Hg^{2+} + 2Cl^- \longrightarrow HgCl_2$$

Calculate (a) the percent Cl in the sample and (b) the percent $C_6H_4Cl_2$.

6-46. The sulfur in a 5.00-g sample of steel was evolved as H_2S which was collected in an ammoniacal solution of $CdCl_2$. The CdS formed was treated with 10 mL of 0.0600 N I_2, and the I_2 back-titrated with 4.82 mL of 0.0510 N sodium thiosulfate. The reaction of CdS with I_2 is

$$CdS + I_2 \longrightarrow S + Cd^{2+} + 2I^-$$

Calculate the percent S in the steel.

precipitation titrations

7

Volumetric methods based upon the formation of sparingly soluble silver salts are among the oldest known; these procedures were and still are routinely employed for the analysis for silver as well as for the determination of such ions as chloride, bromide, iodide, and thiocyanate. Volumetric precipitation methods that do not involve silver as one of the reactants are relatively limited in scope.

7A TITRATION CURVES FOR PRECIPITATION REACTIONS

Titration curves based on p-values are useful for deducing the properties required of an indicator as well as the titration error that its use is likely to cause. Theoretical curves are readily derived from solubility product data and usually show a close resemblance to curves that are obtained experimentally. Three types of calculations are needed to generate such a curve. For the region preceding the end point, the analytical concentration of the excess analyte is obtained; the assumption is then made that this quantity is equal to the equilibrium concentration of the analyte ion. Beyond the equivalence point the analytical concentration of the reagent is computed, and the assumption is made that this value corresponds to its equilibrium concentration as well. Finally, at equivalence, the only source of ions of the two reactants is the sparingly soluble product; their concentrations are readily derived from the solubility product constant.

> **EXAMPLE 7-1.** Derive a curve for the titration of 50.00 mL of 0.00500 M NaBr with 0.01000 M AgNO$_3$.
>
> We shall calculate both pBr and pAg, although only one of the two (the one that affects the indicator behavior) is actually needed.
>
> *Initial point.* At the outset, the solution is 0.00500 M in Br$^-$ and 0.000 M in Ag$^+$. Thus, pBr = $-\log (5.00 \times 10^{-3})$ = 2.301 = 2.30. The pAg is indeterminate.
>
> *After addition of 5.00 mL reagent.* Here, the bromide ion concentration will have been decreased by precipitation as well as by dilution. Thus,

$$M_{NaBr} = \frac{(50.00 \times 0.00500) - (5.00 \times 0.01000)}{50.00 + 5.00} = 3.64 \times 10^{-3}$$

The first term in the numerator is the number of millimoles of NaBr present originally; the second term represents the number of millimoles of $AgNO_3$ added and hence the number of millimoles of AgBr produced. Account is taken of the volume change in the denominator.

Both the unreacted NaBr and the slightly soluble AgBr contribute to the bromide ion concentration of the solution. Thus, the equilibrium concentration of Br^- is larger than the analytical concentration of NaBr by an amount equal to the molar solubility of the precipitate. That is,

$$[Br^-] = 3.64 \times 10^{-3} + [Ag^+]$$

The second term on the right accounts for the contribution of AgBr to the equilibrium bromide concentration because one Ag^+ ion is formed for each Br^- ion from this source. Unless the concentration of NaBr is very small, this second term can be neglected. That is, when $[Ag^+] \ll 3.64 \times 10^{-3}$

$$[Br^-] = 3.64 \times 10^{-3}$$

$$pBr = -\log (3.64 \times 10^{-3}) = 2.439 = 2.44$$

A convenient way to find pAg is to take the negative logarithm of the solubility product expression for AgBr. That is,

$$-\log ([Ag^+][Br^-]) = -\log K_{sp} = -\log (5.2 \times 10^{-13})$$

$$-\log [Ag^+] - \log [Br^-] = -\log K_{sp} = 12.28$$

or

$$pAg + pBr = pK_{sp} = 12.28$$

$$pAg = 12.28 - 2.44 = 9.84$$

This relationship applies to any solution containing silver and bromide ions in contact with solid silver bromide.

Other points up to chemical equivalence can be derived in this same way.

Equivalence point. Here, neither NaBr nor $AgNO_3$ is in excess and the concentrations of silver and bromide ions must be equal. Substitution of this equality into the solubility product expression yields

$$[Ag^+] = [Br^-] = \sqrt{5.2 \times 10^{-13}} = 7.21 \times 10^{-7}$$

$$pAg = pBr = -\log (7.21 \times 10^{-7}) = 6.14$$

After addition of 25.10 mL reagent. The solution now contains an excess of $AgNO_3$; thus,

$$M_{AgNO_3} = \frac{\text{total mmol } AgNO_3 - \text{original mmol NaBr}}{\text{total volume of solution}}$$

$$M_{AgNO_3} = \frac{25.10 \times 0.01000 - 50.00 \times 0.00500}{50.00 + 25.10} = 1.33 \times 10^{-5}$$

and the equilibrium concentration of silver ion is

$$[Ag^+] = 1.33 \times 10^{-5} + [Br^-] \cong 1.33 \times 10^{-5}$$

Here, the second term in the right side of the equation accounts for Ag^+ ions resulting from the slight solubility of AgBr; it can ordinarily be neglected. Thus,

$$pAg = -\log (1.33 \times 10^{-5}) = 4.876 = 4.88$$

$$pBr = 12.28 - 4.88 = 7.40$$

Additional points defining the titration curve beyond the equivalence point can be obtained in an analogous way.

7A-1 SIGNIFICANT FIGURES IN TITRATION-CURVE DERIVATIONS

In deriving titration curves, concentration data associated with the equivalence point region are often of low precision because they are based upon small differences between large numbers. For example, in the calculation for M_{AgNO_3}, following the introduction of 25.10 mL of 0.01000 M AgNO$_3$, the numerator (0.2510 − 0.2500 = 0.0010) contains only two significant figures; at best then, M_{AgNO_3} is known to two significant figures as well. To minimize the rounding error, however, we retained three digits (1.33 × 10^{-5}) in this calculation and rounded after calculating pAg.

In rounding the calculated value for pAg, it is important to recall (p. 79) that it is the *mantissa only of a logarithm (that is, the number to the right of the decimal point) that should be rounded to include only significant figures* because the characteristic serves merely to locate the decimal point. Thus, in the example under consideration we rounded pAg to two figures to the right of the decimal point; that is, pAg = 4.88.

Note that we would have been justified in carrying an additional figure for the points well away from the equivalence point. For example, the initial pBr could have been reported as 2.301. Nothing is gained, however, because it is the equivalence point region that is of primary interest. Thus, here, and in later derivations of titration curves, we will round p-functions to two digits to the right of the decimal. The large changes in p-functions that are characteristic of most equivalence points are not obscured by this limited precision in the calculations.

7A-2 FACTORS INFLUENCING THE SHARPNESS OF END POINTS

A sharp and easily located end point is observed when small additions of reagent cause large changes in p-function. It is therefore of interest to examine those variables that influence the magnitude of such changes during a titration.

Reagent concentration.   Figure 7-1 contains curves for the titration of bromide ion with silver nitrate at three reactant concentrations. These curves were derived from calculations similar to those just considered. It is apparent that an increase in analyte and reagent concentration enhances the change in pAg in the equivalence point region; an analogous effect is observed when pBr is plotted rather than pAg.

These effects have practical significance for the titration of bromide ion. If the analyte concentration is sufficient to permit the use of a silver nitrate solution that is 0.1 M or stronger, easily detected end points are observed, and the titration error is minimal. On the other hand, with solutions that are 0.001 M or less, the change in pAg or pBr is so small that end point detection becomes difficult; a large titration error is thus to be expected.

As will be shown later, the reagent concentration influences the end point sharpness of titrations based upon other reaction types as well.

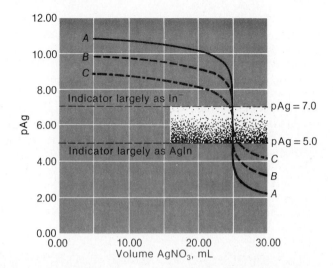

FIGURE 7-1 Effect of reagent concentrations on titrations curves. Curves: A, 50.00 mL of 0.0500 M NaBr with 0.1000 M AgNO$_3$; B, 50.00 mL of 0.00500 M NaBr with 0.01000 M AgNO$_3$; C, 50.00 mL of 0.000500 M NaBr with 0.001000 M AgNO$_3$.

Completeness of reaction. Figure 7-2 shows how the solubility of the species produced influences curves for titrations in which 0.1 M silver nitrate serves as the reagent. Clearly, the greatest change in pAg occurs in the titration of iodide ion which, of all the anions considered, forms the least soluble silver salt and hence reacts most completely with silver ion. The smallest change in pAg is observed for the reaction that is least favorable—that is, in the titration of bromate ion. Reactions that produce silver salts with solubilities intermediate between these ex-

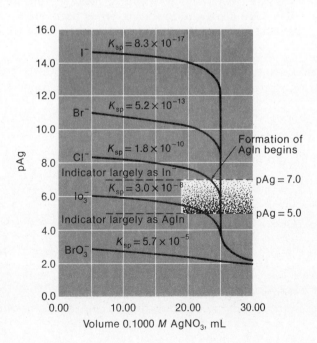

FIGURE 7-2 The effect of completeness of reaction on titration curves. 50.00 mL of a 0.0500 M solution of the anion are titrated with 0.1000 M AgNO$_3$ for each curve.

tremes yield titration curves with end point changes that are also intermediate in magnitude. Again, we shall see that this effect is common to all reaction types.

7A-3 TITRATION CURVES FOR MIXTURES

The methods developed in the previous section can be extended to mixtures that form precipitates of differing solubilities. To illustrate, consider the titration of 50.00 mL of a solution that is 0.0800 M in iodide ion and 0.1000 M in chloride with 0.2000 M silver nitrate.

Because silver iodide has a much smaller solubility than silver chloride, the initial additions of the reagent will result in formation of the iodide exclusively. Here, the titration curve should be similar to that for iodide shown in Figure 7-2. It is of interest, then, to determine the extent to which iodide is precipitated before appreciable formation of silver chloride takes place.

With the first appearance of silver chloride, the solubility product expressions for both precipitates are applicable to the solution; division of one by the other provides a useful relationship.

$$\frac{[Ag^+][I^-]}{[Ag^+][Cl^-]} = \frac{8.3 \times 10^{-17}}{1.82 \times 10^{-10}} = 4.56 \times 10^{-7} = 4.6 \times 10^{-7}$$

or

$$[I^-] = 4.56 \times 10^{-7} [Cl^-]$$

It is apparent that the iodide concentration will have been reduced to a tiny fraction of the chloride ion concentration prior to the onset of precipitation by silver chloride. That is, formation of silver chloride will not occur until very near the equivalence point for iodide or after the addition of, for all practical purposes, 20.00 mL of reagent in this titration. At this point, the chloride ion concentration, because of dilution, will be approximately

$$M_{Cl} \cong [Cl^-] = \frac{50.00 \times 0.1000}{50.00 + 20.00} = 0.0714$$

Substituting into the previous equation yields

$$[I^-] = 4.56 \times 10^{-7} \times 0.0714 = 3.26 \times 10^{-8} = 3.3 \times 10^{-8}$$

The percentage of iodide unprecipitated at this point can be calculated as follows:

$$\text{original mmol } I^- = 50.00 \times 0.0800 = 4.00$$

$$\% \text{ I}^- \text{ unprecipitated} = \frac{3.26 \times 10^{-8} \times 70.00 \times 100}{4.00} = 5.7 \times 10^{-5}$$

Thus, to within about 6×10^{-5} percent of the equivalence point for iodide, no silver chloride should form, and the titration curve will be indistinguishable from that for the iodide alone. The first half of the titration curve shown by the solid line in Figure 7-3 was derived on this basis.

As chloride ion begins to precipitate, the rapid decrease in pAg is terminated abruptly. The pAg is most conveniently calculated from the solubility product constant for silver chloride and the computed chloride concentration. Thus,

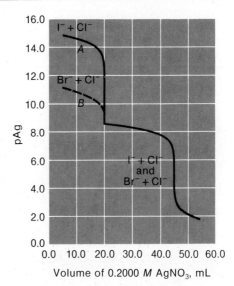

FIGURE 7-3 Titration curves for 50.00 mL of solutions that were 0.100 M in Cl^- and 0.0800 M in Br^- or I^-.

$$[Ag^+] = \frac{1.82 \times 10^{-10}}{0.0714} = 2.55 \times 10^{-9}$$

$$pAg = -\log (2.55 \times 10^{-9}) = 8.59$$

Further additions of silver nitrate decrease the chloride ion concentration, and the curve then becomes that for the titration of chloride by itself. For example, after 25.00 mL of reagent have been added,

$$M_{Cl} \cong [Cl^-] = \frac{50.00 \times 0.1000 + 50.00 \times 0.0800 - 25.00 \times 0.200}{75.00}$$

Here, the first two terms in the numerator give the number of millimoles of chloride and iodide, respectively, and the third is the number of millimoles of titrant added. Thus,

$$[Cl^-] = 0.0533$$

$$[Ag^+] = \frac{1.82 \times 10^{-10}}{0.0533} = 3.41 \times 10^{-9}$$

$$pAg = 8.47$$

The remainder of the curve can be derived in the same way as a curve for chloride by itself.

Curve A in Figure 7-3 is the titration curve for the chloride/iodide mixture just considered; it is a composite of the individual curves for the two anionic species. Two equivalence points are evident. It is to be expected that the change associated with the first equivalence point will become less distinct as the solubilities of the two precipitates approach one another. Curve B demonstrates this effect for a mixture of bromide and chloride ions. In this titration, the initial pAg values are lower because the solubility of silver bromide exceeds that of silver iodide. Beyond the first equivalence point, however, the two titration curves become identical.

It is possible to obtain experimental curves similar to those shown in Figure 7-3 by measuring the potential of a silver electrode immersed in the solution; this technique, which is discussed in Chapter 16, permits the analysis of the individual components in halide ion mixtures.

7A-4 CHEMICAL INDICATORS FOR PRECIPITATION TITRATIONS

A chemical indicator produces a visually detectable change, usually of color or turbidity, in the solution. The indicator functions by reacting competitively with one of the reactants or products of the titration. Thus, in the titration of species X with a reagent R and in the presence of an indicator (In) that can react with R, a chemical description of the solution throughout the titration will be

$$X + R \rightleftharpoons XR(s)$$

$$In + R \rightleftharpoons InR$$

For indicator action, it is necessary, of course, that InR impart a significantly different appearance to the mixture than does In. In addition, the amount of InR required for an observable change should be so small that no appreciable consumption of R occurs when InR is formed. Finally, the equilibrium constant for the indicator reaction must be such that the ratio [InR]/[In] is shifted from a small to a large value as a consequence of the change in [R] (or pR) that occurs in the equivalence point region. This last condition is most likely to be realized where the change in pR is large.

To illustrate, consider the application of a hypothetical indicator which exhibits a full color change as pAg varies from 7 to 5 (as noted on p. 169), a change in p-function of 1 to 2 units is required in order for the color change to be great enough to be detected by the average person) to the three bromide titrations illustrated in Figure 7-1. It is clear that each titration requires a different volume of titrant to encompass this range. Thus, for the most concentrated solution (curve A), less than 0.10 mL of 0.1 M $AgNO_3$ is needed; that is, the color change will begin beyond 24.95 mL and will be completed before 25.05 mL of silver have been added. An abrupt alteration of the color and a minimal titration error can be expected. In contrast, with 0.001 M $AgNO_3$, the color will start to change at about 24.5 mL and will be complete at 25.8 mL (curve C). Location of the end point under these circumstances is impossible. The end point change for a titration with 0.01 M reagent would require somewhat less than 0.2 mL; here, the indicator would be usable, but the uncertainty with respect to the equivalence point would be substantial.

Consider now the applicability and the effectiveness of this same indicator with regard to the titrations represented by the curves in Figure 7-2. During the titration of bromate and iodate ions, the indicator will exist largely as AgIn throughout; thus, no color change is observed after the first addition of silver nitrate. In contrast, the solubilities of silver bromide and silver iodide are small enough to prevent formation of significant amounts of AgIn until the equivalence point region is reached; thus, from Figure 7-2 it is seen that pAg remains above 7 until just before the equivalence point for the bromide titration and just beyond the equivalence point for the iodide. The excess silver ion required to complete the color change corresponds to less than 0.01 mL of reagent in both titrations; the titration error would thus be negligible.

An indicator with a pAg range of 5 to 7 would not be satisfactory for a chloride titration because formation of appreciable amounts of AgIn would begin approximately 1 mL short of the equivalence point and extend over a range of about 1 mL; exact location of the end point would be impossible. On the other hand, an indicator with a pAg range of 4 to 6 would be perfectly satisfactory. No satisfactory chemical indicator exists for the iodate and bromate titrations because the changes in pAg in the equivalence point region are too small.

Examples of indicators employed for precipitation titrations with silver ions are described in the paragraphs that follow.

The formation of a second precipitate; the Mohr method.

The formation of a second, highly colored precipitate is the basis for end point detection with the *Mohr method.* This procedure has been widely applied to the titration of chloride ion and bromide ion with standard silver nitrate. Chromate ion is the indicator, the end point being signaled by the appearance of brick-red silver chromate, Ag_2CrO_4.

The molar solubility of silver chromate is several times greater than that for silver chloride. Thus, the latter precipitate tends to form first in the titration mixture. By adjusting the chromate concentration to a suitable level, formation of silver chromate can be retarded until the silver ion concentration in the mixture is equal to the theoretical equivalence point for the titration of chloride. As we have shown (p. 179), at equivalence we may write

$$[Ag^+] = [Cl^-] = \sqrt{K_{sp}}$$
$$[Ag^+] = \sqrt{1.82 \times 10^{-10}} = 1.35 \times 10^{-5}$$

The chromate concentration required to initiate precipitation of silver chromate under these conditions can be computed with the solubility product constant for silver chromate.

$$[CrO_4^{2-}] = \frac{K_{sp}}{[Ag^+]^2} = \frac{1.1 \times 10^{-12}}{(1.35 \times 10^{-5})^2} = 6 \times 10^{-3}$$

In principle, then, an amount of chromate ion necessary to give this concentration could be added, and the red color of silver chromate would signal the appearance of the first excess of silver ion over its equivalence concentration. In practice, however, end point detection is difficult if not impossible in solutions with chromate ion concentrations as large as 6×10^{-3} M because the yellow color imparted by this concentration of indicator masks the color of the silver chromate. As a result, indicator concentrations smaller than 6×10^{-3} M must be employed (the optimum concentration appears to be about 2.5×10^{-3} M). This decrease in the chromate concentration requires a silver concentration greater than 1.35×10^{-5} M to produce a red precipitate. In addition, a finite amount of silver nitrate must be added to produce a detectable quantity of the precipitate. Both factors cause an overconsumption of reagent. The problem is serious where dilute solutions are involved, but the error is small with 0.1 M solutions. A correction may be made by determining an *indicator blank*—that is, by determining the silver ion consumption for a suspension of chloride-free calcium carbonate in about the same volume and with the same quantity of indicator as the sample. The blank titration mixture serves as a convenient color standard for subsequent titrations. A better method, which largely eliminates the indicator error, is to use the Mohr method to standardize the silver nitrate solution against pure sodium chloride. The "working molarity" obtained for the solution will compensate not only for the overconsumption of reagent but also for the acuity of the analyst in detecting the color change.

Attention must be paid to the acidity of the medium because the equilibrium

$$2CrO_4^{2-} + 2H^+ \rightleftharpoons Cr_2O_7^{2-} + H_2O$$

is displaced to the right as the hydrogen ion concentration is increased. Silver dichromate is considerably more soluble than the chromate; as a consequence, the indicator reaction in acid solution requires substantially larger silver ion concentrations, if indeed it occurs at all. If the medium is made strongly alkaline, there is danger that silver will precipitate as its oxide:

$$2Ag^+ + 2OH^- \rightleftharpoons 2AgOH(s) \rightleftharpoons Ag_2O(s) + H_2O$$

Thus, the determination of chloride by the Mohr method must be carried out in a medium that is neutral or nearly so (pH 7 to 10). The presence of an excess of either sodium hydrogen carbonate or borax in the solution tends to maintain the hydrogen ion concentration within suitable limits.

Formation of a colored complex; the Volhard method. A standard solution of thiocyanate may be used to titrate silver ion by the *Volhard method:*

$$Ag^+ + SCN^- \rightleftharpoons AgSCN(s)$$

Iron(III) ion serves as the indicator, imparting a red coloration to the solution with the first slight excess of thiocyanate:

$$Fe^{3+} + SCN^- \rightleftharpoons \underset{red}{Fe(SCN)^{2+}}$$

The titration must be carried out in acid solution to prevent precipitation of iron(III) as the hydrated oxide. As shown by the following example, the indicator concentration to reduce the titration error to zero is readily obtained.

> **EXAMPLE 7-2.** It has been found from experiment that the average observer can just detect the red color of $Fe(SCN)^{2+}$ when its concentration is 6.4×10^{-6} M. In the titration of 50.0 mL of 0.050 M Ag^+ with 0.100 M KSCN, what concentration of Fe^{3+} should be used to reduce the titration error to zero?
>
> For zero titration error, the $FeSCN^{2+}$ color should appear when the concentration of Ag^+ remaining in the solution is identical to the total concentration of thiocyanate-containing species; that is, at the equivalence point
>
> $$[Ag^+] = [SCN^-] + [Fe(SCN)^{2+}]$$
>
> Since we wish the concentration of the complex to be 6.4×10^{-6} M at the equivalence point,
>
> $$[Ag^+] = [SCN^-] + 6.4 \times 10^{-6}$$
>
> or
>
> $$\frac{K_{sp}}{[SCN^-]} = \frac{1.1 \times 10^{-12}}{[SCN^-]} = [SCN^-] + 6.4 \times 10^{-6}$$

which rearranges to

$$[SCN^-]^2 + 6.4 \times 10^{-6} [SCN^-] - 1.1 \times 10^{-12} = 0$$

Solving this equation yields

$$[SCN^-] = 1.7 \times 10^{-7}$$

The formation constant for $FeSCN^{2+}$ is

$$\frac{[Fe(SCN)^{2+}]}{[Fe^{3+}][SCN^-]} = K_f = 1.4 \times 10^2$$

If we now substitute the $[SCN^-]$ necessary to give a detectable concentration of $FeSCN^{2+}$ at the equivalence point, we obtain

$$\frac{6.4 \times 10^{-6}}{[Fe^{3+}] \, 1.7 \times 10^{-7}} = 1.4 \times 10^2$$

$$[Fe^{3+}] = 0.27$$

The indicator concentration is not critical in the Volhard titration; calculations similar to those just shown demonstrate that a titration error of one part in a thousand or less is, in theory, possible if the iron(III) concentration is held between 0.002 and 1.6 M. In practice, it is found that an indicator concentration greater than 0.2 M imparts a sufficient color to the solution, due to the yellow iron(III) ions, to make detection of the thiocyanate complex difficult. Therefore, lower concentrations (usually about 0.01 M) of iron(III) are employed.

The most important application of the Volhard method is for the indirect determination of halide ions. A measured excess of standard silver nitrate solution is added to the sample, and the excess silver ion is determined by back-titration with a standard thiocyanate solution. The requirement of a strongly acidic environment represents a distinct advantage for the Volhard technique over other methods for halide analysis because such ions as carbonate, oxalate, and arsenate (which form slightly soluble silver salts in neutral but not acidic media) do not interfere.

In contrast to the other silver halides, silver chloride is more soluble than silver thiocyanate. As a consequence, the reaction

$$AgCl(s) + SCN^- \rightleftharpoons AgSCN(s) + Cl^-$$

occurs to a significant extent near the end point during back-titration of the excess silver ion. The result is an end point that fades and an overconsumption of thiocyanate ion, which in turn leads to low values for the chloride analysis.

We can readily estimate the magnitude of the error arising from the reaction by first evaluating the equilibrium constant for the process; to obtain this information, the ratio of the solubility products of silver chloride to silver thiocyanate is computed. That is,

$$\frac{[\cancel{Ag^+}][Cl^-]}{[\cancel{Ag^+}][SCN^-]} = \frac{[Cl^-]}{[SCN^-]} = \frac{1.82 \times 10^{-10}}{1.1 \times 10^{-12}} = 1.7 \times 10^2 = K$$

This equation permits calculation of the chloride ion concentration of the solution if the thiocyanate concentration is known. As was mentioned earlier, the minimum concentration of $Fe(SCN)^{2+}$ that can be detected by the average observer is about 6.4×10^{-6} M. Furthermore, the concentration of Fe^{3+} ordinarily employed is about

0.01 M. Substitution of these quantities into the formation constant expression for the complex gives

$$\frac{[\text{Fe}(\text{SCN})^{2+}]}{[\text{Fe}^{3+}][\text{SCN}^-]} = \frac{6.4 \times 10^{-6}}{0.01\,[\text{SCN}^-]} = 1.4 \times 10^2$$

and

$$[\text{SCN}^-] = 4.6 \times 10^{-6}$$

When this quantity is substituted into the earlier equation, it is found that

$$[\text{Cl}^-] = K[\text{SCN}^-] = 1.7 \times 10^2 \times 4.6 \times 10^{-6}$$
$$= 7.8 \times 10^{-4}$$

Insofar as this chloride concentration results primarily from reaction between the thiocyanate reagent and the silver chloride, we can readily calculate the volume of the overtitration that results. Thus, with a total volume of 100 mL at the end point, the excess of 0.1 M thiocyanate will amount to

$$\frac{7.8 \times 10^{-4} \times 100}{0.1} = 0.8 \text{ mL}$$

In actual practice, the overconsumption of reagent is often greater.

Two general methods are employed to avoid the error resulting from the reaction between thiocyanate and silver chloride. The first involves the use of the maximum allowable indicator concentration [about 0.2 M iron(III) ion].[1] Thus, when an iron(III) concentration of 0.2 M is substituted for 0.01 M in the foregoing calculations, the theoretical titration error is reduced to about 0.04 mL.

A more popular way of performing a Volhard chloride analysis involves isolation of the precipitated silver chloride before back-titration with the thiocyanate. Filtration, followed by titration of an aliquot of the filtrate, yields excellent results provided the precipitated silver chloride is first briefly digested. The time required for filtration is, of course, a disadvantage. Probably the most widely employed modification is that of Caldwell and Moyer,[2] which consists of coating the silver chloride with nitrobenzene, thereby substantially removing it from contact with the solution. The coating is accomplished by shaking the titration mixture with a few milliliters of the organic liquid prior to back-titration.

Adsorption indicators. An adsorption indicator is an organic compound that is adsorbed on or desorbed from the surface of the solid formed during a precipitation titration. Ideally, the adsorption or desorption occurs near the equivalence point and results not only in a color change but also a transfer of color from the solution to the solid (or the reverse). Analytical procedures based upon adsorption indicators are sometimes called *Fajans methods* in honor of the scientist who was active in their development.

A typical adsorption indicator is the organic dye *fluorescein*, which is useful for the titration of chloride ion with silver nitrate. In aqueous solution, fluorescein

---

[1] E. H. Swift, G. M. Arcand, R. Lutwack, and D. J. Meier, *Anal. Chem.*, **22**, 306 (1950).
[2] J. R. Caldwell and H. V. Moyer, *Ind. Eng. Chem., Anal. Ed.*, **7**, 38 (1935).

partially dissociates into hydrogen ions and negatively charged fluoresceinate ions that impart a yellowish green color to the medium. The fluoresceinate ion forms a highly colored silver salt of limited solubility. In its application as an indicator, however, *the concentration of the dye is never large enough to exceed the solubility product for silver fluoresceinate.*

The dye anion is not appreciably adsorbed prior to the equivalence point in a chloride titration; it is in fact repelled from the surface of the silver chloride particles, owing to the negative charge resulting from adsorbed chloride ions (p. 135). Beyond the equivalence point, however, the particles of the precipitate become positively charged by virtue of the strong adsorption by excess silver ions; under these conditions, fluoresceinate ions tend to be held *in the counter-ion layer.* The net result is the appearance of the red color of silver fluoresceinate *on the surface of the precipitate.* It is important to emphasize that the color change is an *adsorption* (not a precipitation) process inasmuch as the solubility product of the silver fluoresceinate is never exceeded. The adsorption is reversible, the dye being desorbed upon back-titration with chloride ion.

The successful application of an adsorption indicator requires that the precipitate and the indicator have the following properties:

1. The particles of the precipitate must be of colloidal dimensions to maximize the quantity of indicator adsorbed.
2. The precipitate must strongly adsorb its own ions. We have seen (Section 5B-2) that this property is characteristic of colloidal precipitates.
3. The indicator dye must be strongly held in the counter-ion layer by the primarily adsorbed ion. In general, this type of adsorption correlates with a low solubility of the salt formed between the dye and the lattice ion. At the same time, this species must have sufficient solubility so that it does not precipitate.
4. The pH of the solution must be such as to assure that the ionic form of the indicator predominates. The active form of most adsorption indicators is an ion that is the conjugate acid or base of the dye molecule; it is thus capable of combining with hydrogen or hydroxide ions to form the inactive parent molecule.

Titrations involving adsorption indicators are rapid, accurate, and reliable. Their application, however, is limited to a relatively few precipitation reactions in which a colloidal precipitate is rapidly formed. End point determinations with adsorption indicators tend to be less satisfactory in the presence of high electrolyte concentrations, owing to coagulation of the precipitate and the consequent decrease in the surface on which adsorption can occur.

Most adsorption indicators are weak acids. Their use is thus confined to basic, neutral, or slightly acidic solutions where the indicator exists predominantly as the anion. A few cationic adsorption indicators are known; these are suitable for titrations in strongly acidic solutions. For such indicators, adsorption of the dye and coloration of the precipitate occur in the presence of an excess of the anion of the precipitate (that is, when the precipitate particles possess a negative charge).

Finally, many adsorption indicators sensitize silver-containing precipitates toward photodecomposition, which may cause difficulties.

7A-5 OTHER METHODS FOR END POINT DETECTION

Electroanalytical methods, which can be applied for detection of end points for some precipitation reactions, are described in Chapters 16 and 18.

7B APPLICATIONS OF PRECIPITATION TITRATIONS

Most applications of precipitation titrations are based upon the use of a standard silver nitrate solution and are sometimes called *argentometric* methods as a consequence. Table 7-1 lists some typical applications of argentometry. Note that many of these analyses are based upon precipitation of the analyte with a measured excess of silver nitrate followed by a Volhard titration with standard potassium thiocyanate. Both of these reagents are obtainable in primary standard quality; however, potassium thiocyanate is somewhat hygroscopic and is ordinarily standardized against silver nitrate. Both silver nitrate and potassium thiocyanate solutions are stable indefinitely.

Table 7-1   Typical Argentometric Precipitation Methods

Substance Determined	End Point	Remarks
AsO_4^{3-}, Br^-, I^-, CNO^-, SCN^-	Volhard	Removal of silver salt not required
CO_3^{2-}, CrO_4^{2-}, CN^-, Cl^-, $C_2O_4^{2-}$, PO_4^{3-}, S^{2-}	Volhard	Removal of silver salt required before back-titration of excess Ag^+
BH_4^-	Modified Volhard	Titration of excess Ag^+ following: $BH_4^- + 8Ag^+ + 8OH^- \rightarrow 8Ag(s) + H_2BO_3^- + 5H_2O$
Epoxide	Volhard	Titration of excess Cl^- following hydrohalogenation
K^+	Modified Volhard	Precipitation of K^+ with known excess of $B(C_6H_5)_4^-$, addition of excess Ag^+ giving $AgB(C_6H_5)_4(s)$, and back-titration of the excess
Br^-, Cl^-	Mohr	
Br^-, Cl^-, I^-, SeO_3^{2-}	Adsorption indicator	
$V(OH)_4^+$, fatty acids, mercaptans	Electroanalytical	Direct titration with Ag^+
Zn^{2+}	Modified Volhard	Precipitate as $ZnHg(SCN)_4$; filter, dissolve in acid, add excess Ag^+; back-titrate excess Ag^+
F^-	Modified Volhard	Precipitate as $PbClF$; filter, dissolve in acid, add excess Ag^+; back-titrate excess Ag^+

Table 7-2 Miscellaneous Volumetric Precipitation Methods

Reagent	Ion Determined	Reaction Product	End Point
$K_4Fe(CN)_6$	Zn^{2+}	$K_2Zn_3[Fe(CN)_6]_2$	Diphenylamine
$Pb(NO_3)_2$	SO_4^{2-}	$PbSO_4$	Erythrosin B
	MoO_4^{2-}	$PbMoO_4$	Eosin A
$Pb(OAc)_2$	PO_4^{3-}	$Pb_3(PO_4)_2$	Dibromofluorescein
	$C_2O_4^{2-}$	PbC_2O_4	Fluorescein
$BaCl_2$	SO_4^{2-}	$BaSO_4$ (50% methanol solvent)	Alizarin red S
$Th(NO_3)_4$	F^-	ThF_4	Alizarin red
$Hg_2(NO_3)_2$	Cl^-, Br^-	Hg_2Cl_2, Hg_2Br_2	Bromophenol blue
$NaCl$	Hg_2^{2+}	Hg_2Cl_2	Bromophenol blue

Silver nitrate is a remarkably useful reagent for both titrimetric and gravimetric methods of analysis. Since about 1978, however, its price has risen precipitously with the advances in silver prices. Thus, when the world price of silver is $15/troy oz, the silver nitrate required to prepare 500 mL of 0.10 M reagent costs in the range of $5 to $15 depending upon its source. Consequently, if silver nitrate is to be used as a reagent, considerable care must be exercised to minimize the quantity used. Husbandry of the reagent may involve the use of smaller samples so that more dilute solutions or smaller volumes of reagent are needed. In addition, the silver salt formed in the analysis and any reagent remaining afterward must be collected for subsequent processing to recover the silver and perhaps converting it to silver nitrate for re-use. Several methods have been recommended for treating silver wastes to yield silver or silver nitrate.[3]

Table 7-2 lists some miscellaneous volumetric methods based on reagents other than silver nitrate.

Specific directions for argentometric titrations are found in Section 2 of Chapter 31.

PROBLEMS *

7-1. Calculate the molar concentration of a silver nitrate solution, 1.00 mL of which will react with 2.00 mg of

* (a) KSCN.
* (b) $MgBr_2$.
* (c) Na_3PO_4.
(d) $K_2Ni(CN)_4$[product: $Ag(CN)_2^-$].
(e) $AlCl_3$.
(f) H_2S.

* 7-2. A Mohr titration of a 0.4137-g sample required 35.82 mL of standard 0.0995 M $AgNO_3$. Calculate the percentage of chloride in the sample.

7-3. The arsenic in an 8.67-g sample of pesticide was converted to arsenate and precipitated as Ag_3AsO_4 with 50.0 mL of 0.02504 M silver nitrate. The excess Ag^+ was then titrated with 3.64 mL of 0.05441 M potassium thiocyanate. Calculate the percentage of As_2O_3 in the sample.

* 7-4. Lead(II) can be titrated with standard potassium chromate (product: $PbCrO_4$); an adsorption indicator signals the end point. A 1.0412-g mineral sample consisting

---

[3] K. J. Bush and H. Diehl, *J. Chem. Educ.* **56**, 54 (1979).

* Answers to problems or parts of problems marked with an asterisk are to be found at the end of the book.

mainly of galena, PbS, was decomposed by treatment with acid and subsequently titrated with 38.34 mL of 0.1033 M K_2CrO_4. Calculate
(a) the PbS titer for the chromate solution.
(b) the percentage of galena in the sample.

7-5. A 50.0-mL aliquot of 0.1124 M $AgNO_3$ was added to an ammoniacal solution containing 0.1589 g of a sample of propargyl alcohol. Reaction:

$$2Ag^+ + NO_3^- + HC\equiv C—CH_2OH \rightleftharpoons AgC\equiv C—CH_2OH \cdot AgNO_3(s) + H^+$$

When reaction was complete, the excess silver ion was titrated with 13.86 mL of 0.0984 M KSCN. Calculate the percentage of propargyl alcohol (fw = 56.06) in the sample.

* 7-6. Gallium(III) is quantitatively precipitated with ferrocyanide ion, according to the reaction

$$4Ga^{3+} + 3Fe(CN)_6^{4-} \longrightarrow Ga_4[Fe(CN)_6]_3(s)$$

Calculate the percentage of gallium(III) chloride if a 1.750-g sample required a 28.85-mL titration with 0.07640 M $K_4Fe(CN)_6$.

7-7. A potassium ferrocyanide solution was standardized against a 0.2147-g sample of pure ZnO; 41.83 mL were required for the titration. The same solution was then used to determine the percentage of zinc carbonate in a 0.2754-g sample of the mineral smithsonite; this titration required 34.72 mL. The same reaction was involved in both the standardization and the titration:

$$2Fe(CN)_6^{4-} + 3Zn^{2+} + 2K^+ \longrightarrow K_2Zn_3[Fe(CN)_6]_2(s)$$

What was the percent $ZnCO_3$ in the sample?

* 7-8. The monochloroacetic acid preservative in a carbonated beverage was extracted from a 100.0-mL sample into diethyl ether. It was then returned to aqueous solution as chloroacetate ion by a second extraction with 1 M NaOH. The extract was acidified and treated with 25.00 mL of 0.08691 M $AgNO_3$. Reaction:

$$ClCH_2COOH + Ag^+ + H_2O \longrightarrow HOCH_2COOH + H^+ + AgCl(s)$$

After filtration of the AgCl, titration of the filtrate and washings required 11.92 mL of an NH_4SCN solution. Titration of a blank taken through the entire process used 23.34 mL of thiocyanate. Calculate the weight (in mg) of chloroacetic acid in the sample.

* 7-9. A 0.1064-g pesticide sample was decomposed by the action of sodium biphenyl in toluene. The liberated chloride was extracted with dilute HNO_3 and titrated with 13.28 mL of 0.0547 M $AgNO_3$ by the Mohr method. Express the results of this analysis in terms of percent aldrin, $C_{12}H_8Cl_6$ (fw = 364.9).

7-10. A carbonate fusion was needed to free the bismuth from a 0.4632-g sample containing the mineral eulytite, $2Bi_2O_3 \cdot 3SiO_2$. The fused mass was dissolved in dilute acid, following which the bismuth(III) was titrated with 36.27 mL of 0.02849 M NaH_2PO_4 solution:

$$Bi^{3+} + H_2PO_4^- \longrightarrow BiPO_4(s) + 2H^+$$

Calculate the percentage purity of eulytite (fw = 1112) in the sample.

* 7-11. A 20-tablet sample of soluble saccharin was treated with 25.00 mL of 0.0624 M $AgNO_3$:

Titration of the filtrate and washings after removal of the solid required 1.81 mL of 0.0583 M KSCN. Calculate the average number of milligrams of saccharin (fw = 205.2) in each tablet.

7-12. The Fajans method is to be used in the routine analysis of solids for their chloride content. It is desired that the volume of standard silver nitrate used in these titrations to be numerically equal to the percentage of chloride when 0.400-g samples are taken for analysis. What will be the molarity of the $AgNO_3$ solution?

* 7-13. The Association of Official Analytical Chemists recommends a Volhard titration for analysis of the insecticide heptachlor, $C_{10}H_5Cl_7$. The percentage of heptachlor is given by the calculation

$$\text{percent heptachlor} = \frac{(\text{mL}_{Ag} \times M_{Ag} - \text{mL}_{SCN} \times M_{SCN}) \times 37.33}{\text{wt of sample}}$$

What does this calculation reveal concerning the stoichiometry of this titration?

7-14. An analysis for borohydride ion is based upon its reaction with silver ion:

$$BH_4^- + 8Ag^+ + 8OH^- \longrightarrow H_2BO_3^- + 8Ag(s) + 5H_2O$$

The purity of a quantity of potassium borohydride for use in an organic synthesis was established by diluting 0.3405 g of the material to exactly 250 mL, treating a 50.0-mL aliquot with 50.0 mL of 0.1978 M $AgNO_3$, and titrating the excess silver ion with 1.36 mL of 0.0512 M KSCN. Calculate the percentage purity of the KBH_4.

7-15. What volume of 0.0512 M KSCN would have been needed if the analysis in Problem 7-14 had been completed by filtering off the metallic silver, dissolving it in acid, diluting to 250 mL, and titrating a 50.0-mL aliquot?

* 7-16. A 100-mL sample of brackish water was made ammoniacal, following which the sulfide it contained was titrated with 6.47 mL of 0.0164 M $AgNO_3$. Net reaction:

$$2Ag^+ + S^{2-} \longrightarrow Ag_2S(s)$$

Calculate the parts per million of H_2S in the water.

7-17. A 1.50-L water sample was evaporated to a small volume and treated with an excess of sodium tetraphenylboron. The precipitated potassium tetraphenylboron was filtered and then redissolved in acetone. The analysis was completed by a Mohr titration, with 24.90 mL of 0.0405 M $AgNO_3$ being used. Net reaction:

$$KB(C_6H_5)_4 + Ag^+ \longrightarrow AgB(C_6H_5)_4 + K^+$$

Express the results of this analysis in terms of parts per million of potassium (that is, mg K/L).

* 7-18. The action of an alkaline iodine solution upon the rodenticide warfarin, $C_{19}H_{16}O_4$ (fw = 308.3), results in formation of one mole of iodoform, CHI_3 (fw = 393.7) from each mole of the parent compound. Analysis for warfarin can then be based upon the reaction between CHI_3 and silver ion:

$$CHI_3 + 3Ag^+ + H_2O \longrightarrow 3AgI(s) + 3H^+ + CO(g)$$

The iodoform produced from a 15.07-g sample was treated with 25.00 mL of 0.0339 M $AgNO_3$, following which the excess Ag^+ was titrated with 2.85 mL of 0.0401 M KSCN. Calculate the percentage of warfarin in the sample.

* 7-19. A 2.050-g sample containing chloride and perchlorate was dissolved in sufficient water to give 250.0 mL of solution. A 50.0-mL aliquot required 14.67 mL of 0.0842 M $AgNO_3$ to titrate the chloride. A second 50.0-mL aliquot was treated with $V_2(SO_4)_3$ to reduce the perchlorate to chloride:

$$ClO_4^- + 4V_2(SO_4)_3 + 4H_2O \longrightarrow Cl^- + 12SO_4^{2-} + 8VO^{2+} + 8H^+$$

Titration of the reduced sample required 39.85 mL of the silver nitrate solution. Calculate the respective percentages of chloride and perchlorate in the sample.

7-20. A 2.336-g sample containing KCl, K_2SO_4, and inert materials was dissolved in sufficient water to give 250.0 mL of solution. A Mohr titration of a 50.00-mL aliquot required 22.10 mL of 0.1174 M $AgNO_3$. A second 50.00-mL aliquot was treated with 40.00 mL of 0.1083 M $(C_6H_5)_4BNa$. Reaction:

$$(C_6H_5)_4BNa + K^+ \longrightarrow (C_6H_5)_4BK(s) + Na^+$$

The solid was filtered, redissolved in acetone, and titrated with 32.37 mL of the $AgNO_3$ solution (see Problem 7-17). Calculate the percentages of KCl and K_2SO_4 in the sample.

* 7-21. A 0.2185-g sample containing only KCl and K_2SO_4 yielded a precipitate of $(C_6H_5)_4BK$ which—after isolation and solution in acetone—required a Fajans titration involving 27.04 mL of 0.1027 M $AgNO_3$. Reaction:

$$(C_6H_5)_4BK + Ag^+ \longrightarrow (C_6H_5)_4BAg(s) + K^+$$

Calculate the respective percentages of KCl and K_2SO_4 in the sample.

7-22. For each of the following precipitation titrations, calculate the concentration of the cation and the anion at reagent volumes corresponding to the equivalence point as well as ± 20.0 mL, ± 10.0 mL, and ± 1.00 mL of equivalence. Construct a titration curve from the data, plotting the p-function of the cation versus reagent volume.

* (a) 20.0 mL of 0.0400 M $AgNO_3$ with 0.0200 M NH_4SCN.
  (b) 30.0 mL of 0.0400 M $AgNO_3$ with 0.0200 M KI.
* (c) 30.0 mL of 0.0800 M $AgNO_3$ with 0.0800 M NaCl.
  (d) 25.0 mL of 0.200 M Na_2SO_4 with 0.100 M $Pb(NO_3)_2$.
* (e) 60.0 mL of 0.0300 M $BaCl_2$ with 0.0600 M Na_2SO_4.
  (f) 50.0 mL of 0.100 M NaI with 0.200 M $TlNO_3$ (K_{sp} for TlI = 6.5×10^{-8}).

7-23. Calculate the silver ion concentration after the addition of 5.00*, 15.0, 25.0, 30.0, 35.0, 39.0, 40.0*, 41.0*, 45.0, and 50.0 mL of 0.100 M $AgNO_3$ to 50.0 mL of 0.0800 M KBr. Construct a titration curve from these data, plotting pAg as a function of titrant volume.

7-24. Calculate the Hg_2^{2+} concentration after the addition of 0, 10.0, 20.0, 30.0*, 35.0, 39.0, 40.0*, 41.0, 45.0, and 50.0* mL of 0.100 M NaCl to 80.0 mL of a solution that is 0.0250 M with respect to Hg_2^{2+}. For the process,

$$Hg_2Cl_2 \rightleftharpoons Hg_2^{2+} + 2Cl^-$$

K_{sp} is equal to 1.3×10^{-18}. Construct a titration curve from these data, plotting pHg_2 as a function of titrant volume. NOTE: No evidence exists for the intermediate species Hg_2Cl^+.

theory of neutralization

titrations for simple systems

8

End point detection in a neutralization titration is ordinarily based upon the abrupt change in pH that occurs in the vicinity of the equivalence point. The pH range within which such a change occurs varies from titration to titration and is determined both by the nature and the concentration of the analyte as well as the titrant. The selection of an appropriate indicator and the estimation of the titration error require knowledge of the pH changes which occur throughout the titration. Thus, we need to know how neutralization titration curves are derived.

This chapter is concerned with simple acids or bases which produce a single hydronium or hydroxide ion per molecule. In Chapter 9 titration curves for more complex systems that contain more than one acidic or basic group will be considered.

In preparation for the discussion that follows, the student may find it helpful to review the material on acids and bases in Sections 2A-2, 2D-3, and 2D-5.

8A REAGENTS AND INDICATORS FOR NEUTRALIZATION TITRATIONS

The standard reagents employed for neutralization titrations are always strong acids or bases because these substances react more completely than their weaker counterparts; in common with precipitation titrations, the more complete the reaction, the greater will be the change in p-value and thus the more satisfactory will be the end point.

The most common reagents for preparation of standard solutions include hydrochloric, perchloric, and sulfuric acids and sodium, potassium, and barium hydroxides.

8A-1 THEORY OF INDICATOR BEHAVIOR

Acid-base indicators are generally weak organic acids or bases which, upon dissociation or association, undergo internal structural changes that give rise to alterations in color. We can symbolize the typical reaction of an acid-base indicator as follows:

$$H_2O + HIn \rightleftharpoons H_3O^+ + In^-$$

(acid color)   (base color)

or

$$In + H_2O \rightleftharpoons InH^+ + OH^-$$

(base color)   (acid color)

For the first indicator, HIn will be the major constituent in strongly acidic solutions and will be responsible for the "acid color" of the indicator, whereas In^- will represent its "base color." For the second indicator, the species In will predominate in basic solutions and thus be responsible for the "base color," whereas InH^+ will constitute the "acid color."

Equilibrium expressions for these processes are:

$$\frac{[H_3O^+][In^-]}{[HIn]} = K_a$$

and

$$\frac{[InH^+][OH^-]}{[In]} = K_b$$

A solution containing an indicator will show continuous changes in color with variations in pH. The human eye is not very sensitive to these changes, however. Typically, a tenfold excess of one form is required before the color of that species appears predominant to the observer; further increases in the ratio have no detectable effect. It is only in the region where the ratio varies from a tenfold excess of one form to a similar excess of the other that the color of the solution appears to change. Thus, the subjective "color change" involves a major alteration in the position of the indicator equilibrium. Using HIn as an example, we may write that the indicator exhibits its pure acid color to the average observer when

$$\frac{[In^-]}{[HIn]} \cong \frac{1}{10}$$

and its basic color when

$$\frac{[In^-]}{[HIn]} \cong \frac{10}{1}$$

The color appears to be intermediate for ratios between these two values. These numerical estimates, of course, represent average behavior only; some indicators require smaller ratio changes and others larger. Furthermore, considerable variation in the ability to judge colors exists among observers; indeed, a color-blind person may be unable to discern any change.

If the two concentration ratios are substituted into the dissociation constant expression for the indicator, the range of hydronium ion concentrations needed to effect the complete indicator color change can be evaluated. Thus, for the full acid color,

$$\frac{[H_3O^+][In^-]}{[HIn]} = \frac{[H_3O^+]1}{10} = K_a$$

$$[H_3O^+] = 10K_a$$

and similarly for the full basic color,

$$\frac{[H_3O^+]10}{1} = K_a$$

$$[H_3O^+] = \frac{1}{10}K_a$$

To obtain the indicator range, we take the negative logarithms of the two expressions. That is,

$$indicator\ pH\ range = -\log 10K_a\ to\ -\log\frac{K_a}{10}$$
$$= -1 + pK_a\ to\ -(-1) + pK_a$$
$$= pK_a \pm 1$$

Thus, the typical indicator with an acid dissociation constant of 1×10^{-5} will show a complete color change when the pH of the solution in which it is dissolved changes from 4 to 6. A similar relationship is easily derived for an indicator of the basic type.

8A-2 TYPES OF ACID-BASE INDICATORS

A list of compounds possessing acid-base indicator properties is large and includes a number of organic structures. An indicator covering almost any desired pH range can ordinarily be found. A few common indicators are given in Table 8-1.

The majority of acid-base indicators possess structural properties that permit classification into perhaps half a dozen categories.[1] Three of these classes are described in the following paragraphs.

Phthalein indicators. Most phthalein indicators are colorless in moderately acidic solutions and exhibit a variety of colors in alkaline media. Their colors tend to fade slowly in strongly alkaline solutions, which is an inconvenience in some applications. As a group, the phthaleins are sparingly soluble in water; ethanol is the ordinary solvent for indicator solutions.

---

[1] See: I. M. Kolthoff and C. Rosenblum, *Acid-Base Indicators*, Chapter 5. Macmillian, 1937.

Table 8-1   Some Important Acid-Base Indicators*

Common Name	Transition Range, pH	pK_a**	Color Change†	Indicator Type‡
Thymol blue	1.2–2.8	1.65	R–Y	1
	8.0–9.6	8.90	Y–B	
Methyl yellow	2.9–4.0		R–Y	2
Methyl orange	3.1–4.4	3.46§	R–O	2
Bromocresol green	3.8–5.4	4.66	Y–B	1
Methyl red	4.2–6.3	5.00§	R–Y	2
Bromocresol purple	5.2–6.8	6.12	Y–P	1
Bromothymol blue	6.2–7.6	7.10	Y–B	1
Phenol red	6.8–8.4	7.81	Y–R	1
Cresol purple	7.6–9.2		Y–P	1
Phenolphthalein	8.3–10.0		C–P	1
Thymolphthalein	9.3–10.5		C–B	1
Alizarin yellow GG	10–12		C–Y	2

* Data from C. A. Streuli, in *Handbook of Analytical Chemistry*, L. Meites, Ed., pp. 3–35 and 3–36. New York: McGraw-Hill, 1963.
** At ionic strength of 0.1.
† B = blue; C = colorless; O = orange; P = purple; R = red; Y = yellow.
‡ (1) acid type: $HIn + H_2O \rightleftharpoons H_3O^+ + In^-$
  (2) base type: $In + H_2O \rightleftharpoons InH^+ + OH^-$
§ For reaction, $InH^+ + H_2O \rightleftharpoons H_3O^+ + In$

     The best-known phthalein indicator is *phenolphthalein,* whose structures may be represented as

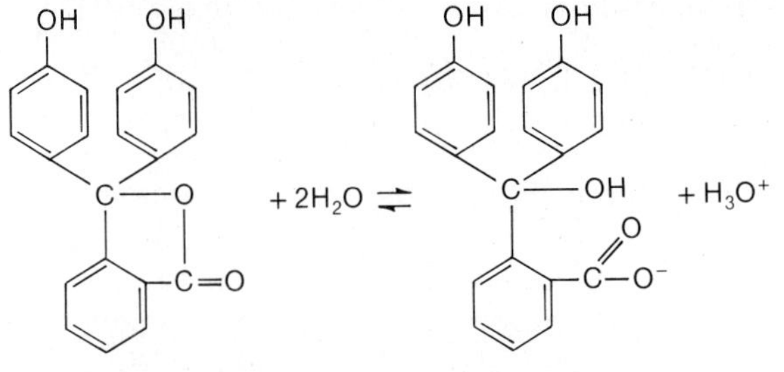

colorless   colorless

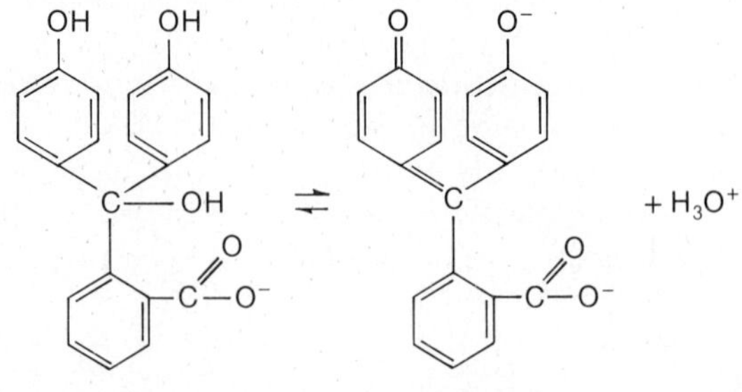

colorless   red

Note that the second equilibrium results in the formation of a quinoid ring, a structure that is often associated with color. Significant concentrations of the colored ion appear in the pH range between 8.0 and 9.8; the pH at which the color is first detectable depends upon the concentration of the indicator and the visual acuity of the observer.

The other phthalein indicators differ only in that the phenolic rings contain additional functional groups; *thymolphthalein*, for example, has two alkyl groups on each ring. The basic structural alterations associated with the color change of this indicator are similar to those of phenolphthalein.

Sulfonphthalein indicators. Many of the sulfonphthaleins exhibit two useful color-change ranges: one occurs in rather acidic solutions and the other in neutral or moderately basic media. In contrast to the phthaleins, the basic color shows good stability toward strong alkali.

The sodium salts of the sulfonphthaleins are ordinarily used for the preparation of indicator solutions owing to the appreciable acidity of the parent molecule. Solutions can be prepared directly from the sodium salt or indirectly by dissolving the sulfonphthalein itself in an appropriate volume of dilute aqueous sodium hydroxide.

The simplest sulfonphthalein indicator is *phenolsulfonphthalein*, known also as *phenol red*. The principal equilibria for this compound are

Only the second of the two color changes, occurring in the pH range between 6.4 and 8.0, is useful.

Substitution of halogens or alkyl groups for the hydrogens in the phenolic rings of the parent compound yields sulfonphthaleins that differ in color and pH range.

Azo indicators. Most azo indicators exhibit a color change from red to yellow with increasing basicity; their transition ranges are generally on the acid side of neutrality. The most commonly encountered examples are *methyl orange* and *methyl red;* the behavior of the former is described by the equations:

red

yellow

Methyl red is similar to methyl orange except that the sulfonic acid group is replaced by a carboxylic acid group. Variations in the substituents on the amino nitrogen and in the rings give rise to a series of indicators with slightly different properties.

8A-3 TITRATION ERRORS WITH ACID-BASE INDICATORS
Two types of titration errors can be distinguished. The first is a determinate error which occurs when the pH at which the indicator changes color differs from the pH at chemical equivalence. This type of error can usually be minimized through judicious indicator selection; often a blank will provide an appropriate correction if such is necessary.

The second type is an indeterminate error that has as its origin the limited ability of the eye to distinguish reproducibly the point at which a color change occurs. The magnitude of this error will depend upon the change in pH per milliliter of reagent at the equivalence point, the concentration of the indicator, and the sensitivity of the eye to the two indicator colors. On the average, the visual uncertainty with an acid-base indicator is in the range between ±0.5 and ±1 pH unit. This uncertainty can often be decreased to as little as ±0.1 pH unit by matching the color of the solution being titrated with that of a reference standard containing a similar amount of indicator at the appropriate pH. It must be understood that these uncertainties are approximations that will vary considerably from indicator to indicator as well as from person to person.

8A-4 VARIABLES THAT INFLUENCE THE BEHAVIOR OF INDICATORS
The pH interval over which a given indicator exhibits a color change is influenced by the temperature, the ionic strength of the medium, and the presence of organic solvents and colloidal particles. Some of these effects, particularly the last two, can cause the transition range to shift by one or more pH units.[2]

---

[2] For a discussion of these effects, see H. A. Laitinen, *Chemical Analysis*, pp. 50–55. New York: McGraw-Hill, 1960.

8B TITRATION CURVES FOR STRONG ACIDS OR STRONG BASES

When both reagent and analyte are strong, the net neutralization reaction can be expressed as

$$H_3O^+ + OH^- = 2H_2O$$

and derivation of a curve for such a titration is analogous to that for a precipitation titration.

8B-1 TITRATION OF STRONG ACID WITH A STRONG BASE

The hydronium ions in an aqueous solution of a strong acid come from two sources, namely, (1) the reaction of the solute with water and (2) the dissociation of water itself. In any but the most dilute solutions, however, the contribution from the solute far exceeds that from the solvent. Thus, for a solution of HCl having a concentration greater than about $1 \times 10^{-6}\ M$, we may write

$$[H_3O^+] = M_{HCl} \qquad (8\text{-}1)$$

For a solution of a strong base such as sodium hydroxide, an analogous situation exists. That is,

$$[OH^-] = M_{NaOH} \qquad (8\text{-}2)$$

Clearly, for the completely dissociated base $Ba(OH)_2$

$$[OH^-] = 2 \times M_{Ba(OH)_2}$$

A useful relationship for calculating the pH of basic solutions can be obtained by taking the negative logarithm of each side of the ion product-constant expression for water. That is,

$$-\log K_w = -\log ([H_3O^+][OH^-]) = -\log [H_3O^+] - \log [OH^-]$$

$$pK_w = pH + pOH$$

At 25°C, pK_w has a value of 14.00.

As will be seen from the following example, the derivation of a curve for the titration of a strong acid with a strong base is simple because the concentration of the hydronium ion or the hydroxide ion is obtained directly from the molar concentration of the acid or base that is present in excess.

EXAMPLE 8-1. Derive a curve for the titration of 50.00 mL of 0.0500 M HCl with 0.1000 M NaOH. Round pH data to two places to the right of the decimal point.

Initial point. The solution is $5.00 \times 10^{-2}\ M$ in HCl. Since HCl is completely dissociated,

$$[H_3O^+] = 5.00 \times 10^{-2}$$

$$pH = -\log (5.00 \times 10^{-2}) = -\log 5.00 - \log 10^{-2}$$

$$= -0.699 + 2 = 1.301 = 1.30$$

pH after addition of 10.00 mL NaOH. The volume of the solution is now 60.00 mL and part of the HCl has been neutralized. Thus,

$$[H_3O^+] = \frac{50.00 \times 0.0500 - 10.00 \times 0.1000}{60.00} = 2.50 \times 10^{-2}$$

$$pH = 2 - \log 2.50 = 1.60$$

Additional data to define the curve in the region short of the equivalence point are derived in the same way. The results of such calculations are given in column 2 of Table 8-2.

pH after addition of 25.00 mL NaOH. Here, the solution contains neither an excess of HCl nor of NaOH; thus, the pH is governed by the dissociation of water

$$[H_3O^+] = [OH^-] = \sqrt{K_w} = 1.00 \times 10^{-7}$$

$$pH = 7.00$$

pH after addition of 25.10 mL NaOH. Here,

$$M_{NaOH} = \frac{25.10 \times 0.1000 - 50.00 \times 0.0500}{75.10} = 1.33 \times 10^{-4}$$

Provided the concentration of OH^- from the dissociation of water is negligible with respect to M_{NaOH},

$$[OH^-] = M_{NaOH} = 1.33 \times 10^{-4}$$

$$pOH = -\log(1.33 \times 10^{-4}) = 3.88$$

$$pH = 14.00 - 3.88 = 10.12$$

Additional data for this titration, calculated in the same way, are given in column 2 of Table 8-2.

Effect of concentration. The effects of reagent and analyte concentrations on neutralization titration curves are shown by the two sets of data in Table 8-2; these data are plotted in Figure 8-1.

For the titration with $0.1\,M$ NaOH (curve A), the change in pH in the equivalence-point region is large. With $0.001\,M$ NaOH, the change is markedly less but still pronounced.

Figure 8-1 shows that the selection of an indicator is not critical when the reagent concentration is approximately $0.1\,M$. Here, the volume differences among titrations with the three indicators are of the same magnitude as the uncertainties associated with the reading of the buret and are thus negligible. On the other hand, bromocresol green would be clearly unsuited for a titration involving the $0.001\,M$

Table 8-2   Changes in pH During the Titration of a Strong Acid With a Strong Base

Volume NaOH, mL	50.00 mL of 0.0500 M HCl With 0.1000 M NaOH pH	50.00 mL of 0.000500 M HCl With 0.001000 M NaOH pH
0.00	1.30	3.30
10.00	1.60	3.60
20.00	2.15	4.15
24.00	2.87	4.87
24.90	3.87	5.87
25.00	7.00	7.00
25.10	10.12	8.12
26.00	11.12	9.12
30.00	11.80	9.80

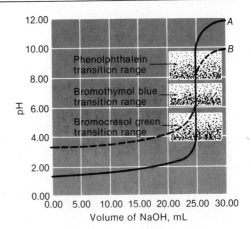

FIGURE 8-1   Curve for the titration of HCl
with standard NaOH. Curve A: 50.00 mL
0.0500 M HCl with 0.1000 M NaOH. Curve B:
50.00 mL 0.000500 M HCl with 0.001000 M
NaOH.

reagent; not only would the color change occur continuously over a substantial
range of titrant volumes, but also the transition to the alkaline form would be essentially complete before the equivalence point was reached; a significant determinate error would result. The use of phenolphthalein would be subject to similar
objections. Of the three indicators, only bromothymol blue would provide a satisfactory end point with a minimal determinate titration error.

8B-2 TITRATION OF A STRONG BASE WITH A STRONG ACID
An analogous situation exists for the titration of strong bases with strong acids. In
the region short of the equivalence point, the solution is highly alkaline, the hydroxide ion concentration being numerically equal to the normality of the base.
The solution is neutral at the equivalence point for precisely the same reason noted
previously. Finally, the solution becomes acidic in the region beyond the equivalence point; here, the pH is computed from the excess of strong acid that has been
introduced. A curve for the titration of a strong base with 0.1 M hydrochloric acid is
shown in Figure 8-5 (p. 216). Indicator selection is based upon the same considerations as noted for the titration of a strong acid with a strong base.

8C PROPERTIES OF WEAK ACID AND WEAK BASE SYSTEMS
In generating a titration curve for a weak acid, it is necessary to evaluate the pH for
a solution containing the acid by itself, its conjugate base by itself, and mixtures of
these two solutes; derivation of a titration curve for a weak base involves analogous
calculations. The first two of these calculations were considered in Section 2D-5;
derivation of the pH for solutions containing appreciable quantities of conjugate
acid-base pairs is treated in the section that follows.

8C-1 THE pH OF SOLUTIONS CONTAINING CONJUGATE ACID-BASE PAIRS
A solution containing a conjugate acid-base pair may be acidic, neutral, or basic depending upon the position of two competitive equilibria. For a solution of the weak
acid HA and its sodium salt NaA, these equilibria are

$$HA + H_2O \rightleftharpoons H_3O^+ + A^- \qquad \frac{[H_3O^+][A^-]}{[HA]} = K_a \tag{8-3}$$

and

$$A^- + H_2O \rightleftharpoons OH^- + HA \qquad \frac{[OH^-][HA]}{[A^-]} = K_b = \frac{K_w}{K_a} \tag{8-4}$$

If the first equilibrium lies farther to the right than the second, the solution will be acidic. Conversely, if the second equilibrium is more favorable, the solution will be basic. From the two equilibrium constant expressions, it is evident that the relative concentrations of the hydronium and hydroxide ions will depend not only upon the magnitudes of K_a and K_b but also upon the ratio between the concentrations of the acid and its conjugate base in the solution.

A similar situation prevails for a weak base and its conjugate acid. Thus, for a solution containing both ammonia and ammonium chloride the equilibria are

$$NH_3 + H_2O \rightleftharpoons NH_4^+ + OH^-$$

$$\frac{[NH_4^+][OH^-]}{[NH_3]} = K_b = 1.76 \times 10^{-5}$$

and

$$NH_4^+ + H_2O \rightleftharpoons NH_3 + H_3O^+$$

$$\frac{[H_3O^+][NH_3]}{[NH_4^+]} = \frac{K_w}{K_b} = K_a$$

$$K_a = \frac{1.00 \times 10^{-14}}{1.76 \times 10^{-5}} = 5.68 \times 10^{-10}$$

We see that K_b is much larger than K_a; thus, the first equilibrium ordinarily predominates, and the solutions are basic. When the ratio of $[NH_4^+]$ to $[NH_3]$ becomes greater than about 200, however, the difference in equilibrium constants is offset, and such solutions are acidic.

Calculation of the pH of a solution containing a weak acid and its conjugate base.   Consider a solution that has a molar concentration of acid M_{HA} and a molar concentration of conjugate base M_{NaA}. The important equilibria and equilibrium constant expressions are given in Equations 8-3 and 8-4.

The reaction shown in Equation 8-3 tends to decrease the concentration of HA by an amount equal to $[H_3O^+]$. Reaction 8-4, conversely, increases the concentration of HA by an amount equal to $[OH^-]$. Thus, the equilibrium concentration of HA is given by

$$[HA] = M_{HA} - [H_3O^+] + [OH^-] \tag{8-5}$$

Similarly, the reaction in Equation 8-3 increases $[A^-]$ by an amount equal to $[H_3O^+]$, whereas reaction 8-4 decreases $[A^-]$ by an amount corresponding to $[OH^-]$. Thus,

$$[A^-] = M_{NaA} + [H_3O^+] - [OH^-] \tag{8-6}$$

Equations 8-5 and 8-6 are also readily derived from mass- and charge-balance considerations.[3]

Owing to the inverse relationship that exists between $[H_3O^+]$ and $[OH^-]$, it will always be possible to eliminate one or the other from Equations 8-5 and 8-6. Moreover, the *difference* in concentration between these two species is usually so small relative to the molar concentrations of acid and conjugate base that Equations 8-5 and 8-6 simplify to

$$[HA] \cong M_{HA} \tag{8-7}$$

$$[A^-] \cong M_{NaA} \tag{8-8}$$

Substitution of these quantities into the dissociation constant expression and rearrangement yield

$$[H_3O^+] = K_a \frac{M_{HA}}{M_{NaA}} \tag{8-9[4]}$$

The assumption leading to Equations 8-7 and 8-8 sometimes breaks down with acids or bases that have dissociation constants greater than about 10^{-3} or where the molar concentration of either the acid or its conjugate base (or both) is very small. In these circumstances, either $[OH^-]$ or $[H_3O^+]$ must be retained in Equations 8-5

---

[3] The [HA] and $[A^-]$ in the solution arise from two sources, the HA and the NaA that have been introduced in known molar amounts. Thus,

$$M_{HA} + M_{NaA} = [HA] + [A^-]$$

Electrical neutrality considerations require that

$$[Na^+] + [H_3O^+] = [A^-] + [OH^-]$$

but

$$[Na^+] = M_{NaA}$$

Therefore,

$$M_{NaA} + [H_3O^+] = [A^-] + [OH^-]$$

which rearranges to Equation 8-6. That is,

$$[A^-] = M_{NaA} + [H_3O^+] - [OH^-]$$

Subtraction of the first equation from the fourth gives upon rearrangement

$$[HA] = M_{HA} - [H_3O^+] + [OH^-]$$

which is identical to Equation 8-5.

[4] An alternative form of Equation 8-9 is frequently encountered in the biological literature and biochemical texts. It is obtained by expressing each term in the form of its negative logarithm and inverting the concentration ratio to keep all signs positive. Thus,

$$-\log [H_3O^+] = -\log K_a - \log \frac{M_{HA}}{M_{NaA}}$$

Therefore,

$$pH = pK_a + \log \frac{M_{NaA}}{M_{HA}} \tag{8-10}$$

This expression is known as the *Henderson-Hasselbalch* equation.

and 8-6 depending upon whether the solution is acidic or basic. In any case, Equations 8-7 and 8-8 should always be used initially. The provisional values for $[H_3O^+]$ and $[OH^-]$ can then be employed to test the assumptions.

Within the limits imposed by the assumptions made in its derivation, Equation 8-9 states that the hydronium ion concentration of a solution containing a weak acid and its conjugate base is dependent only upon the *ratio* between the molar concentrations of these two solutes. Furthermore, this ratio remains *independent of dilution* because the concentration of each component changes in a proportionate manner upon a change in volume. Thus, *the hydronium ion concentration of a solution containing appreciable quantities of a weak acid and its conjugate base tends to be independent of dilution and depends only upon the ratio of molar concentrations between the two solutes.* This independence of pH from dilution is one manifestation of the *buffering* properties of such solutions.

EXAMPLE 8-2. What is the pH of a solution that is $0.400\,M$ in formic acid and $1.00\,M$ in sodium formate?

The equilibrium governing the hydronium ion concentration in this solution is

$$H_2O + HCOOH \rightleftharpoons H_3O^+ + HCOO^-$$

for which

$$K_a = \frac{[H_3O^+][HCOO^-]}{[HCOOH]} = 1.77 \times 10^{-4}$$

$$[HCOO^-] \cong M_{HCOO^-} = 1.00$$

$$[HCOOH] \cong M_{HCOOH} = 0.400$$

Substitution into Equation 8-9 gives

$$[H_3O^+] = 1.77 \times 10^{-4} \times \frac{0.400}{1.00}$$

$$= 7.08 \times 10^{-5}$$

Note that the assumption that $[H_3O^+] \ll$ the molarity of HCOOH and $HCOO^-$ is valid. Thus,

$$pH = -\log(7.08 \times 10^{-5}) = 4.15$$

If we compare the pH of this solution with that calculated for a $0.400\,M$ formic acid solution (Equation 2-18, p. 26), we find that the addition of sodium formate has had the effect of raising the pH from 2.08 to 4.15.

Calculation of the pH of a solution of a weak base and its conjugate acid.

The calculation of pH for a solution of a weak base and its conjugate acid is completely analogous to that discussed in the preceding section.

EXAMPLE 8-3. Calculate the pH of a solution that is $0.280\,M$ in NH_4Cl and $0.0700\,M$ in NH_3.

The equilibrium of interest is

$$NH_3 + H_2O \rightleftharpoons NH_4^+ + OH^-$$

for which, $K_b = 1.76 \times 10^{-5}$.

As before, we assume that

$$[NH_3] = M_{NH_3} = 0.0700$$

$$[NH_4^+] = M_{NH_4^+} = 0.280$$

A provisional value for OH^- is then obtained by substituting these values into the equilibrium constant expression,

$$\frac{0.280 \, [OH^-]}{0.0700} = 1.76 \times 10^{-5}$$

$$[OH^-] = 4.40 \times 10^{-6}$$

Clearly, the approximation is justified and

$$pOH = -\log (4.40 \times 10^{-6}) = 5.36$$

$$pH = 14.00 - 5.36 = 8.64$$

8C-2 THE pH OF VERY DILUTE OR VERY WEAK ACIDS OR BASES

The contribution of the solvent water to the pH of a solution of an acid or base is usually so small that it need not be considered in pH calculations. If the concentration of a strong acid is less than 10^{-6}, however, this source of hydronium ions must be taken into account. Thus, for a very dilute solution of HCl, we must write

$$[H_3O^+] = M_{HCl} + [OH^-]$$

Here, the second term on the right provides a measure of the hydronium ions produced by the dissociation of the solvent. This equation can be rewritten in the form

$$[H_3O^+] = M_{HCl} + \frac{K_w}{[H_3O^+]}$$

which rearranges to

$$[H_3O^+]^2 - [H_3O^+]M_{HCl} - K_w = 0$$

For a dilute solution of a very weak acid or base, an analogous situation prevails. Thus for a very weak acid such as phenol, the hydronium ion concentration is given by

$$[H_3O^+] = [A^-] + [OH^-]$$

where $[A^-]$ provides a measure of the concentration of hydronium ions derived from the weak acid and $[OH^-]$ is a measure of the concentration due to the dissociation of water. Ordinarily, $[A^-]$ can be expressed in terms of the analytical concentration of the weak acid (p. 26) and will be given by

$$[A^-] = \frac{K_a M_{HA}}{[H_3O^+]}$$

Substitution of this and the ion product constant expression into the previous equation gives

$$[H_3O^+] = \frac{K_a M_{HA}}{[H_3O^+]} + \frac{K_w}{[H_3O^+]}$$

which rearranges to

$$[H_3O^+] = \sqrt{K_a M_{HA} + K_w}$$

Note that this equation differs from Equation 2-18 (p. 26) only in that K_w appears as an additional term. Ordinarily, K_w is small with respect to $K_a M_{HA}$ and can be neglected.

An analogous treatment gives the hydroxide ion concentration of very dilute or very weak bases.

EXAMPLE 8-4. Calculate the pH of a 0.0150 M solution of Na_2SO_4. The pertinent equilibria here are

$$SO_4^{2-} + H_2O \rightleftharpoons HSO_4^- + OH^-$$

$$2H_2O \rightleftharpoons H_3O^+ + OH^-$$

and

$$[OH^-] = [HSO_4^-] + [H_3O^+]$$

$$\frac{[HSO_4^-][OH^-]}{[SO_4^{2-}]} = \frac{1.00 \times 10^{-14}}{1.2 \times 10^{-2}} = 8.33 \times 10^{-13}$$

Furthermore,

$$[SO_4^{2-}] = 0.0150 - [HSO_4^-] \cong 0.0150$$

Substituting the latter two equations and the expression for K_w into the previous equation yields

$$[OH^-] = \frac{8.33 \times 10^{-13} \times 0.0150}{[OH^-]} + \frac{1.00 \times 10^{-14}}{[OH^-]}$$

$$[OH^-] = \sqrt{1.25 \times 10^{-14} + 1.00 \times 10^{-14}} = 1.50 \times 10^{-7}$$

$$pH = 14 - (-\log 1.50 \times 10^{-7}) = 7.18$$

8C-3 BUFFER SOLUTIONS

A *buffer solution* is defined as a solution that resists changes in pH as a result of either dilution or small additions of acids or bases. The most effective buffer solutions contain large and approximately equal concentrations of a conjugate acid-base pair.

Effect of dilution. The pH of a buffer solution remains essentially independent of dilution until its concentrations are decreased to the point where the approximations used to develop Equations 8-7 and 8-8 become invalid. The following example illustrates the behavior of a typical buffer during dilution.

EXAMPLE 8-5. Calculate the pH of the buffer described in the example on page 206 (original pH 4.15) upon dilution by a factor of (a) 50 and (b) 10,000.
(a) Upon the dilution by a factor of 50,

$$M_{HCOOH} = \frac{0.400}{50} = 8.00 \times 10^{-3}$$

$$M_{HCOONa} = \frac{1.00}{50} = 2.00 \times 10^{-2}$$

If we assume, as before, that $([H_3O^+] - [OH^-])$ is small with respect to the two molarities (p. 205), we obtain

$$\frac{[H_3O^+] \times 2.00 \times 10^{-2}}{8.00 \times 10^{-3}} = 1.77 \times 10^{-4}$$

$$[H_3O^+] = 7.08 \times 10^{-5}$$

$$\text{pH} = 4.15$$

Here, the error introduced by the assumption $7.08 \times 10^{-5} \ll 8.00 \times 10^{-3}$ and 2.00×10^{-2} is acceptable. (If the assumption is not made, one obtains $[H_3O^+] = 6.99 \times 10^{-5}$ and pH = 4.16.)

(b) Upon dilution by a factor of 10,000,

$$M_{\text{HCOOH}} = 4.00 \times 10^{-5}$$

$$M_{\text{HCOONa}} = 1.00 \times 10^{-4}$$

Here, the solute concentrations are of the same magnitude as $[H_3O^+]$, and the more exact statements given by Equations 8-5 and 8-6 must be used. That is,

$$[\text{HCOOH}] = 4.00 \times 10^{-5} - [H_3O^+] + [\cancel{OH^-}]$$

$$[\text{HCOO}^-] = 1.00 \times 10^{-4} + [H_3O^+] - [\cancel{OH^-}]$$

The solution is acidic; thus, $[OH^-] \ll [H_3O^+]$. Substitution of the foregoing relationships into the dissociation constant expression gives

$$\frac{[H_3O^+](1.00 \times 10^{-4} + [H_3O^+])}{4.00 \times 10^{-5} - [H_3O^+]} = 1.77 \times 10^{-4}$$

This equation rearranges to the quadratic form

$$[H_3O^+]^2 + 2.77 \times 10^{-4} [H_3O^+] - 7.08 \times 10^{-9} = 0$$

The solution is

$$[H_3O^+] = 2.36 \times 10^{-5}$$

$$\text{pH} = 4.63$$

Thus, a 10,000-fold dilution caused the pH to increase from 4.15 to 4.63, whereas a 50-fold dilution had essentially no effect.

Figure 8-2 contrasts the behavior of buffered and unbuffered solutions with dilution. For each the initial solute concentrations are $1.00\ M$. The resistance of the buffered solution to pH changes is clear.

Addition of acids and bases to buffers. The resistance of buffer mixtures to pH changes from added acids or bases is conveniently illustrated with examples.

EXAMPLE 8-6. Calculate the pH change that takes place when 100 mL of (a) $0.0500\ M$ NaOH and (b) $0.0500\ M$ HCl are added to 400 mL of a buffer solution that is $0.200\ M$ in NH_3 and $0.300\ M$ in NH_4Cl.

The initial pH of the solution is obtained by assuming that

$$[NH_3] \cong M_{\text{NH}_3} = 0.200$$

$$[NH_4^+] \cong M_{\text{NH}_4\text{Cl}} = 0.300$$

and substituting into the dissociation constant expression for NH_3

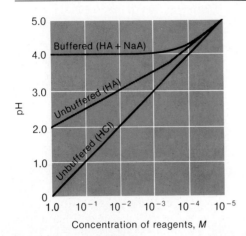

FIGURE 8-2 The effect of dilution on the pH of buffered and unbuffered solutions. The dissociation constant for HA is assumed to be 1.00×10^{-4}. Initial solute concentrations are $1.00\ M$.

$$\frac{0.300\,[\text{OH}^-]}{0.200} = 1.76 \times 10^{-5}$$

$$[\text{OH}^-] = 1.17 \times 10^{-5}$$

$$\text{pH} = 14.00 - (-\log 1.17 \times 10^{-5}) = 9.07$$

(a) Addition of NaOH converts part of the NH_4^+ in the buffer to NH_3:

$$\text{NH}_4^+ + \text{OH}^- \longrightarrow \text{NH}_3 + \text{H}_2\text{O}$$

The analytical concentrations of NH_3 and NH_4Cl then become

$$M_{\text{NH}_3} = \frac{400 \times 0.200 + 100 \times 0.0500}{500} = \frac{85.0}{500} = 0.170$$

$$M_{\text{NH}_4\text{Cl}} = \frac{400 \times 0.300 - 100 \times 0.0500}{500} = \frac{115}{500} = 0.230$$

When substituted into the dissociation constant expression, these values yield

$$[\text{OH}^-] = \frac{1.76 \times 10^{-5} \times 0.170}{0.230} = 1.30 \times 10^{-5}$$

$$\text{pH} = 14.00 - (-\log 1.30 \times 10^{-5}) = 9.11$$

and

$$\Delta\text{pH} = 9.11 - 9.07 = 0.04$$

(b) Addition of HCl converts part of the NH_3 to NH_4^+; thus,

$$\text{NH}_3 + \text{H}_3\text{O}^+ \longrightarrow \text{NH}_4^+ + \text{H}_2\text{O}$$

$$M_{\text{NH}_3} = \frac{400 \times 0.200 - 100 \times 0.0500}{500} = \frac{75}{500} = 0.150$$

$$M_{\text{NH}_4^+} = \frac{400 \times 0.300 + 100 \times 0.0500}{500} = \frac{125}{500} = 0.250$$

$$[\text{OH}^-] = 1.76 \times 10^{-5} \times \frac{0.150}{0.250} = 1.06 \times 10^{-5}$$

$$pH = 14.00 - (-\log 1.06 \times 10^{-5}) = 9.02$$
$$\Delta pH = 9.02 - 9.07 = -0.05$$

It is of interest to contrast the behavior of an unbuffered solution with a pH 9.07 to the buffered ones just considered. It is readily shown that the additions of the same quantity of base would increase the pH to 12.00—a pH change of 2.93 units. The addition of acid, on the other hand, would decrease the pH by slightly over 7 units.

Buffer capacity. The foregoing example demonstrates that a solution containing a conjugate acid-base pair possesses remarkable resistance to changes in pH. The ability of a buffer to prevent a significant change in pH is directly related to the total concentration of the buffering species as well as their concentration ratios. For example, 400-mL portions of a buffer formed by diluting the solutions described in the example by 10 would exhibit changes of about 0.4 to 0.5 pH unit when treated with the same amounts of sodium hydroxide or hydrochloric acid; the change for the more concentrated buffer was only about 0.04 to 0.05.

The *buffer capacity* of a solution is defined as the number of equivalents of strong acid or base needed to cause 1.00 L of the buffer to undergo a 1.00-unit change in pH. As we have seen, the buffer capacity is dependent upon the concentration of the conjugate acid-base pair. It is also dependent upon their concentration ratio and is at a maximum when this ratio is unity.[5]

Preparation of buffers. In principle, a buffer solution of any desired pH can be prepared by combining calculated quantities of a suitable conjugate acid-base pair. In practice, however, buffers prepared in this way are found to differ somewhat in pH from the predicted values. These differences arise in part from the uncertainties that exist in the numerical values for many dissociated constants and the simplifications used in calculations. Moreover, the ionic strength of a buffer is usually so high that good values for the activity coefficients of the ions in the solution cannot be obtained from the Debye-Hückel relationship. Thus, the ionic strength of the NH_3/NH_4^+ buffer considered in the example (p. 209) is about 0.30; the concentration equilibrium constant K_a' (see p. 119) would therefore be significantly larger than 1.76×10^{-5} and quite uncertain (about 4×10^{-5}).

Empirically derived recipes for preparation of buffer solutions of known pH are available in chemical handbooks and reference works.[6] Because of their widespread employment, two buffer systems deserve specific mention. McIlvaine buffers cover a pH range from about 2 to 8 and are prepared by mixing solutions of citric acid with disodium hydrogen phosphate. Clark and Lubs buffers, which encompass a pH range from 2 to 10, make use of three systems: phthalic acid/potassium hydrogen phthalate, potassium dihydrogen phosphate/dipotassium hydrogen phosphate, and boric acid/sodium borate.

---

[5] See: H. A. Laitimen, *Chemical Analysis*, p. 37. New York: McGraw-Hill, 1960.
[6] See, for example, L. Meites, Ed., *Handbook of Analytical Chemistry*, pp. **5**-112 and **11**-5 to **11**-7. New York: McGraw-Hill, 1963.

8C-4 TITRATION CURVES FOR WEAK ACIDS

Derivation of a curve for the titration of a weak acid (or weak base) requires a different calculation for each of four distinct situations. (1) At the outset, the solution contains only a weak acid or a weak base, and the pH is calculated from the concentration of that solute and the corresponding dissociation constant. (2) After various increments of reagent have been added (in quantities up to, but not including, an equivalent amount), the solution consists of a series of buffers; the pH of each can be calculated from the analytical concentrations of the product and the residual weak acid or base. (3) At the equivalence point, the solution contains the conjugate of the weak acid or base being titrated, and the pH is calculated from the concentration of this product. (4) Finally, beyond the equivalence point, the excess of strong acid or base that is used as titrant represses the basic or acidic character of the product to such an extent that the pH is governed largely by the concentration of the excess reagent.

Derivation. To illustrate these calculations, consider derivation of a curve for the titration of 50.00 mL of 0.1000 M acetic acid ($K_a = 1.75 \times 10^{-5}$) with 0.1000 M sodium hydroxide.

EXAMPLE 8-7.

Initial pH. Here, we must calculate the pH of a 0.1000 M solution of HOAc; using the method shown on page 261, a value of 2.88 is obtained.

pH after addition of 10.00 mL of reagent. A buffer solution consisting of NaOAc and HOAc has now been produced. The analytical concentrations of the two constituents are given by

$$M_{HOAc} = \frac{50.00 \times 0.1000 - 10.00 \times 0.1000}{60.00} = \frac{4.000}{60.00}$$

$$M_{NaOAc} = \frac{10.00 \times 0.1000}{60.00} = \frac{1.000}{60.00}$$

Upon substituting these concentrations into the dissociation constant expression for acetic acid, we obtain

$$\frac{[H_3O^+]\,1.000/\cancel{60.00}}{4.000/\cancel{60.00}} = K_a = 1.75 \times 10^{-5}$$

$$[H_3O^+] = 7.00 \times 10^{-5}$$

$$pH = 4.16$$

Calculations similar to this will delineate the curve throughout the buffer region. Data from such calculations are given in column 2 of Table 8-3. Note that when the acid has been 50% neutralized (in this particular titration, after an addition of exactly 25.00 mL of base), the analytical concentrations of acid and conjugate base are identical; within the limits of the usual approximations, so also are their molar concentrations. Thus, these terms cancel one another in the equilibrium constant expression, and the hydronium ion concentration is numerically equal to the dissociation constant; that is, the pH is equal to the pK_a. Likewise, in the titration of a weak base, the hydroxide ion concentration is numerically equal to the dissociation constant of the base at the midpoint.

Table 8-3 Changes in pH During the Titration of a Weak Acid With a Strong Base

Volume NaOH, mL	50.00 mL of 0.1000 M HOAc With 0.1000 M NaOH	50.00 mL of 0.001000 M HOAc With 0.001000 M NaOH
	pH	pH
0.00	2.88	3.91
10.00	4.16	4.30
25.00	4.76	4.80
40.00	5.36	5.38
49.00	6.45	6.46
49.90	7.46	7.47
50.00	8.73	7.73
50.10	10.00	8.09
51.00	11.00	9.00
60.00	11.96	9.96
75.00	12.30	10.30

EXAMPLE 8-8.

Equivalence point pH. At the equivalence point in the titration, the acetic acid has been converted to sodium acetate. The solution is therefore similar to one formed by dissolving that base in water, and the pH calculation is identical to that described on page 29 for a weak base. In the present example, the NaOAc concentration is 0.0500 M. Thus,

$$[OH^-] \cong \sqrt{\frac{K_w}{K_a} \times 0.0500} = \sqrt{\frac{1.00 \times 10^{-14} \times 0.0500}{1.75 \times 10^{-5}}}$$

$$[OH^-] = 5.34 \times 10^{-6}$$

$$pH = 14.00 - (-\log 5.34 \times 10^{-6})$$

$$= 8.73$$

pH after addition of 50.10 mL of base. After 50.10 mL of NaOH have been added, both the excess base and the acetate ion contribute to the hydroxide ion concentration. The dissociation of acetate is vanishingly small, however, because the excess of strong base represses the equilibrium. This fact becomes evident when we consider that the hydroxide ion concentration was only 5.34×10^{-6} M at the equivalence point of the titration; once an excess of strong base has been added, the contribution from the reaction of the acetate will be even smaller. Thus,

$$[OH^-] \cong M_{NaOH} = \frac{50.10 \times 0.1000 - 50.00 \times 0.1000}{100.1} = 1.00 \times 10^{-4}$$

$$pOH = 14.00 - (-\log 1.00 \times 10^{-4}) = 10.00$$

Note that titration curves for a weak acid and a strong base become identical with those for a strong acid with a strong base in the region slightly beyond the equivalence point.

Effect of concentration. The second and third columns of Table 8-3 contain pH data for the titration of 0.1000 M and of 0.001000 M acetic acid, respectively, with sodium hydroxide solutions of the same strengths. Substantial portions of the curve for the dilute solution required the use of more exact expressions for the concentration relationships; an example of this type of calculation appears on page 209.

Figure 8-3 is a plot of the data in Table 8-3. Note that the initial pH values are higher and the equivalence point pH is lower for the more dilute solution than for the 0.1000 M solution. At intermediate titrant volumes, however, the pH values differ only slightly. Acetic acid/sodium acetate buffers exist within this region; Figure 8-2 is graphical confirmation of the fact that the pH of buffers is largely independent of dilution.

Effect of dissociation constants. Titration curves for 0.1000 M solutions of acids with differing dissociation constants are shown in Figure 8-4. Note that the pH change in the equivalence point region becomes smaller as the acid becomes weaker; that is, as the reaction between the acid and the base becomes less complete. The relation between completeness of reaction and reagent concentrations illustrated by Figures 8-3 and 8-4 is analogous to these effects on precipitation titration curves (p. 181).

Indicator choice; feasibility of titration. Figures 8-3 and 8-4 clearly indicate that the choice of indicator for the titration of a weak acid is more limited than that for a strong acid. For example, bromocresol green is totally unsuited for titration of 0.1000 M acetic acid; nor would bromothymol blue be satisfactory because its full color change would occur over a range between about 47 and 50 mL of 0.1000 M base. It might still be possible to employ bromothymol blue by titrating to the full basic color of the indicator rather than to the change in color; this technique would require a comparison standard containing the same concentration of indicator as the solution being titrated. An indicator such as phenolphthalein, which changes color in the basic region, would be most satisfactory for this titration.

The end point pH change associated with the titration of 0.001000 M acetic acid (curve B, Figure 8-3) is so small that a significant titration error is likely to be introduced. By employing an indicator with a transition range between that of phenolphthalein and bromothymol blue and by using a color comparison standard,

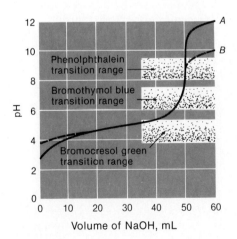

FIGURE 8-3 Curve for the titration of acetic acid with NaOH. Curve A: 0.1000 M acid with 0.1000 M base. Curve B: 0.001000 M acid with 0.001000 M base.

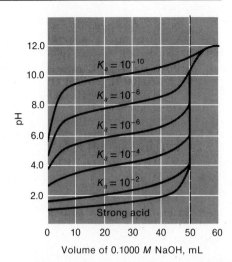

FIGURE 8-4 Influence of acid strength upon titration curves. Each curve represents titration of 50.0 mL of 0.1000 M acid with 0.1000 M NaOH.

it should be possible to establish the end point with a reproducibility of a few percent.

Figure 8-4 illustrates that similar problems exist as the strength of the acid being titrated decreases. Precision on the order of ± 2 ppt can be achieved in the titration of a 0.1000 M acid solution with a dissociation constant of 10^{-8} provided a suitable color comparison standard is available. With more concentrated solutions, somewhat weaker acids can be titrated with reasonable precision.

8C-5 TITRATION CURVES FOR WEAK BASES
The derivation of a curve for the titration of a weak base is analogous to that for a weak acid.

> **EXAMPLE 8-9.** A 50.00-mL aliquot of 0.0500 M NaCN is titrated with 0.1000 M HCl. Calculate the pH after the additions of (a) 0.00, (b) 10.00, (c) 25.00, and (d) 26.00 mL of acid.
>
> (a) *0.00 mL reagent.* Cyanide ion is a simple, weak base; calculation of the pH is performed as on page 291. That is,
>
> $$[OH^-] \cong \sqrt{K_b M_{NaCN}}$$
>
> where
>
> $$K_b = \frac{K_w}{K_{HCN}} = \frac{1.00 \times 10^{-14}}{2.1 \times 10^{-9}} = 4.76 \times 10^{-6}$$
>
> The pH is found to be 10.69.
>
> (b) *10.00 mL reagent.* Addition of acid produces a buffer with a composition given by
>
> $$M_{NaCN} = \frac{50.00 \times 0.0500 - 10.00 \times 0.1000}{60.00} = \frac{1.500}{60.00}$$
>
> $$M_{HCN} = \frac{10.00 \times 0.1000}{60.00} = \frac{1.000}{60.00}$$

These values are then substituted into the acid dissociation constant of HCN to give

$$[H_3O^+] = \frac{2.1 \times 10^{-9} \times 1.000/\cancel{60.00}}{1.500/\cancel{60.00}} = 1.4 \times 10^{-9}$$

$$pH = 8.85$$

(c) *25.00 mL reagent.* This volume corresponds to the equivalence point; the principal solute species is the weak acid HCN. Thus,

$$M_{HCN} = \frac{25.00 \times 0.1000}{75.00} = 0.03333$$

and, as on page 261,

$$[H_3O^+] \cong \sqrt{K_a M_{HCN}} = \sqrt{2.1 \times 10^{-9} \times 0.03333}$$

$$= 8.37 \times 10^{-6}$$

$$pH = 5.08$$

(d) *26.00 mL reagent.* The excess of strong acid now present will repress the dissociation of the HCN to the point where its contribution to the pH is negligible. Thus,

$$[H_3O^+] = M_{HCl} = \frac{26.00 \times 0.1000 - 50.00 \times 0.0500}{76.00}$$

$$= 1.32 \times 10^{-3}$$

$$pH = 2.88$$

Figure 8-5 shows theoretical titration curves for a series of weak bases of differing strengths. Clearly, indicators with *acidic* transition ranges must be employed for weaker bases.

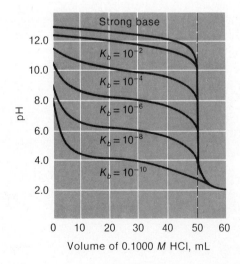

FIGURE 8-5 Influence of base strength upon titration curves. Each curve represents titration of 50.0 mL of 0.1000 M base with 0.1000 M HCl.

PROBLEMS*

All pH or pOH calculations should be rounded to two places beyond the decimal.
* 8-1. Calculate the pH of an aqueous solution that contains
  (a) 4.22% HCl (w/w) and has a density of 1.02 g/mL.
  (b) 2.50% NaOH (w/w) and has a density of 1.02 g/mL.
  (c) 19.00% HNO_3 (w/w) and has a density of 1.10 g/mL.
8-2. Calculate the pH of an aqueous solution that contains
  (a) 8.00% $HClO_4$ (w/w) and has a density of 1.05.
  (b) 4.50% KOH (w/w) and has a density of 1.037.
  (c) 11.0% HNO_3 (w/w) and has a density of 1.08.
* 8-3. Calculate the pH of the solution that results when 15.00 mL of 0.1200 M HCl are mixed with 35.00 mL of
  (a) water.   (d) 0.0150 M $AgNO_3$.
  (b) 0.0400 M NaOH.   (e) 0.0150 M HOCl.
  (c) 0.0400 M $Ba(OH)_2$.   (f) 0.0150 M NH_3.
8-4. Calculate the pH of the solution that results when 25.00 mL of 0.0600 M HBr are mixed with 35.00 mL of
  (a) water.   (d) 0.0300 M $Pb(NO_3)_2$.
  (b) 0.0400 M KOH.   (e) 0.0300 M NH_3.
  (c) 0.0400 M $Ba(OH)_2$.   (f) 0.0200 M CH_3NH_2.
* 8-5. Calculate the pH of the solution that results when 20.00 mL of 0.0800 M NaOH are mixed with 30.00 mL of
  (a) water.   (e) 0.0300 M $MgCl_2$.
  (b) 0.0600 M $HClO_4$. $[Mg^{2+} + 2OH^- \rightleftharpoons Mg(OH)_2(s)]$
  (c) 0.0200 M H_2SO_4.   (f) 0.0600 M $MgCl_2$.
  (d) 0.0150 M NH_3.
8-6. Calculate the pH of the solution that results when 45.00 mL of 0.0250 M $Ba(OH)_2$ are mixed with 25.00 mL of
  (a) water.   (e) 0.0400 M $MgCl_2$.
  (b) 0.0400 M HCl. $[Mg^{2+} + 2OH^- \rightleftharpoons Mg(OH)_2(s)]$
  (c) 0.0400 M H_2SO_4.   (f) 0.0600 M $MgCl_2$.
  (d) 0.0400 M Na_2SO_4.
* 8-7. Calculate the pH of an iodic acid solution that has an analytical concentration of (a) $1.00 \times 10^{-1} M$, (b) $1.00 \times 10^{-2} M$, (c) $1.00 \times 10^{-4} M$.
8-8. Calculate the pH of a lactic acid solution that has an analytical concentration of (a) $1.00 \times 10^{-1} M$, (b) $1.00 \times 10^{-2} M$, (c) $1.00 \times 10^{-4} M$.
* 8-9. Calculate the pH of a hypochlorous acid solution that has an analytical concentration of (a) $1.00 \times 10^{-1} M$, (b) $1.00 \times 10^{-2} M$, (c) $1.00 \times 10^{-4} M$.
8-10. Calculate the pH of an aniline hydrochloride solution that has an analytical concentration of (a) $1.00 \times 10^{-1} M$, (b) $1.00 \times 10^{-2} M$, (c) $1.00 \times 10^{-4} M$.
* 8-11. Calculate the pH of an ammonium chloride solution that has an analytical concentration of (a) $1.00 \times 10^{-1} M$, (b) $1.00 \times 10^{-2} M$, (c) $1.00 \times 10^{-4} M$.
8-12. Calculate the pH of an ethanolamine hydrochloride solution that has an analytical concentration of (a) $1.00 \times 10^{-1} M$, (b) $1.00 \times 10^{-2} M$, (c) $1.00 \times 10^{-4} M$.
* 8-13. Calculate the pH of a piperidine solution that has an analytical concentration of (a) $1.00 \times 10^{-1} M$, (b) $1.00 \times 10^{-2} M$, (c) $1.00 \times 10^{-4} M$.
8-14. Calculate the pH of an aqueous ammonia solution that has an analytical concentration of (a) $1.00 \times 10^{-1} M$, (b) $1.00 \times 10^{-2} M$, (c) $1.00 \times 10^{-4} M$.

* Answers to problems or parts of problems marked with an asterisk are to be found at the end of the book.

* 8-15. Calculate the pH of an hydroxylamine solution that has an analytical concentration of (a) 1.00×10^{-1} M, (b) 1.00×10^{-2} M, (c) 1.00×10^{-4} M.

8-16. Calculate the pH of a sodium formate solution that has an analytical concentration of (a) 1.00×10^{-1} M, (b) 1.00×10^{-2} M, (c) 1.00×10^{-4} M.

* 8-17. Calculate the pH of a sodium azide solution that has an analytical concentration of (a) 1.00×10^{-1} M, (b) 1.00×10^{-2} M, (c) 1.00×10^{-4} M.

8-18. Calculate the pH of a sodium cyanide solution that has an analytical concentration of (a) 1.00×10^{-1} M, (b) 1.00×10^{-2} M, (c) 1.00×10^{-4} M.

* 8-19. Calculate the pH of the solution that results when 15.00 mL of aqueous HIO_3 with an analytical concentration of 0.0800 M are mixed with 45.00 mL of
  (a) water.
  (b) 0.0200 M NaOH.
  (c) 0.0300 M $NaIO_3$.
  (d) 0.0200 M $BaCl_2$.
  (e) 0.0100 M $Ba(OH)_2$.
  (f) 0.0200 M $Ba(OH)_2$.

8-20. Calculate the pH of the solution that results when 20.00 mL of aqueous HNO_2 with an analytical concentration of 0.0600 M are mixed with 30.00 mL of
  (a) water.
  (b) 0.0200 M NaOH.
  (c) 0.0400 M $NaNO_2$.
  (d) 0.0200 M $Ba(OH)_2$.
  (e) 0.0200 M $Ba(NO_2)_2$.
  (f) 0.0400 M HCl.

* 8-21. Calculate the pH of the solution that results when 25.00 mL of aqueous HOCl with an analytical concentration of 0.0600 M are mixed with 25.00 mL of
  (a) water.
  (b) 0.0400 M KOH.
  (c) 0.0300 M NaOCl.
  (d) 0.0300 M $Ca(OCl)_2$.
  (e) 0.0400 M $Ba(OH)_2$.
  (f) 0.0400 M HCl.

8-22. Calculate the pH of the solution that results when 40.00 mL of aqueous piperidine, $C_5H_{11}N$, with an analytical concentration of 0.0150 M are mixed with 20.00 mL of
  (a) water.
  (b) 0.0400 M HCl.
  (c) 0.0400 M $C_5H_{11}NH^+Cl^-$.
  (d) 0.0300 M $C_5H_{11}NH^+Cl^-$.
  (e) 0.0400 M NaOH.

* 8-23. Calculate the pH of the solution that results when 20.00 mL of aqueous ammonia with an analytical concentration of 0.0800 M are mixed with 40.00 mL of
  (a) water.
  (b) 0.0300 M HCl.
  (c) 0.0300 M NH_4Cl.
  (d) 0.0400 M HCl.
  (e) 0.0400 M NH_4Cl.
  (f) 0.0400 M NaOH.

8-24. Calculate the pH of the solution that results when 25.00 mL of aqueous hydroxylamine, $HONH_2$, with an analytical concentration of 0.0800 M are mixed with 40.00 mL of
  (a) water.
  (b) 0.0500 M HCl.
  (c) 0.0400 M $HONH_3^+Cl^-$.
  (d) 0.0400 M $(HONH_3)_2SO_4$.
  (e) 0.0300 M HCl.
  (f) 0.0400 M NaOH.

* 8-25. Calculate the pH of the solution that results when 40.00 mL of aqueous perchloric acid with an analytical concentration of 0.0600 M are mixed with 30.00 mL of
  (a) 0.100 M NaOCl.
  (b) 0.0800 M NH_3.
  (c) 0.0500 M $Ca(HCOO)_2$.
  (d) 0.0400 M $Ca(C_2H_3O_2)_2$.
  (e) 0.100 M $HONH_2$.
  (f) 0.120 M sodium lactate.
  (g) 0.100 M $Mg(NO_2)_2$.
  (h) 0.0500 M $Ce(C_2H_3O_2)_3$.
  (i) 0.125 M $(CH_3)_2NH$.
  (j) 0.120 M H_2NSO_3Na.

8-26. Calculate the pH of the solution that results when 25.00 mL of aqueous sodium hydroxide with an analytical concentration of 0.0400 M are mixed with 20.00 mL of
  (a) 0.0600 M salicylic acid.
  (b) 0.150 M methylammonium chloride.
  (c) 0.100 M hydrazoic acid.
  (d) 0.100 M picric acid.
  (e) 0.125 M $C_5H_{11}NH^+Cl^-$.
  (f) 0.0750 M pyridinium chloride.
  (g) 0.0750 M $(NH_4)_2SO_4$.
  (h) 0.0400 M $[(CH_3)_3NH]_2SO_4$.
  (i) 0.0800 M H_2NSO_3H.
  (j) 0.125 M HOCl.

* 8-27. Calculate the pH of a solution
   (a) containing 0.578 g of salicylic acid and 1.460 g of sodium salicylate in 250 mL of solution.
   (b) prepared by adding 45.0 mL of 0.0250 M HCl to 100.0 mL of 0.0300 M sodium salicylate.
   (c) dissolving 2.42 g of salicylic acid in 75.0 mL of 0.0804 M NaOH.
   (d) dissolving 5.11 g of sodium salicylate in 140 mL of 0.1649 M HCl.
8-28. Calculate the pH of a solution
   (a) containing 14.67 g of mandelic acid and 20.59 g of sodium mandelate in 400 mL of solution.
   (b) prepared by adding 74.4 mL of 0.0968 M NaOH to 95.8 mL of 0.216 M mandelic acid.
   (c) prepared by dissolving 1.885 g of sodium mandelate in 40.0 mL of 0.1461 M HCl.
   (d) prepared by dissolving 2.006 g of mandelic acid in 50.0 mL of 0.0894 M NaOH.
8-29. Calculate the pH of the accompanying buffer solutions:

	Analytical Concentration	
	of Acid	of Conjugate Base
* (a)	0.100 M HNO$_2$	0.0500 M NaNO$_2$
(b)	0.0500 M NH$_4$Cl	0.0600 M NH$_3$
* (c)	0.0333 M (NH$_4$)$_2$SO$_4$	0.0600 M NH$_3$
(d)	0.0475 M HCOOH	0.100 M HCOONa
* (e)	0.0200 M HCOOH	0.0167 M Ca(HCOO)$_2$
(f)	0.0447 M HC$_2$H$_3$O$_2$	0.0500 M NaC$_2$H$_3$O$_2$

8-30. Use activities (Table 4-5) to calculate the pH of the solutions in Problem 8-29 (use for the definition of pH, pH $= -\log a_{H^+}$).
* 8-31. Calculate the molar ratio of acid to conjugate base in a solution that has a pH of 5.40 and contains
   (a) lactic acid and sodium lactate.
   (b) benzoic acid and sodium benzoate.
   (c) hypochlorous acid and potassium hypochlorite.
   (d) formic acid and sodium formate.
   (e) hydroxylamine hydrochloride and hydroxylamine.
8-32. Calculate the molar ratio of acid to conjugate base in a solution that has a pH of 2.17 and contains
   (a) nitrous acid and sodium nitrite.
   (b) acetic acid and sodium acetate.
   (c) formic acid and sodium formate.
   (d) glycolic acid and sodium glycolate.
   (e) aniline hydrochloride and aniline.
* 8-33. Calculate the molar ratio of base to conjugate acid in a solution that has a pH of 9.73 and contains
   (a) ammonia and ammonium chloride.
   (b) sodium hypochlorite and hypochlorous acid.
   (c) pyridine and pyridine hydrochloride.
   (d) sodium phenolate and phenol.
   (e) ethanolamine and ethanolamine hydrochloride.
* 8-34. Calculate the change in pH that results from a 50-fold dilution of a solution that has analytical solute concentration(s) of
   (a) 0.145 M lactic acid.
   (b) 0.259 M sodium lactate.
   (c) 0.145 M lactic acid and 0.259 M sodium lactate.

      (d) $0.0145\ M$ lactic acid and $0.0259\ M$ sodium lactate.

8-35. Calculate the effect of a 100-fold dilution upon the pH of a solution that has analytical concentration(s) of
      (a) $0.165\ M$ NH_3.
      (b) $0.137\ M$ NH_4Cl.
      (c) $0.165\ M$ NH_3 and $0.137\ M$ NH_4Cl.
      (d) $0.0165\ M$ NH_3 and $0.0137\ M$ NH_4Cl.

* 8-36. Calculate the change in pH that occurs when 25.0 mL of $0.0400\ M$ HCl are added to 100 mL of the undiluted solutions in Problem 8-34.

8-37. Calculate the change in pH that occurs when 25.0 mL of $0.0400\ M$ HCl are added to 100 mL of the undiluted solutions in Problem 8-35.

* 8-38. Calculate the change in pH that occurs when 25.0 mL of $0.0400\ M$ NaOH are added to 100 mL of the undiluted solutions in Problem 8-34.

8-39. Calculate the change in pH that occurs when 25.0 mL of $0.0400\ M$ NaOH are added to 100 mL of the undiluted solutions in Problem 8-35.

* 8-40. What weight of sodium lactate must be added to 600 mL of $0.269\ M$ lactic acid in order to produce a buffer with a pH of 4.10?

8-41. What weight of methylamine must be added to 500 mL of $0.316\ M$ methylamine hydrochloride in order to produce a buffer with a pH of 10.00?

* 8-42. What weight of calcium acetate must be added to 200 mL of $0.425\ M$ acetic acid in order to produce a buffer with a pH of 4.90?

8-43. What weight of ammonium sulfate must be added to 400 mL of $0.304\ M$ ammonia in order to produce a buffer with a pH of 9.00?

* 8-44. What volume of $0.200\ M$ sodium hydroxide must be added to 600 mL of $0.269\ M$ lactic acid in order to produce a buffer with a pH of 4.10?

8-45. What volume of $0.250\ M$ sodium hydroxide must be added to 500 mL of $0.316\ M$ methylamine hydrochloride in order to produce a buffer with a pH of 10.00?

* 8-46. What volume of $0.307\ M$ sodium hydroxide must be added to 200 mL of $0.425\ M$ acetic acid in order to produce a buffer with a pH of 4.25?

8-47. What volume of $0.500\ M$ hydrochloric acid must be added to 400 mL of $0.304\ M$ ammonia in order to produce a buffer with a pH of 9.00?

* 8-48. Derive a curve for the titration of 50.0 mL of $0.100\ M$ NaOH with $0.100\ M$ HCl. Calculate the pH of the solution after the addition of 0.0, 10.0, 25.0, 40.0, 45.0, 49.0, 50.0, 51.0, 55.0, and 60.0 mL of acid; prepare a titration curve from the data.

* 8-49. Calculate the pH of the solution after addition of 0.0, 5.0, 15.0, 25.0, 40.0, 45.0, 49.0, 50.0, 51.0, 55.0, and 60.0 mL of $0.100\ M$ NaOH in the titration of 50.0 mL of
      (a) $0.100\ M$ HNO_2.
      (b) $0.100\ M$ lactic acid.
      (c) $0.100\ M$ pyridine hydrochloride.
      Construct titration curves from the data.

8-50. Calculate the pH of the solution after addition of 0.0, 5.0, 15.0, 25.0, 40.0, 45.0, 49.0, 50.0, 51.0, 55.0, and 60.0 mL of $0.100\ M$ HCl in the titration of 50.0 mL of
      (a) $0.100\ M$ ammonia.
      (b) $0.100\ M$ hydrazine.
      (c) $0.100\ M$ sodium hypochlorite.
      Construct titration curves from the data.

8-51. Calculate the pH of the solution after addition of 0.0, 5.0, 15.0, 25.0, 40.0, 49.0, 50.0, 51.0, 55.0, and 60.0 mL of reagent in the titration of 50.0 mL of
      * (a) $0.100\ M$ aniline hydrochloride with $0.100\ M$ NaOH.
      (b) $0.010\ M$ picric acid with $0.010\ M$ NaOH.
      * (c) $0.100\ M$ hypochlorous acid with $0.100\ M$ NaOH.
      (d) $0.100\ M$ hydroxylamine with $0.100\ M$ HCl.
      Construct titration curves from the data.

titration curves for complex

acid-base systems

9

This chapter deals with titration curves for solutions containing two acids or bases, acids or bases that contain or consume two or more protons, and *amphiprotic* substances that act as both acids and bases. A characteristic of all such systems is that two or more equilibria must be considered in describing their behavior; as a consequence, the techniques for derivation of pH data are often more complex than those developed in the previous chapter.

9A TITRATION CURVES FOR MIXTURES OF STRONG AND WEAK ACIDS OR WEAK AND STRONG BASES

To illustrate the derivation of curves for mixtures of strong and weak acids or bases, consider the titration of a solution containing hydrochloric acid and a weak acid HA. The molar hydronium ion concentration in the early stages of this titration is described by the equation

$$[H_3O^+] = M_{HCl} + [A^-] + [OH^-]$$

The first two terms on the right account for the contributions of the two solute acids. The third represents the acidity resulting from dissociation of the solvent; this term is vanishingly small in acidic solutions and can thus be neglected in all calculations.

As shown by the following example, hydrochloric acid represses the dissociation of the weak acid in the early stages of the titration. Thus, $[A^-] \ll M_{HCl}$, and the hydronium ion concentration is simply equal to the molar concentration of the strong acid.

EXAMPLE 9-1. Calculate the pH of a mixture which is 0.1200 M in hydrochloric acid and 0.0800 M in a weak acid HA ($K_a = 1.00 \times 10^{-4}$).

$$[H_3O^+] = 0.1200 + [A^-]$$

Let us assume that $[A^-] \ll 0.1200$. Then $[H_3O^+] = 0.1200$ and the pH is 0.92. To check the assumption, the provisional value for $[H_3O^+]$ can be substituted into the dissociation constant expression for HA to give

$$\frac{[A^-]}{[HA]} = \frac{K_a}{[H_3O^+]} = \frac{1.00 \times 10^{-4}}{0.1200} = 8.33 \times 10^{-4}$$

Since

$$0.0800 = M_{HA} = [HA] + [A^-]$$

then

$$0.0800 = \frac{[A^-]}{8.33 \times 10^{-4}} + [A^-]$$

or

$$[A^-] = 6.67 \times 10^{-5}$$

Thus, the assumption is valid; $[A^-] \ll 0.1200$.

The approximation employed in this example can be shown to apply until most of the hydrochloric acid has been neutralized by the titrant. Thus, the curve in this region is *identical to the titration of a strong acid by itself.*

As shown by the following example, the presence of HA must be taken into account when the concentration of hydrochloric acid becomes small.

EXAMPLE 9-2. Calculate the pH of the solution that results when 29.00 mL of 0.1000 N NaOH have been added to 25.00 mL of the solution described in the earlier example.

Here,

$$M_{HCl} = \frac{25.00 \times 0.1200 - 29.00 \times 0.1000}{54.00} = 1.85 \times 10^{-3}$$

$$M_{HA} = \frac{25.00 \times 0.0800}{54.00} = 3.70 \times 10^{-2}$$

A provisional calculation, based (as in the previous example) on the assumption that $[H_3O^+] = 1.85 \times 10^{-3}$, yields a value of 1.90×10^{-3} for $[A^-]$. Clearly $[A^-]$ is not much smaller than $[H_3O^+]$. Thus,

$$[H_3O^+] = M_{HCl} + [A^-] = 1.85 \times 10^{-3} + [A^-] \tag{9-1}$$

In addition, we may write that

$$[HA] + [A^-] = M_{HA} = 3.70 \times 10^{-2} \tag{9-2}$$

Substitution of [HA] from the dissociation constant expression into 9-2 gives

$$\frac{[H_3O^+][A^-]}{1.00 \times 10^{-4}} + [A^-] = 3.70 \times 10^{-2}$$

which can be rearranged to

$$[A^-] = \frac{3.70 \times 10^{-6}}{[H_3O^+] + 1.00 \times 10^{-4}}$$

Substitution for $[A^-]$ and M_{HCl} in Equation 9-1 yields

$$[H_3O^+] = 1.85 \times 10^{-3} + \frac{3.70 \times 10^{-6}}{[H_3O^+] + 1.00 \times 10^{-4}}$$

or

$$[H_3O^+]^2 - 1.75 \times 10^{-3} [H_3O^+] - 3.885 \times 10^{-6} = 0$$

$$[H_3O^+] = 3.03 \times 10^{-3}$$

$$pH = 2.52$$

Note that the contributions to the hydronium ion concentration from the HCl (1.85×10^{-3}) and from HA ($3.03 \times 10^{-3} - 1.85 \times 10^{-3}$) are of comparable magnitude.

When the amount of base added is equivalent to the amount of hydrochloric acid originally present, the solution is identical in all respects to one prepared by dissolving appropriate quantities of the weak acid and sodium chloride in a suitable amount of water. The latter solute, however, has no effect on the pH (neglecting the influence of increased ionic strength); thus, the remainder of the titration curve is that for a dilute solution of HA.

The shape of the curve for a mixture of weak and strong acids, and hence the information obtainable from it, will depend in large measure upon the strength of the weak acid. Figure 9-1 depicts the pH changes that occur during the titration of mixtures containing hydrochloric acid and several weak acids. Note that a break in the curve at the first equivalence point is small or essentially nonexistent if the weak acid has a relatively large dissociation constant (curves A and B); for such mixtures, only the total acidity (that is, HCl + HA) can be obtained from the titration. Conversely, where the weak acid has a very small dissociation constant, only the

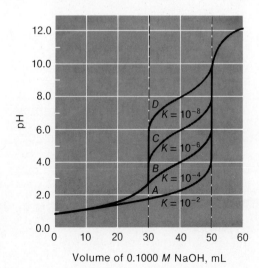

FIGURE 9-1 Curves for the titration of strong acid-weak acid mixtures with 0.1000 M NaOH. Each titration involves 25.00 mL of a solution that is 0.1200 M with respect to HCl and 0.0800 M with respect to HA.

strong acid content can be determined. For weak acids of intermediate strength (K_a somewhat less than 10^{-4} but greater than 10^{-8}), two useful end points usually exist.

The determination of each component in a mixture that contains a strong base and a weak base is also possible, subject to the same constraints that were just described for the strong acid/weak acid system. The derivation of a curve for such a titration is analogous to that for a mixture of acids.

9B EQUILIBRIUM CALCULATIONS FOR COMPLEX ACID-BASE SYSTEMS

This section is concerned with pH calculations for solutions of compounds that contain two acidic functional groups, two basic functional groups, or an acidic and a basic functional group. In addition, the properties of buffer systems prepared from these compounds will be considered.

9B-1 SOLUTIONS OF POLYPROTIC ACIDS

The dibasic weak acid H_2A dissociates in two steps that can be written as:

$$H_2A + H_2O \rightleftharpoons H_3O^+ + HA^-$$

$$HA^- + H_2O \rightleftharpoons H_3O^+ + A^{2-}$$

Application of the mass law yields the corresponding equilibrium constants

$$\frac{[H_3O^+][HA^-]}{[H_2A]} = K_1 \tag{9-3}$$

$$\frac{[H_3O^+][A^{2-}]}{[HA^-]} = K_2 \tag{9-4}$$

If the molar concentration of an H_2A solution is known, mass balance demands that

$$M_{H_2A} = [H_2A] + [HA^-] + [A^{2-}] \tag{9-5}$$

and charge balance requires that

$$[H_3O^+] \cong [HA^-] + 2[A^{2-}] + [OH^-] \tag{9-6}$$

Note that the concentration of hydroxide ions in the acidic solution is assumed to be negligible in Equation 9-6.

These four independent algebraic expressions permit calculation of $[H_3O^+]$ (as well as $[H_2A]$, $[HA^-]$, and $[A^{2-}]$). The mathematical manipulations required are somewhat awkward and tedious without the use of a computer. Fortunately, however, simplification of the relationships is possible provided that the ratio between K_1 and K_2 is sufficiently large ($> 10^3$); under these circumstances, we can ordinarily assume that the hydronium ions produced by the first dissociation step repress the inherently less favorable second step to such an extent that dissociation of the latter can be neglected. This assumption is tantamount to saying that $[A^{2-}]$ is much smaller than $[HA^-]$ and $[H_2A]$ in Equations 9-5 and 9-6. If this assumption can be made, Equation 9-5 becomes

$$M_{H_2A} \cong [H_2A] + [HA^-] \tag{9-7}$$

and Equation 9-6 simplifies to

$$[H_3O^+] = [HA^-] \tag{9-8}$$

Substitution of Equations 9-7 and 9-8 into Equation 9-3 yields

$$\frac{[H_3O^+]^2}{M_{H_2A} - [H_3O^+]} = K_1 \tag{9-9}$$

Equation 9-9 will be recognized as being identical to the one used previously to calculate the hydronium ion concentration for a simple weak acid (p. 25); as before, it may be solved either by the quadratic formula or by making the further assumption that $[H_3O^+]$ is small relative to M_{H_2A}.

The propriety of neglecting the contribution of the second dissociation to the hydronium ion concentration can be judged by substituting Equation 9-8 into Equation 9-4, which gives

$$[A^{2-}] = K_2$$

and provides an approximate concentration for $[A^{2-}]$ against which the trial values for $[H_3O^+]$ or $[HA^-]$ can be compared. In general, it is only where K_1/K_2 is small or where the solution is very dilute that neglect of the second dissociation step will lead to serious error. If the assumption is not justified, a rigorous and tedious solution of the first four equations is required.

EXAMPLE 9-3. Calculate the pH of a 0.100 M maleic acid solution. If we symbolize the acid by H_2M, we may write

$$H_2M + H_2O \rightleftharpoons H_3O^+ + HM^- \qquad \frac{[H_3O^+][HM^-]}{[H_2M]} = K_1 = 1.20 \times 10^{-2}$$

$$HM^- + H_2O \rightleftharpoons H_3O^+ + M^{2-} \qquad \frac{[H_3O^+][M^{2-}]}{[HM^-]} = K_2 = 5.96 \times 10^{-7}$$

If we neglect the effects of the second dissociation and employ Equation 9-9, we obtain

$$\frac{[H_3O^+]^2}{0.100 - [H_3O^+]} = K_1 = 1.20 \times 10^{-2}$$

The first dissociation constant for maleic acid is so large that the quadratic equation must be solved for $[H_3O^+]$. That is,

$$[H_3O^+]^2 + 1.20 \times 10^{-2}\,[H_3O^+] - 1.20 \times 10^{-3} = 0$$

$$[H_3O^+] = 2.92 \times 10^{-2} = [HM^-]$$

$$pH = -\log(2.92 \times 10^{-2}) = 1.54$$

A check of the assumption that $[M^{2-}]$ is much smaller than $[HM^-]$ is done by substituting the relationship $[H_3O^+] = [HM^-]$ into the expression for K_2. That is,

$$\frac{[\cancel{H_3O^+}][M^{2-}]}{[\cancel{HM^-}]} = 5.96 \times 10^{-7}$$

$$[M^{2-}] = 5.96 \times 10^{-7}$$

Clearly, the approximation will not lead to a significant error.

EXAMPLE 9-4. Calculate the hydronium ion concentration of a 0.0400 M H_2SO_4 solution.

One proton in H_2SO_4 is completely dissociated but the second is not. If the dissociation of HSO_4^- is assumed to be negligible, it follows that

$$[H_3O^+] = [HSO_4^-] = 0.0400$$

However, an estimate of $[SO_4^{2-}]$ based upon this approximation and the expression for K_2 reveals that

$$\frac{[\cancel{H_3O^+}][SO_4^{2-}]}{[\cancel{HSO_4^-}]} = 1.20 \times 10^{-2}$$

Clearly, $[SO_4^{2-}]$ is *not* small with respect to $[HSO_4^-]$; a more rigorous solution is thus required.

From stoichiometric considerations, it is necessary that

$$[H_3O^+] = 0.0400 + [SO_4^{2-}]$$

The two terms represent the contributions to $[H_3O^+]$ from the first and second dissociations, respectively. Rearrangement yields

$$[SO_4^{2-}] = [H_3O^+] - 0.0400$$

Mass-balance considerations require that

$$M_{H_2SO_4} = 0.0400 = [HSO_4^-] + [SO_4^{2-}]$$

Combining the last two equations and rearranging yields

$$[HSO_4^-] = 0.0800 - [H_3O^+]$$

Introduction of these equations for $[SO_4^{2-}]$ and $[HSO_4^-]$ into the expression for K_2 yields

$$\frac{[H_3O^+]([H_3O^+] - 0.0400)}{0.0800 - [H_3O^+]} = 1.20 \times 10^{-2}$$

which rearranges to

$$[H_3O^+]^2 - 0.0280\,[H_3O^+] - 9.60 \times 10^{-4} = 0$$
$$[H_3O^+] = 0.0480$$

9B-2 SOLUTIONS OF CONJUGATE BASES OF POLYPROTIC ACIDS

Calculation of the pH for a solution of the conjugate base of a polyprotic acid is analogous to that for the weak acid just considered.

EXAMPLE 9-5. Calculate the pH of a $0.100\ M$ Na_2CO_3 solution. The equilibria that must be considered are

$$CO_3^{2-} + H_2O \rightleftharpoons HCO_3^- + OH^- \qquad \frac{[HCO_3^-][OH^-]}{[CO_3^{2-}]} = \frac{K_w}{K_2} = 2.13 \times 10^{-4}$$

$$HCO_3^- + H_2O \rightleftharpoons H_2CO_3 + OH^- \qquad \frac{[H_2CO_3][OH^-]}{[HCO_3^-]} = \frac{K_w}{K_1} = 2.25 \times 10^{-8}$$

If the first reaction produces enough OH^- to repress the second equilibrium, then $[H_2CO_3] \ll [HCO_3^-]$ or $[CO_3^{2-}]$. Thus,

$$[HCO_3^-] \cong [OH^-]$$
$$[CO_3^{2-}] \cong N_{Na_2CO_3} - [OH^-] \cong M_{Na_2CO_3}$$

Substitution of these quantities into the first equilibrium-constant expression yields

$$[OH^-]^2 = 2.13 \times 10^{-4} \times 0.100 = 2.13 \times 10^{-5}$$

$$[OH^-] = 4.62 \times 10^{-3}$$

$$pH = 14.00 - (-\log 4.62 \times 10^{-3}) = 11.66$$

To test the assumption that $[H_2CO_3]$ is negligible, we turn to the second equilibrium-constant expression and write

$$\frac{[H_2CO_3][\cancel{OH^-}]}{[\cancel{HCO_3^-}]} = 2.25 \times 10^{-8}$$

Clearly, $[H_2CO_3]$ is much smaller than the concentrations of the other two carbonate species. The assumption that $[CO_3^{2-}] = M_{Na_2CO_3}$ introduces an error of about 2.3% in the hydrogen ion concentration.

9B-3 BUFFER SOLUTIONS INVOLVING POLYPROTIC ACID SYSTEMS

Two buffer systems can be prepared from a weak dibasic acid and its salts. The first consists of free acid H_2A and its conjugate base NaHA; the second makes use of the acid NaHA and its conjugate base Na_2A. The pH of the latter system will be the higher, since the acid dissociation constant for HA^- is always less than that for H_2A.

Sufficient independent equations are readily written to permit a rigorous evaluation of the hydronium ion concentration for either of these systems. Ordinarily, however, it is permissible to introduce the simplifying assumption that only one of the equilibria is important in determining the hydronium ion concentration of the solution. Thus, for a buffer prepared from H_2A and NaHA, the dissociation equilibrium for H_2A is considered; the further reaction of HA^- to yield A^{2-} is neglected. With this simplification, the hydronium ion concentration is calculated by the method described in Section 8C-1 for a simple buffer solution. As before, it is an easy matter to check the validity of the assumption by calculating an approximate concentration for A^{2-} and comparing this value with the concentrations of H_2A and HA^-.

EXAMPLE 9-6. Calculate the hydronium ion concentration of a buffer solution that is 2.00 M in phosphoric acid and 1.50 M in potassium dihydrogen phosphate.

The principal equilibrium in this solution is the dissociation of H_3PO_4:

$$H_3PO_4 + H_2O \rightleftharpoons H_3O^+ + H_2PO_4^- \qquad \frac{[H_3O^+][H_2PO_4^-]}{[H_3PO_4]} = K_1 = 7.11 \times 10^{-3}$$

The further dissociation of $H_2PO_4^-$ is assumed to be negligible; that is, $[HPO_4^{2-}] \ll [H_2PO_4^-]$ or $[H_3PO_4]$. Then,

$$[H_3PO_4] \cong M_{H_3PO_4} = 2.00$$

$$[H_2PO_4^-] \cong M_{KH_2PO_4} = 1.50$$

$$[H_3O^+] = \frac{7.11 \times 10^{-3} \times 2.00}{1.50} = 9.48 \times 10^{-3}$$

An approximate value for $[HPO_4^{2-}]$, obtained by substituting values for $[H_3O^+]$ and $[H_2PO_4^-]$ into K_2, verifies that neglect of the second dissociation step was justified.

For a buffer prepared from NaHA and Na_2A, the second dissociation reaction is assumed to predominate, and the reaction

$$HA^- + H_2O \rightleftharpoons H_2A + OH^-$$

is disregarded. The concentration of H_2A is ordinarily negligible compared with that of HA^- and A^{2-}; the hydronium ion concentration can then be calculated from the second dissociation constant, again employing the techniques for a simple buffer solution. To test the assumption, an estimate of the H_2A concentration is compared with the concentrations of HA^- and A^{2-}.

> **EXAMPLE 9-7.** Calculate the hydronium ion concentration of a buffer that is $0.0500\ M$ in potassium hydrogen phthalate (KHP) and $0.150\ M$ in potassium phthalate (K_2P). Here, the principal equilibrium is the dissociation of HP^-.
>
> $$HP^- + H_2O \rightleftharpoons H_3O^+ + P^{2-} \qquad \frac{[H_3O^+][P^{2-}]}{[HP^-]} = K_2 = 3.91 \times 10^{-6}$$
>
> Provided the concentration of H_2P in this solution is negligible, then
>
> $$[HP^-] \cong M_{KHP} = 0.0500$$
>
> $$[P^{2-}] \cong M_{K_2P} = 0.150$$
>
> $$[H_3O^+] = \frac{3.91 \times 10^{-6} \times 0.0500}{0.150} = 1.30 \times 10^{-6}$$
>
> To check the first assumption, an approximate value for $[H_2P]$ is calculated by substituting numerical values for $[H_3O^+]$ and for $[HP^-]$ into K_1.
>
> $$\frac{(1.30 \times 10^{-6})(0.0500)}{[H_2P]} = 1.12 \times 10^{-3}$$
>
> $$[H_2P] = 6 \times 10^{-5}$$
>
> This result justifies the assumption that $[H_2P]$ is much smaller than $[HP^-]$ and $[P^{2-}]$; that is, that the basic dissociation reaction of HP^- can be neglected.

In all but a few situations, the assumption of a single principal equilibrium, as invoked in these examples, will provide a satisfactory estimate for the pH of buffer mixtures derived from polybasic acids. Appreciable errors will occur, however, where the concentration of the acid or the salt is very low or where the two dissociation constants are numerically close to one another. A more laborious and rigorous calculation is then required.

9B-4 SOLUTIONS OF AMPHIPROTIC SUBSTANCES

Many substances exhibit both acidic and basic character when dissolved in water. For example, both acidic and basic dissociation equilibria are established in a solution of the amphiprotic compound ammonium formate. The reactions are

$$NH_4^+ + H_2O \rightleftharpoons H_3O^+ + NH_3 \qquad K_a = \frac{K_w}{K_b} = 5.68 \times 10^{-10}$$

and

$$A^- + H_2O \rightleftharpoons HA + OH^- \qquad K_b = \frac{K_w}{K_a} = 5.65 \times 10^{-11}$$

where A^- represents the formate anion. Note that a solution of ammonium formate will be slightly acidic because the first reaction is somewhat more favorable than the second.

A solute of the type NaHA is a commonly encountered example of an amphiprotic substance. Thus, its acid dissociation reaction can be formulated as

$$HA^- + H_2O \rightleftharpoons A^{2-} + H_3O^+$$

It is also, however, the conjugate base of H_2A. That is,

$$HA^- + H_2O \rightleftharpoons H_2A + OH^-$$

One of these reactions produces hydronium ions and the other hydroxide ions. Whether the solution is acidic or basic will be determined by the relative magnitude of the equilibrium constants for these processes:

$$K_2 = \frac{[H_3O^+][A^{2-}]}{[HA^-]} \tag{9-10}$$

$$K_b = \frac{K_w}{K_1} = \frac{[H_2A][OH^-]}{[HA^-]} \tag{9-11}$$

If K_b is greater than K_2, the solution will be basic; otherwise, it will be acidic (in the unlikely event that K_2 and K_b were equal, the solution would be neutral).

A solution of NaHA can be described in terms of material balance:

$$M_{NaHA} = [HA^-] + [H_2A] + [A^{2-}] \tag{9-12}$$

and charge balance

$$[Na^+] + [H_3O^+] = [HA^-] + 2[A^{2-}] + [OH^-]$$

Since the sodium ion concentration is equal to the molar concentration of the salt, the last equation can be rewritten as

$$M_{NaHA} = [HA^-] + 2[A^{2-}] + [OH^-] - [H_3O^+] \tag{9-13}$$

Frequently neither the hydroxide nor the hydronium ion concentration can be neglected in solutions of this type; therefore, a fifth equation is needed to take care of the five unknowns. The ion-product constant for water will serve:

$$K_w = [H_3O^+][OH^-]$$

The derivation of a rigorous expression for the hydronium ion concentration from these five equations is difficult. However, a reasonable approximation, applicable to solutions of most acid salts, can be obtained. Equating Equation 9-13 with Equation 9-12 yields

$$[H_2A] = [A^{2-}] + [OH^-] - [H_3O^+]$$

With the aid of Equations 9-10 and 9-11, we can express $[H_2A]$ and $[A^{2-}]$ in this equation in terms of $[HA^-]$; that is,

$$\frac{K_w[HA^-]}{K_1[OH^-]} = \frac{K_2[HA^-]}{[H_3O^+]} + [OH^-] - [H_3O^+]$$

Replacement of $[OH^-]$ by the equivalent expression $K_w/[H_3O^+]$ gives

$$\frac{[H_3O^+][HA^-]}{K_1} = \frac{K_2[HA^-]}{[H_3O^+]} + \frac{K_w}{[H_3O^+]} - [H_3O^+]$$

Multiplication of both sides by $[H_3O^+]$ and rearrangement yields

$$[H_3O^+]^2\left(\frac{[HA^-]}{K_1} + 1\right) = K_2[HA^-] + K_w$$

Finally,

$$[H_3O^+] = \sqrt{\frac{K_2[HA^-] + K_w}{1 + [HA^-]/K_1}} \tag{9-14}$$

Under most circumstances, it can be *assumed* that

$$[HA^-] \cong M_{NaHA} \tag{9-15}$$

Introduction of this relationship into Equation 9-14 gives

$$[H_3O^+] = \sqrt{\frac{K_2 M_{NaHA} + K_w}{1 + M_{NaHA}/K_1}} \tag{9-16}$$

Equation 9-16 is a reasonable approximation provided the two equilibrium constants that contain $[HA^-]$ (Equations 9-10 and 9-11) are small and the molar concentration of NaHA is not too low.

Frequently, the ratio M_{NaHA}/K_1 will be much larger than unity; furthermore, $K_2 M_{NaHA}$ is often considerably greater than K_w. Thus, provided $M_{NaHA}/K_1 \gg 1$ and $K_2 M_{NaHA} \gg K_w$, Equation 9-16 simplifies to

$$[H_3O^+] \cong \sqrt{K_1 K_2} \tag{9-17}$$

Note that Equation 9-17 does not contain M_{NaHA}; solutions of this type must then have a substantially constant pH over a considerable range of solute concentrations.

EXAMPLE 9-8. Calculate the hydronium ion concentration of a 0.100 M NaHCO$_3$ solution.

We must first examine the assumptions required for Equation 9-17 to be applicable. The dissociation constants, K_1 and K_2, for H_2CO_3 are 4.45×10^{-7} and 4.7×10^{-11}, respectively. Clearly, M_{NaHA}/K_1 is much larger than 1; in addition, $K_2 M_{NaHA}$ has a value of 4.7×10^{-12}, which is substantially greater than K_w. Thus,

$$[H_3O^+] = \sqrt{4.45 \times 10^{-7} \times 4.7 \times 10^{-11}}$$
$$[H_3O^+] = 4.6 \times 10^{-9}$$

EXAMPLE 9-9. Calculate the hydronium ion concentration of a 1.0×10^{-3} M Na$_2$HPO$_4$ solution.

Here, the pertinent dissociation constants (the two containing $[HPO_4^{2-}]$) are K_2 and K_3 for H_3PO_4 which have values of 6.34×10^{-8} and 4.2×10^{-13}, respectively. Considering again the assumptions implicit in Equation 9-17, we find that $1.0 \times 10^{-3}/6.34 \times 10^{-8}$ is large enough so that the denominator can again be simplified. On the other hand, K_w is by no means much smaller than $K_3 M_{Na_2HPO_4}$. We should therefore use the partially simplified version of Equation 9-16:

$$[H_3O^+] = \sqrt{\frac{4.2 \times 10^{-13} \times 1.0 \times 10^{-3} + 1.0 \times 10^{-14}}{1.0 \times 10^{-3}/6.34 \times 10^{-8}}}$$
$$= 8.1 \times 10^{-10}$$

Use of Equation 9-17 would have yielded a value of $1.6 \times 10^{-10}\ M$.

EXAMPLE 9-10. Find the hydronium ion concentration of a $0.0100\ M$ NaH_2PO_4 solution.

Here, the two dissociation constants of importance (those containing $[H_2PO_4^-]$) are K_1 and K_2 for H_3PO_4; these have values of 7.11×10^{-3} and 6.34×10^{-8}. Here, the denominator of Equation 9-16 cannot be simplified. The numerator however, reduces to $K_2 M_{NaH_2PO_4}$. Thus, Equation 9-16 becomes

$$[H_3O^+] = \sqrt{\frac{6.34 \times 10^{-8} \times 1.0 \times 10^{-2}}{1.00 + 1.0 \times 10^{-2}/7.11 \times 10^{-3}}}$$

$$= 1.6 \times 10^{-5}$$

9C DERIVATION OF TITRATION CURVES FOR POLYPROTIC ACIDS AND THEIR CONJUGATE BASES

Compounds with two or more acidic or basic functional groups will yield multiple end points in a titration provided the acidic or basic groups differ sufficiently in strength. The computation techniques just described permit the derivation of reasonably accurate theoretical titration curves for polyprotic acids or bases provided the ratio K_1/K_2 is somewhat greater than 10^3. If the ratio is smaller than this figure, the error, particularly in the region of the first equivalence point, becomes prohibitive, and a more rigorous treatment of the equilibrium relationships is required.

9C-1 TITRATION CURVES FOR ACIDS

In the example that follows, we shall derive points on the titration curve of maleic acid, a dibasic weak organic acid with the formula $C_2H_2(COOH)_2$. The two dissociation equilibria are

$$H_2M + H_2O \rightleftharpoons H_3O^+ + HM^- \qquad K_1 = 1.20 \times 10^{-2}$$

$$HM^- + H_2O \rightleftharpoons H_3O^+ + M^{2-} \qquad K_2 = 5.96 \times 10^{-7}$$

where H_2M symbolizes the free acid. Because the ratio K_1/K_2 is large (2×10^4), it is feasible to neglect the second dissociation step when deriving points in the early part of the curve; that is, we assume that $[M^{2-}] \ll [HM^-]$ and $[H_2M]$ in this region. It can be shown that, to within a few tenths of a milliliter of the first equivalence point, this assumption does not lead to serious error. Shortly beyond the first equivalence point, the second equilibrium is sufficiently dominant so that the basic reaction of HM^-,

$$HM^- + H_2O \rightleftharpoons OH^- + H_2M$$

does not significantly influence the pH. Here, it can be assumed that $[H_2M] \ll [HM^-]$ and $[M^{2-}]$.

EXAMPLE 9-11. Derive a curve for the titration of 25.00 mL of $0.1000\ M$ maleic acid with $0.1000\ M$ NaOH.

Initial pH. This calculation is shown in the example on page 225. The pH is 1.54.

First buffer region. The addition of 5.00 mL of base results in the formation of a buffer consisting of the weak acid H_2M and its conjugate base HM^-. To the extent that dissociation of HM^- to give M^{2-} is negligible, the solution can be treated as a simple buffer system. Then

$$M_{H_2M} \cong \frac{25.00 \times 0.1000 - 5.00 \times 0.1000}{30.00} = 6.67 \times 10^{-2}$$

$$M_{NaHM} \cong \frac{5.00 \times 0.1000}{30.00} = 1.67 \times 10^{-2}$$

Using Equations 8-7 and 8-8 (p. 205).

$$[H_2M] \cong 6.67 \times 10^{-2}$$

$$[HM^-] \cong 1.67 \times 10^{-2}$$

Substitution of these values into the equilibrium-constant expression for K_1 yields a tentative value of 4.8×10^{-2} mol/L for $[H_3O^+]$. It is clear, however, that the approximation $[H_3O^+] \ll M_{H_2M}$ or M_{HM^-} is not valid; therefore, Equations 8-5 and 8-6 (p. 204) must be used, and

$$[H_2M] = 6.67 \times 10^{-2} - [H_3O^+] + [\cancel{OH^-}]$$

$$[HM^-] = 1.67 \times 10^{-2} + [H_3O^+] - [\cancel{OH^-}]$$

Because the solution is quite acidic, the approximation that $[OH^-]$ is very small is surely valid. Substitution of these expressions into the dissociation-constant relationship gives

$$\frac{[H_3O^+](1.67 \times 10^{-2} + [H_3O^+])}{6.67 \times 10^{-2} - [H_3O^+]} = 1.20 \times 10^{-2} = K_1$$

or $$[H_3O^+]^2 + 2.87 \times 10^{-2}\,[H_3O^+] - 8.00 \times 10^{-4} = 0$$

and $$[H_3O^+] = 1.74 \times 10^{-2}$$

Thus $$pH = 1.76$$

   Additional points in the first buffer region can be computed in a similar way.
First equivalence point.   At the first equivalence point

$$[HM^-] \cong \frac{2.500}{50.00} = 5.00 \times 10^{-2}$$

Simplification of the numerator in Equation 9-16 is clearly justified. On the other hand, the concentration of HM^- is relatively close to the value for K_1. Hence,

$$[H_3O^+] \cong \sqrt{\frac{K_2 M_{HM^-}}{1 + M_{HM^-}/K_1}} = \sqrt{\frac{5.96 \times 10^{-7} \times 5.00 \times 10^{-2}}{1 + 5.00 \times 10^{-2}/1.20 \times 10^{-2}}}$$

$$= 7.60 \times 10^{-5}$$

$$pH = 4.12$$

Second buffer region.   Further additions of base to the solution create a new buffer system consisting of HM^- and M^{2-}. When enough base has been added so that the reaction of HM^- with water to give OH^- may be neglected (a few tenths of a milliliter), the pH of the mixture is readily obtained from K_2. With the introduction of 25.50 mL of NaOH, for example,

$$M_{Na_2M} \cong \frac{(25.50 - 25.00)0.1000}{50.50} = \frac{0.050}{50.50}$$

and the molar concentration of NaHM will be equal to

$$M_{NaHM} \cong \frac{(25.00 \times 0.1000) - (25.50 - 25.00)0.1000}{50.50}$$

$$= \frac{2.45}{50.50}$$

Thus,

$$\frac{[H_3O^+](0.050/50.50)}{(2.45/50.50)} = 5.96 \times 10^{-7}$$

$$[H_3O^+] = 2.92 \times 10^{-5}$$

The assumption that $[H_3O^+]$ is small with respect to the two molar concentrations is valid and

$$pH = 4.54$$

Second equivalence point. After 50.00 mL of 0.1000 M sodium hydroxide have been added, the solution is 0.0333 M in Na_2M. Reaction of the base M^{2-} with water is the predominant equilibrium in the system and the only one that must be taken into account. Thus,

$$M^{2-} + H_2O \rightleftharpoons OH^- + HM^-$$

$$\frac{[OH^-][HM^-]}{[M^{2-}]} = \frac{1.00 \times 10^{-14}}{5.96 \times 10^{-7}} = 1.68 \times 10^{-8}$$

$$[OH^-] \cong [HM^-]$$

$$[M^{2-}] = 0.0333 - [OH^-] \cong 0.0333$$

$$\frac{[OH^-]^2}{0.0333} = \frac{1.00 \times 10^{-14}}{5.96 \times 10^{-7}}$$

$$[OH^-] = 2.36 \times 10^{-5}$$

$$pH = 14.00 - (-\log 2.36 \times 10^{-5}) = 9.37$$

pH beyond the second equivalence point. Further additions of sodium hydroxide repress the basic dissociation of M^{2-}. The pH is calculated from the concentration of NaOH added in excess of that required for the complete neutralization of H_2M.

Figure 9-2 is the curve for the titration of 0.1000 M maleic acid with 0.1000 M NaOH. This curve was derived by use of the calculations shown in the preceding example. Two end points are apparent; in principle, the judicious choice of indicator would permit either the first or both acidic hydrogens of maleic acid to be titrated. The second end point is clearly more satisfactory, however, inasmuch as the pH change is more pronounced.

Figure 9-3 shows titration curves for three other polybasic acids. These curves illustrate that a well-defined end point corresponding to the first equivalence point is observed only where the degree of dissociation of the two acids is sufficiently different.

The ratio of K_1 to K_2 for oxalic acid (curve B) is approximately 1000. The curve for this titration shows an inflection corresponding to the first equivalence point. However, the magnitude of the pH change is too small to permit the location of equivalence by means of an indicator; thus, only the second end point can be used for accurate analyses.

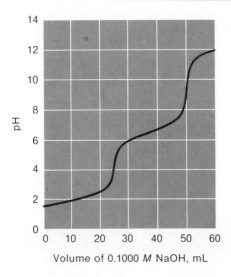

FIGURE 9-2 Titration curve for 25.00 mL of 0.1000 M maleic acid, H_2M.

Curve A illustrates the theoretical titration curve for the tribasic phosphoric acid. Here, the ratio K_1 to K_2 is approximately 10^5, a figure that is about 100 times greater than that for oxalic acid; two well-defined end points are observed, either of which is satisfactory for analytical purposes. If an indicator with an acid transition range is used, one equivalent of base will be consumed per mole of acid. With an indicator exhibiting a color change in the basic region, two equivalents of base will be used. The third hydrogen of phosphoric acid is so slightly dissociated ($K_3 = 4.2 \times 10^{-13}$) that it does not yield an end point of any practical value. The buffering effect of the third dissociation is noticeable, however, causing the pH for curve A to be lower than that for the other two in the region beyond their second equivalence points.

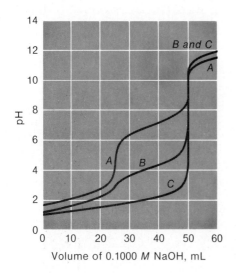

FIGURE 9-3 Curves for the titration of poly-basic acids. A 0.1000 M NaOH solution is used to titrate 25.00 mL of 0.1000 M H_3PO_4 (curve A), 0.1000 M oxalic acid (curve B), and 0.1000 M H_2SO_4 (curve C).

Curve C is for the titration of sulfuric acid, one proton of which is completely dissociated while the dissociation constant for the second is relatively large ($K = 1.2 \times 10^{-2}$). Because of the similarity in strengths of the two acids, only a single end point, corresponding to the titration of both protons, is observed.

In general, the titration of acids or bases with two reactive groups yields individual end points that are of practical value only where the ratio between the two dissociation constants is at least 10^4. If the ratio is much smaller than this, the pH change at the first equivalence point will be too small for accurate detection—only the second end point will prove satisfactory for analysis.

9C-2 TITRATION CURVES FOR SUBSTANCES WITH TWO BASIC FUNCTIONAL GROUPS

The derivation of a titration curve for a polyfunctional base involves no new principles. To illustrate, consider the titration of a sodium carbonate solution with standard hydrochloric acid. The important equilibrium constants are

$$CO_3^{2-} + H_2O \rightleftharpoons OH^- + HCO_3^- \qquad K_{b1} = \frac{K_w}{K_{a2}} = 2.13 \times 10^{-4}$$

$$HCO_3^- + H_2O \rightleftharpoons OH^- + H_2CO_3 \qquad K_{b2} = \frac{K_w}{K_{a1}} = 2.25 \times 10^{-8}$$

The reaction of carbonate ion with water governs the initial pH of the solution; the method shown on page 226 is used for the pH calculation. With the first additions of acid, a carbonate/hydrogen carbonate buffer is established; the hydroxide ion concentration is calculated from K_{b1} (or the hydronium ion concentration from K_{a2}). Sodium hydrogen carbonate is the principal solute species at the first equivalence point; Equation 9-17 will give the hydronium ion concentration of this solution. With the further introductions of acid, a hydrogen carbonate/carbonic acid buffer governs the pH; here, the hydroxide ion concentration is obtained from K_{b2} (or the hydronium ion concentration from K_{a1}). At the second equivalence point, the solution consists of carbonic acid and sodium chloride; the hydronium ion concentration is estimated in the usual way for a simple weak acid, using K_{a1}. Finally, when excess hydrochloric acid has been introduced, the dissociation of the weak acid is repressed to a point where the hydronium ion concentration is essentially that of the molar concentration of the strong acid.

Figure 9-4 illustrates a curve for the titration of sodium carbonate. Two end points are observed, the second being appreciably sharper than the first. It is apparent that mixtures of sodium carbonate and sodium hydrogen carbonate can be analyzed by neutralization methods. Thus, titration to a phenolphthalein end point would yield the number of milliformula weights of carbonate present, while titration to a methyl orange color change would require an amount of acid equal to twice the number of milliformula weights of carbonate plus the number of milliformula weights of bicarbonate in the sample.

9C-3 TITRATION CURVES FOR AMPHIPROTIC SPECIES

As noted on page 228, an amphiprotic substance, when dissolved in a suitable solvent, behaves both as a weak acid and as a weak base. If either its acidic or its basic character predominates sufficiently, titration of the species with a strong base or a

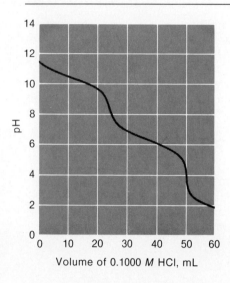

pH

Volume of 0.1000 *M* HCl, mL

FIGURE 9-4 Curve for the titration of 25.00 mL of 0.1000 *M* Na$_2$CO$_3$ with 0.1000 *M* HCl.

strong acid may be feasible. For example, in sodium dihydrogen phosphate solution, the following equilibria exist:

$$H_2PO_4^- + H_2O \rightleftharpoons H_3O^+ + HPO_4^= \qquad K_2 = 6.34 \times 10^{-8}$$

and

$$H_2PO_4^- + H_2O \rightleftharpoons OH^- + H_3PO_4 \qquad K_b = \frac{K_w}{K_1} = 1.41 \times 10^{-12}$$

We see that K_b is too small to permit titration of $H_2PO_4^-$ with an acid; on the other hand, K_2 is large enough for a successful titration of the ion with a standard base solution. A different situation prevails in solutions containing disodium hydrogen phosphate for which the analogous equilibria are

$$HPO_4^{2-} + H_2O \rightleftharpoons H_3O^+ + PO_4^{3-} \qquad K_3 = 4.2 \times 10^{-13}$$

$$HPO_4^{2-} + H_2O \rightleftharpoons OH^- + H_2PO_4^- \qquad K_b = \frac{K_w}{K_2} = 1.58 \times 10^{-7}$$

The magnitude of the constants indicates that HPO_4^{2-} could be titrated with standard hydrochloric acid but not with standard base.

The simple amino acids represent an important class of amphiprotic compounds which owe their acid-base properties to the presence of both a weakly acid and a weakly basic functional group. In an aqueous solution of a typical amino acid, such as glycine, three important equilibria exist:

$$NH_2CH_2COOH \rightleftharpoons NH_3^+CH_2COO^- \qquad (9\text{-}18)$$

$$NH_3^+CH_2COO^- + H_2O \rightleftharpoons NH_2CH_2COO^- + H_3O^+ \qquad (9\text{-}19)$$
$$K_a = 2 \times 10^{-10}$$

$$NH_3^+CH_2COO^- + H_2O \rightleftharpoons NH_3^+CH_2COOH + OH^- \qquad (9\text{-}20)$$
$$K_b = 2 \times 10^{-12}$$

The first reaction constitutes a kind of internal acid-base reaction and is analogous to the reaction one would observe between a carboxylic acid and an amine. That is,

$$R_1NH_2 + R_2COOH \rightleftharpoons R_1NH_3^+ + R_2COO^- \tag{9-21}$$

The typical aliphatic amine has a basic ionization constant of 10^{-4} to 10^{-5} (see Appendix 5) while many carboxylic acids have acidic dissociation constants of about the same magnitude. The consequence is that both reactions 9-18 and 9-21 proceed far to the right with the product or products being the predominant species in the solution.

The amino acid species in Equation 9-18, bearing both a positive and a negative charge, is called a *zwitterion*. As shown by Equations 9-19 and 9-20, the zwitterion of glycine is a slightly stronger acid than a base. Thus, an aqueous solution of glycine is slightly acidic.

The zwitterion of an amino acid, containing as it does a positive and a negative charge, has no tendency to migrate in an electric field; on the other hand, the singly charged anionic or cationic species is attracted to the positive and the negative electrodes, respectively. No *net* migration of the amino acid occurs in an electric field when the pH of the solvent is such that the concentrations of the anionic and cationic forms are identical. The pH at which no net migration occurs is called the *isoelectric point* and is an important physical constant for characterizing amino acids. The isoelectric point is readily related to the ionization constants for the species. Thus, for glycine

$$\frac{[H_3O^+][NH_2CH_2COO^-]}{[NH_3^+CH_3COO^-]} = K_a$$

$$\frac{[OH^-][NH_3^+CH_2COOH]}{[NH_3^+CH_2COO^-]} = K_b$$

At the isoelectric point,

$$[NH_2CH_2COO^-] = [NH_3^+CH_2COOH]$$

Thus, division of K_a by K_b gives

$$\frac{[H_3O^+][\cancel{NH_2CH_2COO^-}]}{[OH^-][\cancel{NH_3^+CH_2COOH}]} = \frac{[H_3O^+]}{[OH^-]} = \frac{K_a}{K_b}$$

Substitution of $K_w/[H_3O^+]$ for $[OH^-]$ and rearrangement yields

$$[H_3O^+] = \sqrt{\frac{K_a K_w}{K_b}}$$

The isoelectric point for glycine occurs at a pH of 6.0. That is,

$$[H_3O^+] = \left[\frac{2 \times 10^{-10}}{2 \times 10^{-12}} \times 1 \times 10^{-14}\right]^{1/2} = 1 \times 10^{-6}$$

The magnitudes of K_a and K_b for simple amino acids are generally so small that determination by direct neutralization titration is impossible. Addition of formaldehyde, however, removes the base functional group and leaves the carboxylic acid available for titration with a standard base. For example, with glycine

$$NH_3^+CH_2COO^- + CH_2O \longrightarrow CH_2{=}NCH_2COOH + H_2O$$

The titration curve for the product is that of a typical carboxylic acid.

9D COMPOSITION OF A POLYBASIC ACID SOLUTION AS A FUNCTION OF pH

In order to understand clearly the compositional changes that occur in the course of a titration involving a polybasic acid, it is instructive to plot the *relative* amount of the free acid as well as each of its anions as a function of the pH of the solution. For this purpose, we shall define so-called α-values for each of the anion-containing species. An α-value is the fraction of the total concentration C_T represented by a particular species. Turning again to the maleic acid system, if we define C_T as the sum of the concentrations of the maleate-containing species, then mass balance requires that

$$C_T = [H_2M] + [HM^-] + [M^{2-}]$$

and by definition,

$$\alpha_0 = \frac{[H_2M]}{C_T}$$

$$\alpha_1 = \frac{[HM^-]}{C_T}$$

$$\alpha_2 = \frac{[M^{2-}]}{C_T}$$

The sum of the α-values for a system must equal unity; that is,

$$\alpha_0 + \alpha_1 + \alpha_2 = 1$$

We can readily express α_0, α_1, and α_2 in terms of $[H_3O^+]$, K_1, and K_2. Rearrangement of the dissociation-constant expressions gives

$$[HM^-] = \frac{K_1[H_2M]}{[H_3O^+]}$$

$$[M^{2-}] = \frac{K_1K_2[H_2M]}{[H_3O^+]^2}$$

After substituting these quantities, the mass-balance equation becomes

$$C_T = [H_2M] + \frac{K_1[H_2M]}{[H_3O^+]} + \frac{K_1K_2[H_2M]}{[H_3O^+]^2}$$

which can be converted to

$$[H_2M] = \frac{C_T[H_3O^+]^2}{[H_3O^+]^2 + K_1[H_3O^+] + K_1K_2}$$

Substituting this value for $[H_2M]$ into the equation defining α_0 gives

$$\alpha_0 = \frac{[H_3O^+]^2}{[H_3O^+]^2 + K_1[H_3O^+] + K_1K_2} \tag{9-22}$$

By similar manipulations, it is easily shown that

$$\alpha_1 = \frac{K_1[H_3O^+]}{[H_3O^+]^2 + K_1[H_3O^+] + K_1K_2} \tag{9-23}$$

$$\alpha_2 = \frac{K_1K_2}{[H_3O^+]^2 + K_1[H_3O^+] + K_1K_2} \tag{9-24}$$

Note that the denominator is the same for each expression; calculation of α-values at any desired pH is thus relatively simple. Furthermore, the equations illustrate that the fractional amount of each species at any fixed pH is *independent* of the total concentration, C_T.

The three curves plotted in Figure 9-5 show the α-values for each maleate-containing species as a function of pH. A consideration of these curves in conjunction with the titration curve for maleic acid (Figure 9-2) gives a clear picture of all concentration changes that occur during the course of the titration. For example, Figure 9-2 reveals that before the addition of any base, the pH of the solution is 1.5. Referring to Figure 9-5, see that at this pH, α_0 for H_2M is roughly 0.7 while α_1 for HM^- is approximately 0.3. For all practical purposes, α_2 is zero. Thus, approximately 70% of the maleic acid exists as H_2M and 30% as HM^-. With addition of base, the pH rises, as does the fraction of HM^-. At the first equivalence point (pH = 4.12), essentially all of the maleate is present as HM^- ($\alpha_1 \to 1$). Beyond the first equivalence point, HM^- decreases and M^{2-} increases. At the second equivalence point (pH = 9.37), it is evident that essentially all of the maleate exists as M^{2-}.

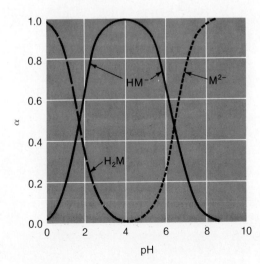

FIGURE 9-5 Composition of H_2M solutions as a function of pH.

PROBLEMS*

* 9-1. Calculate the pH of a solution that has an analytical concentration of 0.0500 M with respect to
  - (a) sulfurous acid.
  - (b) carbonic acid.
  - (c) phosphorous acid.
  - (d) malonic acid.
  - (e) phosphoric acid.
  - (f) oxalic acid.

9-2. Calculate the pH of a solution that has an analytical concentration of 0.0750 M with respect to
  - (a) maleic acid.
  - (b) ethylenediamine.
  - (c) hydrogen sulfide.
  - (d) arsenic acid.
  - (e) sulfuric acid.
  - (f) dihydroxytartaric acid.
    $(K_1 = 1.1 \times 10^{-2}, K_2 = 9.9 \times 10^{-5})$

* 9-3. Calculate the pH of a solution that is 0.0500 M in
  - (a) sodium hydrogen sulfite.
  - (b) sodium hydrogen carbonate.
  - (c) sodium dihydrogen phosphate.
  - (d) disodium hydrogen phosphate.
  - (e) sodium dihydrogen phosphite.

9-4. Calculate the pH of a solution that is 0.0400 M in
  - (a) sodium hydrogen sulfide.
  - (b) sodium hydrogen maleate.
  - (c) disodium hydrogen arsenate.
  - (d) sodium dihydrogen arsenate.
  - (e) ethylenediamine hydrochloride.

* 9-5. Calculate the pH of a solution in which the analytical concentration of
  - (a) Na_2S is 0.06400 M.
  - (b) Na_2CO_3 is 0.0500 M.
  - (c) Na_2SO_3 is 0.0795 M.
  - (d) ethylenediamine dihydrochloride is 0.0378 M.

9-6. Calculate the pH of a solution in which the analytical concentration of
  - (a) Na_3PO_4 is 0.0585 M.
  - (b) tetrasodium ethylenediaminetetraacetate is 0.0328 M.
  - (c) disodium maleate is 0.0705 M.
  - (d) disodium oxalate is 0.0300 M.

* 9-7. Calculate the pH of a solution that has an analytical concentration of 0.0500 M with respect to
  - (a) H_2SO_4.
  - (b) $NaHSO_4$.
  - (c) Na_2SO_4.

9-8. Calculate the pH of a solution that has an analytical concentration of 0.0250 M with respect to
  - (a) p-phenolsulfonic acid (first H strong, $K_2 = 8.8 \times 10^{-10}$).
  - (b) sodium p-phenolsulfonate.
  - (c) disodium p-phenolsulfonate.

* 9-9. Calculate the pH of a solution that has an analytical concentration of 0.0400 M with respect to
  - (a) H_3PO_4.
  - (b) NaH_2PO_4.
  - (c) Na_2HPO_4.
  - (d) Na_3PO_4.

* Answers to problems or parts of problems marked with an asterisk are to be found at the end of the book.

9-10. Calculate the pH of a solution that has an analytical concentration of 0.0600 M with respect to
    (a) H_3AsO_4.   (c) Na_2HAsO_4.
    (b) NaH_2AsO_4.   (d) Na_3AsO_4.

* 9-11. Calculate the pH of a solution in which the analytical concentration of HCl is 0.0100 M and that for
    (a) sulfamic acid is 0.0250 M.
    (b) iodic acid is 0.0250 M.
    (c) maleic acid is 0.0250 M.
    (d) hydrazoic acid is 0.0250 M.

9-12. Calculate the pH of a solution in which the analytical concentration of $HClO_4$ is 0.0100 M and that for
    (a) trichloroacetic acid is 0.0300 M.
    (b) dichloroacetic acid is 0.0300 M ($K_a = 4.3 \times 10^{-2}$).
    (c) chloroacetic acid is 0.0300 M.
    (d) acetic acid is 0.0300 M.

* 9-13. Calculate the pH of a solution in which the analytical concentration of NaOH is 0.0100 M and that for
    (a) disodium arsenite is 0.0300 M.
    (b) trisodium arsenate is 0.0300 M.
    (c) tetrasodium ethylenediaminetetraacetate, Na_4Y, is 0.0300 M.
    (d) aniline is 0.0300 M.

9-14. Calculate the pH of a solution in which the analytical concentration of KOH is 0.0100 M and that for
    (a) disodium sulfide is 0.0200 M.
    (b) trisodium phosphate is 0.0200 M.
    (c) sodium phenolate is 0.0200 M.
    (d) sodium cyanide is 0.0200 M.

* 9-15. Calculate the pH of a solution that is
    (a) 0.1095 M in Na_2CO_3 and 0.1440 M in $NaHCO_3$.
    (b) 0.0624 M in $NH_2CH_2CH_2NH_2$ and 0.0417 M in $NH_2CH_2CH_2NH_3^+Cl^-$.
    (c) 0.0800 M in H_3PO_4 and 0.0460 M in NaH_2PO_4.
    (d) 0.1370 M in Na_2HPO_4 and 0.0893 M in Na_3PO_4.

9-16. Calculate the pH of a solution that is
    (a) 0.0811 M in maleic acid and 0.1040 M in sodium hydrogen maleate.
    (b) 0.0794 M in sodium dihydrogen arsenate and 0.1135 M in disodium hydrogen arsenate.
    (c) 0.1020 M in sodium hydrogen sulfite and 0.1468 M in sodium sulfite.
    (d) 0.0810 M in ethylenediamine hydrochloride and 0.0547 M in ethylenediamine dihydrochloride.

* 9-17. Identify the principal conjugate acid-base pair and calculate the ratio between them in a solution buffered to a pH of 4.00 with species derived from
    (a) dihydroxytartaric acid (see Problem 9-2).
    (b) malic acid.
    (c) oxalic acid.
    (d) succinic acid.
    (e) tartaric acid.

9-18. Identify the principal conjugate acid-base pair and calculate the ratio that exists between them in a solution that is buffered to a pH of 7.00 with species derived from
    (a) H_3AsO_4.   (c) H_3PO_4.
    (b) citric acid.   (d) ethylenediaminetetraacetic acid.

* 9-19. What will be the pH of the solution that results when 40.00 mL of 0.3000 M HCl are

mixed with 20.0 mL of trisodium phosphate that has an analytical concentration of
(a) 0.400 M?
(b) 1.20 M?
(c) 0.300 M?
(d) 0.200 M?
(e) 0.600 M?
(f) 0.240 M?

9-20. What will be the pH of the solution that results when 30.00 mL of 0.4000 M NaOH are mixed with 20.0 mL of phosphoric acid that has an analytical concentration of
(a) 0.200 M?
(b) 0.600 M?
(c) 0.150 M?
(d) 0.400 M?
(e) 0.300 M?
(f) 0.350 M?

* 9-21. Calculate the pH of the solution that results when 25.0 mL of NaH$_2$PO$_4$ with an analytical concentration of 0.120 M are mixed with
(a) 20.0 mL of 0.100 M NaOH.
(b) 40.0 mL of 0.100 M NaOH.
(c) 20.0 mL of 0.100 M HCl.
(d) 60.0 mL of 0.100 M H$_3$PO$_4$.
(e) 60.0 mL of 0.100 M Na$_2$HPO$_4$.
(f) 60.0 mL of 0.100 M Na$_3$PO$_4$.

9-22. Calculate the pH of the solution that results when 25.0 mL of Na$_2$HPO$_4$ with an analytical concentration of 0.120 M are mixed with
(a) 20.0 mL of 0.100 M HCl.
(b) 40.0 mL of 0.100 M HCl.
(c) 20.0 mL of 0.100 M NaOH.
(d) 40.0 mL of 0.100 M H$_3$PO$_4$.
(e) 40.0 mL of 0.100 M NaH$_2$PO$_4$.
(f) 50.0 mL of 0.100 M Na$_3$PO$_4$.

* 9-23. Describe the preparation of a buffer with a pH of
(a) 10.00 by adding sodium carbonate to 1.00 L of 0.450 M sodium hydrogen carbonate.
(b) 10.20 by adding sodium hydrogen carbonate to 500 mL of 0.315 M Na$_2$CO$_3$.
(c) 7.20 by adding Na$_3$PO$_4$ to 500 mL of 0.315 M phosphoric acid.

9-24. Describe the preparation of a buffer with a pH of
(a) 2.40 by adding maleic acid to 600 mL of 0.200 M sodium hydrogen maleate.
(b) 7.00 by adding sodium sulfite to 1.00 L of 0.205 M sulfurous acid.
(c) 3.00 by adding sodium hydrogen malonate to 800 mL of 0.160 M malonic acid.

9-25. Describe the preparation of 2.00 L of a buffer with a pH of
*(a) 10.00 from 0.250 M Na$_2$CO$_3$ and 0.300 M HCl.
*(b) 6.80 from 0.500 M HCl and 0.620 M Na$_2$SO$_3$.
(c) 6.80 from 0.600 M NaOH and 0.400 M H$_2$SO$_3$.

9-26. Calculate α for integer pH values from 1 to 8 for species derived from
* (a) tartaric acid.
(b) maleic acid.
* (c) oxalic acid.
(d) malic acid.
* (e) citric acid.
(f) dihydroxytartaric acid (see Problem 9-2).

9-27 Calculate α for integer pH values from 3 to 12 for species derived from
* (a) sulfurous acid.
(b) ethylenediamine.
* (c) arsenic acid.
(d) carbonic acid.
* (e) phosphoric acid.
(f) glutaconic acid, H$_2$C$_5$H$_4$O$_4$.
$(K_1 = 1.70 \times 10^{-4}, K_2 = 8.38 \times 10^{-6})$

9-28. Calculate α for each
* (a) carbonate-containing species in a solution with a pH of 9.60.
(b) sulfide-containing species in a solution with a pH of 11.40.
(c) citrate-containing species in a solution with a pH of 5.10.
* (d) o-phthalate-containing species in a solution with a pH of 2.30.
(e) phosphate-containing species in a solution with a pH of 10.70.
(f) tartrate-containing species in a solution with a pH of 5.80.

* 9-29. Generate a curve for the titration of 50.0 mL of a solution in which the analytical concentration of HCl is 0.0800 M and that for acetic acid is 0.100 M. Calculate the pH after the addition of 10.0, 18.0, 20.0, 22.0, 35.0, 44.0, 45.0, 46.0, and 50.0 mL of 0.200 M NaOH.

9-30. Generate a curve for the titration of 50.0 mL of a solution in which the analytical concentration of NaOH is 0.1000 M and that for NH_3 is 0.0800 M. Calculate the pH after addition of 10.0, 20.0, 24.0, 25.0, 26.0, 35.0, 44.0, 45.0, 46.0, and 50.0 mL of 0.200 M $HClO_4$.

9-31. Calculate the pH after the addition of 0.00, 5.00, 10.0, 19.0, 20.0, 21.0, 30.0, 39.0, 40.0, 41.0, and 50.0 mL of 0.200 M NaOH to 50.0 mL of 0.0800 M

     * (a) H_2SO_3.   * (c) H_3PO_3.

     (b) H_5IO_6.   (d) H_3PO_4.

applications of

neutralization titrations

10

Volumetric methods based upon neutralization involve the titration of hydronium or hydroxide ions produced directly or indirectly from the analyte. Neutralization methods are extensively employed in chemical analysis. For most applications, water serves conveniently as the solvent; it should be recognized, however, that the acidic or basic character of a solute is determined in part by the nature of the solvent in which it is dissolved and that the substitution of some other solvent for water may be sufficient to permit a titration that cannot be successfully performed in an aqueous environment. Titrations in nonaqueous media are discussed in Chapter 11.

10A REAGENTS FOR NEUTRALIZATION REACTIONS

In Chapter 8, we noted that the most pronounced pH changes in the equivalence-point region occur when strong acids and strong bases are involved in the titration. It is for this reason that standard solutions for neutralization titrations are prepared from such acids and bases.

10A-1 PREPARATION OF STANDARD ACID SOLUTIONS

Hydrochloric acid is the most commonly used standard acid for volumetric analysis. Dilute solutions of the reagent are indefinitely stable and can be used in the presence of most cations without complicating precipitation reactions. It is reported that 0.1 N solutions can be boiled for as long as 1 hr without loss of acid, provided that water lost by evaporation is periodically replaced; 0.5 N solutions can be boiled for at least 10 min without significant loss.

Solutions of perchloric acid and sulfuric acid are also stable and can serve as standard reagents in titrations where the presence of chloride ion would cause precipitation difficulties. Standard solutions of nitric acid are seldom used because of their oxidizing properties.

A standard acid solution is ordinarily prepared by diluting an approximate volume of the concentrated reagent and standardizing it against a primary standard base. Less frequently, the composition of the concentrated acid is established through careful density measurement, following which a weighed quantity is diluted to an exact volume. (Tables relating density of reagents to composition are found in most chemistry or chemical engineering handbooks.) A stock solution with an exactly known hydrochloric acid concentration can also be prepared by distillation of the concentrated reagent; under controlled conditions, the final quarter of the distillate, which is known as *constant boiling* HCl, has a fixed and known composition, its acid content being dependent only upon the atmospheric pressure. For a pressure, P, between 670 and 780 torr, the weight in air of the distillate that contains exactly one equivalent of acid is given by[1]

$$\frac{\text{g constant boiling HCl}}{\text{equivalent}} = 164.673 + 0.02039\ P \tag{10-1}$$

Standard solutions can be prepared by diluting a weighed quantity of this acid to an exactly known volume.

10A-2 STANDARDIZATION OF ACIDS

Sodium carbonate. Sodium carbonate is frequently used as a standard for acid solutions. Primary standard-grade sodium carbonate is available commercially; it can also be prepared by heating purified sodium hydrogen carbonate between 270 to 300°C for 1 hr:

$$2\text{NaHCO}_3(s) \longrightarrow \text{Na}_2\text{CO}_3(s) + \text{H}_2\text{O}(g) + \text{CO}_2(g)$$

As shown in Figure 9-4, two end points are observed in the titration of sodium carbonate. The first, corresponding to conversion of carbonate to hydrogen carbonate, occurs at a pH of about 8.3; the second, involving the formation of carbonic acid, is observed at about pH 3.8; the second end point is always used for standardization because the change in pH is greater. An even sharper end point can be achieved by boiling the solution briefly to eliminate the reaction product, carbonic acid. The sample is titrated to the first appearance of the acid color of the indicator (such as bromocresol green or methyl orange). At this point, the solution contains a large amount of carbonic acid and a small amount of unreacted hydrogen carbonate. Boiling effectively destroys this buffer by eliminating the carbonic acid:

$$\text{H}_2\text{CO}_3(aq) \longrightarrow \text{CO}_2(g) + \text{H}_2\text{O}(l)$$

---

[1] *Official Methods of Analysis of the AOAC*, 12th ed., p. 944. Washington, D.C.: Association of Official Analytical Chemists, 1975.

As a result, the solution again acquires an alkaline pH owing to the residual hydrogen carbonate ion. The titration is completed after the solution has cooled. Now, however, considerably larger changes in pH attend the final additions of acid; a sharper color change is thus observed.

As an alternative, the acid can be introduced in an amount slightly in excess of that needed to convert the sodium carbonate to carbonic acid. The solution is boiled, as before, to remove carbon dioxide; after cooling, the excess acid is back-titrated with a dilute solution of base. Any indicator suitable for a strong acid-strong base titration can be employed. The volume ratio between the acid and the base must, of course, be established by an independent titration.

Directions for the standardization of hydrochloric acid solutions against sodium carbonate are found under Standardization 3-2, Chapter 31.

Other primary standards for acids. *Tris*-(hydroxymethyl)aminomethane, $(HOCH_2)_3CNH_2$, known also as TRIS or THAM, is available in primary standard purity from commercial sources. It possesses the advantage of a substantially larger equivalent weight than sodium carbonate.

Other standards include sodium tetraborate, mercury(II) oxide, and calcium oxalate; details concerning their use can be found in standard reference works.[2]

10A-3 PREPARATION OF STANDARD SOLUTIONS OF BASE

Sodium hydroxide is the most common basic reagent, although potassium hydroxide and barium hydroxide are also encountered. None of these is obtainable in primary standard purity; after preparation to approximate strength, their solutions must be standardized.

Standard base solutions are reasonably stable so long as they are protected from prolonged exposure to glass and contact with the atmosphere. Sodium hydroxide reacts slowly with glass to form silicates; a standard solution that is to be employed for longer than a week or two should be stored in a polyethylene bottle or a glass bottle that has been coated with paraffin.

Effect of carbon dioxide upon standard base solutions. In solution, as well as the solid state, the hydroxides of sodium, potassium, and barium avidly react with atmospheric carbon dioxide to produce the corresponding carbonates:

$$CO_2(g) + 2OH^- \longrightarrow CO_3^{2-} + H_2O$$

The absorption of carbon dioxide by a standardized solution of a base does not necessarily alter the acid titer of the reagent even though hydroxide ion is consumed in the process. For example, if potassium or sodium hydroxide solutions are employed where circumstances permit the use of an indicator with an acid transition range (bromocresol green, for example), each carbonate ion in the reagent will have reacted with two hydronium ions of the analyte (see Figure 9-4). That is,

$$CO_3^{2-} + 2H_3O^+ \longrightarrow H_2CO_3 + 2H_2O$$

---

[2] See, for example, I. M. Kolthoff and V. A. Stenger, *Volumetric Analysis*, vol. 2, pp. 73–93. New York: Interscience, 1947; L. Meites, *Handbook of Analytical Chemistry*, p. 3–34. New York: McGraw-Hill, 1963.

This consumption is chemically equivalent to the amount of base used to form the carbonate, and no error will be incurred.

Unfortunately, most applications of standard base require the use of an indicator with a basic transition range (phenolphthalein, for example). Here, each carbonate ion will have consumed only one hydronium ion when the color change of the indicator is observed:

$$CO_3^{2-} + H_3O^+ \longrightarrow HCO_3^- + H_2O$$

The effective normality of the base is thus diminished, and a determinate error (called a *carbonate error*) will result.

When carbon dioxide is absorbed by standard barium hydroxide, precipitation of barium carbonate occurs:

$$CO_2(g) + Ba^{2+} + 2OH^- \longrightarrow BaCO_3(s) + H_2O$$

The acid titer is thus decreased regardless of the indicator employed in the titration; a carbonate error is the inevitable consequence.

The solid reagents used to prepare standard solutions of base represent a further source of carbonate ion. The extent of contamination is frequently great, owing to the absorption of atmospheric carbon dioxide by the solid. As a result, even freshly prepared solutions of base are likely to contain significant quantities of carbonate. Its presence will not cause a carbonate error provided the analysis is performed with the same indicator that was used for standardization; this restriction, however, causes the reagent to lose much of its versatility.

Several methods exist for the preparation of carbonate-free hydroxide solutions. Barium hydroxide may be employed as the reagent; the concentration of the carbonate can be further diminished by the addition of a neutral barium salt such as the chloride or the nitrate. A barium salt can also be used to eliminate the carbonate from a potassium or sodium hydroxide solution. The presence of barium ion is frequently undesirable, however, owing to its tendency to form slightly soluble salts with anions that may be present in the sample.

Carbonate-free solutions of the alkali-metal hydroxides may be prepared by direct solution of the freshly cleaned metals. Most chemists think that the possibility of fire and explosion during the solution process represents an unacceptable risk.

The preferred method for preparing sodium hydroxide solutions takes advantage of the very low solubility of sodium carbonate in concentrated solutions of the alkali. An approximately 50% aqueous solution of sodium hydroxide is prepared (or purchased from commercial sources); after the sodium carbonate has settled, a portion of the supernatant liquid is decanted and diluted to give the desired concentration. Details for this procedure are given under Preparation 3-3, Chapter 31. Alternatively, the concentrated sodium hydroxide solution can be filtered to eliminate the sodium carbonate.

A carbonate-free base solution must be prepared from water that contains no carbon dioxide. Distilled water, which is sometimes supersaturated with respect to carbon dioxide, should be boiled briefly to eliminate the gas; the water is allowed to cool to room temperature before the introduction of base because hot alkali solutions rapidly absorb carbon dioxide.

Figure 10-1 shows an arrangement for preventing the uptake of atmospheric

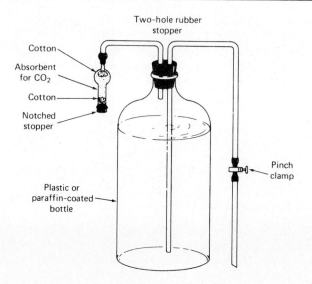

Two-hole rubber stopper

Cotton

Absorbent for CO_2

Cotton

Notched stopper

Plastic or paraffin-coated bottle

Pinch clamp

FIGURE 10-1 Arrangement for the storage of standard base solutions.

carbon dioxide by basic solutions during storage. Air entering the vessel is passed over a solid absorbent for CO_2 such as soda lime or Ascarite II.[3] The contamination that occurs as the solution is transferred from this storage bottle to the buret is negligible. Absorption during a titration can be minimized by covering the open end of the buret with a small test tube or beaker.

If a standard solution of base is to be used for no longer than a week or so, storage in a tightly stoppered polyethylene bottle will usually provide sufficient protection against the uptake of atmospheric carbon dioxide. Before stoppering, the bottle should be squeezed so that little air exists between the solution and stopper. Care should be taken to keep the bottle stoppered except during the brief periods when the contents are being transferred to a buret.

The parts of any ground-glass fitting will freeze upon prolonged exposure to alkaline solutions. For this reason, strong bases should not be stored in glass-stoppered containers. Similarly, burets equipped with glass stopcocks should be promptly drained and thoroughly cleaned after being used to dispense these reagents. A better alternative is the employment of burets equipped with Teflon stopcocks.

10A-4 STANDARDIZATION OF BASES

Several excellent primary standards are available for the standardization of bases. Most are weak organic acids that require the use of an indicator with a basic transition range.

Potassium hydrogen phthalate, $KHC_8H_4O_4$. Potassium hydrogen phthalate possesses many qualities that are desirable in a primary standard. It is a nonhygro-

---

[3] Arthur H. Thomas Company, Philadelphia, PA. Ascarite II consists of sodium hydroxide deposited on a nonfibrous silicate structure.

scopic crystalline solid with a high equivalent weight. For most purposes, the commercial analytical-grade salt can be used without the need for further purification. Potassium hydrogen phthalate of certificated purity is available from the National Bureau of Standards for the most exacting work.

Directions for the standardization of sodium hydroxide solutions are given in Standardization 3-3, Chapter 31.

Other primary standards for bases. Benzoic acid is obtainable in primary standard purity and can be used for the standardization of bases. Because its solubility in water is limited, the reagent is ordinarily dissolved in ethanol prior to dilution with water and titration.

Potassium hydrogen iodate, $KH(IO_3)_2$, is an excellent primary standard with a high equivalent weight. It is also a strong acid; as a result, virtually any indicator with a pH transition range between 4 and 10 can be used.

10B TYPICAL APPLICATIONS OF NEUTRALIZATION TITRATIONS

The most obvious application of neutralization methods is for determining the innumerable inorganic, organic, and biological species that possess inherent acidic or basic properties. Equally important, however, are the many applications that involve conversion of the analyte to an acid or a base by suitable chemical treatment followed by titration with a standard strong base or acid.

Two types of end points are commonly used for neutralization titrations. The first, based upon the color change of an indicator, was discussed in Chapter 8. The second involves the direct measurement of pH throughout the titration by means of a glass/calomel electrode system; here, the potential of the glass electrode is directly proportional to pH. Appropriate plots of the data permit establishment of the end point. The potentiometric procedure is considered in Chapter 16.

10B-1 ELEMENTAL ANALYSIS

Several important elements that occur in organic and biological systems are conveniently determined by methods that involve an acid-base titration as the final step. Generally, the elements susceptible to this type of analysis are nonmetallic. Principal among these are carbon, nitrogen, chlorine, bromine, sulfur, phosphorus, and fluorine; in addition, similar methods exist for several less commonly encountered species. In each instance, the element is converted to an inorganic acid or base that can then be titrated. A few examples follow.

Nitrogen. Nitrogen occurs in many important materials such as proteins, synthetic drugs, fertilizers, explosives, and potable water supplies. The analysis for nitrogen is thus of singular importance to research and to industry.

Two principal methods exist for the determination of organic nitrogen. The *Dumas* method, which is described in Table 25-2, is suitable for the analysis of virtually all organic nitrogen compounds. The *Kjeldahl* method was first described in 1883 and is one of the most widely used of all analytical methods. It requires no special equipment and is readily adapted to the routine analysis of large numbers of samples. The Kjeldahl method (or one of its modifications) is the standard means for determining the protein content of grains, meats, and other biological materials.

Since most proteins contain approximately the same percentage nitrogen, multiplication of the percent nitrogen by a suitable factor (6.25) gives the percent protein in a sample.

In essence, the sample is oxidized in hot, concentrated sulfuric acid during which the bound nitrogen is converted to ammonium ion. The solution is then treated with an excess of strong base; following distillation, the liberated ammonia is titrated.

The critical step in the Kjeldahl method is the oxidation with sulfuric acid. The carbon and hydrogen in the sample are converted to carbon dioxide and water, respectively. The fate of nitrogen, however, depends upon its state of combination in the original sample. If it exists as an amine or an amide, as in proteinaceous matter, conversion to ammonium ion is nearly always quantitative. On the other hand, nitrogen in higher oxidation states (as nitro, azo, and azoxy groups) will be converted to the elemental state or to nitrogen oxides during the oxidation step and will not be retained in the sulfuric acid. Low results due to the existence of nitrogen in these oxidation states can be eliminated by pretreatment of the sample with a reducing agent; complete conversion to ammonium ion during the digestion step is thus assured. One prereduction scheme calls for the addition of salicylic acid and sodium thiosulfate to the concentrated sulfuric acid solution containing the sample; the digestion is then performed in the usual way.

Certain aromatic heterocyclic compounds, such as pyridine and its derivatives, are particularly resistant to complete oxidation by sulfuric acid. Thus, unless special precautions are followed, low results attend the analysis for nitrogen in samples containing these species (see Figure 3-1, p. 42).

The oxidation process is the most time-consuming step in the Kjeldahl method; an hour or more may be needed for refractory samples. Of the many modifications aimed at improving the kinetics of the process, the one proposed by Gunning is now almost universally employed. Here, a neutral salt such as potassium sulfate is added to increase the boiling point of the sulfuric acid solution and thus the temperature at which the oxidation occurs. Care is needed, however, because oxidation of the ammonium ion can occur if the salt concentration is too great. This problem is enhanced if evaporation of the sulfuric acid is excessive during digestion.

Attempts to hasten the Kjeldahl oxidation by introducing such stronger oxidizing agents as perchloric acid, potassium permanganate, and hydrogen peroxide often fail because ammonium ions are partially oxidized to volatile nitrogen oxides.

Many substances catalyze the oxidation of organic compounds by sulfuric acid. Mercury, copper, and selenium, either combined or in the elemental state, are effective. Mercury(II), if present, must be precipitated with hydrogen sulfide prior to the distillation step; otherwise, some ammonia will be retained as an ammine complex.

Figure 10-2 illustrates typical distillation arrangements for a Kjeldahl analysis. The long-necked container, which is used for both oxidation and distillation, is called a *Kjeldahl flask*. After the oxidation is judged complete, the contents of the flask are cooled, diluted with water, and then made basic to liberate the ammonia:

$$NH_4^+ + OH^- \longrightarrow NH_3(g) + H_2O$$

In the apparatus shown in Figure 10-2a, the base is added slowly by partially

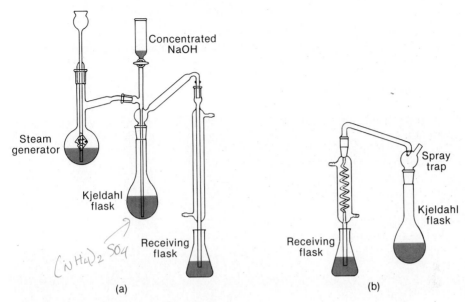

FIGURE 10-2 Kjeldahl distillation apparatus.

opening the stopcock from the storage vessel, and a steam distillation carries liberated ammonia to the receiving flask.

As an alternative, the dense sodium hydroxide solution can be carefully poured down the side of the flask to form a second, lower layer. The flask is then quickly connected to a spray trap (Figure 10-2b) and an ordinary condenser before loss of ammonia can occur; only then are the two layers mixed by gently swirling the flask.

Note that in both types of apparatus shown in Figure 10-2, the end of the condenser extends into the acid in the receiving flask during the distillation. It must be removed as heating is discontinued, however; otherwise, the contents of the flask will be drawn back into the condenser.

Two methods exist for titration of the collected ammonia. In one, the receiver contains a known quantity of standard acid; after distillation is complete, the excess acid is back-titrated with standard base. An indicator with an acidic transition range is required, owing to the presence of ammonium ions at equivalence. A convenient alternative that requires only one standard solution involves use of an unmeasured excess of boric acid in the receiving flask; its reaction with ammonia is

$$HBO_2 + NH_3 \rightleftharpoons NH_4^+ + BO_2^-$$

The borate ion produced is a reasonably strong base and can be titrated with a standard solution of hydrochloric acid:

$$BO_2^- + H_3O^+ \rightleftharpoons HBO_2 + H_2O$$

At the equivalence point, the solution contains boric acid and ammonium ions; an indicator with an acidic transition interval (such as bromocresol green) is again required.

Details of the Kjeldahl method are found in Method 3-7, Chapter 31.

Sulfur. Sulfur in organic and biological materials is conveniently determined by burning the sample in a stream of oxygen. The sulfur dioxide (and sulfur trioxide) formed during the oxidation is collected in a dilute solution of hydrogen peroxide:

$$SO_2(g) + H_2O_2 \longrightarrow H_2SO_4$$

The sulfuric acid is then titrated with standard base.

Other elements. Table 10-1 lists other elements that can be determined by neutralization methods.

10B-2 DETERMINATION OF INORGANIC SUBSTANCES

Numerous inorganic species can be determined by titration with strong acids or bases. A few examples follow.

Ammonium salts. Ammonium salts can be conveniently determined by conversion to ammonia with strong base and distillation in the Kjeldahl apparatus shown in Figure 10-2. The ammonia is collected and titrated as in the Kjeldahl method.

Nitrates and nitrites. The method just considered can also be applied to the determination of inorganic nitrate or nitrite by reducing these species to ammonium ion. Devarda's alloy (50% Cu, 45% Al, 5% Zn) is commonly used as the reducing agent. Granules of the alloy are introduced to a strongly alkaline solution of the sample in a Kjeldahl flask. The ammonia is distilled after reaction is complete. Arnd's alloy (60% Cu, 40% Mg) can also be used as the reducing agent.

Carbonate and carbonate mixtures. The qualitative and quantitative determination of the constituents in a solution containing sodium carbonate, sodium hydrogen carbonate, and sodium hydroxide, alone or admixed, provides interesting examples of neutralization titrations. No more than two of these three constituents can exist in appreciable amount in any solution because reaction will eliminate the third. Thus, the mixing of sodium hydroxide with sodium hydrogen carbonate results in the formation of sodium carbonate until one or the other (or both) of the original reactants is exhausted. If the sodium hydroxide is used up, the solution

Table 10-1 Elemental Analysis Based on Neutralization Titrations

Element	Converted to	Absorption or Precipitation Products	Titration
N	NH_3	$NH_3(g) + H_3O^+ \rightarrow NH_4^+ + H_2O$	Excess HCl with NaOH
S	SO_2	$SO_2(g) + H_2O_2 \rightarrow H_2SO_4$	NaOH
C	CO_2	$CO_2(g) + Ba(OH)_2 \rightarrow BaCO_3(s) + H_2O$	Excess $Ba(OH)_2$ with HCl
Cl(Br)	HCl	$HCl(g) + H_2O \rightarrow Cl^- + H_3O^+$	NaOH
F	SiF_4	$SiF_4(g) + H_2O \rightarrow H_2SiF_6$	NaOH
P	H_3PO_4	$12H_2MoO_4 + 3NH_4^+ + H_3PO_4 \rightarrow$ $(NH_4)_3PO_4 \cdot 12MoO_3(s) + 12H_2O + 3H^+$ $(NH_4)_3PO_4 \cdot 12MoO_3(s) + 26OH^- \rightarrow$ $HPO_4^{2-} + 12MoO_4^{2-} + 14H_2O + 3NH_3(g)$	Excess NaOH with HCl

will contain sodium carbonate and sodium hydrogen carbonate; if sodium hydrogen carbonate is depleted, sodium carbonate and sodium hydroxide will remain. Finally, if equimolar amounts are mixed, the principal solute species will be sodium carbonate.

The analysis of such mixtures requires two titrations with standard acid. An indicator with a transition in the vicinity of pH 8 to 9 is used for one; an acid-range indicator is used for the other. The composition of the solution can be deduced from the relative volumes of acid needed to titrate equal volumes of the sample (see Table 10-2 and Figure 9-4). Once the composition has been established, the volume data can be used to determine the concentration of each component in the sample.

EXAMPLE 10-1. A solution to be analyzed contained $NaHCO_3$, Na_2CO_3, NaOH, alone or in permissible combination. Titration of a 50.0-mL portion to a phenolphthalein end point required 22.1 mL of 0.100 M HCl. A second 50.0-mL aliquot required 48.4 mL of the HCl when titrated to a bromocresol green end point. Deduce the composition, and calculate the molar solute concentrations of the original solution.

Had the solution contained only NaOH, the volume of acid required would have been the same, regardless of indicator (that is, $V_{ph} = V_{bg}$). In fact, however, the second titration required a total of 48.4 mL. Because less than half of this amount was involved in the first titration, the solution must have contained some $NaHCO_3$ in addition to Na_2CO_3. We can now calculate the concentration of the two constituents. When the phenolphthalein end point was reached, the CO_3^{2-} originally present was converted to HCO_3^-. Thus,

$$mmol\ Na_2CO_3 = 22.1\ mL \times 0.100\ mmol/mL = 2.21$$

The titration from the phenolphthalein to the bromocresol green end point (48.4 − 22.1 = 26.3 mL) involved both the hydrogen carbonate originally present as well as that formed by titration of the carbonate. Thus,

$$mmol\ NaHCO_3 + mmol\ Na_2CO_3 = 26.3 \times 0.100$$

Hence,

$$mmol\ NaHCO_3 = 2.63 - 2.21 = 0.42$$

The molar concentrations are readily calculated from these data:

$$M_{Na_2CO_3} = \frac{2.21}{50.0} = 0.0442\ mmol/mL$$

$$M_{NaHCO_3} = \frac{0.42}{50.0} = 0.0084\ mmol/mL$$

In practice, the titration used in this example is not entirely satisfactory because the pH change corresponding to the hydrogen carbonate equivalence point is not sufficient to give a sharp color change with a chemical indicator (see Figure 9-4). Titration to a color match with a solution containing an approximately equivalent amount of sodium hydrogen carbonate is helpful; nevertheless, errors of one percent or more must be expected.

The limited solubility of barium carbonate can be used to improve the titration of carbonate/hydroxide or carbonate/hydrogen carbonate mixtures. The *Winkler* method for the determination of carbonate/hydroxide mixtures involves titration of both components in an aliquot with an acid-range indicator. An excess of

Table 10-2 Volume Relationship in the Analysis of Mixtures Containing Carbonate, Hydrogen Carbonate, and Hydroxide Ions

Constituents Present	Relationship Between Volume of Acid Needed to Reach a Phenolphthalein End Point, V_{ph}, and a Bromocresol Green End Point, V_{bg}
NaOH	$V_{ph} = V_{bg}$
Na_2CO_3	$V_{ph} = \frac{1}{2}V_{bg}$
$NaHCO_3$	$V_{ph} = 0, V_{bg} > 0$
NaOH, Na_2CO_3	$V_{ph} > \frac{1}{2}V_{bg}$
Na_2CO_3, $NaHCO_3$	$V_{ph} < \frac{1}{2}V_{bg}$

neutral barium chloride is then added to a second aliquot to precipitate the carbonate ion, following which the hydroxide ion is titrated to a phenolphthalein end point. Provided the concentration of the excess barium ion is about 0.1 M, the solubility of barium carbonate is too low to interfere with the titration.

An accurate analysis of a carbonate/hydrogen carbonate mixture can be achieved by establishing the total equivalents through titration of an aliquot to the acidic end point with an indicator such as bromocresol green. The hydrogen carbonate in a second aliquot is converted to carbonate by the addition of a known excess of standard base. After a large excess of barium chloride has been introduced, the excess base is titrated with standard acid to a phenolphthalein end point.

The presence of solid barium carbonate does not hamper end point detection in either of these methods.

Directions for the analysis of carbonate/bicarbonate mixtures are given in Method 3-5, Chapter 31.

10B-3 DETERMINATION OF ORGANIC FUNCTIONAL GROUPS

Neutralization titrations are convenient for the direct or indirect determination of several organic functional groups. Brief descriptions of methods for the more common groups follow.

Carboxylic and sulfonic acid groups. Carboxylic and sulfonic acid groups are the two most common structures that impart acidity to organic compounds. Most carboxylic acids have dissociation constants that range between 10^{-4} and 10^{-6} and are thus readily titrated. An indicator that changes color in the basic range is required; phenolphthalein is widely used for this purpose.

Many carboxylic acids are not sufficiently soluble in water to permit a direct titration in this medium. Where this problem exists, the acid can be dissolved in ethanol and titrated with aqueous base. Alternatively, the acid may be dissolved in an excess of standard base; the unreacted base is then back-titrated with standard acid.

Sulfonic acids are generally strong acids and readily dissolve in water. Their titration with a base therefore is straightforward.

Neutralization titrations are often employed to determine equivalent weights of purified organic acids; the data serve as an aid in qualitative identification.

Amine groups. Aliphatic amines generally have basic dissociation constants on the order of 10^{-5} and can thus be titrated directly with a solution of a strong acid. Aromatic amines such as aniline and its derivatives, on the other hand, are usually too weak for titration in aqueous medium ($K_b \cong 10^{-10}$); the same is true for cyclic amines with aromatic character such as pyridine and its derivatives. Many saturated cyclic amines, such as piperidine, behave in the same way as aliphatic amines.

Those amines that are not susceptible to neutralization titration in aqueous media are often readily determined in a nonaqueous solvent, such as glacial acetic acid, that enhances their basicity.

Ester groups. Esters are commonly determined by saponification with a measured quantity of standard base:

$$R_1COOR_2 + OH^- \longrightarrow R_1COO^- + HOR_2$$

The excess base is then titrated with standard acid.

Esters vary widely in their reactivity toward saponification. Some require several hours of heating with a base to complete the saponification. A few react rapidly enough to permit direct titration with standard base. Typically, the ester is refluxed with standard $0.5 N$ base for 1 to 2 hr. After cooling, the excess base is determined with standard acid.

Hydroxyl groups. Hydroxyl groups in organic compounds can be determined by esterification with various carboxylic acid anhydrides or chlorides; the two most common reagents are acetic anhydride and phthalic anhydride. With acetic anhydride, the reaction is

$$(CH_3CO)_2O + ROH \longrightarrow CH_3COOR + CH_3COOH$$

The acetylation is ordinarily carried out by mixing the sample with a carefully measured volume of acetic anhydride in pyridine. After heating, water is added to hydrolyze the unreacted anhydride:

$$(CH_3CO)_2O + H_2O \longrightarrow 2CH_3COOH$$

The acetic acid is then titrated with a standard solution of alcoholic sodium or potassium hydroxide. A blank is carried through the analysis to establish the original amount of anhydride.

Amines, if present, are converted quantitatively to amides by acetic anhydride; a correction for this source of interference is frequently possible by a direct titration of another portion of the sample with standard acid.

Carbonyl groups. Many aldehydes and ketones can be determined with a solution of hydroxylamine hydrochloride. The reaction, which produces an oxime, is

$$\begin{array}{c} R_1 \\ \diagdown \\ \diagup \\ R_2 \end{array} C{=}O + NH_2OH{\cdot}HCl \longrightarrow \begin{array}{c} R_1 \\ \diagdown \\ \diagup \\ R_2 \end{array} C{=}NOH + HCl + H_2O$$

where R_2 may be an atom of hydrogen. The liberated hydrochloric acid is titrated with base. Here, again the conditions necessary for quantitative reaction vary. Typically, 30 min suffice for aldehydes. Many ketones require refluxing with the reagent for an hour or more.

10B-4 DETERMINATION OF SALTS

The total salt content of a solution can be accurately and readily determined by an acid-base titration. Here, the salt is converted to an equivalent amount of an acid or a base by passage through a column packed with an ion-exchange resin. (This application is considered in more detail in Section 26E-3. Standard acid or base solutions can also be prepared with ion-exchange resins. Here, a solution containing a known weight of a pure compound, such as sodium chloride, is washed through the resin column and diluted to a known volume. The salt liberates an equivalent amount of acid or base, permitting calculation of the normality of the reagent in a straightforward way.

PROBLEMS*

10-1. Express the equivalent weight of the acid in the accompanying equations as a fraction or a multiple of its formula weight.
* (a) $H_2SO_3 + KOH \rightarrow KHSO_3 + H_2O$
   (b) $2LiOH + H_4P_2O_7 \rightarrow Li_2H_2P_4O_7 + 2H_2O$
* (c) $Ba(OH)_2 + 2HIO_3 \rightarrow Ba(IO_3)_2(s) + 2H_2O$
   (d) $Na_2CO_3 + 2HCl \rightarrow CO_2 + H_2O + 2NaCl$
* (e) $2H_2CO_3 + Ca(OH)_2 \rightarrow Ca(HCO_3)_2 + 2H_2O$

10-2. Express the equivalent weight of each base in Problem 10-1 in terms of a fraction or multiple of its formula weight.

10-3. Calculate the number of milliequivalents in
* (a) 0.376 g of H_2SO_3 in terms of reaction 10-1(a).
   (b) 21.79 mL of 0.1946 M $H_4P_2O_7$ in terms of reaction 10-1(b).
* (c) 16.84 mL of 0.0240 M $Ba(OH)_2$ in terms of reaction 10-1(c).
   (d) 0.518 g of Na_2CO_3 in terms of reaction 10-1(d).
* (e) 0.854 g of $Ca(OH)_2$ in terms of reaction 10-1(e).

* 10-4. Describe the preparation of
   (a) 2.0 L of approximately 0.12 N HCl from the concentrated reagent (sp gr 1.19, 37% HCl w/w).
   (b) 1.000 L of 0.1200 N reagent from constant-boiling HCl which was distilled at 740 torr.
   (c) 500.0 mL of 0.1200 N reagent from 2.170 N HCl.
   (d) 1.000 L of 0.1500 N reagent from a solution that has a specific gravity of 1.098 and is 20.00% HCl (w/w).

10-5. Describe the preparation of
   (a) 1.0 L of approximately 0.025 N $Ba(OH)_2$ from solid $Ba(OH)_2 \cdot 8H_2O$.
   (b) 2.5 L of approximately 0.12 N NaOH from the solid.
   (c) 250.0 mL of 0.1500 N reagent from 0.825 N NaOH.
   (d) 2.0 L of approximately 0.10 N reagent from concentrated NaOH (sp gr 1.525, 50% NaOH w/w).

---

* Answers to problems or parts of problems marked with an asterisk are to be found at the end of the book.

* 10-6. Calculate the normality of a hydrochloric acid solution if
  (a) a 50.0-mL aliquot yielded 0.572 g of AgCl.
  (b) 37.84 mL were needed to titrate a 0.2056-g sample of primary standard Na_2CO_3 (product: CO_2).
  (c) 19.27 mL were used in the titration of 50.00 mL of 0.04185 N Ba(OH)$_2$.
  (d) 45.88 mL were needed to titrate the Na_2CO_3 (product: CO_2) formed from the ignition of 0.3458 g of primary standard $Na_2C_2O_4$. [$Na_2C_2O_4 \rightarrow Na_2CO_3 + CO(g)$]

10-7. Calculate the normality of a barium hydroxide solution if
  (a) a 50.00-mL aliquot produced 0.2421 g of $BaSO_4$.
  (b) 29.53 mL were needed to titrate a 0.4825-g sample of primary standard potassium hydrogen phthalate.
  (c) addition of a 50.00-mL aliquot to a 0.2841-g sample of primary standard benzoic acid required a 3.19-mL back-titration with 0.0695 N HCl.
  (d) 44.86 mL were needed to titrate a 25.00-mL aliquot of HCl that had a titer of 5.23 mg AgNO$_3$/mL.

* 10-8. Standardization of a sodium hydroxide solution against potassium hydrogen phthalate (KHP) yielded the accompanying results.

| wt KHP, g | 0.7987 | 0.8365 | 0.8104 | 0.8039 |
| volume NaOH, mL | 38.29 | 39.96 | 38.51 | 38.29 |

Calculate
  (a) the average normality of the base.
  (b) the average relative deviation from the mean for the results.
  (c) the standard deviation for the data.

10-9. The normality of a perchloric acid solution was established by titration against primary standard sodium carbonate (product:CO_2); the following data were obtained.

| wt Na_2CO_3, g | 0.2068 | 0.1997 | 0.2245 | 0.2137 |
| volume $HClO_4$, mL | 36.31 | 35.11 | 39.00 | 37.54 |

  (a) Calculate the average normality of the acid and the average relative deviation from the mean for the results.
  (b) Calculate the standard deviation for the data.
  (c) Does statistical justification exist for disregarding the outlying result?

* 10-10. Calculate a range of sample weights that would yield a 35- to 45-mL titration for the standardization of approximately
  (a) 0.10 N HCl against Na_2CO_3.
  (b) 0.12 N NaOH against potassium hydrogen phthalate.
  (c) 0.050 N $HClO_4$ against HgO. Reaction:

$$HgO + 4I^- + 2H_3O^+ \longrightarrow HgI_4^{2-} + 3H_2O$$

  (d) 0.030 N Ba(OH)$_2$ against KH(IO$_3$)$_2$.
  (e) 0.080 N HCl against borax, $Na_2B_4O_7 \cdot 10H_2O$. Reaction:

$$B_4O_7^{2-} + 2H_3O^+ + 3H_2O \longrightarrow 4H_3BO_3$$

  (f) 0.15 N NaOH against benzoic acid, C_6H_5COOH.
  (g) 0.10 N $HClO_4$ against THAM [$tris$(hydroxymethyl)aminomethane]. Reaction:

$$(HOCH_2)_3CNH_2 + H_3O^+ \longrightarrow (HOCH_2)_3CNH_3^+ + H_2O$$

  (h) 0.20 N HCl against a weighed quantity of $Na_2C_2O_4$ which has been ignited to Na_2CO_3 [see Problem 10-6(d) for reaction]. Carbon dioxide and water are the products of the titration.

* 10-11. Select an indicator (Table 8-1) that would be suitable for the titration of
  (a) 0.00100 M HCl with 0.00100 M KOH.

    (b) $0.100\ M$ CH_3NH_2 with $0.100\ M$ $HClO_4$.

    (c) $0.0500\ M$ hydroxylamine hydrochloride with $0.100\ M$ NaOH.

    (d) the sparingly soluble bismuth(III) acetate by titration of the excess HCl used to dissolve the salt with $0.100\ M$ NaOH.

    (e) $0.0500\ M$ H_3AsO_4 to the $H_2AsO_4^-$ end point, with $0.100\ M$ NaOH.

    (f) the strong base in 50 mL of a solution that is $0.0800\ M$ in NaOH and $0.100\ M$ in hydroxylamine with $0.100\ M$ HCl.

    (g) $0.0010\ M$ $HClO_4$ with $0.0010\ M$ KOH.

10-12. Select an indicator (Table 8-1) that would be suitable for the titration of

    (a) $0.0010\ M$ $Ba(OH)_2$ with $0.002\ M$ HCl.

    (b) $0.100\ M$ aniline hydrochloride with $0.0800\ M$ NaOH.

    (c) $0.150\ M$ sodium phenolate with $0.100\ M$ HCl.

    (d) $0.100\ M$ Na_3PO_4 to the HPO_4^{2-} end point with $0.100\ M$ HCl.

    (e) $0.100\ M$ H_3PO_4 to the $H_2PO_4^-$ end point with $0.100\ M$ NaOH.

    (f) $0.100\ M$ Na_2HPO_4 to the $H_2PO_4^-$ end point with $0.100\ M$ HCl.

    (g) the hydrochloric acid in a solution that is $0.0500\ M$ in hydroxylamine hydrochloride and $0.0600\ M$ in HCl with $0.100\ M$ NaOH.

    (h) the total acid content of the solution in (g).

* 10-13. Calculate the equivalent weight of a weak dibasic acid if 47.63 mL of $0.1206\ N$ NaOH are needed to titrate a 0.620-g sample to a phenolphthalein end point.

* 10-14. A 50.0-mL sample of vinegar (density $= 1.06$) was diluted to 500 mL in a volumetric flask. Titration of 50.0-mL aliquots required an average of 42.5 mL of $0.0900\ N$ NaOH to achieve a phenolphthalein end point. Express the acid content of the vinegar as percent acetic acid.

10-15. A 10.00-mL sample of table wine was diluted to 100 mL and titrated to a phenolphthalein end point with 25.62 mL of $0.0409\ N$ NaOH. Express the acidity of the wine in terms of grams tartaric acid ($H_2C_4H_4O_6$, fw $= 150$) per 100 mL, assuming that both of the acidic hydrogens are titrated.

* 10-16. A 0.6216-g sample of impure sodium tetraborate required a 27.73-mL titration with $0.1058\ N$ HCl [see Problem 10-10(e) for reaction]. Express the results of this analysis in terms of percent.

    (a) $Na_2B_4O_7\cdot 10H_2O$.   (c) B_2O_3.

    (b) $Na_2B_4O_7$.   (d) B.

10-17. A 1.883-g sample of impure mercury(II) oxide was brought into solution with an excess of potassium iodide. Reaction:

$$HgO(s) + H_2O + 4I^- \longrightarrow HgI_4^{2-} + 2OH^-$$

Calculate the percentage of HgO in the sample if 41.8 mL of $0.1140\ N$ HCl were needed to titrate the liberated base.

* 10-18. A 0.2947-g specimen containing sodium azide was treated with an unmeasured excess of sodium nitrite and 50.00 mL of $0.1030\ N$ perchloric acid. Upon completion of the reaction

$$2H^+ + N_3^- + NO_2^- \longrightarrow H_2O + N_2O(g) + N_2(g)$$

the excess acid was titrated with 13.49 mL of standard $0.0800\ N$ NaOH. Calculate the percentage of NaN_3 in the sample.

10-19. Specifications for an acid rinse tank in a plating plant call for an operating range between 36 and 48 g of H_2SO_4 per liter. Routine analysis is to be undertaken to assure that this range is maintained. 50.0-mL samples of the bath are diluted to 250 mL in volumetric flasks, and 25.0-mL aliquots of the diluted solutions are taken for titration with standard base.

    (a) What should be the concentration of standard base if a maximum titration no larger than 40.0 mL is desired?

(b) What volume of the standard base solution will be needed when the tank is at its lower tolerable operating limit?

(c) Calculate the normality of a sodium hydroxide solution prepared for this analysis if 29.4 mL were needed to titrate a 0.710-g sample of primary standard potassium hydrogen phthalate.

(d) Compute the titer of the solution in (c) in terms of milligrams of H_2SO_4.

(e) What volume of 40% by weight sodium hydroxide solution ($d = 1.430$ g/mL) should be taken for the preparation of 8.0 L of 0.150 N reagent?

10-20. The formaldehyde in a pesticide preparation was determined by weighing a 1.10-g sample into a flask and treating it with a neutral solution of sodium sulfite. Reaction:

$$\begin{array}{c} \qquad\qquad\qquad\qquad\quad\ \ \text{H} \\ \qquad\qquad\qquad\qquad\quad\ \ | \\ \text{HCHO} + \text{Na}_2\text{SO}_3 + \text{H}_2\text{O} \longrightarrow \text{H}-\text{C}-\text{SO}_3\text{Na} + \text{OH}^- + \text{Na}^+ \\ \qquad\qquad\qquad\qquad\quad\ \ | \\ \qquad\qquad\qquad\qquad\quad\ \ \text{OH} \end{array}$$

Calculate the percentage of formaldehyde in the sample if 37.49 mL of 0.2664 N HCl were needed to titrate the liberated base.

* 10-21. A preliminary titration with 0.1020 N NaOH to a phenolphthalein end point neutralized the acidic contaminants as well as the carboxylic hydrogen of a 0.4186-g aspirin sample. An additional 28.64-mL of the base were then added to cleave the ester linkage:

Calculate the percentage of aspirin (fw = 180.2) in the sample if 15.24 mL of 0.0736 N HCl were needed to titrate the excess base, again to a phenolphthalein end point.

10-22. What is the normality of a hydrochloric acid solution that has a titer of 1.000 mg N/mL with respect to the Kjeldahl method?

10-23. Calculate the percentage of nitrogen in a plant food preparation if a 36.43-mL titration with 0.1241 N HCl were needed to complete the Kjeldahl analysis of a 0.6332-g sample.

* 10-24. A 1.047-g sample of canned tuna was analyzed by the Kjeldahl method; 24.61 mL of 0.1180 N HCl were required to titrate the liberated ammonia. Calculate the percentage of nitrogen in the sample.

* 10-25. The percentage of protein can be estimated by dividing the percentage of nitrogen by 0.160. Calculate the grams of protein in a 6.50-oz can of tuna in Problem 10-24 (1 oz = 28.3 g).

10-26. The *Merck Index* indicates that 10 mg of guanidine, CH_5N_3, may be administered for each kilogram of body weight in the treatment of myasthenia gravis. The nitrogen in a 4-tablet sample that weighed a total of 7.50 g was converted to ammonia by a Kjeldahl digestion, followed by distillation into 100.0 mL of 0.1750 N HCl. The analysis was completed by titrating the excess acid with 11.37 mL of 0.1080 N NaOH. How many of these tablets represent a proper dose for a 48-kg patient?

* 10-27. The atmosphere within a paper mill was tested for compliance with the federal limit of less than 5 ppm SO_2. Air was drawn at the rate of 12 L/min through a trap containing 100 mL of 1% H_2O_2. Reaction:

$$H_2O_2 + SO_2(g) \longrightarrow 2H^+ + SO_4^{2-}$$

The sulfuric acid produced during a 15-min test required 6.49 mL of 0.00537 N NaOH. Calculate the concentration of SO_2 in ppm (i.e., mL $SO_2/10^6$ mL of air), given that the density of sulfur dioxide is 2.85 mg/mL.

10-28. Federal regulations set an upper limit of 50 ppm (i.e., mL $NH_3/10^6$ mL air) for the ammonia content in a work environment (the density of NH_3 is 0.771 g/L). A 300-L sample of air from a manufacturing plant was passed through 40.0 mL of 0.02054 N HCl, following which the excess acid was titrated with 23.79 mL of 0.00618 N NaOH. Is the manufacturer in compliance with the regulation?

* 10-29. A 4.710-g sample that contained ammonium sulfate, potassium nitrate, and inert materials was dissolved in sufficient water to give 250.0 mL of solution. A 50.00-mL aliquot was transferred to a Kjeldahl distillation apparatus and rendered alkaline. The liberated ammonia was collected in an unmeasured excess of boric acid and subsequently titrated with 10.24 mL of 0.0961 N HCl. A second 50.00-mL aliquot was treated with Devarda's alloy to reduce the nitrate. Net reaction:

$$NO_3^- + 6H_2O + 8e \longrightarrow NH_3 + 9OH^-$$

Because this reduction occurs in alkaline solution, all of the ammonia was distilled and again collected in boric acid. Titration of the distillate required 32.07 mL of the standard acid. Calculate the respective percentages of ammonium sulfate and potassium nitrate in the sample.

10-30. The analysis described in Problem 10-29 was used to determine the content of ammonium nitrate and ammonium sulfate in a 3.072-g sample that had been dissolved in sufficient water to give 250.0 mL of solution. Titration of the ammonia liberated from a 50.00-mL aliquot used 34.71 mL of 0.0961 N HCl. A second 50.00-mL aliquot, after treatment with Devarda's alloy, yielded sufficient ammonia to require a 44.83-mL titration with the standard acid. Calculate the percentages of ammonium nitrate and ammonium sulfate in the sample.

* 10-31. Exactly 34.40 mL of 0.1070 N NaOH were needed to titrate a 0.7050-g sample containing $Na_2C_2O_4 \cdot H_2O$, $H_2C_2O_4 \cdot 2H_2O$, and inert materials. After evaporation to dryness, the titrated sample was strongly ignited to convert the product, $Na_2C_2O_4$, to Na_2CO_3 [see Problem 10-6(d) for reaction]. The ignited residue was cooled, dissolved in water, and treated with 50.0 mL of 0.1250 N HCl. Calculate the percentages of $Na_2C_2O_4 \cdot H_2O$ and $H_2C_2O_4 \cdot 2H_2O$ in the sample if a 2.54-mL back-titration with the NaOH solution was needed to neutralize the excess acid.

10-32. A 50.0-mL aliquot of a solution containing Na_3PO_4 and Na_2HPO_4 required 46.9 mL of 0.1080 N HCl when titrated to a bromocresol green end point. A second 50.0-mL aliquot, titrated to a thymolphthalein end point, required 17.6 mL of the acid. Calculate the number of milligrams of the two solutes in each milliliter of solution.

* 10-33. A 5.26-g sample containing sodium carbonate, sodium hydrogen carbonate, and inert materials was dissolved in sufficient water to give 250.0 mL of solution. A 50.0-mL aliquot required 16.70 mL of 0.1204 N HCl when titrated to a phenolphthalein end point; a second 50.0-mL aliquot, titrated to a bromocresol green end point, required 49.40 mL of the acid. Calculate the respective percentages of Na_2CO_3 and $NaHCO_3$ in the sample.

10-34. In addition to inert materials, a solid sample may contain NaOH, Na_2CO_3, $NaHCO_3$, or any of the compatible combinations of these substances. A 5.20-g specimen was dissolved in sufficient water to give 250.0 mL of solution. Titration of 50.0-mL aliquots of this solution required an average of 44.6 mL of 0.1185 N HCl with bromocresol green as indicator and 28.8 mL with phenolphthalein.
(a) What is the composition of the sample?
(b) Calculate the percentage of each basic component that is present.

* 10-35. Calculate the volume of 0.1144 N HCl needed to titrate
  - (a) 24.85 mL of 0.1637 M Na_2HPO_4 to a bromocresol green end point.
  - (b) 19.46 mL of 0.1073 M Na_3PO_4 to a bromocresol green end point.
  - (c) 25.00 mL of a solution that is 0.0614 M with respect to Na_3PO_4 and 0.1005 M with respect to Na_2HPO_4 to a thymolphthalein end point.
  - (d) 25.00 mL of the solution in (c) to a bromocresol green end point.
  - (e) the solution that results upon mixing 30.00 mL of 0.0765 M H_3PO_4 with 20.00 mL of 0.1358 M Na_3PO_4 to a bromocresol green end point.

10-36. Calculate the volume of 0.1374 N NaOH needed to titrate
  - (a) 21.64 mL of 0.1022 M H_3PO_4 to a bromocresol green end point.
  - (b) 34.72 mL of 0.0997 M NaH_2PO_4 to a thymolphthalein end point.
  - (c) 27.66 mL of a solution that is 0.0617 M with respect to H_3PO_4 and 0.1308 M with respect to NaH_2PO_4 to a bromocresol green end point.
  - (d) 20.00 mL of the solution in (c) to a thymolphthalein end point.
  - (e) the solution that results upon mixing 25.00 mL of 0.1047 M H_3PO_4 and 20.00 mL of 0.0862 M Na_3PO_4 to a thymolphthalein end point.

* 10-37. Solutions containing NaOH, Na_2CO_3, and $NaHCO_3$, alone or in compatible combination, were titrated with standard HCl. Use the accompanying data to deduce the composition of the solution and the number of milligrams of solute(s) in each.

| | Volume of 0.1224 N HCl needed for titration to | |
| | bromocresol green | phenolphthalein |
Solution	end point	end point
(a)	40.15	12.66
(b)	17.24	0.00
(c)	36.82	36.80
(d)	43.47	21.74
(e)	41.24	34.66

10-38. A series of solutions can contain HCl, H_3PO_4, and NaH_2PO_4, alone or in any compatible combination. Use the accompanying titration data to establish the composition of the solution and the number of milligrams of solute(s) in each.

| | Volume of 0.1158 N NaOH needed for titration to | |
| | bromocresol green | phenolphthalein |
Solution	end point	end point
(a)	0.00	18.34
(b)	8.48	16.96
(c)	8.48	31.84
(d)	10.77	10.77
(e)	24.56	30.00

acid-base titrations

in nonaqueous media

11

In Chapter 8, we noted that acids or bases with dissociation constants smaller than about 1×10^{-8} cannot be titrated because their reactions are not sufficiently complete to yield a satisfactory end point. Many of these same species become titratable, however, in nonaqueous solvent systems that emphasize the acidic or basic character that they do possess. Nonaqueous solvent systems have the further advantage of permitting the titration of numerous substances that are sparingly soluble in water.[1]

Unfortunately, the quantitative information and data required for the derivation of titration curves are usually lacking for nonaqueous solvents; as a consequence, conclusions regarding the feasibility of titrations must be based on qualitative concepts only. Here, the Brønsted theory (p. 6) is often of considerable help.

11A TYPES OF SOLVENTS FOR NONAQUEOUS TITRATIONS

It is convenient to classify solvents into three categories, depending upon their character. All three find widespread use for titrimetry.

Amphiprotic solvents possess both acidic and basic properties and undergo

[1] For more extensive discussion of nonaqueous neutralization titrations, see: J. S. Fritz, *Titrations in Nonaqueous Solvents*. Boston: Allyn and Bacon, 1973; H. A. Laitinen and W. E. Harris, *Chemical Analysis*, 2d ed., pp. 56–92 and 112–121. New York: McGraw-Hill, 1975; C. A. Streuli, in *Treatise on Analytical Chemistry*, I. M. Kolthoff and P. J. Elving, Eds., Part I, vol. 11, p. 7035. New York: Wiley, 1975.

self-dissociation or *autoprotolysis*. Although water is the most common amphiprotic solvent, many other substances exhibit analogous behavior. Thus,

$$2H_2O \rightleftharpoons H_3O^+ + OH^-$$

$$2C_2H_5OH \rightleftharpoons C_2H_5OH_2^+ + C_2H_5O^-$$

$$2HOAc \rightleftharpoons H_2OAc^+ + OAc^-$$

$$2NH_3 \rightleftharpoons NH_4^+ + NH_2^-$$

or, in general,

$$2SH \rightleftharpoons SH_2^+ + S^-$$

where SH represents the amphiprotic solvent molecule and SH_2^+ the solvated proton; the base thus corresponds to the anion S^-. Table 11-1 lists autoprotolysis constants for several common solvents.

Some amphiprotic solvents, notably water and the alcohols, have approximately equal tendencies to acquire and to donate protons. Others, such as acetic acid, sulfuric acid, and formic acid, have considerably stronger acidic than basic properties. Still others—ammonia and ethylenediamine, for example—are stronger bases than acids.

Aprotic, or inert, solvents have no appreciable acidic or basic character and do not undergo autoprotolysis to any detectable extent. Benzene, carbon tetrachloride, and pentane fall into this category.

Finally, there exist a number of solvents, such as ketones, ethers, esters, and pyridine derivatives, with basic properties but essentially no acidic tendencies. Solvents of this type do not undergo autoprotolysis and are sometimes also classified as inert solvents as a consequence.

11B NEUTRALIZATION REACTIONS IN AMPHIPROTIC SOLVENTS

The completeness of a neutralization reaction in an amphiprotic solvent depends not only upon the strength of the analyte as an acid or base but also upon several

Table 11-1 Properties of Some Common Solvents at 25°C

Solvent	K_s	Dielectric Constant	Boiling Point °C
Water	1.01×10^{-14}	78.5	100
Methanol	2×10^{-17}	32.6	64
Ethanol	8×10^{-20}	24.3	78
Formic acid	6×10^{-7}	58.5	101
Acetic acid	3.5×10^{-15}	6.2	118
Sulfuric acid	1.4×10^{-4}	100	340
Ammonia*	1×10^{-33}	22	−78
Ethylenediamine	5×10^{-16}	14.2	117
Acetonitrile	2.9×10^{-27}	37.4	82
Dimethylformamide	—	37.6	153

* At −50°C.

properties of the solvent as well. This section is devoted to a consideration of those solvent properties that influence the completeness of neutralization reactions.

11B-1 EFFECT OF THE SOLVENT AUTOPROTOLYSIS CONSTANT

In water, the titration of a weak base B with a standard strong acid can be formulated as

$$B + H_3O^+ \rightleftharpoons BH^+ + H_2O \tag{11-1}$$

and the magnitude of the equilibrium constant for this reaction can be used as a measure of its completeness; that is,

$$K_{equil} = \frac{[BH^+]}{[B][H_3O^+]} = \frac{K_b}{K_w} \tag{11-2}$$

Note that the completeness of the reaction is determined in part by the magnitude of the autoprotolysis constant for water.

In an analogous fashion, completeness of the reaction between a weak acid HA and a strong base can be expressed by the equilibrium constant

$$K_{equil} = \frac{[A^-]}{[HA][OH^-]} = \frac{K_a}{K_w} \tag{11-3}$$

Similar relations can be derived for reactions in nonaqueous solvents. For example, when the weak base B is titrated with a strong acid in anhydrous formic acid, we may write

$$B + HCOOH_2^+ \rightleftharpoons BH^+ + HCOOH \tag{11-4}$$

where $HCOOH_2^+$ represents the solvated proton analogous to H_3O^+ in an aqueous solution. Here,

$$K_{equil} = \frac{[BH^+]}{[B][HCOOH_2^+]} = \frac{K_b'}{K_s} \tag{11-5}$$

where K_b' is the dissociation constant for the base *in formic acid;* that is,

$$B + HCOOH \rightleftharpoons BH^+ + HCOO^- \qquad K_b' = \frac{[BH^+][HCOO^-]}{[B]} \tag{11-6}$$

The constant K_s is the autoprotolysis constant for formic acid:

$$2HCOOH \rightleftharpoons HCOOH_2^+ + HCOO^- \qquad K_s = [HCOOH_2^+][HCOO^-] \tag{11-7}$$

As with the ion-product constant for water, the concentration of the solvent HCOOH is essentially invariant and is thus included in K_b' and K_s.

The titration of a weak acid HA with sodium ethoxide in ethanol is readily formulated as

$$HA + C_2H_5O^- \rightleftharpoons A^- + C_2H_5OH \tag{11-8}$$

Here, the standard strong base is a solution of sodium ethoxide (C_2H_5ONa) in ethanol. In common with the previous examples, the completeness of the reaction can be measured with the equilibrium constant

$$K_{equil} = \frac{[A^-]}{[HA][C_2H_5O^-]} = \frac{K_a'}{K_s} \tag{11-9}$$

where K_a' is the dissociation constant for the acid in ethanol,

$$HA + C_2H_5OH \rightleftharpoons C_2H_5OH_2^+ + A^- \qquad K_a' = \frac{[A^-][C_2H_5OH_2^+]}{[HA]} \tag{11-10}$$

and K_s is the autoprotolysis constant for ethanol,

$$2C_2H_5OH \rightleftharpoons C_2H_5OH_2^+ + C_2H_5O^- \qquad K_s = [C_2H_5OH_2^+][C_2H_5O^-] \tag{11-11}$$

These examples demonstrate that completeness of the reaction is a function of both the dissociation constant of the substance being titrated and the autoprotolysis constant of the solvent. The appearance of both constants in the equation can be best understood, perhaps, by considering each neutralization reaction as representing a competition for protons. Thus, for example, the extent of reaction 11-4 is governed by the success with which base molecules B compete with solvent molecules HCOOH for a stoichiometrically limited number of hydrogen ions, H^+. The effectiveness of each participant in this competition is measured by its dissociation constant K_b' and K_s, respectively. Similarly, reaction 11-8 can be thought of as a competition between the ions A^- and $C_2H_5O^-$ for H^+, with the effectiveness of each being measured by K_a' and K_s.

From this discussion it is clearly advantageous to perform acid-base titrations in solvents that have small autoprotolysis constants. Furthermore, acid-base reactions are more complete in those solvents in which K_a' or K_b' is large. These two considerations, which are not entirely independent of one another, govern the choice of an amphiprotic solvent for a particular nonaqueous titration.

11B-2 EFFECT OF ACID OR BASE CHARACTERISTICS OF THE SOLVENT

The acid-base behavior of solutes is greatly affected by the strength of the solvent as an acid or a base. A number of amphiprotic solvents, including formic acid, acetic acid, and sulfuric acid, are considerably better proton donors than proton acceptors and are therefore classed as acidic solvents. In such solvents, the basic properties of a solute are magnified, while its acidic properties are diminished. Thus, for example, aniline, $C_6H_5NH_2$, cannot be titrated in aqueous solution because its basic dissociation constant is only about 10^{-10}. In anhydrous acetic acid, however, aniline is an appreciably stronger base because of the enhanced tendency of the solvent to give up a proton. Thus, the equilibrium constant, K_b', for the reaction

$$C_6H_5NH_2 + HOAc \rightleftharpoons C_6H_5NH_3^+ + OAc^-$$

is significantly larger than K_b for the analogous reaction in water:

$$C_6H_5NH_2 + H_2O \rightleftharpoons C_6H_5NH_3^+ + OH^-$$

A solvent with acid properties, while enhancing the basic properties of a solute, acts as a *discriminating solvent* toward acids. For example, perchloric, hydrochloric, and nitric acids, which have the same strengths in aqueous medium, dissociate to different degrees in acetic acid, with the perchloric acid being stronger than the other two by several orders of magnitude. These differences disappear in aqueous solutions since all react completely with the solvent to form hydronium ions. Water thus acts as a *leveling* solvent with respect to nitric, perchloric, and hydrochloric acids, since each has the same strength in aqueous solutions.

Solvents such as ethylenediamine and liquid ammonia have a strong affinity for protons and are therefore classified as basic solvents. In these media, the acid properties of a solute are enhanced. Thus, phenol, with an acid dissociation constant of about 10^{-10} in water, becomes sufficiently strong in ethylenediamine to permit its titration with a standard base. The strengths of bases are, of course, diminished in solvents of this type and solutes that are strong bases in water may only partially dissociate in solvents with basic properties.

Water and the aliphatic alcohols, such as methanol and ethanol, are examples of neutral amphiprotic solvents. These solvents possess less pronounced capacities as proton donors or acceptors than those just considered. It is important to realize, however, that this name does not necessarily imply an exact equality between acidic and basic character.

11B-3 EFFECT OF SOLVENT DIELECTRIC CONSTANT

The dielectric constant of a solvent measures its capacity for separating particles of opposite charge. In a solvent with a high dielectric constant, such as water (D_{H_2O} = 78.5), a minimum of work is required to separate a positively charged ion from one with a negative charge; for a solvent with a low dielectric constant, such as acetic acid (D_{HOAc} = 6.2), a greater amount of energy is required to accomplish the process. Methanol and ethanol have dielectric constants of 33 and 24, respectively, and are intermediate in their behavior. The dielectric constants for several solvents are shown in Table 11-1.

With solvents of low dielectric constant, dissociation can be considered to be a two-step process: the first step involves ion pair formation and the second step dissociation or separation of the ion pair into individual particles. Thus in anhydrous acetic acid, we may write

$$HClO_4 + CH_3COOH \rightleftharpoons CH_3COOH_2^+ClO_4^- \qquad \text{ion pair formation}$$

$$CH_3COOH_2^+ClO_4^- \rightleftharpoons CH_3COOH_2^+ + ClO_4^- \qquad \text{dissociation}$$

Each component of the ion pair bears a full positive or negative charge; the ion pair, however, is incapable of conducting electricity. A strong acid such as perchloric acid is completely ionized in acetic acid but only partially dissociated ($K_a \cong 10^{-5}$). From the standpoint of acid-base titration, it is the overall dissociation process that is important—that is,

$$HClO_4 + CH_3COOH \rightleftharpoons CH_3COOH_2^+ + ClO_4^-$$

Thus, the dielectric constant of the medium, which influences the dissociation step, plays an important role in determining the strengths of solute acids or bases insofar as the dissociation process produces oppositely charged species. For example, when an uncharged weak acid HA is dissolved in an amphiprotic solvent SH, the dissociation process requires the separation of the charged particles SH_2^+ and A^-:

$$HA + SH \rightleftharpoons SH_2^+ + A^-$$

The same would be true upon solution of an uncharged base B:

$$B + SH \rightleftharpoons BH^+ + S^-$$

Reactions of this sort would be expected to proceed further to the right in a solvent such as water than in methanol or ethanol because less work is required for the dissociation process. The magnitude of this effect can be large; for example, the dissociation constant for a typical carboxylic acid in water is approximately 10^{-5}, whereas it is somewhat smaller than 10^{-10} in ethanol.

The strength of an acid or base is not affected significantly by the dielectric constant of the medium when the dissociation reaction does not involve a charge separation. For example, the following equilibria would not be altered by changes in dielectric constant of the solvent SH:

$$BH^+ + SH \rightleftharpoons SH_2^+ + B$$

$$A^- + SH \rightleftharpoons HA + S^-$$

11B-4 CHOICE OF AMPHIPROTIC SOLVENTS FOR NEUTRALIZATION TITRATIONS

We have shown that the completeness of a neutralization reaction is directly dependent upon the ionization constant of the solute acid or base and inversely related to the autoprotolysis constant of the solvent. Furthermore, the first of these factors, the ionization constant, is dependent upon the acidic or basic properties and the dielectric constant of the solvent. Thus, the most advantageous choice of solvent for a given titration hinges upon three interrelated properties:

1. Its autoprotolysis constant; a numerically small value is desirable.
2. Its properties as a proton donor or acceptor. For the titration of a weak base, a solvent with strong donor tendencies is helpful (that is, an acidic solvent); for the analysis of a weak acid, a solvent that is a good proton acceptor is desirable.
3. Its dielectric constant; a high value is most useful.

In addition, of course, the solute must be reasonably soluble in the solvent.

Glacial acetic acid is often chosen as solvent for the titration of very weak bases because it tends to donate protons and thus enhances the strengths of dissolved bases. Its autoprotolysis constant (3.6×10^{-15}) is also somewhat more favorable than that for water. On the other hand, its low dielectric constant partially offsets these advantages. The two favorable properties outweigh the single disadvantage, however, and acetic acid is a generally superior solvent for the titration of weak bases; it will be clearly inferior to water for titration of weak acids because of its weakness as a proton acceptor.

It is profitable to consider formic acid in the role of an acidic solvent. Like acetic acid, it is a much better proton donor than water; furthermore, in contrast to acetic acid, formic acid has a dielectric constant that is comparable with that of water. On these two counts, then, it would appear to be an ideal solvent for the titration of weak bases. Unfortunately, however, its autoprotolysis constant is much larger than that of water or acetic acid. As a consequence, despite its two very desirable properties, formic acid appears to offer little advantage over water as a solvent.

Methanol and ethanol have been widely applied as solvents for acid-base titrations. Both are classified as neutral solvents because their proton donor and acceptor properties do not differ markedly. Both have advantageous autoprotolysis constants. On the other hand, their low dielectric constants frequently offset the advantage of their small autoprotolysis constants. For example, in ethanol, the dissociation constants of most uncharged acids, such as benzoic acid, are about 10^{-6} as great as in water; at the same time the ratio of autoprotolysis constants is smaller by nearly the same factor (8×10^{-6}). Thus, the ratio K'_a/K_s is only slightly more favorable in ethanol than in water, and the improvement in end points gained by use of this solvent is modest. On the other hand, significant gain is realized by the use of ethanol for the titration of a charged weak acid such as the ammonium ion. Here, no charge separation is involved in the dissociation process:

$$NH_4^+ + C_2H_5OH \rightleftharpoons NH_3 + C_2H_5OH_2^+$$

In contrast to benzoic acid, dissociation of the acid NH_4^+ is not significantly decreased in ethanol. The reaction of NH_4^+ with a strong base is, however, much more complete in ethanol because of the low autoprotolysis constant of the solvent. As a consequence, NH_4^+ can be titrated satisfactorily in ethanol but not in water.

> **EXAMPLE 11-1.** Calculate the percentage of unreacted NH_4^+ at the equivalence point in the titration of $0.20\ M\ NH_4^+$ with (1) $0.20\ M$ NaOH in an aqueous solution and (2) $0.20\ M\ C_2H_5ONa$ in anhydrous ethanol.
>
> The autoprotolysis constants for the solvents are 1×10^{-14} and 8×10^{-20}, respectively; the acid dissociation constant for NH_4^+ is approximately 6×10^{-10} in water and 1×10^{-10} in ethanol.
> For the titration in water,
>
> $$NH_4^+ + OH^- \rightleftharpoons NH_3 + H_2O$$
>
> the equilibrium constant is given by
>
> $$K_{equil} = \frac{[NH_3]}{[NH_4^+][OH^-]} = \frac{K_a}{K_w} = \frac{6 \times 10^{-10}}{1 \times 10^{-14}} = 6 \times 10^4$$
>
> Similarly, for the titration in ethanol,
>
> $$NH_4^+ + C_2H_5O^- \rightleftharpoons NH_3 + C_2H_5OH$$
>
> $$K_{equil} = \frac{[NH_3]}{[NH_4^+][C_2H_5O^-]} = \frac{K'_a}{K_s} = \frac{1 \times 10^{-10}}{8 \times 10^{-20}} = 1.2 \times 10^9$$
>
> At the equivalence point in each titration, the analytical NH_3 concentration will be $0.10\ M$. In the aqueous titration,
>
> $$[NH_4^+] = [OH^-]$$
>
> $$[NH_3] = 0.10 - [NH_4^+]$$
>
> while in the ethanol solution,
>
> $$[NH_4^+] = [C_2H_5O^-]$$
>
> $$[NH_3] = 0.10 - [NH_4^+]$$
>
> If we further assume that in both media $[NH_4^+] \ll 0.10$, then in water,
>
> $$\frac{0.10}{[OH^-]^2} \cong 6 \times 10^4$$

$$[OH^-] = [NH_4^+] = 1.3 \times 10^{-3}$$

$$\% \text{ unreacted } NH_4^+ = \frac{1.3 \times 10^{-3}}{0.10} \times 100 = 1.3$$

and in alcohol,

$$\frac{0.10}{[OH^-]^2} \cong 1.2 \times 10^9$$

$$[OH^-] = [NH_4^+] = 9 \times 10^{-6}$$

$$\% \text{ unreacted } NH_4^+ = \frac{9 \times 10^{-6}}{0.10} \times 100 = 0.009$$

Thus, at the equivalence point in an aqueous titration, approximately 1% of the NH_4^+ remains unreacted; the comparable quantity in anhydrous ethanol is only about 0.01%.

Several amphiprotic basic solvents have been employed for the titration of very weak acids. Among the most basic of these, and thus best from the standpoint of enhancing the acidity of the solute, is ethylenediamine, $NH_2CH_2CH_2NH_2$; its autoprotolysis constant is about 5×10^{-16}. The advantageous acceptor property and autoprotolysis constant of this solvent are partially offset by its low dielectric constant. Dimethylformamide, $HCON(CH_3)_2$, a weaker base than ethylenediamine, has also proved useful; its dielectric constant is 38.

11C NEUTRALIZATION REACTIONS IN APROTIC SOLVENTS AND MIXED SOLVENTS

Solvents without amphiprotic properties offer the advantage that they do not compete for protons with the reactants in a titration; that is, these solvents have autoprotolysis constants that approach zero. Thus, neutralization reactions should be more nearly complete when carried out in solvents of this type.

Inorganic solutes tend to be sparingly soluble in aprotic solvents. As a consequence, many mixtures of aprotic solvents with more polar solvents have been investigated. Examples include benzene/methanol and ethylene glycol/hydrocarbon. Unfortunately, fundamental knowledge regarding the properties and behavior of acids and bases in such systems is meager; the applicability of a mixed solvent to a specific problem can thus be determined only empirically.

11D END-POINT DETECTION IN NONAQUEOUS TITRATIONS

Undoubtedly the most popular method of end-point detection for nonaqueous titrations involves measuring the potential of a glass electrode that responds to the concentration of the solvated proton. The glass electrode is discussed in detail in Chapter 16.

Many acid-base indicators used for aqueous titrations are also applicable in nonaqueous media. To be sure, their behavior in an aqueous environment cannot be extrapolated to predict their properties in nonaqueous solutions; the limited information available with respect to these properties in solvents other than water makes the choice of indicator largely a matter of experience and empirical observation.

11E APPLICATIONS OF NONAQUEOUS ACID-BASE TITRATIONS

For nonaqueous titrations, innumerable combinations of solvents, titrants, and end points exist; the choice among them must be based upon such considerations as solubility, convenience, cost of reagent, toxicity, and availability of suitable equipment. Examples of some typical nonaqueous titrants follow.[2]

11E-1 TITRATION OF BASES IN GLACIAL ACETIC ACID AND OTHER SOLVENTS

Many useful titrations can be performed in glacial acetic acid.[3] The titrant is a standard perchloric acid solution prepared with this solvent; a standard solution of sodium acetate in the same medium can be employed as a base for back-titration if desired.

Coefficient of expansion of glacial acetic acid solutions. In common with most organic solvents, acetic acid has a significantly greater coefficient of expansion than water (0.11% per degree Celsius compared with 0.025%). Therefore, greater care must be exercised to eliminate errors arising from temperature fluctuations during volumetric measurements. A common practice is to note the temperature of the perchloric acid reagent at the time of its standardization; the temperature is then recorded at the time the reagent is used for an analysis, and the volume is corrected to the temperature of standardization with the equation

$$V_{std} = V[1 + 0.0011(T_{std} - T)] \qquad\qquad (11\text{-}12)$$

where T_{std} is the temperature of the reagent at the time of standardization, T is its temperature when used, V is the measured volume of the reagent, and V_{std} is the corrected volume.

Effect of water. Water acts as a weak base in acetic acid and tends to compete with the solvent for protons. As a consequence, its presence leads to smaller pH changes in the equivalence-point region of a neutralization titration and causes less satisfactory end points. The amount of water that can be tolerated during a titration varies; for very weak bases, nearly anhydrous conditions are required. On the other hand, for the titration of bases that are relatively strong in the solvent, as much as 3% of water by volume is not harmful.

Fortunately, nearly anhydrous acetic acid can be obtained by the introduction of acetic anhydride. Any water present reacts with the anhydride to form acetic acid; an excess of the anhydride is avoided because it may interfere with the titration.

End points in acetic acid. Two acid-base indicators, crystal violet and methyl violet, have been found useful for titrations in acetic acid; both exhibit complex color changes. For example, methyl violet changes from violet through green to yellow as the solution becomes more acidic; the disappearance of the violet color signals

---

[2] Reviews of applications of nonaqueous titrations are found in: B. Kratochvil, *Anal. Chem.*, **52**, 151R (1980); **50**, 153R (1978); **48**, 355R (1976).
[3] Glacial acetic acid derives its name from the fact that at about 16°C large ice-like crystals form in the medium. Reagent grade glacial acetic acid may contain up to 0.3% water; thus it is not a truly anhydrous medium.

the end point. The color change is not as obvious as might be desired; with practice, however, significant titration errors can be avoided when stronger bases are titrated. With very weak bases it is impractical to use the visual end point; here, the
potentiometric end point with a glass electrode must be employed (Chapter 16).

Primary standards for solutions of perchloric acid. Potassium hydrogen phthalate is the most commonly used primary standard for the standardization of acetic
acid solutions of perchloric acid. The properties of this substance were discussed
in Section 10A-4 in connection with its use for the standardization of aqueous *base*
solutions. It will be noted that potassium hydrogen phthalate is a sufficiently strong
base in anhydrous acetic acid, so that it can now be employed to standardize an
acid solution. The reaction, of course, involves conversion of the acid salt to the undissociated acid.

Sodium carbonate has also been employed for standardization of perchloric
acid solutions in acetic acid.

Some typical applications. Glacial acetic acid is extensively employed for the titration of aromatic amines, amides, ureas, and other very weak nitrogen bases. Aliphatic amines are generally leveled by this medium and can be determined as a
group in the presence of the aforementioned weak bases. If it is desirable to discriminate among aliphatic amines, a nearly neutral solvent such as acetone or a
weakly acidic solvent such as nitromethane is required. An important application
of acetic acid as a solvent is to the direct titration of most amino acids with a standard acid. It was noted earlier (p. 237) that in aqueous media, these compounds exist
largely as zwitterions which are not strong enough acids or bases for titration. In
glacial acetic acid, however, the dissociation of the carboxylic acid group is essentially completely repressed, leaving the amine group available for titration with
perchloric acid.

The use of glacial acetic acid permits the determination of many inorganic
and organic salts which are not amenable to titrations in aqueous solution because
in this medium they behave as very weak bases or neutral compounds. For example, the sodium salts of inorganic anions such as chloride, bromide, iodide, nitrate, chlorate, and sulfate have all been determined by titration as bases in glacial
acetic acid. The ammonium and alkali metal salts of most carboxylic acids can also
be determined in this medium; typical examples include ammonium benzoate,
sodium salicylate, sodium acetate, potassium tartrate, and sodium citrate. Curve *A*
in Figure 11-1 is a typical titration curve; here, a solution of sodium acetate in glacial acetic acid was titrated with standard perchloric acid solution in the same solvent. A glass electrode (see Chapter 16) was employed to measure the quantity
$(K + pH)$ where K is constant throughout the titration. For comparison, the theoretical curve (B) for the same concentration of sodium acetate in water is also shown.

Laboratory procedures in Chapter 31 (Methods 4-1 and 4-2) illustrate the application of glacial acetic acid solutions to the determination of bases.

11E-2 TITRATION OF ACIDS
Several basic solvents have been employed to determine acids that are too weak to
be titrated in water. Examples include ethylenediamine, dimethylformamide, pyri-

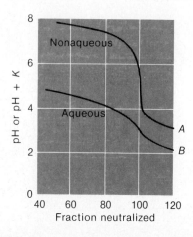

FIGURE 11-1 Titration curves for $0.075\,M$ sodium acetate with 0.100 N $HClO_4$. Curve A: experimental curve with sample and reagent dissolved in glacial acetic acid; ordinate scale is $(pH + K)$ where K is an empirical constant. Curve B: theoretical titration curve for aqueous solution; ordinate is pH.

dine, dimethylsulfoxide, and butylamine. In addition, the alcohols, aprotic solvents, and weakly acidic solvents such as acetone and acetonitrile find use in the determination of acids. Applications of a few of these solvent systems are described in the sections that follow.

Ethylenediamine as a solvent. A variety of inorganic and organic compounds including amine salts, inorganic salts, carboxylic acids, phenols, and imides are soluble in ethylenediamine and exhibit enhanced acidic characteristics therein. Sodium amino ethoxide, $NaOCH_2CH_2NH_2$ can be employed as titrant; solutions are prepared by dissolving sodium in ethanolamine and diluting with ethylenediamine.[4] A standard solution of sodium methoxide in a benzene/methanol mixture has also proved useful.[5] Finally, solutions of tetrabutylammonium hydroxide, $(C_4H_9)_4NOH$ (a strong base in water), may be prepared in isopropyl alcohol, benzene/methanol, or ethanol to give a standard base solution suitable for titrations in ethylenediamine and other basic solvents.

   Phenols, enols, and imides, which are weaker than the carboxylic acids are readily titrated in ethylenediamine with tetrabutylammonium hydroxide as the titrant. Many of the sulfa drugs such as sulfanilamide, sulfathiozole, and sulfathalidine can also be determined in the same way. The acidity of the various sulfa drugs varies with structure; the determination of the individual components of a mixture in some instances is thus feasible.

Dimethylformamide as a solvent. Dimethylformamide is a very weak base with a dielectric constant of about 38. It has found considerable use for the titration of ammonium and amine salts with tetrabutylammonium hydroxide serving as the titrant. In addition, the ability of dimethylformamide to dissolve salts, polymers, and many organic compounds accounts for its popularity as a solvent. The titrants described in the previous section are ordinarily applicable here as well.

[4] M. L. Moss, J. H. Elliott, and R. T. Hall, *Anal. Chem.*, **20**, 784 (1948).
[5] J. S. Fritz and N. M. Lisicki, *Anal. Chem.*, **23**, 589 (1951).

Other solvent systems. Aprotic solvents, ketones, ethers, and esters are often classified as inert solvents even though the latter three have weakly basic properties. Representative inert solvents include acetone for titration of acids, acetonitrile for both acids and bases, and ethyl acetate for titration of amines and mixtures of these compounds with hydrocarbons and chlorinated hydrocarbons. The reagents described in earlier sections are often employed as titrants.

Methanol and ethanol have also had widespread application to nonaqueous titrations for bases and moderately strong acids. They are, however, sufficiently acidic to preclude titration of acids of the strength of phenol or less. *tert*-Butanol, on the other hand, does not suffer from this defect.

Titration of mixtures of acids. The use of a suitable nonaqueous medium makes it possible to discriminate among the various mineral acids that are not leveled by the solvent as they are in water. Figure 11-2 illustrates the discriminatory power of methyl isobutyl ketone toward perchloric and hydrochloric acids as well as other acids. The titrant was a 0.2 N solution of tetrabutylammonium hydroxide in isopropyl alcohol. The ordinate is the potential of a glass electrode system (Chapter 16), which varies linearly with pH. Note that the strengths of perchloric and hydrochloric acids are sufficiently different in this solvent to give two end points. Note also that phenol, which is far too weak to be titrated in water, gives a sharp, well-defined end point in the nonaqueous medium.

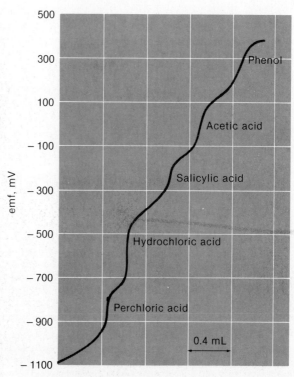

FIGURE 11-2 Titration of a mixture of acids in methyl isobutyl ketone. [D. B. Bruss and G. E. A. Wyld, *Anal. Chem.*, **29**, 234 (1957). With permission of the American Chemical Society.]

Volume of 0.2 N tetrabutylammonium hydroxide

Carboxylic acids with limited solubility in water have been successfully titrated in various nonaqueous media. Most carboxylic acids are leveled and behave as strong acids in the more basic solvents, such as ethylenediamine. A less basic solvent, such as dimethylformamide, or a neutral solvent, such as a ketone, may be preferable if discrimination among such acids is required.

PROBLEMS*

11-1. Write equations depicting the autoprotolysis of
     * (a) water.   (d) ammonia, NH_3.
         (b) ethanol, C_2H_5OH.   * (e) acetic acid, $HC_2H_3O_2$.
     * (c) formic acid, HCOOH.   (f) sulfuric acid.

11-2. Postulate an equilibrium for the autoprotolysis of
     * (a) HCN.   * (c) HF.
         (b) H_2NNH_2.   (d) ethylenediamine, $H_2NCH_2CH_2NH_2$.

* 11-3. The concentration of the species H_2F^+ is about $1.4 \times 10^{-6}\ M$ in liquid hydrogen fluoride. Calculate a value for the autoprotolysis constant, K_s, of HF.

11-4. Evidence exists to suggest that pH_2CN is about 9.3 in liquid hydrogen cyanide. Estimate a value for the autoprotolysis constant, K_s, of HCN.

11-5. Calculate p-values (i.e., pH, pOH, pC_2H_5O, etc.) for the ions in the solvents in Problem 11-1.

* 11-6. Sodium hydroxide is a strong base in aqueous solution because hydroxide ion is the strongest base that can exist in that solvent. By analogy, what sodium compound will be a strong base in the solvents listed in Problem 11-1?

11-7. Write equations depicting the behavior of
     * (a) hydrogen phthalate ion $(HC_8H_4O_4^-)$ as an acid in water.
     * (b) hydrogen phthalate ion as a base in glacial acetic acid.
     * (c) urea (H_2NCONH_2) as a strong base in formic acid.
         (d) urea as an acid in ammonia.
         (e) nitric acid as a base in hydrogen fluoride.

11-8. Briefly account for the observation that
     * (a) the strongest base that can exist in a particular solvent system is the conjugate base of that solvent.
         (b) acetic acid is a more effective leveling solvent for bases than is water.
     * (c) water is a more effective differentiating solvent for acids than is ethylenediamine.
         (d) the acid dissociation constant for ammonium ion is approximately the same in water $(K_b \cong 10^{-9})$ and in ethanol $(K_b' \cong 10^{-10})$.
     * (e) the acid dissociation constant for benzoic acid is approximately 10^{-4} in water and about 10^{-10} in ethanol.
         (f) the acid dissociation constant for benzoic acid is substantially smaller in formic acid than it is in ethylenediamine.

* 11-9. In acetic acid, HCl has a dissociation constant of 2.8×10^{-9}. Calculate pH for
    (a) a $0.0105\ M$ solution of HCl in acetic acid.
    (b) a $0.0284\ M$ solution of $CaCl_2$ in acetic acid.
    (c) a solution that is $0.150\ M$ in HCl and $0.100\ M$ in $CaCl_2$.

* 11-10. Compare the p-values in Problem 11-9 with the pH of the corresponding aqueous solutions.

---

* Answers to problems or parts of problems marked with an asterisk are to be found at the end of the book.

may involve the production of one or more intermediate species. Consider, for example, the equilibrium that exists between the metal ion M with a coordination number of four and the tetradentate ligand D.[1]

$$M + D \rightleftharpoons MD$$

The equilibrium expression for this process is

$$K_f = \frac{[MD]}{[M][D]}$$

where K_f is the *formation constant*.

Similarly, the equilibrium between M and the bidentate ligand B can be represented by

$$M + 2B \rightleftharpoons MB_2$$

This equation, however, is the summation of a two-step process that involves formation of the intermediate MB

$$M + B \rightleftharpoons MB$$

$$MB + B \rightleftharpoons MB_2$$

for which

$$K_1 = \frac{[MB]}{[M][B]} \text{ and } K_2 = \frac{[MB_2]}{[MB][B]}$$

The product of K_1 and K_2 yields the equilibrium constant for the overall process.[2]

$$\beta_2 = K_1 K_2 = \frac{[\cancel{MB}]}{[M][B]} \times \frac{[MB_2]}{[\cancel{MB}][B]} = \frac{[MB_2]}{[M][B]^2}$$

In a like manner, the reaction between M and the unidentate ligand A involves the overall equilibrium

$$M + 4A \rightleftharpoons MA_4$$

and the equilibrium constant β_4 for the formation of MA_4 from M and A is numerically equal to the product of the equilibrium constants for the four constituent processes.

Each of the titration curves depicted in Figure 12-1 is based upon a reaction that has an overall equilibrium constant of 10^{20}. Curve *A* is derived for the formation of MD in a single step. Curve *B* involves the formation of MB_2 in a two-step process for which K_1 is 10^{12} and K_2 is 10^8. Curve *C* represents formation of MA_4 for which the equilibrium constants for the four individual steps are 10^8, 10^6, 10^4, and 10^2, respectively. These curves demonstrate the clear superiority of a ligand that combines with a metal ion in a 1:1 ratio because the change in pM in the equivalence point region is largest with such a system. Thus, polydentate ligands, which

---

[1] The electrostatic charges associated with M and D determine the charge of the product but are not important in terms of the present argument.

[2] It is customary to use the symbol β_i to indicate an overall formation constant. Thus, for example, $\beta_2 = K_1 K_2$, $\beta_3 = K_1 K_2 K_3$, $\beta_4 = K_1 K_2 K_3 K_4$, and so forth.

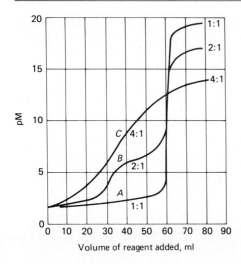

FIGURE 12-1 Curves for complex formation titrations. Titrations of 60.0 mL of 0.020 M M with (curve A) 0.020 M solution of the tetradentate ligand D to MD as produce, (curve B) 0.040 M solution of the bidentate ligand B to give MB_2, and (curve C) 0.080 M solution of the unidentate ligand A to give MA_4. The overall formation constant for each product is 1.0×10^{20}.

generally combine in lower ratios with metal ions, are ordinarily superior reagents for complex formation titrations.

Regardless of reaction type, the error associated with a titration decreases with increasing completeness of the reaction upon which the titration is based. In this respect, polydentate ligands usually offer a distinct advantage over unidentate titrants; formation constants associated with reactions between metal ions and the former tend to be larger than those involving the latter.

12B TITRATIONS WITH INORGANIC COMPLEXING REAGENTS

Complexometric titrations are among the oldest of volumetric methods.[3] For example, the titration of iodide ion with mercury(II)

$$Hg^{2+} + 4I^- \rightleftharpoons HgI_4^{2-}$$

was first reported in 1832; the determination of cyanide based upon formation of the dicyanoargentate(I) ion, $Ag(CN)_2^-$, was described by Liebig in 1851. Table 12-1 lists typical nonchelating complexing reagents as well as some of the uses to which they have been put.

12B-1 TITRATIONS WITH MERCURY(II)

Mercury(II) ion forms neutral complexes with most of the anions that precipitate with silver nitrate. Many of these complexes are sufficiently stable to permit the use of mercuric nitrate for the volumetric determination of these anions. Solutions of mercury(II) have not been used as extensively as silver nitrate even though the reagent appears capable of yielding satisfactory titration data for the determination of such ions as chloride, bromide, thiocyanate, and cyanide. The ever increasing

[3] For further information, see: I. M. Kolthoff and V. A. Stenger, *Volumetric Analysis*, vol. 2, pp. 282, 331. New York: Interscience, 1947.

cost of silver nitrate, however, is likely to cause this situation to change; although not inexpensive, solutions of mercury(II) permit the determination of halide ions at a considerably lower cost. For this reason, a brief discussion of the reagent and its application to the determination of chloride ion is given here.

Reaction of mercury(II) with chloride. The reaction of mercuric ions with chloride and other halide ions is a two-step process, which can be formulated as

$$Hg^{2+} + Cl^- \rightleftharpoons HgCl^+$$

$$HgCl^+ + Cl^- \rightleftharpoons HgCl_2$$

where $HgCl_2$ is a soluble uncharged complex. Formation constants for these reactions are

$$\frac{[HgCl^+]}{[Hg^{2+}][Cl^-]} = K_1 = 5.5 \times 10^6$$

$$\frac{[HgCl_2]}{[HgCl^+][Cl^-]} = K_2 = 3.0 \times 10^6$$

The second constant K_2 is sufficiently large that in the presence of any significant quantity of chloride ion, $HgCl_2$ is the predominant mercury-bearing species in the solution, and the reaction for the titration can be formulated as

$$Hg^{2+} + 2Cl^- \rightleftharpoons HgCl_2$$

The overall formation constant can then be employed for derivation of titration curves. That is,

$$\frac{[HgCl_2]}{[Hg^{2+}][Cl^-]^2} = \beta_2 = 5.5 \times 10^6 \times 3.0 \times 10^6 = 1.6 \times 10^{13}$$

Preparation of standard mercury (II) solutions. Standard solutions of mercuric nitrate can be prepared directly by dissolving primary standard grade mercuric oxide or elemental mercury in nitric acid. Alternatively, solutions can be prepared from

Table 12-1 Typical Inorganic Complex Formation Titrations*

Titrant	Analyte	Remarks
$Hg(NO_3)_2$	Br^-, Cl^-, SCN^-, CN^-, thiourea	Products are neutral mercury(II) complexes; various indicators used
$AgNO_3$	CN^-	Product is $Ag(CN)_2^-$; indicator is I^-; titrate to first turbidity of AgI
$NiSO_4$	CN^-	Product is $Ni(CN)_4^{2-}$; indicator is AgI; titrate to first turbidity of AgI
KCN	Cu^{2+}, Hg^{2+}, Ni^{2+}	Products $Cu(CN)_4^{2-}$, $Hg(CN)_2$, $Ni(CN)_4^{2-}$; various indicators used

* For further applications and selected references see L. Meites, *Handbook of Analytical Chemistry.* New York: McGraw-Hill, 1963, p. 3-226.

reagent grade mercuric nitrate and standardized against primary standard grade sodium chloride.

Indicators for titration with mercury(II). Mercury(II) ions form an intensely colored purple complex with diphenylcarbazide $(C_6H_5NHNH)_2CO$ and diphenylcarbazone $(C_6H_5NHNHCON=NC_6H_5)$; either compound can serve as an indicator for titrations. For the former, the reaction is thought to be

$$2 \; O=C \begin{matrix} \overset{H}{|} \; \overset{C_6H_5}{|} \\ N{-}N{-}H \\[1mm] N{-}N{-}H \\ \overset{|}{H} \; \overset{|}{C_6H_5} \end{matrix} \; + \; Hg^{2+} \longrightarrow \left(O=C \begin{matrix} \overset{H}{|} \; \overset{C_6H_5}{|} \\ N{-}N \\ \overset{H}{|} \\ N{-}N \\ \overset{|}{H} \; \overset{|}{C_6H_5} \end{matrix} \right)_2 Hg \; + \; 2H^+$$

This equilibrium is forced to the left in strongly acidic solution and the color change is not satisfactory. On the other hand, in weakly acidic or neutral solutions, premature end points are observed, probably because the reaction is so favored that the reagent removes mercury(II) ions from the chloride complex. A minimal titration error is observed if a pH of 1.5 to 2.0 is maintained.[4]

12B-2 TITRATION CURVES FOR CHLORIDE ION WITH MERCURY(II)

A titration curve for the reaction between chloride and mercury(II) ions is readily derived by techniques similar to those described in earlier chapters.

EXAMPLE 12-1. Derive a curve for the titration of 50.00 mL of 0.1000 M NaCl with 0.1000 M $Hg(NO_3)_2$.

Because the indicator is responsive to Hg^{2+}, we shall use pHg as the ordinate for the titration curve.

pHg after addition of 10.0 mL of reagent. From the stoichiometry of the reaction we may write

$$M_{Cl^-} = \frac{50.00 \times 0.1000 - 10.00 \times 0.1000 \times 2}{60.00} = 5.00 \times 10^{-2}$$

Note that the second term in the numerator is multiplied by two because each mercury ion consumes two chloride ions.

$$M_{HgCl_2} = \frac{10.00 \times 0.1000}{60.00} = 1.67 \times 10^{-2}$$

The equilibrium chloride ion concentration is given by

$$[Cl^-] = M_{Cl^-} + [HgCl^+] + 2[Hg^{2+}] \cong M_{Cl^-} = 5.00 \times 10^{-2}$$

where the second and third terms on the right account for the chloride ion arising from equilibria involving the complexes; because the formation constants are large, however, the indicated approximation is surely valid.

We may also write that

[4] See: I. Roberts, *Ind. Eng. Chem., Anal. Ed.,* **8,** 365 (1936).

$$[HgCl_2] = M_{HgCl_2} - [HgCl^+] - [Hg^{2+}] \cong M_{HgCl_2} = 1.67 \times 10^{-2}$$

Substituting these values into the overall formation constant expression yields

$$\frac{1.67 \times 10^{-2}}{[Hg^{2+}](5.00 \times 10^{-2})^2} = 1.6 \times 10^{13}$$

$$[Hg^{2+}] = 4.18 \times 10^{-13}$$

$$pHg = -\log 4.18 \times 10^{-13} = 12.38$$

Obviously the assumption that $[Hg^{2+}]$ is much smaller than either of the two molar analytical concentrations is valid. To confirm that $[HgCl^+]$ is also insignificant, we can substitute into the equilibrium expression for K_2. Thus,

$$\frac{1.67 \times 10^{-2}}{[HgCl^+](5.00 \times 10^{-2})} = 3.0 \times 10^6$$

$$[HgCl^+] = 1.1 \times 10^{-7}$$

Other preequivalence point data can be obtained in this same way.

Equivalence point pHg. Here, 25.00 mL of reagent have been added and

$$M_{HgCl_2} = \frac{50.00 \times 0.1000}{75.00} = 6.67 \times 10^{-2}$$

The analytical concentration of chloride ion is now zero. From mass-balance consideration, we may write

$$M_{HgCl_2} = [HgCl_2] + [HgCl^+] + [Hg^{2+}]$$

and from charge-balance (neglecting Na^+ and NO_3^-, whose concentrations cancel)

$$2[Hg^{2+}] + [HgCl^+] = [Cl^-]$$

With the two formation constant expressions shown on page 279, we have four algebraic equations, which would permit exact determination of the concentrations of the four species. The equations are somewhat awkward to solve; therefore, it is worthwhile to try some approximations. First, we will assume that $[HgCl^+] \ll [Hg^{2+}]$. The mass- and charge-balance equations then become

$$6.67 \times 10^{-2} = [HgCl_2] + [Hg^{2+}] \cong [HgCl_2]$$

$$2[Hg^{2+}] \cong [Cl^-]$$

In the first equation we have further assumed $[Hg^{2+}] \ll [HgCl_2]$. Substituting into the overall formation constant expression (p. 279) gives

$$\frac{6.67 \times 10^{-2}}{[Hg^{2+}](2[Hg^{2+}])^2} = 1.6 \times 10^{13}$$

and

$$[Hg^{2+}] = 1.01 \times 10^{-5}$$

To determine whether $[Hg^{2+}]$ is indeed much smaller than $[HgCl^+]$, we substitute this value and $[Cl^-] = 2[Hg^{2+}]$ into the expression for K_1. Thus,

$$\frac{[HgCl^+]}{(1.01 \times 10^{-5})(2 \times 1.01 \times 10^{-5})} = 5.5 \times 10^6$$

$$[HgCl^+] = 1.1 \times 10^{-3}$$

Obviously $[HgCl^+]$ is not smaller than $[Hg^{2+}]$; indeed, the reverse obtains. Therefore, let us perform the calculation with the assumption that $[Hg^{2+}] \ll [HgCl^+]$. The mass- and charge-balance expressions then become

$$M_{HgCl_2} \cong [HgCl_2] + [HgCl^+] \cong [HgCl_2]$$

and

$$[HgCl^+] = [Cl^-]$$

Substitution of the second approximation into the expression for K_1 gives

$$\frac{[\cancel{HgCl^+}]}{[Hg^{2+}][\cancel{HgCl^+}]} = 5.5 \times 10^6$$

$$[Hg^{2+}] = 1.82 \times 10^{-7}$$

We can obtain $[Cl^-]$, and thus $[HgCl^+]$, by substitution into the overall formation constant. That is,

$$\frac{6.67 \times 10^{-2}}{1.82 \times 10^{-7}[Cl^-]^2} = 1.6 \times 10^{13}$$

$$[Cl^-] = [HgCl^+] = 1.51 \times 10^{-4}$$

Thus, the assumption that $[Hg^{2+}] \ll [HgCl^+]$ is valid and

$$pHg = -\log 1.82 \times 10^{-7} = 6.74$$

pHg after addition of 30.0 mL of reagent. Here, we have 5.00 mL excess of the reagent; this excess provides sufficient Hg^{2+} to repress any significant dissociation of $HgCl_2$. Thus,

$$[Hg^{2+}] = \frac{5.00 \times 0.1000}{80.00} = 6.25 \times 10^{-3}$$

and

$$pHg = 2.20$$

Titration curves for the volumetric determination of Cl^- with mercuric nitrate reagent are shown in Figure 12-2.

12C TITRATIONS WITH AMINOPOLYCARBOXYLIC ACIDS

Numerous tertiary amines that also contain carboxylic acid groups form remarkably stable chelates with many metal ions. Their potential as analytical reagents was first recognized by Schwarzenbach in 1945; since that time they have been extensively investigated. The revival of interest in volumetric complex formation methods can be traced to these reagents.[5]

---

[5] These reagents are the subject of several excellent monographs. See, for example, G. Schwarzenbach and H. Flashka, *Complexometric Titrations*, 2d ed., trans. H. M. N. H. Irving. London: Methuen & Co., 1969; A. Johansson and E. Wanninen, in *Treatise on Analytical Chemistry*, I. M. Kolthoff and P. J. Elving, Eds., Part I, vol. 11, Chapter 11. New York: Wiley, 1975; A. Ringbom, *Complexation in Analytical Chemistry*. New York: Interscience, 1963. For annotated summaries of applications, see: L. Meites, *Handbook of Analytical Chemistry*, pp. 3-167 to 3-234. New York: McGraw-Hill, 1963.

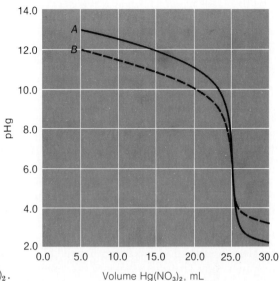

FIGURE 12-2 Titration curves for A 50.00 mL of 0.1000 M Cl$^-$ with 0.1000 M Hg (NO$_3$)$_2$, B 50.00 mL of 0.0100 M Cl$^-$ with 0.01000 M Hg (NO$_3$)$_2$.

The discussion that follows is based largely on ethylenediaminetetraacetic acid, a reagent that has become one of the most widely used titrants in chemical analysis.

12C-1 ETHYLENEDIAMINETETRAACETIC ACID
Ethylenediaminetetraacetic acid (also called ethylenedinitrilotetraacetate and often abbreviated EDTA) has the structure

$$HOOC-CH_2 \qquad\qquad CH_2-COOH$$
$$N-CH_2-CH_2-N$$
$$HOOC-CH_2 \qquad\qquad CH_2-COOH$$

EDTA is the most widely used of the aminopolycarboxylic acids. It is a weak acid for which $K_1 = 1.02 \times 10^{-2}$, $K_2 = 2.14 \times 10^{-3}$, $K_3 = 6.92 \times 10^{-7}$, and $K_4 = 5.50 \times 10^{-11}$. These values indicate that the first two protons are lost much more readily than the remaining two. In addition to the four acidic hydrogens, each nitrogen atom has an unshared pair of electrons; the molecule thus has six potential sites for bonding with a metal ion and is a hexadentate ligand.

The abbreviations H_4Y, H_3Y^-, H_2Y^{2-}, HY^{3-}, and Y^{4-} are often employed in referring to EDTA and its ions.

The free acid H_4Y as well as the dihydrate of the sodium salt $Na_2H_2Y \cdot 2H_2O$ are available in reagent quality. The former can serve as a primary standard after drying for several hours at 130 to 145°C. It is then dissolved in the minimum amount of base required for complete solution.

The dihydrate $Na_2H_2Y \cdot 2H_2O$ contains 0.3% moisture in excess of the stoichiometric amount under normal atmospheric conditions. For all but the most exacting work, this excess is sufficiently reproducible to permit use of a corrected weight for the salt in the direct preparation of a standard solution. If necessary, the pure dihydrate can be prepared by drying at 80°C in an atmosphere of 50% relative humidity for several days.

Another common reagent is nitrilotriacetic acid (abbreviated NTA), which has the structure

$$HOOC-CH_2 \diagdown \diagup CH_2-COOH$$
$$N$$
$$CH_2-COOH$$

Aqueous solutions of this tetradentate ligand are prepared from the free acid.

Other related substances have also been investigated but have not been as widely employed for volumetric analysis. We shall confine our discussion to the applications of EDTA.

12C-2 COMPLEXES OF EDTA AND METAL IONS

A particularly valuable property of EDTA as a titrant is that it combines with metal ions in a $1:1$ ratio regardless of the charge on the cation. For example, with silver and aluminum ions, the complex formation process can be described by the equations

$$Ag^+ + Y^{4-} \rightleftharpoons AgY^{3-}$$

$$Al^{3+} + Y^{4-} \rightleftharpoons AlY^-$$

EDTA is a remarkable reagent, not only because it forms chelates with all cations but also because most of these chelates are sufficiently stable to form the basis for a volumetric analysis. This great stability undoubtedly results from the several complexing sites within the molecule that give rise to structures that effectively surround and isolate the cation from solvent molecules. One form of the complex is depicted in Figure 12-3. Note that all six ligand groups in the EDTA are involved in bonding the divalent metal ion.

Table 12-2 lists formation constants K_{MY} for common EDTA complexes. Note that the constant refers to the equilibrium involving the species Y^{4-} with the metal ion. That is,

$$M^{n+} + Y^{4-} \rightleftharpoons MY^{(n-4)+} \qquad \frac{[MY^{(n-4)+}]}{[M^{n+}][Y^{4-}]} = K_{MY} \tag{12-1}$$

12C-3 EQUILIBRIUM CALCULATIONS INVOLVING EDTA

Most of the indicators used in EDTA titrations are metal ion indicators; that is, they respond to change in the concentration of the uncomplexed analyte cation. The ordinate function for an EDTA titration curve is, therefore, pM; calculation of this quantity is pivotal to the problem of deriving such curves.

Effect of pH on the composition of EDTA solutions. Calculation of equilibrium concentrations of a cation by means of Equation 12-1 is complicated by the fact that the Y^{4-} ion is a conjugate base, and solutions containing this ion will usually have significant concentrations of HY^{3-}, H_2Y^{2-}, H_3Y^-, and H_4Y as well. The ratios

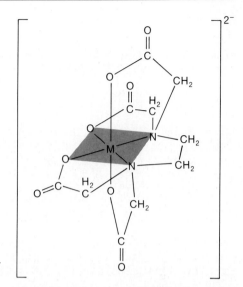

FIGURE 12-3 Structure of a metal-EDTA chelate. (Bond lengths are not to scale.)

among these species are clearly pH dependent. Fortunately, however, EDTA titrations are generally performed in buffered solutions of known pH. Under this circumstance, calculation of the fraction of the uncomplexed EDTA that is present in the form of Y^{4-} is readily accomplished. Here, α_4 is defined by the equation

$$\alpha_4 = \frac{[Y^{4-}]}{C_T} \tag{12-2}$$

where C_T is the total concentration of uncomplexed EDTA or

$$C_T = [Y^{4-}] + [HY^{3-}] + [H_2Y^{2-}] + [H_3Y^-] + [H_4Y]$$

Note that α_4 represents the mole fraction of Y^{4-} in the solution.

Expressions for calculating α values as a function of the hydrogen ion con-

Table 12-2 Formation Constants for EDTA Complexes*

Cation	K_{MY}	log K_{MY}	Cation	K_{MY}	log K_{MY}
Ag^+	2.1×10^7	7.32	Cu^{2+}	6.3×10^{18}	18.80
Mg^{2+}	4.9×10^8	8.69	Zn^{2+}	3.2×10^{16}	16.50
Ca^{2+}	5.0×10^{10}	10.70	Cd^{2+}	2.9×10^{16}	16.46
Sr^{2+}	4.3×10^8	8.63	Hg^{2+}	6.3×10^{21}	21.80
Ba^{2+}	5.8×10^7	7.76	Pb^{2+}	1.1×10^{18}	18.04
Mn^{2+}	6.2×10^{13}	13.79	Al^{3+}	1.3×10^{16}	16.13
Fe^{2+}	2.1×10^{14}	14.33	Fe^{3+}	1×10^{25}	25.1
Co^{2+}	2.0×10^{16}	16.31	V^{3+}	8×10^{25}	25.9
Ni^{2+}	4.2×10^{18}	18.62	Th^{4+}	2×10^{23}	23.2

* Data from G. Schwarzenbach, *Complexometric Titrations*, p. 8. New York: Interscience (London: Chapman & Hall Ltd.), 1957. With permission. (Constants valid at 20°C and an ionic strength of 0.1.)

centration are readily derived by the method demonstrated in Section 9D. For Y^{4-}, it is found that

$$\alpha_4 = \frac{K_1K_2K_3K_4}{[H^+]^4 + K_1[H^+]^3 + K_1K_2[H^+]^2 + K_1K_2K_3[H^+] + K_1K_2K_3K_4} \tag{12-3}$$

$$= \frac{K_1K_2K_3K_4}{D}$$

where K_1, K_2, K_3, and K_4 are the four dissociation constants for H_4Y and D is the denominator of Equation 12-3.[6]

Values for the other species are readily obtained in a similar way and are found to be

$$\alpha_0 = [H^+]^4/D \qquad \alpha_2 = K_1K_2[H^+]^2/D$$

$$\alpha_1 = K_1[H^+]^3/D \qquad \alpha_3 = K_1K_2K_3[H^+]/D$$

Figure 12-4 illustrates how the mole fraction of the five EDTA species varies as a function of pH. It is apparent that H_2Y^{2-} predominates in moderately acid medium pH (3 to 6). Only at pH values greater than 10 does Y^{4-} become a major component of the solution.

Values for α_4 as a function of pH. As will become apparent shortly, values for α_4 are particularly useful in determining the cation concentrations of EDTA solutions. The following example illustrates the method for calculating α_4.

> **EXAMPLE 12-2.** Calculate the percent of EDTA that is present as Y^{4-} in a solution of the reagent that is buffered to a pH of 10.20.
>
> $$[H^+] = \text{antilog}\,(-10.20) = 6.3 \times 10^{-11}$$
>
> From the values for the dissociation constants for H_4Y (p. 283) we obtain
>
> $$K_1 = 1.02 \times 10^{-2} \qquad K_1K_2K_3 = 1.51 \times 10^{-11}$$
>
> $$K_1K_2 = 2.18 \times 10^{-5} \qquad K_1K_2K_3K_4 = 8.31 \times 10^{-22}$$
>
> The various terms in the denominator of Equation 12-3 can then be evaluated. Thus,
>
> $$[H^+]^4 = 1.58 \times 10^{-41} \qquad K_1K_2[H^+]^2 = 8.69 \times 10^{-26}$$
>
> $$K_1[H^+]^3 = 2.56 \times 10^{-33} \qquad K_1K_2K_3[H^+] = 9.53 \times 10^{-22}$$
>
> Substitution into Equation 12-3 gives
>
> $$\alpha_4 = \frac{8.31 \times 10^{-22}}{1.58 \times 10^{-41} + 2.56 \times 10^{-33} + 8.69 \times 10^{-26} + 9.53 \times 10^{-22} + 8.31 \times 10^{-22}}$$
>
> $$= 0.466 = 0.47$$
>
> $$\%\ Y^{4-} = 0.47 \times 100 = 47$$
>
> Note that at pH 10.20, only the last two terms in the denominator are important. At low values for pH, only the first two or three terms are significant.

---

[6] In this chapter and those that follow we shall revert to the use of H^+ as a convenient shorthand notation for H_3O^+; from time to time we shall refer to the species represented by H^+ as the hydrogen ion.

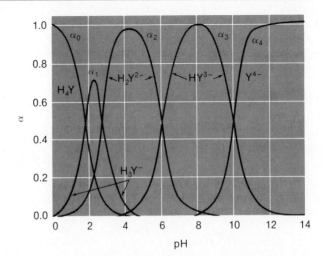

FIGURE 12-4 Composition of EDTA solutions as a function of pH.

Table 12-3 lists values for α_4 at selected pH values. Note that at pH 2.00, only about 4×10^{-12} percent of the EDTA exists as Y^{4-}.

Calculation of the cation concentration of a solution of an EDTA complex. The example that follows demonstrates how the cation concentration of a solution of an EDTA complex can be derived.

EXAMPLE 12-3. Calculate the equilibrium concentration of Ni^{2+} in a solution with an analytical NiY^{2-} concentration of 0.0150 M at a pH of (a) 3.0 and (b) 8.0. From Table 12-2

$$NiY^{2-} \rightleftharpoons Ni^{2+} + Y^{4-} \qquad \frac{[NiY^{4-}]}{[Ni^{2+}][Y^{4-}]} = K_{MY} = 4.2 \times 10^{18}$$

The concentration of the complex is given by the analytical concentration of the complex minus any lost by dissociation. That is,

$$[NiY^{2-}] = 0.0150 - [Ni^{2+}] \cong 0.0150$$

With the very large formation constant, it is unlikely that $[Ni^{2+}]$ is significant with respect to 0.0150.

Since the complex is the only source of both Ni^{2+} and the EDTA species,

$$[Ni^{2+}] = [Y^{4-}] + [HY^{3-}] + [H_2Y^{2-}] + [H_3Y^-] + [H_4Y] = C_T$$

Table 12-3 Values for α_4 for EDTA in Solutions of Various pH

pH	α_4	pH	α_4
2.0	3.7×10^{-14}	7.0	4.8×10^{-4}
3.0	2.5×10^{-11}	8.0	5.4×10^{-3}
4.0	3.6×10^{-9}	9.0	5.2×10^{-2}
5.0	3.5×10^{-7}	10.0	3.5×10^{-1}
6.0	2.2×10^{-5}	11.0	8.5×10^{-1}
		12.0	9.8×10^{-1}

Rearrangement of Equation 12-2 reveals that

$$C_T = \frac{[Y^{4-}]}{\alpha_4}$$

or

$$[Y^{4-}] = \alpha_4[Ni^{2+}]$$

Substitution of this relationship into the formation constant expression gives

$$\frac{[NiY^{2-}]}{[Ni^{2+}]^2} = \alpha_4 K_{MY} = K'_{MY} \tag{12-4}$$

where K'_{MY} is a pH-dependent constant called a *conditional* or *effective formation constant.*

(a) Table 12-3 indicates that α_4 has a value of 2.5×10^{-11} at pH 3.0. Substitution of this value and the concentration of NiY^{2-} into Equation 12-4 gives,

$$\frac{0.0150}{[Ni^{2+}]^2} = 2.5 \times 10^{-11} \times 4.2 \times 10^{18} = 1.05 \times 10^8$$

and

$$[Ni^{2+}] = 1.2 \times 10^{-5}$$

Note that indeed $[Ni^{2+}] \ll 0.0150$ as assumed.

(b) At pH 8.0, the conditional constant is much larger. Thus,

$$K'_{MY} = 5.4 \times 10^{-3} \times 4.2 \times 10^{18} = 2.27 \times 10^{16}$$

and

$$[Ni^{2+}] = \sqrt{0.0150/2.27 \times 10^{16}} = 8.1 \times 10^{-10}$$

Calculation of the cation concentration in the presence of an excess of EDTA.

The following example illustrates the effect of excess EDTA on the concentration of a cation.

EXAMPLE 12-4. Calculate the concentration of Ni^{2+} in a solution that has a pH of 3.00 and was prepared by mixing 50 mL of 0.0300 M Ni^{2+} with 50 mL of 0.0500 M EDTA.

Here, an excess of EDTA is present and the concentration of the complex is determined by the amount of Ni^{2+} originally present. Thus,

$$M_{NiY^{2-}} = \frac{50.0 \times 0.03}{100} = 0.0150$$

$$M_{EDTA} = \frac{50.0 \times 0.0500 - 50.0 \times 0.0300}{100} = 0.0100$$

Again let us assume that $[Ni^{2+}] \ll [NiY^{2-}]$ and

$$[NiY^{2-}] \cong 0.0150$$

Here, the total concentration of uncomplexed EDTA is 0.0100. Thus,

$$C_T = 0.0100$$

Applying Equation 12-2 gives

$$[Y^{4-}] = 0.0100 \times \alpha_4 = 0.0100 \times 2.5 \times 10^{-11}$$

Substitution into the formation constant expression gives

$$\frac{0.0150}{[Ni^{2+}] \times 0.0100} = 2.5 \times 10^{-11} \quad K_{MY} = K'_{MY} = 1.05 \times 10^8$$

$$[Ni^{2+}] = 1.4 \times 10^{-8}$$

Note that in both of the foregoing examples, Equation 12-1 was converted to the form

$$\frac{[MY^{(n-4)+}]}{[M^{n+}]C_T} = \alpha_4 K_{MY} = K'_{MY} \tag{12-5}$$

where K'_{MY}, the conditional formation constant, describes equilibrium conditions *only at the pH for which α_4 is applicable.*

Conditional constants are readily computed and provide a simple means by which the equilibrium concentrations of the metal ion and the complex can be calculated at any point in a titration curve. Note that the expression for the conditional constant differs from that of the formation constant only in that the term C_T replaces the equilibrium concentration of the completely dissociated anion $[Y^{4-}]$. This difference is significant, however, because C_T is more readily determined from the stoichiometry of reactions than is $[Y^{4-}]$.

12C-4 DERIVATION OF TITRATION CURVES FOR EDTA TITRATIONS

EXAMPLE 12-5. Derive a curve (pCa versus volume EDTA) for the titration of 50.0 mL of 0.00500 M Ca^{2+} with 0.0100 M EDTA in a solution buffered to a constant pH of 10.0.

Calculation of conditional constant. The conditional formation constant for the calcium EDTA complex at pH 10 can be obtained from the formation constant of the complex (Table 12-2) and the α_4 value for EDTA at pH 10 (Table 12-3). Thus, substitution into Equation 12-5 gives

$$K'_{CaY} = \alpha_4 K_{CaY} = (3.5 \times 10^{-1})(5.0 \times 10^{10})$$

$$= 1.75 \times 10^{10}$$

At pH 10, then, the formation constant expression for the calcium complex takes the form

$$\frac{[CaY^{2-}]}{[Ca^{2+}]C_T} = 1.75 \times 10^{10}$$

Preequivalence point values for pCa. Before the equivalence point has been reached, the molar concentration of Ca^{2+} will be equal to the sum of the contributions from the untitrated excess of the cation and from dissociation of the complex, the latter being numerically equal to C_T. It is ordinarily reasonable to assume that C_T is small with respect to the analytical concentration of the uncomplexed calcium ion. Thus, for example, after the addition of 10.0 mL of reagent

$$[Ca^{2+}] = \frac{50.0 \times 0.00500 - 10.0 \times 0.0100}{60.0} + C_T \cong 2.50 \times 10^{-3}$$

and

$$pCa = -\log 2.50 \times 10^{-3} = 2.60$$

Equivalence point pCa. Here, the method shown in Example 12-3 is used; the analytical concentration of CaY^{2-} is given by

$$M_{CaY^{2-}} = \frac{50.0 \times 0.00500}{50.0 + 25.0} = 3.33 \times 10^{-3}$$

The only source of Ca^{2+} ions will be from dissociation of this complex. It also follows that the calcium ion concentration must be identical to the sum of the concentrations of the uncomplexed EDTA ions, C_T. Thus,

$$[Ca^{2+}] = C_T$$

$$[CaY^{2-}] = 0.00333 - [Ca^{2+}] \cong 0.00333$$

Substituting into the conditional formation constant expression gives

$$\frac{0.00333}{[Ca^{2+}]^2} = 1.75 \times 10^{10}$$

$$[Ca^{2+}] = 4.36 \times 10^{-7}$$

$$pCa = 6.36$$

Postequivalence point pCa. Beyond the equivalence point, analytical concentrations of CaY^{2-} and EDTA are obtained directly from the stoichiometric data. Here, a calculation similar to that in Example 12-4 is performed. Thus, after 35.0 mL of reagent have been added,

$$M_{CaY^{2-}} = \frac{50.0 \times 0.00500}{85.0} = 2.94 \times 10^{-3}$$

$$M_{EDTA} = \frac{10.0 \times 0.0100}{85.0} = 1.18 \times 10^{-3}$$

As an approximation, we may write

$$[CaY^{2-}] = 2.94 \times 10^{-3} - [Ca^{2+}] \cong 2.94 \times 10^{-3}$$

$$C_T = 1.18 \times 10^{-3} + [Ca^{2+}] \cong 1.18 \times .10^{-3}$$

and substitution into the conditional formation constant expression gives

$$\frac{2.94 \times 10^{-3}}{[Ca^{2+}] \times 1.18 \times 10^{-3}} = 1.75 \times 10^{10} = K'_{CaY}$$

$$[Ca^{2+}] = 1.42 \times 10^{-10}$$

$$pCa = 9.85$$

Curve A in Figure 12-5 is a plot of data derived by calculations illustrated in the foregoing example. Also shown is a titration curve (B) for a solution of magnesium ion under identical conditions. The smaller change in pMg as compared with pCa in the equivalence point region reflects the smaller formation constant for the magnesium EDTA complex and thus the less complete reaction.

Figure 12-6 provides titration curves for calcium ion in solutions buffered to various pH levels. It is apparent that appreciable changes in pCa at the equivalence point can be achieved only if the pH of the solution is maintained at about 8 or greater. As shown in Figure 12-7, however, cations with larger formation constants provide good end points even in acidic solutions. Figure 12-8 shows the minimum permissible pH for satisfactory end points in the titration of various metal ions in the absence of competing complexing agents. Note that a moderately acidic

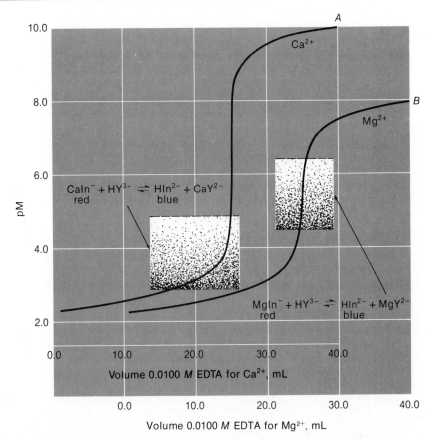

FIGURE 12-5 EDTA titration curves for 50.00 mL of 0.00500 M Ca^{2+} ($K^1CaY^{2-} = 1.75 \times 10^{10}$) and Mg^{2+} ($K^1Mgy^{2-} = 1.7 \times 10^8$) at a pH of 10.0 The shaded areas show the transition range for Eriochrome black T.

environment is satisfactory for many divalent heavy-metal cations and that a strongly acidic medium can be tolerated in the titration of ions such as iron(III) and indium(III).

Effect of other complexing agents on EDTA titrations. Many EDTA titrations are complicated by the tendency on the part of the ion being titrated to precipitate as a basic oxide or hydroxide at the pH required for a satisfactory end point. In order to keep the metal ion in solution, particularly in the early stages of the titration, it is necessary to include an auxiliary complexing agent. Thus, for example, the titration of zinc(II) is ordinarily performed in the presence of relatively high concentrations of ammonia and ammonium chloride. These species buffer the solution to an acceptable pH. In addition, formation of four ammine complexes prevents precipitation of zinc hydroxide. These complexes include $Zn(NH_3)^{2+}$, $Zn(NH_3)_2^{2+}$, $Zn(NH_3)_3^{2+}$, and $Zn(NH_3)_4^{2+}$.

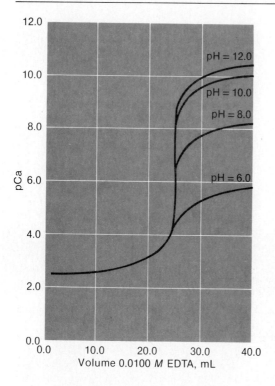

FIGURE 12-6 Effect of pH upon the end point for the titration of 50.0 mL of 0.00500 M Ca^{2+}.

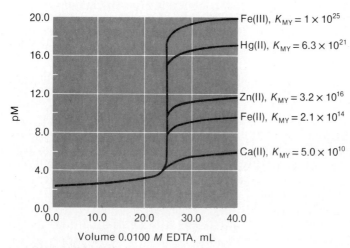

FIGURE 12-7 Titration curves for 50.0 mL of 0.00500 M cation solutions at pH = 6.0.

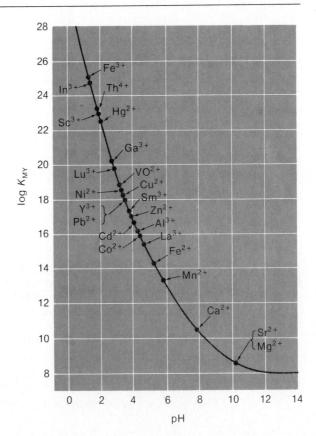

FIGURE 12-8 Minimum pH needed for satisfactory titration of various cations with EDTA. [C. N. Reilley and R. W. Schmid, *Anal. Chem.*, **30**,947 (1958). With permission of the American Chemical Society.]

In a system such as this, the completeness of reaction, and thus the quality of the end point, depends not only upon the pH but also upon the concentration of ammonia present because these ammine complexes compete with the EDTA for zinc ions.

The influence of an auxiliary complexing reagent can be treated in much the same way that account was taken of pH effects on the concentration of EDTA species. A quantity δ is defined that is analogous to α_4. That is,

$$\delta = \frac{[M^{n+}]}{C_M} \tag{12-6}$$

Here, C_M is the sum of the concentrations of species containing the metal ion *exclusive* of that combined with EDTA. For solutions containing zinc(II) and ammonia, then,

$$C_M = [Zn^{2+}] + [Zn(NH_3)^{2+}] + [Zn(NH_3)_2^{2+}] + [Zn(NH_3)_3^{2+}] + [Zn(NH_3)_4^{2+}] \tag{12-7}$$

The value of δ can be readily expressed in terms of the ammonia concentration and the formation constants for the various ammine complexes. Thus, for example,

$$K_1 = \frac{[Zn(NH_3)^{2+}]}{[Zn^{2+}][NH_3]}$$

or

$$[Zn(NH_3)^{2+}] = K_1[Zn^{2+}][NH_3]$$

Similarly, it is readily shown that

$$[Zn(NH_3)_2^{2+}] = K_1K_2[Zn^{2+}][NH_3]^2$$

and so forth.

Substitution of these values into Equation 12-7 gives

$$C_M = [Zn^{2+}](1 + K_1[NH_3] + K_1K_2[NH_3]^2 + K_1K_2K_3[NH_3]^3 + K_1K_2K_3K_4[NH_3]^4)$$

Combining this expression with Equation 12-6 (here, $[M^{n+}] = [Zn^{2+}]$) yields

$$\delta = \frac{1}{1 + K_1[NH_3] + K_1K_2[NH_3]^2 + K_1K_2K_3[NH_3]^3 + K_1K_2K_3K_4[NH_3]^4} \tag{12-8}$$

Finally, a conditional constant for the equilibrium between EDTA and zinc(II) in an ammonia/ammonium chloride buffer is obtained by substituting Equation 12-6 into Equation 12-5 (p. 289) and rearranging

$$\frac{[ZnY^{2-}]}{C_M C_T} = \alpha_4 \delta K_{ZnY} = K''_{ZnY} \tag{12-9}$$

where K''_{ZnY} is a new conditional constant that applies at a single pH as well as a single concentration of ammonia. The following example will show how this conditional constant is employed for derivation of titration curves.

EXAMPLE 12-6. Calculate the pZn of solutions prepared by mixing 20.0, 25.0, and 30.0 mL of 0.0100 M EDTA with 50.0 mL of 0.00500 M Zn^{2+}. Assume that both Zn^{2+} and the EDTA solutions are 0.100 M in NH_3 and 0.176 M in NH_4Cl to provide a constant pH of 9.0.

In Appendix 6, we find that the logarithms of the stepwise formation constants for the four zinc complexes with ammonia are 2.4, 2.4, 2.5, and 2.1, respectively. Thus,

$$K_1 = \text{antilog } 2.4 = 2.51 \times 10^2$$

$$K_1K_2 = \text{antilog } (2.4 + 2.4) = 6.31 \times 10^4$$

$$K_1K_2K_3 = \text{antilog } (2.4 + 2.4 + 2.5) = 2.00 \times 10^7$$

$$K_1K_2K_3K_4 = \text{antilog } (2.4 + 2.4 + 2.5 + 2.1) = 2.51 \times 10^9$$

1. *Calculation of a conditional constant.* A value for δ can be obtained from Equation 12-8 by assuming that the molar and analytical concentrations of ammonia are essentially the same; thus, for $[NH_3] = 0.100$,

$$\delta = \frac{1}{1 + 25 + 631 + 2.00 \times 10^4 + 2.51 \times 10^5} = 3.68 \times 10^{-6}$$

A value for K_{ZnY} is found in Table 12-2, and α_4 for pH 9.0 is given in Table 12-3. Substituting into Equation 12-9, we find

$$K''_{ZnY} = 5.2 \times 10^{-2} \times 3.68 \times 10^{-6} \times 3.2 \times 10^{16} = 6.12 \times 10^9$$

2. *Calculation of pZn after addition of 20.0 mL of EDTA.* The concentration of unreacted Zn^{2+} at this point is given to an excellent approximation by

$$C_M = \frac{50.0 \times 0.00500 - 20.0 \times 0.0100}{70.0} = 7.14 \times 10^{-4}$$

Here, we assume that there is negligible dissociation of the several complexes. The quantity C_M, which is the sum of the equilibrium concentrations of all the Zn species *not* complexed by EDTA, can be employed in Equation 12-6 to give $[Zn^{2+}]$. Thus,

$$[Zn^{2+}] = C_M\delta = (7.14 \times 10^{-4})(3.68 \times 10^{-6})$$

$$= 2.63 \times 10^{-9}$$

$$pZn = 8.58$$

3. *Calculation of pZn after addition of 25.0 mL of EDTA.* At the equivalence point for the titration, the analytical concentration of ZnY^{2-} is given by

$$M_{ZnY^{2-}} = \frac{50.0 \times 0.00500}{50.0 + 25.0} = 3.33 \times 10^{-3}$$

The sum of the concentrations of the various zinc species not combined with EDTA will equal the sum of the concentrations of the uncomplexed EDTA species. That is,

$$C_M = C_T$$

and

$$[ZnY^{2-}] \cong 3.33 \times 10^{-3} - C_M = 3.33 \times 10^{-3}$$

Substituting into Equation 12-9,

$$\frac{3.33 \times 10^{-3}}{C_M^2} = 6.12 \times 10^9 = K''_{ZnY}$$

$$C_M = 7.38 \times 10^{-7}$$

Employing Equation 12-6, we obtain

$$[Zn^{2+}] = C_M\delta = (7.38 \times 10^{-7})(3.68 \times 10^{-6})$$

$$= 2.72 \times 10^{-12}$$

$$pZn = 11.57$$

4. *Calculation of pZn after addition of 30.0 mL of EDTA.* The solution now contains an excess of EDTA; thus,

$$M_{EDTA} = C_T = \frac{30.0 \times 0.0100 - 50.0 \times 0.00500}{80.0} = 6.25 \times 10^{-4}$$

and

$$M_{ZnY^{2-}} = [ZnY^{2-}] = \frac{50.0 \times 0.00500}{80.0} = 3.12 \times 10^{-3}$$

Rearranging Equation 12-9 gives

$$C_M = \frac{[ZnY^{2-}]}{C_T \times K''_{ZnY}}$$

Substituting numerical values yields

$$C_M = \frac{3.12 \times 10^{-3}}{(6.25 \times 10^{-4})(6.12 \times 10^9)} = 8.16 \times 10^{-10}$$

and from Equation 12-6

$$[Zn^{2+}] = C_M\delta = (8.16 \times 10^{-10})(3.68 \times 10^{-6}) = 3.00 \times 10^{-15}$$

$$pZn = 14.52$$

Figure 12-9 shows two theoretical curves for the titration of zinc(II) with EDTA at a pH of 9.00. The equilibrium concentration of ammonia was 0.100 M for the one titration and 0.0100 M for the other. It is seen that the presence of ammonia has the effect of decreasing the change in pZn in the equivalence-point region. Thus, it is desirable to keep the concentration of any auxiliary reagent to the minimum required to prevent hydroxide formation. Note that the magnitude of δ does not affect pZn beyond the equivalence point. On the other hand, it will be recalled (Figure 12-6) that α_4, and thus pH, plays an important role in defining this region of the titration curve.

12C-5 INDICATORS FOR EDTA TITRATIONS

Reilley and Barnard[7] have listed nearly 200 organic compounds that have been suggested as indicators for metal ions in EDTA titrations. In general, these indicators are organic dyes that form colored chelates with metal ions in a pM range that is characteristic of the particular cation and dye. The complexes are often intensely colored, being discernible to the eye in the range of 10^{-6} to 10^{-7} M.

Over half of the compounds listed in reference 7 contain an aromatic azo functional group. One of the first and most widely used of these compounds has the trivial name of *Eriochrome black T*. Its structure is

The proton associated with the sulfonic acid group in this compound is completely dissociated in an aqueous medium. The phenolic protons, on the other hand, are only partially dissociated; that is,

$$H_2O + H_2In^- \rightleftharpoons HIn^{2-} + H_3O^+ \qquad K_1 = 5 \times 10^{-7}$$
$$\text{red} \qquad\qquad \text{blue}$$

$$H_2O + HIn^{2-} \rightleftharpoons In^{3-} + H_3O^+ \qquad K_2 = 2.8 \times 10^{-12}$$
$$\text{blue} \qquad\qquad \text{orange}$$

Note that the acids and their conjugate bases differ in color. Thus, Eriochrome black T also behaves as an acid-base indicator.

The metal complexes of Eriochrome black T are generally red. To observe a color change with this indicator, then, it is necessary to adjust the pH to 7 or above

[7] C. N. Reilley and A. J. Barnard, Jr., in *Handbook of Analytical Chemistry*, L. Meites, Ed., p. 3-77. New York: McGraw-Hill, 1963.

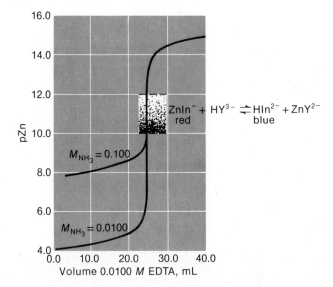

FIGURE 12-9 Influence of ammonia concentration upon the end point for the titration of 50.0 mL of 0.00500 M Zn^{2+}. Solutions buffered to a pH of 9.00. The shaded area shows the transition range for Eriochrome black T.

so that the blue form of the species HIn^{2-} predominates in the absence of a metal ion. The end-point reaction is then

$$MIn^- + HY^{3-} \rightleftharpoons HIn^{2-} + MY^{2-}$$
$$\quad\text{red} \qquad\qquad\qquad \text{blue}$$

Eriochrome black T forms red complexes with more than two dozen different metal ions, but with only certain of these cations are the stabilities of the products appropriate for end point detection. As shown in the following example, the applicability of a given indicator to an EDTA titration can be determined from the change in pM in the equivalence-point region, provided the formation constant for the metal-indicator complex is known.[8]

EXAMPLE 12-7. Derive transition ranges for Eriochrome black T in titrations of Mg^{2+} and Ca^{2+} at a pH of 10.0, given that the second acid dissociation constant for the indicator is

$$HIn^{2-} + H_2O \rightleftharpoons In^{3-} + H_3O^+ \qquad K_2 = 2.8 \times 10^{-12} = \frac{[H_3O^+][In^{3-}]}{[HIn^{2-}]}$$

The formation constant for $MgIn^-$ is

$$Mg^{2+} + In^{3-} \rightleftharpoons MgIn^- \qquad K_f = 1.0 \times 10^7 = \frac{[MgIn^-]}{[Mg^{2+}][In^{3-}]}$$

For Ca^{2+}, the analogous constant has a value of 2.5×10^5.

We will assume, as we have earlier (p. 196), that a detectable color change requires a tenfold excess of one or the other of the colored species; that is, a detectable color change is observed when $[MgIn^-]/[HIn^{2-}]$ changes from 10 to 0.10.

Multiplication of K_2 for the indicator by K_f for $MgIn^-$ gives an expression that contains the foregoing ratio. That is,

---

[8] For a complete discussion of the principles of indicator choice in complex-formation titrations, see C. N. Reilley and R. W. Schmid, *Anal. Chem.*, **31**, 887 (1959).

$$\frac{[MgIn^-][H_3O^+]}{[HIn^{2-}][Mg^{2+}]} = 2.8 \times 10^{-12} \times 1.0 \times 10^7 = 2.8 \times 10^{-5}$$

which rearranges to

$$[Mg^{2+}] = \frac{[MgIn^-]}{[HIn^{2-}]} \times \frac{[H_3O^+]}{2.8 \times 10^{-5}}$$

Substitution of 1.0×10^{-10} for $[H_3O^+]$ and 10 and 0.10 for the ratio yields the range of $[Mg^{2+}]$ over which the color change will occur. Thus,

$$[Mg^{2+}] = 3.6 \times 10^{-5} \text{ to } 3.6 \times 10^{-7}$$

or

$$pMg = 5.4 \pm 1.0$$

Proceeding in the same way, we find the range for pCa to be

$$pCa = 3.8 \pm 1.0$$

The ranges for magnesium and calcium are indicated on the titration curves shown in Figure 12-5. Eriochrome black T is clearly an ideal indicator for magnesium but a totally unsatisfactory one for calcium. Note that the formation constant for $CaIn^{2-}$ is only about one-fortieth of that for magnesium. As a consequence, significant conversion of $CaIn^{2-}$ to HIn^- occurs well before equivalence.

A similar calculation reveals that Eriochrome black T is also well-suited for the titration of zinc with EDTA (see Figure 12-9).

A limitation of Eriochrome black T is that its solutions decompose slowly with standing. It is claimed that solutions of Calmagite, an indicator that for all practical purposes is identical in behavior, do not suffer this disadvantage. The structure of Calmagite is similar to Eriochrome black T.

Many other metal indicators have been developed for EDTA titrations.[9] In contrast to Eriochrome black T, some can be employed in strongly acidic media.

12C-6 TITRATION METHODS EMPLOYING EDTA

EDTA is used in several ways as a titrant for metal ions. The most common of these procedures are considered in the following paragraphs.

Direct titration. Reilley and Barnard[10] list 40 elements that can be determined by direct titration with EDTA using metal-ion indicators for end point detection.

---

[9] See, for example, L. Meites, *Handbook of Analytical Chemistry*, pp. 3-101 to 3-165. New York: McGraw-Hill, 1963.
[10] C. N. Reilley and A. J. Barnard, Jr., in *Handbook of Analytical Chemistry*, L. Meites, Ed., pp. 3-166 to 3-200. New York: McGraw-Hill, 1963.

Direct titration procedures are limited to those reactions for which a method for end point detection exists and to those metal ions that react rapidly with EDTA. When direct methods fail, it is often possible to achieve an analysis by a back titration or a displacement titration.

Back titration. Back-titration procedures are useful for the analysis of cations that form very stable EDTA complexes and for which a satisfactory indicator is not available. In such an analysis, the excess EDTA is determined by back titration with a standard magnesium solution, using Eriochrome black T or Calmagite as the indicator. The metal/EDTA complex must be more stable than the magnesium/EDTA complex; otherwise the back titrant would displace the metal ion.

This technique is also useful for the titration of metals in the presence of anions that would otherwise form slightly soluble precipitates with the cation under the conditions of the analysis; the presence of EDTA prevents their precipitation.

Displacement titration. In displacement titrations, an excess of a solution containing the magnesium (or zinc) complex of EDTA is introduced. If the analyte forms a more stable complex than that of magnesium (or zinc), the following reaction occurs:

$$MgY^{2-} + M^{2+} \rightleftharpoons MY^{2-} + Mg^{2+}$$

The liberated magnesium is then titrated with a standard EDTA solution. This technique is useful where no satisfactory indicator is available for the metal ion being determined.

Alkalimetric titration. An alkalimetric titration involves adding an excess of Na_2H_2Y to a neutral solution of the metallic ion.

$$M^{2+} + H_2Y^{2-} \longrightarrow MY^{2-} + 2H^+$$

The liberated hydrogen ions are then titrated with a standard solution of a base.

12C-7 SCOPE OF EDTA TITRATIONS

Complexometric titrations with EDTA have been reported for the analysis of virtually every cation. Because the reagent tends to form chelates with most cations, EDTA would appear at first glance to be totally lacking in selectivity. In fact, however, considerable control over the behavior of EDTA, as well as other chelating agents, can be achieved through pH regulation. Thus, for example, it is generally possible to determine trivalent cations without interference from divalent species by performing the titration in a medium with a pH of about 1 (see Figure 12-8); under these circumstances, the less stable divalent chelates do not form to any significant extent, whereas the trivalent ions are quantitatively complexed. Similarly, ions such as cadmium and zinc, which form more stable EDTA chelates than magnesium, can be determined in the presence of the latter ion by buffering the mixture to pH 7 before titration. Eriochrome black T serves as an indicator for the cadmium or zinc end points without interference from magnesium because the indicator chelate with magnesium is not formed at this pH. Finally, interference from a particular cation can sometimes be eliminated by adding a suitable *masking agent*, an auxiliary ligand that preferentially forms highly stable complexes with the potential interference.[11] For example, cyanide ion is often employed as a masking agent to permit the titration of magnesium and calcium ions in the presence of ions such as cadmium, cobalt, copper, nickel, zinc, and palladium. All of the latter form sufficiently stable cyanide complexes to prevent reaction with EDTA.

Specific directions for the preparation and use of EDTA solutions are given in Section 5A, Chapter 31.

12C-8 DETERMINATION OF WATER HARDNESS

Historically, "hardness" was defined in terms of the capacity of a water sample to precipitate soap, an undesirable quality. Soap is precipitated by most cations with multiple charges. In natural waters, however, the concentration of calcium and magnesium ions generally far exceeds that of any other metal ion; thus, hardness has come to mean the total concentration of calcium and magnesium expressed in terms of the calcium carbonate equivalent.

The determination of hardness is a useful analytical process for measuring the quality of water for household and industrial uses. The importance for the latter is due to the fact that hard water, upon heating, precipitates calcium carbonate, which then clogs boilers and pipes.

Water hardness is ordinarily determined by an EDTA titration after the sample has been buffered to a pH of 10; Calmagite or Eriochrome black T serves as the indicator. Often, a small concentration of the EDTA complex of magnesium is incorporated in the buffer or the titrant to assure the presence of sufficient magnesium ions for satisfactory indicator action.

Test kits for determining the hardness of household water are available commercially. These consist of a vessel calibrated to contain a known volume of water, a measuring scoop to deliver an appropriate amount of a solid buffer mixture, an indicator solution, and a bottle of standard EDTA, which is equipped with a medicine dropper; a measure of the volume of standard reagent is obtained by counting drops to the end point. The concentration of the EDTA solution is ordinarily such that one drop corresponds to one grain (about 0.065 g) of calcium carbonate per gallon of water.

PROBLEMS*

* 12-1. (a) Describe the preparation of 500 mL of 0.01500 M solution of EDTA from $Na_2H_2Y\cdot2H_2O$. Recall that this compound contains 0.3% excess H_2O (p. 283).
   (b) What is the CaO titer of this solution?
   (c) What is its $Zn_2P_2O_7$ titer?
* 12-2. An EDTA solution was prepared by dissolving 3.842 g $Na_2H_2Y\cdot2H_2O$ in water and diluting to 1.000 L. The starting material contained 0.3% excess water (p. 283).
   (a) Calculate the molarity of the solution.
   (b) Calculate the $MgCO_3$ titer of the solution.
   (c) Calculate the $Ca_3(PO_4)_2$ titer of the solution.

---

[11] For further information, see: D. D. Perrin, *Masking and Demasking of Chemical Reactions.* New York: Wiley-Interscience, 1970; and C. N. Reilley and A. J. Barnard, in *Handbook of Analytical Chemistry*, L. Meites, Ed., pp. 3-208 to 3-225. New York: McGraw-Hill, 1963.
* Answers to problems or parts of problems marked with an asterisk are to be found at the end of the book.

* 12-3. The sulfate in a 0.3734-g sample was determined by precipitation of the anion as $BaSO_4$. After filtration and washing, the precipitate was dissolved in 50.0 mL of 0.02121 M EDTA. The excess EDTA required 11.74 mL of a solution of $MgCl_2$. A blank containing the same volume of EDTA consumed 41.29 mL of the $MgCl_2$ solution.

    (a) Calculate the molarity of the $MgCl_2$ solution.

    (b) Calculate the percent $Al_2(SO_4)_3$ in the sample.

  12-4. A 24-hr urine specimen was diluted to exactly 2.000 L. After buffering to pH 10.0, a 50.00-mL aliquot consumed 22.19 mL of 0.04481 M EDTA. The calcium in a second 50.00-mL aliquot was precipitated as $CaC_2O_4(s)$, filtered, washed, and redissolved in acid. Exactly 25.00 mL of the EDTA were added and the resulting solution was buffered to a pH of 10.0. The excess EDTA was back titrated with 8.31 mL of 0.05053 M $MgCl_2$.

    Calculate the milligrams of Ca^{2+} and Mg^{2+} in the original sample.

* 12-5. The total calcium and magnesium in a hard water was determined by titration of a 100-mL sample with 31.3 mL of 0.0106 M EDTA using Eriochrome black T as the indicator. A second 100-mL sample was made strongly alkaline with NaOH, which precipitated most of the magnesium as magnesium hydroxide. The calcium was then titrated with 19.2 mL of the EDTA using arsenazo indicator, which is sensitive to calcium ion.

    (a) Calculate the total water hardness in terms of mg $CaCO_3/L$.

    (b) Calculate the mg $CaCO_3/L$ and mg $MgCO_3/L$.

  12-6. A 1.296-g sample of a mercuric ointment was suspended in $CHCl_3$ and shaken vigorously with dilute HNO_3 to extract the Hg^{2+}. The extract was titrated with 21.74 mL of 0.04359 M KSCN with Fe^{3+} as the indicator. Reactions:

$$Hg^{2+} + 2SCN^- \longrightarrow Hg(SCN)_2$$

$$Fe^{3+} + SCN^- \longrightarrow FeSCN^+$$

   Calculate the % HgO in the preparation.

* 12-7. Nickel can be titrated in an ammoniacal solution with a standard solution of KCN. The end point is signaled by the disappearance of a turbidity due to a trace of AgI suspended in the solution. Reactions:

$$Ni(NH_3)_4^{2+} + 4CN^- \rightleftharpoons Ni(CN)_4^{2-} + 4NH_3$$

$$AgI(s) + 2CN^- \rightleftharpoons Ag(CN)_2^- + I^-$$

    A 0.4102-g sample of a nickel alloy was dissolved in acid and then made basic with 6 M NH_3. One milliliter of 25% KI was added and the solution was titrated with 0.1164 M KCN until the blue-green color of the $Ni(NH_3)_4^{2+}$ became faint. One or two drops of 0.1 M $AgNO_3$ were added, and the titration was continued until the opalescence due to the AgI disappeared. Calculate the % Ni in the alloy if 37.60 mL of the KCN were required.

* 12-8. A 1.728-g sample of a lead/cadmium alloy was dissolved in acid and diluted to exactly 500 mL in a volumetric flask. Titration of a 25.00-mL aliquot at a pH of 10 in an NH_4^+/NH_3 buffer required 43.60 mL of standard 0.01080 M EDTA and involved both cations. The cadmium ion was masked by using an HCN/NaCN buffer to bring the pH of a second 25.00-mL aliquot to 10; 32.70 mL of the standard EDTA solution were used in this titration. Calculate the respective percentages of lead and cadmium in the sample.

* 12-9. A 2.420-g sample of an alloy containing Bi, Pb, and Cd was dissolved in HNO_3 and diluted to 250.0 mL. A 50.00-mL aliquot was brought to a pH of 2 and titrated with 25.67 mL of 0.02479 M EDTA; at this pH only Bi^{3+} reacted with the reagent. The pH of the solution was then buffered to about 5 by addition of hexamethylenetetra-

mine and the titration continued; 24.76 mL of the EDTA were consumed in reacting with the Pb and Cd. Addition of orthophenanthroline to the solution liberated the Cd^{2+} from the EDTA complex, and the freed EDTA was titrated with 6.76 mL of 0.02174 M $PbNO_3$. Calculate the % Bi, Pb, and Cd in the alloy.

12-10. The Ni^{2+} and Zn^{2+} in a 25.00-mL sample were complexed by addition of 50.00 mL of 0.0125 M EDTA, and the excess chelating agent was titrated with 7.50 mL of 0.0104 M Mg^{2+} solution. Introduction of an excess of 2,3-dimercapto-1-propanol (BAL) displaced the zinc from its EDTA complex; titration of the liberated EDTA required an additional 21.50 mL of the Mg^{2+} solution. Calculate the milligrams of Ni^{2+} and Zn^{2+} in each milliliter of the sample.

* 12-11. A 2.164-g sample of brass was dissolved in nitric acid. The sparingly soluble metastannic acid was removed by filtration, and the combined filtrate and washings were then diluted to exactly 500 mL.

     A 20.0-mL aliquot was suitably buffered; titration of the lead, zinc, and copper in this aliquot required 36.82 mL of 0.02596 M EDTA.

     The copper in a 50.0-mL aliquot was masked with thiosulfate; the lead and zinc were then titrated with 28.13 mL of the EDTA solution.

     Cyanide ion was used to mask the copper and zinc in a 100-mL aliquot; 6.12 mL of the EDTA solution were needed to titrate the lead ion.

     Determine the composition of the brass sample; evaluate the percentage of tin by difference.

12-12. A 25.00-mL sample of a solution of $Al_2(SO_4)_3$ and $NiSO_4$ was diluted to exactly 500.0 mL and the Al^{3+} and Ni^{2+} in a 25.00-mL aliquot were complexed by addition of 40.00 mL of 0.01175 M EDTA after the solution had been buffered to pH 4.8. After boiling the solution for a short time to hasten the formation of the Al complex, the excess EDTA was titrated with 10.07 mL of 0.00993 M Cu^{2+}. An excess of F^- was then added to the hot solution, which displaced the EDTA bound to the Al^{3+}. The liberated EDTA consumed 26.30 mL of the standard Cu^{2+} solution. Calculate the mg/mL of $NiSO_4$ and $Al_2(SO_4)_3$ in the sample.

* 12-13. Calculate conditional constants for the formation of the EDTA complex of Fe^{2+} at a pH of (a) 6.0, (b) 8.0, (c) 10.0.

12-14. Calculate conditional constants for the formation of the EDTA complex of Ba^{2+} at a pH of (a) 7.0, (b) 9.0, (c) 11.0.

* 12-15. Calculate conditional constants for the formation of the EDTA complex of Co^{2+} when
    (a) the concentration of NH_3 is 0.050 M and the pH is 8.0.
    (b) the concentration of NH_3 is 0.050 M and the pH is 10.0.
    (c) the concentration of NH_3 is 0.50 M and the pH is 8.0.
    (d) the concentration of NH_3 is 0.50 M and the pH is 10.0.

12-16. Calculate the conditional constant for the EDTA complex of Cd when
    (a) the concentration of KI is 0.100 M and the pH is 5.0.
    (b) the concentration of KI is 0.100 M and the pH is 7.0.
    (c) the concentration of KI is 0.010 M and the pH is 5.0.
    (d) the concentration of KI is 0.010 M and the pH is 7.0.

* 12-17. Calculate the equilibrium concentration of each of the three silver-containing species in a solution having a total silver concentration of 1.00×10^{-4} M and an equilibrium NH_3 concentration of (a) 0.100 M, (b) 0.0100 M, (c) 0.00100 M.

12-18. Calculate the equilibrium concentration of each of the three lead-containing species in a solution having a total lead concentration of 1.00×10^{-4} M and an equilibrium acetate concentration of (a) 0.10 M, (b) 0.50 M, (c) 1.0 M.

12-19. Assume that for 0.0100 M solutions, the minimum conditional constant for a satisfactory end point for an EDTA titration is 10^8. Calculate the minimum pH (to the nearest integer) for the titration of *(a) Sr^{2+}, (b) Zn^{2+}, *(c) Fe^{3+}, (d) Al^{3+}.

12-20. It can be demonstrated that if the conditional constant for an EDTA reaction is less than about 10^2, no significant consumption of 0.0100 M reagent occurs. On the other hand, if the constant is higher than about 10^8, a satisfactory end point is to be expected. What pH range (approximate) could be used for
  * (a) the titration of Cu^{2+} in the presence of Ba^{2+}?
    (b) the titration of Fe^{3+} in the presence of Fe^{2+}?
  * (c) the titration of Al^{3+} in the presence of Mg^{2+}?
    (d) the titration of Hg^{2+} in the presence of Mn^{2+}?

12-21. Calculate the equivalence point pM in the titration of a 0.0100 M solution of each of the following metal ions with 0.0100 M EDTA.
  * (a) Ba^{2+} at pH 11.0.
    (b) Pb^{2+} at pH 6.0.
  * (c) Al^{3+} at pH 4.0.
    (d) V^{3+} at pH 2.0.

* 12-22. Derive a titration curve for 50.00 mL of 0.00800 M Ba^{2+} with 0.01600 M EDTA in a solution that is buffered to pH 12.0. Derive pBa values after addition of 10.0, 20.0, 24.0, 25.0, 26.0, 30.0, and 35.0 mL of titrant.

12-23. Derive a titration curve for 40.0 mL of 0.01000 M Cu^{2+} with 0.02000 M EDTA in a solution that is buffered to pH 4.0. Derive pCu values after addition of 5.00, 15.0, 19.0, 20.0, 21.0, 25.0, and 30.0 mL of titrant.

* 12-24. Derive a titration curve for the titration of 50.00 mL of 0.00200 M Cu^{2+} with 0.00400 M EDTA in a solution having a pH of 9.0 and an NH_3 concentration of 0.200 M. Derive pCu values after the addition of 10.0, 20.0, 24.0, 25.0, 26.0, 30.0, and 35.0 mL of reagent.

12-25. Derive a titration curve for the titration of 50.0 mL of 0.0100 M Pb^{2+} with 0.0200 M EDTA in a solution having a pH of 9.0 and an acetate ion concentration of 0.40 M. Derive pPb values after the addition of 10.0, 20.0, 24.0, 25.0, 26.0, 30.0, and 35.0 mL of the titrant.

an introduction

to electrochemistry

13

Many important analytical methods are based upon oxidation-reduction equilibria that occur either within a homogeneous solution or at the surface of the electrodes making up an electrochemical cell. This chapter is concerned with certain fundamentals of electrochemistry that will serve as background for the material on oxidation-reduction processes presented in Chapters 14 through 18.

13A OXIDATION-REDUCTION PROCESSES

In an oxidation-reduction (or redox) reaction, one of the reacting species is converted to a higher oxidation state and as a consequence is *oxidized*; the other reactant suffers a decrease in oxidation state and is thus *reduced*. A general chemistry text should be consulted for the arbitrary rules for the assignment of oxidation states for the atomic particles making up an ion or molecule.

13A-1 OXIDIZING AND REDUCING AGENTS

Oxidizing agents or *oxidants* possess a strong tendency to cause oxidation of other species. *Reducing agents* or *reductants*, on the other hand, tend to cause reductions to occur and in the process are themselves oxidized.

The condition of equilibrium in an oxidation-reduction reaction depends upon the relative tendencies of the reactants to increase or decrease in oxidation number. Thus, the mixing of a strong oxidizing agent with a strong reducing agent will result in an equilibrium in which the products are overwhelmingly favored; reactants that are less strong yield correspondingly less favorable equilibria.

13A-2 HALF-REACTIONS

Separation of an oxidation-reduction reaction into its component parts (that is, into *half-reactions*) is a convenient way of indicating clearly the changes that occur in the course of the reaction. Thus, the overall reaction

$$5Sn^{2+} + 2MnO_4^- + 16H^+ \rightleftharpoons 5Sn^{4+} + 2Mn^{2+} + 8H_2O$$

is obtained by combining the half-reaction for the oxidation of tin(II),

$$Sn^{2+} \rightleftharpoons Sn^{4+} + 2e$$

with that for the reduction of permanganate,

$$MnO_4^- + 5e + 8H^+ \rightleftharpoons Mn^{2+} + 4H_2O$$

In balancing half-reactions of this sort, both charge and number of atomic particles on either side of the arrows must be made to agree. A balanced equation for the overall process can then be obtained by multiplication of the half-reactions by integers such that the number of electrons cancel when the two half-reactions are added. For the present reaction, the first half-reaction is multiplied through by five and the second by two. Thus,

$$5Sn^{2+} \rightleftharpoons 5Sn^{4+} + 10e$$

$$2MnO_4^- + 10e + 16H^+ \rightleftharpoons 2Mn^{2+} + 8H_2O$$

Upon addition of the two half-reactions, a balanced oxidation-reduction equation results. It is frequently helpful to separate complicated reactions into their component half-reactions as a way of arriving at a balanced equation for the overall process.

It is important to note that half-reactions provide no information as to the mechanism by which oxidation and reduction occurs. The electrons that appear in half-reactions are needed to account for the changes in oxidation numbers between the participants; the actual mechanism by which this change occurs can only be established by experiment.

13A-3 OXIDATION-REDUCTION REACTIONS IN ELECTROCHEMICAL CELLS

An interesting aspect of oxidation-reduction reactions is that the two constituent half-reactions can often be made to occur in regions that are physically separated. With the arrangement illustrated in Figure 13-1, direct contact between zinc and copper(II) ions is prevented by a porous fritted disk through which the ions of the solution can move. The disk, however, prevents extensive mixing of the two solutions. Although the reactants are physically separated from one another, oxidation of elemental zinc to zinc(II) ions and reduction of copper(II) ions to give the metal occur; the external conductor provides the means by which electrons are transferred from the zinc to copper(II) ions. Electron transfer continues until the copper(II)- and zinc(II)-ion concentrations achieve levels corresponding to equilibrium for the reaction

$$Zn(s) + Cu^{2+} \rightleftharpoons Zn^{2+} + Cu(s)$$

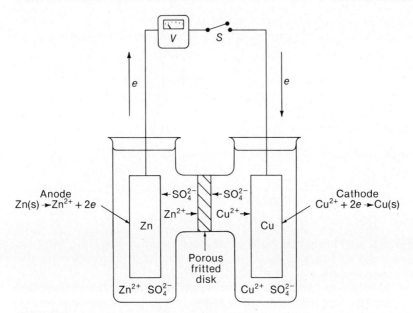

FIGURE 13-1 A galvanic cell with liquid junction. Arrows show direction of flow of charge when S is closed.

When this condition is reached, no further net flow of electrons will occur, and the current drops to zero. It is essential to recognize that the overall process and concentrations that exist at equilibrium are *totally independent of the routes that led to the condition of equilibrium,* be it by direct reaction or by indirect reaction as in Figure 13-1. Thus, studies of the behavior of electrochemical cells provide important quantitative information about oxidation-reduction equilibria as they occur when the reactants are in direct contact.

13B ELECTROCHEMICAL CELLS

Electrochemical cells can be conveniently classified as *galvanic* if they produce electrical energy and *electrolytic* if their operation requires electrical energy from an external source. Both types find use in analytical chemistry. It is important to appreciate that many cells can be operated in either a galvanic or an electrolytic mode by modification of experimental conditions.

An electrochemical cell consists of two conductors called *electrodes,* each immersed in a suitable electrolyte solution. For electricity to flow, it is necessary: (1) that the electrodes be connected externally by means of a metal conductor and (2) that the two electrolyte solutions be in contact to permit movement of ions from one to the other. Figure 13-1 shows an example of a galvanic cell. The fritted glass disk is porous, so that Zn^{2+}, Cu^{2+}, HSO_4^-, SO_4^{2-}, and other ions as well as H_2O molecules can move across the junction between the two electrolyte solutions; the disk minimizes mixing of cell compartment components and thus prevents direct reaction between the elemental zinc and copper ions.

13B-1 CONDUCTION IN AN ELECTROCHEMICAL CELL

Electricity is conducted by three distinct processes in various parts of the galvanic cell shown in Figure 13-1. Within the copper and zinc electrodes, as well as in the external conductor, the current consists of a movement of electrons from the zinc through the conductor to the copper. Within the two solutions, the flow of electricity involves migration of both cations and anions, the former away from the zinc electrode toward the copper and the latter in the reverse direction. *All* ions in the two solutions participate in this process.

A third type of conduction occurs at the two electrode surfaces. Here, an oxidation or a reduction process provides a mechanism whereby the ionic conduction of the solution is coupled with the electron conduction of the electrodes to provide a complete circuit for a current. The two electrode processes are described by the equations

$$Zn(s) \rightleftharpoons Zn^{2+} + 2e \qquad \text{(oxidation)}$$

$$Cu^{2+} + 2e \rightleftharpoons Cu(s) \qquad \text{(reduction)}$$

The net cell reaction is the sum of the two half-cell reactions. That is,

$$Zn(s) + Cu^{2+} \rightleftharpoons Zn^{2+} + Cu(s)$$

This reaction has a strong tendency to proceed to the right; the cell is, therefore, galvanic and produces a potential of about one volt under most conditions.[1]

13B-2 CELL COMPONENTS

Anode and cathode. By definition, the *cathode* of an electrochemical cell is the electrode at which reduction occurs, while the *anode* is the electrode where oxidation takes place. These definitions apply to both galvanic and electrolytic cells.

For the galvanic cell shown in Figure 13-1, the copper electrode is the cathode and the zinc electrode is the anode. Note that this cell could be caused to behave as an electrolytic cell by imposing a sufficiently large potential from an external source. Under these circumstances, the reactions occurring at the electrodes would be

$$Zn^{2+} + 2e \rightleftharpoons Zn(s)$$

$$Cu(s) \rightleftharpoons Cu^{2+} + 2e$$

Now, the roles of the electrodes are reversed; the copper electrode has become the anode and the zinc electrode the cathode.

Reactions at cathodes. Some typical cathodic half-reactions are

$$Cu^{2+} + 2e \rightleftharpoons Cu(s)$$

$$Fe^{3+} + e \rightleftharpoons Fe^{2+}$$

$$2H^+ + 2e \rightleftharpoons H_2(g)$$

---

[1] It is of interest to note that cells based upon the foregoing reactions found widespread use as a source of power for early telegraph installations; these cells were called *Daniell cells.*

$$AgCl(s) + e \rightleftharpoons Ag(s) + Cl^-$$

$$IO_4^- + 2H^+ + 2e \rightleftharpoons IO_3^- + H_2O$$

Electrons are supplied for each of these processes from the external circuit via an electrode that does not participate directly in the chemical reaction. In the first process, copper is deposited on the electrode surface; in the second, only a change in oxidation state of a solution component occurs. The third reaction is frequently observed in aqueous solutions that contain no easily reduced species.

The fourth half-reaction is of interest because it can be considered the result of a two-step process; that is,

$$AgCl(s) \rightleftharpoons Ag^+ + Cl^-$$

$$Ag^+ + e \rightleftharpoons Ag(s)$$

Solution of the sparingly soluble precipitate occurs in the first step to provide the silver ions that are reduced in the second.

The last half-reaction has been included to demonstrate that a cathodic reaction can involve anions as well as cations.

Reactions at anodes.   Examples of typical anodic half-reactions include

$$Cu(s) \rightleftharpoons Cu^{2+} + 2e$$

$$Fe^{2+} \rightleftharpoons Fe^{3+} + e$$

$$2Cl^- \rightleftharpoons Cl_2(g) + 2e$$

$$H_2(g) \rightleftharpoons 2H^+ + 2e$$

$$2H_2O \rightleftharpoons O_2(g) + 4H^+ + 4e$$

The first half-reaction requires a copper electrode to supply Cu^{2+} ions to the solution. The remaining four half-reactions can take place at any of a variety of inert metal surfaces. To cause the fourth half-reaction to occur, it is necessary to replenish the hydrogen in the solution by bubbling the gas across the surface of the electrode (usually platinum). The reaction can then be formulated as

$$H_2(g) \rightleftharpoons H_2(sat'd)$$

$$H_2(sat'd) \rightleftharpoons 2H^+(aq) + 2e$$

The final reaction, giving oxygen as a product, is a common anodic process in aqueous solutions containing no easily oxidized species.

Liquid junctions.   Cells with a liquid junction, such as that shown at the fritted disk in Figure 13-1, are ordinarily employed to avoid direct reaction between the components of the two half-cells. If the two electrolyte solutions in Figure 13-1 were allowed to mix, a decrease in the cell efficiency would occur as a result of the direct deposition of copper on the zinc. As will be shown later, a small potential called a *junction potential* has its origin at the interface between two electrolyte solutions that differ in composition.

Cells without liquid junctions. Occasionally, useful cells can be constructed in which the electrodes share a common electrolyte. An example of a cell without a liquid junction is shown in Figure 13-2. Here, the overall reaction at the silver cathode can be written

$$AgCl(s) + e \rightleftharpoons Ag(s) + Cl^-(aq)$$

As noted earlier (p. 308), the reduction can also be written as a two-step process involving solution of the excess silver chloride that is present and reduction of the resulting silver ions.

Hydrogen is oxidized at the platinum anode:

$$\tfrac{1}{2} H_2(g) \rightleftharpoons H^+(aq) + e$$

The overall cell reaction is then

$$AgCl(s) + \tfrac{1}{2} H_2(g) \rightleftharpoons Ag(s) + H^+(aq) + Cl^-(aq)$$

The direct reaction between hydrogen and solid silver chloride is slow. As a consequence, a common electrolyte can be employed without significant loss of cell efficiency.

The salt bridge. For reasons to be discussed later, electrochemical cells are often equipped with a *salt bridge* to separate the electrolytes in the anode and cathode compartments. This device takes a variety of forms. In Figure 13-4 (p. 313), for example, the bridge consists of a U-shaped tube filled with a saturated solution of potassium chloride. Such a cell has two liquid junctions: one is between the cathode electrolyte and one end of the bridge while the second is between the anode electrolyte and the other end of the bridge.

13B-3 SCHEMATIC REPRESENTATION OF CELLS
Chemists often employ a shorthand notation to simplify the description of cells. For example, the cells shown in Figures 13-1 and 13-2 can be described by

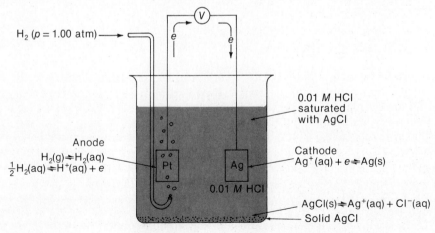

FIGURE 13-2 A galvanic cell without liquid junction.

$$\text{Zn} \mid \text{ZnSO}_4(xM) \mid \text{CuSO}_4(yM) \mid \text{Cu}$$

and

$$\text{Pt,H}_2(p = 1 \text{ atm}) \mid \text{H}^+(0.01 \ M),\text{Cl}^-(0.01 \ M),\text{AgCl(sat'd)} \mid \text{Ag}$$

By convention, *the anode and information with respect to the solution with which it is in contact are always listed on the left.* Single vertical lines represent phase boundaries at which potentials may develop. Thus, in the first example, a part of the cell potential is associated with the phase boundary between the zinc electrode and the zinc sulfate solution. A small potential also develops at liquid junctions; thus, another vertical line is inserted between the zinc and copper sulfate solutions. The cathode is then represented symbolically with another vertical line separating the electrolyte solution from the copper electrode.

In the second cell, only two phase boundaries exist, the electrolyte being common to both electrodes. An equally correct representation of this cell would be

$$\text{Pt} \mid \text{H}_2(\text{sat'd}),\text{HCl}(0.01 \ M),\text{Ag}^+(1.8 \times 10^{-8} \ M) \mid \text{Ag}$$

Here, the molecular hydrogen concentration is that of a saturated solution (in the absence of partial pressure data, 1.00 atm is implied); the indicated molar silver ion concentration was computed from the solubility product constant for silver chloride.

The presence of a salt bridge in a cell is indicated by two vertical lines, implying that a potential difference is associated with each of the two interfaces. Thus, the cell shown in Figure 13-4 (p. 313) would be represented as

$$\text{M} \mid \text{M}^{2+}(yM) \parallel \text{H}^+(xM) \mid \text{H}_2(p \text{ atm}),\text{Pt}$$

13B-4 REVERSIBLE AND IRREVERSIBLE CELLS

The galvanic cell shown in Figure 13-2 would develop a potential of about 0.46 V. If a battery with a potential somewhat greater than 0.46 V were inserted in the circuit with its negative terminal connected to the platinum electrode, a reversal in direction of electron flow would occur; the reactions at the two electrodes would thus become

$$\text{Ag(s)} + \text{Cl}^- \rightleftharpoons \text{AgCl(s)} + e$$

$$\text{H}^+ + e \rightleftharpoons \tfrac{1}{2} \text{H}_2(\text{g})$$

Now the silver electrode is the anode and the platinum electrode the cathode. A cell (or an electrode) for which a change in direction of the current causes a reversal of the electrochemical reaction is said to be *reversible*. Cells in which a current reversal results in different reactions at one or both electrodes are called *irreversible*. The cell shown in Figure 13-1 is also reversible. If, however, a small amount of dilute acid were introduced into the zinc electrode compartment, the cell would tend to become irreversible. Here, zinc would not deposit at the cathode upon application of a potential; instead, hydrogen would be produced.

$$2\text{H}^+ + 2e \rightleftharpoons \text{H}_2(\text{g})$$

Thus, the zinc electrode and the cell tend to be irreversible in the presence of acid.

13C ELECTRODE POTENTIALS

In electroanalytical work, it is often convenient to think of a cell potential as being the difference between two *electrode potentials*, the one being associated with the cathode and the other with the anode. Thus, for the cell just considered, we may write

$$E_{cell} = E_{AgCl} - E_{H_2}$$

where E_{AgCl} is the electrode potential for the silver/silver chloride electrode (which is the cathode of the cell) and E_{H_2} is the electrode potential for the hydrogen gas electrode (here, the anode). Similarly, for the Daniell type cell shown in Figure 13-1,

$$E_{cell} = E_{Cu} - E_{Zn}$$

A more general statement would be

$$E_{cell} = E_{cathode} - E_{anode} \tag{13-1}$$

where $E_{cathode}$ and E_{anode} are *electrode potentials* for the electrodes that are acting as the cathode and anode, respectively. A rigorous definition of the term electrode potential will be forthcoming shortly.

13C-1 NATURE OF ELECTRODE POTENTIALS

At the outset, it should be emphasized that there is *no* way of determining an absolute value for the potential of a single electrode, since all voltage-measuring devices determine only *differences* in potential. One conductor from such a device is connected to the electrode in question; in order to measure a potential difference, however, the second conductor must be brought in contact with the electrolyte solution of the half-cell in question. This latter contact inevitably involves a solid-solution interface and hence acts as a second half-cell at which a chemical reaction *must also take place* if electricity is to flow. A potential will be associated with this second reaction. Thus, an absolute value for the desired half-cell potential is not realized; instead, what is measured is a combination of the potential of interest and the half-cell potential for the second contact between the voltage-measuring device and the solution.

Our inability to measure absolute potentials for half-cell processes turns out not to be a serious handicap because relative half-cell potentials, measured against some reproducible reference half-cell, are just as useful. These relative potentials can be combined to give cell potentials; in addition, they are useful for calculating equilibrium constants for oxidation-reduction processes.

To be useful, relative electrode potentials must all be related to a common reference half-cell.

13C-2 REFERENCE ELECTRODES

The standard hydrogen electrode (SHE) is the universal reference for reporting relative half-cell potentials. It is a special form of the hydrogen gas electrode.

The hydrogen gas electrode. The hydrogen electrode was widely used in early electrochemical studies as a reference electrode and as an indicator electrode for the determination of pH. An example of a hydrogen electrode is shown in the left half of the cell in Figure 13-2 (p. 309). Its composition can be symbolized as

$$Pt,H_2(p \text{ atm}) \mid H^+(xM)$$

or as the reverse as shown on page 310. As suggested by the terms in parentheses, the potential developed at the platinum surface depends upon the hydrogen ion concentration of the solution and the partial pressure of the hydrogen employed to saturate it.

Figure 13-3 illustrates the components of a hydrogen electrode. The conductor is constructed from platinum foil, which has been *platinized,* that is, coated with a finely divided layer of platinum (called platinum black) by rapid chemical or electrochemical reduction of H_2PtCl_6. The platinum black provides a large surface area to assure that the reaction

$$2H^+ + 2e \rightleftharpoons H_2(g)$$

proceeds reversibly at the electrode surface; that is, the reaction is rapid in either direction. As was pointed out earlier, the stream of hydrogen serves simply to keep the solution adjacent to the electrode saturated at all times (and thus constant in concentration) with respect to the gas.

The hydrogen electrode may act as an anode or a cathode, depending upon the half-cell with which it is coupled. Hydrogen is oxidized to hydrogen ions at an anode; the reverse reaction takes place at a cathode. Under proper conditions, then, the hydrogen electrode is electrochemically reversible.

The standard hydrogen electrode. The potential of a hydrogen electrode depends upon the temperature, the hydrogen ion concentration (more correctly, the

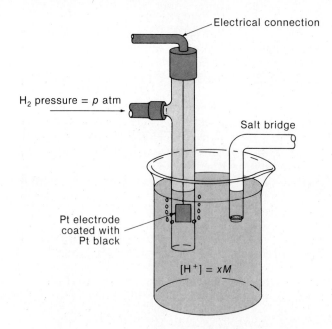

Electrical connection

H_2 pressure $= p$ atm

Salt bridge

Pt electrode coated with Pt black

$[H^+] = xM$

FIGURE 13-3 The hydrogen electrode.

activity) in the solution, and the pressure of the hydrogen at the surface of the electrode. Values for these parameters must be carefully defined in order for the half-cell process to serve as a reference. Specifications for the *standard hydrogen electrode* call for a hydrogen activity of unity and a partial pressure for hydrogen of exactly one atmosphere. *By convention, the potential of this electrode is assigned the value of exactly zero volt at all temperatures.*

Other reference electrodes. Several secondary electrodes, which are more convenient than the standard hydrogen electrode, are also used for electrode potential measurements. The potential data with such electrodes are readily converted by calculation to potentials with respect to the standard hydrogen electrode. Several of these secondary reference electrodes are described at the end of this chapter.

13C-3 DEFINITION OF ELECTRODE POTENTIALS

Electrode potentials are *defined as cell potentials* for a cell consisting of the electrode in question and the standard hydrogen electrode. It must always be borne in mind that electrode potentials are in fact *cell* potentials, all being *relative* to a common reference electrode.

The cell shown in Figure 13-4 can be conveniently employed to illustrate the properties of the electrode potential for the half-reaction

$$M^{2+} + 2e \rightleftharpoons M(s)$$

Here, M represents a metal serving as one of the two electrodes making up the cell.

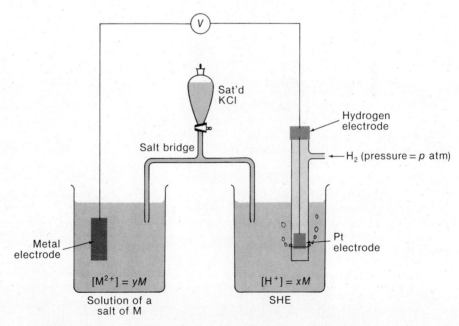

FIGURE 13-4 Schematic diagram of an arrangement for the measurement of electrode potentials against the standard hydrogen electrode.

For the purposes of discussion, we will specify that the electrode on the right is a standard hydrogen electrode; that is, the hydrogen ion activity and the partial pressure of the gas are both unity.

If the metal M in Figure 13-4 is cadmium and the solution has a cadmium ion activity of 1.00, the voltage indicated by the measuring device, V, will be 0.403 V. Moreover, the cadmium will function as the anode; thus, electrons pass from this electrode to the standard hydrogen electrode via the external circuit. That is, the cadmium electrode would be the negative terminal of the galvanic cell. The half-cell reactions for this cell can then be written as

$$Cd(s) \rightleftharpoons Cd^{2+} + 2e \qquad \text{anode}$$

$$2H^+ + 2e \rightleftharpoons H_2(g) \qquad \text{cathode}$$

The overall cell reaction is the sum of these half-reactions, or

$$Cd(s) + 2H^+ \rightleftharpoons Cd^{2+} + H_2(g)$$

If the cadmium electrode is replaced by a zinc electrode immersed in a solution that has a zinc ion activity of 1.00, a potential of 0.763 V will develop. The metal electrode is again the anode in this cell. The larger voltage reflects the greater tendency for zinc to be oxidized. The difference between this potential and the one for cadmium is a quantitative measure of the relative strengths of these two metals as reducing agents.

If the half-cell in Figure 13-4 consisted of a copper electrode in a solution of copper(II) ions with an activity of 1.00, a potential of 0.337 V would develop. However, in distinct contrast to the previous two examples, copper would tend to deposit, and an external electron flow, if allowed, would be from the hydrogen electrode to the copper electrode. That is, the copper electrode would be the positive terminal of this galvanic cell. The spontaneous cell reaction is the reverse of that in the two cells considered earlier:

$$Cu^{2+} + H_2(g) \rightleftharpoons Cu(s) + 2H^+$$

Thus, metallic copper is a much less effective reducing agent than either zinc, cadmium, or *hydrogen*. As before, the observed potential is a quantitative measure of this strength.

It is apparent from the foregoing discussion that a need exists to indicate that the copper electrode behaves as a cathode while the zinc and cadmium electrodes function as anodes when coupled to the hydrogen electrode. Positive and negative signs are used to make this distinction, the potential for half-cells such as copper being provided with one sign and the other two electrodes being assigned the opposite. The choice as to which potential will be positive and which will be negative is *purely arbitrary;* it is vital, however, that whatever sign convention is chosen *must be used consistently.*

Sign convention for electrode potentials.   It is not surprising that the arbitrariness in specifying signs has led to much controversy and confusion in the course of the development of electrochemistry. In 1953, the International Union of Pure and Applied Chemistry (IUPAC), meeting in Stockholm, attempted to resolve these controversies. The sign convention adopted at this meeting is sometimes called the IUPAC or Stockholm convention; there is hope for its general adoption in years to come. We shall always use the IUPAC sign convention.

Any sign convention *must be based upon half-cell processes written in a single way,* that is, entirely as either oxidations or as reductions. According to the IUPAC convention, the term *electrode potential* (or more exactly, *relative electrode potential*) *is reserved exclusively for half-reactions written as reductions.* There is no objection to using the term *oxidation potential* to connote the potential for an electrode process written in the opposite sense, but an oxidation potential should never be called an electrode potential. The sign of an oxidation potential will always be opposite to its corresponding electrode or reduction potential.

The sign of the electrode potential is determined by the actual sign of the electrode of interest when it is coupled with a standard hydrogen electrode in a galvanic cell. Thus, as we have indicated, a zinc or a cadmium electrode will behave as the anode from which electrons flow through the external circuit to the standard hydrogen electrode. These metal electrodes are thus the negative terminals of such galvanic cells, and their electrode potentials are *assigned* negative values. That is,

$$Zn^{2+} + 2e \rightleftharpoons Zn(s) \qquad E^0 = -0.763 \text{ V }[2]$$

$$Cd^{2+} + 2e \rightleftharpoons Cd(s) \qquad E^0 = -0.403 \text{ V}$$

The potential for the copper electrode, on the other hand, is given a positive sign because the copper behaves as a cathode in a galvanic cell constructed from this electrode and the standard hydrogen electrode; electrons flow toward the copper electrode through the exterior circuit. It is thus the positive terminal of the galvanic cell and

$$Cu^{2+} + 2e \rightleftharpoons Cu(s) \qquad E^0 = +0.337 \text{ V}$$

It is important to emphasize that electrode potentials and their signs apply to half-reactions *written as reductions.* Both zinc and cadmium are oxidized by hydrogen ion; the spontaneous reactions are thus oxidations. It is evident, then, that the *sign of the electrode potential will indicate whether or not the reduction is spontaneous with respect to the standard hydrogen electrode.* That is, the positive sign for the copper electrode potential means that the reaction

$$Cu^{2+} + H_2(g) \rightleftharpoons 2H^+ + Cu(s)$$

proceeds toward the right under ordinary conditions. The negative electrode potential for zinc, on the other hand, means that the analogous reaction

$$Zn^{2+} + H_2(g) \rightleftharpoons 2H^+ + Zn(s)$$

does not ordinarily occur; indeed, the equilibrium favors the species on the left.

The IUPAC convention was adopted in 1953, but electrode potential data given in many texts and reference works are not always in accord with it. For example, in a source of oxidation potential data compiled by Latimer,[3] one finds

---

[2] The superscript zero indicates that the electrode potential is a standard potential in which the activity of the reactants and products is 1.00. A detailed discussion of standard electrode potentials is found in Section 13D.

[3] W. M. Latimer, *The Oxidation States of the Elements and Their Potentials in Aqueous Solutions,* 2d ed. Englewood Cliffs, N.J.: Prentice-Hall, 1952.

$$Zn(s) \rightleftharpoons Zn^{2+} + 2e \qquad E^0 = +0.763 \text{ V}$$

$$Cu(s) \rightleftharpoons Cu^{2+} + 2e \qquad E^0 = -0.337 \text{ V}$$

To convert these oxidation potentials to electrode potentials as defined by the IUPAC convention, one must mentally: (1) express the half-reactions as reductions and (2) change the signs of the potentials.

The sign convention employed in a table of electrode potentials may not be explicitly stated. This information is readily ascertained, however, by referring to a half-reaction with which one is familiar and noting the direction of the reaction and the sign of the potential. Whatever changes, if any, are required to convert to the IUPAC convention are then applied to the remainder of the data in the table. For example, all one needs to remember is that strong oxidizing agents such as oxygen have large positive electrode potentials under the IUPAC convention. That is, the reaction

$$O_2(g) + 4H^+ + 4e \rightleftharpoons 2H_2O \qquad E^0 = +1.229 \text{ V}$$

tends to occur spontaneously with respect to the standard hydrogen electrode. The sign and direction of this reaction in a given table can then serve as a key to any changes that may be needed to convert all data to the IUPAC convention.

Effect of concentration on electrode potentials; the Nernst equation. The electrode potential for a half-reaction depends upon the driving force for the half-reaction. Thus, from the Le Châtelier principle, it is to be expected that electrode potentials are concentration dependent. For example, consider the half-reaction

$$Cu^{2+} + 2e \rightleftharpoons Cu$$

As noted earlier, the electrode potential for this process is 0.337 V when the copper(II) ion activity is 1.00. If, however, the copper(II) ion activity is reduced to 0.0100, the potential becomes 0.278 V because the equilibrium tends to be shifted to the left; the lowered driving force for the half-reaction is reflected in the decreased electrode potential.

In the case of the zinc electrode, a reduction of the metal ion activity from 1.00 to 0.0100 changes the potential from -0.763 V to -0.822. The larger negative number again reflects the lesser tendency for the

$$Zn^{2+} + 2e \rightleftharpoons Zn$$

to occur.

The relationship between electrode potential and concentration of electrode species was first enunciated in the nineteenth century by a German chemist by the name of Nernst; his name is attached to this important relationship.

Consider the generalized half-reaction

$$aA + bB + \cdots ne \rightleftharpoons cC + dD + \cdots$$

where the capital letters represent formulas of reacting species (whether charged or uncharged), e represents the electron, and the lower-case italic letters indicate the number of moles of each species (including electrons) participating in the half-cell reaction. According to the Nernst equation, the electrode potential E for this reaction is given by

$$E = E^0 - \frac{RT}{nF} \ln \frac{[C]^c[D]^d \cdots}{[A]^a[B]^b \cdots}$$

where E^0 is a constant called the *standard electrode potential;* its characteristics and significance are discussed in the next section.

At room temperature (298°K), the collection of constants in front of the logarithm has units of joules per coulomb or volts. That is,

$$\frac{RT}{nF} = \frac{8.316 \text{ J mol}^{-1} \text{ deg}^{-1} \times 298 \text{ deg}}{n \text{ equiv mol}^{-1} \times 96491 \text{ C equiv}^{-1}}$$

$$= \frac{2.568 \times 10^{-2} \text{ J C}^{-1}}{n} = \frac{2.568 \times 10^{-2} \text{ V}}{n}$$

Upon converting the natural logarithm to a base ten (by multiplication by 2.303), the foregoing equation becomes at room temperature

$$E = E^0 - \frac{0.0591}{n} \log \frac{[C]^c[D]^d \cdots}{[A]^a[B]^b \cdots} \tag{13-2}$$

To summarize the meaning of the bracketed terms in Equation 13-2, when the substance A is a gas,

[A] = partial pressure in atmospheres

When A is a solute

[A] = activity, a_A, of species A

Ordinarily, however, the approximation will be used that the activity of solute is equal approximately to its molar concentration M_A; that is,

$[A] = a_A \cong M_A$

When A is a pure solid, a pure liquid in excess, or the solvent, [A] is constant and is included in the constant E^0. Formally, then,

[A] = 1.00

Application of the Nernst equation is illustrated in the following examples.

1. $Zn^{2+} + 2e \rightleftharpoons Zn(s) \qquad E = E^0 - \frac{0.0591}{2} \log \frac{1}{[Zn^{2+}]}$

2. $Fe^{3+} + e \rightleftharpoons Fe^{2+} \qquad E = E^0 - \frac{0.0591}{1} \log \frac{[Fe^{2+}]}{[Fe^{3+}]}$

The second electrode potential can be measured with an inert metal electrode immersed in a solution containing iron(II) and iron(III). The potential is dependent upon the *ratio* between the molar concentrations of the two ions.

3. $2H^+ + 2e \rightleftharpoons H_2(g) \qquad E = E^0 - \frac{0.0591}{2} \log \frac{p_{H_2}}{[H^+]^2}$

In this example, p_{H_2} represents the partial pressure of hydrogen, expressed in atmospheres, at the surface of the electrode. Ordinarily, p_{H_2} will be very close to atmospheric pressure.

4. $Cr_2O_7^{2-} + 14H^+ + 6e \rightleftharpoons 2Cr^{3+} + 7H_2O$

$$E = E^0 - \frac{0.0591}{6} \log \frac{[Cr^{3+}]^2}{[Cr_2O_7^{2-}][H^+]^{14}}$$

Here, the potential depends not only upon the concentrations of chromium(III) and dichromate ions but also on the pH of the solution.

13D THE STANDARD ELECTRODE POTENTIAL

An examination of Equation 13-2 reveals that the standard electrode potential E^0 is equal to the electrode potential when the logarithmic term is zero. This condition exists when the activity quotient is equal to unity, one such instance being when the activities or partial pressures of all of the reactants and products are unity. Thus, it is convenient to define the standard electrode potential for a half-reaction as its electrode potential when the reactants and products are all at unit activity or partial pressure.

13D-1 PROPERTIES OF THE STANDARD ELECTRODE POTENTIAL

The standard electrode potential is an important physical property for characterizing oxidation-reduction systems. Several facts regarding this constant should always be kept in mind.

1. Despite its name, the standard electrode potential is, in fact, the potential of a *cell* in which one of the components is always the standard hydrogen electrode. Thus, standard electrode potentials are *relative* potentials.
2. Standard electrode potentials refer only to reactions written as *reductions*. That is, they are relative reduction potentials.
3. Standard electrode potentials bear a sign that is based upon the driving force of the reduction reaction relative to the reduction of hydrogen ion at unit activity. A positive sign indicates that the electrode will behave as the cathode (when the reactants and products are at unit activity) when coupled with the standard hydrogen electrode. A negative sign, on the other hand, indicates that the system will be the anode where oxidation will take place.
4. Standard electrode potentials are temperature dependent.
5. The standard electrode potential is independent of the notation employed to express the half-cell process. Thus, the standard potential for the process

$$Ag^+ + e \rightleftharpoons Ag(s) \qquad E^0 = +0.799 \text{ V}$$

is the same regardless of whether we write the half-reaction as above or as

$$100Ag^+ + 100e \rightleftharpoons 100Ag(s) \qquad E^0 = +0.799 \text{ V}$$

To be sure, the Nernst equation must be consistent with the half-reaction as it has been written. For the first of these, the electrode potential will be

$$E = 0.799 - \frac{0.0591}{1} \log \frac{1}{[Ag^+]}$$

and for the second

$$E = 0.799 - \frac{0.0591}{100} \log \frac{1}{[Ag^+]^{100}}$$

Table 13-1 gives standard electrodes for a few common half-cells. A more complete listing is found in Appendix 2.

13D-2 CALCULATION OF ELECTRODE POTENTIALS FROM STANDARD POTENTIALS

The following examples illustrate the use of standard potential data and the Nernst equation to obtain electrode potentials.

EXAMPLE 13-1. What is the potential for a half-cell consisting of a cadmium electrode immersed in a solution that is 0.0100 M in Cd^{2+}?
From Table 13-1, we find

$$Cd^{2+} + 2e \rightleftharpoons Cd(s) \qquad E^0 = -0.403 \text{ V}$$

Thus,

$$E = E^0 - \frac{0.0591}{2} \log \frac{1}{[Cd^{2+}]}$$

Substituting the Cd^{2+} concentration into this equation gives

$$E = -0.403 - \frac{0.0591}{2} \log \frac{1}{0.0100} = -0.403 - \frac{0.0591}{2} (+2.0)$$

$$= -0.462 \text{ V}$$

Table 13-1   Standard Electrode Potentials*

Reaction	E^0 at 25°C, V
$Cl_2(g) + 2e \rightleftharpoons 2Cl^-$	+1.359
$O_2(g) + 4H^+ + 4e \rightleftharpoons 2H_2O$	+1.229
$Br_2(aq) + 2e \rightleftharpoons 2Br^-$	+1.087
$Br_2(l) + 2e \rightleftharpoons 2Br^-$	+1.065
$Ag^+ + e \rightleftharpoons Ag(s)$	+0.799
$Fe^{3+} + e \rightleftharpoons Fe^{2+}$	+0.771
$I_3^- + 2e \rightleftharpoons 3I^-$	+0.536
$Cu^{2+} + 2e \rightleftharpoons Cu(s)$	+0.337
$Hg_2Cl_2(s) + 2e \rightleftharpoons 2Hg(l) + 2Cl^-$	+0.268
$AgCl(s) + e \rightleftharpoons Ag(s) + Cl^-$	+0.222
$Ag(S_2O_3)_2^{3-} + e \rightleftharpoons Ag(s) + 2S_2O_3^{2-}$	+0.017
$2H^+ + 2e \rightleftharpoons H_2(g)$	0.000
$AgI(s) + e \rightleftharpoons Ag(s) + I^-$	-0.151
$PbSO_4(s) + 2e \rightleftharpoons Pb(s) + SO_4^{2-}$	-0.350
$Cd^{2+} + 2e \rightleftharpoons Cd(s)$	-0.403
$Zn^{2+} + 2e \rightleftharpoons Zn(s)$	-0.763

* See Appendix 2 for a more extensive list.

The sign for the potential permits deduction of the direction of the reaction when this half-cell is coupled with the standard hydrogen electrode. The fact that it is negative shows that the reverse reaction

$$Cd(s) + 2H^+ \rightleftharpoons H_2(g) + Cd^{2+}$$

occurs spontaneously. Note that the calculated potential is a larger negative number than the standard electrode potential itself. This follows from mass-law considerations because the half-reaction, *as written*, has less tendency to occur with the lower cadmium ion concentration.

EXAMPLE 13-2. Calculate the potential for a platinum electrode immersed in a solution prepared by saturating a 0.0100 M solution of KBr with Br_2.
Here, the half-reaction is

$$Br_2(l) + 2e \rightleftharpoons 2Br^- \qquad E^0 = 1.065 \text{ V}$$

Note that the term (l) in the equation indicates that the aqueous solution is kept saturated by the presence of an excess of *liquid* Br_2. Thus, the overall process is the sum of the two equilibria

$$Br_2(l) \rightleftharpoons Br_2(\text{sat'd aq})$$

$$Br_2(\text{sat'd aq}) + 2e \rightleftharpoons 2Br^-$$

The Nernst equation for the overall process is

$$E = 1.065 - \frac{0.0591}{2} \log \frac{[Br^-]^2}{1.00}$$

Here, the concentration of Br_2 in the pure liquid is constant and is incorporated in the standard potential. Thus, its activity is assigned a value of 1.00.

$$E = 1.065 - \frac{0.0591}{2} \log (0.0100)^2 = 1.065 - \frac{0.0591}{2} (-4.00)$$

$$= 1.183 \text{ V}$$

EXAMPLE 13-3. Calculate the potential for a platinum electrode immersed in a solution that is 0.0100 M in KBr and 1.00×10^{-3} M in Br_2.
Here, the half-reaction and standard potential used in the preceding example do not apply *because the solution is no longer saturated in Br_2.* Table 13-1, however, contains a standard potential for the half-reaction

$$Br_2(aq) + 2e \rightleftharpoons 2Br^- \qquad E^0 = 1.087 \text{ V}$$

The term (aq) implies that all of the Br_2 present is in solution; that is, 1.087 is the electrode potential for the half-reaction when the Br^- and Br_2 *solution* activities are 1.00. It turns out, however, that the solubility of Br_2 in water at 25°C is only about 0.18 M. Therefore, the recorded potential of 1.087 is based on a *hypothetical system that cannot be realized experimentally.* Nevertheless, this potential is useful because it provides the means by which potentials can be calculated for real systems that are undersaturated. Thus,

$$E = 1.087 - \frac{0.0591}{2} \log \frac{[Br^-]^2}{[Br_2]}$$

$$= 1.087 - \frac{0.0591}{2} \log \frac{(1.00 \times 10^{-2})^2}{1.00 \times 10^{-3}} = 1.087 - \frac{0.0591}{2} \log 0.100$$

$$= 1.117 \text{ V}$$

Here, the Br_2 activity is approximately 1.00×10^{-3} rather than 1.00, as was the situation when the solution was saturated and in contact with liquid bromine.

13D-3 STANDARD ELECTRODE POTENTIALS FOR HALF-REACTIONS INVOLVING PRECIPITATION AND COMPLEX FORMATION

As shown by the following example, reagents that react with the participants of an electrode process have a marked effect on the potential for that process.

EXAMPLE 13-4. Calculate the potential of a silver electrode in a solution that is saturated with silver iodide and has an iodide ion activity of exactly 1.00 (K_{sp} for AgI = 8.3×10^{-17}). Here, the half-reaction occurring at the silver electrode is

$$Ag^+ + e \rightleftharpoons Ag(s) \qquad E^0 = +0.799 \text{ V}$$

For this reaction, the Nernst equation takes the form

$$E = +0.799 - 0.0591 \log \frac{1}{[Ag^+]}$$

We may obtain $[Ag^+]$ from the solubility product constant

$$[Ag^+] = \frac{K_{sp}}{[I^-]}$$

Substituting into the Nernst equation gives

$$E = +0.799 - \frac{0.0591}{1} \log \frac{[I^-]}{K_{sp}}$$

This equation may be rewritten as

$$E = +0.799 + 0.0591 \log K_{sp} - 0.0591 \log [I^-] \qquad (13\text{-}3)$$

If we substitute 1.00 for $[I^-]$ and use 8.3×10^{-17} for K_{sp}, the solubility product for AgI at 25.0°C, we obtain

$$E = -0.151 \text{ V}$$

This example shows that the half-cell potential for the reduction of silver ion becomes smaller in the presence of iodide ions. Qualitatively this effect is expected because from the Le Châtelier principle, a decrease in the concentration of silver ions diminishes the tendency for their reduction.

Equation 13-3 relates the potential of a silver electrode to the iodide ion concentration of a solution that is also saturated with silver iodide. *When the iodide ion activity is unity,* the potential is the sum of two constants; it can, therefore, be defined as the standard electrode potential for the half-reaction

$$AgI(s) + e \rightleftharpoons Ag(s) + I^- \qquad E^0_{AgI} = -0.151 \text{ V}$$

where

$$E^0_{AgI} = E^0_{Ag^+} + 0.0591 \log K_{sp} = 0.799 + 0.0591 \log K_{sp} \qquad (13\text{-}4)$$

Note that the overall reaction is the sum of the two reactions

$$AgI(s) \rightleftharpoons Ag^+ + I^-$$

$$Ag^+ + e \rightleftharpoons Ag(s)$$

and the standard potential for AgI is a combination of the constants associated with each of the two reactions.

The Nernst relationship for the silver electrode in a solution saturated with silver iodide can then be written as

$$E = E^0 - 0.0591 \log [I^-] = -0.151 - 0.0591 \log [I^-]$$

Thus, when in contact with a solution *saturated with silver iodide*, the potential of a silver electrode can be described *either* in terms of the silver ion concentration (with the standard electrode potential for the simple silver half-reaction) *or* in terms of the iodide ion concentration (with the standard electrode potential for the silver/silver iodide half-reaction). The latter is usually the more convenient because $[I^-]$ can usually be assumed to be the molar analytical concentration of that species.

The potential of a silver electrode in a solution containing an ion that forms a soluble complex with silver ion can be treated in a fashion analogous to the foregoing. For example, a complex ion is produced between thiosulfate and silver ions when the two species are mixed:

$$Ag^+ + 2S_2O_3^{2-} \rightleftharpoons Ag(S_2O_3)_2^{3-} \qquad K_f = \frac{[Ag(S_2O_3)_2^{3-}]}{[Ag^+][S_2O_3^{2-}]^2}$$

where K_f is the formation constant for the complex. The half-reaction for a silver electrode in such a solution can be formulated as

$$Ag(S_2O_3)_2^{3-} + e \rightleftharpoons Ag(s) + 2S_2O_3^{2-} \qquad E^0 = +0.010 \text{ V}$$

The standard electrode potential for this half-reaction will be the electrode potential when both the complex and the complexing anion are at unit activity. Using the same approach as in the previous example, we find that

$$E^0_{cplx} = +0.799 + 0.0591 \log \frac{1}{K_f} = 0.010 \text{ V}$$

13D-4 TABLES OF STANDARD ELECTRODE POTENTIALS

Standard electrode potential data are available for numerous half-reactions. Many have been determined from voltage measurements of cells in which a hydrogen or other reference electrode constituted the other half of the cell. It is possible, however, to calculate E^0 values from equilibrium studies of oxidation-reduction systems and from thermochemical data relating to such reactions. Many of the values found in the literature were so obtained.[4]

For illustrative purposes, a few standard electrode potentials are given in Table 13-1; a more comprehensive table is found in Appendix 2. The species in the upper left part of the equations in Table 13-1 are most easily reduced, as indicated by their large positive E^0 values; they are therefore the most effective oxidizing agents. Proceeding down the left side of the half-reactions, each succeeding species is a less effective oxidant than the one above it. The half-cell reactions at the bottom of the table have little tendency to take place as written. On the other

[4] Two authoritative sources for standard potential data are: L. Meites, *Handbook of Analytical Chemistry*, pp. 5-6 to 5-14. New York: McGraw-Hill, 1963; and G. Milazzo, S. Caroli, and V. K. Sharma, *Tables of Standard Electrode Potentials*. New York: Wiley, 1978.

hand, they do tend to occur in the opposite sense, as oxidations. The most effective reducing agents, then, are those species that appear in the lower right side of the equations in the table.

A compilation of standard potentials provides the chemist with qualitative information regarding the extent and direction of oxidation-reduction reactions between the tabulated species. On the basis of Table 13-1, for example, we see that zinc is more easily oxidized than cadmium, and we conclude that a piece of zinc immersed in a solution of cadmium ions will cause the deposition of metallic cadmium; conversely, cadmium has little tendency to reduce zinc ions. Table 13-1 also shows that iron(III) is a better oxidizing agent than triiodide ion; therefore, in a solution containing an equilibrium mixture of iron(III), iodide, iron(II), and triiodide ions, we can predict that the latter pair will predominate.

13D-5 SOME LIMITATIONS TO THE USE OF STANDARD ELECTRODE POTENTIALS

Standard electrode potentials are of great importance in understanding electroanalytical processes. Certain inherent limitations to the use of these data should be clearly appreciated, however.

Substitution of concentrations for activities. As a matter of convenience, molar concentrations—rather than activities—of reactive species are generally employed in the Nernst equation. Unfortunately, the assumption that these two quantities are identical is valid only in very dilute solutions; with increasing electrolyte concentrations, calculated potentials that are based upon molar concentrations can be expected to depart from those obtained by experiment.

To illustrate, the standard electrode potential for the half-reaction

$$Fe^{3+} + e \rightleftharpoons Fe^{2+}$$

is $+0.771$ V. Neglecting activities, a platinum electrode immersed in a solution in which the analytical concentrations of iron(II), iron(III), and perchloric acid are each 1.00 M would be expected to exhibit a potential numerically equal to this value relative to the standard hydrogen electrode. In fact, however, a potential of $+0.732$ V is observed experimentally. The reason for the discrepancy is seen if we write the Nernst equation in the more exact form

$$E = E^0 - 0.0591 \log \frac{[Fe^{2+}] f_{Fe^{2+}}}{[Fe^{3+}] f_{Fe^{3+}}}$$

where $f_{Fe^{2+}}$ and $f_{Fe^{3+}}$ are the respective activity coefficients. The coefficients of the two species are less than one in this system because of the high ionic strength imparted by the perchloric acid and the iron salts. More important, however, the activity coefficient of the iron(III) ion is smaller than that of the iron(II) ion, inasmuch as the effects of ionic strength on these coefficients increase with the charge on the ion (p. 117). As a consequence, the ratio of the activity coefficients as they appear in the Nernst equation would be larger than one and the potential of the half-cell would be smaller than the standard potential.

Activity coefficient data for ions in solutions of the types commonly encountered in oxidation-reduction titrations and electrochemical work are fairly limited; consequently, molar concentrations rather than activities must be used in many calculations. Appreciable errors may result.

Effect of other equilibria.   The application of standard electrode potentials is further complicated by the occurrence of solvolysis, dissociation, association, and complex-formation reactions involving the species of interest. The equilibrium constants required to correct for these effects are frequently unknown. Lingane[5] cites the ferrocyanide/ferricyanide couple as an excellent example of this problem:

$$Fe(CN)_6^{3-} + e \rightleftharpoons Fe(CN)_6^{4-} \qquad E^0 = +0.356 \text{ V}$$

Although the hydrogen ion does not appear in this half-reaction, the experimentally measured potential is markedly affected by pH. Thus, instead of the expected value of $+0.356$ V, solutions containing equal analytical concentrations of the two species yield potentials of $+0.71$, $+0.56$, and $+0.48$ V with respect to the standard hydrogen electrode when the measurements are made in media that are respectively 1.0 M, 0.1 M, and 0.01 M in hydrochloric acid. These differences are attributable to the difference in the degree of association of ferrocyanide and ferricyanide ions with hydrogen ions. The hydroferrocyanic acids are weaker than the hydroferricyanic acids; thus, the concentration of the ferrocyanide ion is lowered more than that of the ferricyanide ion as the acid concentration increases. This effect tends in turn to shift the oxidation-reduction equilibrium to the right and leads to more positive electrode potentials.

A somewhat analogous effect is encountered in the behavior of the potential of the iron(III)/iron(II) couple. As noted earlier, an equimolar mixture of these two ions in 1 M perchloric acid has an electrode potential of $+0.73$ V. Substitution of hydrochloric acid of the same concentration alters the observed potential to $+0.70$ V; a value of $+0.6$ V is observed in 1 M phosphoric acid. These differences arise because iron(III) forms more stable complexes with chloride and phosphate ions than does iron(II). As a result, the actual concentration of *uncomplexed* iron(III) in such solutions is less than that of *uncomplexed* iron(II), and the net effect is a shift in the observed potential.

Phenomena such as these can be taken into account only if the equilibria involved are known and constants for the processes are available. Such information is frequently lacking, however; the chemist is then forced to neglect such effects and hope that serious errors do not flaw the calculated results.

Formal potentials.   In order to compensate partially for activity effects and errors resulting from side reactions, Swift[6] has proposed the use of a quantity called the *formal potential* in place of the standard electrode potential in oxidation-reduction calculations. The formal potential of a system is the potential of the half-cell with respect to the standard hydrogen electrode when the analytical *concentrations* of reactants and products are 1 M and the concentrations of any other constituents of the solution are carefully specified. Thus, for example, the formal potential for the reduction of iron(III) is $+0.732$ V in 1 M perchloric acid and $+0.700$ V in 1 M hydrochloric acid; similarly, the formal potential for the reduction of ferricyanide ion would be $+0.71$ V in 1 M hydrochloric acid and $+0.48$ V in a 0.01 M solution of this acid. Use of these values in place of the standard electrode potential in the Nernst equation will yield better agreement between calculated and experimental

[5] J. J. Lingane, *Electroanalytical Chemistry*, 2d ed., p. 59. New York: Interscience, 1958.
[6] E. H. Swift, *A System of Chemical Analysis*, p. 50. San Francisco: Freeman, 1939.

potentials, provided the electrolyte concentration of the solution approximates that for which the formal potential was measured. The reader should be aware that application of formal potentials to systems differing greatly as to kind and concentration of electrolyte can lead to errors as great or greater than those encountered with the use of standard potentials. The table in Appendix 2 contains selected formal potentials as well as standard potentials; in subsequent chapters, we shall use whichever is the more appropriate.

Reaction rates. It should be realized that the existence of a half-reaction in a table of electrode potentials does not necessarily imply that there is a real electrode whose potential will respond to the half-reaction. Many of the data in such tables have been obtained by calculations based upon equilibrium or thermochemical measurements rather than from the actual measurement of the potential for an electrode system. For some, no electrode is known; thus, the standard electrode potential for the process,

$$2CO_2 + 2H^+ + 2e \rightleftharpoons H_2C_2O_4 \qquad E^0 = -0.49 \text{ V}$$

has been arrived at indirectly. The reaction is not reversible, and the rate at which carbon dioxide combines to give oxalic acid is negligibly slow. No electrode system is known whose potential varies with the ratio of activities of the reactants and products. Nonetheless, the potential is useful for computational purposes. Many examples of systems of this kind are encountered in biochemical studies.

13D-6 DETERMINATION OF STANDARD POTENTIALS

Although the standard hydrogen electrode is the universal standard of reference and useful standard electrode potentials are available for a wide variety of half-cells, it is important to appreciate that neither the standard hydrogen electrode nor any other electrode in its standard state can be realized in the laboratory. That is, the standard hydrogen electrode is a hypothetical electrode; so also is any electrode system in which the reactants and products are at unit activity or pressure. The reason these electrodes cannot be prepared experimentally is that chemists lack the knowledge to produce solutions having ion activities of exactly one. That is, no adequate theory exists that will permit the calculation of the *concentration* of a compound that must be used in order to achieve unit ionic activity. Here, the ionic strength is so great that the Debye-Hückel relationship (p. 118) is not valid and no independent experimental method exists for the determination of activity coefficients in such solutions. Thus, for example, the *concentration* of HCl or other acids required to give the unit hydrogen ion activity specified in the standard hydrogen electrode *cannot* be calculated or determined experimentally. Notwithstanding, data taken in solutions of low ionic strengths can be extrapolated to give valid measures of standard electrode potentials as theoretically defined. An example of how hypothetical standard electrode potentials can be obtained from experimental data follows.

> **EXAMPLE 13-5.** D. A. MacInnes[7] found that a cell similar to that shown in Figure 13-2 developed a potential of 0.52053 V. The cell is described by

[7] D. A. MacInnes, *The Principles of Electrochemistry*, p. 187. New York: Reinhold, 1939.

$$Pt, H_2(1.00 \text{ atm}) \mid HCl(3.215 \times 10^{-3} M, AgCl(\text{sat'd}) \mid Ag$$

Calculate the standard electrode potential for the half-reaction

$$AgCl(s) + e \rightleftharpoons Ag(s) + Cl^-$$

Here, the electrode potential for the cathode is given by

$$E_{\text{cathode}} = E^0_{\text{AgCl}} - 0.0591 \log M_{\text{HCl}} f_{\text{Cl}^-}$$

where f_{Cl^-} is the activity coefficient of Cl^-. The second half-cell reaction is

$$H^+ + e \rightleftharpoons \tfrac{1}{2} H_2(g)$$

and

$$E_{\text{anode}} = E^0_{H_2} - \frac{0.0591}{1} \log \frac{p_{H_2}^{1/2}}{M_{\text{HCl}} f_{H^+}}$$

The measured potential is the difference between these potentials (p. 311); thus,

$$E_{\text{cell}} = E^0_{\text{AgCl}} - 0.0591 \log M_{\text{HCl}} f_{\text{Cl}^-} - \left(0.000 - 0.0591 \log \frac{p_{H_2}^{1/2}}{M_{\text{HCl}} f_{H^+}} \right)$$

Combining the two logarithmic terms gives

$$E_{\text{cell}} = E^0_{\text{AgCl}} - 0.0591 \log \frac{M_{\text{HCl}}^2 f_{H^+} \times f_{\text{Cl}^-}}{p_{H_2}^{1/2}}$$

The activity coefficients for H^+ and Cl^- can be derived from Equation 4-66 (p. 118) employing 3.215×10^{-3} for the ionic strength μ. These values are 0.9447 and 0.9391 respectively. Substitution of these activity coefficients and the experimental data into the foregoing equation gives upon rearrangement

$$E^0_{\text{AgCl}} = 0.52053 + 0.0591 \log \frac{(3.215 \times 10^{-3})^2 (0.9447)(0.9391)}{1.00^{1/2}}$$

$$= 0.2228 \text{ V}$$

(The mean for this and similar measurements at other concentrations was 0.222 V.)

13E APPLICATIONS OF STANDARD POTENTIAL DATA

Not surprisingly, the standard electrode potentials and the Nernst equation enable the scientist to evaluate the potentials of cells in which the participants are not in their standard states. In addition, electrochemical measurements provide a power-ful means of evaluating equilibrium constants. Examples of each of these applica-tions are included in this section.

13E-1 CALCULATION OF REVERSIBLE CELL POTENTIALS

An important use of standard electrode potentials is the calculation of the potential obtainable from a galvanic cell or for operation of an electrolytic cell. These calcu-lated potentials (sometimes called *reversible* or *thermodynamic potentials*) are theoretical in the sense that they refer to cells in which there is essentially no cur-rent; additional factors must be taken into account where a current is involved.

As shown earlier (Equation 13-1, p. 311), the electromotive force of a cell is obtained by combining half-cell potentials as follows:

$$E_{\text{cell}} = E_{\text{cathode}} - E_{\text{anode}}$$

where $E_{cathode}$ and E_{anode} are the *electrode potentials* for the two half-reactions constituting the cell.

Consider the hypothetical cell

$$Zn \mid ZnSO_4(a_{Zn^{2+}} = 1.00) \parallel CuSO_4(a_{Cu^{2+}} = 1.00) \mid Cu$$

The overall cell process involves the oxidation of elemental zinc to zinc(II) and the reduction of copper(II) to the metallic state. Because the activities of the two ions are specified as unity, the standard potentials are also the electrode potentials. The cell diagram also specifies that the zinc electrode is the anode. Thus, using E^0 data from Table 13-1,

$$E_{cell} = +0.337 - (-0.763) = +1.100 \text{ V}$$

The positive sign for the cell potential indicates that the reaction

$$Zn(s) + Cu^{2+} \longrightarrow Zn^{2+} + Cu(s)$$

occurs spontaneously and that the cell is galvanic.

The foregoing cell, diagrammed as

$$Cu \mid Cu^{2+}(a_{Cu^{2+}} = 1.00) \parallel Zn^{2+}(a_{Zn^{2+}} = 1.00) \mid Zn$$

implies that the copper electrode is now the anode. Thus,

$$E_{cell} = -0.763 - (+0.337) = -1.100 \text{ V}$$

The negative sign indicates the nonspontaneity of the reaction

$$Cu(s) + Zn^{2+} \longrightarrow Cu^{2+} + Zn(s)$$

The application of an external potential greater than 1.100 V would be required to cause this reaction to occur.

EXAMPLE 13-6. Calculate the potentials for the following cell employing (a) concentrations and (b) activities:

$$Zn \mid ZnSO_4(xM), PbSO_4(sat'd) \mid Pb$$

where $x = 5.00 \times 10^{-4}, 2.00 \times 10^{-3}, 1.00 \times 10^{-2}, 2.00 \times 10^{-2}$, and 5.00×10^{-2}.

(a) In a neutral solution, little HSO_4^- will be formed; thus, we may assume that

$$[SO_4^{2-}] = M_{ZnSO_4} = x = 5.00 \times 10^{-4}$$

The half-reactions and standard potentials are

$$PbSO_4(s) + 2e \rightleftharpoons Pb(s) + SO_4^{2-} \qquad E^0 = -0.350 \text{ V}$$

$$Zn^{2+} + 2e \rightleftharpoons Zn \qquad E^0 = -0.763 \text{ V}$$

The potential of the lead electrode is given by

$$E_{Pb} = -0.350 - \frac{0.0591}{2} \log 5.00 \times 10^{-4}$$

$$= -0.252 \text{ V}$$

For the zinc half-reaction,

$$[Zn^{2+}] = 5.00 \times 10^{-4}$$

and

$$E_{Zn} = -0.763 - \frac{0.0591}{2} \log \frac{1}{5.00 \times 10^{-4}}$$

$$= -0.860 \text{ V}$$

Since the Pb electrode is specified as the cathode,

$$E_{cell} = -0.252 - (-0.860) = 0.608 \text{ V}$$

Cell potentials at the other concentrations can be derived in the same way. Their values are tabulated at the end of part (b).

(b) To obtain activity coefficients for Zn^{2+} and SO_4^{2-} we must first calculate the ionic strength with the aid of Equation 4-64 (p. 115).

$$\mu = \tfrac{1}{2}[5.00 \times 10^{-4} \times (2)^2 + 5.00 \times 10^{-4} \times (2)^2]$$

$$= 2.00 \times 10^{-3}$$

In Table 4-5, we find for SO_4^{2-}, $a_A = 4.0$ and for Zn^{2+}, $a_A = 6.0$. Substituting these values into Equation 4-66 gives for sulfate ion

$$-\log f_{SO_4} = \frac{0.5085 \times 2^2 \times \sqrt{2.00 \times 10^{-3}}}{1 + 0.3281 \times 4.0 \sqrt{2.00 \times 10^{-3}}} = 8.59 \times 10^{-2}$$

$$f_{SO_4} = 0.820$$

Repeating the calculations for Zn^{2+} for which $a_A = 6.0$ yields

$$f_{Zn} = 0.825$$

The Nernst equation for the Pb electrode now becomes

$$E_{Pb} = -0.350 - \frac{0.0591}{2} \log 0.820 \times 5.00 \times 10^{-4}$$

$$= -0.250$$

Similarly, for the zinc electrode

$$E_{Zn} = -0.763 - \frac{0.0591}{2} \log \frac{1}{0.825 \times 5.00 \times 10^{-4}}$$

$$= -0.863$$

Thus

$$E_{cell} = -0.250 - (-0.863) = 0.613 \text{ V}$$

Values for other concentrations are found in the accompanying table.

x	μ	(a) E (calc) with Concentrations	(b) E (calc) with Activities	E (exptl)*
5.00×10^{-4}	2.00×10^{-3}	0.608	0.613	0.611
2.00×10^{-3}	8.00×10^{-3}	0.573	0.582	0.583
1.00×10^{-2}	4.00×10^{-2}	0.531	0.550	0.553
2.00×10^{-2}	8.00×10^{-2}	0.513	0.537	0.542
5.00×10^{-2}	2.00×10^{-1}	0.490	0.521	0.529

* Experimental data from: I. A. Cowperthwaite and V. K. LaMer, *J. Amer. Chem. Soc.*, **53**, 4333 (1931).

It is of interest to compare the calculated cell potentials shown in the columns labeled (a) and (b) in the foregoing tabulation with the experimental results shown in the last column. Clearly, the use of activities provides a significant improvement at the higher ionic strengths.

EXAMPLE 13-7. Calculate the potential required to initiate the deposition of copper from a solution that is 0.010 M in $CuSO_4$ and contains sufficient sulfuric acid to give a hydrogen ion concentration of 1.0×10^{-4} M.

The deposition of copper necessarily occurs at the cathode. Because no easily oxidizable species are present, the anode reaction will involve oxidation of H_2O to give O_2. From the table of standard potentials, we find

$$Cu^{2+} + 2e \rightleftharpoons Cu(s) \qquad E^0 = +0.337 \text{ V}$$

$$O_2(g) + 4H^+ + 4e \longrightarrow 2H_2O \qquad E^0 = +1.229 \text{ V}$$

Thus, for the copper electrode

$$E = +0.337 - \frac{0.0591}{2} \log \frac{1}{0.010} = +0.278 \text{ V}$$

Assuming that O_2 is evolved at 1.00 atm, the potential for the oxygen electrode is

$$E = +1.229 - \frac{0.0591}{4} \log \frac{1}{(1.00)(1.0 \times 10^{-4})^4}$$

$$= +0.993 \text{ V}$$

The cell potential is then

$$E_{cell} = +0.278 - 0.993 = -0.715 \text{ V}$$

Thus, to initiate the reaction

$$2Cu^{2+} + 2H_2O \longrightarrow O_2(g) + 4H^+ + 2Cu(s)$$

would require the application of a potential greater than 0.715 V.

13E-2 EQUILIBRIUM CONSTANTS FROM STANDARD ELECTRODE POTENTIALS

Changes in potential during discharge of a cell. Consider again the galvanic cell based upon the reaction

$$Zn(s) + Cu^{2+} \rightleftharpoons Zn^{2+} + Cu(s)$$

At all times, the cell potential is given by

$$E_{cell} = E_{cathode} - E_{anode}$$

As electricity is drawn from this cell, however, the zinc ion concentration increases and the copper(II) concentration decreases. These changes cause the electrode potential for the zinc half-cell to become less negative and that for copper half-cell to be less positive. The net effect then is a decrease in the potential of the cell. Ultimately, the concentrations attain values such that there is no longer a tendency for a net electron transfer to occur. The potential of the cell thus becomes zero, and the system is in equilibrium. That is, at equilibrium

$$E_{cell} = E_{cathode} - E_{anode} = 0$$

or

$$E_{cathode} = E_{anode}$$

This equation expresses an important and general relationship: *for an oxidation-reduction system at chemical equilibrium, the electrode potentials (that is, the reduction potentials) of all half-reactions of the system will be equal.* Note that this generalization applies regardless of the number of half-reactions occurring within the system; interaction among all must take place until all electrode potentials are identical.

Calculation of equilibrium constants. Consider the oxidation-reduction equilibrium

$$a A_{red} + b B_{ox} \rightleftharpoons a A_{ox} + b B_{red}$$

for which the half-reactions may be written

$$a A_{ox} + ne \rightleftharpoons a A_{red}$$
$$b B_{ox} + ne \rightleftharpoons b B_{red}$$

When the components of this system are at chemical equilibrium,

$$E_A = E_B$$

where E_A and E_B are the electrode potentials for the two half-cells. With this equality expressed in terms of the Nernst equation, we find that *at equilibrium*

$$E_A^0 - \frac{0.0591}{n} \log \frac{[A_{red}]^a}{[A_{ox}]^a} = E_B^0 - \frac{0.0591}{n} \log \frac{[B_{red}]^b}{[B_{ox}]^b}$$

which yields upon rearrangement and combination of the log terms

$$E_B^0 - E_A^0 = \frac{0.0591}{n} \log \frac{[A_{ox}]^a [B_{red}]^b}{[A_{red}]^a [B_{ox}]^b} = \frac{0.0591}{n} \log K_{eq}$$

This relationship was derived for equilibrium conditions. The concentration terms are thus *equilibrium concentrations; hence, the quotient is the equilibrium constant for the reaction.* That is,

$$\log K_{eq} = \frac{n(E_B^0 - E_A^0)}{0.0591}$$

EXAMPLE 13-8. Calculate the equilibrium constant for the reaction

$$MnO_4^- + 5Fe^{2+} + 8H^+ \rightleftharpoons Mn^{2+} + 5Fe^{3+} + 4H_2O$$

This reaction can be broken down to the two half-reactions

$$MnO_4^- + 8H^+ + 5e \rightleftharpoons Mn^{2+} + 4H_2O$$
$$5Fe^{3+} + 5e \rightleftharpoons 5Fe^{2+}$$

Since at equilibrium,

$$E_{Fe^{3+}} = E_{MnO_4^-}$$

then application of the Nernst equation to the two half-reactions yields

$$E^0_{Fe^{3+}} - \frac{0.0591}{5} \log \frac{[Fe^{2+}]^5}{[Fe^{3+}]^5} = E^0_{MnO_4^-} - \frac{0.0591}{5} \log \frac{[Mn^{2+}]}{[MnO_4^-][H^+]^8}$$

Rearranging this equation yields

$$\frac{0.0591}{5} \log \frac{[Mn^{2+}][Fe^{3+}]^5}{[MnO_4^-][Fe^{2+}]^5[H^+]^8} = E^0_{MnO_4^-} - E^0_{Fe^{3+}}$$

Substitution of standard potentials from Appendix 2 gives[8]

$$\log K_{eq} = \frac{5(1.51 - 0.771)}{0.0591}$$

$$= 62.5$$

$$K_{eq} = 10^{62.5} = 10^{0.5} \times 10^{62} = 3 \times 10^{62}$$

Here, we are forced to round the result to one significant figure because the first two integers (6 and 2) of the logarithm of K equilibrium simply set the decimal point for the antilogarithm; the mantissa (0.5) then determines the number of significant figures in the result.

EXAMPLE 13-9. A piece of copper is placed in a 0.050 M solution of $AgNO_3$. What is the equilibrium composition of the solution?

The reaction is

$$Cu(s) + 2Ag^+ \rightleftharpoons Cu^{2+} + 2Ag(s)$$

We first calculate the equilibrium constant for the reaction, which is then used to determine the solution composition. The two half-reactions are

$$2Ag^+ + 2e \rightleftharpoons 2Ag(s)$$

$$Cu^{2+} + 2e \rightleftharpoons Cu(s)$$

Since at equilibrium,

$$E_{Cu^{2+}} = E_{Ag^+}$$

then

$$E^0_{Cu^{2+}} - \frac{0.0591}{2} \log \frac{1}{[Cu^{2+}]} = E^0_{Ag^+} - \frac{0.0591}{2} \log \frac{1}{[Ag^+]^2}$$

$$\log \frac{[Cu^{2+}]}{[Ag^+]^2} = \frac{2(E^0_{Ag^+} - E^0_{Cu^{2+}})}{0.0591} = \frac{2(0.799 - 0.337)}{0.0591}$$

$$= 15.63$$

and

$$\frac{[Cu^{2+}]}{[Ag^+]^2} = K_{eq} = 4.3 \times 10^{15}$$

The magnitude of the equilibrium constant suggests that nearly all Ag^+ is precipitated. The molar concentration of Cu^{2+} is thus given by

$$[Cu^{2+}] = \tfrac{1}{2}(0.050 - [Ag^+])$$

---

[8] Recall (p. 318) that the standard potential is the same regardless of whether the reaction is written for $5Fe^{3+}$ or $1Fe^{3+}$.

Because the reaction is nearly complete, it is probably safe to assume that $[Ag^+]$ is small with respect to 0.050. Then

$$[Cu^{2+}] \cong \tfrac{1}{2}(0.050) = 0.025$$

Substituting,

$$\frac{0.025}{[Ag^+]^2} = 4.3 \times 10^{15}$$

$$[Ag^+] = 2.4 \times 10^{-9}$$

$$[Cu^{2+}] = \tfrac{1}{2}(0.050 - 2.4 \times 10^{-9}) \cong 0.025$$

13E-3 DETERMINATION OF DISSOCIATION, SOLUBILITY PRODUCT, AND FORMATION CONSTANTS BY POTENTIAL MEASUREMENTS

We have seen that the potential of an electrode is determined by the concentrations of those species that participate in the electrode reaction; as a consequence, measurement of a half-cell potential often provides a convenient way to determine the concentration of solute species. One important virtue of this approach is that the measurement can be made without affecting appreciably any equilibria that may exist in the solution. For example, the potential of a silver electrode in a solution containing the cyanide complex of silver depends upon the silver ion activity. With suitable equipment, this potential can be measured with a negligible passage of electricity. Thus, the concentration of silver ions in the solution is not sensibly altered during the measurement; the position of the equilibrium

$$Ag^+ + 2CN^- \rightleftharpoons Ag(CN)_2^-$$

is likewise undisturbed.

EXAMPLE 13-10. The solubility product constant for the sparingly soluble CuX_2 can be determined by saturating a 0.0100 M solution of NaX with solid CuX_2. After equilibrium is achieved, this solution is made part of the cell

$$Cu \mid CuX_2(\text{sat'd}),NaX(0.0100\ M) \parallel HCl(0.0104\ M),AgCl(\text{sat'd}) \mid Ag$$

Suppose that the potential of this cell is found to be 0.403 V, with the copper electrode behaving as the anode, as indicated. We may then write

$$E_{cell} = E_{cathode} - E_{anode}$$

$$0.403 = (E^0_{AgCl} - 0.0591 \log [Cl^-]) - \left(E^0_{Cu^{2+}} - \frac{0.0591}{2}\log\frac{1}{[Cu^{2+}]}\right)$$

$$0.403 = 0.222 - 0.0591 \log 0.0104 - 0.337 + \frac{0.0591}{2}\log\frac{1}{[Cu^{2+}]}$$

$$\log [Cu^{2+}] = \frac{2[0.222 - (-0.117) - 0.337 - 0.403]}{0.0591}$$

$$= -13.57 = -13.6$$

$$[Cu^{2+}] = 3 \times 10^{-14}$$

Because $[X^-]$ is 0.0100 M,

$$K_{sp} = (3 \times 10^{-14})(0.0100)^2 = 3 \times 10^{-18}$$

Any electrode system that is sensitive to the hydrogen ion concentration of a solution can, in theory at least, be used to determine the dissociation constants of acids and bases. All half-cells in which hydrogen ion is a participant fall in this category. However, relatively few of these have been applied to the problem.

EXAMPLE 13-11. Calculate the dissociation constant of the acid HP if the cell

$$\text{Pt,H}_2(1.00 \text{ atm}) \mid \text{HP}(0.010 \text{ } M),\text{NaP}(0.030 \text{ } M) \parallel$$
$$\text{NaCl}(5.53 \times 10^{-3} \text{ } M), \text{AgCl(sat'd)} \mid \text{Ag}$$

develops a potential of 0.605 V.

The potential of this cell is given by

$$E_{\text{cell}} = E_{\text{AgCl}} - E_{\text{H}^+}$$

$$0.605 = (0.222 - 0.0591 \log [\text{Cl}^-]) - \left(0.000 - \frac{0.0591}{2} \log \frac{p_{\text{H}_2}}{[\text{H}^+]^2}\right)$$

$$= 0.355 + \frac{0.0591}{2} \log \frac{1.00}{[\text{H}^+]^2}$$

$$\tfrac{1}{2} \log [\text{H}^+]^2 = \frac{0.355 - 0.605}{0.0591} = -4.23$$

$$[\text{H}^+] = 5.89 \times 10^{-5} = 5.9 \times 10^{-5}$$

$$K_a = \frac{[\text{H}^+][\text{P}^-]}{[\text{HP}]}$$

Because the HP present is largely undissociated,

$$K_a \cong \frac{(5.89 \times 10^{-5})(0.030)}{0.010} = 1.77 \times 10^{-4} = 1.8 \times 10^{-4}$$

Formation constants for complex ions can be determined in an analogous fashion. Thus, to evaluate the constant for the Ag(CN)_2^- complex we might measure the potential of the cell

$$\text{Ag} \mid \text{Ag(CN)}_2^-(C_1),\text{CN}^-(C_2) \parallel \text{HCl}(0.0100 \text{ } M),\text{AgCl(sat'd)} \mid \text{Ag}$$

Values for the molar concentrations of the complex and cyanide ion, C_1 and C_2, would be known from the quantities used to prepare the cell. The silver ion concentration could be calculated from the measured potential. Combination of this concentration with C_1 and C_2 would then lead to a value for the desired equilibrium constant.

13F POTENTIALS OF REAL CELLS

The potentials of cells encountered in the laboratory are affected by several other sources in addition to the reversible potentials described in the previous section.

13F-1 LIQUID JUNCTION POTENTIAL

When two electrolyte solutions of different composition are brought in contact with one another, a potential develops at the interface. This *junction potential* arises from differences in rates of migration of anions and cations between the two solutions.

Consider the liquid junction that exists in the system

HCl(1 M) | HCl(0.01 M)

Both hydrogen ions and chloride ions tend to diffuse across this boundary in both directions. The velocity of migration, however, is proportional to concentration. Thus, the net movement of both chloride ions and hydrogen ions tends to be from the more concentrated to the more dilute solution, the driving force for this migration being proportional to the concentration difference. The rate at which various ions move under the influence of a fixed force varies considerably (that is, their *mobilites* are different). In the present example, hydrogen ions are several times more mobile than chloride ions. As a consequence, there is a tendency for the hydrogen ions to outstrip the chloride ions as diffusion takes place; a separation of charge is the net result (see Figure 13-5). The more dilute side of the boundary becomes positively charged owing to the more rapid migration of hydrogen ions; the concentrated side, therefore, acquires a negative charge from the slower-moving chloride ions. The charge that develops tends to counteract the differences in mobilities of the two ions and, as a consequence, an equilibrium condition soon develops. The junction potential difference resulting from this charge separation may amount to 30 mV or more.

In a simple system such as that shown in Figure 13-5, the magnitude of the junction potential is readily calculated from a knowledge of the mobilities of the two ions involved. However, it is seldom that a cell of analytical importance has a sufficiently simple composition to permit such a computation.

It is an experimental fact that the magnitude of the liquid junction potential can be greatly decreased by interposition of a concentrated electrolyte solution (a *salt bridge*) between the two solutions. The effectiveness of this contrivance improves as the mobilities of the ions of the salt approach one another in magnitude and as these concentrations increase. A saturated potassium chloride solution is good from both standpoints, its concentration being somewhat greater than 4 M at room temperature, and the mobility of its ions differing by only 4%. With such a bridge, the junction potential typically amounts to a few millivolts or less, a negligible quantity in many, but not all, analytical measurements.

13F-2 OHMIC POTENTIAL; *IR* DROP

When electricity flows in an electrochemical cell, the overall potential is altered by an amount that corresponds to the driving force necessary to overcome the resistance of ions to movement toward the anode and the cathode. Just as in metallic

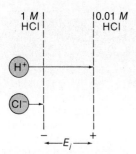

FIGURE 13-5 Schematic representation of a liquid junction showing the source of the junction potential E_j. The length of the arrows corresponds to the relative mobility of the two ions.

conduction, this force follows Ohm's law and is equal to the product of the current in amperes and the resistance of the cell in ohms. The force is generally referred to as the *ohmic potential*, or the *IR drop*.

The net effect of *IR* drop is to increase the potential required to operate an electrolytic cell and to decrease the measured potential of a galvanic cell. Therefore, the *IR* drop is always *subtracted* from the theoretical cell potential. That is,[9]

$$E_{cell} = E_{cathode} - E_{anode} - IR \qquad (13\text{-}5)$$

EXAMPLE 13-12.

1. Calculate the potential when 0.100 A of electricity is drawn from the galvanic cell

$$Cd \mid Cd^{2+}(0.0100\ M) \parallel Cu^{2+}(0.0100\ M) \mid Cu$$

Assume a cell resistance of 4.00 Ω (ohms).

Substitution into the Nernst equation reveals that the electrode potential for the Cu electrode is 0.278 V, while for the Cd electrode it is -0.462 V. Thus, the cell potential is

$$E = E_{Cu} - E_{Cd} = 0.278 - (-0.462) = 0.740\ V$$

$$E_{cell} = 0.740 - IR = 0.740 - (0.100 \times 4.00) = 0.340\ V$$

Note that the emf drops dramatically when electricity is drawn from the cell.

2. Calculate the potential required to generate a current of 0.100 A in the reverse direction in the foregoing cell.

$$E = E_{Cd} - E_{Cu} = -0.462 - 0.278 = -0.740\ V$$

$$E_{cell} = -0.740 - (0.100 \times 4.00) = -1.140\ V$$

Here, an external potential of 1.140 V would be needed to cause Cd^{2+} to deposit and Cu to dissolve at a rate required for a current of 0.100 A.

13F-3 POLARIZATION EFFECTS

In order to describe the current-voltage relationship in an electrolytic cell, it is convenient to rewrite Equation 13-5 in the form

$$I = -\frac{1}{R} E_{applied} + \frac{1}{R} (E_{cathode} - E_{anode})$$

where E_{appl} is the cell potential (E_{cell}) required to generate the current I. For small currents and brief periods of time, $E_{cathode}$ and E_{anode} remain relatively constant. Thus, the cell behavior can be approximated by the relationship

$$I = -\frac{1}{R} E_{applied} + k$$

where k is a constant. As shown in Figure 13-6, a plot of the instantaneous current as a function of applied potential would in theory yield a straight line with a slope that equals the reciprocal of the resistances (note that by convention $E_{applied}$ has a negative sign; thus, the current has a positive sign).

The linear relationship between current and applied potential suggested by

[9] Here and in the subsequent discussion in this chapter we will assume the junction potential is negligible relative to the other potentials. In Chapter 16, it will no longer be possible to make this simplification.

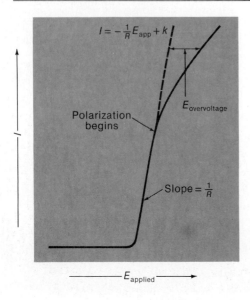

$$I = -\frac{1}{R}E_{app} + k$$

$E_{overvoltage}$

Polarization begins

Slope $= \frac{1}{R}$

$E_{applied}$

FIGURE 13-6   Current-voltage curve (solid line) for an electrolytic cell.

the foregoing equation is frequently observed experimentally when I is small; as shown by the solid line in Figure 13-6, however, marked departures from linearity appear at higher currents. Under these circumstances the cell is said to be *polarized*. Thus, a polarized electrolytic cell requires application of a potential larger than theoretical in order to provide a given current.[10] This added potential is termed an *overvoltage* or an *overpotential*. In order to describe the effects of polarization on a cell behavior, it is necessary to add a term to Equation 13-5. Thus,

$$E_{cell} = E_{anode} - E_{cathode} - IR - E_{overvoltage} \tag{13-6}$$

where $E_{overvoltage}$ is the overpotential whose magnitude is current dependent. The presence of polarization increases (makes more negative) the potential required to operate an electrolytic cell. Thus, $E_{overvoltage}$, like IR, is always subtracted from the thermodynamic cell potential.

Polarization is an electrode phenomenon; either or both electrodes in a cell can be affected. Included among the factors influencing the extent of polarization are the size, shape, and composition of the electrodes; the composition of the electrolyte solution; the temperature and the rate of stirring; the magnitude of the current; and the physical states of the species involved in the cell reaction.

For purposes of discussion, polarization phenomena are conveniently divided into the two categories of *concentration polarization* and *kinetic* or *activation polarization*.

Concentration polarization.   Concentration polarization arises when the current in Equation 13-5 becomes limited by the rate at which the reactant moves from the bulk of the solution to the electrode surface where electron transfer can occur. That

[10] It should be noted that polarization also occurs in galvanic cells; smaller than theoretical currents are produced as a consequence.

is, as the applied potential is increased continuously, ions of the reactant must be brought to the electrode surface at a constantly increasing rate. Ultimately, even with good stirring, a point is reached where the rate of ion transport is insufficient to support the current described by Equation 13-5; the departure from linearity shown in Figure 13-6 occurs at this point.

Concentration polarization increases as the reactant concentration decreases, as the size of the electrode becomes smaller, and as the efficiency of stirring becomes poorer. For some analytical methods, these variables are arranged to limit polarization; in others, effort is made to encourage concentration polarization.

Kinetic polarization. Kinetic polarization results when the rate at which the electrochemical reaction at one or both electrodes is slow; here, the overpotential is required to overcome the energy barrier to the half-reaction. In contrast to concentration polarization, the current is controlled by the *rate of the oxidation or the reduction processes* at the electrode surfaces rather than by the rate of mass transfer.

Kinetic polarization is most pronounced for electrode processes that yield gaseous products; it is often negligible for reactions involving deposition or solution of a metal. Overvoltage effects usually become smaller at elevated temperatures and at lower current densities (current density is defined as A/cm^2 of electrode surface). It also depends upon the composition of the electrode, being most pronounced with softer metals such as lead, zinc, and, particularly, mercury. The magnitude of overvoltage can not be predicted from present theory and can only be crudely approximated from empirical information in the literature.

The overvoltages associated with the formation of hydrogen and oxygen are of particular importance to the chemist because these species are often formed or consumed during an electrolysis. In some instances, the overvoltage associated with formation of these gases may be a volt or more.

13G REFERENCE ELECTRODES

In many electroanalytical applications, it is desirable that the half-cell potential of one electrode be known, constant, and completely insensitive to the composition of the solution under study. An electrode that fits this description is called a *reference electrode*.[11] Employed in conjunction with the reference electrode will be an *indicator electrode*, whose response depends upon the analyte concentration.

A reference electrode should be convenient and easy to assemble and should maintain an essentially constant and reproducible potential in the presence of small currents. The standard hydrogen electrode fails to meet the criterion of convenience among others; several other systems are ordinarily used in its stead.

13G-1 CALOMEL ELECTRODES
Calomel half-cells may be represented as follows:

$$\| \ Hg_2Cl_2(sat'd),KCl(xM) \ | \ Hg$$

where x represents the molar concentration of potassium chloride in the solution. The electrode reaction is given by the equation

[11] For a description of commercially available reference electrodes, see: R. D. Caton, Jr., *J. Chem. Educ.*, **50**, A571 (1973); **51**, A7 (1974).

$$Hg_2Cl_2(s) + 2e \rightleftharpoons 2Hg(l) + 2Cl^-$$

The potential of this cell will vary with the chloride concentration x, and this quantity must be specified in describing the electrode.

Table 13-2 lists the composition and the potentials for the three most commonly encountered calomel electrodes. Note that each solution is saturated with mercury(I) chloride and that the cells differ only with respect to the potassium chloride concentration. Note also that the potential of the normal calomel electrode is greater than the standard potential for the half-reaction ($E^0 = 0.268$ V) because the chloride ion *activity* in a 1 M solution of potassium chloride is significantly smaller than one. The last column in Table 13-2 gives expressions that permit the calculation of electrode potentials for calomel half-cells at temperatures t other than 25°C.

The saturated calomel electrode (SCE) is most commonly used by the analytical chemist because of the ease with which it can be prepared. Compared with the other two, however, its somewhat large temperature coefficient represents a disadvantage.

A simple, easily constructed saturated calomel electrode is shown in Figure 13-7a. The salt bridge, a tube filled with saturated potassium chloride, provides electric contact with the solution in which the indicator electrode is immersed. A fritted disk or wad of cotton at one end of the salt bridge is often employed to prevent siphoning of the cell liquid and contamination of the solutions by foreign ions; alternatively, the tube can be filled with a 5% agar gel that has been saturated with potassium chloride.

Several convenient calomel electrodes are available commercially; typical are the two illustrated in Figure 13-8. The body of each electrode consists of an outer glass or plastic tube that is 5 to 15 cm in length and 0.5 to 1.0 cm in diameter. A mercury/mercury(I) chloride paste is contained in an inner tube that is connected to the saturated potassium chloride solution in the outer tube through a small opening. For electrode (a), contact with the second half-cell is made by means of a fritted disk or a porous fiber sealed in the end of the outer tubing. This type of junction has a relatively high resistance (2000 to 3000 Ω) that limits the current; on the other hand, contamination of the analyte solution due to leakage is minimal. The electrode shown in Figure 13-8b has a much lower electrical resistance but tends to leak small amounts of potassium chloride into the sample. Before it is used, the ground glass collar of this electrode is loosened and turned so that a drop or two of the KCl solution flows from the hole and wets the entire inner ground surface. Better electrical contact to the analyte solution is thus established.

Table 13-2 Specifications of Calomel Electrodes

Name	Concentration of		Electrode Potential (V) vs. Standard Hydrogen Electrode $Hg_2Cl_2(s) + 2e \rightleftharpoons 2Hg(l) + 2Cl^-$
	Hg_2Cl_2	KCl	
Saturated	Saturated	Saturated	$+0.241 - 6.6 \times 10^{-4} (t - 25)$
Normal	Saturated	1.0 M	$+0.280 - 2.8 \times 10^{-4} (t - 25)$
Decinormal	Saturated	0.1 M	$+0.334 - 8.8 \times 10^{-5} (t - 25)$

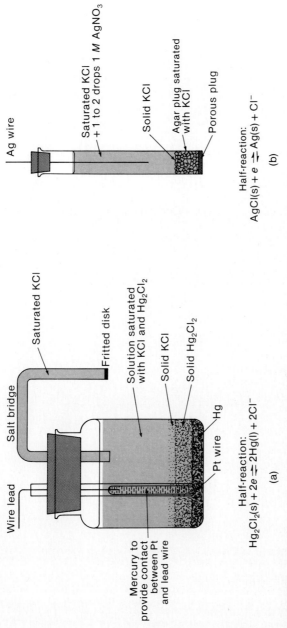

FIGURE 13-7 Two easily constructed reference electrodes. (a) A saturated calomel electrode. (b) A silver/silver chloride electrode.

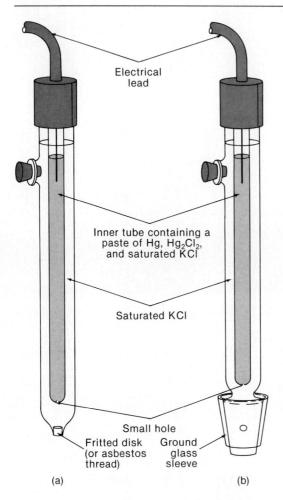

Electrical
lead

Inner tube containing a
paste of Hg, Hg_2Cl_2,
and saturated KCl

Saturated KCl

Small hole
Fritted disk   Ground
(or asbestos   glass
thread)   sleeve

(a)   (b)

FIGURE 13-8   Typical commercial calomel reference electrodes.

13G-2 SILVER/SILVER CHLORIDE ELECTRODES

A reference electrode system analogous to the calomel electrode consists of a silver
electrode immersed in a solution of potassium chloride that has been saturated
with silver chloride

$$\| AgCl(sat'd), KCl(xM) \mid Ag$$

The half-reaction is

$$AgCl(s) + e \rightleftharpoons Ag(s) + Cl^-$$

This electrode is ordinarily prepared with a saturated potassium chloride solution,
the potential at 25°C being 0.197 V with respect to the standard hydrogen electrode.

A simple and easily constructed silver/silver chloride electrode is shown in
Figure 13-7b. The electrode is contained in a Pyrex tube fitted with a 10-mm

fritted-glass disk. A plug of agar gel saturated with potassium chloride is formed on top of the disk to prevent loss of solution from the half-cell. The plug can be prepared by heating 4 to 6 g of pure agar in 100 mL of water until a clear solution is obtained; about 35 g of potassium chloride are then added. A portion of this mixture, while still warm, is poured into the tube; upon cooling, it solidifies to a gel with a relatively low electrical resistance. A layer of solid potassium chloride is placed on the gel, and the tube is filled with a saturated solution of the salt. A drop or two of aqueous silver nitrate is then added and a heavy gauge (1- to 2-mm diameter) silver wire is inserted in the solution.

Silver/silver chloride reference electrodes are also available from commercial sources. Their construction is similar to that of the electrodes shown in Figure 13-8.

PROBLEMS*

* 13-1. Listed below on the left are reactants for an oxidation-reduction reaction; the corresponding products are given on the right (H_2O, H^+, and OH^- are not included). For the reactants, indicate which species is the oxidizing and which is the reducing agent. Write balanced equations for the half-reaction of each.

Reactants	Products
(a) Cl_2, I^-	I_3^-, Cl^-
(b) Zn, Ag^+	Ag, Zn^{2+}
(c) H_2, Mn^{3+}	H^+, Mn^{2+}
(d) $Cr_2O_7^{2-}$, U^{4+}	Cr^{3+}, UO_2^{2+}
(e) $V(OH)_4^+$, V^{3+}	VO^{2+}
(f) HNO_2, MnO_4^-	NO_3^-, Mn^{2+}
(g) I^-, H_2O_2	I_3^-, H_2O
(h) Ce^{4+}, H_2O_2	Ce^{3+}, O_2
(i) IO_3^-, I^-	$I_2(s)$
(j) Ag, I^-, Sn^{4+}	AgI, Sn^{2+}

13-2. Listed below on the left are reactants for an oxidation-reduction reaction; the corresponding products are given on the right (H_2O, H^+, and OH^- are not included). For the reactants, indicate which species is the oxidizing and which is the reducing agent. Write balanced equations for the half-reaction of each.

Reactants	Products
(a) MnO_4^-, Sn^{2+}	Mn^{2+}, Sn^{4+}
(b) O_2, I^-	$I_2(aq)$, H_2O
(c) H_2SO_3, $Br_2(aq)$	SO_4^{2-}, Br^-
(d) BrO_3^-, Br^-	$Br_2(l)$
(e) I_3^-, $S_2O_3^{2-}$	I^-, $S_4O_6^{2-}$
(f) Mn^{2+}, MnO_4^-	MnO_2
(g) I^-, Cu^{2+}	I_3^-, $CuI(s)$
(h) Pb, SO_4^{2-}, H^+	$PbSO_4(s)$, H_2
(i) $PbO_2(s)$, $Ag(s)$, Cl^-	$PbCl_2(s)$, $AgCl(s)$
(j) $H_2C_2O_4$, ClO_3^-	CO_2, Cl_2

* Answers to problems and parts of problems marked with an asterisk are to be found at the end of the book.

* 13-3. Write balanced equations for each of the reactions in Problem 13-1.
  13-4. Write balanced equations for each of the reactions in Problem 13-2.
* 13-5. Express the equivalent weight of each of the reactant oxidizing agents in Problem 13-1 as a fraction or multiple of its formula weight.
  13-6. Express the equivalent weight of each of the reactant reducing agents in Problem 13-2 as a fraction or multiple of its formula weight.
* 13-7. Calculate the equilibrium constant for each of the reactions in Problem 13-3.
  13-8. Calculate the equilibrium constant for each of the reactions in Problem 13-4.
* 13-9. Calculate the electrode potential for a silver electrode immersed in
    (a) $0.0150\ M$ $AgNO_3$.
    (b) $0.0150\ M$ KSCN that is saturated with AgSCN.
    (c) $0.0150\ M$ K_2CrO_4 that is saturated with Ag_2CrO_4.
    (d) $0.0150\ M$ NaCN that is $0.0200\ M$ in $Ag(CN)_2^-$.
    (e) $0.0150\ M$ NH_3 that is $0.0200\ M$ in $Ag(NH_3)_2^+$.

    $$[\beta_2\ \text{for}\ Ag(NH_3)_2^+ = 1.3 \times 10^7]$$

  13-10. Calculate the electrode potential for a lead electrode immersed in
    (a) $0.0350\ M$ $Pb(NO_3)_2$.
    (b) $0.0350\ M$ Na_2SO_4 that is saturated with $PbSO_4$.
    (c) $0.0350\ M$ $Na_2C_2O_4$ that is $0.0640\ M$ in $Pb(C_2O_4)_2^{2-}$.

    $$[\beta_2\ \text{for}\ Pb(C_2O_4)_2^{2-} = 3.2 \times 10^6]$$

    (d) $0.0350\ M$ KI and saturated with PbI_2.
* 13-11. Calculate the electrode potential for a platinum electrode immersed in a solution that is
    (a) $0.0600\ M$ in $V_2(SO_4)_3$ and $0.0250\ M$ in VSO_4.
    (b) $0.0175\ M$ in Na_2SeO_4, $0.0347\ M$ in H_2SeO_3, and $2.10 \times 10^{-4}\ M$ in $HClO_4$.
    (c) $0.0253\ M$ in H_2S, $0.100\ M$ in $HClO_4$, and saturated with S.
    (d) $0.0253\ M$ in H_2S, $1.00 \times 10^{-5}\ M$ in $HClO_4$, and saturated with S.
    (e) $0.00325\ M$ in I_2, $0.0146\ M$ in IO_3^-, and has a pH of 8.00.
    (f) $0.00267\ M$ in $Hg_2(NO_3)_2$ and $0.00932\ M$ in $Hg(NO_3)_2$.
  13-12. Calculate the electrode potential of a platinum electrode immersed in a solution that is
    (a) $0.0315\ M$ in $K_2Cr_2O_7$, $0.0196\ M$ in $Cr_2(SO_4)_3$, and $0.100\ M$ in $HClO_4$.
    (b) $0.0675\ M$ in $U(SO_4)_2$, $0.167\ M$ in UO_2SO_4, and $0.100\ M$ in $HClO_4$.
    (c) $0.0675\ M$ in $U(SO_4)_2$, $0.167\ M$ in UO_2SO_4, and has a pH of 7.00.
    (d) $0.0426\ M$ in Cl^- and saturated with Cl_2 at a partial pressure of 0.0100 atm.
    (e) $0.0456\ M$ in BrO_3^-, saturated in Br_2, and $0.0277\ M$ in $HClO_4$.
    (f) $0.123\ M$ in ICl_2^-, $0.0683\ M$ in KIO_3, and $0.100\ M$ in KCl and has a pH of 3.25.
* 13-13. Indicate whether each of the following half-cells would behave as an anode or a cathode when coupled with a standard hydrogen electrode in a galvanic cell. Calculate the galvanic cell potential.
    (a) $Co\ |\ Co^{2+}\ (0.0644\ M)$
    (b) $Ag\ |\ KI(0.125\ M),AgI(\text{sat'd})$
    (c) $Ag\ |\ KI(1.25 \times 10^{-4}\ M),AgI(\text{sat'd})$
    (d) $Pt\ |\ Ti^{3+}(0.100\ M),TiO^{2+}(0.00613\ M),HClO_4(0.00120\ M)$
    (e) $Pt,H_2(1.00\ \text{atm})\ |\ HClO_4(0.100\ M)$
  13-14. Indicate whether each of the following half-cells would behave as an anode or a cathode when coupled with a standard hydrogen electrode in a galvanic cell. Calculate the galvanic cell potential.
    (a) $Cu\ |\ CuI(\text{sat'd}),KI(0.0440\ M)$
    (b) $Pt\ |\ Tl_2(SO_4)_3(0.0696\ M),Tl_2SO_4(0.0146\ M)$
    (c) $Ag\ |\ S_2O_3^{2-}(0.00212\ M),Ag(S_2O_3)_2^{3-}(0.156\ M)$

(d) $Pt,H_2(1.00 \text{ atm}) \mid HClO_4(2.40 \text{ } M)$

(e) $Pt \mid Hg_2(NO_3)_2(0.0600 \text{ } M),Hg(NO_3)_2(0.0356 \text{ } M)$

* 13-15. Arrange the following substances in order of their decreasing strength as oxidizing agents: $Fe(CN)_6^{3-}$, Fe^{3+}, $Ag(CN)_2^-$ (in 1 M KCN), F_2, H^+ (in 1 M acid), Ce^{4+} (in 1 M H_2SO_4), Cd^{2+}, O_2 (in 1 M acid), Cr^{3+}.

13-16. Arrange the following substances in order of their decreasing strength as reducing agents: Cl^-, Fe^{2+}, Ni, Ag (in 1 M KI), V^{2+}, Pb, Ti^{3+} (in 1 M acid), $Na_2S_2O_3$, KI, H_2 (in 1 M acid).

* 13-17. Indicate in which direction the following reactions will go if all substances are initially at unit activity.

(a) $Fe^{3+} + Ag \rightleftharpoons Fe^{2+} + Ag^+$

(b) $Sn^{4+} + 2Ag + 2Cl^- \rightleftharpoons Sn^{2+} + 2AgCl$

(c) $2Tl + Cd^{2+} \rightleftharpoons 2Tl^+ + Cd$

(d) $2Ce^{4+} + 2Br^- \rightleftharpoons 2Ce^{3+} + Br_2$

(e) $5Cl_2 + I_2 + 6H_2O \rightleftharpoons 2IO_3^- + 10Cl^- + 12H^+$

13-18. Indicate in which direction the following reactions will go if all substances are initially at unit activity.

(a) $Ag(CN)_2^- + Cr^{2+} \rightleftharpoons Ag + Cr^{3+} + 2CN^-$

(b) $Ba + 2Na^+ \rightleftharpoons Ba^{2+} + 2Na$

(c) $2Ag + Hg_2Cl_2 \rightleftharpoons 2Hg + 2AgCl$

(d) $4Ce^{4+} + 2H_2O \rightleftharpoons 4Ce^{3+} + O_2 + 4H^+$

* 13-19. Calculate the theoretical potentials of the following cells as written. Indicate which electrode would act as the anode in a galvanic cell.

(a) $Pb \mid Pb^{2+}(0.100 \text{ } M) \parallel Cd^{2+}(0.00100 \text{ } M) \mid Cd$

(b) $Pt \mid I_3^-(0.0100 \text{ } M),I^-(0.100 \text{ } M),AgI(\text{sat'd}) \mid Ag$

(c) $Pt,H_2(1.00 \text{ atm}) \mid H^+(1.00 \times 10^{-5} \text{ } M),KCl(0.125 \text{ } M),AgCl(\text{sat'd}) \mid Ag$

(d) $Pt \mid Tl^{3+}(1.00 \text{ } M),Tl^+(0.0125 \text{ } M) \parallel Zn^{2+}(0.0175 \text{ } M) \mid Zn$

(e) $Ag \mid AgCl(\text{sat'd}),KCl(1.00 \text{ } M) \parallel KCl(0.00100 \text{ } M),AgCl(\text{sat'd}) \mid Ag$

13-20. Calculate the theoretical cell potentials for the following as written. Indicate which electrode would behave as the negative electrode (anode) of a galvanic cell.

(a) $Ag \mid AgBr(\text{sat'd}),Br^-(0.0125 \text{ } M),H^+(10.0 \text{ } M) \mid H_2(1.00 \text{ atm}),Pt$

(b) $Pt \mid Fe^{3+}(0.250 \text{ } M),Fe^{2+}(0.0135 \text{ } M) \parallel Cl^-(0.115 \text{ } M),Hg_2Cl_2(\text{sat'd}) \mid Hg$

(c) $Cd \mid Cd^{2+}(0.305 \text{ } M) \parallel H^+(0.0180 \text{ } M) \mid O_2(1.00 \text{ atm}),Pt$

(d) $Pt \mid UO_2^{2+}(0.510 \text{ } M),U^{4+}(0.100 \text{ } M),H^+(0.0169 \text{ } M) \parallel Ni^{2+}(0.405 \text{ } M) \mid Ni$

(e) $Pt,H_2(1.00 \text{ atm}) \mid HCl(0.550 \text{ } M) \parallel HCl(1.00 \times 10^{-4} \text{ } M) \mid H_2(1.00 \text{ atm}),Pt$

* 13-21. Assume that each of the half-cells shown in Problem 13-13 was coupled to a saturated calomel electrode. Calculate the cell potential with the SCE as the cathode. Is the cell galvanic or electrolytic?

13-22. Assume that each of the half-cells shown in Problem 13-14 was coupled to a saturated calomel electrode. Calculate the cell potential with the SCE as the cathode. Is the cell galvanic or electrolytic?

* 13-23. In order to determine the solubility product for the salt AgX, the following cell was prepared:

$$Ag \mid AgX(\text{sat'd}),X^-(0.125 \text{ } M),H^+(1.00 \text{ } M) \mid H_2(1.00 \text{ atm}),Pt$$

The cell had a potential of 0.122 V, and the silver electrode behaved as the anode. Calculate the silver ion concentration of the solution and the solubility product of AgX.

13-24. The standard reduction potential for M^{2+} is

$$M^{2+} + 2e \rightleftharpoons M(s) \qquad E^0 = +0.0118 \text{ V}$$

The following cell

$$M \mid MX_2(\text{sat'd}),X^-(0.400 \text{ } M) \parallel \text{standard hydrogen electrode}$$

was found to have a potential of $+0.204$ V with the M electrode as an anode. Calculate the solubility product for MX_2.

* 13-25. From the data in Problem 13-23 calculate E^0 for the reaction

$$AgX(s) + e \rightleftharpoons Ag(s) + X^-$$

13-26. From the data in Problem 13-24 calculate E^0 for the reaction

$$MX_2(s) + 2e \rightleftharpoons M(s) + 2X^-$$

* 13-27. From the standard potentials

$$Tl^+ + e \rightleftharpoons Tl(s) \qquad\qquad E^0 = -0.336 \text{ V}$$

$$TlCl(s) + e \rightleftharpoons Tl(s) + Cl^- \qquad E^0 = -0.557 \text{ V}$$

calculate the solubility product constant for TlCl.

13-28. Calculate the solubility product for $Mn(OH)_2$ from the following data:

$$Mn^{2+} + 2e \rightleftharpoons Mn(s) \qquad\qquad E^0 = -1.180 \text{ V}$$

$$Mn(OH)_2(s) + 2e \rightleftharpoons Mn(s) + 2OH^- \qquad E^0 = -1.556 \text{ V}$$

* 13-29. The solubility product for Tl_2S is 1.2×10^{-22}. Calculate E^0 for the reaction

$$Tl_2S(s) + 2e \rightleftharpoons 2Tl(s) + S^{2-}$$

13-30. The solubility product for $Pb_3(AsO_4)_2$ is 4.1×10^{-36}. Calculate E^0 for the reaction

$$Pb_3(AsO_4)_2(s) + 6e \rightleftharpoons 3Pb(s) + 2AsO_4^{3-}$$

* 13-31. The cation M^{2+} forms a stable complex with anion Y^- having the formula MY_4^{2-}. A solution of the complex was prepared by dissolving 0.0500 of a formula weight of a soluble salt of M^{2+} in 1.00 L of 0.800 M Y^-. A metallic M electrode in this solution behaved as an anode against a standard hydrogen electrode, developing a potential of 0.412 V. The E^0 value for $M^{2+} + 2e \rightleftharpoons M$ is $+0.0118$ V. Calculate the formation constant for the complex MY_4^{2-}.

13-32. The cell

$$Cd \mid CdX_4^{2-}(0.200 \text{ } M), X^{2-}(0.150 \text{ } M) \parallel SCE$$

developed a potential of 0.921 V. Calculate the formation constant for CdX_4^-.

* 13-33. The half-cell

$$Pt, H_2(1.00 \text{ atm}) \mid HA(0.215 \text{ } M), NaA(0.116 \text{ } M)$$

behaves as an anode when coupled to a saturated calomel electrode. Calculate the dissociation constant of the weak acid HA if the cell potential is 0.413 V.

13-34. The half-cell

$$Pt, H_2(1.00 \text{ atm}) \mid HY(0.12 \text{ } M), NaY(0.360 \text{ } M)$$

develops an anode potential of 0.307 V against the standard hydrogen electrode. What is the dissociation constant for HY?

* 13-35. Calculate the theoretical potential required to begin deposition of nickel from a solution buffered to a pH of 7.0 which is 0.0500 M in Ni^{2+}. Assume the anode reaction is evolution of O_2 at a pressure of one atmosphere.

13-36. Calculate the potential necessary to reduce the nickel concentration to 1.0×10^{-4} M in the cell in Problem 13-35.

* 13-37. (a) Calculate the theoretical potential required to begin deposition of cobalt from a solution buffered to a pH of 5.80, which is 0.0675 M in Co^{2+}. Assume the anode reaction is evolution of O_2 at a pressure of one atmosphere.

(b) Calculate the potential required to generate a current of 0.500 A in the cell in (a) assuming the resistance of the cell is 1.25 Ω and that the overvoltage of oxygen on the anode is 0.80 V.

13-38. Perform the calculations in Problem 13-37 assuming that Sn is being deposited from a 0.0675 M solution of Sn^{2+}.

theory of oxidation-reduction

titrations

14

This chapter deals largely with end points and indicators for oxidation-reduction titrations. Theoretically derived titration curves again serve as a basis for determining the feasibility of titrations and the properties required of an indicator for such titrations.

14A TITRATION CURVES

In the titration curves considered thus far, the negative logarithm of the concentration (the p-function) of the analyte or titrant has been plotted against the volume of reagent added. The species chosen for each curve has been the one to which the indicator for the reaction is sensitive. Most of the indicators used for oxidation-reduction titrations are themselves oxidizing or reducing agents that respond to changes in the potential of the system rather than to changes in concentration of any particular reactant or product. For this reason, the usual practice is to plot the electrode potential for the system as the ordinate of the curve for an oxidation-reduction titration rather than a p-function for a reactant.

To develop a clear understanding of what is meant by the term "electrode potential for the system", consider the titration of iron(II) with cerium(IV):

$$Ce^{4+} + Fe^{2+} \rightleftharpoons Ce^{3+} + Fe^{3+}$$

Equilibrium is attained after each addition of titrant. After the first addition of cerium(IV), then, all four species will be present in amounts dictated by the equilibrium constant for the reaction. Recall now (p. 330) that the electrode potentials for the two half-reactions are identical at equilibrium; thus, at any point in the titration

$$E_{Ce^{4+}} = E_{Fe^{3+}} = E_{system}$$

and it is this potential that we call the potential of the system. If the solution contains a reversible oxidation-reduction indicator as well, its potential must also be the same as that for the system. That is, the ratio between the two forms of the indicator changes as a consequence of interaction with the reagent system until

$$E_{In} = E_{Ce^{4+}} = E_{Fe^{3+}} = E_{system}$$

The potential of a system can be measured experimentally by determining the emf of a suitable cell. Thus, for the titration of iron(II) with cerium(IV), the analyte solution could be made part of the cell:

$$SCE \parallel Ce^{4+}, Ce^{3+}, Fe^{3+}, Fe^{2+} \mid Pt$$

Here, the potential of the platinum electrode (versus a saturated calomel electrode)[1] is determined both by the affinity of Fe^{3+} for electrons

$$Fe^{3+} + e \rightleftharpoons Fe^{2+}$$

as well as that of Ce^{4+}

$$Ce^{4+} + e \rightleftharpoons Ce^{3+}$$

At equilibrium the concentration ratios of oxidized and reduced forms of each species are such that these two affinities (and thus their electrode potentials) are identical. Note that the concentration ratios for both species vary continuously as the titration proceeds; so also must E_{system}. It is the characteristic variation in this parameter that provides a means of end point detection.

In deriving E_{system} data for a titration curve, either $E_{Ce^{4+}}$ or $E_{Fe^{3+}}$ can be employed; we choose the more convenient one for any particular calculation. Short of the equivalence point, the concentrations of iron(II), iron(III), and cerium(III) are readily deduced from the amount of titrant that has been added; the concentration of cerium(IV), however, is vanishingly small. Thus, application of the Nernst equation for the iron(III)/iron(II) couple provides a value for the potential of the system directly. The corresponding expression involving the cerium(IV)/cerium(III) couple would give the same answer; however, it would first be necessary to calculate a value for the concentration of cerium(IV), which in turn would require evaluation of the equilibrium constant for the reaction. The situation is reversed after excess titrant has been introduced. Here, the concentrations of cerium(IV), cerium(III), and iron(III) are immediately available, whereas the concentration for iron(II) would require a preliminary calculation involving the equilibrium constant. Thus, the potential of the system is most directly evaluated from the cerium(IV)/cerium(III) couple in this region.

---

[1] If desired, the potential of the system relative to the standard hydrogen electrode could be calculated by adding the potential of the saturated calomel electrode to the measured potential.

14A-1 EQUIVALENCE POINT POTENTIAL

The potential of an oxidation-reduction system at the equivalence point is of particular importance from the standpoint of indicator selection. Calculation of the
equivalence potential is also unique in that there is insufficient stoichiometric information to permit the direct use of the Nernst equation for either half-cell
process. Using the titration of iron(II) with cerium(IV) as an example, values for the
analytical concentrations of cerium(III) and iron(III) at the equivalence point are
easily calculated. On the other hand, we know only that the concentrations of
cerium(IV) and iron(II) are small and numerically equal. As before, the concentrations of the minor constituents can be calculated from the equilibrium constant expression. Alternatively, we may write that, in common with any other point in the
titration, the potential of the system at equivalence, E_{eq}, is given by

$$E_{eq} = E^0_{Ce^{4+}} - 0.0591 \log \frac{[Ce^{3+}]}{[Ce^{4+}]}$$

and also by

$$E_{eq} = E^0_{Fe^{3+}} - 0.0591 \log \frac{[Fe^{2+}]}{[Fe^{3+}]}$$

Upon adding these expressions, we obtain

$$2E_{eq} = E^0_{Ce^{4+}} + E^0_{Fe^{3+}} - 0.0591 \log \frac{[Ce^{3+}][Fe^{2+}]}{[Ce^{4+}][Fe^{3+}]}$$

Note that the concentration quotient in this expression is *not* the equilibrium constant for the titration reaction.

We know from stoichiometric considerations that *at the equivalence point,*

$$[Fe^{3+}] = [Ce^{3+}]$$

$$[Fe^{2+}] = [Ce^{4+}]$$

Thus,

$$2E_{eq} = E^0_{Ce^{4+}} + E^0_{Fe^{3+}} - 0.0591 \log \frac{[\cancel{Ce^{3+}}][\cancel{Fe^{2+}}]}{[\cancel{Ce^{4+}}][\cancel{Fe^{3+}}]}$$

and

$$E_{eq} = \frac{E^0_{Ce^{4+}} + E^0_{Fe^{3+}}}{2}$$

The equilibrium concentrations of the reacting species can be readily calculated from the equivalence point potential.

EXAMPLE 14-1. Calculate the concentration of the various reactants and products at
the equivalence point in the titration of 0.100 N Fe^{2+} with 0.100 N Ce^{4+} at 25°C if
both solutions are 1.0 M in H_2SO_4.

Here, it is convenient to substitute formal potentials (see p. 324) into the
derived expression for the equivalence point potential. That is,

$$E_{eq} = \frac{+1.44 + 0.68}{2} = 1.06 \text{ V}$$

The Nernst equation allows us to evaluate the molar ratio of Fe^{2+} to Fe^{3+} at the equivalence point

$$+1.06 = +0.68 - 0.0591 \log \frac{[Fe^{2+}]}{[Fe^{3+}]}$$

$$\log \frac{[Fe^{2+}]}{[Fe^{3+}]} = -\frac{0.38}{0.0591} = -6.4$$

$$\frac{[Fe^{2+}]}{[Fe^{3+}]} = 4 \times 10^{-7}$$

It is clear that most of the Fe^{2+} has been converted to Fe^{3+} at the equivalence point. As a consequence of dilution, then, the Fe^{3+} concentration will be essentially equal to one-half the original Fe^{2+} concentration. That is,

$$[Fe^{3+}] = \frac{0.100}{2} - [Fe^{2+}] \cong 0.050$$

Thus,

$$[Fe^{2+}] = 4 \times 10^{-7} \times 0.050$$

$$= 2 \times 10^{-8}$$

Finally, the stoichiometry of the reaction requires that

$$[Ce^{4+}] = [Fe^{2+}] \cong 2 \times 10^{-8}$$

$$[Ce^{3+}] = [Fe^{3+}] \cong 0.050$$

Note that identical results would have been obtained if the Nernst equation had been applied to the cerium(IV)/cerium(III) system.

EXAMPLE 14-2. Derive an expression for the equivalence point potential for the somewhat more complicated reaction

$$5Fe^{2+} + MnO_4^- + 8H^+ \rightleftharpoons 5Fe^{3+} + Mn^{2+} + 4H_2O$$

The respective half-reactions may be written as

$$Fe^{3+} + e \rightleftharpoons Fe^{2+}$$

$$MnO_4^- + 8H^+ + 5e \rightleftharpoons Mn^{2+} + 4H_2O$$

The potential of this system is given by either

$$E = E^0_{Fe^{3+}} - \frac{0.0591}{1} \log \frac{[Fe^{2+}]}{[Fe^{3+}]}$$

or

$$E = E^0_{MnO_4^-} - \frac{0.0591}{5} \log \frac{[Mn^{2+}]}{[MnO_4^-][H^+]^8}$$

To combine the logarithmic terms so that the concentrations of various species cancel, it is necessary to multiply the permanganate half-reaction equation by 5.

$$5E_{eq} = 5E^0_{MnO_4^-} - 0.0591 \log \frac{[Mn^{2+}]}{[MnO_4^-][H^+]^8}$$

After addition, we find that

$$6E_{eq} = E^0_{Fe^{3+}} + 5E^0_{MnO_4^-} - 0.0591 \log \frac{[Fe^{2+}][Mn^{2+}]}{[Fe^{3+}][MnO_4^-][H^+]^8}$$

The stoichiometry at the equivalence point requires that

$$[Fe^{3+}] = 5[Mn^{2+}]$$

$$[Fe^{2+}] = 5[MnO_4^-]$$

After substituting and rearranging,

$$E_{eq} = \frac{E^0_{Fe^{3+}} + 5E^0_{MnO_4^-}}{6} - \frac{0.0591}{6} \log \frac{5[MnO_4^-][Mn^{2+}]}{5[Mn^{2+}][MnO_4^-][H^+]^8}$$

$$= \frac{E^0_{Fe^{3+}} + 5E^0_{MnO_4^-}}{6} - \frac{0.0591}{6} \log \frac{1}{[H^+]^8}$$

Note that the equivalence point potential for this titration is dependent upon the pH.

EXAMPLE 14-3. Derive an expression for the equivalence point potential for the reaction

$$6Fe^{2+} + Cr_2O_7^{2-} + 14H^+ \rightleftharpoons 6Fe^{3+} + 2Cr^{3+} + 7H_2O$$

Proceeding as before, we obtain the expression

$$7E_{eq} = E^0_{Fe^{3+}} + 6E^0_{Cr_2O_7^{2-}} - 0.0591 \log \frac{[Fe^{2+}][Cr^{3+}]^2}{[Fe^{3+}][Cr_2O_7^{2-}][H^+]^{14}}$$

At the equivalence point

$$[Fe^{2+}] = 6[Cr_2O_7^{2-}]$$

$$[Fe^{3+}] = 3[Cr^{3+}]$$

Substitution of these quantities into the previous equation reveals that

$$E_{eq} = \frac{E^0_{Fe^{3+}} + 6E^0_{Cr_2O_7^{2-}}}{7} - \frac{0.0591}{7} \log \frac{2[Cr^{3+}]}{[H^+]^{14}}$$

Note that the equivalence point potential in this example is dependent not only upon the concentration of hydrogen ion but also upon that of a product ion (Cr^{3+}). In general, the equivalence point potential will depend upon the concentration of one of the participants in the reaction whenever there exists a molar ratio other than unity between the species containing that participant as a reactant and as a product.

14A-2 VARIATION IN POTENTIAL AS A FUNCTION OF REAGENT VOLUME

The shape of the curve for an oxidation-reduction titration depends upon the nature of the system under consideration. The derivation of typical curves will illustrate the effects of several important variables.

EXAMPLE 14-4. Derive a curve for the titration of 50.00 mL of 0.0500 N iron(II) with 0.1000 N cerium(IV). Assume that both solutions are 1.0 M in H_2SO_4.

Formal potential data for both half-cell processes are available in Appendix 2 and will be used for these calculations.

1. *Initial potential.* The solution contains no cerium ions. It will have a small but unknown concentration of iron(III) owing to air oxidation of the iron(II). Thus, we do not have sufficient information to calculate an initial potential.

2. *Addition of 5.00 mL of cerium(IV).* With the introduction of oxidant, the solution acquires appreciable concentrations of three of the participating ions; that for the fourth, cerium(IV), will be small.

The concentration of cerium(III) is given by its analytical concentration less the equilibrium concentration for the unreacted cerium(IV):

$$[Ce^{3+}] = \frac{5.00 \times 0.1000}{50.00 + 5.00} - [Ce^{4+}] \cong \frac{0.500}{55.00}$$

This approximation appears reasonable because the equilibrium constant for the reaction is large. Similarly,

$$[Fe^{3+}] = \frac{5.00 \times 0.1000}{55.00} - [Ce^{4+}] \cong \frac{0.500}{55.00}$$

$$[Fe^{2+}] = \frac{50.00 \times 0.0500 - 5.00 \times 0.1000}{55.00} + [Ce^{4+}]$$

$$\cong \frac{2.00}{55.00}$$

As we have shown, the potential for the system can be calculated with the aid of *either of the two equations*

$$E = E^0_{Ce^{4+}} - 0.0591 \log \frac{[Ce^{3+}]}{[Ce^{4+}]}$$

$$= E^0_{Fe^{3+}} - 0.0591 \log \frac{[Fe^{2+}]}{[Fe^{3+}]}$$

The second equation is the more convenient for this calculation because the two concentrations that appear in it are known within acceptable limits. Therefore, substituting for the iron(III) and iron(II) concentrations,

$$E = +0.68 - 0.0591 \log \frac{2.00/55.00}{0.500/55.00}$$

$$= +0.64 \text{ V}$$

Had we used the formal potential for the cerium(IV)/cerium(III) system and the equilibrium concentrations of these ions, an identical potential would have been obtained.

Additional values for the potential needed to define the curve short of the equivalence point can be calculated in a fashion strictly analogous to that in step 2. Table 14-1 contains a number of these data; the student should confirm one or two to be sure of knowing how they were obtained.

3. *Equivalence point potential.* We have already seen that the potential at the equivalence point in this titration has a value of 1.06 V.

4. *Addition of 25.10 mL of reagent.* The solution now contains an excess of tetravalent cerium in addition to equivalent quantities of iron(III) and cerium(III) ions. The concentration of iron(II) will be very small. Thus,

$$[Ce^{3+}] = \frac{25.00 \times 0.1000}{75.10} - [Fe^{2+}] \cong \frac{2.500}{75.10} = [Fe^{3+}]$$

$$[Ce^{4+}] = \frac{25.10 \times 0.1000 - 50.00 \times 0.0500}{75.10} + [Fe^{2+}] \cong \frac{0.010}{75.10}$$

These approximations should be reasonable in view of the favorable equilibrium constant. As before, the desired potential could be calculated from the iron(III)/iron(II) system. At this state in the titration, however, it is more convenient to use the cerium(IV)/cerium(III) potential because the concentrations of these species are immediately available. Thus,

$$E = +1.44 - 0.0591 \log \frac{[Ce^{3+}]}{[Ce^{4+}]}$$

$$= +1.44 - 0.0591 \log \frac{2.500/75.10}{0.010/75.10}$$

$$= +1.30 \text{ V}$$

The additional postequivalence point potentials shown in Table 14-1 were calculated in a similar fashion.

The titration for iron(II) with cerium(IV) appears as curve A in Figure 14-1. Its shape is similar to the curves encountered in neutralization, precipitation, and complex formation titrations, the equivalence point being signaled by a large change in the ordinate function. A titration involving more dilute solutions of the analyte and titrant will yield a curve that is, for all practical purposes, identical with the one that was derived because the electrode potentials are independent of dilution.

The titration curve, just considered is symmetric about the equivalence point because the molar ratio of oxidant to reductant is equal to unity. As demonstrated by the following example, an asymmetric curve results if the ratio differs from this value (see Figure 14-1B).

EXAMPLE 14-5. Derive a curve for the titration of 50.00 mL of 0.0500 N iron(II) with 0.1000 N $KMnO_4$. For convenience, assume that the solution contains sufficient H_2SO_4 so that $[H^+] = 1.00$ throughout.

No formal electrode potential is available for MnO_4^-; we must therefore employ its standard potential of 1.51 V.

The reaction is

$$5Fe^{2+} + MnO_4^- + 8H^+ \rightleftharpoons 5Fe^{3+} + Mn^{2+} + 4H_2O$$

Table 14-1 **Electrode Potentials (vs. SHE) During Titrations of 50.0 mL of 0.0500 N Iron(II) Solutions***

Volume of 0.1000 N Reagent, mL	Potential, V	
	Titration with 0.1000 N Ce^{4+}	Titration with 0.1000 N MnO_4^-
5.00	0.64	0.64
15.00	0.69	0.69
20.00	0.72	0.72
24.00	0.76	0.76
24.90	0.82	0.82
25.00	1.06 ← Equivalence point →	1.37
25.10	1.30	1.48
26.00	1.36	1.49
30.00	1.40	1.50

* H_2SO_4 concentration = 1.0 M throughout.

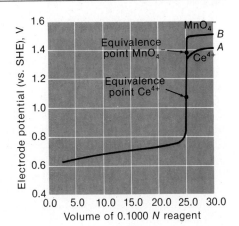

FIGURE 14-1 Titration curves for 50.00 mL of 0.0500 N Fe(II) with (curve A) 0.1000 N Ce(IV) and (curve B) 0.1000 N KMnO$_4$.

It should be stressed that molar rather than normal concentrations are employed in the Nernst equation. For manganese species in the foregoing reaction, the molarity is equal to one-fifth of the normality.

1. *Preequivalence point potentials.* As in the previous example, the preequivalence point potentials are most easily calculated from the concentrations of iron(II) and iron(III) in the solution. Their values will be identical to those computed for the titration with cerium(IV).

2. *Equivalence point potential.* The equivalence point potential for this reaction is given by the equation (p. 350)

$$E = \frac{E^0_{Fe^{3+}} + 5E^0_{MnO_4^-}}{6} - \frac{0.0591}{6} \log \frac{1}{[H^+]^8}$$

It was specified that

$$[H^+] \cong 1.00$$

Therefore,

$$E = \frac{0.68 + 5(+1.51)}{6} - \frac{0.0591}{6} \log \frac{1}{(1.00)^8}$$

$$= 1.37 \text{ V}$$

3. *Postequivalence point potentials.* When 25.10 mL of the 0.1000 N KMnO$_4$ have been added, stoichiometry requires that

$$[Fe^{3+}] = \frac{50.00 \times 0.0500}{75.10} - [Fe^{2+}] \cong \frac{2.500}{75.10}$$

$$[Mn^{2+}] = \frac{1}{5}\left(\frac{50.00 \times 0.0500}{75.10} - [Fe^{2+}]\right) \cong \frac{0.500}{75.10}$$

$$[MnO_4^-] = \frac{1}{5}\left(\frac{25.10 \times 0.1000 - 50.00 \times 0.0500}{75.10} + [Fe^{2+}]\right) \cong \frac{2.0 \times 10^{-3}}{75.10}$$

It is now advantageous to calculate the electrode potential from the standard potential of the manganese system. That is,

$$E = 1.51 - \frac{0.0591}{5} \log \frac{[Mn^{2+}]}{[H^+]^8[MnO_4^-]}$$

$$= 1.51 - \frac{0.0591}{5} \log \frac{0.500/\cancel{75.10}}{(1.00)^8(2.0 \times 10^{-3})/\cancel{75.10}}$$

$$= 1.48 \text{ V}$$

Table 14-1 contains additional data obtained in this way.

Figure 14-1 depicts curves for the titration of iron(II) with permanganate and with cerium(IV). It is of interest to note that the plots for both are alike to within 99.9% of the equivalence point; however, the equivalence point potentials are quite different. Furthermore, the permanganate curve is markedly asymmetric, the potential increasing only slightly beyond the equivalence point. Finally, the total change in potential associated with equivalence is somewhat greater with the permanganate titration, owing to the more favorable equilibrium constant for this reaction.

Effect of concentration on redox titration curves. It is of importance to note that the ordinate function, E_{system}, in the preceding calculations is ordinarily determined by the logarithm of a *ratio* of concentrations; it is, therefore, *independent* of dilution over a considerable range.[2] As a consequence, curves for oxidation-reduction titrations, in distinct contrast to those for other reaction types, tend to be independent of analyte and reagent concentrations.

Effect of the completeness of reaction on redox titration curves. The change in the ordinate function in the equivalence point region of an oxidation-reduction titration becomes larger as the reaction becomes more nearly complete; this effect is identical with that encountered for other reaction types. Figure 14-2 shows curves for titrations involving a hypothetical analyte that has a standard potential of 0.2 V with several reagents that have standard potentials ranging from 0.4 to 1.2 V; the corresponding equilibrium constants for the reaction range from about 2×10^3 to 8×10^{16}. Clearly, the more nearly complete the reaction, the greater is the change in electrode potential for the system.

The curves in Figure 14-2 were derived for reactions in which the oxidant and reductant each exhibit a one-electron change; the change in potential in the region of 24.9 to 25.1 mL is larger by about 0.14 V for a reaction in which the participants both exhibit a two-electron change.

14A-3 TITRATION OF MIXTURES

Solutions containing two oxidizing agents or two reducing agents will yield titration curves that contain two inflection points, provided the standard potentials for the two species are sufficiently different. If this difference is greater than about 0.2 V, the end points are usually distinct enough to permit analysis for each component. This situation is quite comparable to the titration of two acids having dif-

---

[2] When the solution becomes sufficiently dilute so that the reaction is incomplete, E_{system} will in fact vary with further dilution. The magnitude of this effect can be determined by dispensing with the usual approximation that the analytical and equilibrium concentrations of reactants are identical.

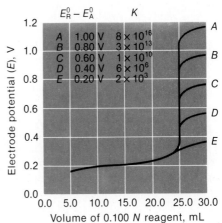

FIGURE 14-2 Titration of 50.0 mL of 0.0500 N A. E_A^0 is assumed to be 0.200 V. From the top, E_R^0 for the reagent is 1.20, 1.00, 0.80, 0.60, 0.40 V, respectively. Both the reagent and analyte are assumed to undergo a one-electron change.

ferent dissociation constants or of two ions forming precipitates of different solubilities with the same reagent.

In addition, the behavior of a few redox systems is analogous to that of polyprotic acids. For example, consider the two half-reactions

$$VO^{2+} + 2H^+ + e \rightleftharpoons V^{3+} + H_2O \qquad E^0 = +0.359 \text{ V}$$

$$V(OH)_4^+ + 2H^+ + e \rightleftharpoons VO^{2+} + 3H_2O \qquad E^0 = +1.00 \text{ V}$$

The curve for the titration of V^{3+} with a strong oxidizing agent such as permanganate will have two inflection points; the first will correspond to oxidation of the V^{3+} to VO^{2+} and the second to oxidation of VO^{2+} to $V(OH)_4^+$. The stepwise oxidation of molybdenum(III), first to the +5 oxidation state and subsequently to the +6 state, is another common example. Here, again, satisfactory inflections occur in the curves because the difference in standard potential between the pertinent half-reactions is about 0.4 V.

Derivation of titration curves for either type of mixture is not difficult if the difference in standard potential is sufficiently great. An example is the titration of a solution containing iron(II) and titanium(III) ions with potassium permanganate. The standard potentials for these systems are

$$TiO^{2+} + 2H^+ + e \rightleftharpoons Ti^{3+} + H_2O \qquad E^0 = +0.099 \text{ V}$$

$$Fe^{3+} + e \rightleftharpoons Fe^{2+} \qquad E^0 = +0.77 \text{ V}$$

The first additions of permanganate are used up by the more readily oxidized titanium(III) ion; as long as an appreciable concentration of this species remains in solution, the potential of the system cannot become high enough to alter greatly the concentration of iron(II) ions. Thus, the first part of the titration curve can be defined from the stoichiometric proportions of titanium(III) and titanium(IV) ion, with the relationship

$$E = +0.099 - 0.0591 \log \frac{[Ti^{3+}]}{[TiO^{2+}][H^+]^2}$$

For all practical purposes, then, the first part of this curve is identical to the titration curve for titanium(III) ion by itself. Beyond the first equivalence point, the solution will contain both iron(II) and iron(III) ions in appreciable concentrations, and the points on the curve can be most conveniently obtained from the relationship

$$E = +0.771 - 0.0591 \log \frac{[Fe^{2+}]}{[Fe^{3+}]}$$

Throughout this region and beyond the second equivalence point, the curve will be essentially identical to that for the titration of iron(II) ion alone (see Figure 14-1). This, then, leaves undefined only the potential at the first equivalence point. A convenient way of estimating its value is to add the Nernst equations for the iron(III) and titanium(III) potentials. Since the potentials for all oxidation-reduction systems in solution will be identical at equilibrium, we can write that

$$2E = +0.099 + 0.771 - 0.0591 \log \frac{[Ti^{3+}][Fe^{2+}]}{[TiO^{2+}][Fe^{3+}][H^+]^2}$$

Iron(II) and titanium(III) ions exist in small and equal amounts as a result of the equilibrium

$$2H^+ + TiO^{2+} + Fe^{2+} \rightleftharpoons Fe^{3+} + Ti^{3+} + H_2O$$

Thus, we may write

$$[Fe^{3+}] \cong [Ti^{3+}]$$

Substitution of this equality into the previous equation for the potential yields

$$E = \frac{+0.87}{2} - \frac{0.0591}{2} \log \frac{[Fe^{2+}]}{[TiO^{2+}][H^+]^2}$$

Finally, if $[TiO^{2+}]$ and $[Fe^{2+}]$ are assumed to be essentially identical to their analytical concentrations, we can compute the equivalence-point potential.

A titration curve for a mixture of iron(II) and titanium(III) ions is shown in Figure 14-3.

14B OXIDATION-REDUCTION INDICATORS

We have seen that the equivalence point in an oxidation-reduction titration is characterized by a marked change in the electrode potential of the system. Several methods exist for detecting such a change; these can serve to signal the end point in the titration.

14B-1 CHEMICAL INDICATORS

Indicators for oxidation-reduction titrations are of two types. *Specific indicators* owe their behavior to a reaction with one of the participants in the titration. *True oxidation-reduction indicators*, on the other hand, respond to the potential of the system rather than to the appearance or disappearance of a particular species during the titration.

Specific indicators. Perhaps the best known specific indicator is starch, which forms a dark-blue complex with triiodide ion. This complex serves to signal the end point in titrations in which iodine is either produced or consumed. Another spe-

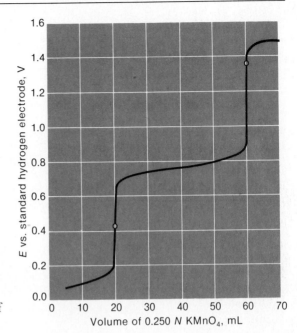

FIGURE 14-3 Curve for the titration of 50.0 mL of a solution that is 0.100 N in Ti^{3+} and 0.200 N in Fe^{2+} with 0.250 N $KMnO_4$. The concentration of H^+ is 1.0 M throughout.

cific indicator is potassium thiocyanate, which may be employed, for example, in the titration of iron(III) with solutions of titanium(III) sulfate; the end point involves the disappearance of the iron(III)/thiocyanate complex, owing to a marked decrease in the iron(III) concentration at the equivalence point.

True oxidation-reduction indicators. True oxidation-reduction indicators enjoy wider application because their behavior is dependent only upon the change in the potential of the system.

The half-reaction responsible for color change in a typical true oxidation-reduction indicator can be written as

$$In_{ox} + ne \rightleftharpoons In_{red}$$

If the indicator reaction is reversible, we may write

$$E = E^0 - \frac{0.0591}{n} \log \frac{[In_{red}]}{[In_{ox}]} \tag{14-1}$$

Typically, a change from the color of the oxidized form of the indicator to the reduced form involves a change in the ratio of reactant concentrations of about 100; that is, a color change can be seen when

$$\frac{[In_{red}]}{[In_{ox}]} \le \frac{1}{10}$$

changes to

$$\frac{[In_{red}]}{[In_{ox}]} \ge 10$$

The conditions for the full color change for a typical indicator can be found by substituting these values into Equation 14-1[3]

$$E = E^0 \pm \frac{0.0591}{n} \qquad\qquad (14\text{-}2)$$

Equation 14-2 suggests that a typical indicator will exhibit a detectable color change when the titrant causes a shift from $(E^0 + 0.0591/n)$ to $(E^0 - 0.0591/n)$ V, or about $(0.118/n)$ V in the potential of the system. With many indicators $n = 2$; a change of 0.059 V is thus sufficient.

The potential at which a color transition will occur depends upon the standard potential for the particular indicator system. Table 14-2 shows that indicators functioning in any desired potential range up to about $+1.25$ V are available.

Structures for and reactions of a few of the indicators listed in Table 14-2 are considered in the paragraphs that follow.

Iron(II) complexes of the orthophenanthrolines. A class of organic compounds known as 1,10-phenanthrolines (or orthophenanthrolines) forms stable complexes with iron(II) and certain other ions. The parent compound has a pair of nitrogen atoms located in such positions that each can form a covalent bond with the iron(II) ion. Three orthophenanthroline molecules combine with each iron ion to yield a complex with the structure

This complex is sometimes called "ferroin"; for convenience its formula will be written as $(Ph)_3Fe^{2+}$.

The complexed iron in the ferroin undergoes a reversible oxidation-reduction reaction that may be written as

$$(Ph)_3Fe^{3+} + e \rightleftharpoons (Ph)_3Fe^{2+} \qquad E^0 = +1.06 \text{ V}$$
$$\text{pale blue} \qquad\qquad\qquad \text{red}$$

The iron(III) complex is pale blue; in practice, the color change associated with the reduction is actually from nearly colorless to red. Because of the difference in color intensity, the end point is usually taken when only about 10% of the indicator is in the iron(II) form. The transition potential is thus approximately $+1.11$ V in one molar sulfuric acid.

Of all the oxidation-reduction indicators, ferroin approaches most closely the ideal substance. Its color change is very sharp, and its solutions are readily prepared and stable. In contrast to many indicators, the oxidized form is remarkably

---

[3] It should be noted that protons are involved in the reduction of many indicators; the transition potentials for these indicators will be pH dependent.

Table 14-2   A Selected List of Oxidation-Reduction Indicators*

| | Color | | Transition | |
Indicator	Oxidized	Reduced	Potential, V	Conditions
5-Nitro-1,10-phenanthroline iron(II) complex	Pale blue	Red-violet	+1.25	1 M H$_2$SO$_4$
2,3'-Diphenylamine dicarboxylic acid	Blue-violet	Colorless	+1.12	7–10 M H$_2$SO$_4$
1,10-Phenanthroline iron(II) complex	Pale blue	Red	+1.11	1 M H$_2$SO$_4$
Erioglaucin A	Bluish red	Yellow-green	+0.98	0.5 M H$_2$SO$_4$
Diphenylamine sulfonic acid	Red-violet	Colorless	+0.85	Dilute acid
Diphenylamine	Violet	Colorless	+0.76	Dilute acid
p-Ethoxychrysoidine	Yellow	Red	+0.76	Dilute acid
Methylene blue	Blue	Colorless	+0.53	1 M acid
Indigo tetrasulfonate	Blue	Colorless	+0.36	1 M acid
Phenosafranine	Red	Colorless	+0.28	1 M acid

* Data taken in part from I. M. Kolthoff and V. A. Stenger, *Volumetric Analysis*, 2d ed., vol. 1, p. 140. New York: Interscience, 1942.

inert toward strong oxidizing agents. The indicator reaction is rapid and reversible. At temperatures above 60°C, ferroin is decomposed.

A number of substituted phenanthrolines have been investigated for their indicator properties, and some have proved to be as useful as the parent compound. Among these, the 5-nitro and the 5-methyl derivatives are noteworthy, with transition potentials of +1.25 V and + 1.02 V, respectively.

Diphenylamine and its derivatives. One of the first redox indicators to be discovered was diphenylamine, C$_{12}$H$_{11}$N. This compound was recommended by Knop in 1924 for the titration of iron(II) with potassium dichromate.

In the presence of a strong oxidizing agent, diphenylamine is believed to undergo the following reactions:

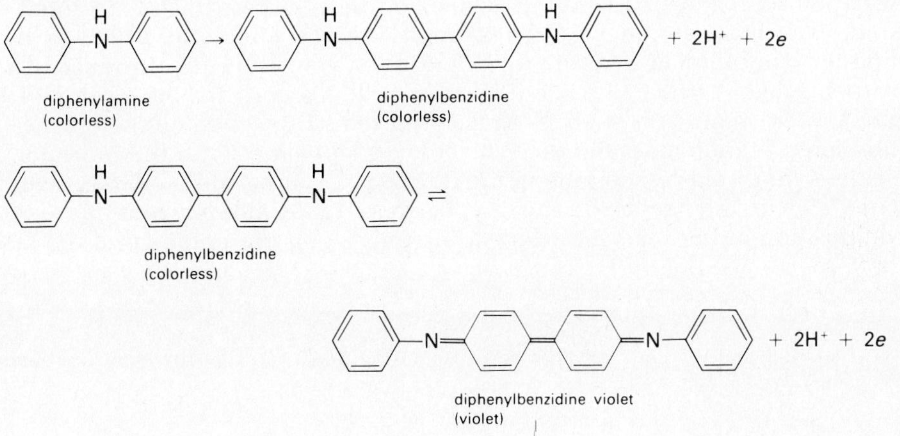

diphenylamine (colorless) → diphenylbenzidine (colorless) + 2H⁺ + 2e

diphenylbenzidine (colorless) ⇌

diphenylbenzidine violet (violet) + 2H⁺ + 2e

The first reaction, involving the formation of the colorless diphenylbenzidine, is nonreversible; the second, however, giving a violet product, can be reversed and constitutes the actual indicator reaction.

The reduction potential for the second reaction is about +0.76 V. Despite the fact that hydrogen ions appear to be involved, variations in acidity have little effect upon the magnitude of this potential, perhaps because of association of hydrogen ions with the colored product.

There are drawbacks in the application of diphenylamine as an indicator. For example, the reagent must be prepared in rather concentrated sulfuric acid solutions because of its low solubility in water. Further, the oxidation product forms a sparingly soluble precipitate with tungstate ion, which precludes its use in the presence of this element. Finally, the indicator reaction is slowed by mercury(II) ions.

The sulfonic acid derivative of diphenylamine does not suffer from these disadvantages:

diphenylamine sulfonic acid

The barium or sodium salt of this acid may be used to prepare aqueous indicator solutions; these salts behave in essentially the same manner as the parent substance. The color change is somewhat sharper, passing from colorless through green to a deep violet. The transition potential is about +0.8 V and again is independent of the acid concentration. The sulfonic acid derivative is now widely used in redox titrations.

It might be surmised from the previous two equations for the indicator reaction of diphenylamine that diphenylbenzidine should behave in an identical fashion and consume less oxidizing agent in its reaction. Unfortunately, the low solubility of diphenylbenzidine in water and sulfuric acid has precluded its widespread use. As might be expected, the sulfonic acid derivative of diphenylbenzidine has proved to be a satisfactory indicator.

Starch/iodine solution. Starch is most commonly used as a specific indicator in oxidation-reduction titrations in which iodine is a reactant. As pointed out by Kolthoff and Stenger,[4] however, a solution of starch containing a little iodine or iodide ion can function as a true redox indicator as well. In the presence of a strong oxidizing agent, the iodine/iodide ratio is high, and the blue color of the iodine/starch complex is seen. With a strong reducing agent, on the other hand, iodide ion predominates, and the blue color disappears. Thus, the indicator system changes from colorless to blue in the titration of many strong reducing agents with various oxidizing agents. This color change is quite independent of the chemical composition of the reactants, depending only upon the potential of the system at the equivalence point.

---

[4] I. M. Kolthoff and V. A. Stenger, *Volumetric Analysis*, 2d ed., vol. 1, p. 105. New York: Interscience Publishers, Inc., 1942.

14B-2 CHOICE OF CHEMICAL INDICATORS
Referring again to Figure 14-2, it is apparent that all of the indicators in Table 14-2 except for the first and the last could be employed with reagent A. On the other hand, only indigo tetrasulfonate could be employed with reagent D. The change in potential with reagent E is too small to be satisfactorily detected by an indicator.

14B-3 POTENTIOMETRIC END POINTS
End points for many oxidation-reduction titrations are readily observed by making the solution of the analyte a part of the cell:

reference electrode ‖ analyte solution | Pt

The potential of this cell, which is measured, will then vary in a way analogous to that shown in Figures 14-1 and 14-2. The end point can be determined from a plot of the measured potential as a function of titrant volume.

The reference electrode could be a standard hydrogen electrode. Usually, however, it is more convenient to use a secondary reference electrode such as the saturated calomel electrode. The experimental titration curves will then take the same form as those in Figures 14-1 and 14-2 but will be displaced on the vertical axis by an amount corresponding to the difference between the potential of the reference electrode and the standard hydrogen electrode. The potentiometric end point is considered in detail in Chapter 16.

14B-4 SUMMARY
Conclusions based upon the calculations in this chapter are helpful to the chemist as guides to the choice of reaction conditions and indicators for oxidation-reduction titrations. Thus, for example, the curves shown in Figures 14-1 and 14-2 clearly define the range of potentials within which an indicator must exhibit a color change for a successful titration. Nevertheless, it is important to emphasize that these calculations are theoretical and that they may not necessarily take into account all factors that determine the applicability and feasibility of a volumetric method. Also to be considered should be the rates at which both the principal and the indicator reactions occur; the effects of electrolyte concentration, pH, and complexing agents; the presence of colored components other than the indicator in the solution; and the variation in color perception among individuals. The state of chemistry has not advanced to the point where the effects of these variables can be completely determined by computation. Theoretical calculations can and will eliminate useless experiments and act as guides to the ones most likely to be profitable. The final test must always come in the laboratory.

*PROBLEMS
* 14-1. Calculate the equivalence point electrode potential of the system for each of the following titrations. Whenever necessary assume that $[H^+] = 0.200$ and the partial pressure of any gas is 1.00 atm.

* Answers to problems and parts of problems marked with an asterisk are to be found at the end of the book.

	Analyte, 0.0400 N	Titrant, 0.100 N
(a)	Fe^{3+}	Sn^{2+}
(b)	U^{4+}	Tl^{3+}
(c)	HNO_2	$KMnO_4$
(d)	H_3AsO_3	$Br_2(aq)$
(e)	Sn^{2+}	H_3AsO_4

14-2. Calculate the equivalence point electrode potential of the system for each of the following titrations. Whenever necessary assume that $[H^+] = 0.200$ and the partial pressure of any gas is 1.00 atm.

	Analyte, 0.0400 N	Titrant, 0.100 N
(a)	Sn^{2+}	Fe^{3+}
(b)	H_2SO_3	MnO_4^-
(c)	Cr^{2+}	$Fe(CN)_6^{3-}$
(d)	$C_6H_4O_2$ (quinone)	Sn^{2+}
(e)	$Fe(CN)_6^{4-}$	$V(OH)_4^+$

* 14-3. Calculate the equilibrium constant for each of the reactions in Problem 14-1.

14-4. Calculate the equilibrium constant for each of the reactions in Problem 14-2.

* 14-5. Calculate the equivalence point concentration of the analyte in Problem 14-1.

14-6. Calculate the equivalence point concentration of the analyte in Problem 14-2.

* 14-7. Construct a titration curve for each of the titrations listed in Problem 14-1. Assume that 50.00 mL of the analyte solution are being titrated and that $[H^+] = 0.200$ throughout the titration; derive data at 5.00, 15.00, 19.00, 20.00, 21.0, 25.0, and 30.0 mL of reagent.

14-8. Construct a titration curve for each of the titrations listed in Problem 14-2. Assume that 50.00 mL of the analyte solution are being titrated and that $[H^+] = 0.200$ throughout the titration; derive data at 5.00, 15.00, 19.00, 20.00, 21.0, 25.0, and 30.0 mL of reagent.

* 14-9. Construct a titration curve for 50.00 mL of 0.0400 N V^{3+} with 0.100 N $KMnO_4$. Assume that $[H^+] = 1.00$ throughout the titration. Derive data at 5.00, 15.00, 19.00, 20.00, 21.00, 26.00, 35.00, 39.00, 40.00, 41.00, 45.00, and 50.00 mL of reagent.

14-10. Construct a titration curve for 50.00 mL of a solution that is 0.0500 N in Sn^{2+} and 0.0400 N in Fe^{2+} with 0.1000 N $Ca(ClO)_2$. Derive data at 10.00, 20.00, 24.00, 25.00, 26.00, 30.00, 40.00, 44.00, 45.00, 46.00, and 50.00 mL. Assume that $[H^+] = 0.100$ throughout the titration.

$$HClO + 2e + H^+ \rightleftharpoons Cl^- + H_2O \qquad E^0 = 1.50 \text{ V}$$

application of

oxidation-reduction titrations

15

This chapter is concerned with the preparation and application of standard solutions of oxidants and reductants. In addition, consideration is given to auxiliary reagents that serve to convert an analyte to a single oxidation state prior to its titration.

Several excellent oxidizing agents are available for the preparation of standard solutions. The number of standard reducing reagents is much more limited because their solutions are susceptible to air oxidation; storage is therefore inconvenient.

15A AUXILIARY REAGENTS

As often as not, the steps that precede an oxidation-reduction titration leave the analyte either in a mixture of oxidation states or else in its highest oxidation state. For example, a mixture of iron(II) and iron(III) results when an iron alloy is dissolved in hydrochloric acid. Before the iron can be titrated, therefore, a reagent must be added that will convert the element quantitatively either to the divalent state for titration with a standard oxidizing agent or to the trivalent state for titration with a reducing agent. The electrode potential for an auxiliary reagent must be such as to assure quantitative conversion of the analyte to the desired oxidation state. Nevertheless, it should not be sufficient to convert other components of the solution into states that will also react with the titrant. Another requirement is that the unused portion of the auxiliary reagent be conveniently removed because it will almost inevitably interfere with the titration by reacting with the standard solution. Thus, a reagent capable of converting iron quantitatively to the divalent state for titration with a standard solution of permanganate would of necessity be a good reducing agent. Any excess remaining after the reduction would surely consume permanganate if it were not removed from the solution.

15A-1 AUXILIARY REDUCING REAGENTS
Reagents that find general application to the prereduction of samples are described in the following paragraphs.

Metals. An examination of standard electrode potential data reveals a number of good reducing agents among the pure metals.[1] Such elements as zinc, cadmium, aluminum, lead, nickel, copper, mercury, and silver have proved useful for prereduction purposes. Where sticks or coils of the metal are used, the excess reductant is simply lifted from the solution and washed thoroughly. Filtration may be required to remove granular or powdered forms of the metal. An alternative is to employ a *reductor* such as that shown in Figure 15-1. Here, the granular or powdered metal is held in a vertical glass tube; the solution to be reduced is then drawn through the column with a moderate vacuum. Normally, this type of a reductor can be employed for several hundred reductions.

The *Jones reductor,* which finds wide application, employs amalgamated zinc as the reducing agent. Amalgamation is accomplished by allowing zinc granules to stand briefly in a solution of mercury(II) chloride:

$$Zn + Hg_x^{2+} \longrightarrow Zn^{2+} + Zn(Hg)_x$$

The zinc amalgam produced is nearly as effective a reducing agent as the metal itself. It has the important additional virtue of inhibiting the reduction of hydrogen ions by zinc, a parasitic reaction that not only consumes the reducing agent but also causes the solutions being analyzed to be heavily contaminated with additional zinc(II) ions. Solutions that are quite acidic can be passed through a Jones reductor without significant formation of hydrogen.

The typical Jones reductor has a diameter of about 2 cm and is packed to a depth of 40 to 50 cm. The packing must be kept covered with liquid at all times to protect against air oxidation and the resultant formation of basic salts which tend to cause clogging.

Table 15-1 lists the principal applications of the Jones reductor.

As seen from Table 15-1, the *Walden reductor* has the advantage of being somewhat more selective than the Jones reductor; here, granular metallic silver is the reductant. The metal is prepared by reducing a solution containing about 30 g of silver nitrate with metallic copper. The resulting suspension of finely divided silver is then poured into a narrow glass column to give about 10 cm of packing. When not in use, the silver is covered with 1 M HCl.

The reducing strength of silver is enhanced by the presence of chloride ion. It is for this reason that hydrochloric acid is always used in conjunction with a

---

[1] For a discussion of metal reductants, the reader should see: I. M. Kolthoff and R. Belcher, *Volumetric Analysis*, vol. 3, pp. 11–23. New York: Interscience, 1957; W. I. Stephen, *Ind. Chemist*, **28**, 13, 55, 197 (1952).

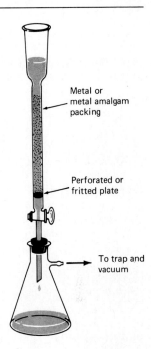

Metal or metal amalgam packing

Perforated or fritted plate

To trap and vacuum

FIGURE 15-1   A metal or metal amalgam reductor.

Walden reductor. The silver chloride that coats the packing with use can be eliminated by dipping a zinc rod into the solution contained in the column.

Table 15-1 lists various uses of the Walden reductor.

Gaseous reductants.   Both hydrogen sulfide and sulfur dioxide are reasonably effective reducing reagents and have been used for prereduction. Excesses of these reagents can be readily eliminated by boiling the acidified solution.

Table 15-1   Uses of the Walden Reductor and the Jones Reductor*

Walden Reductor $Ag(s) + Cl^- \rightleftarrows AgCl(s) + e$	Jones Reductor $Zn(s) \rightleftarrows Zn^{2+} + 2e$
$e + Fe^{3+} \rightarrow Fe^{2+}$	$e + Fe^{3+} \rightarrow Fe^{2+}$
$e + Cu^{2+} \rightarrow Cu^+$	Cu^{2+} reduced to metallic Cu
$e + H_2MoO_4 + 2H^+ \rightarrow MoO_2^+ + 2H_2O$	$3e + H_2MoO_4 + 6H^+ \rightarrow Mo^{3+} + 4H_2O$
$2e + UO_2^{2+} + 4H^+ \rightarrow U^{4+} + 2H_2O$	$2e + UO_2^{2+} + 4H^+ \rightarrow U^{4+} + 2H_2O$
	$3e + UO_2^{2+} + 4H^+ \rightarrow U^{3+} + 2H_2O**$
$e + V(OH)_4^+ + 2H^+ \rightarrow VO^{2+} + 3H_2O$	$3e + V(OH)_4^+ + 4H^+ \rightarrow V^{2+} + 4H_2O$
TiO^{2+} not reduced	$e + TiO^{2+} + 2H^+ \rightarrow Ti^{3+} + H_2O$
Cr^{3+} not reduced	$e + Cr^{3+} \rightarrow Cr^{2+}$

* Taken from I. M. Kolthoff and R. Belcher, *Volumetric Analysis*, vol. 3, p. 12. New York: Interscience, 1957. With permission.
** A mixture of oxidation states is obtained. The Jones reductor may still be used for the analysis of uranium, however, because any U^{3+} formed can be converted to U^{4+} by shaking the solution with air for a few minutes.

The reactions of both gases tend to be slow, with half an hour or more being required to complete the reduction and rid the solution of excess reagent. In addition to this time disadvantage, both gases are noxious and toxic. The employment of other reductants is much preferred.

Tin(II) chloride. The reduction of iron(III) to the divalent state is conveniently accomplished with tin(II) chloride. The chemistry of this application is discussed in Chapter 31, Method 6-1.

15A-2 AUXILIARY OXIDIZING REAGENTS

Sodium bismuthate. Sodium bismuthate is an extremely powerful oxidizing agent capable, for example, of converting manganese(II) quantitatively to permanganate. It exists as a sparingly soluble solid of somewhat uncertain composition; its formula is usually written as $NaBiO_3$. Upon reaction, bismuth(V) is converted to the more common trivalent state. Ordinarily, the solution to be oxidized is boiled in contact with an excess of the solid; the unused reagent is then removed by filtration.

Ammonium peroxodisulfate. In acidic solutions, ammonium peroxodisulfate, $(NH_4)_2S_2O_8$, is a potent oxidizing agent that will convert chromium to dichromate, cerium(III) to the tetravalent state, and manganese(II) ion to permanganate. The half-reaction is

$$S_2O_8^{2-} + 2e \rightleftharpoons 2SO_4^{2-} \qquad E^0 = 2.01 \text{ V}$$

The oxidations are catalyzed by traces of silver ion. The excess reagent is readily decomposed by boiling the solution for a few minutes.

$$2S_2O_8^{2-} + 2H_2O \rightleftharpoons 4SO_4^{2-} + O_2(g) + 4H^+$$

Sodium and hydrogen peroxide. Peroxide is a convenient oxidizing agent. Both the solid sodium salt and dilute solutions of the acid are used. The half-reaction for hydrogen peroxide in acidic solution is

$$H_2O_2 + 2H^+ + 2e \rightleftharpoons 2H_2O \qquad E^0 = 1.78 \text{ V}$$

Any excess reagent is readily decomposed by brief boiling.

$$2H_2O_2 \rightleftharpoons 2H_2O + O_2(g)$$

15B APPLICATIONS OF STANDARD OXIDANTS

Table 15-2 summarizes the characteristics of oxidizing reagents that are commonly employed for titrimetric analysis. These reagents differ considerably in oxidizing strength, with standard potentials that range from 1.6 to 0.5 V. The choice among them depends upon the strength of the analyte (as a reducing agent), the rate of reaction between the oxidant and the analyte, the stability of the standard oxidant solutions, and the availability of a satisfactory indicator for end point detection.

15B-1 POTASSIUM PERMANGANATE

Potassium permanganate, a powerful oxidant, is perhaps the most widely used of all standard oxidizing agents. The color of a permanganate solution is so intense that an indicator is not ordinarily required. The reagent is readily available at modest cost. On the other hand, the tendency of permanganate to oxidize chloride ion is a disadvantage because hydrochloric acid is such a useful solvent. The multiplicity of possible reaction products can at times cause uncertainty regarding the stoichiometry of a permanganate oxidation. Finally, permanganate solutions have limited stability.

Reactions of permanganate ion. Potassium permanganate is most commonly employed with solutions that are 0.1 N or greater in mineral acid; under these conditions, the product is manganese(II):

$$MnO_4^- + 8H^+ + 5e \rightleftharpoons Mn^{2+} + 4H_2O \qquad E^0 = 1.51 \text{ V}$$

Permanganate oxidations typically occur rapidly in acidic solution. Notable exceptions include the reaction with oxalic acid that requires elevated temperatures and with arsenic(III) oxide for which a catalyst such as osmium tetroxide or iodine monochloride is needed.

In solutions that are weakly acidic (above pH 4), neutral, or weakly alkaline, permanganate usually undergoes a three-electron reduction, with a brown precipitate of manganese dioxide, MnO_2, being formed. Titrations of certain species can be carried out to advantage under these conditions. For example, cyanide is oxidized to cyanate; sulfide, sulfite, and thiosulfate are converted to sulfate; manganese(II) is oxidized to manganese dioxide; and hydrazine is oxidized to nitrogen.

Table 15-2   Some Common Oxidants Employed for Standard Solutions

Reagent and Formula	Reduction Product	Standard Potential, V	Primary Standard for	Indicator*	Stability**
Potassium permanganate $KMnO_4$	Mn^{2+}	1.51	$Na_2C_2O_4$, Fe, As_2O_3	MnO_4^-	(b)
Potassium bromate $KBrO_3$	Br^-	1.44	$KBrO_3$	(1)	(a)
Cerium(IV) Ce^{4+}	Ce^{3+}	1.44	$Na_2C_2O_4$, Fe, As_2O_3	(2)	(a)
Potassium dichromate $K_2Cr_2O_7$	Cr^{3+}	1.33	$K_2Cr_2O_7$, Fe	(3)	(a)
Periodic acid H_5IO_6	IO_3^-	1.60	As_2O_3	starch	(b)
Potassium iodate KIO_3	ICl_2^-	1.24	KIO_3	(4)	(a)
Iodine I_2	I^-	0.536	$BaS_2O_3 \cdot H_2O$, As_2O_3	starch	(c)

* (1) α-Naphthoflavone, (2) orthophenanthroline iron(II) complex (ferroin), (3) diphenylamine sulfonic acid, (4) disappearance of I_2 from a layer of chloroform.
** (a) Indefinitely stable; (b) moderately stable, requires periodic standardization; (c) somewhat unstable, requires frequent standardization.

Solutions of manganese(III) are not stable owing to the disproportionation reaction

$$2Mn^{3+} + 2H_2O \rightleftharpoons MnO_2(s) + Mn^{2+} + 4H^+$$

However, manganese(III) ion forms several complexes that are sufficiently stable to permit existence of the $+3$ state in aqueous solution. Lingane has made use of this property to titrate manganese(II) with permanganate in highly concentrated solutions of pyrophosphate ion[2]; the reaction may be expressed as

$$MnO_4^- + 4Mn^{2+} + 15H_2P_2O_7^{2-} + 8H^+ \rightleftharpoons 5Mn(H_2P_2O_7)_3^{3-} + 4H_2O$$

The titration is carried out in a pH range between 4 and 7.

In solutions that are greater than 1 N in sodium hydroxide, permanganate ion undergoes a one-electron reduction to manganate ion, MnO_4^{2-}. Alkaline oxidations with permanganate have proved useful for the determination of certain organic compounds.

End point. One of the obvious properties of potassium permanganate is its intense purple color, which commonly serves as the indicator for titrations. As little as 0.01 to 0.02 mL of a 0.02 M (0.1 N) solution is sufficient to impart a perceptible color to 100 mL of water. For very dilute permanganate solutions, diphenylamine sulfonic acid or the orthophenanthroline/iron(II) complex (Table 14-2) will give a sharper end point.

The permanganate end point is not permanent. Decolorization in acidic solutions results from the reaction of the excess reagent with the relatively large concentration of manganese(II) ion that is present at the end point:

$$2MnO_4^- + 3Mn^{2+} + 2H_2O \rightleftharpoons 5MnO_2(s) + 4H^+$$

The equilibrium constant for this reaction, which is readily calculated from the standard potentials for the two half-reactions, has a numerical value of about 10^{47}. Thus, even in a highly acidic solution, the concentration of permanganate in equilibrium with manganese(II) ion is small. Fortunately, the rate at which this equilibrium is attained is slow, with the result that the end point fades only gradually.

The intense color of a permanganate solution complicates the measurement of reagent volumes in a buret. It is frequently more practical to use the surface of the liquid rather than the bottom of the meniscus as the point of reference.

Stability of permanganate solutions. Aqueous solutions of permanganate are not completely stable because the ion tends to oxidize water. The process may be depicted by the equation

$$4MnO_4^- + 2H_2O \rightleftharpoons 4MnO_2(s) + 3O_2(g) + 4OH^-$$

Although the constant for this equilibrium indicates that the products are favored in neutral solution, the reaction is so slow that a properly prepared solution is moderately stable. The decomposition of permanganate has been shown to be catalyzed by light, heat, acids, bases, manganese(II) ion, and manganese dioxide. To obtain a stable reagent for analysis it is necessary to minimize the influence of these effects.

[2] J. J. Lingane and R. Karplus, *Ind. Eng. Chem., Anal. Ed.*, **18**, 191 (1946).

The decomposition of permanganate solutions is greatly accelerated by the presence of manganese dioxide. Because it is also a product of the decomposition, this solid has an *autocatalytic* effect upon the process.

Photochemical catalysis of the decomposition is often observed when a permanganate solution is allowed to stand in a buret for an extended period. Manganese dioxide forms as a brown stain and serves to show that the concentration of the reagent has undergone a change.

In general, the heating of acidic solutions containing an excess of permanganate should be avoided because of a decomposition error that cannot adequately be compensated for with a blank. At the same time, it is perfectly acceptable to titrate hot, acidic solutions of reductants directly with the reagent because at no time during the titration is the oxidant concentration large enough to cause a measurable uncertainty.

Preparation and storage of permanganate solutions. A permanganate solution possessing reasonable stability can be obtained, provided a number of precautions are observed. Perhaps the most important variable affecting stability is the catalytic influence of manganese dioxide. This compound is an inevitable contaminant in solid potassium permanganate; it is also produced when permanganate oxidizes organic matter in the water used to prepare the solution. Removal of manganese dioxide by filtration markedly enhances the stability of standard permanganate solutions. Sufficient time must be allowed for complete oxidation of contaminants in the water before filtration; the solution may be boiled to hasten the process. Paper cannot be used for the filtration because it reacts with permanganate to form the undesirable dioxide.

Standardized solutions should be stored in the dark. If any solid is detected in the solution, filtration and restandardization are necessary. In any event, restandardization every one to two weeks is a good precautionary measure.[3]

Standardization against sodium oxalate. In acidic solution, permanganate oxidizes oxalic acid to carbon dioxide and water.

$$2MnO_4^- + 5H_2C_2O_4 + 6H^+ \rightleftharpoons 2Mn^{2+} + 10CO_2(g) + 8H_2O$$

This reaction is complex and proceeds slowly at room temperature; even at elevated temperatures, it is not rapid unless catalyzed by manganese(II) ion. Thus, several seconds are required to decolorize a hot oxalic acid solution at the outset of a permanganate titration. Later, when the concentration of manganese(II) ion has become appreciable, decolorization becomes rapid as a consequence of the autocatalysis.

The stoichiometry of the reaction between permanganate ion and oxalic acid has been investigated in great detail by McBride[4] and more recently by Fowler and Bright.[5] The former devised a procedure in which the oxalic acid is titrated slowly at a temperature between 60 and 90°C until the faint pink color of the

[3] It is noteworthy, however, that properly prepared permanganate solutions may have a remarkably long shelf life. For example, Durham has reported that two permanganate solutions changed less than 0.6% in normality over a nine-year period [B. W. Durham, *Anal. Chem.*, **51**, 922A (1979)].

[4] R. S. McBride, *J. Amer. Chem. Soc.*, **34**, 393 (1912).

[5] R. M. Fowler and H. A. Bright, *J. Res. Nat. Bur. Stand.*, **15**, 493 (1935).

permanganate persists. Fowler and Bright have demonstrated, however, that this titration consumes 0.1 to 0.4% too little permanganate, due perhaps to air oxidation of a small part of the oxalic acid.

$$H_2C_2O_4 + O_2(g) \rightleftharpoons H_2O_2 + 2CO_2$$

In the hot solution, the peroxide is postulated to decompose spontaneously to oxygen and water.

Fowler and Bright devised a scheme for standardization in which 90 to 95% of the required permanganate is added rapidly to the cool oxalic acid solution. After all of this reagent has reacted, the solution is heated to 55 to 60°C and titrated as before. Although it minimizes the air oxidation of oxalic acid and gives data that appear to be in exact accord with the theoretical stoichiometry, this method suffers from the disadvantage of requiring a knowledge of the approximate normality of the permanganate solution to permit the proper initial addition of the reagent. The Fowler-Bright procedure is not as convenient as the McBride method in this respect.

The method of McBride will give perfectly adequate data for many purposes (usually 0.2 to 0.3% too high). If a more accurate standardization is required, the approximate normality can be obtained by this procedure, following which a pair of titrations can be performed by the Fowler and Bright method. Directions for both procedures are found in Chapter 31, Standardization 6-1.

Other primary standards. Several other primary standards can be employed to establish the normality of permanganate solutions. These include arsenic(III) oxide, potassium iodide, and metallic iron. Detailed procedures for the use of these standards are described by Kolthoff and Belcher.[6]

Applications of permanganate titrations. Table 15-3 indicates the multiplicity of analyses that make use of standard permanganate solutions in acidic media. Most of these reactions are rapid enough for direct titrations. Directions for two typical applications, the determination of iron and the determination of calcium, are found in Chapter 31 (Methods 6-1 and 6-2).

15B-2 TETRAVALENT CERIUM

A sulfuric acid solution of cerium(IV) is very nearly as potent an oxidizing reagent as permanganate and can be substituted for the latter in most of the applications just described. The reagent is indefinitely stable and does not oxidize chloride ion at a detectable rate. Furthermore, only a single reduction product, trivalent cerium, is possible; thus, the stoichiometry of the reaction is less subject to uncertainty. In these respects cerium(IV) possesses considerable advantages over permanganate. On the other hand, the color of tetravalent cerium solutions is not sufficiently intense to serve as an indicator. In addition, the reagent cannot be used in neutral or basic solutions. A final disadvantage is the relatively high cost of cerium compounds.

[6] I. M. Kolthoff and R. Belcher, *Volumetric Analysis*, vol. 3, pp. 41–59. New York: Interscience, 1957.

Properties of tetravalent cerium solutions. The electrode potential for a cerium(IV) solution depends upon the kind of acid that is used for preparation. Solutions of the reagent prepared with sulfuric acid are roughly comparable in oxidizing power to those of permanganate; solutions containing nitric or perchloric acid are appreciably more potent. To a lesser degree the electrode potential is also influenced by the concentration of the acid.

Composition of cerium(IV) solutions. Acidic solutions of tetravalent cerium have highly complex compositions. The exact nature of the cerium-containing species present has not yet been established. Much of that which is known has come from

Table 15-3 Some Applications of Potassium Permanganate in Acid Solution

Substance Sought	Half-Reaction	Condition
Sn	$Sn^{2+} \rightleftarrows Sn^{4+} + 2e$	Prereduction with Zn
H_2O_2	$H_2O_2 \rightleftarrows O_2(g) + 2H^+ + 2e$	
Fe	$Fe^{2+} \rightleftarrows Fe^{3+} + e$	Prereduction with $SnCl_2$ or with Jones or Walden reductor
$Fe(CN)_6^{4-}$	$Fe(CN)_6^{4-} \rightleftarrows Fe(CN)_6^{3-} + e$	
V	$VO^{2+} + 3H_2O \rightleftarrows V(OH)_4^+ + 2H^+ + e$	Prereduction with Bi amalgam or SO_2
Mo	$Mo^{3+} + 4H_2O \rightleftarrows MoO_4^{2-} + 8H^+ + 3e$	Prereduction with Jones reductor
W	$W^{3+} + 4H_2O \rightleftarrows WO_4^{2-} + 8H^+ + 3e$	Prereduction with Zn or Cd
U	$U^{4+} + 2H_2O \rightleftarrows UO_2^{2+} + 4H^+ + 2e$	Prereduction with Jones reductor
Ti	$Ti^{3+} + H_2O \rightleftarrows TiO^{2+} + 2H^+ + e$	Prereduction with Jones reductor
$H_2C_2O_4$	$H_2C_2O_4 \rightleftarrows 2CO_2 + 2H^+ + 2e$	
Mg, Ca, Zn, Co, Pb, Ag	$H_2C_2O_4 \rightleftarrows 2CO_2 + 2H^+ + 2e$	Sparingly soluble metal oxalates filtered, washed, and dissolved in acid; liberated oxalic acid titrated
HNO_2	$HNO_2 + H_2O \rightleftarrows NO_3^- + 3H^+ + 2e$	15-min reaction time; excess $KMnO_4$ back-titrated
K	$K_2NaCo(NO_2)_6 + 6H_2O \rightleftarrows$ $Co^{2+} + 6NO_3^- + 12H^+ + 2K^+ +$ $Na^+ + 11e$	Precipitated as $K_2NaCo(NO_2)_6$; filtered and dissolved in $KMnO_4$; excess $KMnO_4$ back-titrated
Na	$U^{4+} + 2H_2O \rightleftarrows UO_2^{2+} + 4H^+ + 2e$	Precipitated as $NaZn(UO_2)_3 (OAc)_9$; filtered, washed, dissolved; U determined as above

studies concerned with the effects of various acids and their concentrations upon the potential of the cerium(IV)/cerium(III) couple and upon the color of solutions of the reagent. Table 15-4 provides some typical potential data. Note that stronger oxidizing properties are exhibited by the cerium(IV) species in perchloric acid than by those in nitric or sulfuric acids. In all three media the formal reduction potential varies with acid concentration. These data suggest that cerium(IV) ions form stable complexes with nitrate and sulfate ions. In addition, such species as $Ce(OH)^{3+}$ and $Ce(OH)_2^{2+}$ exist in perchloric acid solution. Finally, the presence of a dimeric cerium(IV) ion, particularly in highly concentrated solutions, has been reported. All evidence indicates that the concentration of the simple hydrated ion, $Ce(H_2O)_x^{4+}$, is small in any of these solutions.

Stability of cerium(IV) solutions. Sulfuric acid solutions of tetravalent cerium are remarkably stable, remaining constant in titer for years. Solutions do not change appreciably when heated to 100°C for considerable periods. Perchloric and nitric acid solutions of the reagent are by no means as stable; these decompose water and decrease in normality by 0.3 to 1% during storage for one month. The decomposition reaction is catalyzed by light.

The oxidation of chloride is so slow that other reducing agents can be titrated without error in the presence of high concentrations of this ion. Hydrochloric acid solutions of cerium(IV), however, are not stable enough for use as standard solutions.

Indicators for cerium(IV) titrations. Several oxidation-reduction indicators are available for use with cerium(IV) solutions. The most widely used of these is the iron(II) complex of orthophenanthroline; indicator solutions of the complex are sometimes given the trivial name "ferroin." Other derivatives of orthophenanthroline (p. 359) are also useful for titration with cerium(IV).

Preparation of solutions. Several cerium(IV) salts are commercially available;[7] the most common are listed in Table 15-5. Cerium(IV) ammonium nitrate of primary standard quality can be purchased; standard solutions can thus be prepared directly by weight. More frequently, solutions of approximately the desired normal-

---

[7] For further information regarding the preparation, standardization, and use of cerium(IV) solutions, see G. Frederick Smith, *Cerate Oxidimetry*, 2d ed. Columbus, Ohio: The G. Frederick Smith Chemical Company, 1964; I. M. Kolthoff and R. Belcher, *Volumetric Analysis*, vol. 3, pp. 121–167. New York: Interscience, 1957.

Table 15-4 Formal Electrode Potentials for Cerium(IV)

	Formal Potential vs. Standard Hydrogen Electrode, V		
Acid Concentration, N	$HClO_4$ Solution	HNO_3 Solution	H_2SO_4 Solution
1	+1.70	+1.61	+1.44
2	1.71	1.62	1.44
4	1.75	1.61	1.43
8	1.87	1.56	1.42

ity are prepared from one of the less expensive reagent-grade salts and then standardized. A stable and entirely satisfactory sulfuric acid solution of tetravalent cerium is obtained from cerium(IV) ammonium nitrate; removal of the ammonium or nitrate ions is unnecessary.

Solutions of cerium(IV) tend to react with water, even in acid solutions, to produce slightly soluble basic salts. The acidity of solutions containing cerium(IV) must be 0.1 N or greater to prevent this precipitation. Neutral or basic solutions cannot be titrated with the reagent.

Standardization against sodium oxalate. Several methods exist for the standardization of sulfuric acid solutions of cerium(IV) against sodium oxalate.[8] The directions in Chapter 31, Standardization 7-1, call for titration at 50°C in hydrochloric acid solution. Iodine monochloride is used as a catalyst; orthophenanthroline is the indicator.

Standardization against arsenic(III) oxide. Arsenic(III) oxide has been perhaps the most satisfactory primary standard for solutions of tetravalent cerium. Unfortunately, this useful analytical reagent has been designated as a carcinogen by OSHA. Thus, special precautions, which may be sufficiently troublesome to restrict its applications, are required for its use.

In the absence of catalysts, the reaction between cerium(IV) and arsenic(III) is so slow that iron(II) can be titrated in the presence of considerable amounts of trivalent arsenic. Fortunately, good catalysts are available that permit standardization with this useful reagent. The best of these is osmium tetroxide, which is effective even at very low concentrations (10^{-5} M). Iodine monochloride also catalyzes the reaction.

Applications of tetravalent cerium solutions. Many applications of cerium(IV) solutions are found in the literature. In general, these parallel the uses of permanganate given in Table 15-3.[9] Directions for the determination of iron are given in Chapter 31, Method 7-1.

15B-3 POTASSIUM DICHROMATE
In its analytical applications, dichromate ion is reduced to the trivalent state:

$$Cr_2O_7^{2-} + 14H^+ + 6e \rightleftharpoons 2Cr^{3+} + 7H_2O \qquad E^0 = 1.33 \text{ V}$$

[8] I. M. Kolthoff and R. Belcher, *Volumetric Analysis*, vol. 3, pp. 132–134. New York: Interscience, 1957.
[9] Information about cerium(IV) methods can be found in I. M. Kolthoff and R. Belcher, *Volumetric Analysis*, vol. 3, pp. 136–167. New York: Interscience, 1957.

Table 15-5 Analytically Useful Cerium(IV) Compounds

Name	Formula	Equivalent Weight
Cerium(IV) ammonium nitrate	$Ce(NO_3)_4 \cdot 2NH_4NO_3$	548.2
Cerium(IV) ammonium sulfate	$Ce(SO_4)_2 \cdot 2(NH_4)_2SO_4 \cdot 2H_2O$	632.6
Cerium(IV) hydroxide	$Ce(OH)_4$	208.1
Cerium(IV) hydrogen sulfate	$Ce(HSO_4)_4$	528.4

Potassium dichromate is more limited in application than either potassium permanganate or tetravalent cerium owing to its lesser oxidizing strength and the slowness of some of its reactions. Despite these handicaps, the reagent is nonetheless useful because its solutions are indefinitely stable and are also inert toward hydrochloric acid. Furthermore, the solid reagent can be obtained in high purity and at modest cost; standard solutions may be prepared directly by weight.

Solutions of potassium dichromate may be boiled for long periods without decomposition.

Preparation and properties of dichromate solutions. For most purposes, commercial reagent-grade or primary standard-grade potassium dichromate can be used to prepare standard solutions with no prior treatment other than drying at 150 to 200°C. If desired, two or three recrystallizations of the solid from water will assure a high-quality, primary standard product.

Although solutions of dichromate are orange, the color is not sufficiently intense for end point determination. Diphenylamine sulfonic acid (Table 14-2) is an excellent indicator for titrations with the reagent; the color change is from the green of the chromium(III) ion to the violet color of the oxidized indicator. An indicator blank is not readily obtained because dichromate oxidizes the indicator only slowly in the absence of other oxidation-reduction systems. Ordinarily, however, the error resulting from inability to correct for the blank is vanishingly small. The reaction of diphenylamine sulfonic acid is reversible, and back-titration of small excesses of dichromate with iron(II) is possible. In the presence of large oxidant concentrations and at low acidities (above pH 2), the indicator is irreversibly oxidized to yellow or red compounds.

Directions for the preparation of a standard dichromate solution are given in Chapter 31, Preparation 8-2.

Determination of iron. The principal use of dichromate involves titration of iron(II):

$$6Fe^{2+} + Cr_2O_7^{2-} + 14H^+ \rightleftharpoons 6Fe^{3+} + 2Cr^{3+} + 7H_2O$$

Moderate amounts of hydrochloric acid do not affect the accuracy of the titration. Detailed procedures for the analysis of iron are found in Chapter 31, Method 8-1.

Other applications. A common method for the determination of oxidizing agents calls for treatment of the sample with a known excess of iron(II) followed by titration of the excess with standard dichromate. This technique has been successfully applied to the determination of nitrate, chlorate, permanganate, dichromate, and organic peroxides, among others.

15B-4 IODIMETRIC METHODS

Many volumetric analyses are based on the half-reaction

$$I_3^- + 2e \rightleftharpoons 3I^- \qquad E^0 = 0.536 \text{ V}$$

These analyses fall into two categories. The first comprises procedures that use a standard solution of iodine to titrate easily oxidized substances. These *direct* or *iodimetric methods* have limited applicability because iodine is a relatively weak oxidizing agent. *Indirect* or *iodometric methods* employ a standard solution of sodium thiosulfate or arsenic(III) to titrate the iodine liberated when an oxidizing substance is allowed to react with an unmeasured excess of potassium iodide. The quantity of iodine formed is chemically equivalent to the amount of the oxidizing agent and thus serves as the basis for the analysis.

Iodine, which is a relatively weak oxidant, is used for the selective determination of strong reducing agents. The availability of a sensitive and reversible indicator for iodine is a great advantage. Disadvantages include the low stability of iodine solutions and the incompleteness of reactions between iodine and many reductants.

Preparation of iodine solutions. A saturated aqueous iodine solution is only about 0.001 M at room temperature. Much higher concentrations can be achieved in the presence of iodide ion, however, owing to formation of the soluble triiodide complex.

$$I_2(s) + I^- \rightleftharpoons I_3^- \qquad K = 7.1 \times 10^2$$

Because a large fraction of the iodine in such solutions is present as I_3^-, it would be more proper to refer to them as *triiodide solutions*. As a practical matter, however, they are usually called *iodine solutions* because of the convenience this terminology affords in writing equations and in describing stoichiometric behavior.

The rate at which iodine dissolves in potassium iodide solution is slow, particularly where the iodide concentration is low. As a consequence, it is necessary to dissolve the solid completely in a small amount of a concentrated iodide solution before diluting to the desired volume. All of the element must be dissolved before dilution; otherwise, the normality of the resulting reagent will increase continuously as the remaining iodine slowly passes into solution.

Stability. Iodine solutions require restandardization every few days. This lack of stability has several sources, one being the volatility of the solute. Even though the excess of iodide is large, a measurable amount of iodine is lost from an open container in a relatively short period of time.

Iodine will slowly attack rubber or cork stoppers as well as other organic substances; reasonable precautions must, therefore, be taken to protect standard solutions of the reagent from contact with these materials. Exposure to organic dust and fumes must also be avoided.

Finally, changes in iodine normality result from air oxidation of iodide ions in the solution:

$$4I^- + O_2 + 4H^+ \rightleftharpoons 2I_2 + 2H_2O$$

This reaction is catalyzed by light, heat, and acids; consequently, it is good practice to store the reagent in a dark, cool place. In contrast to the other effects, air oxidation of iodide causes an increase in normality.

Completeness of iodine oxidations. Because iodine is such a weak oxidizing agent, the chemist frequently must take full advantage of those experimental variables that enhance its reduction to iodide by the analyte. Two effects, pH and the presence of complexing agents, are of particular importance.

In acidic solutions, pH has little influence upon the electrode potential of the iodine/iodide couple because hydrogen ions do not participate in the half-reaction. Many of the substances that react with iodine, however, produce hydrogen ions as they are oxidized. The position of equilibrium may therefore be markedly influenced by pH. The arsenic(III)/arsenic(V) system provides an important example. Its electrode potential differs by only 0.02 V from that for the iodide/iodine half-reaction:

$$H_3AsO_4 + 2H^+ + 2e \rightleftharpoons H_3AsO_3 + H_2O \qquad E^0 = 0.559 \text{ V}$$

Arsenic(V) will quantitatively oxidize iodide to iodine in a strongly acidic medium, however, because of the increase in electrode potential for the system resulting from the common ion effect. On the other hand, in a nearly neutral solution trivalent arsenic can be titrated successfully with iodine; here, the reduction in hydrogen ion concentration is accompanied by a marked decrease in the arsenic(V)/arsenic(III) electrode potential. Although iodine oxidations often become more nearly complete with lowered acidity, care must be taken to prevent the formation of hypoiodite, which takes place in alkaline solutions:

$$I_2 + OH^- \rightleftharpoons HOI + I^-$$

The hypoiodite may subsequently disproportionate to iodate and iodide:

$$3HOI + 3OH^- \rightleftharpoons IO_3^- + 2I^- + 3H_2O$$

The occurrence of these reactions will cause serious errors in an iodimetric analysis. In some titrations, the reaction of iodate and hypoiodite with the reducing reagent is so slow that overconsumption of iodine is observed. In others, the presence of these two species can alter the reaction products; the oxidation of thiosulfate is an important example of such behavior. Thus, solutions to be titrated with iodine cannot have pH values much higher than 9. Occasionally, a pH greater than 7 is detrimental.

Complexing reagents are also used to force certain iodine oxidations toward completion. For example, it is readily seen that the reduction potential of iodine is too low to permit the quantitative oxidation of iron(II) to the trivalent state. In the presence of reagents that strongly complex iron(III), however, complete conversion is achieved (see p. 324 for the effect of complexing agents on the electrode potential for iron); pyrophosphate ion or ethylenediaminetetraacetate are useful for this purpose.

End points for iodine titrations. Several sensitive methods exist for determining the end point in an iodine titration. The color of the triiodide ion itself is often sufficiently intense for the titration of colorless solutions. Thus, a concentration of about 5×10^{-6} M triiodide can just be detected by the eye; in a typical titration, this concentration would be produced by an overtitration of less than one drop of 0.1 N iodine solution.

A greater sensitivity can be obtained, at the sacrifice of convenience, by adding a few milliliters of an immiscible organic solvent such as carbon tetrachloride or chloroform to the solution. Shaking causes the bulk of any iodine present to collect in the organic layer and imparts an intense violet color to it.

The most widely used indicator for iodimetry is an aqueous suspension of starch which imparts an intense blue color to a solution containing a trace of triiodide ion. The nature of the colored species has been the subject of much speculation and controversy.[10] It is now believed that the iodine is held as an adsorption complex within the helical chain of the macromolecule β-amylose, a component of most starches. Another component, α-amylose, is undesirable because it produces a red coloration with iodine that is not readily reversible. Interference from α-amylose, is seldom serious, however, because the substance tends to settle rapidly from aqueous suspension. Other starch fractions do not appear to form colored complexes with iodine. Potato, arrowroot, and rice starches contain large proportions of α- and β-amylose and can be employed as indicators. Corn starch is not suitable because of its high content of the former. The so-called *soluble starch* that is commercially available consists principally of β-amylose, with the α-fraction having been removed. Indicator solutions are readily prepared from this product.

Aqueous starch suspensions decompose within a few days, primarily because of bacterial action. The decomposition products may consume iodine as well as interfere with the indicator properties of the preparation. The rate of decomposition can be greatly inhibited by preparing and storing the indicator under sterile conditions and by the introduction of mercury(II) iodide or chloroform to act as a bacteriostat. Alternatively, a fresh indicator suspension can be prepared each day an iodine titration is to be performed.

Starch added to a solution containing a high concentration of iodine is decomposed to products whose indicator properties are not entirely reversible. Thus, addition of the indicator to a solution containing an excess of iodine should be postponed until most of the iodine has been titrated, as indicated by a light yellow color of the solution.

Standardization of iodine solutions.   Iodine solutions may be standardized against arsenious oxide, barium thiosulfate monohydrate, or potassium antimony(III) tartrate.

Arsenious oxide, As_2O_3, is available commercially in primary standard quality and has been the most widely used primary standard for solutions of iodine. Its popularity may decrease, however, as a consequence of its having been labeled as a carcinogen by OSHA. The oxide dissolves only slowly in water or in the common acids; solution occurs rapidly in 1 N NaOH, however.

$$As_2O_3(s) + 4OH^- \longrightarrow 2HAsO_3^{2-} + H_2O$$

In strongly alkaline solution, arsenic(III) is readily air-oxidized to arsenic(V), whereas neutral or slightly acidic solutions are indefinitely stable toward oxygen. Thus, after the arsenious oxide has been dissolved in base, hydrochloric acid should be added immediately until the solution is slightly acidic. Standard solutions of arsenious acid are useful for periodic standardization of iodine solutions.

The iodimetric titration of arsenious acid must be carried out in a buffered system to use up the hydrogen ions formed in the reaction; otherwise, the pH may decrease to a level where the reaction is not complete. Buffering is conveniently accomplished by acidifying the sample slightly and then saturating with sodium

---

[10] See: R. E. Rundle, J. F. Foster, and R. R. Baldwin, *J. Amer. Chem. Soc.*, **66**, 2116 (1944).

hydrogen carbonate. The carbonic acid/hydrogen carbonate buffer so established will hold the pH in a range between 7 and 8.

Directions for the standardization of iodine solutions against arsenic(III) are given in Chapter 31, Standardization 9-2.

MacNevin and Kriege have demonstrated that barium thiosulfate monohydrate, $BaS_2O_3 \cdot H_2O$ is a suitable substitute for arsenious oxide for the standardization of iodine.[11] The salt is soluble to the extent of only 0.01 M, but the solid reacts so rapidly with iodine solutions that a direct titration is entirely feasible. The reaction is

$$I_2 + 2BaS_2O_3 \cdot H_2O(s) \rightleftharpoons S_4O_6^{2-} + 2Ba^{2+} + 2I^- + H_2O$$

MacNevin and Kriege showed that barium thiosulfate monohydrate with a purity of 99.85% is readily prepared and is stable at room temperature. They also found that the compound from commercial sources assayed 99.5%.[12]

Barium thiosulfate monohydrate begins to lose water at about 50°C; the anhydrous salt is unsuitable for standardization because of its low solubility. The monohydrate should be used without drying.

Directions for standardization of iodine with $BaS_2O_3 \cdot H_2O$ are given in Chapter 31, Standardization 9-1.

Applications of standard iodine. Common analyses that make use of iodine as an oxidizing reagent are summarized in Table 15-6. A procedure for the determination of antimony in a sulfide ore is given in Chapter 31, Method 9-1.

15B-5 POTASSIUM BROMATE SOLUTIONS AS A SOURCE OF BROMINE

Primary standard-grade potassium bromate is available commercially. Its solutions are stable indefinitely.

Direct titrations with potassium bromate are relatively limited. Its principal analytical value is as a convenient and stable source of bromine. In this application, an excess of potassium bromide is added to an acidic solution of the analyte. The

---

[11] W. M. MacNevin and O. H. Kriege, *Anal. Chem.*, **25**, 767 (1953).

[12] Primary standard grade $Ba_2S_2O_3 \cdot H_2O$ assaying 99.9% or better is available from G. Frederick Smith Chemical Co., 867 McKinley Avenue, Columbus, Ohio, 43223.

Table 15-6 Analysis with Standard Iodine Solutions

Substance Analyzed	Half-Reaction
As	$H_3AsO_3 + H_2O \rightleftharpoons H_3AsO_4 + 2H^+ + 2e$
Sb	$H_3SbO_3 + H_2O \rightleftharpoons H_3SbO_4 + 2H^+ + 2e$
Sn	$Sn^{2+} \rightleftharpoons Sn^{4+} + 2e$
H_2S	$H_2S \rightleftharpoons S(s) + 2H^+ + 2e$
SO_2	$SO_3^{2-} + H_2O \rightleftharpoons SO_4^{2-} + 2H^+ + 2e$
$S_2O_3^{2-}$	$2S_2O_3^{2-} \rightleftharpoons S_4O_6^{2-} + 2e$
N_2H_4	$N_2H_4 \rightleftharpoons N_2 + 4H^+ + 4e$
Ascorbic acid*	$C_6H_8O_6 \rightarrow C_6H_6O_6 + 2H^+ + 2e$

* For the structure of ascorbic acid, see Problem 17-29.

addition of standard bromate then releases an equivalent quantity of bromine for reaction with the analyte. That is,

$$BrO_3^- + 5Br^- + 6H^+ \longrightarrow 3Br_2 + 3H_2O$$

standard   excess
solution

This indirect approach eliminates the principal disadvantage associated with the use of standard bromine solutions, namely, lack of stability. Note that each bromate is responsible for the formation of three molecules of bromine, which in turn require six electrons for reduction to bromide. That is,

$$BrO_3^- \equiv 3Br_2 \equiv 6e$$

The equivalent weight of potassium bromate in this application is thus one-sixth of its formula weight.

Indicators for titrations involving bromine. Several organic indicators such as methyl orange and methyl red (Table 8-1) are readily brominated to yield products that differ in color from the original compounds. Unfortunately these reactions are totally nonreversible, which precludes any sort of back-titration. Moreover, direct titrations with these indicators are inconvenient because of the great need to avoid local excesses of the reagent.

Three indicators, α-naphthoflavone, p-ethoxychrysoidine, and quinoline yellow, are reversible with respect to bromine and make employment of the reagent more attractive; they are commercially available.

Applications of standard potassium bromate solutions. Potassium bromate is a convenient source of bromine in organic analysis. Few organic compounds react sufficiently rapidly for a direct titration. Instead, a measured excess of the bromate solution is added to the sample; after the addition of sufficient potassium bromide to convert all of the bromate to bromine, the mixture is acidified and allowed to stand until bromination is judged complete. The excess bromine is then back-titrated with a standard arsenic(III) solution. Alternatively, the analysis can be completed iodometrically by adding an excess of potassium iodide and titrating the iodine liberated with a standard solution of sodium thiosulfate (pp. 368–369). Although the iodometric procedure appears complicated, it is actually quite simple.

In general, bromine is incorporated into an organic molecule either by substitution or by addition.

Substitution reactions. Substitution involves the replacement of hydrogen in an aromatic ring by atoms of the halogen. For example, three hydrogen atoms are replaced when phenol is brominated:

Here, each molecule of phenol requires six atoms of bromine, each of which can be considered to undergo a one-electron change to the -1 state; thus, the equivalent weight of phenol is one-sixth of its formula weight. Table 15-7 lists typical organic compounds that can be determined by bromination.

Substitution methods have been applied successfully to the analysis of aromatic compounds that contain strong *ortho-para* directing groups in the ring, particularly amines and phenols. An important application of the method is the titration of 8-hydroxyquinoline:

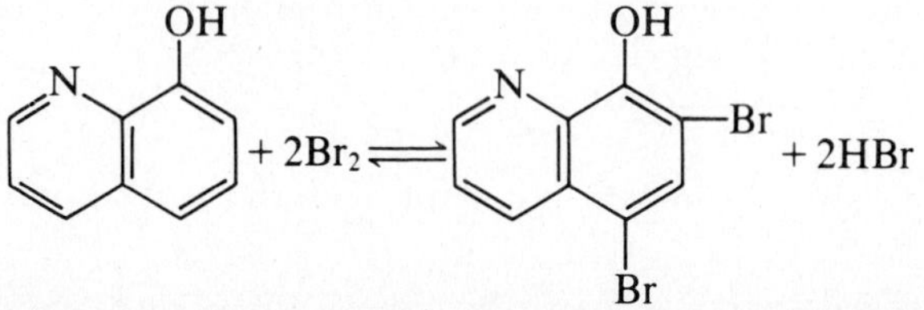

The reaction, which is sufficiently rapid in hydrochloric acid solution to permit direct titration, is of particular interest because 8-hydroxyquinoline is an excellent precipitating reagent for cations (Section 5D-3). Thus, for example, aluminum can be determined by means of the following reaction sequence:

$$Al^{3+} + 3HOC_9H_6N \xrightarrow{pH\ 4-9} Al(OC_9H_6N)_3(s) + 3H^+$$
$$\text{(filter and wash)}$$

$$Al(OC_9H_6N)_3(s) + 3H^+ \xrightarrow[HCl]{hot\ 4\ M} 3HOC_9H_6N + Al^{3+}$$

$$3HOC_9H_6N + 6Br_2 \longrightarrow 3HOC_9H_4NBr_2 + 6HBr$$

Each mole of aluminum is indirectly responsible for the reaction of 6 moles (or 12 equivalents) of molecular bromine; the equivalent weight of the metal is thus one-twelfth of its formula weight.

Table 15-7 Some Organic Compounds That Can Be Analyzed by Bromine Substitution[*]

Compound	Reaction Time, min	Equivalents Br$_2$/mol Compound	Accuracy (percent theoretical)
Phenol	5–30	6	99.89
p-Chlorophenol	30	4	99.87
Salicylic acid	30	6	99.86
Acetylsalicylic acid	30	6	99.84
m-Cresol	1	6	99.75
β-Naphthol	15–20	2	99.89
Aniline	5–10	6	99.92
o-Nitroaniline	30	4	99.86
Sulfanilic acid	30	6	99.98
m-Toluidine	5–10	6	99.87

[*] Data from A. R. Day and W. T. Taggart, *Ind. Eng. Chem.*, **20**, 545 (1928). With permission of the American Chemical Society.

Addition reactions. Addition reactions involve the opening of an olefinic double bond. For example, one mole (or two equivalents) of molecular bromine reacts with each mole of ethylene:

$$\begin{array}{cc} \text{H} & \text{H} \\ | & | \\ \text{H}-\text{C}{=}\text{C}-\text{H} + \text{Br}_2 \longrightarrow \\ \end{array} \quad \begin{array}{cc} \text{H} & \text{H} \\ | & | \\ \text{H}-\text{C}-\text{C}-\text{H} \\ | & | \\ \text{Br} & \text{Br} \end{array}$$

The equivalent weight of ethylene will therefore be one-half of its formula weight.

A variety of methods, many involving the use of bromate/bromide mixtures, are found in the literature[13]; most of these were developed to estimate olefinic unsaturation in fats, oils, and petroleum products.

A procedure for the determination of phenol in waste water by bromine substitution is found in Chapter 31, Method 11-1. In addition, Method 11-2 provides a procedure for the determination of ascorbic acid in vitamin C tablets by bromine oxidation.

15B-6 POTASSIUM IODATE

Potassium iodate is commercially available in a high state of purity and can be used without further treatment, other than drying, for the direct preparation of standard solutions. Iodate solutions, which are stable indefinitely, have a number of interesting and important uses in analytical chemistry.

Reactions of iodate. The reaction of iodate with iodide is analogous to the bromate/bromide reaction. Thus, the position of equilibrium for the reaction

$$\text{IO}_3^- + 5\text{I}^- + 6\text{H}^+ \rightleftharpoons 3\text{I}_2 + 3\text{H}_2\text{O}$$

lies far to the right in acidic solutions and far to the left in basic media.

This reaction is a convenient source for known amounts of iodine (see, for example, p. 388). A measured quantity of iodate is mixed with an excess of iodide in a solution that is 0.1 to 1 N in acid. Exactly six equivalents of iodine are liberated for each mole of iodate; the resulting solution can then be used for the standardization of thiosulfate solutions or for other analytical purposes.

In strongly acidic solutions, iodate will oxidize iodide or iodine to the $+1$ state, provided some anion such as chloride, bromide, or cyanide is present to stabilize this oxidation state. For example, in solutions that are greater than 3 M in hydrochloric acid, the following reaction proceeds essentially to completion:

$$\text{IO}_3^- + 2\text{I}_2 + 10\text{Cl}^- + 6\text{H}^+ \rightleftharpoons 5\text{ICl}_2^- + 3\text{H}_2\text{O}$$

A number of important iodate titrations are performed in concentrated hydrochloric acid, where the iodate is initially reduced to iodine. As the reducing agent is consumed, however, the reaction described in the previous paragraph takes place. Generally, the end point for the process is signaled by the complete disappearance of the iodine. The equivalent weight for the iodate is one-fourth of

[13] For example, see: A. Polgar and J. L. Jungnickel, *Organic Analysis*, vol. 3. New York: Interscience, 1956.

its formula weight because the final reaction product is iodine in the +1 state. When used in this manner, iodate is a less powerful oxidant than either permanganate or cerium(IV).

End points in iodate titrations. The disappearance of iodine from the solution is often sufficient to indicate the end point in an iodate titration. Iodine is nearly always formed in the initial stages of the reaction; only at the equivalence point is it completely oxidized to the +1 state. Starch fails to function as an indicator in the highly acidic medium required for the production of ICl_2^-. Instead, a few milliliters of an immiscible organic solvent such as toluene, carbon tetrachloride, or chloroform are added at the start of the titration. After each addition of iodate the mixture is shaken thoroughly; the organic layer is examined after the phases have separated. The bulk of any unreacted iodine remains in the organic layer and imparts a violet-red color to it. The titration is judged complete when the minimum amount of iodate needed to discharge this color has been added. This excellent method for detecting iodine is quite comparable in sensitivity to the starch/iodine color; it suffers, however, from the disadvantage of being more time-consuming.

Directions for the iodate titration of iodide and iodine in solutions are given in Chapter 31, Method 12-1; further applications can be found in reference sources.[14]

15B-7 PERIODIC ACID

Aqueous solutions of +7 iodine are highly complex.[15] In strongly acidic solutions, paraperiodic acid, H_5IO_6, and its conjugate base predominate, although the metaperiodates HIO_4 and IO_4^- are undoubtedly present as well. Solutions of periodic acid are strong oxidizing agents. Thus,

$$H_5IO_6 + H^+ + 2e \rightleftharpoons IO_3^- + 3H_2O \qquad E^0 = 1.6 \text{ V}$$

Preparation and properties of periodic acid solutions. Several periodates are available for the preparation of standard solutions. Among these is paraperiodic acid itself, a crystalline, readily soluble, hygroscopic solid. An even more useful compound is sodium metaperiodate, $NaIO_4$, which is soluble in water to the extent of 0.06 M at 25°C. Sodium paraperiodate, Na_5IO_6, is not sufficiently soluble for the preparation of standard solutions; however, it is readily converted to the more soluble metaperiodate by recrystallization from hot concentrated nitric acid. Potassium metaperiodate can be used as a primary standard for the preparation of periodate solutions.[16] At room temperature, its solubility is only about 5 g/L. It readily dissolves, however, at elevated temperatures and can be subsequently converted to a more soluble form by the addition of base.

Periodate solutions vary considerably in stability, depending on their mode of preparation and storage. A solution prepared by dissolving sodium metaperiodate in water decomposes at the rate of several percent per week. On the other

[14] For details on the application of iodate titrations, see: I. M. Kolthoff and R. Belcher, *Volumetric Analysis*, vol. 3, pp. 449–473. New York: Interscience, 1957.

[15] For discussion of the composition of acidic solutions of periodate, see: C. E. Crouthamel, A. M. Hayes, and D. S. Martin, *J. Amer. Chem. Soc.*, **73**, 82 (1951).

[16] H. H. Willard and L. H. Greathouse, *J. Amer. Chem. Soc.*, **60**, 2869 (1938).

hand, a solution of potassium metaperiodate in excess alkali was found to change no more than 0.3 to 0.4% in 100 days. The most stable periodate solutions appear to be those containing an excess of sulfuric acid; such solutions decrease in normality by less than 0.1% in four months.[16]

Standardization of periodate solutions. Periodate solutions are most conveniently standardized by adding excess iodide ion and a slightly alkaline buffer, such as borax or hydrogen carbonate, to an aliquot of the reagent; iodine is liberated by the following reaction:

$$H_4IO_6^- + 2I^- \longrightarrow IO_3^- + I_2 + 2OH^- + H_2O$$

As long as the solution is kept neutral, further reduction of iodate does not occur, and the liberated iodine can be titrated directly with a standard arsenite solution.

Applications of periodic acid. The most important applications of periodic acid involve the selective oxidation of organic compounds containing certain combinations of functional groups.[17] Ordinarily, these oxidations are performed at room temperature in the presence of a measured excess of the periodate; most reactions are complete in 30 min to 1 hr. After oxidation, the excess periodate is determined by the method described for standardization. Alternatively, a reaction product such as ammonia, formaldehyde, or a carboxylic acid may be determined; here, the exact quantity of periodate used need not be known.

Periodate oxidations are usually carried out in aqueous solution although solvents such as methanol, ethanol, or dioxane may be added to enhance the solubility of the sample.

Compounds attacked by periodate. At room temperature organic compounds containing aldehyde, ketone, or alcohol functional groups *on adjacent carbon atoms* are rapidly oxidized by periodic acid. Primary and secondary α-hydroxylamines are also readily attacked; α-diamines are not. With a few exceptions, other organic compounds do not react at a significant rate. Thus, compounds containing isolated hydroxyl, carboxyl, or amine groups are not affected by periodic acid; nor are compounds with a carboxylic acid group either isolated from or adjacent to any of the reactive groups. At elevated temperatures, the extraordinary selectivity of periodic acid tends to disappear.

Periodate oxidations of organic compounds follow a regular and predictable pattern; the following rules apply:

1. Attack of adjacent functional groups always results in rupture of the carbon-to-carbon bond between these groups.
2. A carbon atom containing a hydroxyl group is oxidized to an aldehyde or ketone.
3. A carbonyl group is converted to a carboxylic acid group.
4. A carbon atom containing an amine group loses ammonia (or a substituted amine) and is itself converted to an aldehyde.

[17] The use of periodic acid for this purpose was first studied by L. Malaprade, *Compt. rend.*, **186**, 382 (1928).

The following half-reactions will illustrate these rules:

$$\underset{\text{propylene glycol}}{CH_3{-}\overset{\displaystyle OH}{\underset{\displaystyle H}{C}}{-}\overset{\displaystyle OH}{\underset{\displaystyle H}{C}}{-}H} \longrightarrow CH_3\overset{\displaystyle O}{C}{-}H + H{-}\overset{\displaystyle O}{C}{-}H + 2H^+ + 2e$$

$$\underset{\text{glycerol}}{H{-}\overset{\displaystyle OH}{\underset{\displaystyle H}{C}}{-}\overset{\displaystyle OH}{\underset{\displaystyle H}{C}}{-}\overset{\displaystyle OH}{\underset{\displaystyle H}{C}}{-}H} + H_2O \longrightarrow 2H{-}\overset{\displaystyle O}{C}{-}H + H{-}\overset{\displaystyle O}{C}{-}OH + 4H^+ + 4e$$

For purposes of predicting the reaction products, the first step in the glycerol oxidation can be thought of as producing one mole of formaldehyde and one mole of an α-hydroxyaldehyde (glycolic aldehyde). The latter is then further oxidized and by the second rule produces a second mole of formaldehyde plus one mole of formic acid.

$$\underset{\text{biacetyl}}{CH_3{-}\overset{\displaystyle O}{C}{-}\overset{\displaystyle O}{C}{-}CH_3} + 2H_2O \longrightarrow 2CH_3\overset{\displaystyle O}{C}{-}OH + 2H^+ + 2e$$

$$\underset{\text{acetoin}}{CH_3{-}\overset{\displaystyle O}{C}{-}\overset{\displaystyle OH}{C}{-}CH_3} + H_2O \longrightarrow CH_3\overset{\displaystyle O}{C}{-}OH + CH_3\overset{\displaystyle O}{C}{-}H + H^+ + e$$

$$\underset{\text{ethanolamine}}{H{-}\overset{\displaystyle OH}{\underset{\displaystyle H}{C}}{-}\overset{\displaystyle NH_2}{\underset{\displaystyle H}{C}}{-}H} + H_2O \longrightarrow 2H{-}\overset{\displaystyle O}{C}{-}H + NH_3 + 2H^+ + 2e$$

Analysis of glycerol. As noted in the preceding paragraph, glycerol is cleanly oxidized by periodic acid, with two moles of formaldehyde and one mole of formic acid being produced. About one-half hour is required to complete the reaction; the glycerol is readily estimated from the amount of periodate consumed. A procedure for this analysis is found in Chapter 31, Method 13-1.

Determination of α-hydroxylamines. As mentioned earlier, periodate oxidation of compounds having hydroxyl and amino groups on adjacent carbon atoms results in the formation of aldehydes and the liberation of ammonia. The latter is readily distilled from the alkaline oxidation mixture and determined by a neutralization titration. This procedure is particularly useful in the analysis of mixtures of the various amino acids that occur in proteins. Only serine, threonine, β-hydroxyglutamic acid, and hydroxylysine have the requisite structure for liberation of ammonia; the method is thus selective for these compounds.

15C APPLICATION OF REDUCTANTS

Solutions of reducing agents are often troublesome to use because of the readiness with which they react with atmospheric oxygen. As a consequence, titrations ordinarily must be carried out in and the reagents must be stored under an inert atmosphere. Alternatively, a stable standard oxidizing agent such as potassium dichromate is used as the reference substance; typically, then, an aliquot containing an excess of the reductant is added to the sample and the excess is quickly back-titrated with the standard oxidant. The current concentration of the reductant is determined by a similar titration of a blank. Very strong reducing agents such as titanium(III) and chromium(II) react too rapidly with oxygen for use in this manner; for these, a blanket of an inert gas such as N_2 or CO_2 must be employed.

Table 15-8 lists common reductants for the determination of oxidizing substances.

15C-1 SOLUTIONS OF IRON(II)

Solutions of iron(II) are readily prepared from *Mohr's salt*, $Fe(NH_4)_2(SO_4)_2 \cdot 6H_2O$, or *Oesper's salt*, $FeC_2H_4(NH_3)_2(SO_4)_2 \cdot 4H_2O$ [iron(II) ethylenediammonium sulfate].[18] Air oxidation of the iron(II) proceeds rapidly in neutral solution but is inhibited by acid; the most stable solutions are about 0.5 M in H_2SO_4. Such solutions should be standardized daily.

A variety of oxidizing agents such as Cr(VI), Ce(IV), Mo(VI), NO_3^-, NH_2OH,

[18] Both of these standards are available from the source listed in footnote 12.

Table 15-8 Some Common Reductants Employed for Standard Solutions

Reagent and Common Half-Reaction	Oxidation Potential,* V	Primary Standards for	Stability of Solution
Iron(II) $Fe^{2+} \rightarrow Fe^{3+} + e$	−0.77	Fe, $K_2Cr_2O_7$	Unstable unless protected from oxygen
Arsenic(III) $H_3AsO_3 + H_2O \rightarrow$ $H_3AsO_4 + 2H^+ + 2e$	−0.56	As_2O_3, I_2	Indefinitely stable if acidic
Sodium thiosulfate $2S_2O_3^{2-} \rightarrow S_4O_6^{2-} + 2e$	−0.08	KIO_3, I_2	Frequent standardization required
Titanium(III) $Ti^{3+} + H_2O \rightarrow$ $TiO^{2+} + 2H^+ + e$	−0.10	$K_2Cr_2O_7$	Highly unstable unless protected from oxygen

* Note that these are potentials for the reactions written as oxidations; the sign, therefore, is the *opposite* of the corresponding electrode potential. The reaction with the smallest negative potential has the greatest tendency to occur. Thus Ti(III) is a considerably better reducing agent than is iron(II).

and organic peroxides are conveniently determined by reaction with a measured excess of a standard iron(II) solution; as mentioned earlier, standard potassium dichromate is frequently employed for the back-titration.

15C-2 SODIUM THIOSULFATE; IODOMETRIC METHODS

Iodide ion is a moderately effective reducing agent that has been widely employed for the analysis of oxidants. In these applications, a standard solution of sodium thiosulfate (or occasionally arsenious acid) is used to titrate the iodine liberated by reaction of the analyte with an unmeasured excess of potassium iodide.

The reaction of iodine with thiosulfate ion. The reaction between iodine and thiosulfate is described by the equation

$$2S_2O_3^{2-} + I_2 \rightleftharpoons S_4O_6^{2-} + 2I^-$$

The production of the tetrathionate ion requires the loss of two electrons from two thiosulfate ions; the equivalent weight of thiosulfate in this reaction must therefore be equal to its formula weight.

The quantitative conversion of thiosulfate to tetrathionate ion is unique with iodine; other oxidizing reagents tend to carry the oxidation, wholly or in part, to sulfate ion. The reaction with hypoiodous acid provides an important example of this stoichiometry:

$$4HOI + S_2O_3^{2-} + H_2O \rightleftharpoons 2SO_4^{2-} + 4I^- + 6H^+$$

The presence of hypoiodite in slightly alkaline solutions of iodine (p. 376), then, will seriously upset the stoichiometry of the iodine/thiosulfate reaction, causing too little thiosulfate or too much iodine to be used in the titration.

The equilibrium constant for the reaction

$$I_2 + H_2O \rightleftharpoons HOI + I^- + H^+$$

is small (about 3×10^{-13}).[19] It can be shown, however, that hypoiodite formation should become significant in media where the pH is greater than 7. Kolthoff[20] has found that an error of about 4% will occur when a solution containing 25 mL of 0.1 N iodine and 0.5 g of sodium carbonate (pH $\sim$ 11.5) is titrated with 0.1 N thiosulfate; this error becomes 10% or more in solutions containing about 2 g of this salt. Kolthoff recommends that the pH always be less than 7.6 for the titration of 0.1 N solutions, 6.5 or less for 0.01 N solutions, and less than 5 for 0.001 N solutions.

The titration of highly acidic iodine solutions with thiosulfate yields quantitative results, provided care is taken to prevent air oxidation of iodide ion.

The end point in the titration is readily established by means of a starch solution (p. 376). It should be emphasized again that starch is partially decomposed in the presence of a large excess of iodine; it is for this reason that the indicator is not added until the bulk of the iodine has been reduced. The proper moment for addition is signaled by a color change of the solution from red-brown to faint yellow.

[19] W. C. Bray and E. L. Connolly, *J. Amer. Chem. Soc.*, **33**, 1485 (1911).
[20] I. M. Kolthoff and R. Belcher, *Volumetric Analysis*, vol. 3, pp. 214–215. New York: Interscience, 1957.

Stability of thiosulfate solutions. Principal among the variables affecting the stability of thiosulfate solutions are the pH, the presence of microorganisms and impurities, the concentration of the solution, the presence of atmospheric oxygen, and exposure to sunlight. The change in iodine titer may amount to as much as several percent in a few weeks. Proper attention to detail will yield standard thiosulfate solutions that need only occasional restandardization.

The following reaction occurs at an appreciable rate when the pH of a thiosulfate solution is 5 or less:

$$S_2O_3^{2-} + H^+ \rightleftharpoons HS_2O_3^- \longrightarrow HSO_3^- + S(s)$$

The velocity of this disproportionation increases with the hydrogen ion concentration; in a strongly acidic solution, elemental sulfur forms within a few seconds. This decomposition may result in either a decrease or an increase in iodine titer. The hydrogen sulfite ion produced is also oxidized by iodine, reacting with twice the quantity of that reagent as the thiosulfate from which it was derived. On the other hand, sulfite ion is readily air oxidized to the unreactive sulfate ion; when this reaction occurs to a significant extent, a decrease in normality results.

Clearly, thiosulfate solutions cannot be allowed to stand in contact with acid. On the other hand, iodine solutions that are 3 to 4 M in acid may be titrated without error as long as care is taken to introduce the thiosulfate slowly and with good mixing. Under these conditions, the thiosulfate is oxidized so rapidly by the iodine that the slower acid decomposition cannot occur to any measurable extent.

Experiments indicate that the stability of thiosulfate solutions is at a maximum in the pH range between 9 and 10, although for most purposes a pH of 7 is adequate. The addition of a small amount of a base such as sodium carbonate, borax, or disodium hydrogen phosphate is frequently recommended to preserve standard solutions of the reagent. If this procedure is followed, the iodine solutions to be titrated must be sufficiently acidic to neutralize the added base. Otherwise, hypoiodite formation may occur before the equivalence point is attained and cause the partial oxidation of the thiosulfate to sulfate.

The most important single cause of instability can be traced to certain bacteria that metabolize the thiosulfate ion, converting it to sulfite, sulfate, and elemental sulfur.[21] Solutions that are free of bacteria are remarkably stable; it is common practice, therefore, to impose reasonably sterile conditions in the preparation of standard solutions. Substances such as chloroform, sodium benzoate, or mercury(II) iodide can be added to inhibit bacterial growth. Bacterial activity appears to be at a minimum at a pH between 9 and 10, which accounts, at least in part, for the maximum stability of thiosulfate solutions in this range.

Many other variables affect the stability of thiosulfate solutions. Decomposition is reported to be catalyzed by copper(II) ions as well as by the decomposition products themselves. Solutions that have become turbid from the formation of sulfur should be discarded. Exposure to sunlight increases the rate of decomposition as does atmospheric oxygen. Finally, the decomposition rate is greater in more dilute solutions.

[21] M. Kilpatrick, Jr., and M. L. Kilpatrick, *J. Amer. Chem. Soc.*, **45**, 2132 (1923); F. O. Rice, M. Kilpatrick, Jr., and W. Lemkin, *J. Amer. Chem. Soc.*, **45**, 1361 (1923).

Standardization of thiosulfate solutions. Potassium iodate is an excellent primary standard for thiosulfate solutions. Iodate ion reacts rapidly with iodide in slightly acidic solution to give iodine.

$$IO_3^- + 5I^- + 6H^+ \rightleftharpoons 3I_2 + 3H_2O$$

$$3I_2 + 6e \rightleftharpoons 6I^-$$

Three moles of iodine are furnished by each formula weight of potassium iodate; its equivalent weight is thus one-sixth its formula weight because a six-electron change is associated with the reduction of three iodine molecules, the species actually titrated. That is,

$$IO_3^- \equiv 3I_2 \equiv 6I^- \equiv 6e$$

The sole disadvantage of potassium iodate as a primary standard is its low equivalent weight (35.67). Only slightly more than 0.1 g can be used for standardization of a 0.1 N thiosulfate solution; the relative error to be expected in weighing this quantity may, under some circumstances, be somewhat greater than that desirable for a standardization. As an alternative, a larger quantity of the solid can be dissolved in a known volume; aliquots of the resulting solution can then be taken. This procedure suffers from the disadvantage that it provides no duplicate check on the precision of the solution preparation process.

Other primary standards for sodium thiosulfate include potassium dichromate, potassium bromate, potassium hydrogen iodate [$KH(IO_3)_2$], potassium ferricyanide, and metallic copper. Details for the use of these may be found in various reference works.[22]

Directions for the preparation and standardization of a thiosulfate solution are given in Chapter 31, Standardizations 10-1 and 10-2.

Errors in iodometric methods. Three sources of errors in iodometric methods have already been mentioned, namely, the decomposition of thiosulfate solutions, the alteration of the stoichiometric relationship between iodine and thiosulfate ion in the presence of base, and the premature addition of starch. In addition, care must be taken to avoid loss of iodine by volatilization; significant losses can generally be prevented by using stoppered containers when solutions must stand, by maintaining a goodly excess of iodide ion, and by avoiding elevated temperatures.

Air oxidation of iodide ion can also be a serious source of error in iodometric analyses (p. 375). Clearly, the reaction is favored by hydrogen ions; from the standpoint of minimizing this source of error, reactions should be carried out at low acidities. Fortunately, the air oxidation is slow under many circumstances. It is, however, catalyzed by acid, light, traces of copper(I), and nitrogen oxides. The latter two are thus potential interferences in any iodometric procedure.

Applications of the indirect iodometric methods. Numerous substances can be determined iodometrically; some of the more common applications are summarized in Table 15-9. Detailed instructions for the iodometric determination of copper, a typical example of the indirect procedure, are found in Chapter 31,

---

[22] For example, see: I. M. Kolthoff and R. Belcher, *Volumetric Analysis*, vol. 3, pp. 234–243. New York: Interscience, 1957.

Methods 10-1 and 10-2. Method 10-3 provides instructions for a widely used method for the iodometric determination of oxygen in natural waters.

15C-3 THE KARL FISCHER REAGENT FOR WATER DETERMINATION
A number of chemical methods for the determination of water in solids and organic solvents have been devised. Unquestionably the most important of these involves the use of Karl Fischer reagent, which is relatively specific for water.[23]

Reaction and stoichiometry. Karl Fischer reagent is composed of iodine, sulfur dioxide, pyridine, and methanol. Upon addition of this reagent to water, the following reactions occur:

$$C_5H_5N \cdot I_2 + C_5H_5N \cdot SO_2 + C_5H_5N + H_2O \longrightarrow$$
$$2C_5H_5N \cdot HI + C_5H_5N \cdot SO_3 \quad (15\text{-}1)$$

$$C_5H_5N \cdot SO_3 + CH_3OH \longrightarrow C_5H_5N(H)SO_4CH_3 \quad (15\text{-}2)$$

Only the first step, which involves the oxidation of sulfur dioxide by iodine to give sulfur trioxide and hydrogen iodide, consumes water. In the presence of a large amount of pyridine, C_5H_5N, all reactants and products exist as complexes, as indicated in the equations.

The second step in the reaction occurs when an excess of methanol is present and is important to the success of the titration, for the pyridine/sulfur trioxide complex is also capable of consuming water:

$$C_5H_5N \cdot SO_3 + H_2O \longrightarrow C_5H_5NHSO_4H \quad (15\text{-}3)$$

This last reaction is undesirable because it is not as specific for water as the reaction shown in Equation 15-1; it can be prevented completely by having a large excess of methanol present.

Equation 15-1 indicates that the stoichiometry of the Karl Fischer titration involves one mole of iodine, one mole of sulfur dioxide, and three moles of pyri-

[23] For reviews on this subject, see: J. Mitchell, *Anal. Chem.*, **23**, 1069 (1951); J. Mitchell and D. M. Smith, *Aquametry*, 2d ed. New York: Interscience, 1977.

Table 15-9   Some Applications of the Iodometric Method

Substance	Half-Reaction	Special Condition
IO_4^-	$IO_4^- + 8H^+ + 7e \rightarrow \frac{1}{2}I_2 + 4H_2O$	Acidic solution
	$IO_4^- + 2H^+ + 2e \rightarrow IO_3^- + H_2O$	Neutral solution
IO_3^-	$IO_3^- + 6H^+ + 5e \rightarrow \frac{1}{2}I_2 + 3H^+$	Strong acid
BrO_3^-, ClO_3^-	$XO_3^- + 6H^+ + 6e \rightarrow X^- + 3H_2O$	Strong acid
Br_2, Cl_2	$X_2 + 2I^- \rightarrow I_2 + 2X^-$	
NO_2^-	$HNO_2 + H^+ + e \rightarrow NO + H_2O$	
Cu^{2+}	$Cu^{2+} + I^- + e \rightarrow CuI(s)$	
O_2	$O_2 + 4Mn(OH)_2(s) + 2H_2O \rightarrow 4Mn(OH)_3(s)$	Basic solution
	$Mn(OH)_3(s) + 3H^+ + e \rightarrow Mn^{2+} + 3H_2O$	Acidic solution
O_3	$O_3 + 2H^+ + 2e \rightarrow O_2 + H_2O$	
Organic peroxide	$ROOH + 2H^+ + 2e \rightarrow ROH + H_2O$	

dine for each mole of water. In practice, excesses of both sulfur dioxide and pyridine are employed so that the combining capacity of the reagent for water is determined by its iodine content.

End-point detection. The end point in the Karl Fischer titration is signaled by the appearance of the first excess of the pyridine/iodine complex when all water has been consumed. The color of the reagent is intense enough for a visual end point; the change is from the yellow of the reaction products to the brown of the excess reagent. With some practice, and in the absence of other colored materials, the end point can be established with a reasonable degree of certainty (that is, to perhaps ±0.2 mL).

Various electrometric end points are also employed for the Karl Fischer titration, the most widely used being the "dead stop" technique described in Section 18C.

Stability of the reagent. As mentioned earlier, the Karl Fischer reagent is prepared so that its combining capacity for water is determined by the concentration of iodine in the solution. For typical application, the titer is about 3.5 mg of water per milliliter of reagent; a twofold excess of sulfur dioxide and a threefold to fourfold excess of pyridine are provided. Stabilized Karl Fischer reagent can be purchased from commercial sources.

The titer of the Karl Fischer reagent decreases with standing. Decomposition is particularly rapid immediately after preparation; it is therefore good practice to prepare the reagent a day or two before it is to be used. Ordinarily, its titer should be established at least daily against a standard solution of water in methanol.

It is obvious that great care must be exercised to prevent contamination of the Karl Fischer reagent and the sample by atmospheric moisture. All glassware must be carefully dried before use, and the standard solution must be stored out of contact with air. It is also necessary to minimize contact between the atmosphere and the solution during the titration.

Applications. The Karl Fischer reagent has been applied to the determination of water in numerous substances.[24] The techniques employed vary considerably, depending upon the solubility of the material, the state in which the water is retained, and the physical state of the sample. If the sample can be dissolved completely in methanol, a direct and rapid titration is feasible. This method has been applied to the analysis of water in many organic acids, alcohols, esters, ethers, anhydrides, and halides. Hydrated salts of most organic acids, as well as hydrates of a number of inorganic salts that are soluble in methanol, can also be analyzed by direct titration.

If it is impossible to produce a solution of the sample and reagent, direct titration ordinarily results in an incomplete reaction. Satisfactory results can frequently be obtained, however, by addition of an excess of reagent and back-titration with a standard solution of water in methanol after a suitable reaction time. An often effective alternative is to extract the water from the sample with an anhydrous solution of methanol or perhaps some other organic solvent; the rate of

---

[24] For a complete discussion of the applications of the reagent, see: J. Mitchell and D. M. Smith, *Aquametry*, 2d ed. New York: Interscience, 1977.

transfer of moisture is increased by refluxing the mixture of sample and solvent. The methanol can then be titrated directly with the Karl Fischer solution.

Difficulty is also encountered in the analysis of sorbed moisture and tightly bound hydrate water. For these, the preceding extraction techniques are frequently effective.

Certain substances interfere with the Fischer method. Among these are compounds that react with one of the components of the reagent to produce water. For example, carbonyl compounds combine with methanol to give acetals:

$$RCHO + 2CH_3OH \longrightarrow R-CH\begin{smallmatrix}OCH_3 \\ \\ OCH_3\end{smallmatrix} + H_2O$$

The result is a fading end point in the titration. Many metal oxides will react with the hydrogen iodide formed in the titration to give water:

$$MO + 2HI \rightleftharpoons MI_2 + H_2O$$

Again, erroneous data result. Preliminary treatment of the sample can sometimes prevent these interferences.

Oxidizing or reducing substances frequently interfere with the Karl Fischer water titration by reoxidizing the iodide produced or reducing the iodine in the reagent.

PROBLEMS*

* 15-1. Write balanced equations for the following processes:
  (a) the oxidation of oxalate ions to carbon dioxide by a solution of vanadium(V). Assume VO^{2+} is the reduction product.
  (b) the reduction of iron(III) to iron(II) with hydrogen sulfide (elemental S is formed).
  (c) the oxidation of Mn^{2+} to MnO_4^- with sodium bismuthate, $NaBiO_3$ (reduction product: BiO^+).
  (d) the oxidation of uranium(IV) to uranium(VI) by permanganate ion in neutral solution (reduction product: MnO_2).
  (e) the air oxidation of iron(II) in acidic solution.
  (f) the oxidation of hydrogen peroxide by chlorine.
  (g) the reduction of molybdenum(VI) in a Walden reductor.

15-2. Write balanced equations for the following processes:
  (a) the oxidation of elemental bismuth to bismuth(III) by potassium dichromate.
  (b) the air oxidation of a basic solution of $HAsO_3^{2-}$.
  (c) the reaction of p-chlorophenol with an acidified solution containing bromide and bromate ions.
  (d) the reaction of hydrogen peroxide with iodide ion.
  (e) the reduction of molybdenum(VI) in a Jones reductor.
  (f) the oxidation of manganese(II) with permanganate ion to produce MnO_2.
  (g) the oxidation of arsenic(III) by iodate in concentrated hydrochloric acid solution.

* Answers to problems and parts of problems marked with an asterisk are to be found at the end of the book.

15-3. Write balanced equations for the oxidation of each of the following with periodic acid:

   * (a) ethylene glycol, CH_2OHCH_2OH.

   (b) glyoxal, CHOCHO.

   * (c) mannitol, $CH_2OH(CHOH)_4CH_2OH$.

   (d) tartaric acid, $COOH(CHOH)_2COOH$.

   * (e) glucose, $CH_2OH(CHOH)_4CHO$.

   (f) serine, $CH_2OHCH(NH_2)COOH$.

* 15-4. A solution of permanganate was standardized against pure As_2O_3. Exactly 0.200 g of the standard required 38.1 mL of $KMnO_4$. What was the normality of the solution? The molarity?

15-5. A solution of $KMnO_4$ was standardized against 0.179 g of pure KI. The titration was carried out in the presence of HCN, where the reaction was $2MnO_4^- + 5I^- + 5HCN + 11H^+ \rightleftharpoons 2Mn^{2+} + 5ICN + 8H_2O$. Exactly 37.9 mL of reagent were used. What was the normality? The molarity?

* 15-6. Describe the preparation of 2.000 L of 0.05000 N solution of arsenious acid from primary standard arsenious oxide.

15-7. Describe the preparation of 500 mL of a 0.0750 N KIO_3 solution from the primary standard reagent. Assume the solution is to be used for titration of solutions containing an excess of cyanide ion.

* 15-8. What is the normality of a ceric sulfate solution having a titer of 26.0 mg of Mohr's salt $[FeSO_4 \cdot (NH_4)_2SO_4 \cdot 6H_2O]$ per mL?

15-9. A 0.770-g sample of KIO_3 consumed 41.0 mL of a $KMnO_4$ solution after the IO_3^- was reduced to I^- and titrated as in Problem 15-5. What was the normality of the $KMnO_4$?

* 15-10. A 0.814-g sample of a stibnite ore was decomposed in acid and the Sb(III) was oxidized to the +5 state with 40.0 mL of 0.119 N $KMnO_4$. The excess $KMnO_4$ was back-titrated with 3.82 mL of 0.0961 N Fe^{2+}. Calculate the percent Sb_2S_3 in the sample.

15-11. A solution of KNO_2 was analyzed by treating exactly 25.0 mL with 50.0 mL of 0.0880 N $KMnO_4$. After reaction was complete, the excess $KMnO_4$ was back-titrated with 1.20 mL of 0.100 N Fe^{2+}. What is the normality of the KNO_2 solution? How many grams KNO_2 are in 1.000 L of the solution?

* 15-12. A 2.02-g sample of La_2O_3 (fw = 325.8) was brought into solution and the La was precipitated as $La_2(C_2O_4)_3$. The precipitate was filtered, washed, and dissolved in acid and the oxalic acid titrated with 43.2 mL of 0.120 N ceric sulfate. Calculate the percent La_2O_3 present.

15-13. What is the normality of a 0.0500 M $KMnO_4$ solution when

   (a) it is used for titrations in strong acid?

   (b) it is used for titrations in neutral solutions where MnO_2 is the product?

   (c) it is used for titrations in strongly basic solutions?

* 15-14. A solution of $KMnO_4$ was found to be 0.0761 N when standardized against oxalate in the usual way. The solution was used to determine manganese by the Volhard procedure ($2MnO_4^- + 3Mn^{2+} + 4OH^- \rightleftharpoons 5MnO_2 + 2H_2O$). A 0.543-g sample required 29.2 mL of the $KMnO_4$. What percent Mn was contained in the sample?

15-15. In the presence of a high concentration of fluoride ion, Mn^{2+} can be titrated with MnO_4^-, with both reactants being converted to a complex of Mn^{3+}. A 0.312-g sample containing Mn_3O_4 was dissolved in such a way as to convert all of the manganese to Mn^{2+}. Titration in the presence of fluoride ion consumed 22.7 mL of $KMnO_4$ which was 0.121 N against oxalate. Calculate the percent Mn_3O_4.

* 15-16. A sample containing both iron and titanium was analyzed by solution of a 3.00-g sample and dilution to exactly 500 mL in a volumetric flask. A 50.0-mL aliquot of the solution was passed through a silver reductor, which reduced the iron to the divalent state but left the titanium in the +4 state; titration with 0.0750 N Ce^{4+} required 18.2 mL. A 100-mL aliquot was passed through a Jones reductor which reduced both Fe and Ti (TiO^{2+} → Ti^{3+}). The reduced solution consumed 46.3 mL of the Ce^{4+} solution. Calculate the percent Fe$_2$O$_3$ and the percent TiO$_2$ in the sample.

15-17. A sample of alkali metal chlorides was analyzed for sodium by dissolving a 0.800-g sample in water and diluting to exactly 500 mL. A 25.0-mL aliquot of this solution was treated in such a way as to precipitate the sodium as NaZn (UO$_2$)$_3$(OAc)$_9$·6H$_2$O. The precipitate was filtered, dissolved in acid, and passed through a lead reductor, which converted the uranium to U^{4+}. Oxidation of this ion to UO$_2^{2+}$ required 19.9 mL of 0.100 N K$_2$Cr$_2$O$_7$. Calculate the percent NaCl in the sample.

* 15-18. A 0.320-g sample was analyzed for Cr by solution and oxidation of the Cr^{3+} to Cr$_2$O$_7^{2-}$ with an excess of ammonium persulfate. After the excess persulfate was destroyed by boiling, exactly 1.00 g of Mohr's salt (fw = 392.2) was added and the excess Fe^{2+} was titrated with 8.77 mL of 0.0500 N K$_2$Cr$_2$O$_7$. Calculate the percent Cr$_2$O$_3$ in the sample.

15-19. A 25.0-mL sample containing ClO$_3^-$ was made strongly acidic and the ClO$_3^-$ was reduced to Cl$^-$ by the addition of 10.0 mL of 0.250 N ferrous ammonium sulfate. After the reaction was complete, the excess Fe^{2+} was titrated with 9.12 mL of 0.100 N K$_2$Cr$_2$O$_7$. Calculate the percent Ca(ClO$_3$)$_2$ in the sample.

* 15-20. Under suitable conditions, thiourea is oxidized to sulfate by solutions of bromate

$$3CS(NH_2)_2 + 4BrO_3^- + 3H_2O \rightleftharpoons 3CO(NH_2)_2 + 3SO_4^{2-} + 4Br^- + 6H^+$$

A 0.0715-g sample of a material was found to consume 14.1 mL of 0.0500 N KBrO$_3$. What was the percent purity of the thiourea sample?

15-21. The H$_2$S concentration of a sample of air was obtained by bubbling 10.0 L of the air through an absorption tower containing a solution of Cd^{2+}, which retained the S^{2-} as CdS. The mixture was treated with 20.0 mL of 0.0107 N I$_2$ and acidified, which resulted in the sulfide being oxidized to elemental sulfur. The excess iodine consumed 10.1 mL of 0.0120 N thiosulfate. Calculate the parts per million H$_2$S present in the air assuming a gas density of 12 × 10^{-4} g/mL.

* 15-22. An aqueous solution of phenol was analyzed by addition of 25.0 mL of 0.100 N KBrO$_3$, an excess of KBr, and several milliliters of strong acid. After the bromination was complete, 10.0 mL of 0.120 N arsenious acid were added which reduced the excess Br$_2$. Finally, the excess H$_3$AsO$_3$ was titrated with 1.47 mL of the standard bromate solution. Calculate the number of milligrams of phenol in the sample.

15-23. What is the normality of a 0.0400 M solution of KH(IO$_3$)$_2$ when used for the following purposes:
(a) as an oxidizing reagent in strong HCN solution?
(b) as a source of iodine for a redox titration wherein an excess of iodide and acid are present?
(c) as a standard for the titration of a strong acid?

* 15-24. Calculate parts per million of SO$_2$ in a gas stream (density = 0.0012 g/mL) that was analyzed by absorption of the compound from a 6.0-L sample in sodium hydroxide followed by acidification with HCl and titration of the sulfite with 4.98 mL of 0.0125 N KIO$_3$. The IO$_3^-$ was converted to ICl$_2^-$ in the titration.

15-25. Strong acids may be standardized against KIO$_3$. The iodate is dissolved in water containing an excess of KI and sodium thiosulfate. As acid is added, the KIO$_3$ consumes the protons, liberating iodine which reacts with the S$_2$O$_3^{2-}$. Calculate the normality of a H$_2$SO$_4$ solution if 29.10 mL is used in the titration of 0.150 g of KIO$_3$.

* 15-26. A sample of ethylene glycol, CH$_2$OHCH$_2$OH, was analyzed by treatment with 50.0 mL of a solution of periodic acid. After the reaction was complete, the mixture was buffered to pH 7.5 and an excess of iodide was added. The liberated iodine reacted with 14.3 mL of a standard 0.100 N arsenite solution. A blank solution con-

taining all but the glycol was found to consume 40.1 mL of the standard arsenite. Calculate the weight of ethylene glycol in the sample if its reaction product is formaldehyde, HCHO.

15-27. What is the titer of a 0.100 M solution of periodic acid in terms of milligrams of each of the following compounds: (a) propylene glycol, (b) glycerol, and (c) biacetyl.

* 15-28. A 0.6935-g sample of a zinc oxide ointment, used as a sun screen, was ignited in a crucible to destroy the organic material. The residue was then dissolved in dilute sulfuric acid. After dilution, the zinc was precipitated as ZnC_2O_4 which was filtered, washed, and dissolved in dilute sulfuric acid. Titration of the resulting oxalate species required 32.64 mL of 0.1042 N $KMnO_4$. Calculate the percent ZnO in the sample.

15-29. The CO concentration of a gas was obtained by passing a 22.1-L sample over iodine pentoxide heated to 150°C.

$$I_2O_5 + 5CO \longrightarrow 5CO_2 + I_2$$

The iodine distilled at this temperature and was collected in an absorber containing 8.25 mL of 0.01097 N $Na_2S_2O_2$. The excess thiosulfate consumed 2.16 mL of 0.00947 N I_2. Calculate the parts per million CO in the gas, assuming a gas density of 1.25 g/L.

* 15-30. What is the titer of a 0.0250 M KIO_3 solution in terms of milligrams of each of the following? Assume in each case the oxidation is carried out in HCl solution where the product is ICl_2^-: As(III), I_2, N_2H_4 (product: N_2), KSCN (products: CN^-, SO_4^{2-}).

15-31. What is the titer of a 0.200 N I_2 solution in terms of milligrams of As_2O_3, Sb, SO_2, H_2S, As?

* 15-32. Nitroglycerin tablets are used for the treatment of angina pectoris. Ten tablets weighing 2.230 g were ground to a fine powder. A 0.1906-g sample of the powder was dissolved in a solution of sulfuric acid containing an excess of salicylic acid. Reaction between the nitroglycerin and the salicylic acid resulted in conversion of the nitrate group to a nitro salicylic acid derivative. Exactly 25.00 mL of a 0.01276 N titanium(III) chloride were added, which resulted in the reduction of the nitro group to an amine:

$$RNO_2 + 6H^+ + 6e \longrightarrow RNH_2 + 2H_2O$$

The excess Ti^{3+} was titrated with 10.63 mL of 0.01045 N Fe^{3+}, with SCN^- serving as the indicator. Calculate the milligrams nitroglycerin ($CH_2ONO_2CHONO_2CH_2ONO_2$, fw = 227.1) per tablet.

15-33. Acetanilide $C_6H_5NHCOCH_3$ is sometimes used in combination with aspirin and caffeine in headache remedies. It can be determined by hydrolysis to give aniline, which is then brominated. Eight tablets of a headache remedy weighing 4.17 g were crushed and a 0.521-g sample of the resulting powder was treated to remove the other constituents of the preparation and then hydrolyzed in dilute sulfuric acid. Exactly 40.00 mL of 0.1105 N $KBrO_3$ were added to the solution as well as an excess of KBr. After 10 min, bromination of the aniline was considered to be complete (Table 15-7) and an excess of KI was added to the mixture. The liberated iodine required 7.67 mL of 0.0949 N $Na_2S_2O_3$. Calculate the grams acetanilide (fw = 135) in each tablet.

* 15-34. The lanthanum from a 2.761-g sample was precipitated as $La(IO_3)_3$ by the addition of 50.00 mL of 0.06730 N KIO_3 to a solution of the sample. The filtrate and washings from the precipitate were acidified and treated with an excess of KI. The liberated iodine required 6.42 mL of 0.05152 N thiosulfate. Calculate the percent $La_2(SO_4)_3$ in the sample.

15-35. Ethyl mercaptan, which often serves as the odorant in household gas supplies, was determined by bubbling 5.75 L of a gas through two absorber solutions. The first contained 25.00 mL of 0.01063 N I_2. The second contained 5.00 mL of 0.0113 N $Na_2S_2O_3$; its purpose was to retain any iodine lost from the first absorber by volatilization. The analysis was completed by combining the contents of the two absorbers and titrating the residual I_2 with 4.17 mL of the thiosulfate solution. Calculate the concentration in terms of mg/L of ethyl mercaptan in the sample. Reaction:

$$2C_2H_5SH + I_2 \longrightarrow C_2H_5SSC_2H_5 + 2H^+ + 2I^-$$

* 15-36. A 0.2785-g sample of a rocket fuel was dissolved and diluted to 250.0 mL in a volumetric flask. A 25.00-mL aliquot was treated with 50.00 mL of 0.0932 N $KBrO_3$, which oxidized the hydrazine to N_2 and converted the BrO_3^- to Br^-. The solution was then acidified after adding additional Br^-, which converted the excess BrO_3^- to Br_2. The bromine was determined by adding an excess of KI and titrating the liberated I_2 with 10.76 mL of 0.1115 N thiosulfate. Calculate the percent hydrazine, N_2H_4, in the sample.

15-37. A 10.0-mL sample of a brandy was diluted to 1.000 L. The ethanol in a 10.0-mL aliquot was distilled into 50.00 mL of 0.1246 N $K_2Cr_2O_7$ in sulfuric acid where oxidation of the ethanol to acetic acid occurred. The excess dichromate was titrated with 26.08 mL of 0.1025 N iron(II). Calculate the weight-volume percent alcohol in the brandy.

* 15-38. Calculate the numerical value of the equilibrium constant for the reaction

$$2MnO_4^- + 3Mn^{2+} + 2H_2O \rightleftharpoons 5MnO_2 + 4H^+$$

15-39. Calculate a value for the equilibrium constant for the air oxidation of iodide ion to give triiodide ion.

* 15-40. Air slowly oxidizes solutions of arsenic(III) to arsenic(V). Calculate the equilibrium constant for this reaction.

* 15-41. Calculate the concentration of iron(II) in a solution having a pH of 4.50 that is in equilibrium with oxygen at a partial pressure of 0.80 atm. Assume that the solution was initially 0.1000 M in iron(II).

* 15-42. A sample containing KI, KBr, and other inert substances was analyzed as follows: A 1.00-g sample was dissolved in water and diluted to exactly 200 mL. A 50.0-mL aliquot was treated with Br_2 in neutral solution which converted the I^- to IO_3^-. The excess Br_2 was removed by boiling and KI added. After acidification, the liberated iodine was titrated with 40.8 mL of 0.0500 N thiosulfate. A 50.0-mL aliquot was oxidized with strongly acidic $K_2Cr_2O_7$; the liberated I_2 and Br_2 were distilled and collected in a strong KI solution. The iodine present after complete reaction required 29.8 mL of the thiosulfate. Calculate the percent KI and KBr in the sample.

15-43. A solution containing IO_3^- and IO_4^- was analyzed as follows: A 50.0-mL aliquot was buffered with a borax buffer and treated with an excess of KI, which resulted in reduction of the IO_4^- to IO_3^-. The liberated iodine consumed 18.4 mL of 0.100 N $Na_2S_2O_3$. A 10.0-mL aliquot was then acidified strongly and an excess of KI was added. This required 48.7 mL of the thiosulfate. Calculate the molar concentration of IO_3^- and IO_4^- in the solution.

potentiometric methods

16

In the preceding chapters, it was shown that the potential of an electrode is determined by the concentration (or, more correctly, the activity) of one or more species in a solution. This chapter considers how this phenomenon is applied to the quantitative determination of ions or molecules.[1]

The equipment required for a potentiometric measurement includes a *reference electrode*, an *indicator electrode*, and a *potential measuring device*. The preparation and properties of some common reference electrodes were considered in Section 13G; the latter two components will be discussed in this chapter, in addition to applications of potentiometric measurements to quantitative analyses.

16A INDICATOR ELECTRODES

Indicator electrodes for potentiometric measurements are of two basic types, namely, *metallic* and *membrane*. The latter are also referred to as *specific* or *selective ion* electrodes.

16A-1 METALLIC INDICATOR ELECTRODES

First-order electrodes for cations. A first-order electrode serves to determine the concentration of the cation derived from the electrode metal. Several metals, such as silver, copper, mercury, lead, and cadmium, exhibit reversible half-reactions with their ions and are satisfactory as first-order electrodes. In contrast, other metals are less suitable because they tend to develop nonreproducible potentials that are influenced by strains or crystal deformations in their structures and by

---

[1] For a detailed review of potentiometric methods, see: R. P. Buck, in *Physical Methods of Chemistry*, A. Weissberger and B. W. Rossiter, Eds., Part IIA, vol. 1, Chapter 2. New York: Wiley-Interscience, 1971.

oxide coatings on their surfaces. Metals in this category include iron, nickel, cobalt, tungsten, and chromium.

Second-order electrodes for anions. A metal electrode can also be indirectly responsive to anions that form slightly soluble precipitates or stable complexes with its cation. For the former, it is only necessary to saturate the solution under study with the sparingly soluble salt. Thus, the potential of a silver electrode will accurately reflect the concentration of iodide ion in a solution that is saturated with silver iodide. Here, the electrode behavior can be described by

$$AgI(s) + e \rightleftharpoons Ag(s) + I^- \qquad E^0_{AgI} = -0.151 \text{ V}$$

Application of the Nernst equation to this half-reaction gives the relationship between the electrode potential and the anion concentration. Thus,

$$\begin{aligned} E &= -0.151 - 0.0591 \log [I^-] \\ &= -0.151 + 0.0591 \text{ pI} \end{aligned}$$

where pI is the negative logarithm of the iodide ion concentration. A silver electrode serving as an indicator for iodide is an example of an *electrode of the second-order* because it measures the concentration of an ion that is not directly involved in the electron transfer process.

An important second-order electrode for measuring the concentration of the EDTA anion Y^{4-} is based upon the response of a mercury electrode in the presence of a small concentration of the stable EDTA complex of mercury(II). The half-reaction for the electrode process can be written as

$$HgY^{2-} + 2e \rightleftharpoons Hg(l) + Y^{4-} \qquad E^0 = 0.21 \text{ V}$$

for which

$$E = 0.21 - \frac{0.0591}{2} \log \frac{[Y^{4-}]}{[HgY^{2-}]}$$

To employ this electrode system, it is necessary to introduce a small concentration of HgY^{2-} into the analyte solution at the outset. The complex is very stable (for HgY^{2-}, $K_f = 6.3 \times 10^{21}$); thus, its concentration remains essentially constant over a wide range of Y^{4-} concentrations because dissociation to form Hg^{2+} is slight. The foregoing equation can then be written in the form

$$E = K - \frac{0.0591}{2} \log [Y^{4-}] = K + \frac{0.0591}{2} \text{ pY}$$

where the constant K is equal to

$$K = 0.21 - \frac{0.0591}{2} \log \frac{1}{[HgY^{2-}]}$$

This second-order electrode is useful for establishing end points of EDTA titrations.

Indicator electrodes for redox systems. Electrodes fashioned from platinum, gold, palladium, or carbon serve as indicator electrodes for oxidation-reduction systems. Of itself, such an electrode is inert; the potential it develops depends solely upon the potential of the oxidation-reduction systems of the solution in which it is immersed. For example, the potential of a platinum electrode in a solution containing cerium(III) and cerium(IV) ions is given by

$$E = E^0 - 0.0591 \log \frac{[Ce^{3+}]}{[Ce^{4+}]}$$

Thus, a platinum electrode can serve as the indicator electrode in a titration in which cerium(IV) serves as the standard reagent.

16A-2 MEMBRANE INDICATOR ELECTRODES[2]

For many years, the most convenient method for determining pH has involved measurement of the potential that develops across a thin glass membrane that separates two solutions with different hydrogen ion concentrations. The phenomenon, first reported by Cremer,[3] has been extensively studied by many investigators; as a result, the sensitivity and selectivity of glass membranes to pH are reasonably well understood. Furthermore, membrane electrodes have now been developed for the direct potentiometric determination of two dozen or more ions such as K^+, Na^+, F^-, and Ca^{2+}.[4]

It is convenient to divide membrane electrodes into four categories based upon membrane composition. These include: (1) glass electrodes, (2) liquid-membrane electrodes, (3) solid-state or precipitate electrodes, and (4) gas-sensing membrane electrodes. We shall consider the properties and behavior of the glass electrode in particular detail both because of its historical and its current importance.

16A-3 THE GLASS ELECTRODE FOR pH MEASUREMENTS

Figure 16-1 shows a modern cell for the measurement of pH. It consists of commercially available calomel and glass electrodes immersed in a solution whose pH is to be measured. The calomel reference electrode is similar to the ones described in Section 13G-1. The glass electrode is manufactured by sealing a thin, pH-sensitive glass tip to the end of a heavy-walled glass tubing. The resulting bulb is filled with a solution of hydrochloric acid (often 0.1 M) that is saturated with silver chloride. A silver wire is immersed in the solution and is connected via an external lead to one terminal of a potential-measuring device; the calomel electrode is connected to the other terminal. Note that the cell contains *two* reference electrodes, each with a potential that is constant and *independent of pH*. One reference electrode is the external calomel electrode; the other is the internal silver/silver chloride electrode (Section 13G-2), which is a *part* of the glass electrode but is *not* the pH-sensitive component. In fact, it is the *thin membrane at the tip of the electrode* that responds to pH changes.

[2] For further information on this topic, see: R. A. Durst, Ed., *Ion-Selective Electrodes.* National Bureau of Standards Special Publication 314. Washington, D.C.: U.S. Government Printing Office, 1969; K. Cammann, *Working With Ion-Selective Electrodes.* New York: Springer-Verlag, 1979; J. Vesely, D. Weiss, and K. Stulik, *Analysis With Ion-Selective Electrodes.* New York: Wiley, 1979; and H. Freiser, Ed., *Ion Selective Electrodes in Analytical Chemistry.* New York: Plenum Press, 1978.
[3] M. Cremer, Z. *Biol.,* **47**, 562 (1906).
[4] See: J. Vesely, D. Weiss, and K. Stulik, *Analysis with Ion-Selective Electrodes*, pp. 125–203. New York: Wiley, 1979.

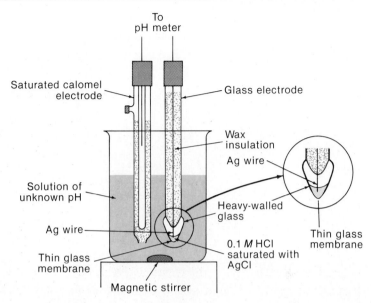

FIGURE 16-1 Typical electrode system for measuring pH.

Figure 16-2 is a schematic representation of the cell shown in Figure 16-1. It is found experimentally that the potential of this cell can be resolved into several components. That is,

$$E_{cell} = E_{Ag,AgCl} - E_{SCE} + E_j + E_{asy} + E_b \qquad (16\text{-}1)$$

Here, the first two terms represent the difference in potential between the saturated calomel and silver/silver chloride reference electrodes, and E_j is the junction potential associated with the salt bridge that joins the calomel electrode to the analyte solution. The *asymmetry potential*, E_{asy}, is a constant potential (over limited periods of time), which is independent of pH; its source and properties will be considered later.

The first four terms on the right side of the foregoing equation are constants and can be combined to form a single constant Q. The equation then reduces to

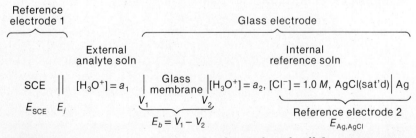

FIGURE 16-2 Schematic diagram of a glass/calomel cell for measurement of pH.

$$E_{cell} = Q + E_b \tag{16-2}$$

The *boundary potential*, E_b, is the difference between two potentials, each of which is associated with one surface of the glass membrane (see Figure 16-2). That is,

$$E_b = V_1 - V_2$$

Empirically, it is observed that each of the potentials, V_1 and V_2, is dependent upon the hydrogen ion activity of the solution with which it interfaces. At 25°C, this dependence is found to be

$$V_1 = \frac{RT}{F} \ln a_1 = 0.0591 \log a_1$$

$$V_2 = \frac{RT}{F} \ln a_2 = 0.0591 \log a_2$$

Then,

$$E_b = 0.0591 \log a_1 - 0.0591 \log a_2$$

In practice, the hydrogen ion activity of the internal solution a_2 is fixed. Thus, Equation 16-2 can be written in the form

$$E_{cell} = Q + 0.0591 \log a_1 - 0.0591 \log a_2$$

or

$$E_{cell} = L + 0.0591 \log a_1 \tag{16-3}$$

where L is a new constant made up of Q and the logarithmic function of a_2.

It is important to emphasize that, in principal, the potentials making up Q remain constant during a pH measurement. Thus, the source of the pH-dependent variation in E must lie *across the glass membrane*. That is, when a_1 and a_2 differ, the two surfaces of the membrane must differ by some potential $V_1 - V_2$ (see Figure 16-2). The *only* function of the two reference electrodes is to make observation of this difference possible.

In order to understand the source of the boundary potential ($V_1 - V_2$), it is necessary to consider the properties of pH-sensitive glass membranes in some detail.

Composition of glass membranes. Much systematic investigation has been devoted to the effects of glass composition on the sensitivity of membranes to protons and other cations, and a variety of compositions are now used commercially.[5] For many years, Corning 015 glass (consisting of approximately 22% Na_2O, 6% CaO, and 72% SiO_2) has been widely used. This glass shows an excellent specificity toward hydrogen ions up to a pH of about 9. At higher pH values, however, the membrane becomes somewhat sensitive to sodium and other singly charged cations.

[5] For a summary of this work, see: J. O. Isard, "The Dependence of Glass-Electrode Properties on Composition," in *Glass Electrodes for Hydrogen and Other Cations*, G. Eisenman, Ed., Chapter 3. New York: Marcel Dekker, 1967.

Hygroscopicity of glass membranes. It has been shown that the surfaces of a glass membrane must be hydrated in order to have pH activity; nonhygroscopic glasses such as Pyrex and quartz show no pH function. Even Corning 015 glass shows little pH response after dehydration by storage over a desiccant; its sensitivity is restored, however, after immersion in water for a few hours. Hydration involves absorption of approximately 50 mg of water per cubic centimeter of glass.

It has also been demonstrated experimentally that hydration of a pH-sensitive glass membrane is accompanied by a chemical reaction in which singly charged cations of the glass are exchanged for protons of the solution. The divalent and trivalent cations in the silicate structure are much more strongly bonded and thus do not exchange. The ion-exchange reaction can be written as

$$H^+ + Na^+Gl^- \rightleftharpoons Na^+ + H^+Gl^- \qquad (16\text{-}4)$$

soln   solid   soln   solid

where Gl^- represents a cation bonding site *in the glass.* The equilibrium constant for this process favors incorporation of hydrogen ions into the silicate lattice; as a result, the surface of a well-soaked membrane will ordinarily consist of a layer of silicic acid gel that is 10^{-5} to 10^{-4} mm thick.

In all but highly alkaline media—where alkali metal ions may occupy an appreciable number of bonding sites—the predominant univalent cation in the outer surface of the gel is the proton. There is a continuous decrease in the number of protons from the surface to the interior of the gel and a corresponding increase in the number of sodium ions. A schematic representation of the two surfaces of a glass membrane is shown in Figure 16-3.

Conductivity of glass membranes. The membrane in a typical commercial glass electrode has a thickness between 0.03 and 0.1 mm and an electrical resistance of 50 to 500 MΩ (megohm; 1 MΩ = 10^6 Ω). Electrical conduction through the membrane involves migration of singly charged cations. Across each solution interface, the current involves a transfer of protons; the direction of migration is from glass to solution at one interface and from solution to glass at the other. That is,

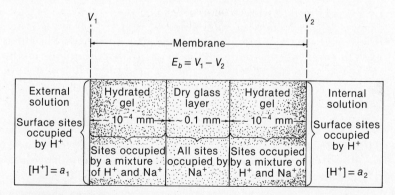

FIGURE 16-3 Schematic representation of a well-soaked glass membrane. Note that the dimensions of the three inner layers are not to scale.

$$H^+Gl^- \rightleftharpoons Gl^- + H^+ \qquad \text{(16-5)}$$
$$\text{solid} \qquad\quad \text{solid} \quad\; \text{soln}$$

and

$$H^+ + Gl^- \rightleftharpoons H^+Gl^- \qquad \text{(16-6)}$$
$$\text{soln} \quad\; \text{solid} \qquad\;\; \text{solid}$$

The positions of these two equilibria are determined by the hydrogen ion concentrations in the two solutions. When these positions differ, the surface at which the greater dissociation has occurred will be negative with respect to the other surface. Thus, a potential develops whose magnitude depends upon the *difference* in hydrogen ion concentration on the two sides of the membrane. It is this potential that serves as the analytical parameter in a potentiometric pH measurement.

Conduction within the two silicic gel layers is due to migration of hydrogen and sodium ions. In the dry center region of the membrane, sodium ions take over this function.

Theory of the glass electrode potential.[6] The potential across a glass membrane consists of a boundary potential and a *diffusion potential*. Ideally, only the former is affected by pH.

The two potentials V_1 and V_2 that make up the boundary potential can be thought of as measures of the driving forces of the two reactions shown by Equations 16-5 and 16-6. From thermodynamic considerations it can be shown that these driving forces are related to the hydrogen ion activities of the solutions and the gel surfaces as follows:

$$V_1 = j_1 + \frac{RT}{F} \ln \frac{a_1}{a_1'} \qquad \text{(16-7)}$$

$$V_2 = j_2 + \frac{RT}{F} \ln \frac{a_2}{a_2'} \qquad \text{(16-8)}$$

where R, T, and F have their usual meanings; a_1 and a_2 are activities of the hydrogen ion in the *solutions* on either side of the membrane; and a_1' and a_2' are the hydrogen ion activities in each of the *gel layers* contacting the two solutions. If the two gel surfaces have the same number of sites from which protons can leave, then the two constants j_1 and j_2 will be identical; so also will be the two activities a_1' and a_2' in the gel layers, provided that all original sodium ions on the surface have been replaced by protons (that is, insofar as the equilibrium shown in Equation 16-4 lies far to the right). With these equalities the boundary potential simplifies to

$$E_b = V_1 - V_2 = \frac{RT}{F} \ln \frac{a_1}{a_2} \qquad \text{(16-9)}$$

Thus, *provided the two gel surfaces are identical,* the boundary potential E_b depends only upon the activities of the hydrogen ion in the *solutions* on the two sides of the membrane.

---

[6] For a comprehensive discussion of this topic, see: G. Eisenman, *Glass Electrodes for Hydrogen and Other Cations,* Chapters 4–6. New York: Marcel Dekker, 1967.

Asymmetry potential. If identical solutions and identical reference electrodes are placed on both sides of the membrane shown in Figure 16-3, $V_1 - V_2$ should be zero. In fact, however, it is often found that a small potential, called the *asymmetry potential*, does develop when this experiment is performed. Moreover, the asymmetry potential associated with a given glass electrode changes slowly with time.

The causes for the asymmetry potential are obscure; they undoubtedly include such factors as differences in strains established within the two surfaces during manufacture of the membrane, mechanical and chemical attack of the surfaces, and contamination of the outer face during use. The effect of the asymmetry potential on a pH measurement is eliminated by frequent calibration of the electrode against a standard buffer of known pH.

The alkaline error. Glass electrodes respond to the concentrations of both the hydrogen ion and alkali metal ions in basic solution, where the former is small relative to the latter. The magnitude of the resulting error for four types of glass membranes is indicated on the right side of Figure 16-4. To obtain each of these curves, the sodium ion concentration was maintained at 1 M and the pH was varied. Note that the pH error is negative at high pH values (that is, the observed values for pH were lower than the true values), which suggests that the electrode is responding to sodium ions as well as to protons. This observation is confirmed by data obtained for solutions with different sodium ion concentrations. Thus, at pH 12, an electrode fashioned from Corning 015 glass gave a pH reading of 11.3 when the sodium ion concentration was 1 M (Figure 16-4) but 11.7 in solutions that were 0.1 M in this ion.

All singly charged cations cause alkaline errors; the magnitude of the error varies according to the kind of metallic ion and the composition of the glass.

The alkaline error can be satisfactorily explained by assuming an exchange equilibrium between the hydrogen ions on the surface of the glass and the cations in the solution. This process, which is the reverse of the one described by Equation 16-4, can be formulated as

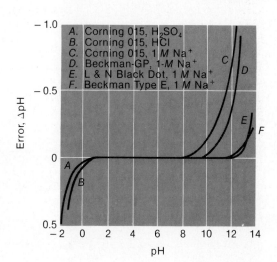

FIGURE 16-4 Acid and alkaline error of selected glass electrodes at 25°C. [R. G. Bates, *Determination of pH: Theory and Practice*, p. 316. New York: Wiley, 1964. With permission.]

$$H^+Gl^- + B^+ \rightleftharpoons B^+Gl^- + H^+ \tag{16-10}$$
$$\text{gel} \quad\quad \text{soln} \quad\quad \text{gel} \quad\quad \text{soln}$$

where B^+ represents a singly charged cation, such as sodium ion. The equilibrium constant for this reaction is

$$K_{ex} = \frac{a_1 b_1'}{a_1' b_1}$$

where a_1 and b_1 are activities of H^+ and B^+ in solution, and a_1' and b_1' are the activities of the same ions in the surface of the gel. The constant K_{ex} depends upon the composition of the glass membrane and is ordinarily small. Thus, the fraction of the surface sites populated by cations other than hydrogen is small except when the hydrogen ion concentration is very low and the concentration of B^+ is high.

Eisenman has demonstrated that the effect of an alkali metal ion on the potential across a membrane can be described by a modification of Equation 16-3.[7] That is,

$$E_{cell} = L + \frac{RT}{F} \ln \left[a_1 + K_{ex} \left(\frac{U_B}{U_H} \right) b_1 \right] \tag{16-11}$$

where U_B and U_H are constants that depend upon the mobilities of B^+ and H^+ in the gel. This equation may be rewritten in the form

$$E_{cell} = L + \frac{RT}{F} \ln (a_1 + K_{H,B} b_1) \tag{16-12}$$

where $K_{H,B}$ is the *selectivity constant* of the electrode.

The selectivity constant is small for many glasses so that $K_{H,B} b_1$ is negligible compared with hydrogen ion activity a_1, provided the pH of the solution is less than about 9. Under these conditions, Equation 16-12 simplifies to Equation 16-3. With high concentrations of singly charged cations and at higher pH levels, however, this second term plays an increasingly important part in determining E. The composition of the glass membrane determines the magnitude of $K_{H,B}$; this parameter is relatively small for glasses that have been designed for work in strongly basic solutions. The membrane of the Beckman Type E glass electrode, illustrated in Figure 16-4, is of this type.

The acid error. As shown in Figure 16-4, the typical glass electrode exhibits an error, opposite in sign to the alkaline error, in solutions of pH less than about 0.5; pH readings tend to be too high in this region. The magnitude of the error depends upon a variety of factors and is generally not very reproducible. The causes of the acid error are not well understood.

16A-4 GLASS ELECTRODES FOR THE DETERMINATION OF OTHER CATIONS
The existence of the alkaline error in early glass electrodes led to studies concerning the effect of glass composition on the magnitude of this error. One consequence of such work has been the development of glasses for which the alkaline error is negligible below a pH of about 12. Other studies have been directed

[7] See: G. Eisenman, *Glass Electrodes for Hydrogen and Other Cations*, Chapters 4 and 5. New York: Marcel Dekker, 1967.

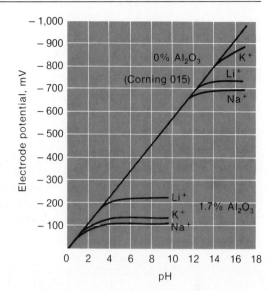

FIGURE 16-5 Response of two glass membranes in the presence of alkali-metal ions (0.1 M in each case). [Curves from G. Eisenman, in *Advances in Analytical Chemistry and Instrumentation*, C. N. Reilley, Ed. vol. 4, p. 213. New York: Wiley. With permission.]

toward finding compositions in which $K_{H,B}$ is greatly enhanced, in the interest of developing glass electrodes for the determination of cations other than hydrogen. This application requires that the hydrogen ion activity a_1 in Equation 16-12 be negligible with respect to the second term containing the activity b_1 of the other cation; under these circumstances, the potential of the electrode would be *independent* of pH but would vary with pB in the typical way.

A number of investigators have demonstrated that the presence of Al_2O_3 or B_2O_3 in glass causes the desired effect. Eisenman and co-workers have carried out a systematic study of glasses containing Na_2O, Al_2O_3, and SiO_2 in various proportions; they have demonstrated clearly that it is practical to prepare membranes for the selective measurement of several cations in the presence of others.[8]

Figure 16-5 illustrates how two types of glass electrodes respond to pH changes in solutions that are 0.1 M in various alkali-metal ions. The one electrode was fabricated from Corning 015 glass, a composition that contains no Al_2O_3; at pH levels below 9, the term $K_{H,B}b_1$ in Equation 16-12 is small with respect to a_1. At higher pH values, the second term becomes important; note that its magnitude also depends upon the type of alkali ion present.

With the glass containing Al_2O_3, both potential-determining terms in Equation 16-12 are clearly important at low pH values; at levels above about pH 5, however, the potential becomes *independent* of pH. Thus, $a_1 \ll K_{H,B}b_1$ in this region, with the result that the potential can be expected to vary linearly with pNa, pK, or pLi. Experimentally, this prediction is verified; such a glass can be employed to measure the concentration of these ions provided the pH is maintained at a level greater than 5.

Glass membranes that are suitable for direct potentiometric measurement of

[8] For a summary of this work, see: G. Eisenman, in *Advances in Analytical Chemistry and Instrumentation*, C. E. Reilley, Ed., vol. 4, p. 213. New York: Wiley, 1965.

the concentrations of such singly charged species as Na^+, K^+, NH_4^+, Rb^+, Cs^+, Li^+, and Ag^+ have now been developed. Some of these glasses are reasonably selective toward other singly charged cations. Glass electrodes for potassium and for sodium ions are available commercially.

16A-5 LIQUID MEMBRANE ELECTRODES

Liquid membranes owe their response to the potential that is established across the interface between the solution to be analyzed and an immiscible liquid that selectively bonds with the ion being determined. Liquid membrane electrodes are particularly important because they permit the direct potentiometric determination of the activities of several polyvalent cations and certain anions as well.

A liquid membrane electrode differs from a glass electrode only in that the solution of known and fixed activity is separated from the analyte solution by a thin layer of an immiscible organic *liquid* instead of a thin glass membrane. Such an electrode for the determination of calcium ion activities is shown schematically in Figure 16-6. A porous, hydrophobic (that is, water-repelling), plastic disk serves to hold the organic layer between the two aqueous solutions. Wick action causes the pores of the disk or membrane to stay filled with the organic liquid from the reservoir in the outer of the two concentric tubes. The inner tube contains an aqueous standard solution of $CaCl_2$ that is also saturated with AgCl to form a Ag/AgCl reference electrode with the silver lead wire.

The organic liquid is a nonvolatile, water-immiscible, organic ion exchanger that reacts selectively with calcium ions. In one commercially available calcium ion electrode, the ion exchanger is an aliphatic diester of phosphoric acid dissolved in a polar solvent. The chain length of the aliphatic groups of the ester ranges from 8 to 16 carbon atoms. The diester contains a single acidic proton; thus, two molecules are required to bond the divalent calcium ion. The equilibrium established at each interface of the porous disk can be represented as

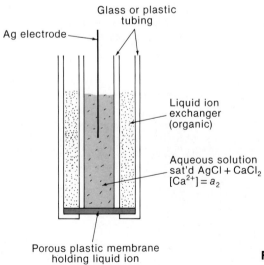

Glass or plastic
tubing

Ag electrode

Liquid ion
exchanger
(organic)

Aqueous solution
sat'd AgCl + CaCl$_2$
[Ca^{2+}] = a_2

Porous plastic membrane
holding liquid ion
exchanger

FIGURE 16-6   Liquid membrane electrode sensitive to Ca^{2+}.

$$[(RO)_2POO]_2Ca \rightleftharpoons 2(RO)_2POO^- + Ca^{2+}$$

organic   organic   aqueous

Note the similarity of this equation to Equation 16-5 for the glass electrode. Selectivity here is the result of the special affinity possessed by the ion exchanger for calcium ions.

As with the glass electrode, a liquid membrane electrode is used in conjunction with an external reference—usually a saturated calomel electrode. The potential of this cell is given by an equation analogous to Equation 16-3 for the glass electrode; that is,

$$E_{cell} = L + \frac{0.0591}{2} \log a_1 \tag{16-13}$$

where a_1 here is the activity of Ca^{2+} and L is a constant whose value is determined by the measurement of E when the electrode is immersed in a standard solution of calcium ion.

Figure 16-7 shows construction details of a commercial liquid-membrane electrode that is selective for calcium ion and compares its structural features with those of a glass electrode.

The sensitivity of the electrode just described is reported to be 50 times greater for calcium ion than for magnesium ion and 1000 times greater than that for sodium or potassium ions. Calcium ion activities as low as $5 \times 10^{-7} M$ can be measured. The performance of the electrode is said to be independent of pH in the range between 5.5 and 11. At lower pH levels, hydrogen ions undoubtedly replace some of the calcium ions on the exchanger to a significant extent; the electrode then becomes pH- as well as pCa-sensitive.

The calcium ion membrane electrode has proved to be a valuable tool for physiological studies because this ion plays important roles in nerve conduction, bone formation, muscle contraction, cardiac expansion and contraction, and renal tubular function. At least some of these processes are influenced more by calcium ion *activity* than calcium concentration; activity, of course, is the parameter measured by the electrode.

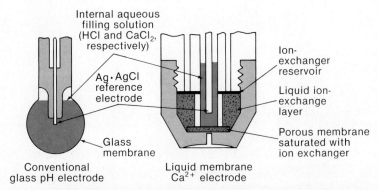

FIGURE 16-7 Comparison of a liquid membrane calcium ion electrode with a glass electrode. (Courtesy of Orion Research, Inc.)

Another liquid specific-ion electrode of great value for physiological studies is that for potassium, because the transport of nerve signals appears to involve movement of this ion across nerve membranes. Study of the process requires an electrode that can detect small concentrations of potassium ion in the presence of much larger concentrations of sodium. A number of liquid membrane electrodes show promise of meeting these needs; one is based upon the antibiotic valinomycin, a cyclic ether that has a strong affinity for potassium ion. Of equal importance is the observation that a liquid membrane consisting of valinomycin in diphenyl ether is about 10^4 times as responsive to potassium ion as to sodium ion.[9]

Table 16-1 lists commercially available liquid membrane electrodes. Here, the selectivity factor is defined by a more general form of Equation 16-12. That is, for the measurement of A^{n+} in the presence of a second ion B^{m+} at 25°C,

$$E_{\text{cell}} = L + \frac{0.0591}{n} \log\left(a_A + K_{A,B} a_B^{n/m}\right) \tag{16-14}$$

EXAMPLE 16-1. Calculate the relative error in the determination of pCa in a solution that was $1.00 \times 10^{-3}\ M$ in Ca^{2+} and $0.200\ M$ in Na^+ if the calcium ion electrode described in Table 16-1 were employed.

In the presence of sodium ion, Equation 16-14 would apply with n/m having a value of 2. Thus, the actual measured cell potential minus L would be

$$E_{\text{cell}} - L = \frac{0.0591}{2} \log\left(1.00 \times 10^{-3} + 0.003 \times 0.200^2\right)$$

$$= -8.720 \times 10^{-2}$$

[9] M. S. Frant and J. W. Ross, Jr., *Science*, **167**, 987 (1970).

Table 16-1   Commercial Liquid-Membrane Electrodes*

Analyte Ion	Concentration Range, M	Preferred pH	Selectivity Constant for Interferences, $K_{A,B}$
Ca^{2+}	10^0–5×10^{-7}	6–8	Zn^{2+}, 3.2; Fe^{2+}, 0.8; Pb^{2+}, 0.63; Mg^{2+}, 0.014; Na^+, 0.003
Cl^-	10^0–8×10^{-6}	2–11	I^-, 17; NO_3^-, 4.2; Br^-, 1.6; HCO_3^-, 0.19; SO_4^{2-}, 0.14; F^-, 0.10
BF_4^-	10^{-1}–3×10^{-6}	3–10	NO_3^-, 0.1; Br^-, 0.04; OAc^-, 0.004; HCO_3^-, 0.004; Cl^-, 0.001
NO_3^-	10^0–7×10^{-6}	3–10	I^-, 20; Br^-, 0.1; NO_2^-, 0.04; Cl^-, 0.004; SO_4^{2-}, 0.00003; CO_3^{2-}, 0.0002; ClO_4^-, 1000; F^-, 0.00006
ClO_4^-	10^0–2×10^{-6}	3–10	I^-, 0.012; NO_3^-, 0.0015; Br^-, 0.00056; F^-, 0.00025; Cl^-, 0.00022
K^+	10^0–10^{-6}	3–10	Cs^+, 1.0; NH_4^+, 0.03; H^+, 0.01; Ag^+, 0.001; Na^+, 0.002; Li^+, 0.001
$Ca^{2+} + Mg^{2+}$	10^{-2}–6×10^{-6}	5–8	Zn^{2+}, 3.5; Fe^{2+}, 3.5; Cu^{2+}, 3.1; Ni^{2+}, 1.35; Ba^{2+}, 0.94; Na^+, 0.015

* *Analytical Methods Guide*, 6th ed. Cambridge, Mass.: Orion Research, Inc., August, 1973.

If sodium were neglected, however, Equation 16-13 would be employed to obtain pCa. That is,

$$E_{cell} - L = \frac{0.0591}{2} \log a_1 \cong \frac{0.0591}{2} \log [Ca^{2+}]$$

and

$$-8.720 \times 10^{-2} = \frac{0.0591}{2} \log [Ca^{2+}] = -\frac{0.0591}{2} pCa$$

$$pCa = 2.95$$

$$\text{rel error} = \frac{2.95 - 3.00}{3.00} \times 100 = -1.6 = -2\%$$

The anion-sensitive electrodes shown in Table 16-1 employ a solution of an anion exchange resin in an organic solvent. Liquid membrane electrodes for Ca^{2+}, K^+, NO_3^-, and BF_4^- have recently become available in which the liquid is held in the form of a polyvinylchloride gel. These electrodes have the appearance of a solid-state electrode described in the next section.

16A-6 SOLID-STATE AND PRECIPITATE ELECTRODES

Considerable work has been devoted to the development of solid membranes that are selective toward anions in the way that some glasses behave toward specific cations. We have seen that the selectivity of a glass membrane is due to the presence of anionic sites on its surface that show particular affinity toward certain positively charged ions. By analogy, a membrane having similar cationic sites might be expected to respond selectively toward anions. To exploit this possibility, attempts have been made to prepare membranes of salts containing the anion of interest and a cation that selectively precipitates that anion from aqueous solutions; for example, barium sulfate has been proposed for sulfate ion and silver halides for the various halide ions. The problem encountered in this approach has been in finding methods for fabricating membranes from the desired salts that possess adequate physical strength, conductivity, and resistance to abrasion and corrosion.

Membranes prepared from cast pellets of silver halides have been successfully used in electrodes for the selective determination of chloride, bromide, and iodide ions. An electrode employing a polycrystalline Ag_2S membrane is offered by one manufacturer for the determination of sulfide ion. Silver ions are sufficiently mobile to conduct electricity through the membrane. Mixtures of PbS, CdS, and CuS with silver sulfide provide membranes which are selective for Pb^{2+}, Cd^{2+}, and Cu^{2+}, respectively; in these, silver ion serves to transport electricity within the solid membrane.

A solid-state electrode selective for fluoride ion is also available from commercial sources. The membrane consists of a single crystal of lanthanum fluoride that has been doped with europium(II) to increase its electrical conductivity. The membrane, supported between a reference solution and the solution to be measured, shows the theoretical response to changes in fluoride ion activity from 10^0 to 10^{-6} M; that is,

$$E_{cell} = L - 0.0591 \log a_{F^-}$$

The electrode is reported to be selective for fluoride over other common anions by several orders of magnitude. Only hydroxide ion appears to offer serious interference.

Table 16-2 lists some of the commercially available solid-state electrodes.

16A-7 GAS-SENSING ELECTRODES

Figure 16-8 is a schematic diagram of a so-called gas-sensing electrode, which consists of a reference electrode, a specific ion electrode, and an electrolyte solution housed in a cylindrical plastic tube. A thin, replaceable, gas-permeable membrane is attached to one end of the tube and serves to separate the internal electrolyte solution from the analyte solution. The membrane is described by its manufacturers as being a thin microporous film fabricated from a hydrophobic plastic through which water and electrolytes cannot pass. Thus, the pores of the film contain only air or other gases to which the membrane is exposed. When a solution containing a gaseous analyte such as carbon dioxide is in contact with the membrane, the CO_2 distills into the pores, as shown by the reaction

$$CO_2(aq) \rightleftharpoons CO_2(g)$$

external solution   membrane pores

Because the pores are numerous, a state of equilibrium is rapidly approached. The

Table 16-2 Commercial Solid-State Electrodes*

Analyte Ion	Concentration Range, M	Preferred pH	Inferences
Br^-	10^0–5×10^{-6}	2–12	Max: $[S^{2-}] = 10^{-7}\ M$; $[I^-] = 2 \times 10^{-4}[Br^-]$
Cd^{2+}	10^{-1}–10^{-7}	3–7	Max: $[Ag^+]$, $[Hg^{2+}]$, $[Cu^{2+}] = 10^{-7}\ M$
Cl^-	10^0–5×10^{-5}	2–11	Max: $[S^{2-}] = 10^{-7}\ M$; traces of Br^-, I^-, CN^- do not interfere
Cu^{2+}	10^0–10^{-8}	3–7	Max: $[S^{2-}]$, $[Ag^+]$, $[Hg^{2+}] = 10^{-7}\ M$; high levels of Cl^-, Br^-, Fe^{3+}, Cd^{2+} interfere
CN^-	10^{-2}–10^{-6}	11–13	Max: $[S^{2-}] = 10^{-7}\ M$; $[I^-] = 0.1\ [CN^-]$; $[Br^-] = 5 \times 10^3[CN^-]$; $[Cl^-] = 10^6[CN^-]$
F^-	10^0–10^{-6}	5–8	Max: $[OH^-] = 0.1[F^-]$
I^-	10^0–5×10^{-8}	3–12	Max: $[S^{2-}] = 10^{-7}\ M$
Pb^{2+}	10^0–10^{-6}	4–7	Max: $[Ag^+]$, $[Hg^{2+}]$, $[Cu^{2+}] = 10^{-7}\ M$; high levels of Cd^{2+} and Fe^{3+} interfere
Ag^+	10^0–10^{-7}	2–9	Max: $[Hg^{2+}] = 10^{-7}\ M$
S^{2-}	10^0–10^{-7}	13–14	Max: $[Hg^{2+}] = 10^{-7}\ M$
Na^+	10^0–10^{-6}	9–10	Selectivity constants: Li^+, 0.002; K^+, 0.001; Rb^+, 0.00003; Cs^+, 0.0015; NH_4^+, 0.00003; Tl^+, 0.0002; Ag^+, 350; H^+, 100
SCN^-	10^0–5×10^{-6}	2–12	Max: $[OH^-] = [SCN^-]$; $[Br^-] = 0.003[SCN^-]$; $[Cl^-] = 20[SCN^-]$; $[NH_3] = 0.13[SCN^-]$; $[S_2O_3^{2-}] = 0.01[SCN^-]$; $[CN^-] = 0.007[SCN^-]$; $[I^-]$, $[S^{2-}] = 10^{-7}\ M$

* From *Analytical Methods Guide*, 9th ed. Cambridge, Mass.: Orion Research, Inc., December, 1978.

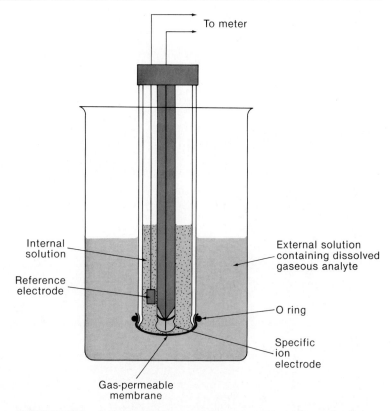

FIGURE 16-8 Schematic diagram of a gas-sensing electrode. Specific ion electrode is shown as a glass electrode. The reference electrode is a Ag/AgCl electrode. Other electrode combinations are possible.

CO_2 in the pores, however, is also in contact with the internal solution and a second equilibrium reaction is readily established, namely

$$\underset{\substack{\text{membrane} \\ \text{pores}}}{CO_2(g)} \rightleftharpoons \underset{\substack{\text{internal} \\ \text{solution}}}{CO_2(aq)}$$

As a consequence of the two reactions, the external solution rapidly (in a few seconds to minutes) equilibrates with the film of internal solution adjacent to the membrane. Here, another equilibrium causes the pH of the internal surface film to change, namely,

$$\underset{\substack{\text{internal} \\ \text{solution}}}{CO_2(aq) + 2H_2O} \rightleftharpoons \underset{\substack{\text{internal} \\ \text{solution}}}{HCO_3^- + H_3O^+}$$

A glass-reference electrode pair immersed in the film of internal solution (see Figure 16-8) then detects the pH change.

The overall reaction for the process just described is obtained by adding the three chemical equations to give

$$CO_2(aq) + 2H_2O \rightleftharpoons H_3O^+ + HCO_3^-$$

external solution   internal solution

The equilibrium constant for the reaction is given by

$$\frac{[H_3O^+][HCO_3^-]}{[CO_2(aq)]_{ext}} = K$$

If the concentration of HCO_3^- in the internal solution is made relatively high so that its concentration is not altered significantly by the CO_2 which distills, then

$$\frac{[H_3O^+]}{[CO_2(aq)]_{ext}} = \frac{K}{[HCO_3^-]} = K_g$$

which may be rewritten as

$$a_1 = [H_3O^+] = K_g[CO_2(aq)]_{ext} \tag{16-15}$$

where a_1 is the internal hydrogen ion activity.

The potential of the electrode system in the internal solution is dependent upon a_1 as described by Equation 16-3. Substitution of Equation 16-15 into 16-3 yields

$$E_{cell} = L + 0.0591 \log K_g[CO_2(aq)]_{ext}$$

or

$$E_{cell} = L' + 0.0591 \log [CO_2(aq)]_{ext}$$

where

$$L' = L + 0.0591 \log K_g$$

Thus, the potential of the cell consisting of the internal reference and indicator electrodes is determined by the CO_2 concentration of the external solution. Note that *no electrode comes directly in contact* with the analyte solution; for this reason, it would be better to call the device a gas-sensing *cell* rather than a gas-sensing electrode. Note also that the only species that will interfere with the measurement are dissolved gases that can pass through the membrane and can additionally affect the pH of the internal solution.

The possibility exists for increasing the selectivity of the gas-sensing electrode by employing an internal electrode that is sensitive to some species other than hydrogen ion. For example a nitrate-sensing electrode could be used to provide a cell that would be sensitive to nitrogen dioxide. Here, the equilibrium would be

$$2NO_2(aq) + H_2O \rightleftharpoons NO_2^- + NO_3^- + 2H^+$$

external solution   internal solution

This electrode should permit the determination of NO_2 in the presence of gases such as SO_2, CO_2, and NH_3, which would also alter the pH of the internal solution.

Gas-sensing electrode systems are commercially available for CO_2, NO_2, H_2S, SO_2, HF, HCN, and NH_3. An oxygen-sensitive cell system is also on the market; it is based on a voltammetric measurement and is discussed in Chapter 17.

16B INSTRUMENTS FOR CELL POTENTIAL MEASUREMENTS

An instrument for potentiometric measurements must draw essentially no electricity from the galvanic cell being studied. One reason for this requirement is that a current causes changes in reactant concentrations and thus changes in potential. More important is the dependency of cell potentials on current owing to the effects of IR drop and polarization phenomena (p. 334). As shown by the example that follows, the influence of IR drop is particularly significant with specific ion electrodes, which may have resistances of 100 MΩ or more. With such electrodes, currents must be limited to 10^{-12} A or smaller; this limitation requires that the potential measuring device have an internal resistance of 10^{12} Ω or more.

EXAMPLE 16-2. The true potential of a glass/calomel electrode system is 0.800 V; its internal resistance is 120 MΩ. What would be the relative error in the measured potential if the measuring device has a resistance of 600 MΩ?

Here, the circuit can be considered to consist of a potential source E_S and two resistors in series, R_S being that of the source and R_M that of the measuring device. That is,

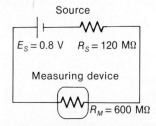

Source

$E_S = 0.8$ V $R_S = 120$ MΩ

Measuring device

$R_M = 600$ MΩ

From Ohm's law, we may write

$$E_S = IR_S + IR_M$$

where I is the current in this circuit consisting of the cell and the measuring device. The current is then given by

$$I = \frac{0.800}{(120 + 600) \times 10^6} = 1.11 \times 10^{-9} \text{ A}$$

The potential drop across the measuring device (which is the potential indicated by the device) is IR_M. Thus,

indicated potential $= 1.11 \times 10^{-9} \times 600 \times 10^6 = 0.667$ V

and

$$\text{rel error} = \frac{0.667 - 0.800}{0.800} \times 100 = -17\%$$

It is important to appreciate that an error in potential, such as that shown in Example 16-2 (0.133 V), would have an enormous effect on the accuracy of a concentration measurement based upon that potential. Thus, as shown in Section 16C-2, a 0.001-V uncertainty in potential leads to a relative error of about 4% in the determination of the hydrogen ion concentration of a solution by potential measurement with a glass electrode. An error of the size found in Example 16-2 would result in a concentration uncertainty of two orders of magnitude or more.

Historically, potential measurements were performed with a potentiometer, a null-point instrument in which the unknown potential was just offset or nulled by means of a standard reference potential. At null, zero electricity is drawn from the source being measured.

The potentiometer has by now been almost completely displaced by electronic voltmeters with very high internal resistances. These high-resistance instruments are usually called *pH meters*, although with the widespread development of membrane electrodes for other ions, they could be called more appropriately *pIon* or *ion meters*.

Numerous direct-reading pH meters are available commercially. Generally, these are solid-state devices employing an operational amplifier circuit which is designed to have an internal resistance of greater than 10^{12} Ω. The readout is either a digital meter or a meter with a 5- to 10-in. scale that covers a range of 0 to 14 pH units. Many of the latter instruments are also equipped with scale expansion capabilities which provide full scale ranges of 0.5 to 2 pH units; a precision of ± 0.001 to 0.005 pH unit can then be realized, depending upon amplifier noise level. It should be appreciated, however, that it is seldom, if ever, possible to measure pH with this kind of *accuracy;* indeed, uncertainties of ± 0.02 to 0.03 pH unit are typical (see p. 416).

The price of pH meters ranges from perhaps \$100 to \$1000 or more.

16C DIRECT POTENTIOMETRIC MEASUREMENTS

Direct potentiometric measurements can be used to complete chemical analyses of those species for which an indicator electrode is available. The technique is simple, requiring only a comparison of the potential developed by the indicator electrode in the test solution with its potential when immersed in a standard solution of the analyte; insofar as the response of the electrode is specific for the analyte and independent of matrix effects, no preliminary separation steps are required. In addition, direct potentiometric measurements are readily adapted to the continuous and automatic monitoring of analytical parameters.

Notwithstanding these attractive advantages, the user of direct potentiometric measurements must be alert to limitations that are inherent to the method. An important example is the existence of a *liquid junction* potential (p. 333) in most potentiometric measurements. For most electroanalytical methods, this junction potential is inconsequential and can be neglected. Unfortunately, however, its existence places a limitation on the accuracy that can be attained from a direct potentiometric measurement.

16C-1 EQUATION FOR DIRECT POTENTIOMETRY

The observed potential of a cell employed for a direct potentiometric measurement can be expressed in terms of the potentials developed by the indicator electrode, the reference electrode, and a junction potential. That is,

$$E_{obs} = E_{ref} - E_{ind} + E_j \tag{16-16}^{10}$$

Typically, the junction potential E_j has two components, the first being located at the junction of the analyte solution and one end of a salt bridge and the second where the bridge interfaces with the reference electrode solution. The two potentials tend to cancel one another but seldom do so completely. Thus, E_j in Equation 16-16 may be as large as 1 mV or more.

Ideally, the potential of the indicator electrode is related to the activity a_1 of M^{n+}, the ion of interest, by a Nernst-type equation

$$E_{ind} = L' + \frac{0.0591}{n} \log a_1 = L' - \frac{0.0591}{n} \, \text{pM} \tag{16-17}$$

where L' is a constant. For metallic electrodes, L' is often the standard potential for the indicator electrode; with membrane electrodes, however, L' may also include an unknown, time-dependent asymmetry potential (p. 403).

Combination of Equation 16-17 with Equation 16-16 and rearrangement yields

$$\text{pM} = -\log a_1 = \frac{E_{obs} - (E_{ref} + E_j - L')}{0.0591/n} \tag{16-18}$$

$$\text{pM} = -\log a_1 = \frac{E_{obs} - K}{0.0591/n} \tag{16-19}$$

The new constant K consists of three constants, of which at least one (E_j) has a magnitude that *cannot be evaluated from theory*. Thus, K must be determined experimentally with the aid of a standard solution of M before Equation 16-19 can be used for the calculation of pM.

Several methods for performing analyses by direct potentiometry have been developed; all are based, directly or indirectly, upon Equation 16-19.

16C-2 ELECTRODE CALIBRATION METHOD

In the electrode calibration method, K in Equation 16-19 is determined by measuring E_{obs} for one or more standard solutions of known pM. The assumption is then made that K does not change during measurement of the analyte solution. Generally, the calibration operation is performed at the time that pM for the unknown is determined; recalibration may be required if measurements extend over several hours.

The electrode calibration method offers the advantages of simplicity, speed, and applicability to the continuous monitoring of pM. Two important disadvantages may attend its use, however. One of these is that the results of an analysis are in terms of activities rather than concentrations (in some situations, this property may be an advantage rather than a disadvantage); the other is that the accuracy of a measurement obtained by this procedure is limited by the inherent uncertainty caused by the junction potential; unfortunately, this uncertainty can never be totally eliminated.

---

[10] As written, Equation 16-16 has the reference electrode acting as cathode and the indicator electrode acting as anode. If, in a particular cell, the roles are reversed, the signs of the two electrodes are likewise reversed.

Activity versus concentration. Electrode response is related to activity rather than to analyte concentration. Ordinarily, however, the scientist is interested in concentration, and the determination of this quantity from a potentiometric measurement requires activity coefficient data. More often than not, activity coefficients will be unavailable because the ionic strength of the solution is either unknown or so high that the Debye-Hückel equation is not applicable. Unfortunately, the assumption that activity and concentration are identical may lead to serious errors, particularly if the analyte is polyvalent.

The difference between activity and concentration is illustrated by Figure 16-9, where the lower curve gives the change in potential of a calcium electrode as a function of calcium chloride concentration (note that the activity or concentration scale is logarithmic). The nonlinearity of the curve is due to the increase in ionic strength—and the consequent decrease in the activity coefficient of the calcium ion—as the electrolyte concentration becomes larger. When these concentrations are converted to activities, the upper curve is obtained; note that this straight line has the theoretical slope of 0.0296 (0.0591/2).

Activity coefficients for singly charged ions are less affected by changes in ionic strength than are those for species with multiple charges. Thus, the effect shown in Figure 16-9 will be less pronounced for electrodes that respond to H^+, Na^+, and other univalent ions.

In potentiometric pH measurements, the pH of the standard buffer employed for calibration is generally based on the activity of hydronium ions. Thus, the resulting hydrogen ion analysis is also on an activity scale. If the unknown sample has a high ionic strength, the hydrogen ion *concentration* will differ appreciably from the activity measured.

Many physiological functions tend to be dependent upon the activity of metallic ions rather than their concentration. When this situation prevails, an indicator electrode provides an ideal analytical tool.

Inherent error in the electrode calibration procedure. A serious disadvantage of the electrode calibration method is the existence of an inherent uncertainty that results from the assumption that K in Equation 16-19 remains constant after calibration. This assumption can seldom, if ever, be exactly true because the electrolyte

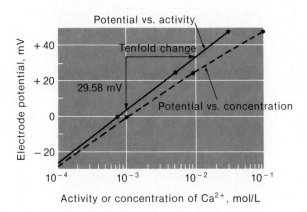

FIGURE 16-9 Response of a calcium ion electrode to variations in the calcium ion concentration and activity of solutions prepared from pure calcium chloride. (Courtesy of Orion Research, Inc.)

composition of the unknown will almost inevitably differ from that of the solution employed for calibration. The junction potential contained in K will vary slightly as a consequence, even when a salt bridge is used. This uncertainty will frequently be of the order of 1 mV or more; unfortunately, because of the nature of the potential-activity relationship, such an uncertainty has an amplified effect on the inherent accuracy of the analysis. The magnitude of the uncertainty in analyte concentration can be estimated by differentiating Equation 16-19, holding E_{obs} constant:

$$-\log_{10} e \, \frac{da_1}{a_1} = -0.434 \, \frac{da_1}{a_1} = -\frac{dK}{0.0591/n}$$

$$\frac{da_1}{a_1} = \frac{ndK}{0.0256}$$

Upon replacing da_1 and dK with finite increments and multiplying both sides of the equation by 100, we obtain

$$\frac{\Delta a_1}{a_1} \times 100 = 3.90 \times 10^3 \, n\Delta K = \% \text{ relative error}$$

The quantity $\Delta a_1/a_1$ is the relative error in a_1 associated with an absolute uncertainty ΔK in K. If, for example, ΔK is ± 0.001 V, a relative error in activity of about $\pm 4n\%$ can be expected. *It is important to appreciate that this uncertainty is characteristic of all measurements involving cells that contain a salt bridge and that this error cannot be eliminated by even the most careful measurements of cell potentials or the most sensitive and precise measuring devices;* nor does it appear possible to devise a method for completely eliminating the uncertainty in K that is the source of this error.

16C-3 CALIBRATION CURVES FOR DIRECT POTENTIOMETRY

An obvious way of correcting potentiometric measurements to give results in terms of concentration is to make use of an empirical calibration such as the lower curve in Figure 16-9. Success of this approach, however, is initially dependent upon the ionic composition of the standard being essentially the same as that of the analyte—a condition that is difficult to realize experimentally for complex samples.

Where electrolyte concentrations are not too great, it is often helpful to swamp both the samples and the calibration standards with a measured excess of an inert electrolyte. Under these circumstances, the added effect of the electrolyte in the sample becomes negligible, and the empirical calibration curve yields results in terms of concentration. This approach has been employed, for example, in the potentiometric determination of fluoride in public water supplies. Here, both samples and standards are diluted on a 1:1 basis with a solution containing sodium chloride, a citrate buffer, and an acetate buffer; the diluent is sufficiently concentrated so that the samples and standards do not differ significantly in ionic strength. The procedure permits rapid measurement of fluoride ion concentrations on the order of 1 ppm with an accuracy of about 5% relative.

Procedures for the determination of fluoride ion in water and in toothpaste samples are found in Chapter 31, Method 14-4.

16C-4 STANDARD ADDITION METHOD

In the standard addition method, the potential of the electrode system is measured before and after addition of a small volume of a standard to a known volume of the sample. The assumption is made that this addition does not alter the ionic strength significantly and, thus, the activity coefficient f of the analyte. It is further assumed that the added standard does not significantly alter the junction potential.

If the potentials before and after the addition are E_1 and E_2, respectively, Equation 16-19 can be written as

$$-\log a_1 = -\log C_x f = \frac{E_1 - K}{0.0591/n}$$

where a_1 is the activity of the analyte in the sample, f is its activity coefficient, and C_x is its molar concentration. Addition of V_s mL of a standard with a molar concentration of C_s to V_x mL of the sample causes the potential to acquire a value E_2; Equation 16-19 then becomes

$$-\log \left[\frac{C_x V_x + C_s V_s}{V_x + V_s} f \right] = \frac{E_2 - K}{0.0591/n}$$

Subtraction of the first equation from the second and rearrangement gives,

$$\log \left[\frac{C_x(V_x + V_s)}{C_x V_x + C_s V_s} \right] = \frac{E_2 - E_1}{0.0591/n} = \frac{\Delta E}{0.0591/n}$$

This equation in turn rearranges to

$$C_x = \frac{C_s V_s}{(V_x + V_s) \, 10^{-n\Delta E/0.0591} - V_x} \tag{16-20}$$

Thus, C_x is readily calculated from the concentration of the standard, the two volumes, and the potential difference ΔE.

The standard addition method has been applied to the determination of chloride and fluoride in samples of commercial phosphors.[11] Here, two solid-state indicator electrodes (one for fluoride and the other for chloride) and a reference electrode were used; the added standard contained known quantities of the two anions. The relative standard deviations for the measurement of replicate samples were found to be 0.7% for fluoride and 0.4% for chloride. When the standard addition method was not used, relative errors for the analyses appeared to range between 1 and 2%.

16C-5 POTENTIOMETRIC pH MEASUREMENTS WITH A GLASS ELECTRODE[12]

The glass electrode is unquestionably the most important indicator electrode for hydrogen ion. It is convenient to use and is subject to few of the interferences that affect other pH-sensing electrodes.

Glass electrodes are available at relatively low cost and in many shapes and sizes. A common variety is illustrated in Figure 16-1 (p. 399); the reference electrode is usually a commercial saturated calomel electrode.

[11] L. G. Bruton, *Anal. Chem.*, **43**, 579 (1971).
[12] For a detailed discussion of potentiometric pH measurements, see: R. G. Bates, *Determination of pH, Theory and Practice*. New York: Wiley, 1964.

The glass/calomel electrode system is a remarkably versatile tool for the measurement of pH under many conditions. The electrode can be used without interference in solutions containing strong oxidants, reductants, proteins, and gases; the pH of viscous or even semisolid fluids can be determined. Electrodes for special applications are available. Included among these are small electrodes for pH measurements in a drop (or less) of solution, in a cavity of a tooth, or in the sweat on the skin; microelectrodes which permit the measurement of pH inside a living cell, systems for insertion in a flowing liquid stream to provide a continuous monitoring of pH; and a small glass electrode that can be swallowed to indicate the acidity of the stomach contents (the calomel electrode is kept in the mouth).

Summary of errors affecting pH measurements with the glass electrode. The ubiquity of the pH meter and the general applicability of the glass electrode tend to lull the chemist into the attitude that any measurement obtained with such an instrument is surely correct. It is well to guard against this false sense of security since there are distinct limitations to the electrode system. Some of these have been discussed in earlier sections and include the following:

1. **The alkaline error.** The ordinary glass electrode becomes somewhat sensitive to alkali-metal ions at pH values greater than 9; low pH readings result.
2. **The acid error.** At a pH less than 0.5, values obtained with a glass electrode tend to be somewhat high.
3. **Dehydration.** Dehydration of the electrode may cause unstable performance and errors.
4. **Errors in unbuffered neutral solutions.** Equilibrium between the electrode surface layer and the solution is achieved only slowly in poorly buffered, approximately neutral solutions. Errors will arise unless time is allowed for this equilibrium to be established. In determining the pH of poorly buffered solutions, the glass electrode should first be thoroughly rinsed with water. Then, if the unknown is plentiful, the electrodes should be placed in successive portions until a constant pH reading is obtained. Good stirring is also helpful; several minutes should be allowed for the attainment of steady readings.
5. **Variation in junction potential.** It should be reemphasized that junction potential variations represent a fundamental uncertainty in the measurement of pH, for which a correction cannot be applied. Absolute values more reliable than 0.01 pH unit are generally unobtainable. Even reliability to 0.03 pH unit requires considerable care. On the other hand, it is often possible to detect pH *differences* between similar solutions or pH *changes* in a single solution that are as small as 0.001 unit. For this reason, many pH meters are designed to permit readings in smaller increments than 0.01 pH unit.
6. **Error in the pH of the standard buffer.** Any inaccuracies in the preparation of the buffer used for calibration or changes in its composition during storage will be propagated as errors in pH measurements. A common cause of deterioration is the action of bacteria on organic components of buffers.

16D POTENTIOMETRIC TITRATIONS

The measurement of the potential of a suitable indicator electrode permits the establishment of the equivalence point for a titration (a *potentiometric titration*). A potentiometric titration provides different information from a direct potentiometric measurement. For example, the direct measurement of 0.100 *M* acetic and hydrochloric acid solutions with a pH-sensitive electrode would yield widely different hydrogen ion concentrations because the former is only partially dissociated. On the other hand, potentiometric titrations of equal volumes of the two acids would require the same amount of standard base for neutralization.

The potentiometric end point is widely applicable and provides inherently more accurate data than the corresponding method that makes use of indicators. It is particularly useful for titration of colored or turbid solutions and for detecting the presence of unsuspected species in a solution. Unfortunately, it is more time-consuming than a conventional titration. On the other hand, it is readily applied with automatic titration devices.

Figure 16-10 shows a typical apparatus for performing a potentiometric titration. Ordinarily, the titration involves measuring and recording a cell potential (in units of mV or pH) after each addition of reagent. The titrant is added in large increments at the outset; as the end point is approached (as indicated by larger potential changes per addition), the increments are made smaller.

Sufficient time must be allowed for the attainment of equilibrium after each addition of reagent. Precipitation reactions may require several minutes for equilibration, particularly in the vicinity of the equivalence point. A close approach to equilibrium is indicated when the measured potential ceases to drift by more than

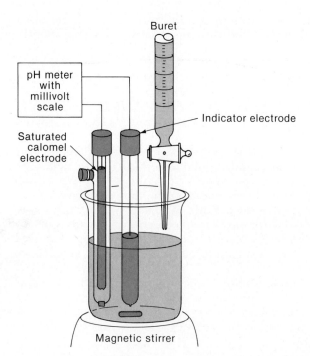

FIGURE 16-10 Apparatus for a potentiometric titration.

one or two millivolts per minute. Good stirring is frequently effective in hastening the achievement of equilibirum.

The first two columns of Table 16-3 consist of typical potentiometric titration data obtained with the apparatus illustrated in Figure 16-10. The data near the end point are plotted in Figure 16-11a. Note that this experimental plot closely resembles titration curves derived from theoretical considerations.

16D-1 END POINT DETERMINATION

Several methods can be used to determine the end point for a potentiometric titration. The most straightforward involves a direct plot of potential versus reagent volume, as in Figure 16-11a. The midpoint in the steeply rising portion of the curve is then estimated visually and taken as the end point. Various mechanical methods to aid in the establishment of the midpoint have been proposed; it is doubtful, however, that these significantly improve the accuracy.

A second approach is to calculate the change in potential per unit change in volume of reagent (that is, $\Delta E / \Delta V$), as has been done in column 3 of Table 16-3. A plot of this parameter as a function of average volume leads to a sharp maximum at the end point (see Figure 16-11b). Alternatively, the ratio can be evaluated during the titration and recorded directly in lieu of the potential itself. Thus, in column 3 of Table 16-3, it is seen that the maximum is located between 24.3 and 24.4 mL; selection of 24.35 mL would be adequate for most purposes.

Column 4 in Table 16-3 and Figure 16-11c show that the second derivative of the data changes sign at the point of inflection in the titration curve. This change is often used as the analytical signal in automatic titrators.

Table 16-3 Potentiometric Titration Data for 2.433 mmol of Chloride with 0.1000 M Silver Nitrate

Vol AgNO$_3$, mL	E vs. SCE, V	$\Delta E/\Delta V$, V/mL	$\Delta^2 E/\Delta V^2$, V/mL^2
5.0	0.062		
15.0	0.085	0.002	
20.0	0.107	0.004	
22.0	0.123	0.008	
23.0	0.138	0.015	
23.50	0.146	0.016	
23.80	0.161	0.050	
24.00	0.174	0.065	
24.10	0.183	0.09	
24.20	0.194	0.11	2.8
24.30	0.233	0.39	4.4
24.40	0.316	0.83	-5.9
24.50	0.340	0.24	-1.3
24.60	0.351	0.11	-0.4
24.70	0.358	0.07	
25.00	0.373	0.050	
25.5	0.385	0.024	
26.0	0.396	0.022	
28.0	0.426	0.015	

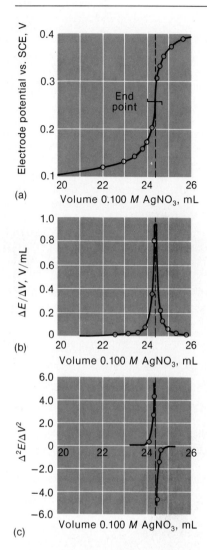

(a)

(b)

(c)

FIGURE 16-11 (a) Potentiometric titration curve of 2.433 mmol of Cl^- with 0.100 M $AgNO_3$. (b) First derivative curve. (c) Second derivative curve.

Gran's plots. Gran[13] has suggested a method to permit establishment of the end point with data from a region of the titration where the reaction is forced nearly to completion by the common ion effect. Such a procedure has the advantage of requiring fewer data points than a conventional plot. Furthermore, it may yield more accurate end points for titrations where the rate of change in p-function is small in the equivalence-point region.

As an example, we shall apply Gran's method to the data shown in Table 16-3 for the potentiometric titration of chloride ion with silver nitrate. Short of the

[13] G. Gran, *Analyst*, **77**, 661 (1952).

equivalence point in this titration, the silver electrode behaves as second-order electrode for chloride ion, and the cell potential E_{obs} is given by the equation

$$E_{obs} = E^0_{AgCl} - 0.0591 \log [Cl^-] - E_{SCE}$$

where E_{SCE} is the potential of a saturated calomel reference electrode (0.241 V). We have neglected the junction potential because it is of no significance in a potentiometric titration. It is convenient to rewrite this equation in the form

$$\log [Cl^-] = -16.9 E_{obs} + K'$$

where 16.9 is the reciprocal of 0.0591 and K' is equal to $(E^0_{AgCl} - E_{SCE})/0.0591$. Taking the antilogarithm of both sides of this equation leads to

$$[Cl^-] = \text{antilog}\,(-16.9 E_{obs} + K')$$

Short of the end point in the titration, $[Cl^-]$ is given by

$$[Cl^-] = \frac{V_{Cl}M_{Cl} - V_{Ag}M_{Ag}}{V_{Cl} + V_{Ag}}$$

where V_{Cl} and V_{Ag} are the volumes of the analyte and the silver nitrate solutions; M_{Cl} and M_{Ag} are the initial molar concentrations of the two solutions. Combining the two equations and rearranging gives

$$(V_{Cl}M_{Cl} - V_{Ag}M_{Ag}) = (V_{Cl} + V_{Ag})\,\text{antilog}\,(-16.9 E_{obs} + K')$$

For a given titration, all of the terms in this equation are constant except V_{Ag} and E_{obs}. Thus, a plot of the quantity $(V_{Cl} + V_{Ag})\,\text{antilog}\,(-16.9 E_{obs})$ against V_{Ag} should be linear. In addition, at equivalence, the left side of the equation will equal zero. Thus, an extrapolation of the plot to zero on the antilog axis will give the equivalence-point volume. A Gran's plot of the data in Table 16-3 is shown in Figure 16-12. Note that the first four points lie on a good straight line. A departure from linearity is observed, however, for the data within about 2 mL of the equivalence point. For less complete reactions, the curvature occurs earlier in the titration.

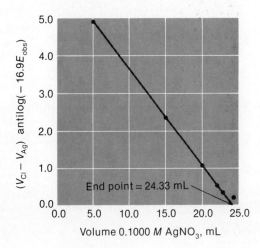

FIGURE 16-12 Gran's plot of the data in Table 16-3.

Clearly, a satisfactory end point can be obtained from 3 or 4 data points, in contrast to the 19 that were recorded in Table 16-3.

When the concentration of the reagent is made large with respect to the analyte, $(V_{Cl} + V_{Ag})$ is equal to approximately V_{Cl} throughout the titration; the quantity antilog $(-16.9 E_{obs})$ can then serve as the ordinate.

Semiantilog paper is available[14] that makes it possible to obtain Gran's plots without computations. The vertical axis of this paper is skewed to eliminate the need to correct for the volume change, provided the titrant volume is less than 10% of the total at the end point.

Titration to a fixed potential. Another procedure consists of titrating to a predetermined end point potential. The value chosen may be the theoretical equivalence-point potential calculated from formal potentials or an empirical potential obtained by the titration of standards. Such a method demands that the equivalence-point behavior of the system be entirely reproducible.

16D-2 PRECIPITATION TITRATIONS

Electrode systems. The indicator electrode for a precipitation titration is often the metal from which the reacting cation is derived. Membrane electrodes that are sensitive to one of the ions involved in the titration process may also be employed. For example, a fluoride-sensitive membrane electrode has been employed for the determination of the fluoride content of toothpastes; here, a standard solution of lanthanum(III) served as the precipitant.

Silver nitrate is without question the most versatile reagent for precipitation titrations; argentometric methods exist for the determination of halides (see Section 7B), halogenoids, mercaptans, sulfides, arsenates, phosphates, and oxalates. A silver electrode serves as indicator for the potentiometric titration of all of these ions. For reagent and analyte concentrations that are 0.1 M or greater, a calomel reference electrode (Figure 13-8a) can be located directly in the titration mixture; the slight leakage of chloride ions from the salt bridge is negligible. If the analyte solution is very dilute or if the highest precision is required, this leakage can be the source of significant positive error. The difficulty can be eliminated by immersing the electrode in a concentrated potassium nitrate solution that in turn is connected to the analyte solution by a salt bridge consisting of an agar gel that contains about 3% potassium nitrate.

Titration curves for mixtures. Multiple end points are often revealed by a potentiometric titration when the sample contains more than one reacting species. The titration of halide mixtures with silver nitrate is an important example. Figure 7-3 (p. 183) is a theoretically derived titration curve for a mixture of chloride and iodide with silver nitrate. Because the measured potential is directly proportional to the negative logarithm of the silver ion concentration, the experimental potential curve has the same shape although the ordinate units are different.

---

[14] Orion Research Inc., 380 Putnam Ave., Cambridge, MA 02139.

16D-3 COMPLEX-FORMATION TITRATIONS

Both metal electrodes and membrane electrodes have been applied to the detection of end points in reactions that involve formation of a soluble complex. The mercury electrode, illustrated in Figure 16-13, is particularly useful for EDTA titrations.[15] It will function as an indicator electrode for the titration of cations forming complexes that are less stable than HgY^{2-} (see p. 397).

Reilley and co-workers have made a systematic theoretical and experimental study of the application of the mercury electrode to the potentiometric determination of 29 divalent, trivalent, and tetravalent cations with EDTA.[16] In these studies, 5- to 500-mg quantities of the cation were titrated with 0.1 or 0.005 M reagent solutions. One drop of a 10^{-3} M solution of HgY^{2-} was used in each titration.

Figure 16-14 illustrates an application of their procedure to the determination of bismuth, cadmium, and calcium in a mixture. Bismuth(III) is first titrated at a pH of 1.2. At this acidity, neither cadmium nor calcium ions react to any significant extent. Bismuth, on the other hand, forms a complex that is sufficiently stable to provide a satisfactory end point. After bismuth has been titrated, the solution is brought to a pH of 4 by addition of an acetate/acetic acid buffer, and the titration is continued to an end point for cadmium. Calcium ions do not react appreciably at this pH but can be subsequently titrated in a basic solution obtained by addition of an ammonia/ammonium chloride buffer.

---

[15] These electrodes can be obtained from Kontes Manufacturing Corp., Vineland, NJ 08360.
[16] See: C. N. Reilley and R. W. Schmid, *Anal. Chem.*, **30**, 947 (1958); and C. N. Reilley, R. W. Schmid, and D. W. Lamson, *Ibid.*, 953 (1958).

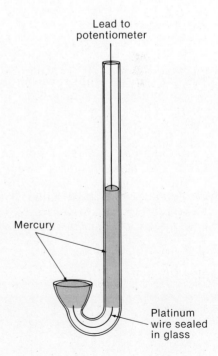

Lead to
potentiometer

Mercury

Platinum
wire sealed
in glass

FIGURE 16-13 A typical mercury electrode.

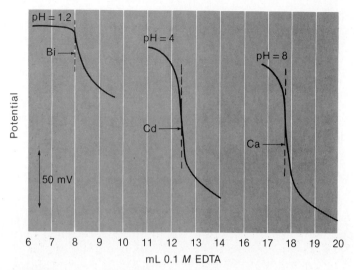

FIGURE 16-14   Potentiometric EDTA titration of a mixture
of three cations. [C. N. Reilley, R. W. Schmid, and D. W.
Lamson, *Anal. Chem.*, **30**, 957 (1958). With permission of the
American Chemical Society.]

16D-4 NEUTRALIZATION TITRATIONS

In Chapters 8 and 9 we considered theoretical curves for various neutralization titrations in some detail. The shapes of these curves can be closely approximated
experimentally. Often, however, the experimental curves will be displaced from
the theoretical curves because concentrations rather than activities were used in
their derivation. A study of the theoretical curves will show that the small error
inherent in the potentiometric measurement of pH is of no consequence insofar as
locating the end point is concerned.

  Potentiometric acid-base titrations are particularly useful for the analysis of
mixtures of acids or polyprotic acids (or bases) since discrimination between the
end points can often be made.

Determination of dissociation constants.   An approximate numerical value for
the dissociation constant of a weak acid or base can be estimated from potentiometric titration curves. In theory, this quantity can be obtained from any point along
the curve; as a practical matter, it is most easily derived from the pH at the point of
half-neutralization. For example, in the titration of the weak acid HA, we may ordinarily assume that at the midpoint,

$$[HA] \cong [A^-]$$

and, therefore,

$$K_a = \frac{[H^+][A^-]}{[HA]} = [H^+]$$

or

$$pK_a = pH$$

It is important to note that a dissociation constant determined in this way may differ from that shown in a table of dissociation constants by a factor of 2 or more because the latter is based upon activities while the former is not. Thus, if we write the dissociation-constant expression in its more exact form we obtain

$$K_a = \frac{a_{H_3O^+} \cdot a_{A^-}}{a_{HA}} = \frac{a_{H_3O^+} \cdot [A^-] \cdot f_{A^-}}{[HA] \cdot f_{HA}}$$

The assumption that $[A^-]$ and $[HA]$ are approximately equal is still valid; thus we may write

$$K_a = \frac{a_{H_3O^+} \cdot \cancel{[A^-]} \cdot f_{A^-}}{\cancel{[HA]} \cdot f_{HA}}$$

$$= \frac{a_{H_3O^+} \cdot f_{A^-}}{f_{HA}} \tag{16-21}$$

but measurement with the glass electrode provides a good approximation of $a_{H_3O^+}$. Thus, the measured K_a will differ from the thermodynamic value found in the literature by the ratio of the two activities.

EXAMPLE 16-3. What would be the experimental value for K_a for H_3PO_4 and $H_2PO_4^-$ if it were assumed that the two dissociation constants were equal to the measured hydrogen ion activities at the midpoint in their titration curves and that the ionic strength at this point was 0.1.

The activity coefficient for H_3PO_4 is approximately 1 since it bears no charge while the coefficients for $H_2PO_4^-$ and HPO_4^{2-} are 0.78 and 0.36, respectively (see Table 4-5, p. 119). Rearranging Equation 16-21 gives

$$K_a(\text{exptl}) = a_{H_3O^+} = \frac{K_a \cdot f_{HA}}{f_{A^-}}$$

and for H_3PO_4

$$K_1 \text{ (exptl)} = 7.11 \times 10^{-3} \times 1.0/0.78 = 9.1 \times 10^{-3}$$

$$K_2 \text{ (exptl)} = 6.34 \times 10^{-8} \times 0.78/0.36 = 1.37 \times 10^{-7}$$

Note that the experimental value for K_1 would be about 30% high while that for K_2 somewhat greater than 100%.

A value of the equivalent weight and the approximate dissociation constant of a pure sample of an unknown acid can be obtained from a single potentiometric titration; this information is frequently sufficient to identify the acid.

End points in nonaqueous titrations. The potentiometric method has proved particularly useful for signaling end points of titrations in nonaqueous solvents. The ordinary glass/calomel electrode system can be used; the electrodes must be stored in water between titrations to prevent dehydration of the glass and precipitation of potassium chloride in the salt bridge. The millivolt scale rather than the pH scale of potentiometer should be employed because the potentials in nonaqueous solvents may exceed the pH scale. Furthermore, the pH scale based upon aqueous buffers has no significance in a nonaqueous environment. The titration curves are thus empirical; however, they provide a useful and satisfactory means of end point detection.

16D-5 OXIDATION-REDUCTION TITRATIONS

The derivation of theoretical titration curves for oxidation-reduction processes was considered in Chapter 14. In each example, an electrode potential related to the concentration ratio between the oxidized and reduced forms of either of the reactants was determined as a function of the titrant volume. These curves can be duplicated experimentally (here again the curves may be displaced on the ordinate scale because of the effect of ionic strength), provided an indicator electrode responsive to at least one of the couples involved in the reaction is available. Such electrodes exist for most, but not all, of the reagents described in Chapter 14.

Indicator electrodes for oxidation-reduction titrations are generally constructed from platinum, gold, mercury, or palladium. The metal chosen must be unreactive with respect to the components of the reaction—it is merely a medium for electron transfer. Without question, the platinum electrode is most widely used for oxidation-reduction titrations. Curves similar to those shown in Figure 14-1 can be obtained experimentally with a platinum/calomel electrode system. The end point can be established by the methods already discussed.

16D-6 DIFFERENTIAL TITRATIONS

We have seen that a derivative curve generated from the data of a conventional potentiometric titration curve (Figure 16-11b, p. 422) reaches a distinct maximum in the vicinity of the equivalence point. It is also possible to acquire titration data directly in derivative form by means of a relatively simple apparatus.

A differential titration requires the use of two *identical* indicator electrodes, one of which is well shielded from the bulk of the solution. Figure 16-15 illustrates a typical arrangement. Here, one of the electrodes is contained in a small sidearm test tube. Contact with the bulk of the solution is made through a small (~1 mm) hole in the bottom of the tube. Because of this restricted access, the composition of the solution surrounding the shielded electrode will not be immediately affected by an addition of titrant to the bulk of the solution. The resulting difference in solution composition gives rise to a difference in potential, ΔE, between the electrodes. After each potential measurement, the solution is homogenized by squeezing the rubber bulb several times, whereupon ΔE again becomes zero. If the volume of solution in the tube that shields the electrode is kept small (say, 1 to 5 mL), the error arising from failure of the final addition of reagent to react with this portion of the solution can be shown to be negligibly small.

For oxidation-reduction titrations, the indicator electrodes can be two platinum wires with one enclosed in an ordinary medicine dropper.

The main advantage of a differential method is the elimination of the need for the reference electrode and salt bridge.

16D-7 AUTOMATIC TITRATORS

In recent years, several automatic titrators based on the potentiometric principle have become available from commercial sources. These instruments are useful

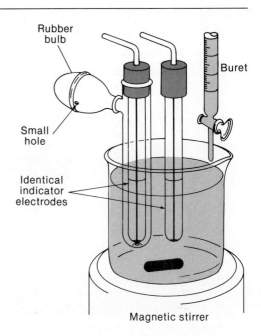

FIGURE 16-15 Apparatus for differential potentiometric titrations.

where many routine analyses are required.[17] Automatic titrators cannot yield results that are more accurate than those obtained by manual potentiometric techniques; however, they do decrease the operator time needed to perform the titrations and, thus, may offer a significant economic advantage.

Flow control and measurement of reagent volume. Several methods exist for automatic control of the addition of a reagent and the measurement of its volume. The simplest employs an ordinary buret in which the stopcock is replaced with an electromagnetic pincer device; an elastic plastic tube is inserted between the buret body and the tip. Flow is prevented by pinching the tube between a spring-loaded soft iron piece and a metal wedge. Titrant is introduced by passage of electricity through a solenoid that surrounds the pinching device. In another type of valve, a small piece of iron is sealed into a glass or plastic tube that fits inside the outflow tube of a buret. The two surfaces are ground to form a stopper. Current in a solenoid unseats the stopper and allows a flow of reagent.

The most widely used apparatus for automatic reagent addition consists of a calibrated syringe that is activated by a motor-driven micrometer screw. The volume is determined from the number of turns that the screw makes during the titration.

Preset end-point titrators. Figure 16-16 is a schematic diagram of the simplest and least expensive type of automatic titrator. Here, a preset equivalence point po-

[17] For a comprehensive discussion of automatic titrators, see: G. Svehla, *Automatic Potentiometric Titrations*. New York: Pergamon Press, 1978; and J. K. Foreman and P. B. Stockwell, *Automatic Chemical Analysis*, pp. 44–62. New York: Wiley, 1975.

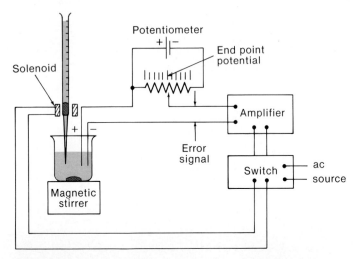

FIGURE 16-16   Automatic preset end-point titrator.

tential is applied across the electrodes by means of a calibrated potentiometer. If a difference exists between this potential and that of the electrodes, an "error" signal results. This signal is amplified and closes an electronic switch that permits a flow of electricity through the solenoid-operated valve of the buret. As the error signal approaches zero, current to the solenoid is switched off, and flow of titrant ceases.

Second-derivative titrators.   Second-derivative titrators can also be relatively simple devices; they have the advantage that no preknowledge of the equivalence-point potential is required. The signal processor of these devices contains two electronic derivative circuits in series to convert the amplified signal from the electrode to a voltage that is proportional to the second derivative of the electrode potential of the indicator electrode. The output is then similar in form to that shown in Figure 16-12c, where the sign of the signal changes at the equivalence point. This change in sign then causes a switching device to turn off the flow of titrant.

Recording titrators.   Recording titrators carry a titration beyond the equivalence point while recording a curve for the analyte. In some instruments, the rate of reagent addition is held constant and is synchronized with the chart drive of a millivolt recorder. The pen then records the amplified output potential of the cell as a function of time, which is proportional to the reagent volume.

Recording titrators are frequently equipped to perform a titration the way a skilled chemist does; that is, to add titrant rapidly before and after the end point but in small increments as the end point is approached and passed.

Fully automatic titrators.   A fully automatic titrator is equipped with a turntable that holds a series of samples for titration. After a titration has been completed, the solution is discarded, the titration vessel and electrodes are rinsed, the buret is refilled, the sample table is rotated so that a measured volume of a new sample is introduced into the system, and the titration process is resumed. Such instruments are controlled by microprocessors and usually have computing facilities for calculating and printing out the analytical results. Instruments of this kind are expensive, but their costs can be easily justified for laboratories that must perform large numbers of routine titrations on a regular basis.

PROBLEMS*

* 16-1. (a) Calculate the standard potential for the reaction

$$AgVO_3(s) + e \rightleftharpoons Ag(s) + VO_3^-$$

For $AgVO_3$, $K_{sp} = 5.2 \times 10^{-7}$.

(b) Give a schematic representation of a cell with a silver indicator electrode as a cathode and a saturated calomel electrode as an anode that could be used for the determination of VO_3^-.

(c) Derive an equation that relates the measured potential of the cell in (b) to pVO_3 (assume that the junction potential is zero).

(d) Calculate the pVO_3 of a vanadate-containing solution that is saturated with $AgVO_3$ and contained in the cell described in (b) if the resulting potential is 0.525 V.

16-2. (a) Calculate the standard potential for the reaction

$$ZnCO_3(s) + 2e \rightleftharpoons Zn(s) + CO_3^{2-}$$

For $ZnCO_3$, $K_{sp} = 1.4 \times 10^{-11}$.

(b) Give a schematic representation of a cell with a zinc indicator electrode as the anode and a saturated calomel electrode as the cathode that could be used for determining CO_3^{2-}.

(c) Derive an equation that relates the measured potential of the cell in (b) to pCO_3 (assume that the junction potential is zero).

(d) Calculate the pCO_3 of a solution that is saturated with $ZnCO_3$ and contained in the cell described in (b) if the resulting potential is 0.989 V.

* 16-3. Give a schematic representation of each of the following cells and an equation for the relationship between the cell potential and the desired quantity. Assume that the junction potential is negligible, and specify any necessary concentrations as 1.00×10^{-4} M. Treat the indicator electrode as the cathode in each case.

(a) A cell with a Cu indicator electrode for the determination of $pFe(CN)_6$. [$Cu_2Fe(CN)_6$ has a low solubility.]

(b) A cell with a Hg indicator electrode for the determination of pIO_3. [$Hg_2(IO_3)_2$ has a low solubility.]

(c) A cell with a platinum electrode for the determination of $pTl(I)$.

16-4. Give a schematic representation of each of the following cells and an equation for the relationship between the cell potential and the desired quantity. Assume that the junction potential is negligible, and specify any necessary concentrations as 1.00×10^{-4} M. Treat the indicator electrode as the anode in each case.

(a) A cell with a Pb indicator electrode for the determination of PbC_2O_4. (PbC_2O_4 has a low solubility.)

(b) A cell with a Ag indicator electrode for the determination of pPO_4. (Ag_3PO_4 has a low solubility.)

* Answers to problems and parts of problems marked with an asterisk are to be found at the end of the book.

(c) A cell with a platinum electrode for the determination of pTi(II).

* 16-5. The following cell was employed for the determination of $pCrO_4$:

$$SCE\|Ag_2CrO_4(sat'd),CrO_4^{2-}(xM)|Ag$$

Calculate $pCrO_4$ if the cell potential is 0.476 V.

16-6. The following cell was employed to determine the pSO_4 of a solution:

$$SCE\|Hg_2SO_4(sat'd),SO_4^{2-}(xM)|Hg$$

Calculate the pSO_4 if the potential was 0.506 V.

16-7. Calculate the potential of the cell (neglecting the junction potential)

$$\text{indicator electrode}\|SCE$$

where the indicator electrode is mercury immersed in a solution that is

* (a) $4.00 \times 10^{-6} M$ Hg^{2+}.
  (b) $4.00 \times 10^{-6} M$ Hg_2^{2+}.
* (c) saturated with Hg_2Br_2 ($K_{sp} = 5.8 \times 10^{-23}$) and is $3.00 \times 10^{-3} M$ in Br^-.
  (d) $2.50 \times 10^{-5} M$ in $Hg(NO_3)_2$ and $0.0400 M$ in KCl.

$$Hg^{2+} + 2Cl^- \rightleftharpoons HgCl_2(aq) \qquad K_f = 1.6 \times 10^{13}$$

* (e) $2.50 \times 10^{-5} M$ in $Hg(NO_3)_2$ and $0.00400 M$ in KCl.

* 16-8. The formation constant for the mercury(II) acetate complex is

$$Hg^{2+} + 2OAc^- \rightleftharpoons Hg(OAc)_2(aq) \qquad K_f = 2.7 \times 10^8$$

Calculate the standard potential for the half-reaction

$$Hg(OAc)_2(aq) + 2e \rightleftharpoons Hg(l) + 2OAc^-$$

16-9. The standard electrode potential for the reduction of the mercury complex of EDTA is given by

$$HgY^{2-} + 2e \rightleftharpoons Hg(l) + Y^{4-} \qquad E^0 = 0.21 V$$

Calculate the formation constant for the reaction

$$Hg^{2+} + Y^{4-} \rightleftharpoons HgY^{2-}$$

16-10. Calculate the potential of the cell (neglecting the junction potential)

$$Hg|HgY^{2-}(4.50 \times 10^{-5} M),Y^{4-}(xM)\|SCE$$

where Y^{4-} is the EDTA anion, and the concentration of Y^{4-} is

* (a) $3.33 \times 10^{-1} M$, (b) $3.33 \times 10^{-3} M$, and * (c) $3.33 \times 10^{-5} M$.

$$HgY^{2-} + 2e \rightleftharpoons Hg(l) + Y^{4-} \qquad E^0 = 0.21 V$$

* 16-11. The following cell was found to have a potential of 0.209 V when the solution in the left compartment was a buffer of pH 6.34:

$$\text{glass electrode}|H^+(a = x)\|SCE$$

The following potentials were obtained when the buffered solution was replaced with unknowns. Calculate the pH and the hydrogen ion activity of each unknown.

(a) 0.064 V   (c) 0.510 V
(b) 0.329 V   (d) 0.677 V

16-12. The following cell was found to have a potential of 0.411 V:

$$\text{membrane electrode for } Mg^{2+}|Mg^{2+}(a = 1.77 \times 10^{-3} M)\|SCE$$

When the solution of known magnesium activity was replaced with an unknown solution, the potential was found to be 0.439 V. What was the pMg of this unknown solution? Neglect the junction potential.

* 16-13. The following cell was found to have a potential of 0.972 V:

$$Cd|CdX_2(sat'd),X^-(0.0200\ M)||SCE$$

Calculate the solubility product of CdX_2, neglecting the junction potential.

16-14. The following cell was found to have a potential of 0.512 V:

$$Pt,H_2(1.00\ atm)|HA(0.200\ M)||SCE$$

Calculate the dissociation constant of HA, neglecting the junction potential.

* 16-15. A 40.00-mL aliquot of $0.1000\ N$ U^{4+} is diluted to 75.0 mL and titrated with $0.0800\ N$ Ce^{4+}. The pH of the solution is maintained at 1.00 throughout the titration. (Use 1.44 V for the formal potential of the cerium system.)
    (a) Calculate the potential of the indicator cathode with respect to a saturated calomel reference electrode after the addition of 5.00, 10.00, 15.00, 25.00, 40.00, 49.00, 50.00, 51.00, 55.00, and 60.00 mL of cerium(IV).
    (b) Draw a titration curve for these data.

16-16. Calculate the potential of a mercury cathode (vs. SCE) after the addition of 5.00, 15.00, 25.00, 30.00, 35.00, 39.00, 40.00, 41.00, 45.00, and 50.00 mL of 0.0500 M $Hg_2(NO_3)_2$ to 50.00 mL of 0.0800 M NaCl. Construct a titration curve from these data (K_{sp} Hg_2Cl_2, 1.3×10^{-18}).

* 16-17. Calculate the potential (vs. SCE) of a lead anode after the addition of 0.00, 10.00, 20.00, 24.00, 24.90, 25.00, 25.10, 26.00, and 30.00 mL of 0.2000 M $NaIO_3$ to 50.00 mL of 0.0500 M $Pb(NO_3)_2$. For $Pb(IO_3)_2$, $K_{sp} = 3.2 \times 10^{-13}$.

* 16-18. A glass/calomel electrode system was found to develop a potential of 0.0620 V when used with a buffer of pH 7.00; with an unknown solution the potential was observed to be 0.2794 V.
    (a) Calculate the pH and $[H^+]$ of the unknown.
    (b) Assume that K is uncertain by ± 0.001 V as a consequence of a difference in the junction potential between standardization and measurement. What is the range of $[H^+]$ associated with this uncertainty?
    (c) What is the relative error in $[H^+]$ associated with the uncertainty in E_j?

16-19. The following cell was found to have a potential of 0.3674 V:

$$\text{membrane electrode for } Mg^{2+}|Mg^{2+}(a = 6.87 \times 10^{-3}\ M)||SCE$$

    (a) When the solution of known magnesium activity was replaced with an unknown solution, the potential was found to be 0.4464 V. What was the pMg of this unknown solution?
    (b) Assuming an uncertainty of ± 0.002 V in the junction potential, what is the range of Mg^{2+} activities within which the true value might be expected?
    (c) What is the relative error in $[Mg^{2+}]$ associated with the uncertainty in E_j?

* 16-20. The sodium ion concentration of a solution was determined by measurements with a glass-membrane electrode. The electrode system developed a potential of 0.2331 V when immersed in 10.0 mL of the unknown. After addition of 1.00 mL of a standard solution that was 2.00×10^{-2} M in Na^+, the potential decreased to 0.1846 V. Calculate the sodium ion concentration and the pNa of the original solution.

16-21. The calcium ion concentration of a solution was determined by measurements with a liquid-membrane electrode. The electrode system developed a potential of 0.4965 V when immersed in 25.0 mL of the sample. After addition of 2.00 mL of 5.45×10^{-2} M $CaCl_2$, the potential changed to 0.4117 V. Calculate the calcium concentration and the pCa of the sample.

electrogravimetric and

coulometric methods

17

Three related electroanalytical methods, namely, *electrogravimetric analysis,* *constant-potential coulometry,* and *coulometric titrations,* are discussed in this chapter. Each involves an electrolysis that is carried on for a sufficient length of time to assure quantitative oxidation or reduction of the analyte. In electrogravimetric methods, the product of the electrolysis is weighed as a deposit on one of the electrodes (the *working electrode*). In the two coulometric procedures, on the other hand, the quantity of electricity needed to complete the electrolysis serves as a measure of the amount of analyte present.

   The three methods have moderate sensitivity and speed; for many applications, they are among the most precise and accurate methods available to the chemist, with attainable uncertainties of the order of a few tenths percent. In common with gravimetric methods, but in contrast to all other methods discussed in this text, these procedures require no calibration against standards; that is, the functional relationship between the quantity measured and the weight of analyte can be derived from theory.

17A SELECTIVITY OF ELECTROLYTIC DEPOSITION METHODS

In principle, electrolytic methods appear to offer a reasonably selective means for carrying out the separation and determination of a number of ions. The feasibility of and the theoretical conditions for accomplishing a given separation can be readily derived from the standard electrode potentials for the species of interest.

EXAMPLE 17-1. In principle, is a quantitative separation of Ni^{2+} and Cd^{2+} by electrolytic deposition feasible? If so, what conditions must be maintained? Assume that the sample solution is initially 0.1000 M in each ion and that the quantitative removal of an ion has been achieved when only 1 part in 10,000 remains undeposited.

In Appendix 2, we find

$$Ni^{2+} + 2e \rightleftharpoons Ni(s) \qquad E^0 = -0.250 \text{ V}$$

$$Cd^{2+} + 2e \rightleftharpoons Cd(s) \qquad E^0 = -0.403 \text{ V}$$

It is apparent that Ni^{2+} will begin to deposit at a less negative electrode potential than the Cd^{2+}. Let us first calculate that potential required to reduce the Ni^{2+} concentration to 10^{-4} of its original concentration (that is, to 1.00×10^{-5} M). Substituting into the Nernst equation, we obtain

$$E = -0.250 - \frac{0.0591}{2} \log \frac{1}{1.00 \times 10^{-5}} = -0.398 \text{ V}$$

Similarly, we can then derive the cathode potential to initiate deposition of Cd. Thus,

$$E = -0.403 - \frac{0.0591}{2} \log \frac{1}{0.100} = -0.433 \text{ V}$$

Therefore if the cathode potential can be maintained between -0.398 and -0.433 V, a quantitative separation could in theory be accomplished.

Calculations such as the foregoing make it possible to determine the differences in standard electrode potentials theoretically needed to permit the deposition of one metal without interference from another; these differences range from about 0.04 V for triply charged ions to about 0.24 V for singly charged. It is important to appreciate that an approach to these theoretical limits can be realized only through control of the potential of the *working electrode* (ordinarily the cathode). The potential of this electrode is in turn determined by the cell potential which includes not only the cathode potential but also the anode potential, the IR drop, and any overvoltages associated with the electrode processes (Equation 13-6, p. 336). All of these potentials vary continuously as the electrolysis proceeds, and it is seldom feasible to calculate the fraction of the applied potential that is attributable to the working electrode. As a consequence, the only practical way of achieving separation of species whose electrode potentials differ by only a few tenths of a volt is to measure the cathode potential continuously against a reference electrode whose potential is known; the cell potential can then be adjusted to maintain the potential of the working electrode within the desired range. An analysis performed in this way is called a *controlled* or *constant cathode* (or *anode*) *potential electrolysis*.

Both uncontrolled and controlled electrode potential methods are used in electrogravimetric and coulometric procedures. In the uncontrolled procedure, the applied cell potential is generally maintained at some more or less constant level that provides a large enough current to complete the electrolysis in a reasonable period. In the controlled method, the applied potential must be decreased continuously as the electrolysis proceeds in order to keep the working electrode at the desired potential.

17A-1 CURRENT AND POTENTIAL CHANGES DURING AN ELECTROLYSIS WITHOUT CATHODE POTENTIAL CONTROL

Electrolytic procedures in which the applied potential to a cell is held approximately constant throughout the deposition offer the advantage that the equipment required is simple and inexpensive. To understand the limitations of this procedure, it is necessary to consider how the current and the potential of the working electrode vary as a function of time when a constant potential is applied to the cell.

In order to illustrate current-voltage relationships during an electrolysis at constant cell potential, it is convenient to consider the deposition of copper from a solution that is 0.1 M in copper(II) ions and 1 M in sulfuric acid. Here, the electrode reactions are:

cathode $Cu^{2+} + 2e \longrightarrow Cu(s)$

anode $H_2O \longrightarrow \frac{1}{2}O_2(g) + 2H^+ + 2e$

The reversible potential of this cell can be shown to be -0.92 V. We will assume that an initial current of 1.5 A is desired and that the cell resistance is 0.5 Ω.

To obtain the relationship between current and applied potential, it is necessary to take into account both kinetic and concentration overvoltage at the two electrodes. At the anode, concentration polarization should be negligible throughout the electrolysis because an enormous excess of the reactant (H_2O) is always present. As indicated earlier, however, kinetic polarization often accompanies formation of gases at electrodes. We will assume for illustrative purposes that the resulting overvoltage required to bring about the deposition of oxygen at the desired rate is 0.8 V.

As with the deposition of most metals, kinetic polarization is negligible during the reduction of copper. Before the electrolysis is completed, however, concentration polarization will undoubtedly set in. Thus, the applied potential for the cell is given by (see Equation 13-6)

$$E_{applied} = E_{cathode} - E_{anode} - IR - \Pi_a - \Pi_c \tag{17-1}$$

where Π_a and Π_c are the overvoltages at the anode and cathode, respectively.

At the beginning of the electrolysis, the copper ion concentration at the electrode surface will be large enough so that concentration overpotential is negligible. Thus, the potential required to produce a current of 1.5 A is given by

$$E_{applied} = -0.92 - 1.5 \times 0.5 - 0.8 - 0.0 = -2.5 \text{ V}$$

The variation in current as a function of time in the cell under consideration is shown in Figure 17-1a. The exponential decrease in current arises because concentration polarization begins to occur at the cathode almost immediately after application of the potential. That is, a 1.5-A current is sufficiently great that copper ions cannot be brought to the electrode surface rapidly enough to sustain this current for more than a few seconds. Thus, the current is limited by the rate of transport of copper ions, and this rate decreases rapidly as the copper concentration becomes smaller.

Figure 17-1b shows that the decrease in current brought about by concentration polarization results in a positive shift in both IR and the kinetic overpotential,

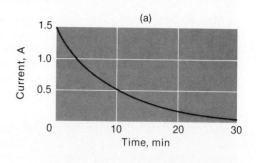

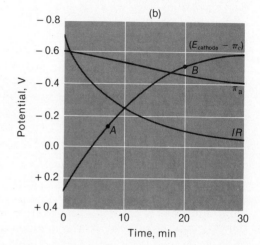

FIGURE 17-1 Changes in (a) current and (b) potentials during electrolytic deposition of Cu and O_2 at an applied potential of -2.5 V.

Π_a. Since $E_{applied}$ in Equation 17-1 is fixed, however, the remaining terms, $E_{cathode}$, E_{anode}, and Π_c must become more negative to offset these changes. The reversible anode potential is given by

$$E_{anode} = +1.229 - \frac{0.0591}{2} \log \frac{1}{p_{O_2}^{1/2} \, [H^+]^2}$$

Because the initial hydrogen ion concentration is large and the partial pressure of oxygen over the solution is constant, E_{anode} changes only slightly as the reaction proceeds. Thus, as shown in Figure 17-1b, the positive shift in the IR drop results in a corresponding negative shift in the potential of the working electrode $(E_{cathode} - \Pi_c)$.

The rapid change in cathode potential that accompanies concentration polarization often leads to codeposition of other species and loss of selectivity. For example, lead, if present in a concentration roughly equal to that of copper, would begin to codeposit at point A in Figure 17-1b. In the absence of lead, the evolution of hydrogen would commence at about point B (this process was not taken into account in deriving the curve in Figure 17-1a).

The interference just described could be avoided by a decrease in the initial applied potential by several tenths of a volt to limit the negative drift of the cathode potential. The consequence, however, would be a diminution in the initial currents and, ordinarily, an enormous increase in the time required for the analysis.

In practice, an electrolysis at constant cell potential is limited to a separation of an easily reduced cation from those that are more difficult to reduce than hydrogen ion. Evolution of hydrogen can also be expected near the end of such an electrolysis unless precautions are taken to prevent it.

17A-2 CONSTANT CATHODE POTENTIAL ELECTROLYSIS

An approach to theoretical separations within a reasonable period requires a considerably more sophisticated technique than the one just described because concentration polarization at the cathode, if unchecked, will prevent all but the crudest of separations. In a constant cathode potential electrolysis, the potential of the cathode is measured continuously by insertion of a third electrode into the solution. This electrode is a reference electrode against which the potential of the working electrode can be monitored. The overall potential applied to the cell is then controlled with a voltage divider so that the cathode potential is maintained at a level suitable for the separation. Figure 17-2 is a schematic diagram of a simple apparatus that would permit deposition at a constant cathode potential. This apparatus can be operated at relatively high initial applied potentials to give high currents. As the electrolysis progresses, however, a lowering of the applied potential across *AC* is required. This decrease, in turn, diminishes the current. Completion of the electrolysis will be indicated by the approach of the current to zero. The changes that occur in a typical constant cathode potential electrolysis are depicted in Figure 17-3. In contrast to an electrolysis without cathode potential control, this technique, when employed manually, demands constant attention during operation. Usually, some provision is made for automatic control; otherwise, the operator time required represents a major disadvantage to the controlled cathode potential method.

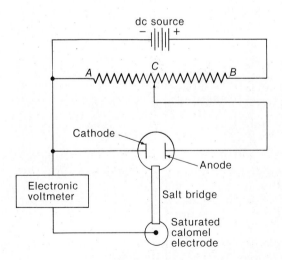

FIGURE 17-2 Apparatus for electrolysis at a controlled cathode potential. Contact *C* is continuously adjusted to maintain the cathode potential at the desired level.

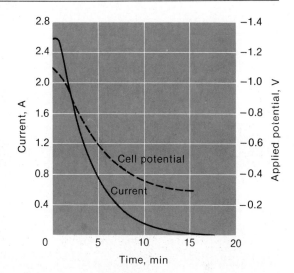

FIGURE 17-3 Changes in applied potential and current during a controlled cathode potential electrolysis. Deposition of copper upon a cathode maintained at $-0.36V$ vs. SCE. [Experimental data from J. J. Lingane, *Anal. Chim. Acta,* **2,** 589 (1949). With permission.]

17B ELECTROGRAVIMETRIC METHODS OF ANALYSIS

Electrolytic precipitation has been used for over a century for the gravimetric determination of metals. In most applications, the metal is deposited on a weighed platinum cathode, and the increase in weight is determined. Important exceptions to this procedure include the anodic depositions of lead as lead dioxide on platinum and chloride as silver chloride on silver.

17B-1 PHYSICAL PROPERTIES OF ELECTROLYTIC PRECIPITATES

Ideally, an electrolytically deposited metal should be strongly adherent, dense, and smooth so that the processes of washing, drying, and weighing can be performed without mechanical loss or reaction with the atmosphere. Good metallic deposits are fine grained and have a metallic luster; spongy, powdery, or flaky precipitates are likely to be less pure and less adherent.

The principal factors that influence the physical characteristics of deposits include current density, temperature, and the presence of complexing agents. Ordinarily, the best deposits are formed at current densities that are less than 0.1 A/cm². Stirring generally improves the quality of a deposit. The effects of temperature are unpredictable and must be determined empirically.

It is also found that many metals form smoother and more adherent films when deposited from solutions in which their ions exist primarily as complexes. Cyanide and ammonia complexes often provide the best deposits. The reasons for this effect are not obvious.

Codeposition of hydrogen during electrolysis is likely to cause the formation of nonadherent deposits, which are unsatisfactory for analytical purposes. The evolution of hydrogen can be avoided by introduction of a *cathode depolarizer*—a substance that is reduced at a lower potential than hydrogen ion. Nitrate functions in this manner, being reduced to ammonium ion

$$NO_3^- + 10H^+ + 8e \rightleftharpoons NH_4^+ + 3H_2O$$

17B-2 INSTRUMENTATION

The apparatus for an analytical electrodeposition consists of a suitable cell and a direct-current power supply.

Cells. Figure 17-4 shows a typical cell for the deposition of a metal on a solid electrode. Tall-form beakers are ordinarily employed, and mechanical stirring is provided to minimize concentration polarization; frequently, the anode is rotated with an electric motor.

Electrodes. Electrodes are usually constructed of platinum, although copper, brass, and other metals find occasional use. Platinum electrodes have the advantage of being relatively nonreactive; moreover, they can be ignited to remove any grease, organic matter, or gases that could have a deleterious effect on the physical properties of the deposit. Certain metals (notably bismuth, zinc, and gallium) cannot be deposited directly onto platinum without causing permanent damage; a protective coating of copper should always be deposited on a platinum electrode before the electrolysis of these metals is undertaken.

The mercury cathode. A mercury cathode is particularly useful in removing easily reduced elements as a preliminary step in an analysis. For example, copper, nickel, cobalt, silver, and cadmium are readily separated from ions such as aluminum, titanium, the alkali metals, sulfates, and phosphates. The precipitated elements dissolve in the mercury; little hydrogen evolution occurs even at high applied potentials because of large overvoltage effects. Ordinarily, no attempt is

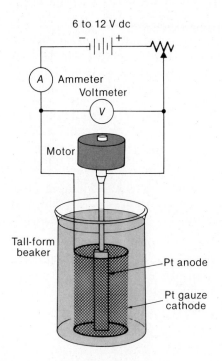

FIGURE 17-4 Apparatus for electrodeposition of metals without cathode potential control.

made to determine the elements deposited in the mercury; the goal is simply removing them from solution. A cell such as that shown in Figure 17-5 is used.

Power supplies. The apparatus shown in Figure 17-4 is typical of that employed for most electrolytic analyses. The dc power supply may consist of a storage battery, a generator, or an alternating-current rectifier. A rheostat is used to control the applied potential; an ammeter and a voltmeter are provided to indicate the approximate current and applied voltage.

The apparatus required for a controlled potential electrolysis does not need to be elaborate. A schematic diagram of the essential components is given in Figure 17-2. The cathode potential is measured against a saturated calomel or silver/silver chloride reference electrode with a simple potentiometer or electronic voltmeter; an ordinary moving-coil voltmeter draws sufficient current to introduce an error in the voltage measurement and is therefore unsatisfactory.

The power supply can be a storage battery or a rectifier unit with a well-filtered, direct-current output. The voltage divider AB should have a high current rating and a resistance of no more than 20 or 30 Ω. A tubular type of rheostat, found in most laboratories, is quite suitable. The working electrodes can be similar to those described in previous sections.

Employment of such a simple manual device demands the constant attention of the operator. Initially, the applied potential can be quite high and the currents, therefore, large. As the electrolysis proceeds, however, continuous decreases in the applied potential are required to maintain a constant cathode potential. During this period the chemist can do little else but adjust the equipment. Fortunately, however, automatic instruments, called *potentiostats*, have been designed to maintain the cathode potential at some constant, preset value throughout the electrolysis. Several commercial models are available.

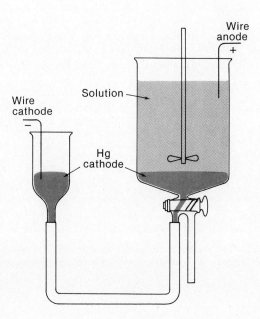

FIGURE 17-5 Mercury cathode for the electrolytic removal of metal ions from solution.

17B-3 APPLICATIONS OF CONSTANT-CURRENT ELECTROLYSIS

Without control of the cathode potential, electrolytic methods suffer from a lack of specificity. Despite this limitation, several applications of practical importance make use of this relatively unrefined technique. In general, the analyte must be the only component in the solution that is more readily reduced than hydrogen ions. Any interfering species should be eliminated by a chemical precipitation or prevented from depositing by complexation with a ligand that does not appreciably influence the electrochemical behavior of the analyte.

Constant-current deposition with a mercury cathode is also useful in removing easily reduced ions from solution prior to completion of an analysis by some other method. The deposition of interfering heavy metals prior to the quantitative determination of the alkali metals is an example of this application.

Table 17-1 lists the common elements that can be determined by electrogravimetric procedures for which control over the cathode potential is not required.

17B-4 APPLICATION OF CONTROLLED ELECTRODE POTENTIAL ELECTROLYSIS[1]

The controlled cathode potential method is a potent tool for the direct analysis of solutions containing a mixture of metallic elements. Such control permits quantitative separation of elements with standard potentials that differ by only a few tenths of a volt. For example, Lingane and Jones[2] developed a method for the successive determination of copper, bismuth, lead, and tin. The first three can be deposited from a nearly neutral tartrate solution. Copper is first reduced quantitatively by maintaining the cathode potential at -0.2 V with respect to a saturated calomel electrode. After weighing, the copper-plated cathode is returned to the solution, and bismuth is removed at a potential of -0.4 V. Lead is then deposited quantita-

---

[1] This method was first suggested by H. J. S. Sand, *Trans. Chem. Soc.*, **91**, 373 (1907). For many of its applications, see: H. J. S. Sand, *Electrochemistry and Electrochemical Analysis*, vol. 2. Glasgow: Blackie & Son, Ltd., 1940. An excellent discussion of applications of automatic control to the method can be found in J. J. Lingane, *Electroanalytical Chemistry*, 2d ed., Chapters 13–16. New York: Interscience, 1958. See also: G. A. Rechnitz, *Controlled-Potential Analysis*. New York: Macmillan, 1963.

[2] J. J. Lingane and S. L. Jones, *Anal. Chem.*, **23**, 1798 (1951).

Table 17-1   Common Elements That Can Be Determined by Electrogravimetric Methods Without Cathode Potential Control

Ion	Weighed As	Conditions
Cd^{2+}	Cd	Alkaline cyanide solution
Co^{2+}	Co	Ammoniacal sulfate solution
Cu^{2+}	Cu	HNO_3-H_2SO_4 solution
Fe^{3+}	Fe	$(NH_4)_2C_2O_4$ solution
Pb^{2+}	PbO_2	HNO_3 solution
Ni^{2+}	Ni	Ammoniacal sulfate solution
Ag^+	Ag	Cyanide solution
Sn^{2+}	Sn	$(NH_4)_2C_2O_4$-$H_2C_2O_4$ solution
Zn^{2+}	Zn	Ammoniacal or strong NaOH solution

tively by increasing the cathode potential to -0.6 V. Throughout these depositions, tin is retained in solution as a very stable tartrate complex. Acidification of the solution after deposition of the lead is sufficient to decompose the complex by converting tartrate ion to the undissociated acid; tin can then be readily deposited at a potential of -0.65 V. This method can be extended to include zinc and cadmium as well. Here, the solution is made ammoniacal after removal of the copper, bismuth, and lead. Cadmium and zinc are then successively deposited and weighed. Finally, the tin is determined after acidification, as before.

A procedure such as this is particularly attractive for use with a potentiostat because of the small operator time required for the complete analysis.

Table 17-2 lists some other separations that have been performed by the controlled-cathode method.

17C COULOMETRIC METHODS OF ANALYSIS

Coulometry encompasses a group of methods which involve measuring the quantity of electricity (in coulombs) needed to convert the analyte quantitatively to a different oxidation state. In common with gravimetric methods, coulometry offers the advantage that the proportionality constant between coulombs and the weight of analyte can be derived from known physical constants; thus, a calibration or standardization step is not ordinarily required. Coulometric methods are often as accurate as gravimetric or volumetric procedures; they are usually faster and more convenient than the former. Finally, coulometric procedures are readily adapted to automation.[3]

17C-1 QUANTITY OF ELECTRICITY

Units for the quantity of electricity include the *coulomb* (C) and the *faraday* (F). *The coulomb is the quantity of electricity that is transported in one second by a*

[3] For summaries of coulometric methods, see: H. L. Kies, *J. Electroanal. Chem.*, **4**, 257 (1962); J. J. Lingane, *Electroanalytical Chemistry*, 2d ed., Chapters 19–21. New York: Interscience, 1958; G. W. C. Milner and G. Phillips, *Coulometry in Analytical Chemistry*. New York: Pergamon Press, 1967; and J. T. Stock, *Anal. Chem.*, **52**, 1R (1980); **50**, 1R (1978); **48**, 1R (1976).

Table 17-2   Some Applications of Controlled Cathode Potential Electrolysis

Element Determined	Other Elements That May Be Present
Ag	Cu and heavy metals
Cu	Bi, Sb, Pb, Sn, Ni, Cd, Zn
Bi	Cu, Pb, Zn, Sb, Cd, Sn
Sb	Pb, Sn
Sn	Cd, Zn, Mn, Fe
Pb	Cd, Sn, Ni, Zn, Mn, Al, Fe
Cd	Zn
Ni	Zn, Al, Fe

constant current of one ampere. For a constant current of I amperes operating for t seconds, the number of coulombs Q is given by the expression

$$Q = It \tag{17-2}$$

For a variable current, the number of coulombs is given by the integral

$$Q = \int_0^t I \, dt \tag{17-3}$$

The faraday is the quantity of electricity that will produce one equivalent of chemical change at an electrode. Since the equivalent in an oxidation-reduction reaction corresponds to the change brought about by one mole of electrons, the faraday is equal to 6.02×10^{23} electrons. One faraday is also equal to 96,491 coulombs.

> **EXAMPLE 17-2.** A constant current of 0.800 A was maintained in a solution for 15.2 min. Calculate the grams of copper deposited at the cathode and the grams of O_2 evolved at the anode, assuming that these are the only products formed.
>   The equivalent weights are determined from consideration of the changes in oxidation numbers for the two species. The half-reactions are

$$Cu^{2+} + 2e \longrightarrow Cu(s)$$

$$2H_2O \longrightarrow 4e + O_2(g) + 4H^+$$

Thus, the equivalent weights of copper and molecular oxygen are one-half and one-fourth their respective formula weights.
    From Equation 17-2, we find

$$\text{quantity of electricity} = 0.800 \text{ A} \times 15.2 \text{ min} \times 60 \text{ s/min}$$

$$= 729.6 \text{ A s} = 729.6 \text{ C}$$

or

$$729.6 \text{ C} \times \frac{1 \text{ F}}{96491 \text{ C}} = 7.56 \times 10^{-3} \text{ F} \equiv 7.56 \times 10^{-3} \text{ eq}$$

From the definition of the faraday, 7.56×10^{-3} equivalent of copper is deposited on the cathode; a similar quantity of oxygen is evolved at the anode. Therefore,

$$\text{wt Cu} = 7.56 \times 10^{-3} \text{ eq Cu} \times 63.5 \frac{\text{g Cu}}{\text{mol Cu}} \times \frac{1 \text{ mol Cu}}{2 \text{ eq Cu}}$$

$$= 0.240 \text{ g}$$

and

$$\text{wt O}_2 = 7.56 \times 10^{-3} \text{ eq O}_2 \times \frac{32.0 \text{ g O}_2}{\text{mol O}_2} \times \frac{1 \text{ mol O}_2}{4 \text{ eq O}_2}$$

$$= 0.0605 \text{ g}$$

17C-2 TYPES OF COULOMETRIC METHODS

Coulometric analyses are performed in either of two general ways. The first involves maintaining the potential of the working electrode at a constant level by means of a potentiostat such that quantitative oxidation or reduction of the analyte occurs without involvement of less reactive species in the sample or solvent. Here, the current is initially high but decreases rapidly and approaches zero as the analyte is removed from the solution (see Figure 17-3). The quantity of electricity required is measured by integration of the current-time curve. A second coulometric technique makes use of a constant current that is continued until an indicator signals completion of the reaction. The quantity of electricity required to attain the end point is then calculated from the magnitude of the current and the time of its passage. The latter method has enjoyed wider application than the former; it is frequently called a *coulometric titration*.

A fundamental requirement of all coulometric methods is that the species determined interact with 100% current efficiency. That is, each faraday of electricity must bring about a chemical change corresponding to one equivalent of the analyte. This requirement does not imply that the analyte must necessarily participate directly in the electron-transfer process at the electrode. Indeed, more often than not, the substance being determined is involved wholly or in part in a reaction that is secondary to the electrode reaction. For example, at the outset of the oxidation of iron(II) at a platinum anode, all of the current results from the reaction

$$Fe^{2+} \rightleftharpoons Fe^{3+} + e$$

As the concentration of iron(II) decreases, however, concentration polarization will ultimately cause the anode potential to rise until decomposition of water occurs as a competing process. That is,

$$2H_2O \rightleftharpoons O_2(g) + 4H^+ + 4e$$

The quantity of electricity required to complete the oxidation of iron(II) would then exceed that demanded by theory. To avoid the consequent error, an unmeasured excess of cerium(III) can be introduced at the start of the electrolysis. This ion is oxidized at a lower anode potential than is water:

$$Ce^{3+} \rightleftharpoons Ce^{4+} + e$$

The cerium(IV) produced diffuses rapidly from the electrode surface, where it then oxidizes an equivalent amount of iron(II):

$$Ce^{4+} + Fe^{2+} \longrightarrow Ce^{3+} + Fe^{3+}$$

The net effect is an electrochemical oxidation of iron(II) with 100% current efficiency even though only a fraction of the iron(II) ions are directly oxidized at the electrode surface.

The coulometric determination of chloride provides another example of an indirect process. Here, a silver electrode serves as the anode, and silver ions are produced by the current. These cations diffuse into the solution and precipitate the chloride. A current efficiency of 100% with respect to the chloride ion is achieved even though this ion is neither oxidized nor reduced in the cell.

17C-3 COULOMETRIC METHODS WITH CONSTANT ELECTRODE POTENTIALS

In variable-current coulometry, the potential of the working electrode is maintained at a constant level that will cause the analyte to react quantitatively without involvement of other components in the sample. A constant electrode potential or variable-current analysis of this kind possesses all the advantages of an electrogravimetric method but is not subject to the limitation imposed by the need for a weighable product. The technique can therefore be applied to systems that yield deposits with poor physical properties as well as to reactions that yield no solid product at all. For example, arsenic may be determined coulometrically by the electrolytic oxidation of arsenious acid (H_3AsO_3) to arsenic acid (H_3AsO_4) at a platinum anode. Similarly, the analytical conversion of iron(II) to iron(III) can be accomplished with suitable control of the anode potential.

Apparatus. A constant-electrode-potential coulometric analysis requires a potentiostat, an instrument for measuring current as a function of time, and an integrating device for deriving the quantity of electricity by means of Equation 17-3.

Modern integrators are generally electronic. Such devices are frequently a part of a recorder, which provides a plot of current as a function of time. Several controlled-potential coulometric instruments are available from commercial sources.

Application.[4] Controlled-potential coulometric methods have been widely used for the determination of various metal ions. Mercury appears to be favored as the cathode (Figure 17-5), and methods for the deposition of two dozen metals at this electrode have been described. The method has found widespread use in the nuclear energy field for the relatively interference-free determination of uranium and plutonium.

Controlled-potential coulometric methods also offer possibilities for the electrolytic determination (and synthesis) of organic compounds. For example, Meites and Meites[5] have demonstrated that trichloroacetic acid and picric acid are quantitatively reduced at a mercury cathode whose potential is suitably controlled:

$$Cl_3CCOO^- + H^+ + 2e \longrightarrow Cl_2HCCOO^- + Cl^-$$

Coulometric measurements permit the analysis of these compounds with an accuracy of a few tenths of a percent.

Variable-current coulometric methods are frequently used to monitor the concentration of constituents in gas or liquid streams continuously and automatically. An important example is the determination of small concentrations of oxygen.[6] A schematic diagram of the apparatus is shown in Figure 17-6. The porous silver cathode serves to break up the incoming gas into small bubbles; the reduction of oxygen takes place quantitatively within the pores. That is,

$$O_2(g) + 2H_2O + 4e \rightleftharpoons 4OH^-$$

[4] For a summary of the applications, see: J. E. Harrar, *Electroanalytical Chemistry*, A. J. Bard, Ed., vol. 8. New York: Marcel Dekker, 1975.
[5] T. Meites and L. Meites, *Anal. Chem.*, **27**, 1531 (1955); **28**, 103 (1956).
[6] For further details, see: F. A. Keidel, *Ind. Eng. Chem.*, **52**, 491 (1960).

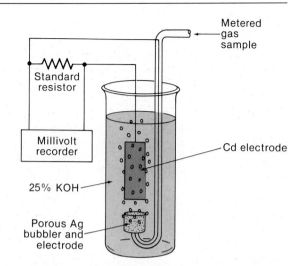

FIGURE 17-6 An instrument for continuously recording the O$_2$ content of a gas stream.

The anode is a heavy cadmium sheet; here, the half-cell reaction is

$$Cd(s) + 2OH^- \rightleftharpoons Cd(OH)_2(s) + 2e$$

Note that a galvanic cell is formed so that no external power supply is required. The electricity produced is passed through a standard resistor and the potential drop is recorded with a millivolt recorder. The oxygen concentration is proportional to the recorded potential, and the chart paper can be made to display the instantaneous oxygen concentration directly. The instrument is reported to provide oxygen concentration data in the range from 1 ppm to 1%.

17C-4 COULOMETRIC TITRATIONS

A coulometric titration employs an electrolytically generated titrant for reaction with the analyte. In some analyses, the active electrode process involves only generation of the reagent; an example is the titration of a halide by electrolytically generated silver ions. In other titrations, the analyte may also be directly involved at the generator electrode; an example of the latter is the coulometric oxidation of iron(II)—in part by electrolytically generated cerium(IV) and in part by direct electrode reaction (p. 445). The net process must, under any circumstance, approach 100% current efficiency with respect to a single chemical change in the analyte.

The current in a coulometric titration is carefully maintained at a constant and accurately known level; the product of this current in amperes and the time in seconds required to reach an end point yields the number of coulombs and thus the number of equivalents involved in the electrolysis. The constant current aspect of this operation precludes the quantitative oxidation or reduction of the unknown species entirely at the generator electrode. As the solution is depleted of analyte, concentration polarization is inevitable; the electrode potential must then change if a constant current is to be maintained (p. 436). Unless this change produces a reagent that can react with the analyte, the current efficiency will be less than 100%. In a coulometric titration, then, *at least part* (and frequently all) of the analytical reaction occurs away from the surface of the working electrode.

A coulometric titration, like a more conventional volumetric titration, requires some means of detecting the point of chemical equivalence. Most of the end points applicable to volumetric analysis are equally satisfactory here; color changes of indicators, potentiometric, amperometric (Chapter 18), and conductance measurements have all been successfully applied.

The analogy between a volumetric and a coulometric titration extends well beyond the common requirement of an observable end point. In both, the amount of analyte is determined through evaluation of its combining capacity—in the one case, for a standard solution and, in the other, for a quantity of electricity. Similar demands are made of the reactions; that is, they must be rapid, essentially complete, and free of side reactions.

Electrical apparatus. Coulometric titrators are available from several laboratory supply houses. In addition, they can be readily assembled from components available in most laboratories.

Figure 17-7 depicts the principal components of a simple coulometric titrator. Included are a source of constant current and a switch that simultaneously initiates the current and starts an electric stopclock (position 1). Note that the circuit is so arranged that electricity is drawn from the source when the switch is moved from position 1 to 2. Ordinarily R_1 will have a resistance about the same as the titration cell. With this arrangement, electricity is drawn continuously from the source throughout the titration (and also between titrations), a condition that generally leads to a more constant current.

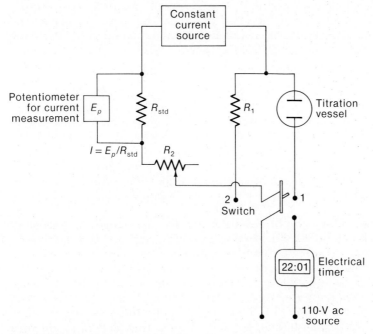

FIGURE 17-7 Schematic diagram of a coulometric titration apparatus.

A coulometric instrument must include a means for accurately measuring the current. In the apparatus shown in Figure 17-7, the potential drop across the standard resistor, R_{std}, is used for this measurement.

Many electronic or electromechanical constant-current sources are described in the literature. The ready availability of inexpensive operational amplifiers makes their construction a relatively simple matter.

An ordinary electric stopclock is inadequate for determining the total electrolysis time because the motor tends to coast when stopped and lag when started; the resulting accumulated error can be appreciable. Stopclocks with solenoid-operated brakes or electronic timers eliminate this problem.

It is useful to point out the close analogy between the various components of the apparatus shown in Figure 17-7 and the apparatus and solutions employed in a conventional volumetric analysis. The constant-current source of known magnitude serves the same function as the standard solution in a volumetric method. The electric clock and switch correspond closely to the buret: the switch performs the same function as a stopcock. During the early phases of a coulometric titration, the switch is kept closed for extended periods; as the end point is approached, however, small additions of "reagent" are achieved by switching from position 2 to position 1 for shorter and shorter intervals. The similarity to the operation of a buret is obvious.

Cells for coulometric titrations. A typical coulometric titration cell is shown in Figure 17-8. It consists of a generator electrode at which the reagent is formed and an auxiliary electrode to complete the circuit. The generator electrode, which should have a relatively large surface area, is often a rectangular strip or a wire coil of platinum; a gauze electrode such as that shown on page 440 can also be employed.

The products formed at the second electrode are frequently a potential

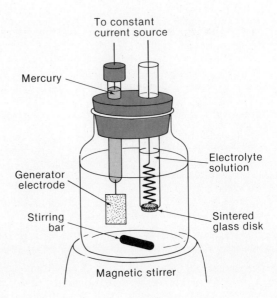

FIGURE 17-8 A typical coulometric titration cell.

source of interference. For example, the anodic generation of oxidizing agents is often accompanied by the evolution of hydrogen from the cathode; unless this gas is allowed to escape from the solution, reaction with the oxidizing agent becomes a likelihood. To eliminate this type of difficulty, the second electrode is isolated by a sintered disk or some other porous medium.

As an alternative to isolation of the auxiliary electrode, a device such as that shown in Figure 17-9 may be employed to generate the reagent externally. The apparatus is so arranged that flow of the electrolyte continues briefly after the current is discontinued, thus flushing the residual reagent into the titration vessel. Note that the apparatus shown in Figure 17-9 provides either hydrogen or hydroxide ions depending upon which arm is used. The apparatus has also been used for generation of other reagents such as iodine produced by oxidation of iodide.

17C-5 APPLICATIONS OF COULOMETRIC TITRATIONS[7]

Coulometric titrations have been developed for all types of volumetric reactions. Selected applications are described in the following paragraphs.

Neutralization titrations. Both weak and strong acids can be titrated with a high degree of accuracy using hydroxide ions generated by the reaction

$$2H_2O + 2e \longrightarrow 2OH^- + H_2(g)$$

The apparatus shown in Figures 17-8 and 17-9 can be employed. A convenient alternative to disposing of the H^+ involves substitution of a silver wire as the anode and the addition of chloride or bromide ions to the solution of the analyte. The anode reaction then becomes

$$Ag(s) + Br^- \rightleftharpoons AgBr(s) + e$$

---

[7] Applications of the coulometric procedure are summarized in J. J. Lingane, *Electroanalytical Chemistry*, 2d ed., pp. 536–613. New York: Interscience, 1958; H. L. Kies, *J. Electroanal. Chem.*, **4**, 257 (1962); and J. T. Stock, *Anal. Chem.*, **52**, 1R (1980); **50**, 1R (1978); **48**, 1R (1976).

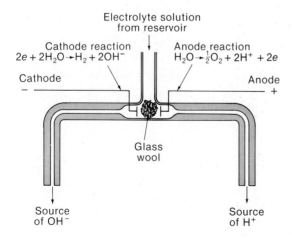

FIGURE 17-9 A cell for external generation of acid and base.

Clearly, the silver bromide will not interfere with the neutralization reaction as would the hydrogen ions that are formed at most anodes.

Both potentiometric and indicator end points can be employed for these titrations. The problems associated with the estimation of the equivalence point are identical with those encountered in a conventional volumetric analysis. A real advantage to the coulometric method, however, is that interference by carbonate ion is far less troublesome; it is only necessary to eliminate carbon dioxide from the solution containing the analyte by aeration with a carbon dioxide-free gas before beginning the analysis.

The coulometric titration of strong and weak bases can be performed with hydrogen ions generated at a platinum anode.

$$H_2O \rightleftharpoons \tfrac{1}{2}O_2(g) + 2H^+ + 2e$$

Here, the cathode must be isolated from the solution or external generation must be employed to prevent interference from the hydroxide ions produced at that electrode.

Precipitation and complex-formation titrations. A variety of coulometric titrations involving anodically generated silver ions have been developed (see Table 17-3). A cell, such as that shown in Figure 17-8, can be employed with a generator electrode constructed from a length of heavy silver wire. End points are detected potentiometrically, amperometrically (Chapter 18), or with chemical indicators. Similar analyses, based upon the generation of mercury(I) ion at a mercury anode, have also been described.

An interesting coulometric titration makes use of a solution of the ammine mercury(II) complex of ethylenediaminetetraacetic acid (H_4Y).[8] The complexing agent is released to the solution as a result of the following reaction at a mercury cathode:

$$HgNH_3Y^{2-} + NH_4^+ + 2e \rightleftharpoons Hg(l) + 2NH_3 + HY^{3-} \tag{17-4}$$

[8] C. N. Reilley and W. W. Porterfield, *Anal. Chem.*, **28**, 443 (1956).

Table 17-3   Summary of Applications of Coulometric Titrations Involving Neutralization, Precipitation, and Complex-Formation Reactions

Species Determined	Generator Electrode Reaction	Secondary Analytical Reaction
Acids	$2H_2O + 2e \rightleftharpoons 2OH^- + H_2$	$OH^- + H^+ \rightleftharpoons H_2O$
Bases	$H_2O \rightleftharpoons 2H^+ + \tfrac{1}{2}O_2 + 2e$	$H^+ + OH^- \rightleftharpoons H_2O$
Cl^-, Br^-, I^-	$Ag \rightleftharpoons Ag^+ + e$	$Ag^+ + Cl^- \rightleftharpoons AgCl(s)$, etc.
Mercaptans	$Ag \rightleftharpoons Ag^+ + e$	$Ag^+ + RSH \rightleftharpoons AgSR(s) + H^+$
Cl^-, Br^-, I^-	$2Hg \rightleftharpoons Hg_2^{2+} + 2e$	$Hg_2^{2+} + 2Cl^- \rightleftharpoons Hg_2Cl_2(s)$, etc.
Zn^{2+}	$Fe(CN)_6^{3-} + e \rightleftharpoons Fe(CN)_6^{4-}$	$3Zn^{2+} + 2K^+ + 2Fe(CN)_6^{4-} \rightleftharpoons$ $K_2Zn_3[Fe(CN)_6]_2(s)$
$Ca^{2+}, Cu^{2+}, Zn^{2+},$ and Pb^{2+}	See Equation 17-4	$HY^{3-} + Ca^{2+} \rightleftharpoons CaY^{2-} + H^+$, etc.

Because the mercury chelate is more stable than the corresponding complexes with calcium, zinc, lead, or copper, complexation of these ions will not occur until the electrode process frees the ligand.

Oxidation-reduction titrations. Table 17-4 indicates the variety of reagents that can be generated coulometrically and some of the analyses to which they have been applied. Electrogenerated bromine has proved to be particularly useful among the oxidizing agents and forms the basis for a host of methods. Of interest also are some of the unusual reagents not ordinarily encountered in volumetric analysis because of the instability of their solutions; these include dipositive silver ion, tripositive manganese, and the chloride complex of unipositive copper.

Comparison of coulometric and volumetric titrations. Some real advantages can be claimed for a coulometric titration in comparison with the classical volumetric process. Principal among these is the elimination of problems associated with the preparation, standardization, and storage of standard solutions. This advantage is particularly important with labile reagents such as chlorine, bromine, or titanium(III) ion; owing to their instability, these species are inconvenient as volumetric reagents. Their utilization in coulometric analysis is straightforward, however, because they undergo reaction with the analyte immediately after being generated.

A coulometric titration offers a considerable advantage where small quantities of reagent are required. By proper choice of current, micro quantities of a substance can be introduced with ease and accuracy; the equivalent volumetric process requires small volumes of very dilute solutions, a recourse that is always difficult.

A single constant-current source can be employed to generate precipitation, complex formation, oxidation-reduction, or neutralization reagents. Furthermore, the coulometric method is readily adapted to automatic titrations, because current control is easily accomplished.

Table 17-4 Summary of Applications of Coulometric Titrations Involving Oxidation-Reduction Reactions

Reagent	Generator Electrode Reaction	Substance Determined
Br_2	$2Br^- \rightleftharpoons Br_2 + 2e$	As(III), Sb(III), U(IV), Tl(I), I⁻, SCN⁻, NH_3, N_2H_4, NH_2OH, phenol, aniline, mustard gas, mercaptans, 8-hydroxyquinoline, olefins
Cl_2	$2Cl^- \rightleftharpoons Cl_2 + 2e$	As(III), I⁻, styrene, fatty acids
I_2	$2I^- \rightleftharpoons I_2 + 2e$	As(III), Sb(III), $S_2O_3^{2-}$, H_2S, ascorbic acid
Ce^{4+}	$Ce^{3+} \rightleftharpoons Ce^{4+} + e$	Fe(II), Ti(III), U(IV), As(III), I⁻, $Fe(CN)_6^{4-}$
Mn^{3+}	$Mn^{2+} \rightleftharpoons Mn^{3+} + e$	$H_2C_2O_4$, Fe(II), As(III)
Ag^{2+}	$Ag^+ \rightleftharpoons Ag^{2+} + e$	Ce(III), V(IV), $H_2C_2O_4$, As(III)
Fe^{2+}	$Fe^{3+} + e \rightleftharpoons Fe^{2+}$	Cr(VI), Mn(VII), V(V), Ce(IV)
Ti^{3+}	$TiO^{2+} + 2H^+ + e \rightleftharpoons Ti^{3+} + H_2O$	Fe(III), V(V), Ce(IV), U(VI)
$CuCl_3^{2-}$	$Cu^{2+} + 3Cl^- + e \rightleftharpoons CuCl_3^{2-}$	V(V), Cr(VI), IO_3^-
U^{4+}	$UO_2^{2+} + 4H^+ + 2e \rightleftharpoons U^{4+} + 2H_2O$	Cr(VI), Ce(IV)

Coulometric titrations are subject to five potential sources of error: (1) variation in the current during electrolysis; (2) departure of the process from 100% current efficiency; (3) error in the measurement of current; (4) error in the measurement of time; and (5) titration error due to the difference between the equivalence point and the end point. The last of these difficulties is common to conventional volumetric methods as well; where the indicator error is the limiting factor, the two methods are likely to have comparable reliability.

With simple instrumentation, currents constant to 0.2% relative are easily achieved; with somewhat more sophisticated apparatus, control to 0.01% is obtainable. In general, then, errors due to current fluctuations are seldom of importance.

Generalizations concerning the magnitude of uncertainty associated with the electrode process are difficult; nevertheless, current efficiencies of 99.5% to better than 99.9% are often reported in the literature. Currents are readily measured to ±0.1% relative or better. Similarly, a good quality timer permits measurement of time to within ±0.1% relative.

To summarize, then, the current-time measurements required for coulometric titration are inherently as accurate or more accurate than the comparable volume-normality measurements of a classical volumetric analysis, particularly where small quantities of reagent are involved. Often, however, the accuracy of a titration is not limited by these measurements but by the sensitivity of the end point; in this respect, the two procedures are equivalent.

17C-6 AUTOMATIC COULOMETRIC TITRATORS

A number of instrument manufacturers offer automatic coulometric titrators. Most of these employ the potentiometric end point and are similar in construction to the automatic titrators discussed in the previous chapter (p. 428). Here, however, the error signal controls a flow of electricity rather than a flow of liquid reagent. Some of the commercial instruments are multipurpose and can be used for the determination of a variety of species. Others are designed for a single analysis. Examples of the latter include titrators designed specifically for the determination of chloride in biological fluids such as blood serum, in which silver ion is generated coulometrically; sulfur dioxide monitors, where anodically generated bromine oxidizes the analyte to sulfate ions; and water titrators in which the iodine for the Karl Fischer reagent is generated electrolytically.

PROBLEMS*

* 17-1. Copper is to be deposited from a solution that is 0.200 M in Cu^{2+} and buffered to a pH of 3.00. Oxygen is evolved at a partial pressure of 1.00 atm at a platinum anode. The cell has a resistance of 2.80 Ω; the temperature is 25°C. Calculate
  (a) the theoretical potential needed to initiate deposition of copper in this cell.
  (b) the IR drop for a current of 0.15 A.
  (c) the initial applied potential, given that the oxygen overvoltage is 0.85 V.
  (d) the applied potential that will be needed when [Cu^{2+}] is 0.0010, assuming that all other variables remain unchanged.

17-2. Silver is to be deposited from a solution in which the analytical concentration of

---

* Answers to problems and parts of problems marked with an asterisk are to be found at the end of the book.

$Ag(CN)_2^-$ is 0.0400 M and that for KCN is 0.300 M. Oxygen is evolved at the anode at a partial pressure of 740 torr. The cell has a resistance of 1.80 Ω; the temperature is 25°C. Calculate
(a) the theoretical potential needed to initiate deposition of silver in this cell.
(b) the IR drop for a current of 0.12 A.
(c) the initial applied potential, given that the oxygen overvoltage is 0.72 V.
(d) the theoretical potential when the analytical concentration of the undeposited silver is 1.00×10^{-5} M.

17-3. Calculate the minimum difference in standard electrode potentials needed to lower the concentration of the metal M_1 to 1.00×10^{-5} M in a solution that is 1.00×10^{-1} M in the less reducible metal M_2, where
* (a) M_1 is univalent and M_2 is divalent.
  (b) M_1 and M_2 are both divalent.
* (c) M_1 is trivalent and M_2 is univalent.
  (d) M_1 is divalent and M_2 is univalent.
  (e) M_1 is divalent and M_2 is trivalent.

* 17-4. A solution is 0.100 M in Pb^{2+} and 0.0750 M in Ni^{2+}. Calculate
  (a) the lead-ion concentration in the solution as the first nickel starts to deposit.
  (b) the cathode potential needed to lower the lead-ion concentration to 1.00×10^{-5} M.

17-5. A solution is 0.0500 M in Co(II) and 0.0400 M in Cd(II).
  (a) What will be the concentration of the more readily reduced cation at the onset of deposition of the less reducible one?
  (b) What will be the potential of the cathode when the concentration of the more easily reduced species is 1.00×10^{-6} M?

* 17-6. Electrodeposition is to be used to separate the cations in a solution that is buffered to a pH of 8.00 and is 5.00×10^{-2} M with respect to Zn(II) and 8.00×10^{-3} M with respect to Cd(II). Oxygen is evolved at a platinum anode at a pressure of 1.00 atm; the oxygen overvoltage is 0.80 V. The cell has a resistance of 2.40 Ω.
  (a) Which cation will deposit first?
  (b) Estimate the initial potential that must be applied in order to operate the cell at 0.50 A.
  (c) Taking 1.00×10^{-6} M as a reasonable estimate for quantitative removal, calculate the range (vs. SCE) within which it will be necessary to maintain the potential of the cathode.

17-7. Electrodeposition is proposed as the means for separating the cations in a solution that is buffered to a pH of 5.00 and is 1.00×10^{-3} M with respect to Pb^{2+} and 5.00×10^{-2} M with respect to Tl^+. Oxygen is evolved at a platinum anode at a pressure of 750 torr. The cell has a resistance of 1.50 Ω; the oxygen overvoltage is 0.75 V.
  (a) Estimate the initial potential that must be applied in order to operate the cell at 0.20 A.
  (b) Within what range (vs. SCE) should the cathode potential be maintained in order to lower the lead-ion concentration to at least 1.00×10^{-6} M without interference from Tl^+?

17-8. Electrogravimetric analysis involving control of the cathode potential is proposed as a means for separating Bi(III) and Pb(II) in a solution that is 0.100 M with respect to each ion and is buffered to a pH of 2.00.
  (a) Calculate the theoretical potential of the cathode at the start of deposition of the more readily reduced ion.
  (b) Calculate the residual concentration of the more readily reduced species at the outset of deposition by the less readily reduced species.
  (c) Propose a range (vs. SCE), if such exists, between which the potential of the cathode should be maintained; consider a residual concentration less than 10^{-6} M as constituting quantitative removal.

* 17-9. Halide ions can be deposited on a silver anode. Reaction:

$$Ag(s) + X^- \longrightarrow AgX(s) + e$$

(a) Using 1.00×10^{-6} M as the criterion for quantitative removal, is it theoretically feasible to effect a separation of Br^- and I^- through control of a silver anode in a solution that is initially 0.0500 M with respect to each ion?

(b) Is a separation of Cl^- and I^- theoretically feasible in a solution that is initially 0.0400 M in each ion?

(c) If a separation is feasible in either (a) or (b), what range of anode potentials (vs. SCE) should be used?

* 17-10. What cathode potential (vs. SCE) would be needed to lower the analytical concentration of silver(I) containing species to 1.00×10^{-6} M in a solution that is

(a) 0.0200 M in $HClO_4$?

(b) 0.100 M in CN^-?

(c) 0.150 M in NH_3 (analytical concentration) and has a pH of 9.00?

$$Ag(NH_3)_2^+ + e \longrightarrow Ag(s) + 2NH_3 \qquad E^0 = 0.120 \text{ V}$$

(d) 4.00×10^{-2} M in EDTA and is buffered to a pH of 10.0?

17-11. What cathode potential (vs. SCE) would be needed to lower the analytical concentration of a nickel(II) containing species to 1.00×10^{-6} M in a solution that is

(a) 0.0010 M in HCl (assume no chloride complex forms)?

(b) 0.108 M in CN^-?

$$Ni^{2+} + 4CN^- \longrightarrow Ni(CN)_4^{2-} \qquad \beta_4 = 1.0 \times 10^{22}$$

(c) 0.100 M in NH_3 (analytic concentration) and has a pH of 10.0?

$$Ni(NH_3)_4^{2+} + 2e \longrightarrow Ni(s) + 4NH_3 \qquad E^0 = -0.530 \text{ V}$$

(d) 0.0380 M in EDTA and is buffered to a pH of 6.00?

* 17-12. Calculate the time needed for a constant current of 1.20 A to deposit 0.400 g of cobalt(II) as

(a) the element on the surface of a cathode.

(b) Co_2O_3 on an anode.

17-13. Calculate the time needed for a constant current of 0.800 A to deposit 0.100 g of

(a) Tl(III) as the element on a cathode.

(b) Tl(I) as Tl_2O_3 on an anode.

(c) Tl(I) as the element on a cathode.

* 17-14. Calculate the percentage of lead in a 0.148-g sample if quantitative deposition required 32.25 C.

17-15. The quantitative oxidation of the iron(II) in a 25.0-mL aliquot required 36.4 C. Calculate the weight of iron(II) in each milliliter of the sample.

* 17-16. An excess of $HgNH_3Y^{2-}$ was introduced to a 50.00-mL aliquot of well water. Express the hardness of the water in terms of ppm $CaCO_3$ if the EDTA needed for the titration (Equation 17-4) was generated at a mercury cathode in 3.91 min by a constant current of 30.4 mA.

17-17. A 0.1516-g sample of purified organic acid was neutralized by the hydroxide ion produced in 330 s by a constant current of 0.384 A. Calculate the equivalent weight of the acid.

* 17-18. The iodide in a 7.58-g sample of table salt was titrated with silver ion in an ammoniacal solution. Calculate the percentage of KI in the sample if the anodic generation of Ag^+ required a constant current of 0.0294 A for a total of 59.9 s.

17-19. Electrolytically generated iodine was used to determine the hydrogen sulfide in a 100.0-mL sample of brackish water. Following addition of excess potassium iodide, titration required a constant current of 47.9 mA for a total of 8.53 min. Reaction:

$$H_2S + I_2 \longrightarrow S(s) + 2H^+ + 2I^-$$

Express the results of the analysis in terms of ppm H_2S.

17-20. The oxygen in a gas stream with a density of 0.00205 g/mL was monitored with the cell shown in Figure 17-6. Use the accompanying data to calculate the ppm of O_2 in the gas.

sample	* (1)	(2)	(3)	(4)
volume of gas, L	25.0	21.4	30.2	27.5
coulombs	3.34	2.89	4.00	3.66

17-21. Calculate the standard deviation of the data in Problem 17-20.

17-22. The chromium deposited on one surface of a 10.0-cm² test plate was dissolved in acid and oxidized to the +6 state with ammonium peroxodisulfate. Reaction:

$$3S_2O_8^{2-} + 2Cr^{3+} + 7H_2O \longrightarrow Cr_2O_7^{2-} + 14H^+ + 6SO_4^{2-}$$

The solution was boiled to eliminate the excess peroxodisulfate, cooled, and then subjected to coulometric titration with Cu(I) generated from 25.0 mL of 0.0800 M Cu^{2+}. Reaction:

$$Cr_2O_7^{2-} + 6CuCl_3^{2-} + 14H^+ \longrightarrow 2Cr^{3+} + 7H_2O + 6Cu^{2+} + 18Cl^-$$

Calculate the average thickness of chromium on the test plate if the titration required a constant current of 36.4 mA for a period of 6 min and 45 s (sp gr Cr = 7.1).

* 17-23. A 0.8774-g mixture consisting of KCl, KBr, and inert materials caused the weight of a silver anode to increase by 0.3050 g as a consequence of formation of AgBr and AgCl. Calculate the respective analyte percentages if the total electrical requirement amounted to 187.4 C.

17-24. The combined deposit of zinc and cadmium from a 0.1967-g alloy sample caused the weight of a mercury cathode to increase by 0.1048 g. Calculate the respective concentrations of these metals in the sample if the deposition required 255.8 C.

* 17-25. An applied potential of -1.0 V (vs. SCE) causes the reduction of carbon tetrachloride to chloroform at the surface of a mercury cathode:

$$2CCl_4 + 2H^+ + 2e + 2Hg(l) \longrightarrow 2CHCl_3 + Hg_2Cl_2(s)$$

A 0.1037-g sample containing CCl_4 was dissolved in methanol and electrolyzed at a fixed potential of -1.0 V. A coulometer in series with the electrolysis cell indicated that 22.87 C were needed to cause the current to approach zero. Calculate the percentage of carbon tetrachloride in the sample.

17-26. Chloroform is reduced to methane at -1.80 V (vs. SCE):

$$2CHCl_3 + 6H^+ + 6e + 6Hg(l) \longrightarrow 2CH_4 + 3Hg_2Cl_2(s)$$

Calculate the percentage of $CHCl_3$ in a 0.0914-g sample if 61.23 C were needed for the reaction.

* 17-27. A 0.1002-g sample containing CCl_4, $CHCl_3$, and inert materials was dissolved in methanol and electrolyzed at the surface of a mercury electrode at -1.0 V, with a total of 10.04 C being required (see Problem 17-25). The potential of the cathode was then held constant at -1.80 V. Completion of the titration at this potential required an additional 76.47 C (see Problem 17-26). Calculate the respective percentages of CCl_4 and $CHCl_3$ in the sample.

17-28. A 0.1076-g sample containing only CCl_4 and $CHCl_3$ was dissolved in methanol. Electrolysis at a mercury cathode at -1.80 V (vs. SCE) resulted in the reduction of both compounds to methane (see Problems 17-25 and 17-26) and required a total of 266.6 C. Calculate the percentage composition of the sample.

* 17-29. Ascorbic acid is oxidized to dehydroascorbic acid with Br_2:

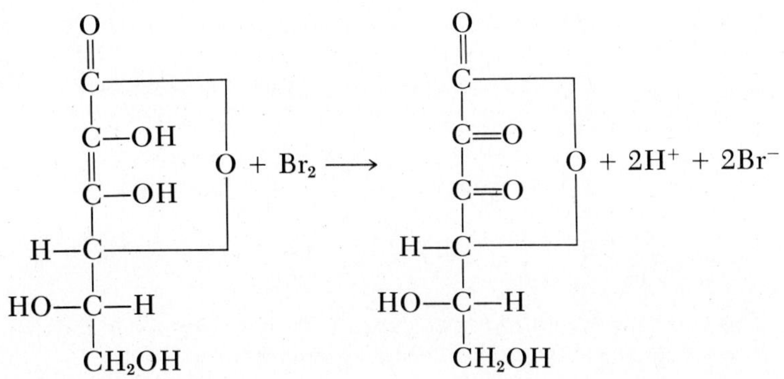

A vitamin C tablet was dissolved in sufficient water to give 250 mL of solution. A 50.0-mL aliquot was then mixed with an equal volume of 0.100 M KBr. Calculate the weight of ascorbic acid in the tablet if reaction required the bromine generated by a constant current of 0.0500 A for a total of 7.53 minutes.

17-30. Phenol undergoes a quantitative reaction with bromine

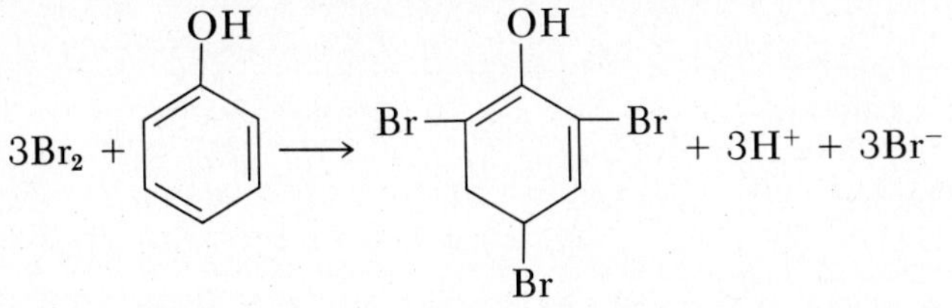

The bromine needed to react with the phenol in a 50.0-mL sample of a disinfectant preparation was generated anodically with a constant current of 80.0 mA for a period of 10.40 min. Express the results of this analysis in terms of mg phenol/mL of sample.

* 17-31. Traces of aniline can be determined by reaction with an excess of electrolytically generated Br_2:

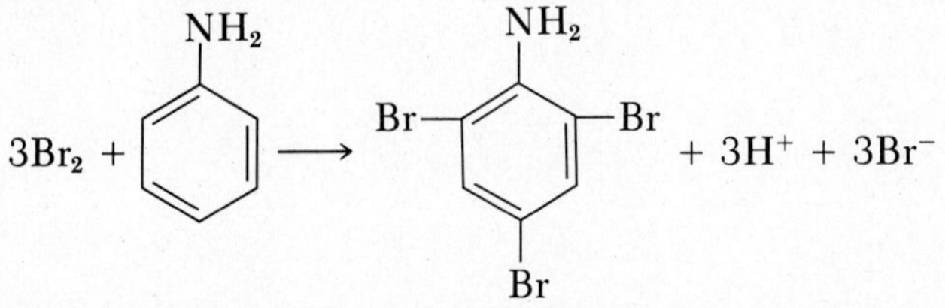

The polarity of the working electrode is then reversed, and the excess bromine is determined by a coulometric titration involving the generation of Cu(I):

$$Br_2 + 2Cu^+ \longrightarrow 2Br^- + 2Cu^{2+}$$

Suitable quantities of KBr and copper(II) sulfate were added to a 25.0-mL sample containing aniline. Calculate the micrograms of $C_6H_5NH_2$ in the sample from the accompanying data:

Working electrode functioning as	Generation time (min) with a constant current of 1.00 mA
anode	3.46
cathode	0.41

17-32. Quinone can be reduced to hydroquinone with an excess of electrolytically generated Sn(II):

The polarity of the working electrode is then reversed, and the excess Sn(II) is oxidized with Br_2 generated in a coulometric titration:

$$Sn^{2+} + Br_2 \rightleftharpoons Sn^{4+} + 2Br^-$$

Appropriate quantities of $SnCl_4$ and KBr were added to a 50.0-mL aliquot of sample. Calculate the weight of quinone in the sample from the accompanying data:

Working electrode functioning as	Generation time (min) with a constant current of 1.30 mA
cathode	7.05
anode	0.75

polarography and

amperometric titrations

18

The polarographic method of analysis was first described by the Czechoslovakian chemist Jaroslav Heyrovsky in 1922. His discovery proved to be of sufficient importance to merit the 1959 Nobel Prize in chemistry.

Several variants of the original polarographic procedure have been developed, and the term *voltammetry* is often used to describe these as well as the original polarographic method. A discussion of one of these variants, *amperometric titrations,* is included near the end of this chapter.

18A POLAROGRAPHIC MEASUREMENTS

In a polarographic analysis, the solution of the analyte is made a part of a special type of electrolytic cell containing a mercury microelectrode at which the analyte reacts. Both qualitative and quantitative information is obtained from plots of the current generated in the cell as a function of applied potential.[1]

---

[1] The principles and applications of polarography are considered in detail in a number of monographs. See, for example: I. M. Kolthoff and J. J. Lingane, *Polarography*, 2d ed. New York: Interscience, 1952; J. Heyrovsky and J. Kŭta, *Principles of Polarography*, New York: Academic Press, 1966; P. Zuman, *Organic Polarographic Analysis*. Oxford: Pergamon Press, 1964; P. Zuman, *The Elucidation of Organic Electrode Processes*. New York: Academic Press, 1969; H. W. Nurnberg, Ed., *Electroanalytical Chemistry*, Chapters 1–5. New York: Wiley, 1974; A. M. Bond, *Modern Polarographic Methods in Analytical Chemistry*. New York: Marcel Dekker, 1980.

18A-1 POLAROGRAPHIC APPARATUS

For polarography, as well as the other voltammetric techniques, the working electrode at which the analyte reacts must be small; typically, its surface area will range from 1 to 10 mm². In addition, the electrode must be chemically inert. Thus, microelectrodes are fabricated of conducting materials such as mercury, platinum, gold, silver, and graphite. With the exception of mercury, the electrodes usually are fine wires or disks that are sealed into glass tubing.

The most important microelectrode for voltammetry, and the one used in Heyrovsky's early work, is the *dropping mercury electrode*. The discussion in this chapter will be principally concerned with this electrode, which has unique properties that make it particularly well suited for polarographic analyses.

The dropping mercury electrode. A typical dropping mercury electrode is shown as part of the polarographic cell in Figure 18-1. Here, mercury is forced through a 5- to 10-cm capillary tubing with an internal diameter of approximately 0.05 mm (capillaries of this type are available commercially). Under a head of about 50 cm of mercury, a continuous flow of identical droplets with diameters of 0.5 to 1 mm results; typically, the lifetime of an individual drop is 2 to 6 seconds.

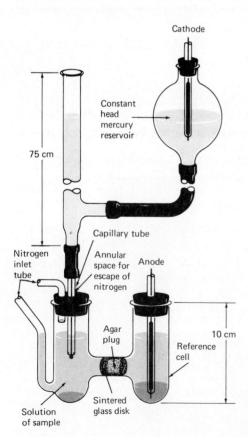

FIGURE 18-1 A dropping mercury electrode and cell. [J. J. Lingane and H. A. Laitinen, *Ind. Eng. Chem., Anal. Ed.,* **11**, 504 (1939). With permission of the American Chemical Society.]

The success of a polarographic experiment is critically dependent upon the reproducible behavior of the dropping electrode. Care must therefore be taken to be sure that the tip is cut at as nearly a right angle as possible and that the electrode is mounted in a vertical position. Furthermore, the mounting must be as free of vibrations as possible.

With reasonable care, a capillary will last for several months or even years. The precautions needed to ensure such performance involve the use of scrupulously clean mercury and the maintenance of a slight mercury flow under all circumstances. Malfunction of the electrode can be expected if solution comes in contact with the inner surface of the capillary. Therefore, a sufficient head of mercury should always be provided before the electrode is immersed in a solution.

To prevent its becoming contaminated when not in use, the electrode should be rinsed thoroughly with water and dried. The mercury head should then be decreased until the flow in air just ceases. Care must be taken to avoid lowering the mercury too far. Before use, the head is increased. The tip is immersed in 1:1 nitric acid for a minute or so and then washed with distilled water.

Reference electrodes. Any of the reference electrodes discussed in Section 13-G (p. 337) can be employed for polarographic measurements. The cell shown in Figure 18-1 employs a saturated calomel electrode separated from the analyte solution by means of a sintered glass disk or an agar plug.

For some applications, a pool of mercury in the bottom of the cell serves as an adequate reference electrode.

Electrical apparatus. Polarographic measurements require means for varying the voltage continuously over the range from 0 to 3.0 V. The applied potential should usually be known to about 0.01 V. In addition, it must be possible to measure cell currents over the range between 0.01 and perhaps 100 μA with an accuracy of about 0.01 μA. A manual apparatus meeting these requirements is easily constructed from equipment available in most laboratories. More elaborate instruments that record polarograms automatically are commercially available.

Figure 18-2 shows a circuit diagram of a simple instrument for polarographic work. Two 1.5-V batteries provide a voltage across the 100-Ω potential divider R_1 to permit variation of the potential that is applied to the cell. The magnitude of this voltage can be determined by means of the potentiometer with the switch of the instrument shifted to position 2. The current is measured by determining the potential drop across a precision of 10,000-Ω resistance R_2 with the same potentiometer and the switch in position 1.

18A-2 POLAROGRAMS

A polarogram is a plot of current as a function of the potential applied to a polarographic cell. The microelectrode is ordinarily connected to the negative terminal of the power supply; *by convention* the applied potential is given a negative sign under these circumstances. By convention also, the currents are designated as positive when the flow of electrons is from the power supply into the microelectrode—that is, when that electrode behaves as a cathode.

Figure 18-3 shows two polarograms. The lower one is for a solution that is 0.1 M in potassium chloride; the upper one is for a solution that is additionally

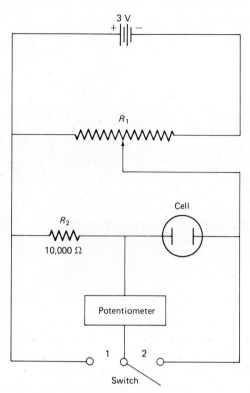

FIGURE 18-2 A simple polarographic circuit. [J. J. Lingane, *Anal. Chem.*, **21**, 47 (1949). With permission of the American Chemical Society.]

1×10^{-3} *M* in cadmium chloride. A step-shaped current-voltage curve, called a *polarographic wave*, is produced as a result of the reaction

$$Cd^{2+} + 2e + Hg \rightleftharpoons Cd(Hg)$$

where Cd(Hg) represents elementary cadmium dissolved in mercury. The sharp increase in current at about -2 V in both plots is associated with reduction of potassium ions to give a potassium amalgam.

For reasons to be considered presently, a polarographic wave suitable for analysis is obtained only in the presence of a large excess of a *supporting electrolyte;* potassium chloride serves this function in the present example. Examination of the polarogram for the supporting electrolyte alone reveals that a small current, called the *residual current*, exists even in the absence of cadmium ions.

A characteristic feature of a polarographic wave is the region in which the current, after increasing sharply, becomes essentially independent of the applied voltage; this constant current is called a *limiting current*. In the limiting current region, the microelectrode is completely polarized because a maximum has been reached in the rate at which reactant can be supplied to the electrode surface for reduction. In general, transport of reactant to an electrode surface occurs as a result of electrostatic attraction, thermal or mechanical convection, and diffusion. When these forces are no longer sufficient to bring reactant to the surface at an increasing rate, a limiting current is observed. In polarography, every effort is made to mini-

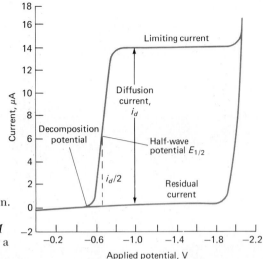

FIGURE 18-3 Polarogram for cadmium ion. The upper curve is for a solution that is $1 \times 10^{-3}\ M$ with respect to Cd^{2+} and $0.1\ M$ with respect to KCl. The lower curve is for a solution that is $0.1\ M$ in KCl only.

mize electrostatic and mechanical forces so that only diffusion is responsible for the transport of the reactant. The electrostatic attraction between the analyte and the electrode is reduced to nearly zero by the large excess of supporting electrolyte (a 50-fold or greater excess is required); insofar as possible, mechanical and thermal convection is avoided by insulating the cell from external sources of vibration and heat. Under these circumstances, the current is almost entirely limited by the diffusion rate and is given the special name *diffusion current* (symbolized by i_d). Ordinarily, the diffusion current is directly proportional to the concentration of the reactive constituent and is thus of prime importance from the standpoint of analysis. As shown in Figure 18-3, measurement of the diffusion current requires a correction for the residual current.

One other important quantity, the *half-wave potential,* is the potential at which the current is equal to one-half the diffusion current. The half-wave potential is usually given the symbol $E_{1/2}$; it may permit qualitative identification of the reactant.

18A-3 INTERPRETATION OF POLAROGRAPHIC WAVES
This section provides a qualitative description of the phenomena that are responsible for the characteristic, step-shaped polarographic wave. The reduction of cadmium ion to yield cadmium amalgam at a dropping mercury electrode will be used as a specific example; the conclusions, however, are applicable to other types of electrodes, oxidation processes, and reactions that yield other types of products.

The half-reaction

$$Cd^{2+} + Hg(l) + 2e \rightleftharpoons Cd(Hg)(l)$$

is reversible. In the context of polarography, *reversibility* implies that the electron transfer process is sufficiently rapid that the activities of reactants and products in the liquid film at the interface between the solution and the mercury electrode are determined by the electrode potential alone; thus, for the reversible reduction of cadmium ion, it may be assumed that the activities of the reactant and product at this interface are always those needed to satisfy the equation

$$E_{\text{applied}} = E_A^0 - \frac{0.0591}{2} \log \frac{[\text{Cd}]_0}{[\text{Cd}^{2+}]_0} - E_{\text{ref}} \tag{18-1}$$

Here, $[\text{Cd}]_0$ is the activity of metallic cadmium dissolved in the surface film of the mercury, and $[\text{Cd}^{2+}]_0$ is the activity of the ion in the aqueous phase. Note that the subscript zero has been employed for the activity terms to emphasize that this relationship *applies to the surface films of the two media only;* the activity of cadmium ion in the bulk of the solution and of elemental cadmium in the interior of the mercury drop *will ordinarily be quite different from the surface activities.* The films we are concerned with are no more than a few atoms or molecules thick.

The term E_{applied} in Equation 18-1 is the potential applied to the cell consisting of the dropping electrode and a reference electrode whose potential is E_{ref}; E_A^0 is the standard potential for the half-reaction in which a saturated cadmium amalgam is the product. The difference between E_A^0 and the standard potential when the product is elemental cadmium is about $+0.05$ V.

Consider now what occurs when E_{applied} is sufficiently negative to cause appreciable reduction of cadmium ion. Because the reaction is reversible, the activity of cadmium ion in the film surrounding the electrode decreases, and the activity of cadmium in the outer layer of the mercury drop increases instantaneously to the levels demanded by Equation 18-1; a surge of current results. This current would rapidly decay to zero were it not for the fact that cadmium ions are mobile in the aqueous medium and can migrate to the surface of the mercury as a consequence. Because the reduction reaction is essentially instantaneous, *the magnitude of this current depends upon the rate at which the cadmium ions move from the bulk of the solution to the surface where reaction occurs.* That is,

$$i = k' \times v_{\text{Cd}^{2+}}$$

where i is the current at an applied potential E_{applied}, $v_{\text{Cd}^{2+}}$ is the rate of migration of cadmium ions, and k' is a proportionality constant.

As mentioned earlier, ions or molecules migrate as a consequence of diffusion, thermal or mechanical convection, and electrostatic attraction. In polarography every effort is made to eliminate the latter two effects. Under these circumstances, diffusion becomes the only force responsible for the transport of cadmium ions to the electrode surface. Because the rate of diffusion is directly proportional to the concentration (strictly activity) difference between two parts of a solution, we may write

$$v_{\text{Cd}^{2+}} = k''([\text{Cd}^{2+}] - [\text{Cd}^{2+}]_0)$$

where $[\text{Cd}^{2+}]$ is the concentration *in the bulk of the solution* from which ions are diffusing and $[\text{Cd}^{2+}]_0$ is the concentration in the aqueous film surrounding the electrode. As long as diffusion is the only process bringing cadmium ions to the surface, it follows that

$$i = k'v_{\text{Cd}^{2+}} = k'k''([\text{Cd}^{2+}] - [\text{Cd}^{2+}]_0)$$

$$= k([\text{Cd}^{2+}] - [\text{Cd}^{2+}]_0)$$

Note that $[\text{Cd}^{2+}]_0$ becomes smaller as E_{applied} is made more negative (Equation 18-1). Thus, the rate of diffusion as well as the current increases with increases in applied potential. This potential ultimately becomes so negative that essentially every cadmium ion reaching the drop is reduced, and the activity of that ion in the surface film approaches zero; under this circumstance, the rate of diffusion and, thus, the current become constant, and the expression for current becomes

$$i_d = k[Cd^{2+}]$$

where i_d is the potential-independent diffusion current. Note that *the magnitude of the diffusion current is directly proportional to the concentration of the reactant in the bulk of the solution.* Quantitative polarography is based upon this fact.

A state of *complete concentration polarization* is said to exist when the current in a cell is limited by the rate at which a reactant can be brought to the surface of an electrode. With a microelectrode the current required to reach this condition is small—typically 3 to 10 μA for a 10^{-3} M solution. It is important to note that deposition by such current levels does not significantly alter the reactant concentration in the time required to obtain a polarogram.

Half-wave potential. An equation relating the applied potential, $E_{applied}$, and current, i, is readily derived.[2] For the reduction of cadmium ion to cadmium amalgam the equation takes the form

$$E_{applied} = E_{1/2} - \frac{0.0591}{n} \log \frac{i}{i_d - i} \tag{18-2}$$

where

$$E_{1/2} = E_A^0 - \frac{0.0591}{n} \log \frac{f_{Cd}k_{Cd}}{f_{Cd^{2+}}k_{Cd^{2+}}} - E_{ref} \tag{18-3}$$

Here, f_{Cd} and $f_{Cd^{2+}}$ are activity coefficients for the metal in the amalgam and the ion in the solution; k_{Cd} and $k_{Cd^{2+}}$ are proportionality constants related to the rates at which the two species diffuse in the respective media.

Examination of Equation 18-3 reveals that the half-wave potential is a reference point on a polarographic wave; it is independent of the reactant concentration but directly related to the standard potential for the half-reaction. In practice, the half-wave potential can be a useful quantity for identification of the species responsible for a given polarographic wave.

It is important to note that the half-wave potential may vary considerably with concentration for electrode reactions that are not rapid and reversible; in addition, Equation 18-2 no longer describes such waves.

Effect of complex formation on polarographic waves. We have already seen (Section 13D-3) that the potential for the oxidation or reduction of a metallic ion is greatly affected by the presence of species that form complexes with that ion. It is not surprising, therefore, that similar effects are observed with polarographic half-wave potentials. The data in Table 18-1 indicate that the half-wave potential for the reduction of a metal complex is generally more negative than that for reduction of the corresponding simple metal ion.

Lingane[3] has shown that the shift in half-wave potential as a function of the

[2] I. M. Kolthoff and J. J. Lingane, *Polarography*, vol. 1, p. 190. New York: Interscience, 1952.
[3] J. J. Lingane, *Chem. Rev.*, **29**, 1 (1941).

Table 18-1   Effect of Complexing Agents on Polarographic Half-Wave Potentials at the Dropping Mercury Electrode

Ion	Noncomplexing Media	1 M KCN	1 M KCl	1 M NH$_3$, 1 M NH$_4$Cl
Cd^{2+}	−0.59	−1.18	−0.64	−0.81
Zn^{2+}	−1.00	NR*	−1.00	−1.35
Pb^{2+}	−0.40	−0.72	−0.44	−0.67
Ni^{2+}	—	−1.36	−1.20	−1.10
Co^{2+}	—	−1.45	−1.20	−1.29
Cu^{2+}	+0.02	NR*	+0.04 and −0.22**	−0.24 and −0.51**

* No reduction occurs before involvement of the supporting electrolyte.
** Here, the reduction occurs in two steps having different electrode potentials. That is,

$$Cu^{2+} + 2Cl^- + e \rightleftharpoons CuCl_2^-$$
$$CuCl_2^- + Hg + e \rightleftharpoons Cu(Hg) + 2Cl^-$$

concentration of a complexing agent can be employed to determine the formula and the formation constant for the complex, provided that the cation involved reacts reversibly at the dropping electrode.

Polarograms for mixtures of reactants. The reactants of a mixture will ordinarily behave independently of one another at a microelectrode; a polarogram for a mixture is thus simply the summation of the waves for the individual components. Figure 18-4 shows the polarogram of a five-component cation mixture. Clearly, a single polarogram may permit the quantitative determination of several elements. Success depends upon the existence of a sufficient difference between succeeding half-wave potentials to permit evaluation of individual diffusion currents. Approximately 0.2 V is required if the more reducible species undergoes a two-electron reduction; a minimum of about 0.3 V is needed if the first reduction is a one-electron process.

Polarograms for irreversible reactions. Many polarographic electrode processes, particularly those associated with organic systems, are irreversible; drawn-out and less well-defined waves result. Although half-wave potentials for irreversible reactions ordinarily show a dependence upon concentration, diffusion currents remain linearly related to this variable, and such processes are readily adapted to quantitative analysis.

18A-4 PROPERTIES OF THE DROPPING MERCURY ELECTRODE
The dropping mercury electrode has certain unique features that make it particularly well suited for voltammetric studies. Some of these features are discussed briefly in the paragraphs that follow.

Current variations during the lifetime of a drop. The current in a cell containing a dropping electrode undergoes periodic fluctuations corresponding in frequency to the drop rate. As a drop breaks, the current falls to zero; it then increases rapidly as the new drop develops because of the greater surface to which diffusion can occur

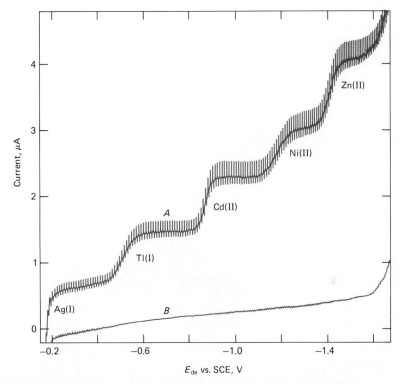

FIGURE 18-4   Polarograms of A: approximately 0.1 mM each of silver (I), thallium (I), cadmium (II), nickel (II), and zinc (II), listed in the order in which their waves appear, in 1 M ammonia, 1 M ammonium chloride containing 0.002% Triton X-100; B: the supporting electrolyte alone. [L. Meites, *Polarographic Techniques*, 2nd ed., p. 164. New York: Wiley, 1967. With permission.]

(see Figure 18-5a). A well-damped electrical system is needed to measure such rapidly fluctuating currents. As shown in Figure 18-5b, the oscillations under these circumstances are limited to a reasonable magnitude, and the average current is readily determined provided the drop rate is reproducible. Note the effect of irregular drops in the center of the limiting current region, which are probably caused by vibration of the apparatus.

When currents are measured by means of a recorder with a reasonably fast response, it is better to use the maximum current rather than the average current as the analytical parameter.

Advantages and limitations of the dropping mercury electrode. The dropping mercury electrode offers several advantages over other types of microelectrodes. The first is the large overvoltage for the formation of hydrogen from hydrogen ions. As a consequence, the reduction of many substances from acidic solutions can be studied without interference. Second, because a new metal surface is continuously generated, the behavior of the electrode is independent of its history. Thus, repro-

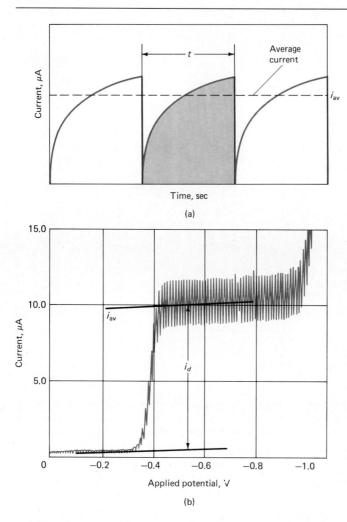

FIGURE 18-5 Effect of drop growth on polarographic currents. (a) Current-time relationship during the lifetime t of drops. The shaded area represents the microcoulombs of electricity associated with each drop. Note that the current increases by only about 20% during the second half of the drop lifetime. (b) A recorded polarogram in which the current fluctuations have been damped, thus permitting the measurement of the average current i_{av}.

ducible current-voltage curves are obtained regardless of how the electrode has been used previously. The third useful feature is that reproducible average currents are immediately achieved at any given applied potential.

The most serious limitation to the dropping electrode is the ease with which mercury is oxidized; this property severely restricts the use of mercury as an anode. At applied potentials much above $+0.4$ V (versus the saturated calomel electrode), formation of mercury(I) occurs; the resulting current masks the polarographic waves of other oxidizable species in the solution. Thus, the dropping mercury electrode can be employed only for the analysis of reducible or very easily oxidizable substances. Other disadvantages are that it is cumbersome to use and tends to malfunction by clogging.

18A-5 SCOPE OF POLAROGRAPHY

Virtually every element in one form or another is amenable to polarographic analysis. In addition, the method can be extended to the determination of numerous organic functional groups. Because the polarographic behavior of any species is unique for a given set of experimental conditions, the technique offers attractive possibilities for selective analysis.

Most polarographic analyses are performed in aqueous solution, but other solvent systems may be substituted if necessary. The optimum concentration range for quantitative analyses lies between 10^{-2} and 10^{-4} M; by suitably modifying the basic polarographic procedure, however, concentration determinations in the parts-per-billion range become possible. An analysis can be easily performed on 1 to 2 mL of solution; with a little effort, a volume as small as one drop is sufficient. The polarographic method is thus particularly useful for the determination of quantities in the milligram to microgram range.

Relative errors ranging between 2 and 3% are to be expected in routine polarographic work. This order of uncertainty is comparable to or smaller than errors affecting other methods for the analysis of small quantities.

18A-6 ANALYTICAL DETAILS

Several important variables that affect the applicability and the accuracy of polarography are considered in this section.

Temperature control. The diffusion currents for most substances are found to increase by about 2.5% per degree Celsius. Thus, solutions must be thermostated to a few tenths of a degree for accurate polarographic analysis.

Oxygen removal. Dissolved oxygen undergoes a two-step irreversible reduction at the dropping electrode; the H_2O_2 produced in the first step is reduced to H_2O in the second. Two waves of equal size result, the first with a half-wave potential at about -0.14 V and the second at about -0.9 V (vs. SCE). The two half-reactions are somewhat slow. As a consequence, the waves are drawn out over a considerable potential range.

While these polarographic waves are convenient for the determination of oxygen in solutions, the presence of this element often interferes with the accurate determination of other species. Thus, oxygen removal is ordinarily the first step in a polarographic analysis. Deaeration of the solution for several minutes with an inert gas accomplishes this end. A stream of the same gas, usually nitrogen, is passed over the surface during the analysis to prevent reabsorption.

Current maxima. Polarographic waves are frequently distorted by so-called current maxima (see Figure 18-6). Maxima are troublesome because they interfere with the accurate evaluation of diffusion currents and half-wave potentials. The cause or causes of maxima are not fully understood; fortunately, however, considerable empirical knowledge exists for their elimination. The addition of traces of such high-molecular-weight substances as gelatin, Triton X-100 (a commercial surface-active agent), methyl red, other dyes, or carpenter's glue is effective. The first two of these additives are particularly useful.

Determination of diffusion currents. In measuring currents obtained with the dropping electrode, it is common practice to use the *average* value of the galvanometer or recorder oscillations rather than the maximum or minimum values (see Figure 18-5a); the measurement is thus less dependent upon the damping employed.

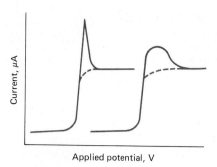

Applied potential, V

FIGURE 18-6 Typical current maxima.

For analytical work, limiting currents must always be corrected for the residual current. A residual current curve is ordinarily obtained immediately before or after acquiring the polarogram for the analyte. The diffusion current is taken as the difference between the two at some potential in the limiting current region (see Figure 18-3).

Because the residual current usually increases nearly linearly with applied voltage, it is often possible to dispense with a separate residual-current curve and obtain the correction by extrapolation (see Figure 18-5).

Concentration determination. The best and most straightforward method for quantitative polarographic analysis involves preliminary calibration with a series of standard solutions. As nearly as possible, these standards should be identical with the samples being analyzed and cover a concentration range within which the unknown samples will likely fall. The linearity of the current-concentration relationship can be assessed from such data; if this relationship is nonlinear, the analysis can be based upon the calibration curve.

The standard addition method (p. 418) is also applicable to polarographic analysis. This procedure is particularly effective when the diffusion current is sensitive to other constituents that are present in the sample.

18A-7 INORGANIC POLAROGRAPHIC ANALYSIS

The polarographic method is generally applicable to the analysis of inorganic substances. Most metallic cations, for example, are reduced at the dropping electrode to form a metal amalgam or an ion of lower oxidation state. Even the alkali- and alkaline-earth metals are reducible, provided the supporting electrolyte used does not decompose at the high potentials required. The tetraalkyl ammonium halides serve this function well.

18A-8 ORGANIC POLAROGRAPHIC ANALYSIS

Almost from its inception the polarographic method has been used for the study and analysis of organic compounds, and a large number of papers have been devoted to this subject.[4] Organic compounds containing any of the following functional groups can be expected to react at the dropping mercury electrode and produce one or more polarographic waves.

[4] See: P. Zuman, *Topics in Organic Polarography*. New York: Pergamon Press, 1970.

1. *The carbonyl group* includes aldehydes, ketones, and quinones. Aldehydes tend to be reduced at lower potentials than ketones. Conjugation of the carbonyl double bond also leads to lower half-wave potentials.
2. *Certain carboxylic acids* are reduced polarographically, although the simple aliphatic and aromatic monocarboxylic acids are not. Dicarboxylic acids such as fumaric, maleic, or phthalic acid, in which the carboxyl groups are conjugated with one another, give characteristic polarograms. The same is true of certain keto and aldehydo acids.
3. *Most peroxides and epoxides* yield useful polarograms.
4. *Nitro, nitroso, amine oxide,* and *azo groups* are generally reduced at the dropping electrode.
5. *Most organic halogen groups* produce a polarographic wave as the halogen is replaced with an atom of hydrogen.
6. *The carbon/carbon double bond* is reduced when it is conjugated with another double bond, an aromatic ring, or an unsaturated group.
7. *Hydroquinones* and *mercaptans* produce anodic waves.

In addition, numerous applications to biological systems have been reported.[5]

18B AMPEROMETRIC TITRATIONS WITH ONE MICROELECTRODE

The polarographic method can be used to estimate the equivalence point of a reaction, provided at least one of the participants or products of the titration is oxidized or reduced at a microelectrode. Here, the current in a polarographic cell at some fixed potential is measured as a function of titrant volume (or of time in a coulometric titration). Plots of the data on either side of the equivalence point are straight lines with differing slopes; the end point can be fixed by extrapolation to their intersection (see curve *B*, Figure 6-3, p. 170).

The amperometric method is inherently more accurate than the polarographic method and less dependent upon the characteristics of the electrode and the supporting electrolyte. Furthermore, the temperature need not be fixed accurately, although it must be kept constant during the titration. Finally, the substance being determined need not be reactive at the electrode; a reactive reagent or product will suffice as well.

18B-1 TITRATION CURVES

Amperometric titration curves typically take one of the forms shown in Figure 18-7. Figure 18-7a represents a titration in which the analyte reacts at the electrode while the reagent does not. The titration of lead with sulfate or oxalate ions may be cited as an example. Here, a sufficiently high potential is applied to give a diffusion current for lead; a linear decrease in current is observed as lead ions are removed from the solution by precipitation. The curvature near the equivalence point reflects the incompleteness of the analytical reaction in this region. The end point is obtained by extrapolation of the linear portions, as shown.

The curve of Figure 18-7b is typical of a titration in which the reagent reacts

[5] M. Brezina and P. Zuman, *Polarography in Medicine, Biochemistry and Pharmacy.* New York: Interscience, 1958; *Polarography of Molecules of Biological Significance,* W. F. Smyth, Ed. New York: Academic Press, 1979.

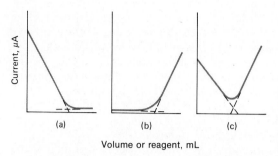

FIGURE 18-7 Typical amperometric titration curves. (a) Analyte is reduced, reagent is not. (b) Reagent is reduced, analyte is not. (c) Both reagent and analyte are reduced.

at the microelectrode and the analyte does not. An example is the titration of magnesium with 8-hydroxyquinoline. A diffusion current for the latter is obtained at −1.6 V (vs. SCE), whereas magnesium ion is inert at this potential.

The curve of Figure 18-7c corresponds to the titration of lead ion with a chromate solution at an applied potential greater than −1.0 V. Both lead and chromate ions give diffusion currents at this potential, and a minimum in the curve signals the end point. The same titration would yield a curve like that of Figure 18-7b at an applied potential of 0.0 V because only chromate ions are reduced under these conditions.

In order to obtain plots with linear regions before and after the equivalence point (Figure 18-7), it is necessary to apply corrections for the volume change that results from the added titrant. This correction is made by multiplying the measured diffusion current by $(V + v)/V$, where V is the original volume and v is the volume of reagent; thus, all measured currents are corrected back to the original volume. An alternative, which is often satisfactory, is to make v negligibly small through use of a titrant solution that is 20 or more times as concentrated as the solution being titrated. This approach, however, does require the use of a microburet so that a total reagent volume of 1 or 2 mL can be measured with suitable accuracy.

18B-2 EQUIPMENT

A simple manual polarograph is entirely adequate for amperometric titrations. Ordinarily the applied voltage does not have to be known any closer than about ±0.05 V because it is only necessary to select a potential within the diffusion current region of at least one reactant or product in the titration.

Many amperometric titrations can be carried out conveniently with a dropping mercury electrode. For reactions involving oxidizing agents that attack mercury, e.g., bromine, silver ion, and iron(III), a rotating platinum electrode is preferable. This microelectrode consists of a short length of platinum wire sealed into the side of a glass tube. Mercury inside the tube provides electrical contact between the wire and the lead to the polarograph. The tube is held in the hollow chuck of a synchronous motor and rotated at a constant speed in excess of 600 rpm. Commercial models of the rotating electrode are available. A typical apparatus is shown in Figure 18-8.

Polarographic waves, similar to those observed with the dropping mercury electrode, can be obtained with the rotating platinum electrode. Here, however, the reactive species is brought to the electrode surface not only by diffusion but also by mechanical mixing. As a consequence, the limiting currents are as much as 20 times larger than those obtained with a microelectrode that is supplied by diffu-

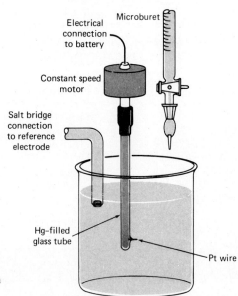

FIGURE 18-8 Typical cell arrangement for amperometric titrations with a rotating platinum electrode.

sion only. With a rotating electrode, steady currents are instantaneously obtained. This behavior is in distinct contrast to that of a solid microelectrode in an unstirred solution.

Several limitations restrict the widespread application of the rotating platinum electrode to polarography. The low hydrogen overvoltage prevents its use as a cathode in acidic solutions. In addition, the high currents obtained with the electrode make it particularly sensitive to traces of oxygen in the solution. These two factors have largely confined its employment to anodic reactions. Limiting currents are often influenced by the history of the electrode and are seldom as reproducible as the diffusion currents obtained with a dropping mercury electrode. This drawback, however, does not seriously restrict the use of the rotating electrode for amperometric titrations.

18B-3 APPLICATION OF AMPEROMETRIC TITRATIONS
The amperometric end point has been largely confined to titrations in which a slightly soluble precipitate or a stable complex is the reaction product. Selected applications are listed in Table 18-2. Some organic precipitants that are reduced at the dropping mercury electrode appear in this table.

A few applications of the amperometric method to oxidation-reduction reactions exist. For example, the technique has been applied to various titrations involving iodine and bromine (in the form of bromate) as reagents.

18C AMPEROMETRIC TITRATIONS WITH TWIN POLARIZED MICROELECTRODES
A convenient modification of the amperometric method involves the use of two identical, stationary microelectrodes immersed in a well-stirred solution of the sample. A small potential (say, 0.1 to 0.2 V) is applied between these electrodes, and the current is followed as a function of the volume of added reagent. The end

Table 18-2   Applications of Amperometric Titrations

Reagent	Reaction Product	Type Electrode*	Substance Determined
K_2CrO_4	Precipitate	DME	Pb^{2+}, Ba^{2+}
$Pb(NO_3)_2$	Precipitate	DME	SO_4^{2-}, MoO_4^{2-}, F^-, Cl^-
8-Hydroxyquinoline	Precipitate	DME	Mg^{2+}, Zn^{2+}, Cu^{2+}, Cd^{2+}, Al^{3+}, Bi^{3+}, Fe^{3+}
Cupferron	Precipitate	DME	Cu^{2+}, Fe^{3+}
Dimethylglyoxime	Precipitate	DME	Ni^{2+}
α-Nitroso-β-naphthol	Precipitate	DME	Co^{2+}, Cu^{2+}, Pd^{2+}
$K_4Fe(CN)_6$	Precipitate	DME	Zn^{2+}
$AgNO_3$	Precipitate	RP	Cl^-, Br^-, I^-, CN^-, RSH
EDTA	Complex	DME	Bi^{3+}, Cd^{2+}, Cu^{2+}, Ca^{2+}, and so on
$KBrO_3$, KBr	Substitution, addition, or oxidation	RP	Certain phenols, aromatic amines, olefins; N_2H_4, As(III), Sb(III)

* DME = dropping mercury electrode; RP = rotating platinum electrode.

point is marked by a sudden current rise from zero, a decrease in the current to zero, or a minimum (at zero) in a V-shaped curve.

Although the use of two polarized electrodes for end-point detection was first proposed before 1900, almost 30 years passed before chemists came to appreciate the potentialities of the method.[6] The name *dead-stop end point* was used to describe the technique, and this term is still occasionally encountered. It was not until about 1950 that a clear interpretation of dead-stop titration curves was made.[7]

18C-1 APPLICATIONS TO TITRATION WITH SILVER IONS

Twin silver microelectrodes permit observation of the end point for the various titrations employing silver nitrate shown in Table 7-1 (p. 190). Consider what happens, for example, when a small potential is applied between two such electrodes during the titration of bromide with silver ions. Short of the equivalence point essentially no current can exist because no easily reduced species is present in the solution. Consequently, electron transfer at the cathode is precluded and that electrode is completely polarized. Note that the anode is not polarized because the reaction

$$Ag \rightleftharpoons Ag^+ + e$$

could occur in the presence of a suitable cathodic reactant or depolarizer.

After equivalence has been passed, the cathode becomes depolarized owing to the presence of significant amount of silver ions which can react to give silver. That is,

$$Ag^+ + e \rightleftharpoons Ag$$

[6] C. W. Foulk and A. T. Bawden, *J. Amer. Chem. Soc.*, **48**, 2045 (1926).
[7] For an excellent analysis of this type of end point, see; J. J. Lingane, *Electroanalytical Chemistry*, 2d ed, pp. 280–294. New York: Interscience, 1958.

Current is permitted as a result of this half-reaction and the corresponding oxidation of silver at the anode. The magnitude of the current is, as in other amperometric methods, directly proportional to the concentration of the excess reagent. Thus, the titration curve is similar to that shown in Figure 18-7b.

18C-2 APPLICATION TO TITRATIONS WITH BROMINE AND IODINE

An amperometric titration with twin platinum microelectrodes has also been applied to titrations in which iodine or bromine is the titrant. When this technique is employed for the titration of arsenious acid, for example, a curve similar to Figure 18-7b is again obtained. No current is observed in the early stages of the titration because of cathodic polarization. In contrast to the previous example, however, polarization here is of the kinetic type. That is, even though a cathodic half-reaction involving H_3AsO_4, the titration product, can be written

$$H_3AsO_4 + 2e + 2H^+ \rightleftharpoons H_3AsO_3 + H_2O$$

this process occurs so slowly at the electrode surface that no current can be detected.

The principal advantage of the twin microelectrode procedure is its simplicity. One can dispense with a reference electrode; the only instrumentation needed is a simple voltage divider, powered by a dry cell, and a galvanometer or microammeter for current detection.

PROBLEMS*

* 18-1. As the terms pertain to a polarographic wave, make a distinction between
  - (a) decomposition potential and half-wave potential.
  - (b) limiting current and diffusion current.
* 18-2. How are limiting currents affected in a series of solutions in which
  - (a) the concentration of the analyte is increased while the concentration of the supporting electrolyte is held constant?
  - (b) the concentration of the supporting electrolyte is decreased while the concentration of the analyte is held constant?
  18-3. The accompanying data pertain to generation of a calibration curve for the polarographic determination of lead. Supply the missing numbers.

Concentration, C, of Pb^{2+}, mol/L	Limiting Current, i, μA	Diffusion Current, i_d, μA
0	2.4	0
* 3.00×10^{-4}	6.4	
5.00×10^{-4}		6.7
*	15.6	13.2
2.50×10^{-3}	35.4	
* 4.00×10^{-3}		53.2
	84.7	82.3
7.50×10^{-3}	101	

18-4. A linear relationship between the diffusion current and concentration requires that

$$i_d = kC$$

Evaluate i_d/C for the data in Problem 18-3. Calculate the standard deviation in this quantity.

18-5. Solutions were analyzed for their lead content by the polarographic method in Problem 18-3. Calculate their respective concentrations, based upon the accompanying limiting-current data:

Solution	Limiting Current, i, μA	Solution	Limiting Current, i, μA
* (a)	61.7	(c)	53.8
* (b)	24.9	(d)	80.0

18-6. A standard-addition method for the polarographic analysis of zinc calls for introduction of 10.0 mL of 4.00×10^{-3} M Zn^{2+} to one of two 25.0-mL aliquots of the sample. Both then receive 10.0 mL of a tartrate solution that acts as supporting electrolyte and buffer; finally, both are diluted to 50.0 mL. Calculate the weight (mg) of zinc in 100 mL of the following solutions:

	Diffusion Current, μA	
Solution	Aliquot Only	Aliquot Plus Standard Addition
* (a)	29.4	36.1
(b)	41.6	48.3
* (c)	82.9	89.6
(d)	63.4	70.1
(e)	17.3	24.0

18-7. Calculate the milligrams of cadmium in each milliliter of sample, based upon the following data (corrected for residual current):

	Volumes Used, mL				
Solution	Sample	0.400 M KCl	2.00×10^{-3} M Cd^{2+}	H_2O	Current, μA
* (a)	15.0	20.0	0.00	15.0	79.7
	15.0	20.0	5.00	10.0	95.9
(b)	10.0	20.0	0.00	20.0	49.9
	10.0	20.0	10.0	10.0	82.3
* (c)	20.0	20.0	0.00	10.0	41.4
	20.0	20.0	5.00	5.00	57.6
(d)	15.0	20.0	0.00	15.0	67.9
	15.0	20.0	10.0	5.00	100.3
(e)	5.00	20.0	0.00	25.0	56.4
	5.00	20.0	10.0	15.0	88.8

18-8. The sulfate in a 50.0-mL aliquot of solution was determined amperometrically by titration with 0.0160 M $Pb(NO_3)_2$ at -1.20 V (vs. SCE). Use the accompanying data to

(a) construct a curve relating the current (corrected for volume change) observed to the volume of reagent.

(b) calculate the molar concentration of sulfate in the solution.

0.0160 M Pb^{2+}, mL	i_{obs}, μA
5.00	1.0
10.0	1.4
15.0	2.1
18.0	2.9
21.0	5.0
24.0	20.8
27.0	122
30.0	262
33.0	386

18-9. Sketch the general features of the curve to be expected from the amperometric titration of

(a) chloride ion with standard AgNO$_3$ at 0.0 V (vs. SCE); silver ion is reduced at this potential.

(b) bismuth(III) with standard EDTA at -0.18 V (vs. SCE); at this potential, the titrant is not involved at the dropping mercury electrode.

(c) nickel(II) with dimethylglyoxime at an applied potential of -1.85 V (vs. SCE); both participants are reduced at this potential.

(d) tetraphenylboron with standard AgNO$_3$ at -0.10 V (vs. SCE); only Ag$^+$ is reduced at the dropping mercury electrode at this potential.

(e) molybdate with a standard solution of lead at -0.80 V (vs. SCE). Reaction:

$$Pb^{2+} + MoO_4^{2+} \longrightarrow PbMoO_4(s)$$

Both participants are reduced at this potential.

(f) a mixture of Pb^{2+} and Ba^{2+} with a standard chromate solution at an applied potential of -1.2 V (vs. SCE); see appendix for K_{sp} data. Chromate and lead ions are reduced at this potential.

an introduction to spectroscopic

methods of analysis

19

Spectroscopic methods of analysis are based upon the production or interaction of electromagnetic radiation with matter. *Emission methods* depend upon the electromagnetic radiation produced when the analyte is *excited* by thermal, electrical, or radiant energy. *Absorption methods,* on the other hand, are based upon the *attenuation* (the weakening) of a beam of electromagnetic radiation as a consequence of its interaction with and partial absorption by the analyte. Spectroscopic methods are among the most widespread and powerful analytical tools for both qualitative and quantitative analysis.

It is convenient to characterize spectroscopic methods according to the type of electromagnetic radiation involved. The categories include X-ray, ultraviolet, visible, infrared, microwave, and radio-frequency methods. We will be concerned mainly with the absorption and emission of ultraviolet, visible, and infrared radiation but will make occasional reference to other types as well.[1]

19A PROPERTIES OF ELECTROMAGNETIC RADIATION
Electromagnetic radiation is a type of energy that is transmitted through space at enormous velocity. Many of the properties of electromagnetic radiation are conven-

---

[1] For a further review of these topics, see: E. J. Meehan, in *Treatise on Analytical Chemistry*, Part I, vol. 5, Chapter 53, I. M. Kolthoff and P. J. Elving, Eds. New York: Wiley, 1964; F. Grum, in *Physical Methods of Chemistry*, Vol. I, Part IIIB, pp. 214–305, A. Weissberger and B. W. Rossiter, Eds. New York: Wiley-Interscience, 1972; and J. E. Crooks, *The Spectrum in Chemistry*. New York: Academic Press, 1978.

iently described by means of a classical wave model that employs such parameters as wavelength, frequency, velocity, and amplitude. In contrast to other wave phenomena, such as sound, electromagnetic radiation requires no supporting medium for its transmission; thus, it readily passes through a vacuum.

Phenomena associated with the absorption or emission of radiant energy cannot be explained adequately by treating radiation as waves; here, it is necessary to view electromagnetic radiation as a stream of discrete particles of energy called *photons* with energies that are proportional to the frequency of the radiation. These dual views of radiation as particles and waves are not mutually exclusive. Indeed, the duality is found to apply to the behavior of streams of electrons and other elementary particles as well and is rationalized by wave mechanics.

19A-1 WAVE PROPERTIES

For many purposes, electromagnetic radiation is conveniently treated as an oscillating electrical force field in space; associated with the electrical field, and at right angles to it, is a magnetic force field.

The electrical and magnetic fields associated with radiation are vector quantities; at any instant, they can be represented by an arrow whose length is proportional to the magnitude of the force and whose direction is parallel to that of the force. A graphic representation of a beam of radiation can be obtained by plotting one of these vector quantities as a function of time as the radiation passes a fixed point in space. Alternatively, the vector can be plotted as a function of distance, with time held constant.

In Figure 19-1, the electrical vector serves as the ordinate; a plot of the magnetic vector would be identical in every regard except that the ordinate would be rotated 90 deg around the zero axis. Normally, only the electrical vector is of concern because it is the electrical force that is responsible for such phenomena as

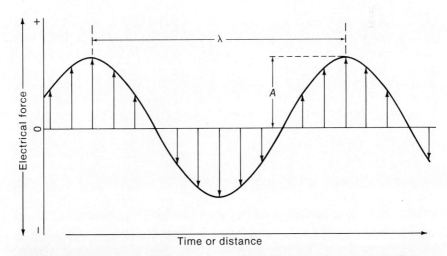

FIGURE 19-1   Representation of a beam of monochromatic radiation of wavelength λ and amplitude A. The arrows represent the electrical vector of the radiation.

transmission, reflection, refraction, and absorption of radiation. Figure 19-1 is a two-dimensional representation of monochromatic (that is, a single wavelength) radiation. Viewed end on, it would appear as a circular cross section that periodically fluctuates in radius from zero to the maximum amplitude A.

Wave parameters.   The time required for the passage of successive maxima through a fixed point in space is called the *period, p,* of the radiation. The *frequency, ν,* is the number of oscillations of the field that occurs per second[2] and is equal to $1/p$. It is important to realize that *the frequency is determined by the source and remains invariant* regardless of the media traversed by the radiation. In contrast, the *velocity* of propagation, v_i, the rate at which the wave front moves through a medium, is *dependent* upon both the medium and the frequency; the subscript i is employed to indicate this frequency dependence. Another parameter of interest is the *wavelength, λ_i,* which is the linear distance between successive maxima or minima of a wave.[3] Multiplication of the frequency in cycles per second by the wavelength in centimeters per cycle gives the velocity of propagation in centimeters per second; that is,

$$v_i = \nu\lambda_i \tag{19-1}$$

In a vacuum, the velocity of propagation of radiation becomes independent of frequency and is at its maximum. This velocity, given the symbol c, has been accurately determined to be 2.99792×10^{10} cm/s. Thus, for a vacuum,

$$c = \nu\lambda = 3.00 \times 10^{10} \text{ cm/s} \tag{19-2}$$

The rate of propagation in any other medium is less because of interactions between the electromagnetic field of the radiation and the bound electrons in the atoms or molecules of the medium. Since the radiant frequency is invariant and

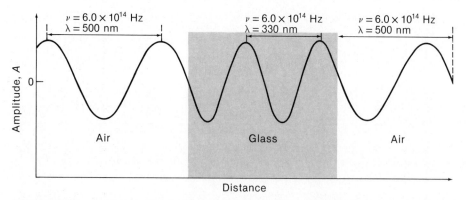

FIGURE 19-2   Wavelength of radiation in air and in a dense glass.

[2] The common unit of frequency is the *hertz,* Hz, which is equal to one cycle per second.
[3] The commonly used units for wavelength depend upon the spectral region in question. For example, the ångström unit, Å (10^{-10} m), is convenient for X-ray and short ultraviolet radiation; the nanometer, nm (10^{-9} m), is employed with visible and ultraviolet radiation; the micrometer, μm (10^{-6} m), is useful for the infrared region.

fixed by the source, the *wavelength must decrease* as radiation passes from a vacuum to a medium containing matter (Equation 19-1). This effect is illustrated in Figure 19-2.

It should be noted that the velocity of radiation in air differs only slightly from c (about 0.03% less); for most purposes, then, Equation 19-2 is usually applicable to air as well as to a vacuum.

The wavenumber σ is defined as the number of waves per centimeter and is yet another way of describing electromagnetic radiation. When the wavelength *in vacuo* is expressed in centimeters, the wavenumber is equal to $(1/\lambda)$ cm^{-1}.

Radiant power or intensity. The *power, P,* of radiation is the energy of the beam that reaches a given area per second; the *intensity, I,* is the power per unit solid angle. These quantities are related to the square of the amplitude A (see Figure 19-1). Although it is not strictly correct to do so, power and intensity are often used synonymously.

19A-2 PARTICLE PROPERTIES OF RADIATION

Energy of electromagnetic radiation. An understanding of certain interactions between radiation and matter requires that the radiation be treated as packets of

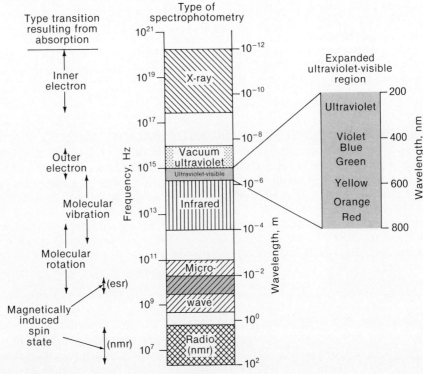

FIGURE 19-3 Parts of the electromagnetic spectrum which are employed for spectrophotometry.

energy called *photons* or *quanta*. The energy of a photon depends upon the frequency of the radiation and is given by

$$E = h\nu \tag{19-3}$$

where h is Planck's constant (6.63×10^{-27} erg sec). In terms of wavelength and wavenumber,

$$E = \frac{hc}{\lambda} = hc\sigma \tag{19-4}$$

Note that the wavenumber, σ, like the frequency, is directly proportional to energy.

19B THE ELECTROMAGNETIC SPECTRUM

The electromagnetic spectrum covers an immense range of wavelengths and energies. For example, an X-ray photon ($\lambda \sim 10^{-10}$ cm) is approximately 10,000 times more energetic than a photon emitted by an incandescent tungsten wire ($\lambda \sim 10^{-6}$ m) and 10^{11} times that for one in the radio-frequency range.

Figure 19-3 depicts qualitatively the major divisions of the electromagnetic spectrum. Note that the frequency (or wavelength) scale is logarithmic; note also that the region to which the human eye is perceptive (the *visible spectrum*) is but a minute part of the whole spectrum. Such diverse radiations as gamma rays or radio waves differ from visible light only in the matter of frequency and, hence, energy (Equation 19-3).

Figure 19-3 also indicates the regions of the spectrum that are useful for analytical purposes and the molecular or atomic transitions responsible for absorption or emission of radiation in each region.

19C EMISSION OF ELECTROMAGNETIC RADIATION

Atoms, ions, and molecules have a limited number of discrete and quantized energy levels. At room temperature, most species exist primarily in the lowest of these energy levels, called the *ground state*. A sample of matter can be *excited* to one or more higher energy levels by any of several processes, including bombardment with electrons or other elementary particles, exposure to a high-voltage alternating-current spark, heat treatment in a flame or arc, or exposure to an intense source of electromagnetic radiation. The lifetime of an excited species is generally transitory (10^{-6} to 10^{-9} s), and *relaxation* to a lower energy level or the ground state may take place with a release of the excess energy in the form of electromagnetic radiation, heat, or perhaps both.

19C-1 LINE AND BAND EMISSION SPECTRA

Radiation from a source is conveniently characterized by means of an *emission spectrum*, which usually takes the form of a plot of relative power of the radiation as a function of wavelength or frequency. Three types of spectra can be distinguished, namely, *line*, *band*, and *continuous*. All are of importance in analytical chemistry.

Source of line spectra. Line spectra are encountered when the radiating species are individual atomic particles that are well separated, as in a gas. Under this circumstance the individual particles behave independently of one another, and the spectrum consists of a series of sharp lines with widths of about 10^{-4} Å. Figure 19-4 illustrates the appearance of a typical emission line spectrum. Note that the sharp lines for sodium, potassium, strontium, and calcium are superimposed upon a broad band spectrum, which will be discussed in the next section.

The energy level diagram in Figure 19-5a shows the source of two of the lines in a typical spectrum of an element. The horizontal line labeled G corre-

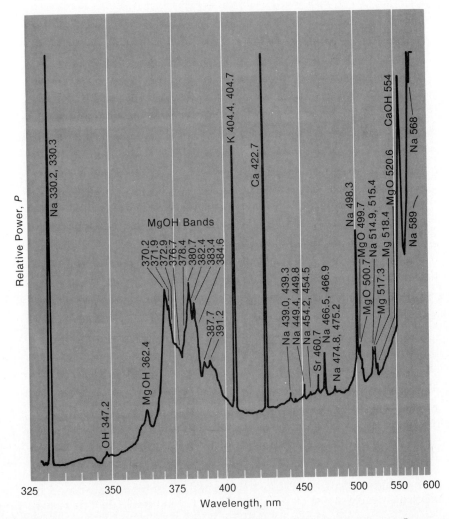

FIGURE 19-4   Emission spectrum of a brine obtained with an oxyhydrogen flame. [R. Hermann and C. T. J. Alkemade, *Chemical Analysis by Flame Photometry*, 2nd ed., p. 484. New York: Interscience, 1963. With permission.]

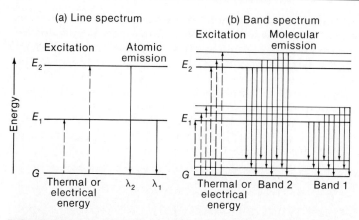

FIGURE 19-5 Energy level diagrams for an atom and a simple molecule showing the source of (a) a line spectrum and (b) a band spectrum.

sponds to the lowest or ground state energy of the atom. Also shown by heavy horizontal lines labeled E_1 and E_2 are two higher energy electronic levels of the species. For example, the single outer electron in the ground state G for a sodium atom is located in the $3s$ orbital. Energy level E_1 could then represent the energy of the atom when this electron has been promoted to the $3p$ state by absorption of thermal, electrical, or radiant energy. This process is depicted by the dashed arrow on the left in Figure 19-5a. After perhaps 10^{-8} s, the atom returns to the ground state, emitting radiation whose frequency is

$$\nu_1 = (E_1 - G)/h$$

or

$$\lambda_1 = hc/(E_1 - G)$$

This process is illustrated by the solid arrow on the right side of Figure 19-5a. For the sodium atom, E_2 could correspond to the more energetic $4p$ state; the resulting radiation, λ_2, would then appear at a shorter wavelength. The line appearing at about 330 nm in Figure 19-4 results from this transition; the $3p$ to $3s$ transition provides a line at about 590 nm.[4]

Source of band spectra.   For a *molecule*, the energy of various electronic states depends not only upon the electronic energy but also upon how its atomic components are vibrating with respect to one another and how the molecule is rotating about its center of gravity. The energy differences among the various quantized vibrational and rotational states are much smaller than among the electronic states. The energies of two of the many excited vibrational states associated with each electronic state are shown by the lighter horizontal lines in Figure 19-5b. Typically, many, many more vibrational energy levels would exist. In addition, super-

[4] In fact, both these lines are doublets that differ in wavelength by only a few tenths of a nanometer. The cause of the splitting is beyond the scope of this text.

imposed on each of the vibrational states are a series of rotational states that differ by yet smaller energies; these states are not shown in the figure. The consequence of this multitude of energy states is depicted by the solid lines in Figure 19-5b. Here, two bands of radiation are produced, each of which consists of a large number of closely spaced lines. Unless a very high-resolution instrument is employed, the discrete nature of the component lines of the band may not be detected.

Figure 19-4 contains band spectra for such polyatomic species as CaOH, MgOH, and OH radicals. The bands between 365 and 400 nm, for example, are the result of vibrational and rotational transitions superimposed on electronic transitions of the species MgOH. Note that the resolving power of the instrument is not sufficient to reveal the discrete nature of the bands very well.

Effect of concentration on line and band spectra. The radiant power, P, of a line or band depends directly upon the number of excited atoms or molecules, which in turn is proportional to the total concentration of the species present in the source. Thus, we may write

$$P = kC \tag{19-5}$$

where k is a proportionality constant. This relationship is the basis of quantitative emission spectroscopy.

19C-2 CONTINUOUS EMISSION SPECTRA

Truly continuous radiation is produced when solids are heated to incandescence. Thermal radiation of this kind, which is called *black-body radiation*, is more characteristic of the temperature of the emitting surface than the material of which that surface is composed. Black-body radiation is produced by the innumerable atomic and molecular oscillations excited in the condensed solid by the thermal energy. Theoretical treatment of black-body radiation leads to the following conclusions: (1) the radiation exhibits a maximum emission at a wavelength that varies inversely with the absolute temperature; (2) the total energy emitted by a black body (per

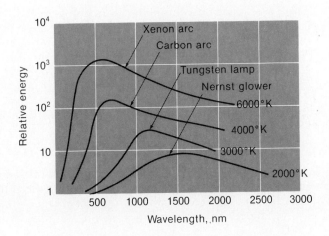

FIGURE 19-6 Black-body radiation curves.

unit of time and area) varies as the fourth power of temperature; and (3) the emissive power at a given temperature varies inversely as the fifth power of wavelength. These relationships are reflected in the behavior of several experimental radiation sources shown in Figure 19-6; the emission from these sources approaches that of the ideal black body. Note that the energy peaks in Figure 19-6 shift to shorter wavelengths with increasing temperature. It is clear that very high temperatures are needed to cause a thermally excited source to emit a substantial fraction of its energy as ultraviolet radiation.

Heated solids are important sources of infrared, visible, and longer wavelength ultraviolet radiation for analytical instruments.

19D ABSORPTION OF RADIATION

Absorption refers to a process by which a chemical species in a transparent medium selectively removes certain frequencies of electromagnetic radiation. In this process, the energy of the photon converts the absorbing atom, molecule, or ion M to a more energetic or excited form, M^*; excitation can be depicted by the equation

$$M + h\nu \longrightarrow M^*$$

After a brief period (10^{-6} to 10^{-9} s), the excitation energy is lost, most often as heat, and the species relaxes to its former state; that is,

$$M^* \longrightarrow M + heat$$

Relaxation may also result from decomposition of M^* to form new species; such a process is called a *photochemical decomposition*. Alternatively, relaxation may involve the *fluorescent* or *phosphorescent* reemission of radiation. It is important to note that the lifetime of M^* is so very short that its concentration at any instant is ordinarily negligible. Furthermore, the amount of thermal energy created is usually so small as to be undetectable. Thus, absorption measurements have the advantage of creating minimal disturbance of the system under study.

19D-1 ATOMIC ABSORPTION

The passage of polychromatic ultraviolet or visible radiation through a medium consisting of monatomic particles, such as gaseous mercury or sodium, results in the absorption of but a few well-defined frequencies. The relative simplicity of such spectra is due to the small number of possible energy states for the particles. Excitation can occur only by an electronic process in which one or more of the electrons of the atom is raised to a higher energy level. Thus, with sodium, excitation of the $3s$ electron to the $3p$ state requires energy corresponding to a wavenumber of 1.697×10^4 cm^{-1}. As a result, sodium vapor exhibits a sharp absorption peak at 589.3 nm (yellow light). It is important to understand that the wavelengths for absorption peaks will be identical to those of the lines emitted by the gaseous element because the transitions involved in the two processes are identical. Thus, excited sodium atoms emit radiation at 589.3 while sodium atoms in the ground state absorb this wavelength. Furthermore, the element depicted in Figure 19-5a would absorb radiation having wavelengths of precisely λ_1 and λ_2.

The width of absorption lines in the ultraviolet and visible regions is about 0.005 nm.

Ultraviolet and visible radiation have sufficient energy to cause transitions of the outermost or bonding electrons only. X-ray frequencies, on the other hand, are several orders of magnitude more energetic and are capable of interacting with electrons closest to the nuclei of atoms. Absorption peaks corresponding to electronic transitions of these innermost electrons are thus observed in the X-ray region.

19D-2 MOLECULAR ABSORPTION

Figure 19-7a is a partial energy diagram that depicts some of the processes that occur when a polyatomic species absorbs infrared, visible, and ultraviolet radiation. As in the previous energy-level diagram, two excited electronic states, E_1 and E_2, are shown as well as the ground state. In addition, a few of the many vibrational states associated with each electronic state are indicated as lighter horizontal lines. The superimposed rotational states are not shown.

Infrared absorption. Infrared radiation generally is not sufficiently energetic to cause electronic transitions; it can, however, cause transition among vibrational and rotational states associated with *the ground electronic state* of the molecule. Three of these transitions are depicted in the lower left corner of Figure 19-7a. For absorption to occur, the source would have to emit frequencies corresponding exactly to the energies indicated by the lengths of the three arrows; infrared radiation of only those three frequencies would be absorbed.

Absorption of ultraviolet and visible radiation. The center arrows in Figure 19-7a suggest that visible radiation of four wavelengths is absorbed, thereby promoting electrons to the four vibrational levels of electronic level E_1. More energetic ultraviolet photons are required to produce the absorption indicated by the four arrows to the right.

As suggested by Figure 19-7a, molecular absorption consists of bands made up of closely spaced lines. In a solution in which the absorbing species is surrounded by solvent, the band nature becomes blurred because collisions tend to

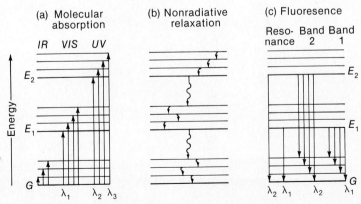

FIGURE 19-7 Energy level diagram for absorption, nonradiative relaxation, and fluorescence.

spread the energies of a given quantum state; a smooth and continuous absorption peak is often the result.

19D-3 TERMS EMPLOYED IN ABSORPTION SPECTROSCOPY

Table 19-1 lists the common terms and symbols found in absorption spectroscopy. In recent years, a considerable effort has been made by the American Society for Testing Materials to develop a standard nomenclature; the terms and symbols listed in the first two columns of Table 19-1 are based on ASTM recommendations. Column 3 contains alternative symbols that will be encountered in the older literature. A standard nomenclature seems most worthwhile in order to avoid ambiguities; the reader is, therefore, urged to learn and use the recommended terms and symbols.

Transmittance. Figure 19-8 depicts a beam of parallel radiation before and after it has passed through a layer of solution with a thickness of b cm and a concentration of c of an absorbing species. As a consequence of interactions between the photons and absorbing particles, the power of the beam is attenuated from P_0 to P. The *transmittance*, T, of the solution is the fraction of incident radiation transmitted by the solution. That is,

$$T = P/P_0 \tag{19-6}$$

Transmittance is often expressed as a percentage.

Table 19-1 Important Terms and Symbols Employed in Absorption Measurement

Term and Symbol*	Definition	Alternative Name and Symbol
Radiant power, P, P_0	Energy of radiation (in ergs) impinging on a 1-cm² area of a detector per second	Radiation intensity, I, I_0
Absorbance, A	$\log \dfrac{P_0}{P}$	Optical density, D; extinction, E
Transmittance, T	$\dfrac{P}{P_0}$	Transmission, T
Path length of radiation,** b	—	l, d
Absorptivity,** a	$\dfrac{A}{bc}$	Extinction coefficient, k
Molar absorptivity,† ϵ	$\dfrac{A}{bc}$	Molar extinction coefficient

* Terminology recommended by the American Chemical Society [*Anal. Chem.*, **24**, 1349 (1952); **48**, 2298 (1976)].
** c may be expressed in g/L or other specified concentration units; b may be expressed in cm or in other units of length.
† c is expressed in units of mol/L; b is expressed in cm.

Absorbance. The absorbance of a solution is defined by the equation

$$A = -\log_{10} T = \log \frac{P_0}{P} \qquad (19\text{-}7)$$

Note that, in contrast to transmittance, the absorbance of a solution increases as the attenuation of the beam becomes greater.

19D-4 RELATIONSHIP BETWEEN ABSORPTION AND CONCENTRATION

The functional relationship between the quantity measured in an absorption analysis (A) and the quantity sought (the concentration, c) is known as *Beer's law* which can be written as

$$A = abc \qquad (19\text{-}8)$$

Here, a is a proportionality constant called the *absorptivity*, and b is the path length of the radiation through the absorbing medium. Absorbance is a unitless quantity; the absorptivity will thus require units that likewise render the right side of the equation dimensionless.

When the concentration is expressed in moles per liter and b is in centimeters, the proportionality constant is called the *molar absorptivity* and is given the special symbol ϵ. Thus,

$$A = \epsilon bc \qquad (19\text{-}9)$$

where ϵ has the units of L cm^{-1} mol^{-1} [liter/(cm mol)].

Beer's law can be rationalized as follows.[5] Consider the block of absorbing matter (solid, liquid, or gas) shown in Figure 19-9. A beam of parallel monochromatic radiation with power P_0 strikes the block perpendicular to a surface; after passing through a length b of the material, which contains n absorbing particles (atoms, ions, or molecules), its power is decreased to P as a result of absorption. Consider now a cross section of the block having an area S and an infinitesimal thickness dx. Within this section, there are dn absorbing particles; associated with each particle, we can imagine a surface at which photon capture will occur. That is, if a photon reaches one of these areas by chance, absorption will follow immediately. The total projected area of these capture surfaces within the section is designated as dS; the ratio of the capture area to the total area, then, is dS/S. On a statistical average, this ratio represents the probability for the capture of photons within the section.

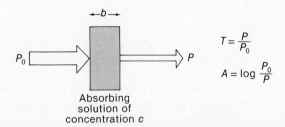

FIGURE 19-8 Attenuation of a beam of radiation by an absorbing solution.

Absorbing solution of concentration c

$$T = \frac{P}{P_0}$$

$$A = \log \frac{P_0}{P}$$

[5] The discussion that follows is based on a paper by F. C. Strong, *Anal. Chem.*, **24**, 338 (1952). For a rigorous derivation of the law, see: D. J. Swinehart, *J. Chem. Educ.*, **39**, 333 (1972).

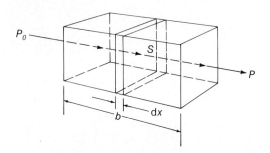

P_0

S

P

b dx

FIGURE 19-9   Attenuation of radiation with initial power P_0 by a solution containing c moles per liter of absorbing solute and a path length of b cm. $P < P_0$.

The power of the beam entering the section, P_x, is proportional to the number of photons per square centimeter per second, and dP_x represents the quantity removed per second within the section; the fraction absorbed is then $-dP_x/P_x$, and this ratio also equals the average probability for capture. The term is given a minus sign to indicate that P undergoes a decrease. Thus,

$$-\frac{dP_x}{P_x} = \frac{dS}{S} \tag{19-10}$$

Recall, now, that dS is the sum of the capture areas for particles within the section; it must therefore be proportional to the number of particles, or

$$dS = adn \tag{19-11}$$

where dn is the number of particles and a is a proportionality constant, which can be called the *capture cross section*. Combining Equations 19-10 and 19-11 and summing over the interval between zero and n, we obtain

$$-\int_{P_0}^{P} \frac{dP_x}{P_x} = \int_{0}^{n} \frac{adn}{S}$$

which, upon integration, gives

$$-\ln \frac{P}{P_0} = \frac{an}{S}$$

Upon converting to base 10 logarithms and inverting the fraction to change the sign, we obtain

$$\log \frac{P_0}{P} = \frac{an}{2.303S} \tag{19-12}$$

where n is the total number of particles within the block shown in Figure 19-9. The cross-sectional area S can be expressed in terms of the volume of the block V and its length b. Thus,

$$S = \frac{V}{b} \text{ cm}^2$$

Substitution of this quantity into Equation 19-12 yields

$$\log \frac{P_0}{P} = \frac{anb}{2.303V} \tag{19-13}$$

Note that n/V has the units of concentration (that is, number of particles per cubic centimeter); we can readily convert n/V to moles per liter. Thus,

$$c = \frac{n \text{ particles}}{6.02 \times 10^{23} \text{ particles/mol}} \times \frac{1000 \text{ cm}^3/\text{L}}{V \text{ cm}^3} = \frac{1000n}{6.02 \times 10^{23} \, V} \text{ mol/L}$$

Combining this relationship with Equation 19-13 yields

$$\log \frac{P_0}{P} = \frac{6.02 \times 10^{23} \, abc}{2.303 \times 1000}$$

Finally, the constants in this equation can be collected into a single term ϵ to give

$$\log \frac{P_0}{P} = \epsilon bc = A$$

which is a statement of Beer's law.

19D-5 EXPERIMENTAL MEASUREMENT OF T AND A

The relationships given by Equation 19-8 or 19-9 are not directly applicable to chemical analysis. Neither P nor P_0 as defined can be conveniently measured in the laboratory because the sample to be studied must be held in some sort of container. Interaction between the radiation and the walls is inevitable, leading to a loss by reflection at each interface; moreover, a significant absorption may occur within the walls themselves. Finally, the beam may suffer a diminution in power during its passage through the solution as a result of scattering by large molecules or inhomogeneities. Reflection losses can be appreciable; for example, about 4% of a beam is reflected upon vertical passage of visible radiation across an air-to-glass or glass-to-air interface.

In order to compensate for these effects, the power of the beam transmitted through a cell containing an absorbing solution is generally compared with that which passes through an identical cell containing the solvent. An experimental absorbance that closely approximates the true absorbance of the solution can then be obtained; that is,

$$A = \log \frac{P_{\text{solvent}}}{P_{\text{solution}}} = \log \frac{P_0}{P} \tag{19-14}$$

The terms P_0 and P, when used henceforth, refer to the power of a beam of radiation after it has passed through a cell containing the solvent and the solution of the analyte, respectively.

19D-6 APPLICATION OF BEER'S LAW TO MIXTURES

Beer's law also applies to a solution containing more than one kind of absorbing substance. Provided there is no interaction among the various species, the total absorbance for a multicomponent system is given by

$$A_{\text{total}} = A_1 + A_2 + \cdots + A_n = \epsilon_1 bc_1 + \epsilon_2 bc_2 + \cdots + \epsilon_n bc_n \tag{19-15}$$

where the subscripts refer to absorbing components $1, 2, \cdots, n$.

19D-7 LIMITATIONS TO THE APPLICABILITY OF BEER'S LAW

The linear relationship between absorbance and path length at a fixed concentration of an absorbing substance is a generalization for which no exceptions are known. On the other hand, deviations from the direct proportionality between the measured absorbance and concentration when b is constant are frequently encountered. Some of these deviations are fundamental and represent real limitations to the law. Others occur as a consequence of the manner in which the absorbance measurements are made or as a result of chemical changes associated with concentration changes; the latter two are sometimes known, respectively, as *instrumental deviations* and *chemical deviations*.

Real limitations to Beer's law. Beer's law is successful in describing the absorption behavior of dilute solutions only; in this sense, it is a limiting law. At high concentrations (usually $>0.01\ M$), the average distance between the species responsible for absorption is diminished to the point where each affects the charge distribution of its neighbors. This interaction, in turn, can alter their ability to absorb a given wavelength of radiation. Because the extent of interaction depends upon concentration, the occurrence of this phenomenon causes deviations from the linear relationship between absorbance and concentration. A similar effect is sometimes encountered in solutions containing low absorber concentrations and high concentrations of other species, particularly electrolytes. The close proximity of ions to the absorber alters the molar absorptivity of the latter by electrostatic interactions; the effect is lessened by dilution.

While the effect of molecular interactions is ordinarily not significant at concentrations below $0.01\ M$, some exceptions are encountered among certain large organic ions or molecules. For example, the molar absorptivity for the cation of methylene blue at 436 nm is reported to increase by 88% as the dye concentration is increased from 10^{-5} to $10^{-2}\ M$; even below $10^{-6}\ M$, strict adherence to Beer's law is not observed.

Deviations from Beer's law also arise because ϵ is dependent upon the refractive index of the solution.[6] Thus, if concentration changes cause significant alterations in the refractive index, n, of a solution, departures from Beer's law are observed. In general, this effect is small and is rarely significant at concentrations less than $0.01\ M$.

Chemical deviations. Apparent deviations from Beer's law are frequently encountered as a consequence of association, dissociation, or reaction of the absorbing species with the solvent. These deviations are, in fact, more apparent than real because they result from shifts in chemical equilibria and not from changes in the molar absorptivities of the species present. As shown by the following example, apparent departures from Beer's law are readily predicted from the equilibrium constants for the reactions and the molar absorptivities of the solutes.

> **EXAMPLE 19-1.** A series of solutions containing various concentrations of the acidic indicator HIn ($K_a = 1.42 \times 10^{-5}$) were prepared in $0.1\ M$ HCl and $0.1\ M$ NaOH. In both media, a perfectly linear relationship between absorbance and concentration was observed at 430 and 570 nm. From the magnitude of the acid dissociation constant, it was apparent that, for all practical purposes, the indicator was entirely in the undissociated form (HIn) in the $0.1\ M$ HCl solution and completely dissociated as

[6] G. Kortum and M. Seiler, *Angew. Chem.*, **52**, 687 (1939).

In$^-$ in the 0.1 M NaOH. The molar absorptivities at the two wavelengths were found to be

	ϵ_{430}	ϵ_{570}
HIn (HCl solution)	6.30×10^2	7.12×10^3
In$^-$ (NaOH solution)	2.06×10^4	9.60×10^2

Derive absorbance data (1.00-cm cell) at the two wavelengths for unbuffered solutions with indicator concentrations ranging from 2×10^{-5} to 16×10^{-5} M. Plot the data.

Let us calculate the concentration of HIn and In$^-$ in an unbuffered 2.00×10^{-5} M solution of HIn. From the equation for the dissociation reaction, it is apparent that

$$[H^+] = [In^-]$$

Furthermore,

$$[In^-] + [HIn] = 2.00 \times 10^{-5}$$

Substitution of these relationships into the expression for K_a gives

$$\frac{[In^-]^2}{2.00 \times 10^{-5} - [In^-]} = 1.42 \times 10^{-5}$$

Rearrangement yields the quadratic expression

$$[In^-]^2 + 1.42 \times 10^{-5} [In^-] - 2.84 \times 10^{-10} = 0$$

which gives

$$[In^-] = 1.12 \times 10^{-5}$$

$$[HIn] = 2.00 \times 10^{-5} - 1.12 \times 10^{-5} = 0.88 \times 10^{-5}$$

The absorbances at the two wavelengths are given by

$$A_{430} = 6.30 \times 10^2 \times 1.00 \times 0.88 \times 10^{-5} + 2.06 \times 10^4 \times 1.00 \times 1.12 \times 10^{-5}$$

$$= 0.236$$

$$A_{570} = 7.12 \times 10^3 \times 1.00 \times 0.88 \times 10^{-5} + 9.60 \times 10^2 \times 1.00 \times 1.12 \times 10^{-5}$$

$$= 0.073$$

The accompanying data were derived in a similar way and are plotted in Figure 19-10.

M_{HIn}	[HIn]	[In$^-$]	A_{430}	A_{570}
2.00×10^{-5}	0.88×10^{-5}	1.12×10^{-5}	0.236	0.073
4.00×10^{-5}	2.22×10^{-5}	1.78×10^{-5}	0.381	0.175
8.00×10^{-5}	5.27×10^{-5}	2.73×10^{-5}	0.596	0.401
12.00×10^{-5}	8.52×10^{-5}	3.48×10^{-5}	0.771	0.640
16.00×10^{-5}	11.9×10^{-5}	4.11×10^{-5}	0.922	0.887

Figure 19-10 is a plot of the data derived in the foregoing example and illustrates the departures from Beer's law that occur when the absorber is a participant

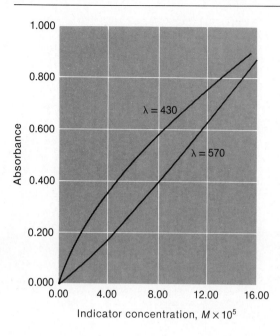

FIGURE 19-10 Chemical deviations from Beer's law for unbuffered solutions of the indicator HIn. For data, see example on page 492.

in an association or dissociation equilibrium. Note that the direction of curvature is opposite for the two wavelengths.

Instrumental deviations with polychromatic radiation. Strict adherence to Beer's law is observed only when the radiation employed is truly monochromatic; this observation is yet another manifestation of the limiting character of the law. Unfortunately, the use of radiation that is restricted to a single wavelength is seldom practical; devices that isolate portions of the output from a continuous source produce a more or less symmetric band of wavelengths around the desired one (p. 505).

The following derivation illustrates the effect of polychromatic radiation on Beer's law.

Consider a beam comprised of just two wavelengths, λ' and λ''. Assuming that Beer's law applies strictly for each of these individually, we may write for radiation λ'

$$A' = \log \frac{P_0'}{P'} = \epsilon' bc$$

or

$$\frac{P_0'}{P'} = 10^{\epsilon' bc}$$

Similarly, for λ''

$$\frac{P_0''}{P''} = 10^{\epsilon'' bc}$$

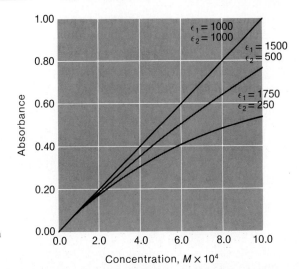

FIGURE 19-11 Derivations from Beer's law with polychromatic light. Here, two wavelengths of radiation λ_1 and λ_2 have been assumed for which the absorber has the indicated molar absorptivities.

When an absorbance measurement is made with radiation composed of both wavelengths, the power of the beam emerging from the solution is given by $(P' + P'')$ and that of the beam from the solvent by $(P_0' + P_0'')$. Therefore, the measured absorbance is

$$A_M = \log \frac{(P_0' + P_0'')}{(P' + P'')}$$

which can be rewritten as

$$A_M = \log \frac{(P_0' + P_0'')}{(P_0' \, 10^{-\epsilon' bc} + P_0'' \, 10^{-\epsilon'' bc})}$$

or

$$A_M = \log (P_0' + P_0'') - \log (P_0' \, 10^{-\epsilon' bc} + P_0'' \, 10^{-\epsilon'' bc})$$

Now, when $\epsilon' = \epsilon''$, this equation simplifies to

$$A_M = \epsilon' bc$$

and Beer's law is followed. As shown in Figure 19-11, however, the relationship between A_M and concentration is no longer linear when the molar absorptivities differ; moreover, greater departures from linearity can be expected with increasing differences between ϵ' and ϵ''. This derivation can be expanded to include additional wavelengths; the effect remains the same.

It is an experimental fact that deviations from Beer's law resulting from the use of a polychromatic beam are not appreciable, provided the radiation used does not encompass a spectral region in which the absorber exhibits large changes in absorbance as a function of wavelength. This observation is illustrated in Figure 19-12.

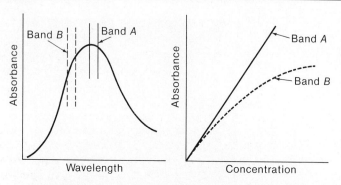

FIGURE 19-12 The effect of polychromatic radiation upon the Beer's law relationship. Band A shows little deviation since ϵ does not change greatly throughout the band. Band B shows marked deviations since ϵ undergoes significant changes in this region.

Instrumental deviations in the presence of stray radiation. Generally, the devices used for providing a limited band of wavelengths for absorption measurements are seldom perfect. As a consequence, the radiation produced is likely to be contaminated with small amounts of scattered or stray radiation owing to reflections from various internal surfaces (see p. 570). Stray radiation often differs greatly in wavelength from that of the principal radiation and, in addition, may not have passed through the sample or solvent.

When measurements are made in the presence of stray radiation, the observed absorbance is given by

$$A' = \log \frac{P_0 + P_s}{P + P_s}$$

where P_s is the power of the stray radiation. Figure 19-13 shows a plot of A' versus concentration for various levels of P_s as compared to P_0.

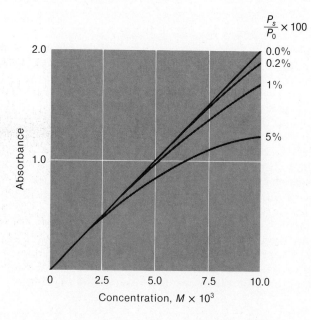

$$\frac{P_s}{P_0} \times 100$$

0.0%
0.2%

1%

5%

FIGURE 19-13 Apparent deviation from Beer's law caused by various amounts of stray radiation.

Note that the instrumental deviations illustrated in Figures 19-12 and 19-13 result in absorbances that are smaller than theoretical. It can be shown that instrumental deviations always lead to negative absorbance errors.[7]

19E FLUORESCENCE AND PHOSPHORESCENCE

Fluorescence and phosphorescence are analytically important emission processes in which atoms or molecules are excited by absorption of a beam of electromagnetic radiation; radiant emission then occurs as the excited species return to the ground state. The resulting emission spectrum serves as the basis for analysis.

Fluorescence differs from phosphorescence in the respect that it occurs much more rapidly and is generally complete after about 10^{-5} s (or less) from the time of excitation. Phosphorescence emission takes place over periods longer than 10^{-5} s and may indeed continue for minutes or even hours after irradiation has ceased. Generally, fluorescence and phosphorescence are observed at a 90-deg angle to the excitation beam. Of the two, fluorescence has found much more widespread application.

19E-1 SOURCES OF FLUORESCENCE AND PHOSPHORESCENCE

Resonance fluorescence describes a process in which the emitted radiation is identical in frequency to the radiation employed for excitation. The lines labeled λ_1 and λ_2 in Figures 19-7a and 19-7c (p. 487) illustrate this type of fluorescence. Here, the species is excited to the energy states E_1 or E_2 by radiation with an energy of $(E_1 - G)$ or $(E_2 - G)$. After a brief period, emission of radiation of identical energy occurs, as depicted in Figure 19-7c. Resonance fluorescence is most commonly produced by atoms in the gaseous state which do not have vibrational energy states superimposed on electronic energy levels.

Nonresonance fluorescence is brought about by irradiation of *molecules* in solution or in the gaseous state. As shown in Figure 19-7a, absorption of radiation promotes the molecules into any of the several vibrational levels associated with the two excited electronic levels. The lifetimes of the excited *vibrational* states are momentary, however ($\sim 10^{-15}$ s), and, as shown in Figure 19-7b, relaxation to the lowest vibrational level of a given electronic state takes place as a result of collision with other molecules. As a consequence, fluorescence radiation is often of lower energy than the excitation energy. For example, suppose that excitation were carried out with radiation of λ_3 which promotes the species to an upper vibrational level of E_2. Radiationless relaxation to the lowest vibrational level of this state would occur essentially instantaneously and well before fluorescence could take place. The fluorescent radiant energy would then correspond to λ_2 or one of the less energetic wavelengths of band 2 (Figure 19-7c). Furthermore, as shown in Figure 19-7b, relaxation from state E_2 to E_1 can also take place without emission. Here, fluorescence excited by λ_3 would consist of band 1, which is much lower in energy or longer in wavelength. This shift in wavelength to lower frequencies is sometimes called the *Stokes shift*. Clearly, both resonance and nonresonance radiation

---

[7] I. M. Kolthoff and P. J. Elving, Eds., *Treatise on Analytical Chemistry*, Part I, vol. 5, pp. 2767–2773. New York: Interscience, 1964.

can accompany fluorescence of molecules; the latter tends to predominate, however, because of the much larger number of vibrational excited states.

Phosphorescence occurs when an excited molecule relaxes to a metastable
excited electronic state which has an average lifetime of greater than about 10^{-5} s.

Relationship between fluorescence and concentration.   If the absorption by the
sample is small, it can be shown that the power of fluorescent radiation F is given
by

$$F = k'abcP_0$$

where k' is a constant and the other terms are from Equations 19-8 and 19-9. When
the power of the incident beam is held constant, this equation reduces to

$$F = kc \tag{19-16}$$

Equation 19-16 is the basis for quantitative methods based upon fluorescence.

PROBLEMS*

19-1. Complete the accompanying tabulation:

	λ, nm	λ, Å	λ, cm	λ, μm	ν, Hz	σ, cm^{-1}
* (a)	280					
(b)		9130				
* (c)			3×10^{-9}			
(d)				2.75		
* (e)					5×10^9	
(f)						6.4×10^7

19-2. Calculate the frequency in hertz of
   * (a) the calcium emission line at 422.7 nm.
   (b) an infrared absorption peak at 3.00 μm.
   * (c) the line in the X-ray emission spectrum of potassium at 3.742 Å.
   (d) a microwave beam with a wavelength of 250 cm.
19-3. Calculate the wavenumber, σ, for the radiative processes described in Problem
   19-2.
19-4. Calculate
   (a) the wavenumber of the sodium emission line at 330.2 nm.
   (b) the frequency (Hz) of a carbonyl absorption maximum at 280 nm.
   (c) the wavelength (Å) of the X-ray emission line for sodium that has a frequency of
      1.57×10^{17} Hz.
   (d) the wavelength (cm) of a microwave emission with a wavenumber of 1.5 cm^{-1}.
19-5. Calculate the energy (in erg/photon) of the radiation in Problem 19-1.
19-6. Convert the absorbance data for the accompanying solutions into percent transmittance:
   * (a) 0.168   (d) 0.413
   * (b) 0.936   (e) 0.798
   * (c) 0.024   (f) 0.561

* Answers to problems or parts of problems marked with an asterisk are to be found at the end of the
book.

19-7. Convert the percent transmittance data for the accompanying solutions to absorbance:

* (a) 36.7   (d) 4.39
* (b) 29.4   (e) 64.3
* (c) 81.0   (f) 50.0

* 19-8. Calculate the percent transmittance of solutions with half the absorbance of those in Problem 19-6.

* 19-9. Calculate the absorbance of solutions with half the percent transmittance of those in Problem 19-7.

* 19-10. Calculate the transmittance after the solutions in Problem 19-7 have been diluted to twice their original volume.

19-11. Calculate the missing quantity:

	Absorbance, A	Molar Absorptivity, ϵ	Path Length, cm	Concentration (As Indicated)
* (a)	0.244	3.39×10^4	2.00	M
(b)		6.81×10^3	1.00	$7.49 \times 10^{-5}\ M$
* (c)	0.865	3.24×10^3		$1.78 \times 10^{-4}\ M$
(d)	0.473		0.10	$5.61 \times 10^{-3}\ M$
* (e)		5.09×10^3	1.00	17.65 ppm (fw = 360)
(f)	0.602	1.14×10^4	1.00	ppm (fw = 200)
* (g)	0.539	2.18×10^4	1.00	g/L (fw = 250)
(h)	0.749	1.35×10^4	1.00	μg/L (fw = 320)

19-12. Calculate the missing quantity:

	Absorbance, A	Molar Absorptivity, ϵ	Path Length, cm	Concentration (As Indicated)
(a)	0.827	1.98×10^4	1.00	M
(b)		7.70×10^3	1.50	$6.43 \times 10^{-5}\ M$
(c)	0.199	2.37×10^4		$3.36 \times 10^{-6}\ M$
(d)	0.904		0.50	$8.89 \times 10^{-5}\ M$
(e)		6.58×10^3	1.50	22.61 ppm (fw = 420)
(f)	0.765	2.36×10^4	1.00	ppm (fw = 295)
(g)	0.470	1.08×10^4	1.50	g/L (fw = 360)
(h)	0.173	6.92×10^3	1.00	μg/L (fw = 275)

19-13. What are the units for absorptivity when the concentration is expressed as

* (a) parts per million?   * (c) grams/liter?
    (b) weight/volume percent?   (d) milligrams/milliliter?

* 19-14. An 8.64 ppm solution of $FeSCN^{2+}$ has a transmittance of 0.295 when measured in a 1.00-cm cell at 580 nm. Calculate the molar absorptivity for the complex at this wavelength.

* 19-15. The complex that is formed between gallium(III) and 8-hydroxyquinoline has an absorption maximum at 393 nm. A $1.29 \times 10^{-4}\ M$ solution of the complex has a transmittance of 14.6% when measured in a 1.00-cm cell at this wavelength. Calculate the molar absorptivity of the complex.

19-16. The molar absorptivity of the Bi(III)/thiourea complex is 9.3×10^3 L cm^{-1}mol^{-1} at 470 nm. What concentration of this complex would be needed to produce a solution with a transmittance of 6.85% when measured in a 1.00-cm cell at this wavelength?

19-17. A 25.0-mL aliquot of a solution that contains 5.64 ppm of Bi(III) is treated with an appropriate excess of thiourea and diluted to 50.0 mL. Calculate the absorbance of the resulting solution at 470 nm (1.50-cm cell); see Problem 19-16 for absorptivity data.

* 19-18. A 50.0-mL aliquot of well water is treated with an excess of KSCN and diluted to 100.0 mL. Calculate the ppm of iron(III) in the sample if the diluted solution has an absorbance of 0.394 at 580 nm when measured in a 2.50-cm cell; see Problem 19-14.

* 19-19. A 2.83×10^{-4} M solution of potassium permanganate has a molar absorbance of 0.510 when measured in a 1.00-cm cell at 520 nm. Calculate
  (a) the molar absorptivity for $KMnO_4$ at this wavelength.
  (b) the absorptivity when the concentration is expressed in ppm.
  (c) the concentration of permanganate in a solution that has an absorbance of 0.697 when measured in a 1.50-cm cell at 520 nm.
  (d) the transmittance of the solution in (c).
  (e) the absorbance of a solution that has twice the transmittance of the solution in (c).

19-20. Chromium(III) forms a complex with diphenylcarbazide whose molar absorptivity is 4.17×10^4 at 540 mm. Calculate
  (a) the absorbance of a 7.68×10^{-6} M solution of the complex at 540 nm when measured in a 1.00-cm cell.
  (b) the transmittance of the solution described in (a).
  (c) the path length needed to cause a 2.56×10^{-6} M solution of the complex to have the same absorbance as the solution described in (a).
  (d) the concentration of the complex in a solution that has an absorbance of 0.649 at 540 nm when measured in a 1.00-cm cell.
  (e) the concentration of the complex in a solution that has a transmittance of 0.649 at 540 nm when measured in a 1.00-cm cell.

optical spectroscopic instruments

20

The first spectroscopic instruments were developed for use in the visible region and were thus called *optical instruments*. This term has by now been extended to include instruments designed for use in the ultraviolet and infrared regions as well; while not strictly correct, the terminology is nevertheless useful in that it emphasizes the many features that are common to the instruments used for studies in these three important spectral regions.

The early sections of this chapter describe the nature and properties of components of instruments employed for optical spectroscopy. The sections that follow describe how these several components are assembled to produce optical instruments for various purposes. The instruments and techniques for spectroscopic studies in regions more energetic than the ultraviolet and less energetic than the infrared have characteristics that differ substantially from optical instruments and will not be considered in this text.

Optical spectroscopic methods are based upon the phenomena of *emission, absorption, fluorescence, phosphorescence,* and *scattering.* Only methods based upon the first three phenomena are treated here. While the instruments for each of these methods differ somewhat in configuration, their basic components are remarkably similar. Furthermore, the required properties of these components are the same regardless of the spectral regions for which they are designed.

Spectroscopic instruments contain five components, including: (1) a stable source of radiant energy; (2) a wavelength selector that permits isolation of a restricted wavelength region; (3) a transparent container for holding the sample; (4) a radiation detector or transducer that converts radiant energy to a usable signal (usually electrical); and (5) a signal processor and *readout.* Figure 20-1 shows the arrangement of these components for the three common types of spectroscopic measurements mentioned earlier. As can be seen in the figure, the configuration of components (4) and (5) is the same for each type of instrument.

Emission spectroscopy differs from the other two types in that no external radiation source is required; the sample itself is the emitter. Here, the sample con-

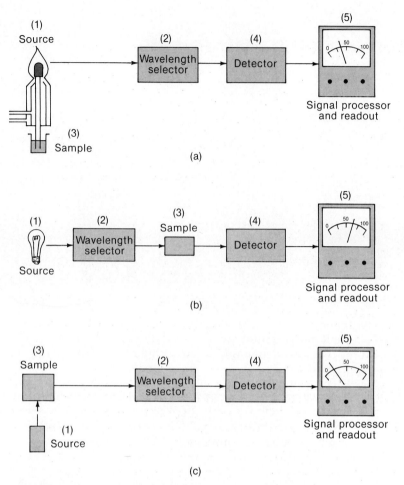

FIGURE 20-1   Components for various types of instruments for optical spectroscopy: (a) emission spectroscopy, (b) absorption spectroscopy, and (c) fluorescence and scattering spectroscopy.

tainer is an arc, a spark, a heated surface, or a flame, which both holds the sample and causes it to emit characteristic radiation.

Absorption as well as fluorescence spectroscopy require an external source of radiant energy. In the former, the beam from the source passes through the sample after leaving the wavelength selector (in some instruments, the position of the sample and selector is reversed). For fluorescence, the source induces the sample, held in a container, to emit characteristic radiation, which is measured at an angle (usually 90 deg) with respect to the source.

Figure 20-2 summarizes the characteristics of the first four components shown in Figure 20-1. It is clear that instrument components differ in detail, depending upon the wavelength region within which they are to be used. Their de-

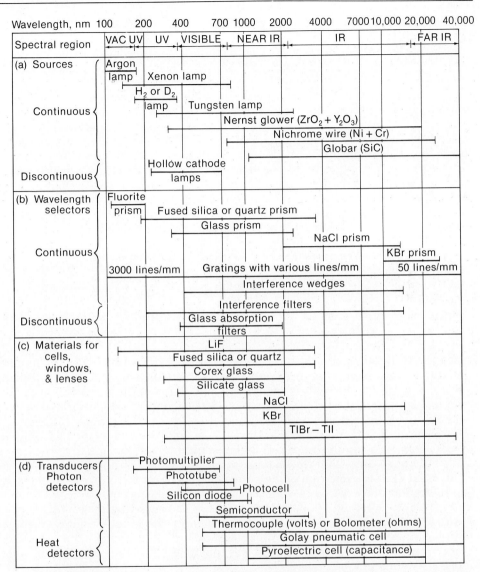

FIGURE 20-2 Components and materials for spectroscopic instruments. [Adapted from a figure by Professor A. R. Armstrong, College of William and Mary. With permission.]

sign also depends on the primary use of the instrument; that is, whether it is to be employed for qualitative or quantitative analysis and whether it is to be applied to atomic or molecular spectroscopy. Nevertheless, the general function and performance requirements of each type of component are similar, regardless of wavelength region and application.

20A RADIATION SOURCES FOR ABSORPTION AND FLUORESCENCE SPECTROSCOPY

In order to be suitable for spectroscopic studies, a source must generate a stable beam of radiation with sufficient power for ready detection and measurement. In addition, its output should be stable. Typically, the radiant power of a source varies exponentially with the electrical power supplied to it. Thus, a regulated power supply is ordinarily needed to provide the required stability. The problem of source stability is circumvented in instruments in which the output from the source is split into a reference beam and a sample beam. The first passes directly to a transducer; the second interacts with the sample before being focused on a matched transducer (or perhaps a single one that alternately receives the signal from both beams). The ratio of the two responses serves as the analytical parameter. All but the most transitory fluctuations of the source are canceled by this arrangement.

Both continuous and line sources are used in optical spectroscopy. The former find wide application in molecular absorption methods; the latter are employed in fluorescence and atomic absorption spectroscopy.

Figure 20-2a lists the most widely used spectroscopic sources.

20A-1 CONTINUOUS SOURCES OF ULTRAVIOLET, VISIBLE, AND NEAR-INFRARED RADIATION

Continuous sources provide radiation whose power does not change sharply among adjacent wavelengths.

Hydrogen or deuterium lamps. Two types of hydrogen lamps are encountered. The high-voltage variety requires potentials of 2000 to 6000 V to cause a discharge between aluminum electrodes; water cooling of the lamp is required if high radiation intensities are to be produced. In low-voltage lamps, an arc is formed between a heated, oxide-coated filament and a metal electrode. The heated filament provides electrons to maintain a dc current when a voltage of about 40 V is applied; a regulated power supply is required for constant intensities.

Hydrogen and deuterium lamps produce a truly continuous spectrum as a consequence of the excitation of the gaseous molecules to quantized excited states. The relaxation process involves dissociation of the excited molecule to produce a photon of ultraviolet radiation and two atoms of hydrogen in the ground state. The absorbed energy is released in two forms, namely the kinetic energy of the two hydrogen atoms and the energy of the ultraviolet photon. Inasmuch as the kinetic energy imparted to the two atoms is not quantized, a broad spectrum of photon energies is obtained.

Both high- and low-voltage lamps produce continuous spectrum in the region of 160 to 375 nm. Quartz windows must be employed in the tubes since glass absorbs strongly in this wavelength region. Deuterium lamps produce higher intensities than hydrogen lamps under the same operating conditions.

Tungsten filament lamps. The most common source of visible and near-infrared radiation is the tungsten filament lamp. The energy distribution of this source approximates that of a black body and is thus temperature dependent. Figure 19-6 (p. 485) illustrates the behavior of the tungsten filament lamp at 3000°K. In most absorption instruments, the operating filament temperature is about 2900°K; the bulk of the energy is thus emitted in the infrared region. A tungsten filament lamp is useful for the wavelength region between 320 and 2500 nm.

The energy output of a tungsten lamp in the visible region varies approximately as the fourth power of the operating voltage. Close voltage control is thus required for a stable radiation course. Constant voltage transformers or electronic voltage regulators are often employed for this purpose. As an alternative, the lamp can be operated from a 6-V storage battery, which provides a remarkably stable voltage source if it is maintained in good condition.

20A-2 INFRARED SOURCES

Common infrared sources include the Nernst glower, which consists of a cylinder of rare earth oxides having dimensions of 2 by 20 mm. Electricity is forced through the cylinder to provide the radiant energy. The Globar source is a silicon carbide rod with dimensions of 5 by 50 mm. It also is heated by an electrical current. Tightly wound spirals of nichrome wire also serve as infrared sources.

20B WAVELENGTH SELECTORS

For most spectroscopic analyses, radiation consisting of a limited, narrow, continuous group of wavelengths called a band is required.[1] A narrow bandwidth tends to enhance the sensitivity of the absorbance measurement, may provide selectivity to a method, and is frequently a requirement from the standpoint of obtaining a linear relationship between the optical signal and concentration (p. 496). Two types of wavelength selectors are encountered, *filters* and *monochromators*. The latter

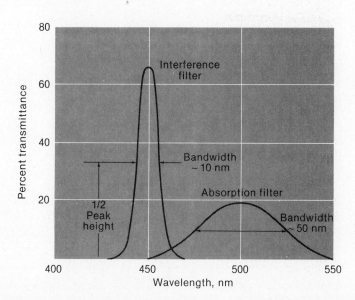

FIGURE 20-3 Bandwidths for two types of filters.

---

[1] Note that the term *band* in this context has a somewhat different meaning from that used in describing types of spectra.

permit continuous variation in the wavelength whereas the former provide but a single wavelength region.

Ideally, the output from a wavelength selector would be radiation of a single wavelength or frequency. No existing wavelength selector even approaches this ideal; instead, a Gaussian-shaped distribution of wavelengths such as those shown in Figure 20-3 is obtained. Here, the percent of incident radiation of a given wavelength that is transmitted by the selector is plotted as a function of wavelength. The *bandwidth* or *effective bandwidth*, which is defined in Figure 20-3, is an inverse measure of the quality of the device, a narrower bandwidth representing better performance. Figure 20-3 shows the characteristics of typical filters; the output from a monochromator has the same shape but ordinarily has a considerably narrower bandwidth.

20B-1 FILTERS

Absorption and interference filters are employed for wavelength selection. The former are restricted to the visible region of the spectrum. Interference filters, on the other hand, are available for ultraviolet, visible, and up to about 14 μm in the infrared region.

Interference filters. As the name implies, an interference filter relies on optical interference to provide a relatively narrow band of radiation. An interference filter consists of a transparent material (frequently calcium fluoride or magnesium fluoride) that occupies the space between two semitransparent metallic films coated on the inside surfaces of two glass plates. The thickness of the center layer is carefully controlled and determines the wavelength of the transmitted radiation. When a perpendicular beam of collimated radiation strikes this array, a fraction passes through the first metallic layer while the remainder is reflected. The portion that is passed undergoes a similar partition upon striking the second metallic film. If the reflected portion from this second interaction is of the proper wavelength, it is partially reflected from the inner side of the first layer in phase with incoming light of the same wavelength. The result is that this particular wavelength is reinforced, while most other wavelengths, being out of phase, suffer destructive interference.

Figure 20-3 illustrates the performance characteristics of a typical interference filter. Typically, bandwidths are about 1.5% of the wavelength at peak transmittance, although this figure is reduced to 0.15% in some narrow-band filters; these have maximum transmittances of about 10%.

Absorption filters. Absorption filters, which are generally less expensive than interference filters, have been widely used for band selection in the visible region. Absorption filters limit radiation by absorbing certain portions of the spectrum. The most common type consists of colored glass or of a dye suspended in gelatin and sandwiched between glass plates. The former have the advantage of greater thermal stability.

Absorption filters have effective bandwidths that range from perhaps 30 to 250 nm (see Figure 20-3). Filter that provide the narrowest bandwidths also absorb a significant fraction of the desired radiation and may have a transmittance of 1% or less at their band peaks. Glass filters with transmittance maxima throughout the entire visible region are available commercially.

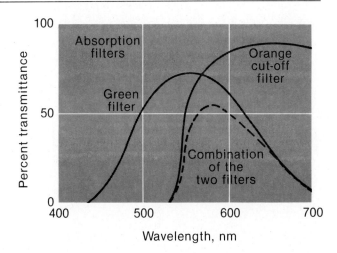

FIGURE 20-4 Comparison of various types of filters for visible radiation.

Cut-off filters have transmittances of nearly 100% over a portion of the visible spectrum but then rapidly decrease to zero transmittance over the remainder. A narrow spectral band can be isolated by coupling a cut-off filter with a second filter (see Figure 20-4).

The performance characteristics of absorption filters are significantly inferior to those of interference filters (see Figure 20-3). Not only are their bandwidths greater, but the fraction of light transmitted is often less. Nevertheless, absorption filters are totally adequate for many applications.

20B-2 MONOCHROMATORS

Monochromators for ultraviolet, visible, and infrared radiation are all similar in mechanical construction in the sense that they employ slits, lenses, mirrors, windows, and dispersing devices. To be sure, the materials from which these components are fabricated will depend upon the wavelength region of intended use (see Figure 20-2b and c).

Components of a monochromator. All monochromators contain an entrance slit, a collimating lens or mirror to produce a parallel beam of radiation, a prism or grating to disperse the radiation into its component wavelengths, and a focusing element which projects a series of rectangular images of the entrance slit upon a plane surface (the *focal plane*). In addition, most monochromators have entrance and exit *windows*, which are designed to protect the components from dust and corrosive laboratory fumes.

Figure 20-5 shows the optical designs of two typical monochromators, one employing a prism and the other a grating to disperse radiation. A source of radiation containing but two wavelengths, λ_1 and λ_2, is shown for purposes of illustration. This radiation enters the monochromators via a narrow rectangular opening or slit, is collimated, and then strikes the surface of the dispersing element at an angle. In the prism monochromator, refraction at the two faces results in angular dispersion of the radiation, as shown; for the grating, angular dispersion results

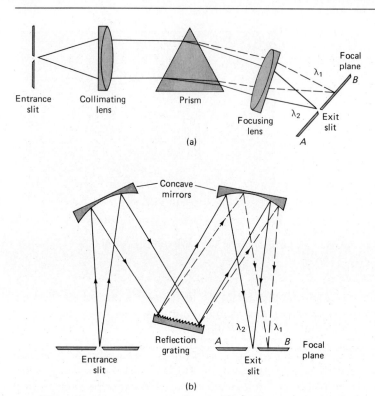

FIGURE 20-5   Two types of monochromators: (a) Bunsen prism monochromator, and (b) Czerney-Turner grating monochromator. (In both instances, $\lambda_1 > \lambda_2$.)

from diffraction, which occurs at the reflective surface. In either case, the dispersed radiation is focused on the focal plane, AB, where it appears as two images of the entrance slit (one for each wavelength).

Prism monochromators.   Prisms can be used to disperse ultraviolet, visible, and infrared radiation. The material used for their construction will differ, depending upon the wavelength region (see Figure 20-2b). Figure 20-5a shows one of the two most common types of prism configurations; here, the three angles of the prism are exactly 60 deg. Figure 20-6 depicts a *Littrow prism*, which is a 30-deg prism with a

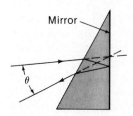

FIGURE 20-6   A Littrow prism.

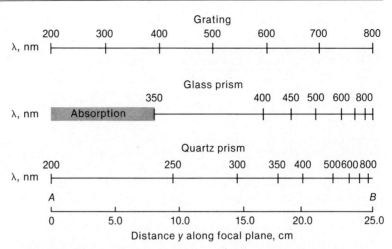

FIGURE 20-7 Dispersion for three types of monochromators. The points A and B on the scale correspond to the points shown in Figure 20-5. The two prisms were Littrow types having heights of 57 mm.

mirrored back. It is seen that refraction occurs twice at the same interface; the performance characteristics of this prism are thus similar to those of the 60-deg prism. The Littrow prism permits somewhat more compact monochromator designs.

The dispersion of a monochromator is often described in terms of the variation of wavelength as a function of the linear distance along the line AB of the focal plane of the instrument (see Figure 20-5a and b). Figure 20-7 shows the dispersion in these terms for two Littrow prism monochromators and one grating instrument. Note that the dispersion of the prism monochromators is highly nonlinear, with the longer wavelengths being bunched at one end of the focal plane. Note also the superior dispersion exhibited by the glass monochromator in the region of 350 to 800 nm.

Grating monochromators. Dispersion of ultraviolet, visible, and infrared radiation can be brought about by passage of a beam through a *transmission grating* or by reflection from a *reflection grating*. A transmission grating consists of a series of parallel and closely spaced grooves ruled on a piece of glass or other transparent material. A grating suitable for use in the ultraviolet and visible region has between 2000 and 6000 lines per millimeter. An infrared grating requires considerably fewer lines; thus, for the far infrared region, gratings with 20 to 30 lines per millimeter may suffice.

Reflection gratings are used more extensively in instrument construction than are transmission gratings. Such gratings are made by ruling a polished metal surface or by evaporating a thin film of aluminum onto a ruled surface. Incident radiation is reflected from one of the faces of each groove; interference among the host of individual reflected beams results in radiation of differing wavelengths being reflected at different angles.

An important advantage of a grating monochromator is the linear dispersion of radiation along the focal plane of the instrument. Figure 20-7 shows the contrast between a grating and a prism monochromator in this regard. The main disadvantages of grating monochromators include a greater amount of stray radiation (see below) and the presence of higher order radiation along the focal plane. These problems have been largely eliminated in modern instruments, and grating monochromators are now more widely used than their prism counterparts.

Monochromator slits. The slits of a monochromator play an important part in determining its quality. Slit jaws are formed by carefully machining two pieces of metal to give sharp edges. Care must be taken to assure that these edges are parallel to one another and in the same plane.

The effective bandwidth of a monochromator depends upon both the dispersion of the prism or grating as well as the width of the entrance and/or exit slit. Most monochromators are equipped with variable slits so that the effective bandwidth can be changed. The use of minimal slit widths is desirable where the resolution of narrow absorption or emission bands is needed. On the other hand, the marked decrease in the available radiant power that accompanies narrowing of slits makes the accurate measurement of this power more difficult. Generally, wider slit widths are used for quantitative analysis where accurate absorbance measurements are required; narrower slits are employed for qualitative work where spectral detail is important.

Stray radiation in monochromators. The exit beam of a monochromator is usually contaminated with small amounts of radiation having wavelengths far different from that of the instrument setting. Sources of this unwanted radiation include reflection from various optical parts and the monochromator housing and scattering by dust particles in the atmosphere or on the surfaces of optical parts. Generally, the effects of spurious radiation are minimized by introducing baffles in appropriate spots in the monochromator and by coating interior surfaces with flat black paint. In addition, the monochromator is sealed with windows over the slits to prevent entrance of dust and fumes. Despite these precautions, however, some spurious radiation is still emitted; we shall find that its presence can have serious effects on spectroscopic measurements under certain conditions (p. 531).

20C SAMPLE CONTAINERS

Sample containers are required for many types of spectroscopic studies. In common with the optical elements of monochromators, the *cells* or *cuvettes* that hold the samples must be made of material that passes radiation in the spectral region of interest. Thus, as shown in Figure 20-2c, quartz or fused silica is required for work in the ultraviolet region (below 350 nm); both of these substances are transparent in the visible region and to about 3000 nm in the infrared as well. Silicate glasses can be employed in the region between 350 and 2000 nm. Plastic containers have also found application in the visible region. Crystalline sodium chloride is the most common substance employed for cell windows in the infrared region; the other infrared-transparent materials shown in Figure 20-2 also find application.

The best cells have windows that are perfectly normal to the direction of the beam in order to minimize reflection losses. The most common cell length for studies in the ultraviolet and visible regions is 1 cm; matched, calibrated cells of this size are available from several commercial sources. Other path lengths, from 0.1 cm (and shorter) to 10 cm, can also be purchased. Transparent spacers for shortening the path length of 1-cm cells to 0.1 cm are also available.

Cells for infrared studies of liquids and solutions generally have path lengths of less than 1 mm; for gases, cells up to several meters in length are used.

For reasons of economy, cylindrical cells are sometimes employed in the ultraviolet and visible regions. Particular care must be taken to duplicate the position of such cells with respect to the beam; otherwise, variations in path length and reflection losses at the curved surfaces can cause significant errors.

The quality of spectroscopic data is critically dependent upon the way the matched cells are used and maintained. Fingerprints, grease, or other deposits on the walls markedly alter the transmission characteristics of a cell. Thus, thorough cleaning before and after use is imperative; the surface of the windows must not be touched during the handling. Matched cells should never be dried by heating in an oven or over a flame—such treatment may cause physical damage or a change in path length. The cells should be calibrated against each other regularly with an absorbing solution.

20D RADIATION DETECTORS

A *transducer* is a device that converts one type of signal to another. For example, the pointer on the beam of a balance transduces a mass signal to a signal consisting of a visually observable position on a scale. Early instruments for the measurement of emission or absorption employed the eye or a photographic plate. In the first of these, the radiation signal was transduced to an optic nerve signal; in the second, the signal appeared as a blackening of the photographic emulsion. Most modern transducers convert radiant energy into an electrical signal; our discussion will be confined to detectors of this kind.

20D-1 PROPERTIES OF PHOTOELECTRIC DETECTORS

To be useful, a detector must respond to radiant energy over a broad wavelength range. It should, in addition, be sensitive to low levels of radiant power, respond rapidly to the radiation it receives, produce an electrical signal that can be readily amplified, and have a relatively low noise level.[2] Finally, it is essential that the signal produced be directly proportional to the beam power, P; that is,

$$G = KP + K' \tag{20-1}$$

where G is the electrical response of the detector in units of current, resistance, or emf. The proportionality constant K measures the sensitivity of the detector in terms of electrical response per unit of radiant power. Many detectors exhibit a small constant response, known as a *dark current*, K', even when no radiation im-

---

[2] Generally, the output from analytical instruments fluctuates in a random way as a consequence of the operation of a large number of uncontrolled variables. These fluctuations, which limit the sensitivity of an instrument, are called noise. The terminology is derived from radio engineering, where the presence of unwanted signal fluctuations was recognizable to the ear as static or noise.

pinges on their surfaces. Instruments with detectors that have a dark-current response are ordinarily equipped with a compensating circuit that permits application of a countersignal to reduce K' to zero. Thus, under ordinary circumstances, we may write

$$P = KG \qquad\qquad (20\text{-}2)$$

Types of photoelectric detectors. As indicated in Figure 20-2d, two general types of radiation transducers exist: one responds to photons, the other to heat. All photon detectors are based upon interaction of radiation with a reactive surface to produce electrons (photoemission) or to promote electrons to energy states in which they can conduct electricity (photoconduction). Only ultraviolet, visible, and near-infrared (180 to 2000 nm) radiation has sufficient energy to cause these processes to occur.

20D-2 PHOTON DETECTORS

Photon detectors can be classified into several categories, including: (1) photovoltaic cells, (2) phototubes, (3) photomultiplier tubes, (4) semiconductor detectors, and (5) silicon diodes. The discussion that follows will be concerned with the first three of these.

Photovoltaic or barrier-layer cells. The photovoltaic cell is used primarily to detect and measure radiation in the visible region. The typical cell has a maximum sensitivity at about 550 nm; the response falls off to perhaps 10% of the maximum at 350 and 750 nm. Its range approximates that of the human eye.

    The photovoltaic cell consists of a flat copper or iron electrode upon which is deposited a layer of semiconducting material, such as selenium or copper(I) oxide (see Figure 20-8). The outer surface of the semiconductor is coated with a thin, transparent metallic film of gold, silver, or lead, which serves as the second or collector electrode; the entire array is protected by a transparent envelope. When radiation of sufficient energy reaches the semiconductor, covalent bonds are broken, with the result that conduction electrons are formed that are free to migrate through the semiconductor. Promotion of electrons to the conduction bands leaves behind positive charges, which are termed *holes*, which can migrate in the opposite direction. The result is an electrical current whose magnitude is proportional to the number of photons striking the semiconductor surface. Ordinarily, these currents are large enough to be measured with a galvanometer or microammeter; if the resistance of the external circuit is kept small, the magnitude of the photocurrent is directly proportional to the power of the radiation striking the cell. Currents on the order of 10 to 100 μA are typical.

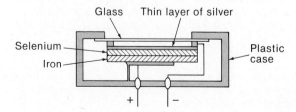

FIGURE 20-8 Schematic of a typical barrier-layer cell.

The barrier-layer cell constitutes a rugged, low-cost means for measuring radiant power. No external source of electrical energy is required. On the other hand, the low internal resistance of the cell makes the amplification of its output less convenient. Thus, although the barrier-layer cell provides a readily measured response at high levels of illumination, it suffers from lack of sensitivity at low levels.

Another disadvantage of the barrier-type cell is that it exhibits *fatigue;* that is, its current output decreases gradually during continual illumination. Proper circuit design and choice of experimental conditions minimize this effect.

Barrier-type cells are widely used in simple, portable instruments where ruggedness and low cost are important. For routine analyses, these instruments often provide reliable analytical data.

Phototubes. A second type of photoelectric device is the phototube, which consists of a semicylindrical cathode and a wire anode sealed inside an evacuated transparent envelope. The concave surface of the electrode supports a layer of photoemissive material that tends to emit electrons upon being irradiated. When a potential is applied across the electrodes, the emitted electrons flow to the wire anode, generating a photocurrent. The currents produced are generally about one-tenth as great as those from a photovoltaic cell for a given radiant intensity. In contrast, however, amplification is easily accomplished since the phototube has a high electrical resistance. Figure 20-9 is a schematic diagram of a typical phototube arrangement.

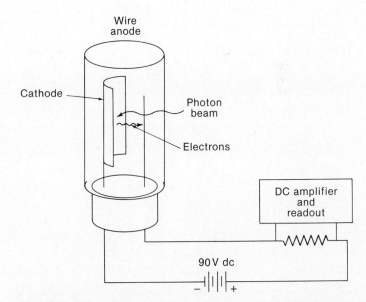

FIGURE 20-9 A phototube and accessory circuit. The photocurrent induced by the radiation causes a potential drop in the resistor, which is then amplified to drive a meter or recorder.

The number of electrons ejected from a photoemissive surface is directly proportional to the radiant power of the beam striking that surface. With an applied potential of about 90 V, all of these electrons reach the anode; a current that is proportional to radiant power results. Phototubes frequently produce a small current even though no radiation is striking their emissive surfaces; this *dark current* (see Equation 20-1) results from thermally induced electron emission.

Photomultiplier tubes.   For the measurement of low radiant power, the *photomultiplier* tube offers advantages over the ordinary phototube. Figure 20-10 is a sche-

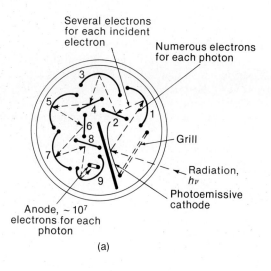

(a)

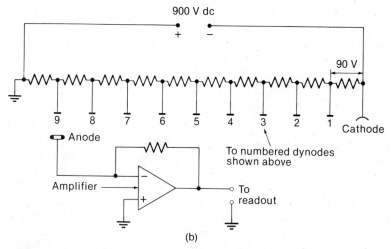

(b)

FIGURE 20-10   Photomultiplier tube: (a) cross section of the tube; (b) electrical circuit.

matic diagram of such a device. The cathode surface is similar in composition to that of a phototube, with electrons being emitted upon exposure to radiation. The tube also contains additional electrodes (nine in Figure 20-10) called *dynodes*. Dynode 1 is maintained at a potential 90 V more positive than the cathode, and electrons are accelerated toward it as a consequence. Upon striking the dynode, each photoelectron causes emission of several additional electrons; these, in turn, are accelerated toward dynode 2, which is 90 V more positive than dynode 1. Again, several electrons are emitted for each electron striking this surface. By the time this process has been repeated nine times, 10^6 to 10^7 electrons have been formed for each photon; this cascade is finally collected at the anode. The resulting current is then electronically amplified and measured.

20D-3 HEAT DETECTORS

The measurement of infrared radiation is difficult owing to the low intensity of available sources and the low energy of the infrared photon. As a consequence of these properties, the electrical signal from an infrared detector is weak, and its measurement requires large amplification. It is usually the detector system that limits the sensitivity and the precision of an infrared instrument.

Generally, infrared radiation is detected by measuring the temperature rise of a blackened material placed in the path of the beam. The temperature changes resulting from absorption of the radiant energy are minute; thus, close control of the ambient temperature is required if large errors are to be avoided.

One method of determining the temperature change involves use of a tiny thermocouple or a group of thermocouples called a *thermopile*. With this device, the temperature-dependent electromotive force developed across a dissimilar metal junction is measured.

A *bolometer* is a second type of temperature detector; it consists of a resistance wire or a thermistor whose resistance varies as a function of temperature. Here, it is the change in electrical resistance of the detector that is measured.

20E SIGNAL PROCESSORS AND READOUTS

The signal processor is ordinarily an electronic device that amplifies the electrical signal from the detector; in addition, it may alter the signal from dc to ac (or the reverse), change the phase of the signal, and filter it to remove unwanted components. The signal processor may also be called upon to perform such mathematical operations on the signal as differentiation, integration, or conversion to a logarithm.

Several types of readout devices are found in modern instruments. Some of these include the d'Arsonval meter, digital meters, the scales of potentiometers, recorders, and cathode-ray tubes.

20F SPECTROSCOPIC INSTRUMENTS

The instrumental components discussed in the previous section have been combined in various ways to produce dozens of commercial instruments for emission, absorption, and fluorescence analysis. Some of the simplest instruments can be purchased for $100 to $200, whereas the most sophisticated cost a hundred or even a thousand times this amount. No single instrument is best for all purposes. Selection is governed by the type of work for which the instrument is to be used and the economics of the situation to which it is to be applied.

In this section, a few simple instruments that are typical of ones the student is likely to encounter will be described.

20F-1 TYPES OF INSTRUMENTS

A *spectroscope* consists of a monochromator such as that shown in Figure 20-5a. The focal plane containing the slit is replaced, however, by a movable eyepiece that permits visual detection of the emission lines from a sample. The wavelength is determined by measurement of the angle between the incident beam and the dispersed beam being observed by means of the eyepiece.

A *colorimeter* is an instrument for absorption measurements in which the human eye serves as the detector. One or more color-comparison standards are required.

A *photometer* consists of a source, a filter, and a photoelectric detector plus a signal processor and readout. Filter photometers are commercially available for absorption measurements in the ultraviolet, visible, and infrared regions as well as emission and fluorescence in the first two wavelength regions. Photometers designed for fluorescence measurements are also called *fluorometers*.

In a *spectrograph*, radiation is detected by a photographic film or plate located along the focal plane of the monochromator (see Figure 20-5). Spectrographs are ordinarily employed for qualitative and quantitative analyses based on ultraviolet and visible atomic line spectra.

Monochromators with a fixed slit in the focal plane (such as those in Figure 20-5) are called *spectrometers*. A spectrometer equipped with a photoelectric detector is called a *spectrophotometer*. A spectrophotometer for fluorescence analysis is sometimes called a *spectrofluorometer*. Spectrophotometers are employed for absorbance measurements in the ultraviolet, visible, and infrared regions and for emission and fluorescence measurements in the former two.

20F-2 SINGLE-BEAM AND DOUBLE-BEAM DESIGN

Photometers and spectrophotometers are designed either with one or with two light paths. The beam of a double-beam absorption instrument is split to permit approximately half to traverse a cell containing the sample while the other half passes through a second cell containing the solvent; every effort is made to be sure the path lengths in the two media are identical. The power of the two beams is then compared to give the transmittance of the solution (Equation 19-6). Because comparison of the beam passing through the solvent with that passing through the sample is made simultaneously or nearly simultaneously, a split-beam instrument compensates for all but the most short-term electrical fluctuations, as well as for irregular performance in the source, the detector, and the amplifier.

Single-beam instruments are particularly suited for quantitative analyses that involve measurements at a single wavelength. Here, simplicity of instrumentation and the concomitant ease of maintenance offer real advantages. The greater speed and convenience of measurement, on the other hand, make the double-beam instrument particularly useful for qualitative analyses, where measurements must be made at numerous wavelengths in order to obtain a spectrum. Furthermore, the double-beam device is readily adapted to continuous monitoring of optical signals;

it is for these reasons that most modern recording spectrophotometers employ twin beams.

Both single-beam and double-beam instruments are available for ultraviolet and visible radiation. Commercial infrared spectrophotometers are always double-beam because they are ordinarily employed to scan and record a large spectral region.

20F-3 COLORIMETERS

Colorimeters employ the human eye as a detector and the brain as a transducer and signal readout. The eye and brain, however, can only match colors. As a consequence, colorimetric methods always require the use of one or more standards for color matching with the analyte solutions.

The simplest colorimetric methods involve the comparison of the sample with a set of standards until a match is found. Flat-bottomed *Nessler tubes* are frequently employed for this purpose. These tubes are calibrated so that a uniform light path is achieved. Daylight commonly serves as a radiation source; ordinarily no attempt is made to restrict the portion of the spectrum employed.

Colorimetric methods suffer from several disadvantages. A standard or a series of standards must always be available. The human eye responds to a limited spectral region (400 to 700 nm) and additionally is unable to match absorbances if the analyte solution contains a second colored substance. Finally, the eye is not as sensitive to small differences in absorbance as a photoelectric device; as a consequence, concentration differences smaller than about 5% relative cannot be detected.

Despite their limitations, visual comparison methods find extensive application for routine analyses in which the requirements for accuracy are modest. For example, simple but useful colorimetric test kits are sold for determining the pH and the chlorine content of swimming pool water; kits are also available for the analysis of soils. Filtration plants commonly employ color comparison tests for the estimation of iron, silicon, fluorine, and chlorine in city water supplies. For such analyses a colorimetric reagent is introduced to the sample, and the resulting color is compared with permanent standard solutions or with colored glass disks. Accuracies of perhaps 10 to 50% relative are to be expected and suffice for the purposes intended.

20F-4 PHOTOMETERS

The photometer provides a simple, relatively inexpensive tool for performing emission, absorption, and fluorescence analyses. Convenience, ease of maintenance, and ruggedness are properties of a filter photometer that may not exist in the more sophisticated spectrophotometer. Moreover, where high spectral purity is not important to a method (and often it is not), analyses can be performed as accurately with a photometer as with more complex instrumentation.

Visible absorption photometers. Figure 20-11 presents schematic diagrams of two photometers for absorption measurements in the visible region. The first is a single-beam, direct-reading instrument consisting of a tungsten-filament lamp, a lens to provide a parallel beam of light, a filter, and a photovoltaic cell. The current

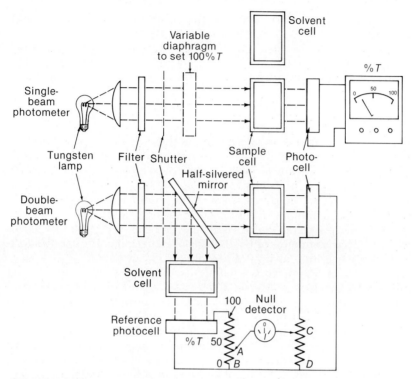

FIGURE 20-11   A single- and double-beam absorption photometer.

produced is indicated with a microammeter, the face of which may be scribed with a linear scale from 0 to 100.

The measurement of %T involves three steps: (1) the *0% T adjustment*, (2) the *100% T adjustment*, and (3) the *determination of % T for the analyte.* The 0% T adjustment is carried out with a shutter imposed between the source and the photodetector. The needle of the meter is then made to read 0 by either a mechanical or electrical adjustment.

The shutter is then opened and the instrument is adjusted to give a reading of 100% T with the blank in the light path. This adjustment involves changing the voltage applied to the lamp in some instruments or altering the aperture of a diaphragm in the light path of others. Since the signal from a photovoltaic cell is linear with respect to the radiation it receives, the scale reading with the sample in the light path will be the percent transmittance (that is, the percent of full scale). Clearly, a logarithmic scale could be substituted to give the absorbance of the solution directly.

Also shown in Figure 20-11 is a schematic representation of a double-beam, null-type photometer. Here, the light beam is split by a half-silvered mirror that transmits about 50% of the radiation striking it and reflects the other 50% . One part passes through the sample and thence to a photovoltaic cell; the other part passes through the solvent to a similar detector. The currents from the two photovoltaic cells are passed through variable resistances; one of these is calibrated as a transmittance scale in linear units from 0 to 100. A sensitive galvanometer, which serves as a null detector, is connected across the two resistances. When the potential drop across AB is equal to that across CD, no electricity passes through the galvanometer; under all other circumstances, a current is indicated. At the outset, the contact on the left is set to 0% T, the shutter is closed, and the pointer of the detector is centered mechanically; the center mark then corresponds to zero current. Next, the solvent is placed in both cells, and contact A is set at 100 (the 100% T setting); with the shutter open, contact C is then adjusted until zero current is indicated. Replacement of the solvent with the sample in the one cell results in a decrease in radiant power that reaches the working phototube and a corresponding decrease in its output potential; a current is then indicated by the galvanometer. The lack of balance is compensated for by moving A to a lower value. At balance, the percent transmittance is read directly from the scale.

Commercial photometers usually cost a few hundred dollars. The majority employ the double-beam principle because this design largely compensates for fluctuations in the source intensity due to voltage variations.

Filter selection for absorption analysis. Photometers are generally supplied with several filters, each of which transmits a different portion of the visible spectrum. Selection of the proper filter for a given application is important inasmuch as the sensitivity of the measurement is directly dependent upon this choice. The color of the light absorbed is the complement of the color of the solution itself. A liquid appears red, for example, because it transmits the red portion of the spectrum but absorbs the green. It is the intensity of green radiation that varies with concentration; a green filter should thus be employed. In general, then, the most suitable filter for a photometric analysis will be the color complement of the solution being analyzed. If several filters possessing the same general hue are available, the one that causes the sample to exhibit the greatest absorbance (or least transmittance) should be used.

Ultraviolet absorption photometers. The most important application of ultraviolet photometry is as a detector in high-performance liquid chromatography. In this application, a mercury-vapor lamp serves as a source and the emission line at 254 nm is isolated by filters. This type of detector is described briefly in Section 28A-1.

Infrared absorption photometers. Figure 20-12 is a schematic diagram of a portable (weight = 18 lb), infrared filter photometer designed for quantitative analysis of various organic substances in the atmosphere. A variety of interference filters, which transmit in the range between about 3 and 14 μm (3000 to 750 cm^{-1}), are available; each is designed for the analysis of a specific compound. The filters are readily interchangeable.

The gaseous sample is introduced into the cell by means of a battery-operated pump. The path length of the cell as shown is 0.5 m; a series of reflecting mirrors (not shown in Figure 20-12) permit increases in the effective cell length to 20 m in increments of 1.5 m. This feature greatly enhances the concentration range of the instrument.

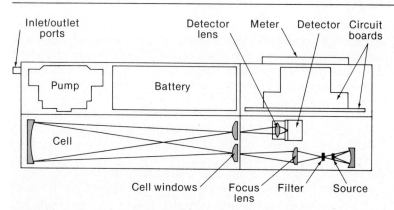

FIGURE 20-12 A portable, infrared photometer designed for gas analysis. (Courtesy of Foxboro Analytical, Wilks Infrared Center, Burlington, Massachusetts, 01803.)

The photometer is reported to be sensitive to a few tenths parts per million of such substances as acrylonitrile, chlorinated hydrocarbons, carbon monoxide, phosgene, and hydrogen cyanide.

20F-5 SPECTROPHOTOMETERS

A large variety of spectrophotometers is available from commercial sources. Some have been designed for the visible region only; others are applicable in the ultraviolet and visible regions. Still others have been designed exclusively for the infrared region.

Ultraviolet-visible range. Several spectrophotometers designed to operate within the visible region of 380 to 800 nm are available; within these limits, the range varies somewhat from instrument to instrument.

Spectrophotometers designed for the visible region are frequently simple, single-beam grating instruments that are relatively inexpensive (less than $1000), rugged, and readily portable. At least one is battery operated and light enough to be hand-held. The most common application of these instruments is to quantitative analysis, although several produce surprisingly good absorption spectra as well.

Figure 20-13 depicts two typical visible spectrophotometers.

The Bausch and Lomb Spectronic 20 spectrophotometer shown in Figure 20-13a employs a reference phototube, which serves to compensate for fluctuations in the output of the tungsten-filament light source; this design eliminates the need for a stabilized lamp power supply. The amplified difference signal from the two phototubes powers a meter with a $5\frac{1}{2}$-in. scale calibrated in transmittance and absorbance.

The Spectronic 20 is equipped with an occluder, which is a vane that automatically falls between the beam and the detector whenever the cuvette is removed from its holder; the 0% T adjustment can then be made. The light control device shown in Figure 20-13a consists of a V-shaped slot that can be moved in or out of the beam in order to set the meter to 100% T.

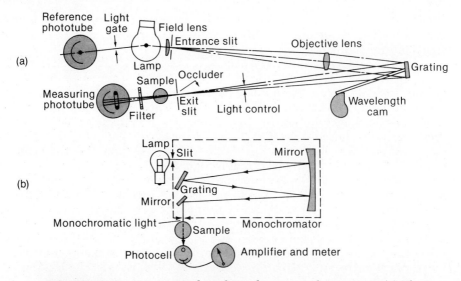

FIGURE 20-13   Two examples of simple spectrophotometers. (a) The
Spectronic 20; diagram courtesy of Bausch and Lomb, Rochester, New York.
(b) The Turner 350; diagram courtesy of Amsco Instrument Company
(formerly Turner Associates), Carpinteria, California.

   The range of the Spectronic 20 is from 340 to 625 nm; an accessory phototube
extends this range to 950 nm. Other specifications for the instrument include a
bandwidth of 20 nm and a wavelength accuracy of ± 2.5 nm.
   The Turner instrument, shown schematically in Figure 20-13b, makes use of
a tungsten-filament bulb as a source, a plane reflection grating in an *Ebert mounting* for dispersion, and a phototube detector that is sensitive in the range between
210 and 710 nm (the low wavelength limit of the instrument is about 350 nm unless
a deuterium source is used). The readout device is a meter calibrated both in transmittance and absorbance; instruments with 4- or 7-in. scales are offered. The transmittance is first set to zero by adjustment of the amplifier output while a shutter
screens the lamp from the detector. With the solvent in the light path, the 100% T
adjustment is then accomplished by varying the output of the stabilized lamp
power supply. Finally, the transmittance or absorbance is read with the sample in
the beam. Instrument specifications include a bandwidth of 9 nm, a wavelength
accuracy of ± 2 nm, and a photometric accuracy of 0.5% A.
   Several accessories are offered with the Turner instrument. One, which includes a deuterium lamp, extends the range of the instrument to 210 nm; another
provides an additional phototube that permits measurements to 1000 nm.

Infrared double-beam spectrophotometers for qualitative analysis.   Because of
the complexity of infrared spectra, a recording instrument is required for qualitative work; as a consequence, all commercially available instruments of this type are
of double-beam design. An added reason for employing two beams is that this design is less demanding in terms of performance of source and detector than the

single-beam arrangement. This property is important because of the low energy of infrared radiation, the low stability of sources and detectors, and the need for large signal amplification.

All commercial infrared spectrophotometers incorporate a low-frequency chopper (5 to 13 cycles per minute) to modulate the source output. This feature permits the detector to discriminate between the signal from the source and signals from extraneous radiation, such as infrared emission from various bodies surrounding the detector. Low chopping rates are demanded by the slow response times of most infrared detectors.

Infrared instruments are generally of the null type with the beam being attenuated by a comb or an absorbing wedge.

Several dozen models of infrared instruments, ranging in cost from $3000 to more than $30,000, are available from various instrument manufacturers. In general, the optical design of these instruments is similar to that of double-beam ultraviolet-visible spectrophotometers except that the sample and reference compartment is always located between the source and the monochromator in infrared instruments. This arrangement is desirable because the monochromator then removes most of the stray radiation generated by the sample, the reference, and the cells. In ultraviolet-visible spectroscopy, stray radiation from these sources is seldom serious. Location of the cell compartment between the monochromator and the detector of such instruments, however, has the advantage of protecting the sample from the full power and wavelength range of the ultraviolet source; thus, photochemical decomposition is less likely.

Figure 20-14 shows schematically the arrangement of components in a typical infrared spectrophotometer. Note that three types of systems link the components: (1) a radiation linkage indicated by dashed lines; (2) a mechanical linkage shown by thick dark lines; and (3) an electrical linkage shown by narrow solid lines.

Radiation from the source is split into two beams, half passing into the sample-cell compartment and the other half into the reference area. The reference beam then passes through an attenuator and on to a chopper. The attenuator consists of a wedge whose absorbance depends upon its vertical position with respect to the beam; movement of the wedge has the effect of increasing or decreasing the power of the reference beam. The chopper consists of a motor-driven disk that alternately reflects the reference beam or transmits the sample beam into the monochromator. After dispersion by a prism or grating, the alternating beams fall on a detector and are converted to an electrical signal. The signal is amplified and passed to the synchronous rectifier, a device that is mechanically or electrically coupled to the chopper to cause the rectifier switch and the beam leaving the chopper to change simultaneously. If the two beams are identical in power, the signal from the rectifier is an unfluctuating direct current. If, on the other hand, the two beams differ in power, a fluctuating or ac current is produced, the phase of which is determined by which beam is the more intense. The current from the rectifier is filtered and further amplifed to drive a synchronous motor in one direction or the other, depending upon the phase of the input current. The synchronous motor is mechanically linked to both the attenuator and the pen drive of the recorder and causes both to move until a null is achieved. A second synchronous motor drives the chart and varies the wavelength simultaneously. There is fre-

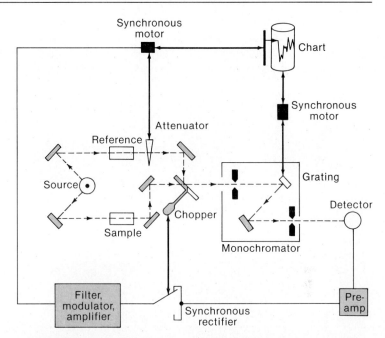

FIGURE 20-14 Schematic diagram of a double-beam spectro-photometer. Heavy dark line indicates mechanical linkage; light line indicates electrical linkage; dashed line indicates radiation path.

quently a mechanical linkage between the wavelength and slit drives to keep the radiant power reaching the detector approximately constant.

PROBLEMS*

20-1. Describe the differences between the following and list any particular advantages possessed by one over the other:

* (a) hydrogen- and deuterium-discharge lamps as sources for ultraviolet radiation.
(b) filters and monochromators as wavelength selectors.
* (c) grating and prism monochromators.
(d) quartz and glass prisms as dispersing elements for absorbance measurements.
* (e) photovoltaic cells and phototubes as detectors for electromagnetic radiation.
(f) phototubes and photomultiplier tubes.
* (g) photometers and colorimeters.
(h) spectrophotometers and photometers.
* (i) single-beam and double-beam instruments for absorbance measurements.
* 20-2. A portable photometer with a linear response to radiation registered $73.6\mu A$ with a blank solution in the light path. Replacement of the blank with an absorbing solution yielded a response of $24.9\mu A$. Calculate

* Answers to problems and parts of problems marked with an asterisk are to be found at the end of the book.

(a) the percent transmittance of the sample solution.
(b) the absorbance of the sample solution.
(c) the transmittance to be expected for a solution in which the concentration of the absorber is one-third that of the original sample solution.
(d) the transmittance to be expected for a solution that has twice the concentration of the sample solution.

20-3. A photometer with a linear response to radiation gave a reading of 685 mV with a blank in the light path and 179 mV when the blank was replaced by an absorbing solution. Calculate
(a) the percent transmittance and absorbance of the absorbing solution.
(b) the expected transmittance if the concentration of absorber is one-half that of the original solution.
(c) the transmittance to be expected if the light path through the original solution is doubled.

applications of

molecular spectroscopy

21

Molecular spectroscopy involving ultraviolet, visible, and infrared radiation finds widespread application to qualitative and quantitative analyses. Infrared absorption spectroscopy is particularly useful for qualitative analysis because these spectra generally contain a host of narrow peaks that serve to characterize the absorbing species. Ultraviolet and visible spectra, which usually consist of broad absorption bands, have less value for identification but great utility for quantitative analysis. Molecular fluorescence spectroscopy is particularly useful for quantitative work because of its high sensitivity and good selectivity.

21A MOLECULAR ABSORPTION SPECTRA

Absorption by polyatomic molecules, particularly in the condensed state, is a complex process because the number of possible energy states is large. Here, the total energy of a molecule is given by

$$E = E_{\text{electronic}} + E_{\text{vibrational}} + E_{\text{rotational}} \tag{21-1}$$

where $E_{\text{electronic}}$ describes the energy associated with the various orbitals of the outer electrons of the molecule, while $E_{\text{vibrational}}$ refers to the energy of the molecule as a whole due to interatomic vibrations. The third term in Equation 21-1 accounts for the energy associated with the rotation of the molecule around its center of gravity.

21A-1 ABSORPTION IN THE INFRARED AND MICROWAVE REGIONS

The three terms on the right in Equation 21-1 are ordered according to decreasing energy, with the average value for each term differing from the next by roughly two orders of magnitude. Pure rotational absorption spectra, which are free from electronic and vibrational transitions, can be brought about by microwave radiation, which is less energetic than infrared. Spectroscopic studies of gaseous species in this region are important in gaining fundamental information concerning molecular behavior; applications of microwave absorption to analytical problems, however, have been limited.

Vibrational absorption occurs in the infrared region, where the energy of radiation is insufficient for electronic transitions. Here, spectra exhibit narrow, closely spaced absorption peaks resulting from transitions among the various vibrational quantum levels (see Figure 21-1). Variations in rotational levels may also give rise to a series of peaks for each vibrational state; with liquid or solid samples, however, rotation is often hindered or prevented, and the effects of these small energy differences are not detected. Thus, a typical infrared spectrum for a liquid such as that in Figure 21-1 consists of a series of vibrational peaks.

The number of individual ways a molecule can vibrate is largely related to the number of atoms, and thus the number of bonds, it contains. Even for a simple molecule, the number is large. Thus, n-butanal ($CH_3CH_2CH_2CHO$) has 33 vibrational modes, most differing from each other in energy. Not all of these vibrations give rise to infrared peaks; nevertheless, as shown in Figure 21-1, the spectrum for n-butanal is relatively complex.

Infrared absorption is not confined to organic molecules. Covalent metal-ligand bonds are also infrared active in the longer wavelength region. Infrared spectrophotometric studies have thus provided much useful information about complex metal ions.

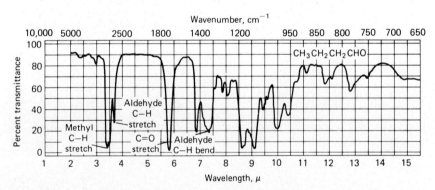

FIGURE 21-1   Infrared spectrum for n-butanal (n-butyraldehyde). Note that transmittance rather than absorbance is plotted. [*Catalog of Selected Infrared Spectral Data*, Serial No. 225, Thermodynamics Research Center Data Project Thermodynamics Research Center, Texas A & M University, College Station, Texas (loose-leaf data sheets extant, 1964).]

21A-2 ABSORPTION IN THE ULTRAVIOLET AND VISIBLE REGION

The first term in Equation 21-1 is ordinarily larger than the other two, and electronic transitions of outer electrons generally require energies corresponding to ultraviolet or visible radiation. In contrast to atomic absorption spectra, molecular spectra are often characterized by absorption bands that encompass a wide range of wavelengths (see Figure 21-2). Here, numerous vibrational and rotational energy states exist for each electronic state. Thus, for a given value of $E_{electronic}$ in Equation 21-1, values for E will exist that differ only slightly due to variations in $E_{vibrational}$ and/or $E_{rotational}$. As a consequence, the spectrum for a molecule often consists of a series of closely spaced absorption bands, such as that shown for benzene vapor in curve A of Figure 21-2. Individual bands may be undetected unless a high-resolution instrument is employed; as a result, the spectra appear as smooth curves. Finally, in the condensed state and in the presence of solvent molecules, the individual bands tend to broaden to give the type of spectra shown in the upper two curves in Figure 21-2.

21A-3 ABSORPTION INDUCED BY A MAGNETIC FIELD

When electrons or the nuclei of certain elements are subjected to a strong magnetic field, additional quantized energy levels are produced; these owe their origins to magnetic properties possessed by such elementary particles. The difference in energy between the induced states is small, and transitions between them are brought about only by absorption of radiation of long wavelengths or low frequency. For nuclei, radio waves ranging from 10 to 100 MHz are generally employed; for electrons, microwaves with frequencies of 1000 to 25,000 MHz are absorbed.

Absorption by nuclei and electrons in magnetic fields is studied by *nuclear magnetic resonance* (NMR) and *electron spin resonance* (ESR) techniques, respectively. This type of absorption is not considered in this text.

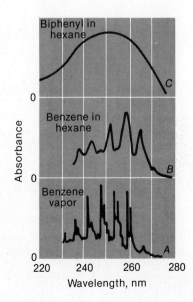

FIGURE 21-2 Some typical molecular spectra.

21B ABSORBING SPECIES

Absorption in the visible and ultraviolet regions can yield both quantitative as well as qualitative information about both inorganic and organic absorbing species.

21B-1 ABSORPTION OF ULTRAVIOLET AND VISIBLE RADIATION BY ORGANIC COMPOUNDS

The electrons responsible for absorption of ultraviolet and visible radiation by organic molecules are of two types: (1) those that participate directly in bond formation and are thus associated with more than one atom and (2) unshared outer electrons that are largely localized about such atoms as oxygen, the halogens, sulfur, and nitrogen.

The wavelength at which an organic molecule absorbs depends upon how tightly its various electrons are bound. Thus, the shared electrons in single bonds such as carbon/carbon or carbon/hydrogen are so firmly held that their excitations require energies corresponding to wavelengths in the vacuum ultraviolet region ($\lambda < 180$ nm). This region is not readily accessible because components of the atmosphere also absorb; as a result, absorption by single bonds of these types has not been important for analytical purposes.

The unshared electrons in sulfur, bromine, and iodine are less strongly held

Table 21-1 Absorption Characteristics of Some Common Organic Chromophores

Chromophore	Example	Solvent	λ_{max}(nm)	ϵ_{max}
Alkene	$C_6H_{13}CH{=}CH_2$	n-Heptane	177	13,000
Conjugated alkene	$CH_2{=}CHCH{=}CH_2$	n-Heptane	217	21,000
Alkyne	$C_5H_{11}C{\equiv}C{-}CH_3$	n-Heptane	178	10,000
			196	2000
			225	160
Carbonyl	$CH_3\overset{\overset{O}{\|}}{C}CH_3$	n-Hexane	186	1000
			280	16
	$CH_3\overset{\overset{O}{\|}}{C}H$	n-Hexane	180	Large
			293	12
Carboxyl	$CH_3\overset{\overset{O}{\|}}{C}OH$	Ethanol	204	41
Amido	$CH_3\overset{\overset{O}{\|}}{C}NH_2$	Water	214	60
Azo	$CH_3N{=}NCH_3$	Ethanol	339	5
Nitro	CH_3NO_2	Isooctane	280	22
Nitroso	C_4H_9NO	Ethyl Ether	300	100
			665	20
Nitrate	$C_2H_5ONO_2$	Dioxane	270	12
Aromatic	Benzene	n-Hexane	204	7900
			256	200

than the shared electrons of a saturated bond. Organic molecules containing these elements frequently have absorption peaks in the ultraviolet region as a consequence.

Electrons of double and triple bonds in organic molecules are relatively easily excited by radiation; thus, species containing unsaturated bonds generally exhibit useful absorption peaks. Unsaturated organic functional groups that absorb in the ultraviolet and visible regions are termed *chromophores*. Table 21-1 lists common chromophores and the approximate location of their absorption maxima. The data for position and peak intensity can serve only as a rough guide for identification purposes because the position of a maximum is also affected by solvent as well as structural details. Furthermore, when two chromophores are conjugated, shifts in peak maxima to longer wavelengths usually occur. Finally, peaks in the ultraviolet and visible regions are ordinarily broad because of vibrational effects; it is thus difficult to locate the position of the maximum with precision.

21B-2 ABSORPTION OF ULTRAVIOLET AND VISIBLE RADIATION BY INORGANIC SPECIES

The spectra for most absorbing inorganic ions or molecules resemble those for organic compounds, with broad absorption maxima and little fine structure. The spectra for ions of the lanthanide and actinide series represent an important exception. The electrons responsible for absorption by these elements ($4f$ and $5f$, respectively) are shielded from external influences by electrons that occupy orbitals with larger principal quantum numbers. As a consequence, the bands are narrow and relatively unaffected by the nature of the species bonded by the outer electrons.

With few exceptions, the ions and complexes of the 18 elements in the first two transition series are colored in one if not all of their oxidation states. Here, absorption involves transitions between filled and unfilled d orbitals which differ in energy as a consequence of ligands bonded to the metal ions. The energy differences between d orbitals (and thus the position of the corresponding absorption peak) depend upon the oxidation state of the element, its position in the periodic table, and the kind of ligand bonded to its ion.

21B-3 CHARGE-TRANSFER ABSORPTION[1]

For analytical purposes, *charge-transfer absorption* by inorganic species is of particular importance because the molar absorptivities of the band peaks are very large ($\epsilon_{max} > 10,000$). Thus, a highly sensitive means for detecting and determining the absorbing species is provided. Many inorganic and organic complexes exhibit charge-transfer absorption and are therefore called *charge-transfer complexes*. Common examples include the thiocyanate and phenolic complexes of iron(III), the *o*-phenanthroline complex of iron(II), the iodide complex of molecular iodine, and the ferro-ferricyanide complex responsible for the color of Prussian blue.

A requirement for a charge-transfer spectrum is the existence of an electron-donor group as well as an electron acceptor within the complex. Absorption of radiation involves transition of an electron from the donor group to an orbital that is largely associated with the acceptor. The excited state is thus the product of an internal oxidation-reduction process.

[1] For a brief discussion of this type of absorption, see C. N. R. Rao, *Ultra-Violet and Visible Spectroscopy*, 3d ed. London: Butterworths, 1975.

The iron(III)/thiocyanate complex is an example of charge-transfer absorption. Absorption of a photon results in the transition of an electron from the thiocyanate ion to an orbital that is largely associated with the iron(III) ion. The product is an excited species involving predominantly iron(II) and the thiocyanate radical SCN. As with an intramolecular excitation, the electron, under ordinary circumstances, returns to its original state after a brief period. Occasionally, however, dissociation of an excited complex may occur to produce a photochemical oxidation-reduction process.

In most charge-transfer complexes involving a metal ion, the metal serves as the electron acceptor. An exception is observed in the o-phenanthroline complex of iron(II) or copper(I), in which the ligand is the acceptor and the metal ion is the donor. A few other examples of this type of complex are known.

21B-4 ABSORPTION OF INFRARED RADIATION

The relative positions of atoms in a molecule are not fixed; instead, they fluctuate continuously as a consequence of a multitude of different types of vibrations. These vibrations are quantized in the sense that their frequencies can assume only certain values. Vibrational absorption requires that the radiation frequency exactly match the vibrational frequency of a bond. Thus, infrared absorption typically consists of narrow peaks, each one of which corresponds to a vibrational frequency of a bond in the molecule. The energy transferred to the bond by the absorption of the radiation increases the amplitude of the vibration.

Another requirement for infrared absorption is that the vibrational motions about a bond must cause a change in dipole moment. Only then can the alternating field of the radiation interact with the bond and cause a change in amplitude of the vibration. For example, the charge distribution around a molecule such as hydrochloric acid is not symmetric; the chlorine has a greater electron density than the hydrogen. As the hydrogen and chlorine atoms move toward or away from one another, the dipole moment changes and an electrical field is established that can interact with the electric vector of the radiation. In contrast, vibrations in homonuclear species such as O_2, N_2, or Cl_2 cause no change in dipole moment; thus, these molecules do not absorb in the infrared region.

Most organic and inorganic molecules contain bonds between atoms of differing charge density. Therefore, most molecules, both inorganic and organic, exhibit infrared absorption peaks.

21C APPLICATIONS OF ABSORPTION SPECTROSCOPY TO QUALITATIVE ANALYSIS

Absorption spectroscopy is a useful tool for qualitative analysis. Identification of a pure compound by this method involves an empirical comparison of the spectral details of the unknown (maxima, minima, and inflection points) with those of authentic compounds; a close match is considered good evidence of chemical identity, particularly if the spectrum of the unknown contains a number of sharp and well-defined features. Absorption in the infrared region is particularly useful for qualitative purposes because of the wealth of fine structure that exists in most spectra (see Figure 21-1). The application of ultraviolet and visible spectrophotometry to qualitative analysis is more limited because the absorption bands tend to be broad and, hence, lacking in detail. Nevertheless, spectral investigations in this region frequently provide useful qualitative information concerning the presence or absence of certain functional groups (such as carbonyl, aromatic, nitro, or conjugated diene) in organic compounds. A further important application involves the detection of highly absorbing impurities in nonabsorbing media. If an absorption peak for the contaminant has a sufficiently high absorptivity, its presence in trace amounts can be readily established.

21C-1 QUALITATIVE TECHNIQUES

Solvents. Factors to be considered in choosing a solvent include not only its transparency in the spectral region of interest but also its possible effects upon the absorbing system. Quite generally, polar solvents such as water, alcohols, esters, and ketones tend to obliterate spectral fine structure arising from vibrational effects; spectra that more closely approach those of the gas phase are most likely to be observed in nonpolar solvents such as hydrocarbons. In addition, the positions of absorption maxima are influenced by the nature of the solvent. Clearly, a spectrum prepared for identification purposes should involve the same solvent as the spectrum with which it is to be compared.

Table 21-2 lists common solvents for studies in the ultraviolet and visible regions and the approximate wavelengths below which they absorb and cannot be used. These lower wavelength limits are strongly dependent upon the purity of the solvent.

No single solvent is transparent throughout the infrared region. Water is generally avoided not only because it absorbs broadly but also because it attacks the infrared-transparent materials required for windows and optics. Carbon tetrachloride and carbon disulfide are the two most satisfactory solvents; the former is useful in the region up to about 7.6 μm and the latter from this wavelength to 15 μm. Many compounds are not soluble in these solvents and must be examined in the pure (or neat) form; here, very thin films of the sample are required.

Effect of slit width. Absorption spectra that are to be used for qualitative comparison should be generated with the narrowest possible slit width; otherwise, significant details of the spectrum may be lost. This effect is demonstrated in Figure 21-3.

Effect of scattered radiation at wavelength extremes of a spectrophotometer. We have already noted (p. 496) that scattered radiation may cause instrumental de-

Table 21-2 Solvents for the Ultraviolet and the Visible Regions

Solvent	Lower Wavelength Limit (nm)	Solvent	Lower Wavelength Limit (nm)
Water	180	Carbon tetrachloride	260
Ethanol	220	Diethyl ether	210
Hexane	200	Acetone	330
Cyclohexane	200	Dioxane	220
Benzene	280	Cellosolve	320

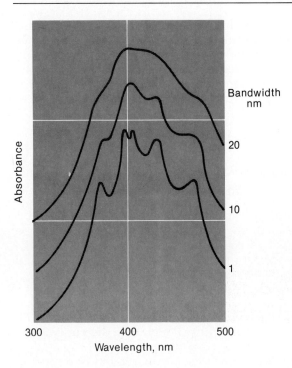

Bandwidth
nm

20

10

1

Absorbance

300   400   500

Wavelength, nm

FIGURE 21-3 Effect of bandwidth
upon spectra of identical solutions.
Note that spectra have been displaced
from one another by 0.1 in absorbance.

viations from Beer's law. When measurements are attempted at the wavelength extremes of an instrument, the effects of stray radiation may be even more serious
and, on occasion, lead to the appearance of false absorption peaks. For example,
consider a spectrophotometer for the visible region equipped with glass optics, a
tungsten source, and a photocell detector. At wavelengths below about 380 nm, the
windows, cells, and prism begin to absorb radiation, diminishing the radiant energy reaching the transducer. The output of the source falls off rapidly in this
region; so also does the sensitivity of the photoelectric device. Thus, the radiant
power for the 100% T adjustment may be as low as 1 to 2% of that in the region
between 500 and 650 nm. Scattered radiation, however, often consists of wavelengths to which the instrument is highly sensitive. Thus, its effects can be enormously magnified. Indeed, the output signal produced by the stray radiation may
even exceed that from the monochromator beam; under these circumstances, the
measured absorbance is as much affected by the stray radiation as by the radiation
to which the instrument is set.

   An example of a false peak appearing at the wavelength extremes of a
visible-region spectrophotometer is shown in Figure 21-4. Curve B is the spectrum
for a solution of cerium(IV) that was acquired with an ultraviolet-visible spectrometer sensitive in the region of 200 to 750 nm. Curve A is a spectrum for the same
solution obtained with a simple visible spectrophotometer. The apparent maximum shown in curve A arises from the instrument responding to stray wavelengths longer than 400 nm, which (as can be seen from the spectrum) are not absorbed by the cerium(IV) ions.

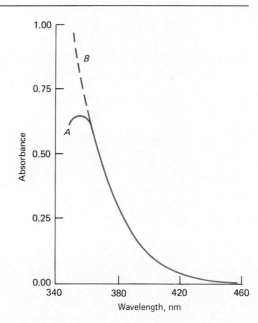

FIGURE 21-4 Spectra of cerium(IV) obtained with a spectrophotometer having glass optics (A) and quartz optics (B). The false peak in A arises from transmission of stray radiation of longer wavelengths.

    This same effect is sometimes observed with ultraviolet-visible instruments when attempts are made to measure absorbances at wavelengths lower than about 200 nm.

21C-2 IDENTIFICATION OF COMPOUNDS

The certainty with which an identification can be made by absorption measurement is directly related to the number of similar spectral features (peaks, minima, and inflection points) that can be observed between the unknown and an authentic reference standard. For the ultraviolet and visible regions, the number of these characteristic features is often limited; identification based on a single pair of spectra may thus be ambiguous. Occasionally, the identity of unknown and standard can be confirmed by comparing their spectra in other solvents, at different pH values, or following suitable chemical treatment. Unfortunately, however, the electronic absorption peaks of many chromophores are not greatly influenced by structural features of attached nonabsorbing groups; as a consequence, only the absorbing functional groups can be identified, and other means must be used to determine the remaining features of the compound under study.

Ultraviolet and visible regions. Catalogs of ultraviolet and visible absorption data for organic compounds are available in several publications.[2] These compilations frequently are of help in organic qualitative work.

---

[2] American Petroleum Institute, *Ultraviolet Spectral Data, A.P.I. Research Project 44.* Pittsburgh: Carnegie Institute of Technology; *Sadtler Ultraviolet Spectra.* Philadelphia: Sadtler Research Laboratories; American Society for Testing Materials, Committee E-13, Philadelphia. Ultraviolet spectra are coded on IBM cards.

Table 21-3 Some Characteristic Infrared Absorption Peaks

| | Functional Group | Absorption Peaks | |
		Wavenumber (cm^{-1})	Wavelength (μm)
O—H	Aliphatic and aromatic	3600–3000	2.8–3.3
NH$_2$	Also secondary and tertiary	3600–3100	2.8–3.2
C—H	Aromatic	3150–3000	3.2–3.3
C—H	Aliphatic	3000–2850	3.3–3.5
C≡N	Nitrile	2400–2200	4.2–4.6
C≡C—	Alkyne	2260–2100	4.4–4.8
COOR	Ester	1750–1700	5.7–5.9
COOH	Carboxylic acid	1740–1670	5.7–6.0
C=O	Aldehydes and ketones	1740–1660	5.7–6.0
CONH$_2$	Amides	1720–1640	5.8–6.1
C=C—	Alkene	1670–1610	6.0–6.2
ϕ—O—R	Aromatic ethers	1300–1180	7.7–8.5
R—O—R	Aliphatic ethers	1160–1060	8.6–9.4

Infrared region. An infrared absorption spectrum, even for relatively simple compounds, often contains a bewildering array of sharp peaks and minima. It is this multiplicity of peaks, however, that imparts specificity to the spectrum; no two compounds give identical spectra. Thus, identity between the spectra of an unknown and an authentic sample is widely accepted as proof of identity of composition.

Peaks useful for the identification of functional groups are located in the shorter wavelength region of the infrared (below about 8.5 μm); the positions of the maxima in this region are only slightly affected by the carbon skeleton to which the groups are attached. Investigation of this region of the spectrum thus provides considerable information regarding the overall constitution of the molecule under investigation. Table 21-3 gives the positions of characteristic maxima for some common functional groups.[3]

Identification of the functional groups in a molecule is seldom sufficient to permit positive identification of the compound, and the entire spectrum from 2.5 to 15 μm must be compared with that of known compounds. Collections of spectra are available for this purpose.[4]

21D QUANTITATIVE ANALYSIS BY ULTRAVIOLET AND VISIBLE ABSORPTION MEASUREMENTS

Absorption spectroscopy based upon ultraviolet and visible radiation is one of the most useful tools available to the chemist for quantitative analysis. Important characteristics of spectrophotometric and photometric methods include:

1. *Wide applicability.* As we have pointed out, numerous inorganic and

[3] For more detailed information, see N. B. Colthup, *J. Opt. Soc. Amer.*, **40**, 397 (1950); R. M. Silverstein, G. W. Bassler, and T. C. Morrill, *Spectrometric Identification of Organic Compounds*, 4th ed. New York: Wiley, 1978.

[4] American Petroleum Institute, *Infrared Spectral Data, A.P.I. Research Project 44.* Pittsburgh: Carnegie Institute of Technology; *Sadtler Standard Spectra.* Philadelphia: Sadtler Research Laboratories.

organic species absorb in the ultraviolet and visible ranges and are thus susceptible to quantitative determination. In addition, many nonabsorbing species can be analyzed after conversion to absorbing species by suitable chemical treatment.

2. *High sensitivity.* Molar absorptivities in the range of 10,000 to 40,000 are common, particularly among the charge-transfer complexes of inorganic species. Thus, analyses for concentrations in the range of 10^{-4} to 10^{-5} M are ordinary; the range can often be extended to 10^{-6} or even 10^{-7} M with suitable procedural modifications.

3. *Moderate to high selectivity.* It may be possible to locate a wavelength region in which the only absorbing component in a sample is the substance being determined. Furthermore, where overlapping absorption bands do occur, corrections based on additional measurements at other wavelengths are sometimes possible. As a consequence, the separation step can be omitted.

4. *Good accuracy.* For the typical spectrophotometric or photometric procedure employing ultraviolet and visible radiation, the relative errors in concentration measurements lie in the range of 1 to 3%. Such errors can often be decreased to a few tenths of a percent with special precautions.

5. *Ease and convenience.* Spectrophotometric or photometric measurements are easily and rapidly performed with modern instruments.

21D-1 SCOPE OF ULTRAVIOLET-VISIBLE ABSORPTION SPECTROSCOPY
The applications of quantitative absorption methods not only are numerous but also touch upon every field in which quantitative chemical information is required. The reader can obtain a notion of the scope of spectrophotometry by consulting a series of review articles published periodically in *Analytical Chemistry*[5] and from monographs on the subject.[6]

Applications to absorbing species. Table 21-1 lists many of the common organic chromophoric groups. Spectrophotometric determination of any organic compound containing one or more of these groups is potentially feasible; many examples of this type of analysis are found in the literature.

A number of inorganic species also absorb and are thus susceptible to direct determination; we have already mentioned the various transition metals. In addition, numerous other species also show characteristic absorption. Examples include nitrite, nitrate, and chromate ions; osmium and ruthenium tetroxides; molecular iodine; and ozone.

Applications to nonabsorbing species. Numerous reagents react selectively with various nonabsorbing species to yield products that absorb strongly in the ultravio-

[5] See, for example, D. F. Boltz and M. G. Mellon, *Anal. Chem.*, **48**, 216R (1976); J. A. Howell and L. G. Hargis, *Anal. Chem.*, **50**, 243R (1978); **52**, 306R (1980).

[6] See, for example, E. B. Sandell and H. Onishi, *Colorimetric Determination of Traces of Metals*, 4th ed. New York: Interscience, 1978; *Colorimetric Determination of Nonmetals*, 2d ed., D. F. Boltz and J. A. Howell, Eds. New York: Wiley, 1978; Z. Marczenko, *Spectrophotometric Determination of Elements*. New York: Halsted Press, 1975; and M. Pisez and J. Bartos, *Colorimetric and Fluorometric Analysis of Organic Compounds and Drugs*. New York: Marcel Dekker, 1974.

let or visible regions. The successful application of such reagents to quantitative analysis usually requires that the color-forming reaction be forced to near completion. It should be noted that these reagents are frequently employed as well for the determination of an absorbing species such as a transition-metal ion; the molar absorptivity of the product will frequently be orders of magnitude greater than that of the uncombined species.

A host of complexing agents have been employed for the determination of inorganic species. Typical inorganic reagents include thiocyanate ion for iron, cobalt, and molybdenum; the anion of hydrogen perioxide for titanium, vanadium, and chromium; and iodide ion for bismuth, palladium, and tellurium. Of even more importance are organic chelating agents which form stable, colored complexes with cations. Examples include o-phenanthroline for the determination of iron, dimethylglyoxime for nickel, diethyldithiocarbamate for copper, and diphenyldithiocarbazone for lead.

21D-2 PROCEDURAL DETAILS

The first steps in a photometric or spectrophotometric analysis involve the establishment of working conditions and the preparation of a calibration curve relating concentration to absorbance.

Selection of wavelength. Spectrophotometric absorbance measurements are ordinarily made at a wavelength corresponding to an absorption peak because the change in absorbance per unit of concentration is greatest at this point; the maximum sensitivity is thus realized. In addition, the absorption curve is often flat in this region; under these circumstances, good adherence to Beer's law can be expected (p. 495). Finally, the measurements are less sensitive to uncertainties arising from failure to reproduce precisely the wavelength setting of the instrument.

The absorption spectrum, if available, aids in choosing the most suitable filter for a photometric analysis; if this information is lacking, the alternative method for selection, given on page 519, may be used.

Variables that influence absorbance. Common variables that influence the absorption spectrum of a substance include the nature of the solvent, the pH of the solution, the temperature, the electrolyte concentration, and the presence of interfering substances. The effects of these variables must be known; conditions for the analysis can then be chosen such that the absorbance will not be materially influenced by small, uncontrolled variations in their magnitudes.

Cleaning and handling of cells. It is apparent that accurate spectrophotometric analysis requires the use of good quality, matched cells. These should be regularly calibrated against one another to detect differences that can arise from scratches, etching, and wear. Equally important is the use of proper cell cleaning and drying techniques. Erickson and Surles[7] recommend the following cleaning sequence for the outside windows of cells: Prior to measurement, the cell surfaces are cleaned with a lens paper soaked in spectrograde methanol. The paper is held with a hemostat; after wiping, the methanol is allowed to evaporate, leaving the cell surfaces

[7] J. O. Erickson and T. Surles, *American Laboratory*, 8 (6), 50 (1976).

free of contaminants. The authors showed that this method was far superior to the usual procedure of wiping the cell surfaces with a dry lens paper, which apparently leaves lint and films on the surface.

Determination of the relationship between absorbance and concentration. After deciding upon the conditions for the analysis, it is necessary to prepare a calibration curve from a series of standard solutions. These standards should approximate the overall composition of the actual samples and should cover a reasonable analyte concentration range. Seldom, if ever, is it safe to assume adherence to Beer's law and use only a single standard to determine the molar absorptivity. The results of an analysis should *never* be based on a literature value for the molar absorptivity.

The difficulties that attend production of a set of standards whose overall composition closely resembles that of the sample can be formidable if not insurmountable. Under these circumstances, a *standard addition* approach may prove useful. Here, a known quantity of standard is added to an aliquot of the sample, with the absorbance being measured before and after the addition. Provided Beer's law is obeyed, the analyte concentration can be calculated from the two absorbances.

EXAMPLE 21-1. A 2.00-mL urine specimen was diluted to 100 mL. Photometric analysis for the phosphate in a 25.0-mL aliquot yielded an absorbance of 0.428. Addition of 1.00 mL of a solution containing 0.0500 mg of phosphate to a second 25.0-mL aliquot resulted in an absorbance of 0.517. Calculate the milligrams phosphate per milliliter of sample.

We must first correct the second measurement for dilution. Thus,

$$\text{corrected absorbance} = 0.517 \times \frac{26.0}{25.0} = 0.538$$

$$\text{absorbance due to } 0.0500 \text{ mg phosphate} = 0.538 - 0.428 = 0.110$$

$$\text{mg phosphate in aliquot of specimen} = \frac{0.428}{0.110} \times 0.0500 = 0.195$$

Thus,

$$\text{mg phosphate/mL of specimen} = \frac{100}{25.0} \times 0.195 \times \frac{1}{2.00} = 0.390$$

Analysis of mixtures of absorbing substances. The total absorbance of a solution at a given wavelength is equal to the sum of the absorbances of the individual components present. This relationship makes it possible to analyze for the individual components of a mixture even if an overlap in their spectra exists. Consider, for example, the spectra in Figure 21-5. There is obviously no wavelength at which the absorbance of this mixture is dependent upon only one of the components. Nevertheless, absorbance measurements at two wavelengths make resolution of this mixture possible.

EXAMPLE 21-2. Curves *A* and *B* of Figure 21-5 are absorption spectra for alcoholic solutions of sulfanilamide and sulfathiazole. Beer's law plots for a series of standard solutions of the two drugs provided the following information:

λ, nm	a_A	a_B
260.0	0.1367	0.07333
287.5	0.01583	0.09500

where a_A and a_B are the absorptivities for sulfanilamide and sulfathiazole, respectively, when the concentration unit was mg/L.

A 0.126-g sample of a bactericidal ointment was dissolved in alcohol and diluted to 100 mL. The absorbance of this solution (cell length 0.976 cm) was found to be 1.360 at 260 nm and 0.763 at 287.5 nm. Calculate the milligrams of sulfanilamide and sulfathiazole contained in 1.00 g of the sample.

The absorbance of the two wavelengths is given by

$$1.360 = 0.1367 \times 0.976 \times C_A + 0.07333 \times 0.976 \times C_B$$

$$0.763 = 0.01583 \times 0.976 \times C_A + 0.09500 \times 0.976 \times C_B$$

The solution to the two simultaneous equations is

$$C_A = 6.35 \text{ mg/L} \qquad C_B = 7.17 \text{ mg/L}$$

$$\text{mg sulfanilamide/g} = 6.35 \, \frac{\text{mg}}{\text{L}} \times \frac{100 \text{ mL}}{1000 \text{ mL/L}} \times \frac{1}{0.126 \text{ g sample}} = 5.04$$

$$\text{mg sulfathiazole/g} = 7.17 \times 0.100 \times \frac{1}{0.126} = 5.69$$

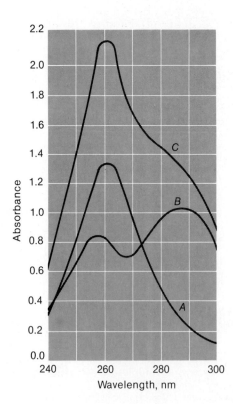

FIGURE 21-5   Absorption spectra for 95% ethanol solutions which contained; A 6.0 mg/L sulfanilamide; B 10.5 mg/L sulfathiazole; and C a mixture of these concentrations of the drugs.

21D-3 THE EFFECT OF INSTRUMENTAL UNCERTAINTIES ON THE PRECISION OF SPECTROPHOTOMETRIC ANALYSES[8]

The accuracy and precision of spectrophotometric analyses are often limited by the random uncertainties, or noise, associated with the instrument.[9] As was pointed out earlier (Section 20F-4), a spectrophotometric measurement entails three steps: a $0\% \, T$ adjustment, a $100\% \, T$ adjustment, and a measurement of $\% \, T$ with the sample in the beam of radiation. The uncertainty associated with each of these steps combines to give a net uncertainty for the final value obtained for T. The relationship between the noise encountered in the measurement of T and the uncertainty in concentration can be derived by writing Beer's law in the form

$$c = -\frac{1}{\epsilon b} \log T = \frac{-0.434}{\epsilon b} \ln T$$

The partial derivative of this equation, holding b and c constant, is given by

$$\frac{\delta c}{c} = \frac{-0.434}{\epsilon bT} \delta T$$

where δc is the uncertainty in c that results from the noise or uncertainty δT in T. Dividing by the first equation yields

$$\frac{\delta c}{c} = \frac{0.434}{\log T} \times \frac{\delta T}{T} \tag{21-2}$$

Here, $\delta T/T$ is the net *relative* uncertainty or noise in T that arises from the three measurement steps and $\delta c/c$ is the resulting relative uncertainty in concentration.

The best and most useful measure of the uncertainty δT is the standard deviation σ_T (Section 3C-2). Thus, it is useful to write Equation 21-2 in the form

$$\frac{\sigma_c}{c} = \frac{0.434}{\log T} \times \frac{\sigma_T}{T} \tag{21-3}$$

where (σ_c/c) and (σ_T/T) are relative standard deviations.

It is clear from an examination of Equation 21-3 that the uncertainty in a photometric concentration measurement varies in a complex way with the magnitude of the transmittance. The situation is even more complicated than is suggested by the equation, however, because the uncertainty σ_T is, under many circumstances, also *dependent upon T*.

In a detailed theoretical and experimental study, Rothman, Crouch and Ingle[8] have described several sources of instrumental uncertainties and shown their net effect on the precision of absorbance or transmittance measurements. These uncertainties fall into three categories; those for which the magnitude of σ_T is (1) proportional to T, (2) proportional to $\sqrt{T^2 + T}$, and (3) independent of T.

---

[8] See: L. D. Rothman, S. R. Crouch, and J. D. Ingle, Jr., *Anal. Chem.*, **47**, 1226 (1975); J. D. Ingle, Jr. and S. R. Crouch, *Anal. Chem.*, **44**, 1375 (1972); J. O. Erickson and T. Surles, *American Laboratory*, **8** (6), 41 (1976); *Optimum Parameters for Spectrophotometry*. Varian Instrument Division, Palo Alto, CA.

[9] In the context of this discussion, "noise" refers to random variations in the output of instruments due not only to electrical fluctuations but also to other variables such as the way the operator reads the meter, the position of the cell in the light beam, the temperature of the solution, and the output of the source.

Table 21-4   Sources of Instrumental Uncertainties in Transmittance Measurements

Category	Sources
$\sigma_T = k_1$	Dark current and amplifier uncertainties (noise)
	Thermal detector uncertainties in infrared detectors
	Limited resolution of the readout device
$\sigma_T = k_2 T$	Variation in output of the source
	Uncertainties in cell positioning
$\sigma_T = k_3\sqrt{T^2 + T}$	Photon detector uncertainties (shot noise) in ultraviolet and visible
	detectors

Table 21-4 summarizes information about these sources of uncertainty. Clearly, each source has a different effect upon the magnitude of the concentration error. An error analysis for a given spectrophotometric method thus requires an understanding of the likely type (or types) that predominate in a given set of circumstances.

Concentration errors when $\sigma_T = k_1$. For many photometers and spectrophotometers, the standard deviation in the measurement of T is constant and independent of the magnitude of T. The meter of a direct-reading instrument is a particularly common source of this type of error. The resolution of such a meter is ordinarily a few tenths percent of its 5- to 7-in. scale, and the absolute uncertainty in T is the same from one end of the scale to the other. A similar limitation occurs in some digital panel meters.

  The infrared spectrophotometer is another instrument which may exhibit an uncertainty that is independent of transmittance. Here, the limiting indeterminate

Table 21-5   Relative Concentration Error as a Function of Transmittance and Absorbance for Various Types of Uncertainties

		Relative Error, $\frac{\sigma_c}{c} \times 100$, for σ_T Equal to		
Transmittance, T	Absorbance, A	k_1	$k_2 T$	$k_3\sqrt{T^2 + T}$
0.95	0.022	±6.2*	±25.3**	±8.4†
0.90	0.046	±3.2	±12.3	±4.1
0.80	0.097	±1.7	±5.8	±2.0
0.60	0.222	±0.98	±2.5	±0.96
0.40	0.398	±0.82	±1.4	±0.61
0.20	0.699	±0.93	±0.81	±0.46
0.10	1.00	±1.3	±0.56	±0.43
0.032	1.50	±2.7	±0.38	±0.50
0.010	2.00	±6.5	±0.28	±0.65
0.0032	2.50	±16.3	±0.23	±0.92
0.0010	3.00	±43.4	±0.19	±1.4

* From Equation 21-3 employing $k_1 = \sigma_T = \pm 0.003$.
** From Equation 21-3 employing $k_2 = \pm 0.013$.
† From Equation 21-3 employing $k_3 = \pm 0.003$.

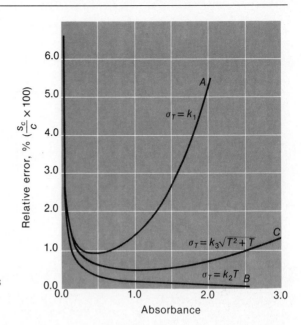

FIGURE 21-6 Error curves for various categories of instrumental uncertainties.

error lies in the thermal detector. Fluctuations in the output of this type of transducer are independent of the output; indeed, fluctuations are observed even in the absence of radiation and, therefore, net current.

The precision of instruments for which $\sigma_T = k_1$ can be readily obtained by measuring the transmittance of 20 or 30 portions of a solution and calculating the standard deviation σ_T of the measurements. The precision of analyses made with the instrument can then be derived from Equation 21-3. Clearly, the precision for a particular analysis depends upon the magnitude of T. The third column of Table 21-5 shows data obtained with this equation when an absolute standard deviation σ_T of ±0.003 or $\pm0.3\%$ T was assumed. A plot of the data is shown by curve A in Figure 21-6.

An indeterminate uncertainty of 0.3% T is typical of many moderately priced spectrophotometers or photometers. Clearly, concentration errors of 1 to 2% relative are to be expected with these instruments. It is also evident that indeterminate errors at this level can only be realized if the absorbance of the sample lies between about 0.1 and 1.

Concentration errors when $\sigma_T = k_2T$. Uncertainties, which are proportional to T, commonly result from failure to position the sample cell reproducibly with respect to the beam during replicate measurements. All cells have minor imperfections. As a consequence, reflection and scattering losses vary as different sections of the cell window are exposed to the beam. Small variations in transmittance result. Rothman, Crouch, and Ingle[8] have shown that this uncertainty is probably the most common limitation to the accuracy of high-quality spectrophotometers.

Fluctuation in source intensity also yields standard deviations that are proportional to transmittance.

Substitution of $\sigma_T = k_2 T$ into Equation 21-3 yields

$$\frac{\sigma_c}{c} = -\frac{0.434\ k_2}{\log T} \qquad (21\text{-}4)$$

Column 4 of Table 21-5 contains data obtained from Equation 21-4 with k_2 assumed to be 0.013. The data are plotted as curve B in Figure 21-6.

The value of k_2 used in the foregoing calculation was obtained experimentally by Rothman, Crouch, and Ingle[8] with a high-quality spectrophotometer. They were able to demonstrate that the source of this uncertainty was a variation in cell positioning. The uncertainty could be eliminated by leaving the cell in position at all times; samples, standards, and wash liquid were then introduced by means of a syringe.

Concentration errors when $\sigma_T = k_3 \sqrt{T^2 + T}$. This type of uncertainty often limits the accuracy of the highest quality instruments. It has its origin in the so-called *shot noise* that causes the output of photomultipliers and phototubes to fluctuate in a random way about a mean value. Substitution of this relationship between σ_T and T into Equation 21-3 gives

$$\frac{\sigma_c}{c} = -\frac{0.434\ k_3}{T \log T} \sqrt{T^2 + T} = -\frac{0.434\ k_3}{\log T} \sqrt{1 + \frac{1}{T}} \qquad (21\text{-}5)$$

The last column in Table 21-5 demonstrates the effect of shot noise on the indeterminate errors associated with an analysis. Here, k_3 was assumed to have a value of 0.003. The data are plotted as curve C in Figure 21-6.

The value used for k_3 in the foregoing calculations is typical of high-quality spectrophotometers. Note that with such instruments the most accurate analyses are obtained in an absorbance range of about 0.6 to 2.5, in contrast to 0.1 to 1.0 for moderate quality spectrophotometers.

21D-4 DIFFERENTIAL ABSORPTION METHODS[10]
In the previous section, it was pointed out that the accuracy of many photometers and spectrophotometers is limited by the sensitivities of their readout devices. With such instruments, the uncertainty σ_T in the measurement of T is *constant*, and the resulting relative concentration errors are given by Equation 21-3.

Curve O in Figure 21-7 gives a plot of the relative concentration errors to be expected from a *readout-limited* uncertainty of $\sigma_T = \pm 0.005$. This uncertainty would be typical for an instrument equipped with a 5-in. meter. Clearly, serious analytical errors are encountered when the analyte concentration is such that $\% T$ is smaller than 10% ($A = 1.0$) or greater than 70% ($A = 0.15$).

Differential methods provide a means of expanding the scale of a readout-limited instrument and thus decreasing this type of error significantly. These methods employ standard solutions of the analyte to adjust the zero and/or the 100% transmittance setting of the photometer or spectrophotometer rather than the shutter and the solvent. The three types of differential methods are compared with the ordinary method in Table 21-6.

[10] For a complete analysis of these methods, see: C. N. Reilley and C. M. Crawford, *Anal. Chem.*, **27**, 716 (1955); C. F. Hiskey, *Anal. Chem.*, **21**, 1440 (1949).

FIGURE 21-7 Relative errors in spectrophotometric analysis when photometric uncertainties are independent of T. Curve O: ordinary method; reference solution has transmittance of 100%. Curve A: high-absorbance method; reference solution has transmittance of 10%. Curve B: low-absorbance method; reference solution has transmittance of 90%. Curve C: method of ultimate precision; reference solutions have transmittances of 45% and 55%, respectively. s_T is 0.5% for each curve. See Figure 21-8 for the type of scale expansion corresponding to curves A, B, C.

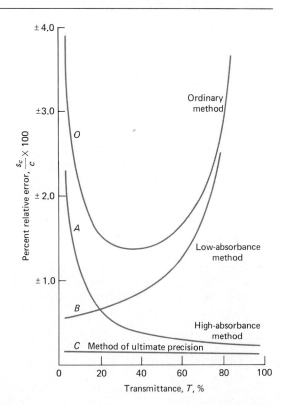

High-absorbance method. In the high-absorbance method, the zero adjustment is carried out in the usual way, with the shutter imposed between the source and the detector. The 100% transmittance adjustment, however, is made with a standard solution of the analyte, which is less concentrated than the sample, in the light path.

Table 21-6 Comparison of Methods for Absorption Measurements

		Imposed in Beam for Indicator Setting of:	
Method	Designation*	0% T	100% T
Ordinary	O	Shutter	Solvent
High absorbance	A	Shutter	Standard solution less concentrated than sample
Low absorbance	B	Standard solution more concentrated than sample	Solvent
Ultimate precision	C	Standard solution more concentrated than sample	Standard solution less concentrated than sample

* See Figures 21-7 and 21-8.

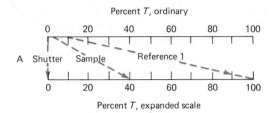

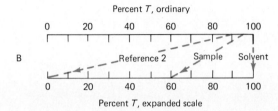

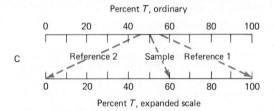

FIGURE 21-8 Scale expansion by various methods: (A) high-absorbance method; (B) low-absorbance method; and (C) method of ultimate precision.

Finally, the standard is replaced by the sample, and a relative transmittance is read directly. As shown in Figure 21-8A, the effect of this modification is to expand a small portion of the transmittance scale to a full 100%; thus, as shown by the upper scale of Figure 21-8A, the transmittance of the reference standard employed is 10% *when compared with pure solvent*, while the sample exhibits a 4% transmittance against the same solvent. That is, the sample transmittance is four-tenths that of the standard when compared with a common solvent. When the standard is substituted as the reference, however, its transmittance becomes 100% (lower scale of Figure 21-8A). The transmittance of the sample remains four-tenths that of the reference standard; here, then, it is 40%.

   If the instrumental uncertainty in the measurement of absorbance is not affected by the modification, use of the standard as a reference has the effect of bringing the transmittance of the sample into the middle of the scale, where instrumental uncertainties have a minimal effect on the relative concentration error.

   It can be shown that a linear relationship exists between concentration and the relative absorbance measured by the high-absorbance technique.[11] The method is particularly useful for the analysis of samples with absorbances greater than unity.

---

[11] See: C. N. Reilley and C. M. Crawford, *Anal. Chem.*, **27**, 716 (1955).

Low-absorbance method. In the low-absorbance procedure, a standard solution somewhat more concentrated than the sample is employed in lieu of the shutter for the 0% T adjustment of the indicator scale; the 100% T adjustment is made in the usual way, with the solvent in the light path. The sample transmittance is then obtained by replacing the solvent with the sample. The effect of this procedure is shown in Figure 21-8B; note that, again, a small segment of the scale is expanded to 100% transmittance and that the transmittance of the sample now lies near the center of the expanded scale.

The low-absorbance method is particularly applicable to samples having absorbances of less than 0.1. Here, a nonlinear relationship exists between relative absorbance and concentration. As a consequence, a number of standard solutions must be employed to establish accurately the calibration curve for the analysis.

The method of ultimate precision. The two techniques just described can be combined to give the method of ultimate precision. As shown in Figure 21-8C, a scale expansion is again achieved; here two reference solutions, one having a smaller and one having a greater transmittance than the sample, are employed to adjust the 100% and the 0% scale readings. A nonlinear relationship between relative absorbance and concentration is again observed.

Precision gain by differential methods. It is possible to derive relationships analogous to Equation 21-3 for the three differential methods. Figure 21-7 compares the corresponding error curves with that of the ordinary method. From curve A, it is apparent that a significant gain in precision results when the high-absorbance method elevates the transmittances of solutions to values that are greater than 10%. The plot was based upon the reference standard employed in Figure 21-8A, that is, for a reference having a transmittance of 10% against the solvent. The error curve approaches that for the ordinary procedure (see curve O, Figure 21-7) as the percent transmittance of the *reference* increases and approaches 100. Error curves for the low-absorbance method (curve B) and the method of ultimate precision (curve C) are based on the corresponding data in the caption accompanying Figure 21-7.

Instrumental requirements for precision methods. For the low-absorbance method, it is clearly necessary that the spectrophotometer possess a dark-current compensating circuit capable of offsetting larger currents than are normally produced when no radiation strikes the photoelectric detector. The high-absorbance method, on the other hand, requires an instrument with a sufficient reserve capacity to permit setting the indicator at 100% transmittance when an absorbing solution is placed in the light path. Here, the full-scale reading may be realized either by increasing the radiation intensity (most often by widening the slits) or by increasing the amplification of the photoelectric current. The method of ultimate precision requires instruments with both of these qualities.

The capability of a spectrophotometer to be set to full scale with an absorbing solution in the radiation path will depend both upon the quality of its monochromator and the stability of its electronic circuit. Furthermore, this capacity will be wavelength dependent, since the intensity of the source and the sensitivity of the detector both change with wavelength. In regions where the intensity and sensitivity are low, an increase in slit width may be necessary to realize a full-scale setting; under these circumstances, scattered radiation may lead to errors unless the quality of the monochromator is high. Alternatively, a very high current amplification may be required; again, unless the electronic stability is good, significant photometric error may result.

21E QUANTITATIVE APPLICATIONS OF INFRARED ABSORPTION

Quantitative infrared-absorption methods differ somewhat from those discussed in the previous section because of the greater complexity of the spectra, the narrowness of the absorption bands, and the instrumental limitations of infrared instruments.

21E-1 DEVIATIONS FROM BEER'S LAW

Instrumental deviations from Beer's law are more common in the infrared than in the ultraviolet and visible regions because infrared absorption bands are relatively narrow. Moreover, the low intensity of sources and low sensitivities of detectors in this region require the use of relatively wide monochromator slit widths; thus, the bandwidths employed are frequently of the same order of magnitude as the widths of absorption peaks. We have pointed out (see Figure 19-12, p. 496) that this combination of circumstances usually leads to a nonlinear relationship between absorbance and concentration. Calibration curves, having significant curvature are therefore often encountered in quantitative work.

21E-2 ABSORBANCE MEASUREMENT

Matched absorption cells for solvent and solution are ordinarily employed in the ultraviolet and visible regions. This technique is seldom practical for measurements in the infrared region because of the difficulty in obtaining cells whose transmission characteristics are identical. Most infrared cells have very short path lengths that are difficult to duplicate exactly. In addition, the cell windows are readily attacked by contaminants in the atmosphere and the solvent; thus, their transmission characteristics change continually with use. For these reasons, a reference absorber is often dispensed with entirely in infrared work, and the intensity of the radiation passing through the sample is simply compared with that of the unobstructed beam; alternatively, a salt plate may be placed in the reference beam. Either way, the resulting transmittance is ordinarily less than 100%, even in regions of the spectrum where the sample is not absorbing; this effect is readily seen by examining the spectrum in Figure 21-1.

For quantitative work, it is necessary to correct for the scattering and absorption by the solvent and the cell. Two methods are employed. In the so-called *cell in-cell out* procedure, spectra of the solvent and sample are obtained successively with respect to the unobstructed reference beam. The same cell is used for both measurements. The transmittance of each solution versus the reference beam is then determined at an absorption maximum of the analyte. These transmittances can be written as

$$T_0 = P_0/P_r$$

and

$$T_s = P/P_r$$

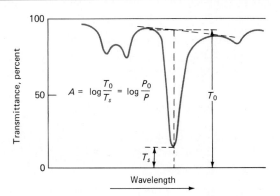

Figure 21-9 Base-line method for determination of absorbance.

where P_r is the power of the reference beam and T_0 and T_s are the transmittances of the solvent and sample, respectively, against this reference. If P_r remains constant during the two measurements, then the transmittance of the sample with respect to the solvent can be obtained by division of the two equations. That is,

$$T = T_s/T_0 = P/P_0$$

An alternative way of obtaining P_0 and T is the *base-line* method, in which the solvent transmittance is assumed to be constant or at least to change linearly between the shoulders of the absorption peak. This technique is demonstrated in Figure 21-9.

21E-3 APPLICATIONS OF QUANTITATIVE INFRARED SPECTROSCOPY

With the exception of homonuclear molecules, all organic and inorganic molecular species absorb in the infrared region; thus, infrared spectrophotometry offers the potential for determining an unusually large number of substances. Moreover, the uniqueness of an infrared spectrum provides a degree of specificity that is matched or exceeded by relatively few other analytical methods. This specificity has particular application to analysis of mixtures of closely related organic compounds. Two examples that typify these applications follow.

Analysis of a mixture of aromatic hydrocarbons. A typical application of quantitative infrared spectroscopy involves the resolution of C_8H_{10} isomers in a mixture which includes *o*-xylene, *m*-xylene, *p*-xylene, and ethylbenzene. The infrared absorption spectra of the individual components in the 12 to 15 μm range are shown in Figure 21-10; cyclohexane is the solvent. Useful absorption peaks for determination of the individual compounds occur at 13.47, 13.01, 12.58, and 14.36 μm, respectively. Unfortunately, however, the absorbance of a mixture at any one of these wavelengths is not entirely determined by the concentration of just one component, because of overlapping absorption bands. Thus, molar absorptivities for each of the four compounds must be determined at the four wavelengths. Then four simultaneous equations can be written, which permits the calculation of the concentration of each species from four absorbance measurements. Such calculations are most easily performed with a computer.

When the relationship between absorbance and concentration is nonlinear

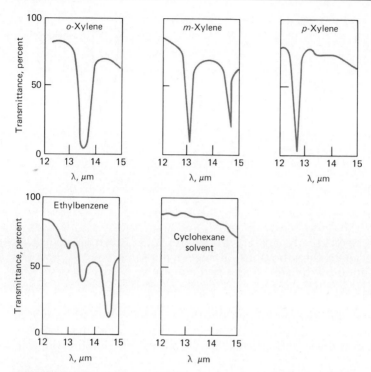

FIGURE 21-10   Spectra of C_8H_{10} isomers in cyclohexane. [R. P. Bauman, *Absorption Spectroscopy*, p. 406. New York: Wiley & Sons. With permission.]

(as frequently occurs in the infrared region), the algebraic manipulations associated with an analysis of several components having overlapping absorption peaks are considerably more complex.

Analysis of air contaminants.   The recent proliferation of government regulations with respect to atmospheric contaminants has demanded the development of sensitive, rapid, and highly specific methods for a variety of chemical compounds. Infrared absorption procedures appear to meet this need better than any other single analytical tool.

Table 21-7 demonstrates the potential of infrared spectroscopy for the analy-

Table 21-7   An Example of Infrared Analysis of Air Contaminants*

Contaminants	Concn (ppm)	Found (ppm)	Relative Error (%)
Carbon monoxide	50	49.1	1.8
Methylethyl ketone	100	98.3	1.7
Methyl alcohol	100	99.0	1.0
Ethylene oxide	50	49.9	0.2
Chloroform	100	99.5	0.5

* (Courtesy of Foxboro Analytical, Burlington, MA. 01803.)

Table 21-8 Some Examples of Infrared Vapor Analysis for OSHA Compliance*

Compound	Allowable Exposure (ppm)**	λ (μm)	Minimum Detectable Concentration (ppm)†
Carbon disulfide	20	4.54	0.5
Chloroprene	25	11.4	4
Diborane	0.1	3.9	0.05
Ethylenediamine	10	13.0	0.4
Hydrogen cyanide	10	3.04	0.4
Methyl mercaptan	10	3.38	0.4
Nitrobenzene	1	11.8	0.2
Pyridine	5	14.2	0.2
Sulfur dioxide	5	8.6	0.5
Vinyl chloride	1	10.9	0.3

* (Courtesy of Foxboro Analytical, Burlington, MA 01803.)
** 1977 OSHA exposure limits for 8-hour weighted average.
† For 20.25-m cell.

sis of gas mixtures. The standard sample of air containing five species in known concentration was analyzed with an automated, computerized infrared spectrophotometer equipped with a 20-m gas cell. Table 21-8 shows potential applications of the filter photometer shown in Figure 20-12 for the quantitative determination of various chemicals in the atmosphere for the purpose of assuring compliance with regulations established by the Occupational Safety and Health Administration (OSHA). Of the more than 400 chemicals for which maximum tolerable limits have been set by OSHA, more than half appear to have absorption characteristics suitable for determination by means of infrared filter photometers or spectrophotometers. Obviously, among all of these absorbing compounds, peak overlaps are to be expected; yet the method should provide a moderately high degree of selectivity.

21F PHOTOMETRIC TITRATIONS
Photometric or spectrophotometric measurements can be employed to advantage in locating the equivalence point of a titration.[12] The end point in a direct photometric titration is the result of a change in the concentration of a reactant or a product or both; clearly, at least one of these species must absorb radiation at the wavelength selected. In the indirect method, the absorbance of an indicator is observed as a function of titrant volume.

21F-1 TITRATION CURVES
A photometric titration curve is a plot of absorbance (corrected for volume change) as a function of the volume of titrant. If conditions are chosen properly, the curve will consist of two straight-line portions with differing slopes, one before and the other well beyond the equivalence-point region; the end point is taken as the intersection of extrapolated linear portions. Figure 21-11 shows some typical titration

---

[12] For further information concerning this technique, see: J. B. Headridge, *Photometric Titrations*. New York: Pergamon Press, 1961.

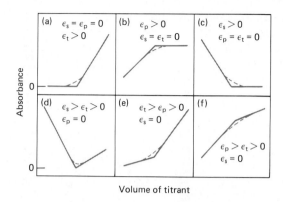

FIGURE 21-11 Typical photometric titration curves. Molar absorptivities of the substance titrated, the product, and the titrant are given by ϵ_s, ϵ_p, ϵ_t, respectively.

curves. Titration of a nonabsorbing species with an absorbing titrant that is decolorized by the reaction produces a horizontal line in the initial stages, followed by a rapid rise in absorbance beyond the equivalence point (Figure 21-11a). The formation of an absorbing product from colorless reactants, on the other hand, initially produces a linear rise in the absorbance, followed by a region in which the absorbance is independent of reagent volume (Figure 21-11b). Depending upon the absorption characteristics of the reactants and the products, the other curve forms shown in Figure 21-11 are also possible.

In order to obtain a satisfactory photometric end point, it is necessary that the absorbing system(s) obey Beer's law; otherwise, the titration curve will lack the linear portions needed for end-point extrapolation. Further, it is necessary to correct the absorbance for volume changes. The observed values are multiplied by $(V + v)/V$, where V is the original volume of the solution and v is the volume of added titrant.

21F-2 INSTRUMENTATION

Photometric titrations are ordinarily performed with a spectrophotometer or a photometer that has been modified to permit insertion of the titration vessel into the light path.[13] After the zero adjustment of the meter scale has been made, radiation is allowed to pass through the solution of the analyte, and the instrument is adjusted by varying the source intensity or the detector sensitivity to obtain a convenient absorbance reading. Ordinarily, no attempt is made to measure the true absorbance, since relative values are perfectly adequate for the purpose of end-point detection. Data for the titration are then collected without alteration of the instrument setting.

The power of the radiation source and the response of the detector must be reasonably constant during the period required for a photometric titration. Cylindrical containers are ordinarily used, and care must be taken to avoid any movement of the vessel that might alter the length of the radiation path.

Both filter photometers and spectrophotometers have been employed for photometric titrations. The latter are preferred, however, because their narrower bandwidths enhance the probability of adherence to Beer's law.

---

[13] Titration flasks and cells for use in a Spectronic 20 spectrophotometer are available from the Kontes Manufacturing Corp., Vineland, NJ, 08360.

21F-3 APPLICATION OF PHOTOMETRIC TITRATIONS

Photometric titrations often provide more accurate results than a direct photometric analysis because the data from several measurements are pooled in determining the end point. Furthermore, the presence of other absorbing species may not interfere, since only a change in absorbance is being measured.

The photometric end point possesses the advantage over many conventional end points in that the experimental data are taken well away from the equivalence-point region. Thus, the reactions need not have as favorable equilibrium constants as those required for a titration that depends upon observations near the equivalence point (for example, potentiometric or indicator end points). For the same reason, more dilute solutions may be titrated.

The photometric end point has been applied to all types of reactions.[14] Most of the reagents used in oxidation-reduction titrations have characteristic absorption spectra and thus produce photometrically detectable end points. Acid-base indicators have been employed for photometric neutralization titrations. The photometric end point has also been used to great advantage in titrations with EDTA and other complexing agents. Figure 21-12 illustrates the application of this end point to the successive titration of bismuth(III) and copper(II). At 745 nm, neither cation nor the reagent absorbs, nor does the more stable bismuth complex, which is formed in the first part of the titration; the copper complex, however, does absorb. Thus, the solution exhibits no absorbance until essentially all of the bismuth has been titrated. With the first formation of the copper complex, an increase in absorbance occurs. The increase continues until the copper equivalence point is reached. Further reagent additions cause no further absorbance change. Clearly, two well-defined end points result.

The photometric end point has also been adapted to precipitation titrations; here, the suspended solid product has the effect of diminishing the radiant power by scattering; titrations are carried to a condition of constant turbidity.

---

[14] See, for example, the review: A. L. Underwood, *Advances in Analytical Chemistry and Instrumentation*, C. N. Reilley, Ed., vol. 3, pp. 31–104. New York: Interscience, 1964.

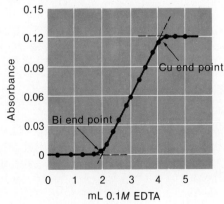

FIGURE 21-12 Photometric titration curve of 100 mL of a solution that was $2.0 \times 10^{-3} M$ in Bi^{3+} and Cu^{2+}. Wavelength: 745 nm. [A. L. Underwood, *Anal. Chem.*, **26**, 1322 (1954). With permission of the American Chemical Society.]

21G SPECTROPHOTOMETRIC STUDIES OF COMPLEX IONS

Spectrophotometry is a valuable tool for elucidating the composition of complex ions in solution and for determining their formation constants. The power of the technique lies in the fact that quantitative absorption measurements can be performed without disturbing the equilibria under consideration. Although most spectrophotometric studies of complexes involve systems in which a reactant or a product absorbs, this condition is not a necessity. It is also possible to cause one of the components to participate in a competing equilibrium that does produce an absorbing species. Thus, for example, complexes involving iron(II) and a nonabsorbing ligand might be studied by investigating the effect of this ligand on the color of the iron(II)/orthophenanthroline complex (p. 358). The formation constant and the composition of the nonabsorbing species can then be evaluated, provided the corresponding data are available for the phenanthroline complex.

The three most common techniques employed for complex-ion studies are (1) the method of continuous variations, (2) the mole-ratio method, and (3) the slope-ratio method. Each of these is examined briefly.

21G-1 THE METHOD OF CONTINUOUS VARIATIONS[15]

In the method of continuous variations, volumes of solutions with identical analytical concentrations of the cation and the ligand are mixed in such a way that the total volume of each mixture is the same (for example, 1:9, 8:2, 7:3, and so forth). The absorbance of each solution is then measured at a suitable wavelength and corrected for any absorbance the mixture might possess if no reaction had occurred. The corrected absorbance is plotted against the volume fraction (which is equal to the mole fraction) of one reactant; that is, $V_M/(V_M + V_L)$ where V_M is the volume of the cation solution and V_L that of the ligand. A typical plot is shown in Figure 21-13.

[15] See: W. C. Vosburgh and G. R. Cooper, *J. Amer. Chem. Soc.*, **63**, 437 (1941).

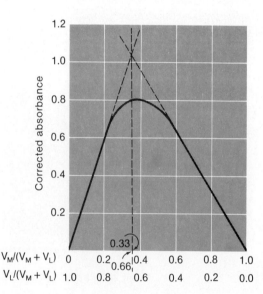

FIGURE 21-13 Continuous variation plot for the 1:2 complex, ML_2.

A maximum (or a minimum if the complex absorbs less than the reactants) occurs at a volume ratio, V_M/V_L, corresponding to the combining ratio of cation and ligand in the complex. In Figure 21-13 $V_M/(V_M + V_L)$ is 0.33 and $V_L/(V_M + V_L)$ is 0.66; thus, V_M/V_L is 0.33/0.66, which suggests that the complex has the formula ML_2.

The curvature of the experimental lines in Figure 21-13 is the result of incompleteness of the complex-formation reaction. A formation constant for the complex can be evaluated from measurement of the deviations from the theoretical straight lines.

To determine whether more than one complex forms between the reactants, the experiment is ordinarily repeated with different reactant concentrations and at several wavelengths.

21G-2 THE MOLE-RATIO METHOD

In this method a series of solutions is prepared in which the analytical concentration of one reactant (usually the cation) is held constant while that of the other is varied. A plot of the absorbance versus the mole ratio of the reactants is then prepared. If the formation constant is reasonably favorable, two straight lines of different slope are obtained; the intersection occurs at a mole ratio that corresponds to the combining ratio in the complex. Typical mole-ratio plots are shown in Figure 21-14. Note that the ligand of the 1:2 complex absorbs at the wavelength selected; as a result, the slope beyond the equivalence point is greater than zero. The uncomplexed cation involved in the 1:1 complex absorbs, since the initial point has an absorbance greater than zero.

The formation constant can be evaluated from the data in the curved portion of the mole-ratio plots.

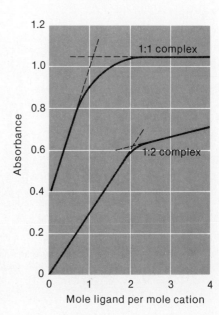

FIGURE 21-14 Mole-ratio plots for a 1:1 and a 1:2 complex. The 1:2 complex is the more stable, as indicated by less curvature near stoichiometric ratio.

EXAMPLE 21-3. Derive sufficient equations to permit the calculation of the equilibrium concentrations for all of the species involved in the $1:2$ complex formation reaction illustrated in Figure 21-14.

Two mass-balance expressions can be written that are based upon the preparatory data. Thus, for the reaction

$$M + 2L \rightleftharpoons ML_2$$

we may write

$$M_M = [M] + [ML_2]$$
$$M_L = [L] + 2[ML_2]$$

where M_M and M_L are the molar concentrations of M and L before reaction occurred for 1-cm cells, the absorbance of the solution is given by

$$A = \epsilon_M[M] + \epsilon_L[L] + \epsilon_{ML_2}[ML_2]$$

From the mole-ratio plot we see that $\epsilon_M = 0$. Values for ϵ_L and ϵ_{ML_2} can be obtained from the two straight-line portions of the curve. With one or more measurements of A in the curved portion of the plot, sufficient data are available for the calculation of the three equilibrium concentrations and, thus, the formation constant.

A mole-ratio plot may reveal the stepwise formation of two or more complexes as successive slope changes, provided the complexes have different molar absorptivities and provided that the formation constants are sufficiently different.

21G-3 THE SLOPE-RATIO METHOD
This procedure is particularly useful for weak complexes; it is applicable only to systems in which a single complex is formed. The method assumes that the complex-formation reaction can be forced to completion in the presence of a large excess of either reactant and that Beer's law is followed under these circumstances. For the reaction

$$mM + lL \rightleftharpoons M_mL_l$$

the following equation can be written when L is present in very large excess:

$$[M_mL_l] \cong M_M/m$$

where M_M is the molarity of M before complex formation took place. If Beer's law is obeyed

$$A_m = \epsilon b[M_mL_l] = \epsilon b M_M/m$$

and a plot of A with respect to M_M will be linear. When M is very large with respect to L, a similar relationship exists. That is,

$$A_l = \epsilon b[M_mL_l] = \epsilon b M_L/l$$

The slopes of the straight lines (A/M_m and A/M_l) are obtained under these conditions; their ratio is the desired combining ratio. That is,

$$\frac{A_m/M_m}{A_l/M_l} = \frac{\epsilon b/m}{\epsilon b/l} = l/m$$

21H AUTOMATIC PHOTOMETRIC AND SPECTROPHOTOMETRIC ANALYSIS

One of the major instrumental developments during the past two decades has been the appearance, from commerical sources, of automatic analysis systems which provide analytical data with a minimum of operator intervention. The need for these systems has been greatest in clinical laboratories, where perhaps 30 or more different analyses are routinely used in large number for diagnostic and screening purposes. The number of such analyses required by modern medicine is enormous; the need to keep their costs at a reasonable level is obvious. These two considerations have led to a focus of effort toward using automatic instruments for clinical laboratories.[16] As an outgrowth, automatic instruments are now beginning to find application in such diverse fields as analyses for the control of industrial processes and the routine analyses of air, water, soil, and agricultural products.

Most clinical analyses are based upon photometric or spectrophotometric measurements in the ultraviolet or visible regions.[17] Thus, an important component of most automatic instruments is a photometer or spectrophotometer. It should be pointed out, however, that some automatic systems make use of atomic spectroscopy and electroanalytical methods to measure analyte concentrations.

An example of an automatic system is discussed briefly in the paragraphs that follow.

21H-1 CONTINUOUS-FLOW ANALYZER

A continuous analyzer is designed so that successive samples pass through the same system of tubes and chambers. Thus, samples must be isolated from one another to avoid cross contamination, and means must be provided for rinsing the system between samples. In the Technicon AutoAnalyzer® (the earliest of the commercial instruments), movement of sample, reagents, and diluent through plastic tubing is accomplished by means of a peristaltic pump. Successive samples are isolated from one another by introducing bubbles of air into the tubing.

Figure 21-15 is a schematic diagram of a single-channel autoanalyzer used for the analysis of a constituent of blood. Here, samples are removed automatically and successively from containers held in the rotating table sampler and are mixed with diluent and air bubbles. The latter promote mixing and serve to separate the sample from earlier and later ones. The diluted sample then passes into the dialyzer, which contains membranes through which the small analyte ions or molecules are free to diffuse into the reagent stream. The residual large-protein molecules of the blood remain in the diluent stream and pass from the system to waste. The remainder of the system shown in Figure 21-15 is self-explanatory.

A system for routine blood analysis includes a module for diluting and partitioning the sample into several aliquots, each of which passes through a separate channel similar to the one shown in Figure 21-15. The final output from a typical

---

[16] For a description of some commerically available instruments, see: R. H. Laessig, *Anal. Chem.*, **43** (8), 18A (1971); and J. K. Foreman and P. B. Stockwell, *Automatic Chemical Analysis*, Chapter 4. New York: Wiley, 1975.

[17] For example, a breakdown on the basis of instruments used for analyses in a typical large chemical laboratory in 1971 was as follows: filter photometers, 42%; spectrophotometers, 27%, flame photometers, 18%; atomic absorption spectrometers, 3%; fluorometers, 1%; electroanalytical instruments, 4%; and other instruments, 5%.

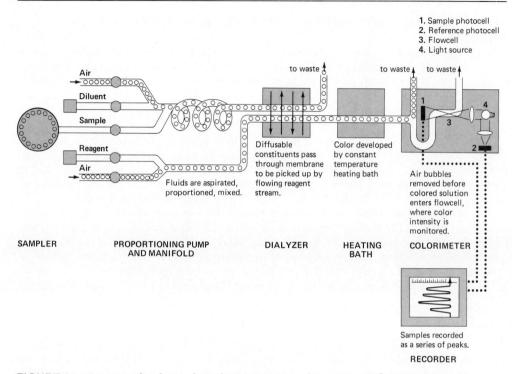

FIGURE 21-15   A single-channel Technicon AutoAnalyzer System®. (Reproduced by permission of Technicon Instruments Corporation, Tarrytown, New York. Technicon and Auto-Analyzer are trademarks of Technicon Instruments Corporation.)

system is shown in Figure 21-16. The shaded areas show the range of concentrations which are considered to be normal for the population.

21I MOLECULAR FLUORESCENCE ANALYSIS
Measurement of fluorescent intensity permits the quantitative determination of many inorganic and organic species in trace amounts.[18] One of the most attractive features of fluorometry is its inherent sensitivity, which is generally greater than absorptiometric methods and often lies in the parts per billion range. In addition, selectivity is usually as good and often better than other spectroscopic methods. Fluorometry, however, is less widely applicable than absorption methods because of the relatively limited number of systems that can be made to fluoresce.

21I-1 FLUORESCENT SPECIES
As was pointed out in Section 19E-1, one mechanism by which excited molecules can relax to lower energy states is by fluorescence. Several other competing relax-

---

[18] For detailed discussions of fluorescence methods, see: J. D. Winefordner, S. G. Schulman, and T. C. O'Haver, *Luminescence Spectrometry in Analytical Chemistry.* New York: Wiley-Interscience, 1972; G. G. Guilbault, *Practical Fluorescence: Theory, Methods and Technique.* New York: Marcel Dekker, 1973; S. G. Schulman, *Fluorescence and Phosphorescence Spectroscopy.* New York: Pergamon Press, 1977; and *Modern Fluorescence Spectroscopy,* E. L. Wehry, Ed. New York: Plenum Press, 1966.

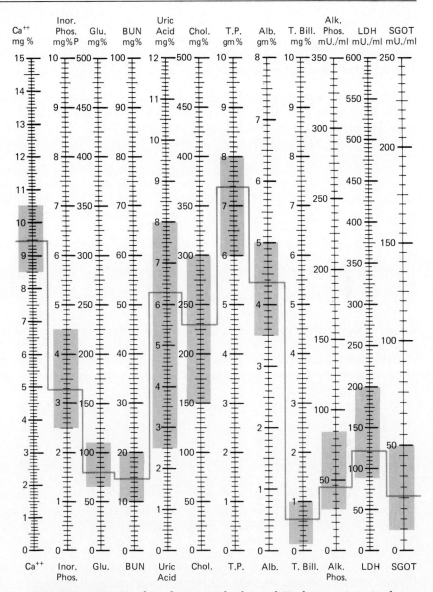

FIGURE 21-16 Readout from a multichannel Technicon AutoAnalyzer System®. (Reproduced by permission of Technicon Instruments Corporation, Tarrytown, New York. Technicon and AutoAnalyzer are trademarks of Technicon Instruments Corporation.)

ation mechanisms exist, however, whose rates are often greater than the rate of fluorescence. As a consequence, although all absorbing species are potentially capable of fluorescent behavior, the phenomenon is observed only in those compounds for which the other relaxation processes are slow.

The *quantum yield*, or quantum efficiency, for a fluorescent process is simply the ratio of molecules that fluoresce to the total number of excited molecules. Highly fluorescent molecules, such as fluorescein, have quantum efficiencies that may approach unity under some conditions. Nonfluorescent species have efficiencies that are essentially zero.

Fluorescence and structure.   The most intense and most useful fluorescent behavior is found in compounds containing aromatic rings. Aliphatic and alicyclic carbonyl compounds as well as highly conjugated double-bond structures may also fluoresce; their numbers are small, however, in comparison with those in aromatic systems.

Most unsubstituted aromatic hydrocarbons fluoresce in solution, with the quantum efficiency usually increasing with the number or rings and their degree of condensation. The simplest heterocyclics, such as pyridine, furan, thiophene, and pyrrole, do not exhibit fluorescent behavior; on the other hand, fused-ring structures containing these rings often do.

Substitution on an aromatic ring causes shifts in the wavelength of absorption maxima and corresponding changes in the fluorescence peaks. In addition, substitution frequently affects the fluorescent efficiency. For example, the relative fluorescent intensity for aniline, phenol, benzene, and bromobenzene is 20, 18, 10, and 5, respectively.

Effect of structural rigidity.   It is found experimentally that fluorescence is particularly favored in molecules that possess rigid structures. For example, the quantum efficiencies for fluorene and biphenyl are nearly 1.0 and 0.2, respectively, under similar conditions of measurement. The difference in behavior appears to be

fluorene   biphenyl

largely a result of the increased rigidity furnished by the bridging methylene group in fluorene. Many similar examples can be cited. In addition, enhanced emission frequently results when fluorescing dyes are adsorbed on a solid surface; here again, the added rigidity provided by the solid surface may account for the observed effect.

The influence of rigidity has also been invoked to account for the increase in fluorescence of certain organic chelating agents when they are complexed with a metal ion. For example, the fluorescent intensity of 8-hydroxyquinoline is much less than that of the zinc complex:

21I-2 TEMPERATURE AND SOLVENT EFFECTS

The quantum efficiency of fluorescence by most molecules decreases with increasing temperature because the increased frequency of collisions at elevated temperatures improves the probability for alternative deactivation. A decrease in solvent viscosity leads to the same result.

21I-3 EFFECT OF CONCENTRATION ON FLUORESCENT INTENSITY

The power of fluorescent radiation F is proportional to the radiant power of the excitation beam that is absorbed by the system. That is,

$$F = K'(P_0 - P) \tag{21-6}$$

where P_0 is the power of the beam incident upon the solution and P is its power after traversing a length b of the medium. The constant K' depends upon the quantum efficiency of the fluorescence process. In order to relate F to the concentration c of the fluorescing particle, we write Beer's law in the form

$$\frac{P}{P_0} = 10^{-\epsilon bc} \tag{21-7}$$

where ϵ is the molar absorptivity of the fluorescent species and ϵbc is the absorbance A. By substitution of Equation 21-7 into Equation 21-6, we obtain

$$F = K'P_0(1 - 10^{-\epsilon bc}) \tag{21-8}$$

Expansion of the exponential term in Equation 21-8 leads to

$$F = K'P_0 \times \left[2.3\epsilon bc - \frac{(-2.3\epsilon bc)^2}{2!} - \frac{(-2.3\epsilon bc)^3}{3!} - \cdots \right] \tag{21-9}$$

Provided $\epsilon bc = A < 0.05$, all of the subsequent terms in the brackets become small with respect to the first; under these conditions, we may write

$$F = 2.3K'\epsilon bcP_0 \tag{21-10}$$

or at constant P_0,

$$F = Kc \tag{21-11}$$

Thus, a plot of the fluorescent power of a solution versus concentration of the emitting species should be linear at low concentrations, c. When c becomes great enough so that the absorbance is larger than about 0.05 or the transmittance is smaller than about 90%, linearity is lost and F lies below an extrapolation of the straight-line plot.

Two other factors responsible for further negative departures from linearity at high concentration are *self-quenching* and *self-absorption*. The former is the result of collisions between excited molecules. Radiationless transfer of energy occurs. Self-quenching can be expected to increase with concentration.

Self-absorption occurs when the wavelength of emission overlaps an absorption peak; fluorescence is then decreased as the radiation beam traverses the solution.

The effects of these phenomena are such that a plot relating fluorescent power to concentration often exhibits a maximum.

21J APPLICATIONS OF FLUOROMETRY

Fluorometric methods are inherently applicable to lower concentration ranges than are spectrophotometric determinations and are thus among the most sensitive analytical techniques available to the scientist. The basic difference in the sensitivity between the two methods arises from the fact that the concentration-related parameter for fluorometry F can be measured independently of the power of the source P_0. In contrast, a spectrophotometric measurement requires evaluation of both P_0 and P because the concentration-dependent parameter A depends upon the *ratio* between these two quantities. The sensitivity of a fluorometric method can be improved by increasing P_0 or by further amplifying the fluorescent signal. In spectrophotometry, an increase in P_0 results in a proportionate change in P and therefore fails to affect A; thus, no improvement in sensitivity results. Similarly, amplification of the detector signal has the same effect on both P and P_0 and results in no net gain with respect to A. As a consequence, fluorometric methods generally have sensitivities that are two to four orders of magnitude better than the corresponding spectrophotometric procedures.

21J-1 INORGANIC ANALYSIS

Inorganic fluorometric methods are of two types. Direct methods involve the formation of a fluorescent chelate and the measurement of its emission. A second group is based upon the diminution of fluorescence in a reagent resulting from the quenching action of the analyte. The latter technique has been most widely used for anion analysis.

Cations that form fluorescent chelates. Two factors greatly limit the number of transition-metal ions that form fluorescent chelates. First, many of these ions are paramagnetic; second, transition-metal complexes are characterized by many

Table 21-9   Selected Fluorometric Methods for Inorganic Species*

| | | Wavelength (nm) | | | |
| | | Absorption | Fluorescence | Sensitivity ($\mu g/ml$) | Interference |
Ion	Reagent				
Al^{3+}	Alizarin garnet R	470	500	0.007	Be, Co, Cr, Cu, F^-, NO_3^-, Ni, PO_4^{3-}, Th, Zr
F^-	Al complex of Alizarin garnet R (quenching)	470	500	0.001	Be, Co, Cr, Cu, Fe, Ni, PO_4^{3-}, Th, Zr
$B_4O_7^{2-}$	Benzoin	370	450	0.04	Be, Sb
Cd^{2+}	2-(o-Hydroxyphenyl)-benzoxazole	365	Blue	2	NH_3
Li^+	8-Hydroxyquinoline	370	580	0.2	Mg
Sn^{4+}	Flavanol	400	470	0.1	F^-, PO_4^{3-}, Zr
Zn^{2+}	Benzoin	—	Green	10	B, Be, Sb, Colored ions

* A. Weissler and C. E. White, *Handbook of Analytical Chemistry*, L. Meites, Ed., pp. 6-178 to 6-181. New York: McGraw-Hill Book Company, Inc., 1963. With permission.

closely spaced energy levels. Both of these characteristics tend to enhance the rate of relaxation by nonradiative processes. Nontransition-metal ions are less susceptible to these deactivation processes; it is for such elements that the principal applications of fluorometry are to be found. It is noteworthy that few nontransition-metal cations absorb in the visible region and tend to form chelates which have characteristic absorption spectra. Thus, fluorometry often complements spectrophotometry.

Fluorometric reagents. The most successful fluorometric reagents for cation analyses have aromatic structures with two or more donor functional groups that permit chelate formation with the metal ion. The structures of four common reagents follow:

Flavanol
(Reagent for Zr and Sn)

8-Hydroxyquinoline
(Reagent for Al, Be, and other metal ions)

Benzoin
(Reagent for B, Zn, Ge, and Si)

Alizarin garnet R
(Reagent for Al, F⁻)

Selected fluorometric reagents and their applications are presented in Table 21-9. For a more complete summary see Meites,[19] St. John,[20] or the review articles in *Analytical Chemistry*.[21]

21J-2 ORGANIC SPECIES

The number of applications of fluorometric analysis to organic problems is impressive. Weissler and White have summarized the most important of these in several tables.[22] Over 100 entries are found under a heading of *Organic and General Biochemical Substances*, including such diverse compounds as adenine, anthranilic acid, aromatic polycyclic hydrocarbons, cysteine, guanidine, indole, naphthols, certain nerve gases, proteins, salicylic acid, skatole, tryptophan, uric acid, and warfarin. Some 50 medicinal agents that can be determined fluorometrically are listed. Included among these are adrenaline, alkylmorphine, chloroquin, digitalis princieples, lysergic acid diethylamide (LSD), penicillin, phenobarbital, procaine, and re-

[19] L. Meites, *Handbook of Analytical Chemistry*, pp. 6-178 to 6-181.New York: McGraw-Hill, 1963.

[20] P. A. St. John, in *Trace Analysis*, J. D. Winefordner, Ed., pp. 263–271. New York: Wiley, 1976.

[21] E. L. Wehry, *Anal. Chem.*, **52**, 75R (1980); C. M. O'Donnell and T. N. Solle, *Anal. Chem.*, **50**, 189R (1978); **48**, 175R (1976).

[22] A. Weissler and C. E. White, in *Handbook of Analytical Chemistry*, L. Meites, Ed., pp. **6**-182 to **6**-196. New York: McGraw-Hill, 1963.

serpine. Methods for the analysis of ten steroids and an equal number of enzymes and coenzymes are also listed in these tables. Some of the plant products listed include chlorophyll, ergot alkaloids, rauwolfia serpentina alkaloids, flavonoids, and rotenone. Eighteen listings for vitamins and vitamin products are also included; among these are ascorbic acid, folic acid, nicotinamide, pyridoxal, riboflavin, thiamin, vitamin A, and vitamin B_{12}.

Without question, the most important applications of fluorometry are in the analyses of food products, pharmaceuticals, clinical samples, and natural products. The sensitivity and selectivity of the method make it a particularly valuable tool in these fields.

PROBLEMS*

21-1. Briefly account for the following observations:
* (a) the filter selected for a photometric analysis is ordinarily the complement of the color of the absorbing species.
  (b) the narrowest practical slit width is used in the generation of an absorption spectrum.
* (c) the slope-ratio method is applicable to systems in which the reactants combine to form just one absorbing species.
  (d) fluorescence measurements are potentially capable of greater sensitivity than absorption measurements.
* (e) a photometric titration of picric acid with sodium hydroxide based upon absorption by picrate ion is not practical.

21-2. Combination of the cation M^{2+} with the ligand L^- results in formation of an adduct that has an absorption maximum at 605 nm. Neither the ligand nor the cation absorbs at this wavelength. Given the further information that L^- is the conjugate base of the weak acid HL ($K_a = 10^{-7}$) and the K_{sp} for $M(OH)_2$ is about 10^{-18}, make a *qualitative* prediction regarding the effect of pH upon development of color in the M^{2+}/L^- system.

21-3. Estimate the color transmitted by (see Figure 21-17)
* (a) filter A.
  (b) filter D.
* (c) filter G.

21-4. Estimate the color transmitted through a combination of filters shown in Figure 21-17:
* (a) A and B.
  (b) C and D.
* (c) C and E.
  (d) B and C.
* (e) F and G.
  (f) D and E.

* Answers to problems and parts of problems marked with an asterisk are to be found at the end of the book.

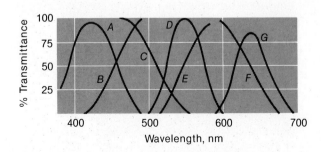

FIGURE 21-17

21-5. From Figure 21-17, select a filter (or combination of filters) suitable for the photometric analysis of an analyte that has an absorption maximum at
* (a) 550 nm.   (d) 650 nm.
   (b) 415 nm.   * (e) 500 nm.
* (c) 450 nm.   (f) 620 nm.

21-6. From Figure 21-7, suggest a filter (or combination of filters) that would be suitable
for a photometric analysis based upon
* (a) the red species formed by the addition of excess thiocyanate to Fe(III).
   (b) the deep blue tetraammine complex of copper(II).
* (c) the yellow complex formed between H_2O_2 and Ti(IV).

21-7. From Figure 21-17, estimate the effective bandwidth of
* (a) filter A.
   (b) filter D.
* (c) filter G.

* 21-8. The iron content of a sample of well water was determined by treating a 25.0-mL
aliquot with nitric acid and an excess of thiocyanate and then diluting to 50.0 mL. A
10.0-mL portion of 5.97×10^{-4} M Fe^{3+} was treated in an identical fashion. Color
match between the two solutions was observed when the light path through the
standard was 3.04 cm and that through the sample was 2.61 cm. Calculate the parts
per million of iron in the sample.

21-9. A 1.374-g alloy sample was dissolved in acid, treated with an excess of potassium
periodate, and heated to oxidize any manganese present to the +7 state. Reaction:

$$5IO_4^- + 2Mn^{2+} + 3H_2O \longrightarrow 2MnO_4^- + 5IO_3^- + 6H^+$$

The resulting solution was diluted to 500 mL in a volumetric flask. Color match was
achieved between the diluted solution and a 7.61×10^{-4} M $KMnO_4$ solution when
the light path through the standard was 5.84 cm and that through the sample solution was 3.67 cm. Calculate the percentage of manganese in the alloy.

* 21-10. The molar absorptivity for acetone in ethanol is 2.75 L cm^{-1} mol^{-1} at 366 nm. Calculate the range of acetone concentrations that can be used if the percent transmittance is to be greater than 10% and less than 90% with a 1.00-cm cell.

21-11. The logarithm for the molar absorptivity of phenol in aqueous solution is 3.79 at
211 nm. Calculate the range of phenol concentrations that can be used if the absorbance is to be greater than 0.100 and less than 1.500 with a 1.00-cm cell.

21-12. The molar absorptivity of the bismuth(III) thiourea complex is 9.3×10^3 L cm^{-1}
mol^{-1} at 470 nm. Calculate the range of Bi(III) concentrations that correspond to
absorbances of 0.15 and 0.80 (with a 1.00-cm cell) if the measurements are to be
made upon solutions with a total volume of 50.0 mL, of which the sample taken for
analysis is
(a) 5.00 mL.
(b) 15.00 mL.
(c) 25.00 mL.

* 21-13. A standard solution was put through appropriate dilutions to give the indicated concentrations of iron. The iron(II)-1,10-phenanthroline complex was then developed
in 25.0-mL aliquots of these solutions, following which each was diluted to
50.0 mL. The following absorbances were recorded at 510 nm:

Concentration of Fe(II) in the original solutions, ppm	Absorbance, A (1.00-cm cells)	Concentration of Fe(II) in the original solutions, ppm	Absorbance, A (1.00-cm cells)
2.00	0.080	12.0	0.475
5.00	0.195	16.0	0.630
8.00	0.315	20.0	0.790

(a) Produce a calibration curve from these data.
(b) By the method of least squares (p. 69), derive an equation relating absorbance and concentration of iron(II).
(c) Calculate the standard deviation of the residuals.
(d) Calculate the standard deviation of the slope b.

21-14. The method developed in Problem 21-13 was used for the routine determination of iron in 25.0-mL aliquots of ground water. Express the concentration (as ppm Fe) in samples that yielded the accompanying absorbance data (1.00-cm cell). Estimate standard deviations for the derived concentrations. Repeat the calculations assuming the absorbance data are means of three measurements.

* (a) 0.143 (d) 0.384
  (b) 0.675 * (e) 0.722
* (c) 0.068 (f) 0.546

21-15. Calculate the ppm of chromium in the solutions that result upon diluting volumes of 5.00×10^{-4} M Cr(III) as indicated:

	Volume of 5.00×10^{-4} M Cr(III) solution taken, mL	Final volume, mL
* (a)	10.00	500.0
(b)	10.00	250.0
* (c)	20.00	250.0
(d)	12.00	100.0
* (e)	16.00	100.0
(f)	10.00	50.00

* 21-16. The solutions in Problem 21-15 were used to prepare a calibration curve, with 10.00-mL aliquots being mixed with 1,5-diphenylcarbazide solution and subsequently diluted to 50.0 mL.
(a) Plot a calibration curve from the data, given below.
(b) By the method of least squares (p. 69), derive an equation relating absorbance to concentration of chromium(III).
(c) Calculate the standard deviation of the residuals.
(d) Calculate the standard deviation of the slope b.

Solution	Absorbance, A (1.00-cm cells)	Solution	Absorbance, A (1.00-cm cells)
(1)	0.083	(4)	0.496
(2)	0.163	(5)	0.664
(3)	0.335	(6)	0.832

21-17. The analysis developed in Problem 21-16 was used to determine the chromium in 10.00-mL aliquots of treated waste water from an electroplating plant. Calculate the ppm of Cr, based upon the accompanying data: Estimate standard deviations for

Sample	Absorbance, A (1.00-cm cells)	Sample	Absorbance, A (1.00-cm cells)
* (a)	0.214	(d)	0.529
(b)	0.375	* (e)	0.110
* (c)	0.476	(f)	0.700

the concentrations, assuming the absorbance data are for a single measurement and for the mean of three.

21-18. A 25.0-mL aliquot of an aqueous quinine solution was diluted to 50.0 mL and found to have an absorbance of 0.528 at 348 nm when measured in a 1.00-cm cell. A second 25.0-mL aliquot was mixed with 10.00 mL of a solution containing 23.4 ppm of quinine; after dilution to 50.0 mL, this solution had an absorbance of 0.610 (1.00-cm cell). Calculate the ppm of quinine (fw = 324) in the sample.

* 21-19. A 7.94-g pesticide sample was decomposed by wet ashing and then diluted to 250.0 mL in a volumetric flask. The analysis was completed by treating aliquots of this solution as indicated.

Volume of sample taken, mL	Reagent volumes used, mL			Absorbance, A, 545 nm (1.00-cm cells)
	3.00 ppm Cu^{2+}	Ligand	H_2O	
50.0	0.00	20.0	30.0	0.364
50.0	4.00	20.0	26.0	0.688

Calculate the percentage of copper in the sample.

21-20. A standard addition method for the routine analysis of copper is based upon dilution of the sample, after suitable pretreatment, to 500.0 mL. The copper(I) complex with 1,10-phenanthroline is developed in aliquots with and without addition of known volumes of 5.00×10^{-3} M Cu^{2+}. The absorbance is measured at 508 nm in 1.00-cm cells. Calculate the percentage of copper in the following samples.

Sample	Wt of sample, g (in 500 mL)	Volumes taken, mL				
		Sample	0.00500 M Cu^{2+}	Ligand	H_2O	A
* (a)	3.858	25.00	0.00	15.00	10.00	0.374
		25.00	5.00	15.00	5.00	0.555
(b)	3.969	20.00	0.00	15.00	15.00	0.339
		20.00	4.00	15.00	11.00	0.484
* (c)	4.007	25.00	0.00	15.00	10.00	0.176
		25.00	8.00	15.00	2.00	0.466
(d)	3.564	20.00	0.00	15.00	15.00	0.191
		20.00	6.00	15.00	9.00	0.409

21-21. A. J. Mukhedkar and N. V. Deshpande [*Anal. Chem.*, **35**, 47 (1963)] report on a simultaneous determination for cobalt and nickel based upon absorption by their respective 8-hydroxyquinolinol complexes. Molar absorptivities corresponding to their absorption maxima are

	Molar absorptivity, ϵ, at	
	365 nm	700 nm
Co	3529	428.9
Ni	3228	0.00

Calculate the ppm of nickel and cobalt in each of the following solutions, based upon the accompanying data:

Solution	Absorbance, A, 1.00-cm cells	
	365 nm	700 nm
* (a)	0.537	0.044
(b)	0.684	0.050
* (c)	0.802	0.027
(d)	0.873	0.081
* (e)	0.729	0.062

21-22. The equilibrium constant for the reaction

$$2CrO_4^{2-} + 2H^+ \rightleftharpoons Cr_2O_7^{2-} + H_2O$$

has a value of 4.2×10^{14}. Molar absorptivities for Cr(VI) species in a $K_2Cr_2O_7$ solution are

Wavelength, nm	ϵCrO_4^{2-}	$\epsilon Cr_2O_7^{2-}$
345	1.84×10^3	1.07×10^1
370	4.81×10^3	7.27×10^2
400	1.88×10^3	1.89×10^2

Calculate the absorbance (1.00-cm cell) at 345, 370, and 400 nm for a solution that is buffered to a pH of 5.40 in which the analytical concentration of $K_2Cr_2O_7$ is
* (a) 4.00×10^{-4}.   * (c) 2.00×10^{-4}.
   (b) 3.00×10^{-4}.   (d) 1.00×10^{-4}.

21-23. Absorptivity data for cobalt and nickel complexes with 2,3-quinoxalinedithiol at their respective absorption peaks are

	Wavelength	
	510 nm	656 nm
ϵ_{Co}	36,400	1240
ϵ_{Ni}	5250	17,500

A 0.524-g soil sample was dissolved and subsequently diluted to 50.0 mL. A 25.0-mL aliquot was treated to eliminate interferences, after which 2,3-quinoxalinedithiol was added along with sufficient water to give a total volume of 50.0 mL. This solution had an absorbance of 0.467 at 510 nm and 0.347 at 656 nm when measured in a 1.00-cm cell. Calculate the respective percentages of cobalt and nickel in the soil sample.

21-24. Solutions of substances A and B individually exhibit adherence to Beer's law over wide concentration ranges. From the spectrophotometric data that follow, generate an absorption spectrum (1.00-cm cell) for a solution in which the concentrations of A and B are 8.50×10^{-5} M and 4.65×10^{-5} M, respectively.

λ, nm	Absorbance, A (1.00-cm cells)		λ, nm	Absorbance, A (1.00-cm cells)	
	8.50×10^{-5} M A	4.65×10^{-5} M B		8.50×10^{-5} M A	4.65×10^{-5} M B
400	0.200	0.270	575	0.354	0.336
425	0.148	0.423	600	0.470	0.269
450	0.131	0.542	625	0.591	0.208
475	0.129	0.567	650	0.703	0.163
500	0.147	0.536	675	0.752	0.120
525	0.193	0.475	700	0.764	0.083
550	0.262	0.403	725	0.755	0.054

21-25. Use the data in Problem 21-24 to calculate the absorbance (1.00-cm cell) at 475 nm and at 700 nm for solutions with the following concentrations of A and B:

	Concentration, mol/L	
Solution	A	B
* (a)	6.85×10^{-5}	5.83×10^{-5}
(b)	2.94×10^{-5}	7.56×10^{-5}
* (c)	6.39×10^{-5}	6.24×10^{-5}
(d)	4.78×10^{-5}	3.85×10^{-5}
* (e)	8.17×10^{-5}	4.25×10^{-5}

21-26. Use the data in Problem 21-24 to calculate the concentrations of A and B in solutions that yielded the accompanying absorbance data.

	Absorbance, A (1.00-cm cell)	
Solution	475 nm	700 nm
* (a)	0.466	0.918
(b)	0.882	0.468
* (c)	0.710	0.729
(d)	0.726	0.576
* (e)	0.470	0.552

* 21-27. The acid dissociation constant for the indicator HIn has a value of 5.40×10^{-7} at ordinary temperatures. Absorbance data (1.00-cm cells) for 5.00×10^{-4} M solutions of the indicator in strongly acidic and strongly alkaline media appear on page 568.
(a) Predict the color of the acid form of the indicator.
(b) What color filter would be suitable for the photometric analysis of the indicator in a strongly acidic medium?
(c) What wavelength would be suitable for the spectrophotometric analysis of the indicator in its alkaline form?
(d) What would be the absorbance of a 1.00×10^{-4} M solution of the indicator in its alkaline form when measured at 590 nm in a 2.50-cm cell?
(e) At what wavelength is the absorbance of the indicator independent of pH?

	Absorbance, A			Absorbance, A	
λ, nm	pH = 1.00	pH = 13.00	λ, nm	pH = 1.00	pH = 13.00
440	0.401	0.067	570	0.303	0.515
470	0.447	0.050	585	0.263	0.648
480	0.453	0.050	600	0.226	0.764
485	0.454	0.052	615	0.195	0.816
490	0.452	0.054	625	0.176	0.823
505	0.443	0.073	635	0.160	0.816
535	0.390	0.170	650	0.137	0.763
555	0.342	0.342	680	0.097	0.588

* 21-28. A solution that is 5.00×10^{-4} M with respect to the indicator in Problem 21-27 has an absorbance of 0.309 at 485 nm and 0.410 at 625 nm when measured in a 1.00-cm cell.
   (a) What is the pH of the solution?
   (b) What will be the absorbance of this solution at 555 nm?

21-29. Calculate the absorbance of a solution in which the analytical concentrations of Na_2HPO_4 and NaH_2PO_4 are 4.00×10^{-3} M and 2.50×10^{-2} M, respectively, and that for the indicator in Problem 21-27 is 5.00×10^{-4}, at 440 nm and 680 nm (1.00-cm cell).

21-30. A 25.00-mL aliquot of a solution containing a purified weak acid of unknown composition required 38.50 mL of 0.1180 N NaOH for neutralization. This amount of base was then precisely introduced to a 50.00-mL aliquot of the acid. Sufficient HIn (Problem 21-27) was introduced to make the solution 5.00×10^{-4} M with respect to the indicator. The absorbance of the resulting solution, measured in a 1.00-cm cell, was found to be 0.223 at 485 nm and 0.547 at 625 nm.
   (a) Calculate the pH of the solution.
   (b) Calculate K_a for the weak acid.

21-31. Construct absorption curves for solutions in which the analytical concentration of the indicator in Problem 21-27 is constant at 5.00×10^{-4} M, 1.00-cm cells are used, and the pH is
   * (a) 6.268.   * (c) 7.000.
      (b) 5.842.   (d) 5.293.

21-32. Construct absorption spectra for solutions in which the analytical concentration of the indicator in Problem 21-27 is 5.00×10^{-4} M, the measurements are made in 1.00-cm cells, and

   * (a) $\dfrac{[HIn]}{[In^-]} = 4.00.$ * (c) $\dfrac{[HIn]}{[In^-]} = \dfrac{2.00}{5.00}.$

      (b) $\dfrac{[HIn]}{[In^-]} = \dfrac{5.00}{2.00}.$ (d) $\dfrac{[HIn]}{[In^-]} = \dfrac{1.00}{4.00}.$

21-33. The absolute error in transmittance for a particular photometer is 0.005 and independent of the magnitude of T. Calculate the percent relative error in concentration that is caused by this source when
   * (a) $A = 0.633.$ (f) $A = 0.500.$
   * (b) $T = 45.8\%.$ (g) $T = 79.4\%.$
   * (c) $A = 0.961.$ (h) $A = 0.433.$
   * (d) $T = 0.100.$ (i) $T = 0.500.$
   * (e) $A = 0.247.$ (j) $T = 18.8\%.$

* 21-34. An instrument for which the absolute error in transmittance does not depend upon

the magnitude of T registered values of 0.294, 0.302. 0.297, and 0.305 for the transmittance of a particular solution. Calculate the relative error in concentration for a solution that has a transmittance of 48.0%, based upon the assumption that the absolute indeterminate error in transmittance is
(a) the standard deviation of the four data.
(b) the 90% confidence interval for a single measurement.

* 21-35. Maxima exist at 470 nm in the absorption spectrum for the bismuth(III)/thiourea complex and at 265 nm in the spectrum for the bismuth(III)/EDTA complex. Predict the shape of a curve for the photometric titration of
(a) bismuth(III) with thiourea at 470 nm.
(b) bismuth(III) with EDTA at 265 nm.
(c) the bismuth(III)/thiourea complex with EDTA at 470 nm. Reaction:

$$Bi(tu)_6^{3+} + H_2Y^{2-} \longrightarrow BiY^- + 6tu + 2H^+$$

(d) the reaction in (c) at 265 nm.

21-36. Ethylenediaminetetraacetic acid will abstract the cation from the magnesium/Eriochrome black T complex at a pH of 10:

$$\underset{\text{red}}{MgIn^-} + HY^{3-} \longrightarrow \underset{\text{blue}}{HIn^{2-}} + MgY^{2-}$$

The magnesium/EDTA chelate absorbs at 225 nm. Predict the shape of the curve for the photometric titration of
(a) magnesium ion with EDTA at 225 nm.
(b) a magnesium solution containing a small amount of Eriochrome black T, with EDTA at 640 nm, which corresponds to the absorption maximum for the indicator.
(c) the solution in (b) at a wavelength corresponding to the maximum in the absorption spectrum for the magnesium(II)/Eriochrome black T chelate.

* 21-37. Given the information that

$$Fe^{3+} + Y^{4-} \rightleftharpoons FeY^- \qquad K_f = 1.0 \times 10^{25}$$
$$Cu^{2+} + Y^{4-} \rightleftharpoons CuY^{2-} \qquad K_f = 6.3 \times 10^{18}$$

and the further information that, among the several reactants and products, only CuY^{2-} absorbs at 750 nm, describe how Cu(II) could be used as indicator for the photometric titration of Fe(III) with H_2Y^{2-}. Reaction:

$$Fe^{3+} + H_2Y^{2-} \longrightarrow FeY^- + 2H^+$$

21-38. A solution containing small amounts of chromium can be determined by a preliminary oxidation to the +6 state, followed by a photometric titration with Fe(II):

$$6Fe^{2+} + Cr_2O_7^{2-} + 14H^+ \longrightarrow 6Fe^{3+} + 2Cr^{3+} + 7H_2O$$

(a) Indicate whether it would be preferable to base this analysis on the disappearance of the orange Cr(VI) or the formation of green Cr(III) if the sample also contains appreciable quantities of green Ni(II) ions.
(b) Which would be the preferable process to monitor if the analyte solution contains a high concentration of chloride ion instead of Ni(II)? (Fe^{3+} forms orange chloride complexes)

* 21-39. A 15.0-mL aliquot of a solution containing nickel(II) was diluted to 25.0 mL with ethanol and titrated with 5.68×10^{-3} M dimethylglyoxime (HDMG). Reaction:

$$Ni^{2+} + 2HDMG \longrightarrow Ni(DMG)_2 + 2H^+$$

Calculate the concentration of nickel in the sample, based upon the accompanying data:

Volume of dimethylglyoxime, mL	A, 410 nm	Volume of dimethylglyoxime, mL	A, 410 nm
0.00	0.000	4.20	0.422
1.10	0.126	5.15	0.444
2.06	0.231	6.24	0.432
2.85	0.311	7.56	0.416
3.50	0.375	8.50	0.415

* 21-40. The method of continuous variations was used to determine the combining ratio between cation and ligand in the species responsible for the absorption peak of solutions containing iron(III) and thiocyanate ion. The accompanying data were obtained upon mixing the indicated volumes of 2.50×10^{-3} M Fe(III) with sufficient 2.50×10^{-3} M SCN$^-$ to give a total of 10.00 mL; both stock solutions were 0.20 M with respect to HNO$_3$.

Volume of solution taken, mL	A, 480 nm (1.00-cm cells)	Volume of solution taken, mL	A, 480 nm (1.00-cm cells)
0.00	0.000	6.00	0.649
1.00	0.221	7.00	0.572
2.00	0.447	8.00	0.442
3.00	0.579	9.00	0.222
4.00	0.659	10.00	0.002
5.00	0.691		

(a) Determine the composition of the complex.
(b) Evaluate the molar absorptivity of the complex, based upon the assumption that the reactant in lesser amount is completely incorporated in the complex in the linear portion of the plot (i.e., at 1.00 mL).
(c) Calculate a value of K_f for the complex, based upon the stoichiometric relationships that exist under conditions of maximum absorption.

21-41. The following data were collected in an experiment to determine the combining ratio in the complex that is formed with the cation M and the ligand L:

Solution	Reactant volumes, mL		Absorbance, 580 nm (1.00-cm cells)
	6.00×10^{-5} M M	6.00×10^{-5} M L	
0	10.00	0.00	0.00
1	9.00	1.00	0.095
2	8.00	2.00	0.186
3	7.00	3.00	0.284
4	6.00	4.00	0.364
5	5.00	5.00	0.448
6	4.00	6.00	0.527
7	3.00	7.00	0.596
8	2.00	8.00	0.545
9	1.00	9.00	0.268
10	0.00	10.00	0.00

(a) Determine the composition of the complex.
(b) Calculate an average value for the molar absorptivity; assume that the reactant

in stoichiometrically lesser amount is totally incorporated in the complex in linear regions of the curve.

(c) Calculate K_f for the complex, using the stoichiometric relationships that exist under conditions of maximum absorption.

* 21-42. A slope-ratio investigation of the chelate formed between Cu(II) and the ligand Q at 364 nm yielded the accompanying data:

$C_Q = 8.00 \times 10^{-3}\ M$		$C_{Cu} = 8.00 \times 10^{-3}\ M$	
Concentration of Cu(II), mol/L	Absorbance, A (1.00-cm cells)	Concentration of Q, mol/L	Absorbance, A (1.00-cm cells)
8.00×10^{-6}	0.081	1.20×10^{-5}	0.058
2.30×10^{-5}	0.233	2.30×10^{-5}	0.113
3.90×10^{-5}	0.386	4.10×10^{-5}	0.204
5.10×10^{-5}	0.508	5.80×10^{-5}	0.287
6.60×10^{-5}	0.656	7.60×10^{-5}	0.389

Calculate the ligand to cation ratio in this complex.

21-43. Menis, Manning, and Goldstein [*Anal. Chem.*, **29**, 1426 (1957)] investigated the yellow complex formed between thorium(IV) and quercetin (3,3',4',5,7-pentahydroxyflavone) at its absorption maximum of 422 nm. Evaluate the combining ratio between the two reactants, based upon the accompanying absorbance data (1.00-cm cells).

$C_{Th} = 6.0 \times 10^{-4}\ M$		$C_{quercetin} = 6.0 \times 10^{-4}\ M$	
Concentration of quercetin, mol/L	Absorbance, A 422 nm	Concentration of Th(IV), mol/L	Absorbance, A 422 nm
6.0×10^{-6}	0.101	4.0×10^{-6}	0.134
1.1×10^{-5}	0.185	9.0×10^{-6}	0.302
1.5×10^{-5}	0.253	1.6×10^{-5}	0.537
2.0×10^{-5}	0.338	2.0×10^{-5}	0.675
2.5×10^{-5}	0.422	2.3×10^{-5}	0.778

* 21-44. Use the accompanying data to evaluate the cation-to-ligand ratio of the chelate that is formed between Fe(II) and 1,10-phenanthroline (Ph).

$C_{phenanthroline} = 4.00 \times 10^{-3}\ M$		$C_{Fe} = 1.00 \times 10^{-2}\ M$	
Concentration of Fe(II), mol/L	Absorbance, 510 nm (1.00-cm cells)	Concentration of Ph, mol/L	Absorbance, 510 nm (1.00-cm cells)
8.50×10^{-6}	0.094	2.10×10^{-5}	0.077
2.60×10^{-5}	0.286	5.22×10^{-5}	0.191
4.20×10^{-5}	0.462	7.80×10^{-5}	0.286
5.45×10^{-5}	0.600	9.45×10^{-5}	0.346
7.20×10^{-5}	0.792	1.08×10^{-4}	0.396

21-45. The complex that is formed between zinc(II) and the chelating reagent X was studied by measuring the absorbance of solutions in which the analytical concentration of Zn(II) was held constant at 1.38×10^{-4} mol/L, while the concentration of X was varied. Measurements were made in 1.00-cm cells at 645 nm.

Concentration of X, mol/L	Absorbance, A	Concentration of X, mol/L	Absorbance, A
2.00×10^{-5}	0.058	3.50×10^{-4}	0.720
5.00×10^{-5}	0.134	4.50×10^{-4}	0.765
9.00×10^{-5}	0.245	5.50×10^{-4}	0.781
1.50×10^{-4}	0.405	7.00×10^{-4}	0.779
2.50×10^{-4}	0.612	9.00×10^{-4}	0.780

(a) Elucidate the composition of the complex from a plot of these data.

(b) Calculate a value for the formation constant, K_f, using the stoichiometric relationships that exist where the extrapolated lines intersect.

* 21-46. The complex between Co(II) and the ligand R was investigated spectrophotometrically at 550 nm, the wavelength of its absorption maximum. The cation concentration was maintained at 2.50×10^{-5} M in solutions with differing concentrations of R. Absorbance data (1.00-cm cells) follow:

Concentration of R, mol/L	Absorbance, A	Concentration of R, mol/L	Absorbance, A
1.50×10^{-5}	0.106	9.50×10^{-5}	0.523
3.25×10^{-5}	0.232	1.15×10^{-4}	0.529
4.75×10^{-5}	0.339	1.25×10^{-4}	0.531
6.25×10^{-5}	0.441	1.65×10^{-4}	0.529
7.75×10^{-5}	0.500	2.00×10^{-4}	0.530

(a) Determine the ligand-to-cation ratio for the complex, based upon a plot of these data.

(b) Calculate a value of K_f for the complex, using the stoichiometric relationships that exist where the extrapolated lines intersect.

21-47. The chelate CuA_2^{2-} exhibits maximum absorption at 480 nm. When the chelating reagent is present in at least a 10-fold excess, the absorbance is dependent only upon the analytical concentration of Cu(II) and conforms to Beer's law over a wide range. A solution in which the analytical concentration of Cu^{2+} is 2.30×10^{-4} M and that for A^{2-} is 8.60×10^{-3} M has an absorbance of 0.690 when measured in a 1.00-cm cell at 480 nm. A solution in which the analytical concentrations of Cu^{2+} and A^{2-} are 2.30×10^{-4} M and 5.00×10^{-4} M, respectively, has an absorbance of 0.540 when measured under the same conditions. Use this information to calculate a value of K_f for the process

$$Cu^{2+} + 2A^{2-} \rightleftharpoons CuA_2^{2-}$$

* 21-48. Mixture of the chelating reagent B with Ni(II) gives rise to formation of the highly colored NiB_2^{2+}, solutions of which obey Beer's law over a wide range. Provided the analytical concentration of the chelating reagent exceeds that of Ni(II) by a factor of 5 (or more), the cation exists, within the limits of observation, entirely in the form of the complex. Use the accompanying data to evaluate K_f for the process

$$Ni^{2+} + 2B \rightleftharpoons NiB_2^{2+}$$

Analytical concentration, M		Absorbance, A, 395 nm (1.00-cm cells)
Ni^{2+}	B	
2.50×10^{-4}	2.20×10^{-1}	0.765
2.50×10^{-4}	1.00×10^{-3}	0.360

atomic spectroscopy

22

Atomic spectroscopy is based upon the absorption, emission, or fluorescence of electromagnetic radiation by atomic particles. Two regions of the spectrum provide atomic spectral data—the ultraviolet-visible and the X-ray. Only the former will be discussed here.

22A TYPES OF ATOMIC SPECTROSCOPY

A requirement for the generation of ultraviolet or visible spectra for elements is that the samples be *atomized.* In this process, the constituents of a sample are decomposed and converted to gaseous elementary particles (atoms or ions). The emission, absorption, or fluorescence spectrum of the resulting atoms or ions then serves as the basis for qualitative and quantitative analysis for the elements contained therein.

It is convenient to classify atomic spectral methods according to the type of spectra and the method of atomization (see Table 22-1). Of the various methods listed, four will be considered in this chapter—flame atomic absorption, electrothermal atomic absorption, flame emission, and inductively coupled plasma emission spectroscopy.

22A-1 EMISSION METHODS

Atomization in an arc, spark, plasma, or flame is ordinarily accompanied by electronic excitation of a small fraction of the particles formed in the process. Relaxation of the atomized particles to their ground states produces emission spectra that are useful for analysis. Emission arises from several types of particles, including atoms, ions, small molecules, and radicals.

Emission by atoms and elementary ions. Most emission methods are based upon the line spectrum produced by an excited atom or its ion. For a given element, the spectra for the two differ significantly. For example, the pattern of lines produced

Table 22-1 Classification of Atomic Spectral Methods

	Type of Spectroscopy	Method of Atomization	Radiation Source
Emission	Spark	Sample excited in a high voltage spark	Sample
	Inductively coupled plasma (ICP)	Sample heated in an inductively heated argon plasma	Sample
	Arc	Sample heated in an electric arc	Sample
	Flame emission	Sample solution aspirated into a flame	Sample
Absorption	Atomic absorption (flame)	Sample solution aspirated into a flame	Hollow cathode tube
	Atomic absorption (electrothermal)	Sample solution evaporated and ignited on a hot surface	Hollow cathode tube
Fluorescence	Atomic fluorescence (flame)	Sample solution aspirated into a flame	Sample (excited by radiation from a pulsed lamp)
	Atomic fluorescence (electrothermal)	Sample solution evaporated and ignited on a hot surface	Sample (excited by radiation from a pulsed lamp)

by excitation of one of the two $3s$ electrons of atomic magnesium differs markedly from that for elemental sodium with its single $3s$ electron. On the other hand, the patterns for Mg^+ and elemental sodium, both of which contain but a single outer electron, are remarkably alike. To be sure, the lines for Mg^+ are displaced to shorter wavelengths because the greater charge of the magnesium nucleus causes the excited states to have higher energies than the corresponding ones for sodium. Nonetheless, a strong similarity exists between the two while the spectrum for magnesium atoms is totally different from either.

The spectrum formed by atomization of the sample in a high-voltage spark or in an inductively coupled argon plasma is generally rich in lines from excited ions. In contrast, atomization by heating in an electric arc provides spectra consisting largely of lines from neutral atoms because this source is less energetic than the other two. A flame provides even fewer lines for ions because it is the least energetic of all of the sources shown in Table 22-1. Figures 19-4 (p. 483) and 22-1 illustrate the relatively simple line spectra produced by flame atomization. Note that in both figures, the line spectra are superimposed on band spectra that result from excitation of polyatomic species. In many instances these species are products of the combustion such as carbon monoxide, oxides of nitrogen, and cyanogen. At other times, however, the bands are produced by incompletely atomized sample components.

Molecular spectra. It is noteworthy that some alkaline earth and rare earth ele-

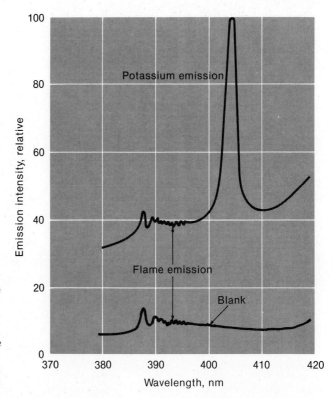

FIGURE 22-1 A portion of the emission spectrum for potassium in a hydrogen/oxygen flame. Note that the potassium band is superimposed upon the background emission from the flame. (The blank spectrum has been displaced downward.)

ments form volatile oxide and hydroxide species that are not completely dissociated in some atomization sources such as low-temperature flames. Consequently, molecular band spectra are observed, some of which are useful for quantitative analytical purposes. An example is the emission band for CaOH shown in Figure 22-2. The peak at about 5540 Å has been employed for the quantitative determination of calcium. When bands of this kind are used, the process would be more properly referred to as "molecular emission" rather than "atomic emission."

22A-2 ABSORPTION METHODS
In atomic absorption methods, atomization is carried out in a flame or on an electrically heated surface. The vapor from the sample is then interposed into the light path of a spectrophotometer similar to those discussed in Chapter 20. As with flame emission, the location of absorption lines serves to identify the components of a sample; the absorbance of a line, as in molecular spectrophotometry, is generally proportional to the concentration of the absorbing species.

22A-3 FLUORESCENCE METHODS
The unexcited atoms in a flame or above a heated surface serve as the basis for *atomic fluorescence spectroscopy*. In this approach, a fraction of the atoms in atom-

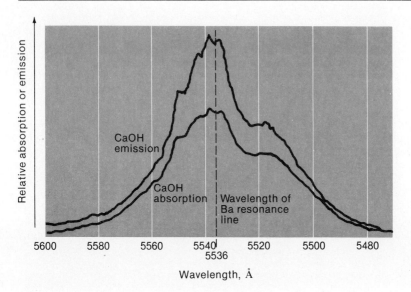

FIGURE 22-2   Molecular flame and flame absorption spectra for CaOH. The wavelength of an important line for barium is also shown. [Adapted from L. Capacho-Delgado and S. Sprague, *Atomic Absorption Newsletter*, **4**, 363 (1965). Courtesy of Perkin-Elmer Corporation, Norwalk, Connecticut.]

ized samples are excited by suitable radiation; the resonance fluorescence is then observed at 90 deg to the excitation beams.

22A-4 RELATIONSHIP BETWEEN ATOMIC EMISSION AND ATOMIC ABSORPTION MEASUREMENTS

Atomic emission spectroscopy is based upon the intensity of radiation emitted by atoms that are excited in the atomization process, whereas the unexcited atoms serve as the basis for atomic absorption measurements (and atomic fluorescence measurements as well). The unexcited species predominate by a large margin at the temperatures employed in atomic spectroscopy. For example, at 2500°K, only about 0.02% of the sodium atoms in a flame are excited to the $3p$ state at any instant; the percentages in higher electronic states are even less. At 3000°K, the corresponding percentage is about 0.09.

This large ratio of unexcited to excited species in a flame is important in comparing absorption and emission procedures. Because atomic absorption methods are based upon a much larger population of particles, it might be expected to be the more sensitive procedure. This apparent advantage is offset, however, by the fact that an absorbance measurement involves a difference measurement ($A = \log P_0 - \log P$); when the two numbers are nearly alike, large relative errors in the difference result. As a consequence, the two procedures are often complementary in sensitivity, the one being advantageous for one group of elements and the other for a different group.

An important advantage of atomic absorption spectroscopy is that the absolute number of unexcited particles is far less dependent upon temperature than is the number of excited species. For example, at about 2500°K, a 20°K fluctuation causes a change of about 8% in the number of sodium atoms excited to the $3p$ state. Because so many more unexcited atoms exist at this temperature, however, their population changes by only about 0.02% for the same temperature variation.

It should be noted that temperature fluctuations do exert an indirect influence on atomic absorption by increasing the total number of atoms that are available for absorption. In addition, line broadening and a consequent decrease in peak height occur. These indirect effects necessitate a reasonable control of the flame temperature for quantitative atomic absorption measurements.

22B ATOMIZERS FOR ATOMIC SPECTROSCOPY

Five types of atomizers are listed in Table 22-1. Three of these are discussed in this text—flames, electrothermal devices, and the inductively coupled plasma. Arc and spark atomizers, which find widespread use for certain industrial analytical applications, involve highly specialized equipment and techniques and, thus, are not considered here.

22B-1 FLAME ATOMIZERS

The most common atomization device for atomic spectroscopy consists of a nebulizer and a burner.[1] The nebulizer produces a fine spray or aerosol from the liquid sample, which is then fed into the flame. Both *total consumption (turbulent flow)* and *premixed (laminar flow) burners* are encountered.

Turbulent flow burner. Figure 22-3 is a schematic diagram of a commercially available turbulent flow burner. Here, the nebulizer and burner are combined into a single unit. Venturi action of the gases around the tip of the burner causes the sample to be drawn up the capillary and nebulized. Typical sample flow rates are 1 to 3 mL/min.

[1] For a detailed discussion of flame atomizers, see: R. D. Dresser, R. A. Mooney, E. M. Heithmar, and F. W. Plankey, *J. Chem. Educ.*, **52**, A403 (1975).

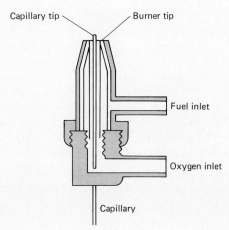

FIGURE 22-3 A turbulent flow burner. (Courtesy of Beckman Instrument, Inc., Fullerton, California.

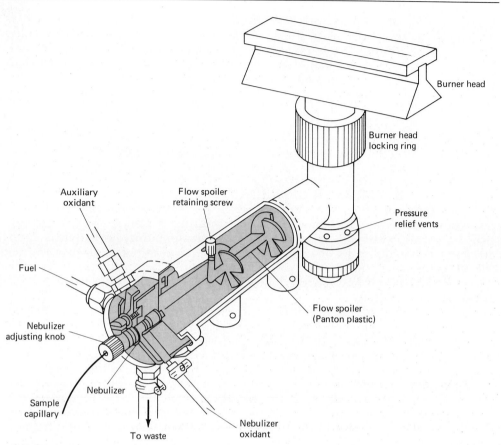

Auxiliary
oxidant

Flow spoiler
retaining screw

Burner head

Burner head
locking ring

Pressure
relief vents

Fuel

Flow spoiler
(Panton plastic)

Nebulizer
adjusting knob

Nebulizer

Sample
capillary

To waste

Nebulizer
oxidant

FIGURE 22-4 A laminar flow burner. (Courtesy of Perkin-Elmer Corporation, Norwalk, Connecticut.)

Turbulent flow burners offer the advantage of letting a relatively large and representative sample reach the flame. Furthermore, no possibility of flashback and explosion exists. The disadvantages of such burners include a relatively short flame path length and problems with clogging of the tip. In addition, turbulent flow burners are noisy both from the electronic and auditory standpoints. Although sometimes used for emission and fluorescence analyses, turbulent flow burners find little use in present-day absorption work.

Laminar flow burner. Figure 22-4 is a diagram of a typical commercial laminar flow burner. The sample is nebulized by the flow of oxidant past a capillary tip. The resulting aerosol is then mixed with fuel and flows past a series of baffles that remove all but the finest droplets. As a result of the baffles, the majority of the sample collects in the bottom of the mixing chamber where it is drained to a waste container. The aerosol, oxidant, and fuel are then burned in a slotted burner that provides a flame which is usually 5 or 10 cm in length.

Laminar flow burners provide a significantly longer path length; this property tends to enhance sensitivity and reproducibility for absorption measurements. Furthermore, clogging is seldom a problem. Disadvantages include a lower rate of sample introduction (which may offset the longer path length advantage) and the possibility of selective evaporation of mixed solvents in the mixing chamber, which may lead to analytical uncertainties. Furthermore, the mixing chamber contains a potentially explosive mixture which can be ignited by a flashback. Note that the burner in Figure 22-4 is equipped with pressure relief vents for this reason.

Fuel and oxidant regulators. An important variable that requires close control in flame spectroscopy is the flow rate of both oxidant and fuel. It is desirable to be able to vary each over a considerable range so that ideal atomization conditions can be found experimentally. Fuel and oxidant are ordinarily combined in approximately stoichiometric amounts. For the analysis of metals that form stable oxides, however, a reducing flame, which contains an excess of fuel, may prove more desirable. Flow rates are ordinarily controlled by means of double-diaphragm pressure regulators followed by needle valves in the instrument housing.

A requirement for reproducible analytical conditions is that both the fuel and oxidant systems be equipped with some type of flowmeter. The most widely used of these is the rotameter, which consists of a tapered transparent tube that is mounted vertically with the smaller end down. A lightweight conical or spherical float is lifted by the gas flow; its vertical position is determined by the flow rate.

Fuels and oxidants. Fuels used for flame production include natural gas, propane, butane, hydrogen, and acetylene. The last is perhaps most widely employed. The common oxidants are air, oxygen-enriched air, oxygen, and nitrous oxide. The acetylene/nitrous oxide mixture is advantageous when a hot flame is required.

Low-temperature flames (natural gas/air, for example) are used to advantage for elements that are readily converted to the atomic state, such as the alkali metals, copper, lead, zinc, and cadmium. Elements such as the alkaline earths, on the other hand, form refractory oxides which require somewhat higher temperatures for decomposition; for these, an acetylene/air mixture often produces the most sensitive results. Aluminum, beryllium, the rare earths, and certain other elements form unusually stable oxides; a reasonable concentration of their atoms can be obtained only at the high temperatures developed in an acetylene/oxygen or an acetylene/nitrous oxide flame.

22B-2 ELECTROTHERMAL ATOMIZERS

In terms of sampling efficiency (and thus sensitivity), a flame suffers in comparison with other atomization methods. Two reasons for the lower sampling efficiency of the flame can be cited. First, a large portion of the sample flows down the drain (laminar burner) or is not completely atomized (turbulent burner). Second, the residence time of individual atoms in the optical path in the flame is brief ($\sim 10^{-4}$ s).

Since 1970, electrothermal atomizers have appeared on the market.[2] These

[2] See: R. D. Dresser, R. A. Mooney, E. M. Heithmar, and F. W. Plankey, *J. Chem. Educ.*, **52**, A451, A503 (1975); R. E. Sturgeon, *Anal. Chem.*, **49**, 1255A (1977); and C. W. Fuller, *Electrothermal Atomization for Atomic Absorption Spectroscopy*. London: The Chemical Society, 1978.

devices generally provide enhanced sensitivity because the entire sample is atomized in a short period, and the average residence time of the atoms in the optical path is a second or more.

In an electrothermal atomizer, a few microliters of sample are evaporated and ashed at low temperatures on an electrically heated surface of carbon, tantalum, or other conducting material. The conductor can be a hollow tube, a strip or rod, a boat, or a trough. After ashing, the current is increased to 100 A or more, which causes the temperature to soar to perhaps 2000 to 3000°C; atomization of the sample occurs in a few seconds. The absorption or fluorescence (but not the emission) of the atomized particles can then be measured in the region immediately above the heated conductor. At a wavelength where absorption or fluorescence takes place, the signal is observed to rise to a maximum after a few seconds and then decay to zero, corresponding to the atomization and subsequent escape of the volatilized sample; analyses are based upon peak height or area. Atomization is ordinarily performed in an inert gas atmosphere to prevent oxidation of the conductor.

Nonflame atomizers offer the advantage of unusually high sensitivity for small volumes of sample. Typically, sample volumes between 0.5 and 10 μL are employed; absolute limits thus lie in the range of 10^{-10} to 10^{-13} g of analyte. These sensitivities are 1000 times or greater than those obtained with flames.

The relative precision of electrothermal methods is generally in the range of 5 to 10% compared with the 1 to 2% that can be attained with flame atomization.

22B-3 THE INDUCTIVELY COUPLED PLASMA (ICP)

The inductively coupled plasma source, which became commercially available in the mid 1970s, offers several advantages over flame sources for emission spectroscopy.[3] Figure 22-5 is a schematic drawing of an inductively coupled plasma source. It consists of three concentric quartz tubes through which streams of argon flow at a total rate between 11 and 17 L/min. The diameter of the largest tube is about 2.5 cm. Surrounding the top of this tube is a water-cooled induction coil powered by a radio-frequency generator, which produces 2 kW of energy at about 27 MHz. Ionization of the flowing argon is initiated by a spark from a Tesla coil. The resulting ions, and their associated electrons, then interact with the fluctuating magnetic field (labeled "H" in Figure 22-5) produced by the induction coil. This interaction causes the ions and electrons within the coil to flow in the closed annular paths depicted in the figure; ohmic heating is the consequence of their resistance to this movement.

The temperature of the plasma formed in this way is high enough (8000 to 10,000°K) to require thermal isolation from the outer quartz cylinder. This isolation is achieved by flowing argon tangentially around the walls of the tube, as indicated by the arrows in Figure 22-5; the flow rate of this stream is 10 to 15 L/min. The tangential flow cools the inside walls of the center tube and centers the plasma radially.

[3] For a more complete discussion, see: V. A. Fassel, *Science*, **202**, 183 (1978); A. F. Ward, *Amer. Lab.*, **10** (11), 79 (1978); R. M. Ajhar, P. D. Dalager, and A. L. Davison, *Amer. Lab.*, **8** (3), 71 (1976); *Applications of Inductively Coupled Plasmas to Emission Spectroscopy*, R. M. Barnes, Ed. Philadelphia: The Franklin Institute Press, 1978; and *Applications of Plasma Emission Spectrochemistry*, R. M. Barnes, Ed. Philadelphia: Heyden, 1979.

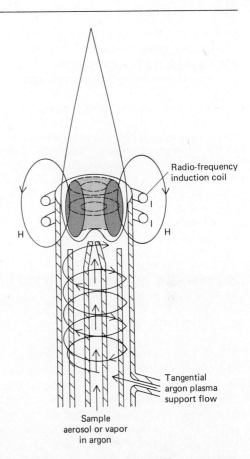

FIGURE 22-5 A typical inductively coupled plasma source. [V. A. Fassel, *Science*, **202**, 185 (1978). With permission.]

Sample injection. The sample is carried into the hot plasma at the head of the tubes by argon flowing at about 1 L/min through the central quartz tube. The sample may be an aerosol, a thermally generated vapor, or a fine powder. The most widely used device for sample injection is similar in construction to the nebulizer employed for flame methods.

Plasma appearance and spectra. The typical plasma has a very intense, brilliant white, nontransparent core topped by a flamelike tail. The core, which extends a few millimeters above the tube, is a continuum upon which is superimposed the atomic spectrum for argon. The continuum is apparently the result of recombination of argon and other ions with electrons. In the region 10 to 30 mm above the core, the continuum fades, and the plasma is optically transparent. Spectral observations are generally made at a height of 15 to 20 mm above the induction coil. Here, the background radiation is remarkably free of argon lines and is well suited for analysis. Many of the most sensitive emission lines in this region of the plasma are from ions such as Ca^+, Ca^{2+}, Cd^+, Cr^{2+}, and Mn^{2+}.

Analyte atomization and ionization. By the time the sample atoms have reached the observation point in the plasma, they will have had a residence time of about 2 ms at temperatures ranging from 6000 to 8000°K. These times and temperatures are roughly twice as great as those attainable in the hotter combustion flames (acetylene/nitrous oxide). As a consequence, atomization is more complete, and fewer interference problems due to dissociation reactions (p. 588) are encountered. Surprisingly, ionization interference effects (p. 588) are small or nonexistent, perhaps because the large concentration of electrons from ionization of the argon represses the ionization of the sample components.

Several other advantages are associated with the plasma source. First, atomization occurs in a chemically inert environment, which should also enhance the lifetime of the analyte. In addition, and in contrast to arc, spark, and flame sources, the temperature cross section of the plasma is relatively uniform; self-absorption and self-reversal effects (p. 592) are not encountered as a consequence. Calibration curves thus tend to remain linear over several orders of magnitude of concentration. The plasma source has been used as a source for emission methods only.

22C ATOMIC ABSORPTION SPECTROSCOPY

Atomic absorption spectroscopy finds considerably wider use than atomic emission with flame sources because it is best suited to routine analyses in the hands of relatively unskilled operators.[4]

22C-1 RADIATION SOURCES FOR ATOMIC ABSORPTION METHODS

Atomic absorption lines are remarkably narrow and are unique for each element; the potential for specificity is thus great. On the other hand, the limited line widths create a measurement problem not encountered in molecular absorption. Recall that Beer's law applies only for monochromatic radiation; a linear relationship between absorbance and concentration can be expected, however, only if the bandwidth of the source is narrow with respect to the width of the absorption peak (p. 496).

Line widths. Atomic emission and absorption peaks are much narrower than the bands resulting from emission or absorption by molecules. The natural width of atomic lines can be shown to be about 10^{-4} Å. Two effects, however, combine to broaden the lines to a range from 0.02 to 0.05 Å.

Doppler broadening arises from the rapid motion of the atomic particles in the flame plasma. Atoms that are moving toward the monochromator emit shorter wavelengths due to the well-known Doppler shift; the effect is reversed for atoms moving away from the monochromator.

Doppler broadening is also observed for absorption lines. Those atoms traveling toward the source absorb radiation of slightly shorter wavelength than that absorbed by particles traveling perpendicular to the incident beam. The reverse is true of atoms moving away from the source.

Pressure broadening also occurs; here, collisions among atoms cause small

[4] Reference books on atomic absorption spectroscopy include: G. F. Kirkbright and M. Sargent, *Atomic Absorption and Fluorescence Spectroscopy*. New York: Academic Press, 1974; J. W. Robinson, *Atomic Absorption Spectroscopy*, 2d ed. New York: Marcel Dekker, 1975; and W. Slavin, *Atomic Absorption Spectroscopy*, 2d ed. New York: Interscience, 1978.

changes in the ground-state energy levels and a consequent broadening of peaks.

No ordinary monochromator is capable of yielding a band of radiation as narrow as the peak width of an atomic absorption line. Thus, only a small fraction of the total power of the beam emerging from the monochromator is absorbed by the sample. Under these conditions, the sensitivity of the measurement is diminished, and Beer's law is not followed.

This problem has been overcome by making use of a radiation source that emits a line of the same wavelength as the one to be used for absorption analysis. Thus, for example, an analysis for sodium based upon absorption at 589.6 nm would involve the use of a sodium vapor lamp as the source for this radiation. Gaseous sodium atoms are excited by electrical discharge in such a lamp; the excited atoms then emit characteristic radiation as they return to lower energy levels. An

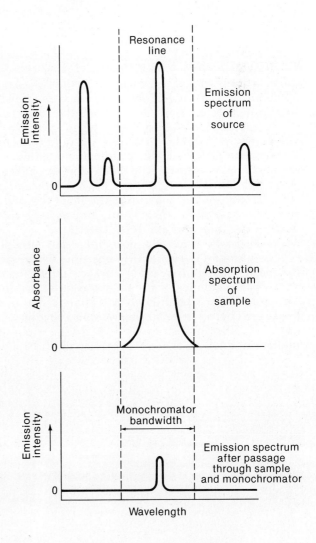

FIGURE 22-6 Absorption of a resonance line by atoms.

emitted line will thus have the same wavelength as the resonance absorption line. With a properly designed source (one that operates at a lower temperature than the flame to minimize Doppler broadening), the emission lines will have bandwidths significantly narrower than the absorption bandwidths. It is only necessary, then, for the monochromator to have the capability of isolating a suitable emission line for the absorption measurement; the process is depicted in Figure 22-6. Here it is seen that the radiation employed in the analysis is sufficiently limited in bandwidth to permit measurements at the absorption peak. Greater sensitivity and better adherence to Beer's law result.

    A separate lamp source is needed for each element (or sometimes, group of elements). To avoid this inconvenience, attempts have been made to employ a continuous source with a very high resolution monochromator or, alternatively, to produce a line source by introducing a compound of the element to be determined into a high-temperature flame. Neither of these alternatives is as satisfactory as individual lamps.

Hollow cathode lamps. The most common source for atomic absorption measurements is the *hollow cathode lamp*, which consists of a tungsten anode and a cylindrical cathode sealed in a glass tube that is filled with neon or argon at a pressure of 1 to 5 torr (see Figure 22-7). The cathode is constructed of the metal whose spectrum is desired or serves to support a layer of that metal.

    Ionization of the gas occurs when a potential is applied across the electrodes, and a current of about 5 to 10 mA is generated as the ions migrate to the electrodes. If the potential is sufficiently large, the gaseous cations acquire enough kinetic energy to dislodge some of the metal atoms from the cathode surface and produce an atomic cloud; this process is called *sputtering*. Portions of the sputtered metal atoms are in excited states and, thus, emit their characteristic radiation in the usual way. Eventually, the metal atoms diffuse back to the cathode surface or to the glass walls of the tube and are redeposited.

    The cylindrical configuration of the cathode tends to concentrate the radiation in a limited region of the tube; this design also enhances the probability that redeposition will occur at the cathode rather than on the glass walls.

    The efficiency of the hollow cathode lamp depends upon its geometry and the operating potential. High potentials, and thus high currents, lead to greater intensities. This advantage is offset somewhat by an increase in Doppler broadening of the emission lines. Furthermore, higher currents result in an increase in the number of unexcited atoms in the cloud; the unexcited atoms, in turn, are capable

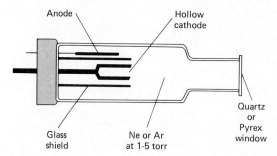

FIGURE 22-7   Schematic cross section of a hollow cathode lamp.

of absorbing the radiation emitted by the excited ones. This *self-absorption* leads to lowered intensities, particularly at the center of the emission band.

A variety of hollow cathode tubes are available commercially. The cathodes of some consist of a mixture of several metals; such lamps permit the analysis of more than a single element.

Gaseous discharge lamps.   Gas discharge lamps produce a line spectrum when electricity is passed through a vapor of metal atoms; the familiar sodium and mercury vapor lamps are examples. Sources of this kind are particularly useful for producing spectra of the alkali metals.

Source modulation.   In the typical atomic absorption instrument, it is necessary to eliminate interferences caused by emission of radiation from the flame. Most of the emitted radiation can be removed by locating the monochromator between the flame and the detector; nevertheless, this arrangement does not remove the flame radiation corresponding to the wavelength selected for the analysis. The flame will likely contain such radiation due to excitation and radiant emission by some atoms of the analyte. This difficulty is overcome by modulating the output of the source so that its intensity fluctuates at a constant frequency. The detector then receives two types of signal: an alternating one from the source and a continuous one from the flame. These signals are converted to the corresponding electrical responses. A simple electronic filter is then employed to remove the unmodulated dc signal and pass the ac signal for amplification.

A simple and entirely satisfactory way of modulating radiation from the source is to interpose a circular disk in the beam between the source and the flame. Alternate quadrants of this disk are removed to permit passage of light. Rotation of the disk at constant speed provides a beam that is chopped to the desired frequency. As an alternative, the power supply for the source can be designed for intermittent or ac operation.

22C-2 INSTRUMENTS FOR ATOMIC ABSORPTION SPECTROSCOPY
Instruments for atomic absorption work are offered by numerous manufacturers; both single- and double-beam designs are available. The range of sophistication and cost (upward from a few thousand dollars) is substantial.

In general, the instrument must be capable of providing a sufficiently narrow bandwidth to isolate the line chosen for the measurement from other lines that may interfere with or diminish the sensitivity of the analysis. A glass filter suffices for some of the alkali metals, which have only a few widely spaced resonance lines in the visible region. An instrument equipped with readily interchangeable interference filters is available commercially. A separate filter (and light source) is used for each element. Satisfactory results for the analysis of 22 metals are claimed. Most instruments, however, incorporate a good-quality ultraviolet and visible monochromator.

The detector-indicator components for an atomic absorption spectrophotometer are fundamentally the same as those for the typical ultraviolet and visible spectrophotometer for molecular spectroscopy (see Section 20D). Photomultiplier tubes are generally used to convert the radiant energy received to an electrical signal. As we have pointed out, the electronic system must be capable of discriminating between the modulated signal from the source and the continuous signal from the flame. Both null-point and direct-reading meters, calibrated in terms of absorbance or transmittance, are used. Many instruments are equipped with recorders to provide spectral plots. Some instruments give a digital readout of analyte concentration.

22C-3 APPLICATIONS OF ATOMIC ABSORPTION SPECTROSCOPY

Atomic absorption spectroscopy is a sensitive means for the determination of more than 60 elements. Details concerning methods of sample preparation and quantitative determination of individual elements have been compiled in the several reference works listed in footnote 4 (p. 582).

Sensitivity and detection limits. Two terms are employed in characterizing atomic absorption methods. The *sensitivity* is defined as the concentration of an element in $\mu g/mL$ (or ppm) which produces a transmittance signal of 0.99 or a corresponding absorbance signal of 0.0044. Modern atomic instruments have adequate precision to discriminate between absorbance signals that differ by less than 0.0044. A second term, which is used to characterize not only atomic absorption methods but other procedures as well, is the *detection limit*. This quantity is defined as the concentration of the element that produces an analytical signal equal to twice the standard deviation of the background signal. (For flame atomization, the standard deviation of the background signal is obtained by observing the signal variation while a blank is aspirated into the flame.) Both the sensitivity and the detection limits are affected by such variables as temperature, spectral bandwidth, detector sensitivity, and type of signal processing. Small differences among quoted

Table 22-2 Sensitivity and Detection Limits for Selected Elements by Various Atomic Spectroscopy Methods*

Element	AA (flame)** Sensitivity (ppm)	AA (electrothermal)** Sensitivity (ppm)	AE (ICP)** Detection Limit (ppm)	AE (flame)† Detection Limit (ppm)
As	0.1	0.0002	0.002	—
Ba	0.2	0.00006	0.00002	0.001
Ca	0.02	0.0004	0.000001	0.0001
Cd	0.01	0.000003	0.00006	2
Co	0.07	0.00006	0.0001	0.05
Cu	0.04	0.00004	0.0004	0.01
Fe	0.06	0.000008	0.0005	0.05
Hg	2.2	0.0005	0.001	—
K	0.01	0.0001	—	0.0005
Mg	0.003	0.000007	0.000003	0.005
Na	0.003	0.000004	0.0002	0.0005
Pb	0.1	0.00005	0.001	0.2
Zn	0.009	0.000002	0.0001	—

* AA = atomic absorption; AE = atomic emission
** From G. H. Morrison, *Crit. Rev. Anal. Chem.*, **8**, 296 (1979).
† From E. E. Pickett and S. R. Koirtyohann, *Anal. Chem.*, **41**(14), 28A (1969).

values of the two parameters are not significant. For example, a factor of 2 or 3 probably is not meaningful; an order of magnitude difference certainly is.

Flame atomic absorption methods yield sensitivities ranging from about 3×10^{-4} ppm to 20 ppm for various metallic elements. Electrothermal atomization often enhances this limit by a factor of 10^{-1} to 10^{-3}.

Columns 2 and 3 of Table 22-2 list sensitivities for several common elements by flame and electrothermal atomization.

22C-4 ACCURACY

Under usual conditions, the relative error associated with a flame absorption analysis is of the order of 1 to 2%. With special precautions, this figure can be lowered to a few tenths of 1%. For electrothermal methods, uncertainties in the range of 5 to 10% relative are encountered.

22C-5 SPECTRAL INTERFERENCES

Atomic absorption methods are subject to two types of interference. *Spectral interferences* occur when the absorption of an interfering species either overlaps or lies so close to the analyte absorption that resolution by the monochromator becomes impossible. *Chemical interferences* result from various chemical processes occurring during atomization that alter the absorption characteristics of the analyte. Both of these problems are considered briefly in the paragraphs that follow.

Because the emission lines of hollow cathode sources are so very narrow, interference due to overlap of atomic spectral lines is rare. More common are spectral interferences resulting from the presence of combustion products that exhibit broad band absorption or particulate products that scatter radiation. Both diminish the power of the transmitted beam and lead to positive analytical errors. Where the source of these products is the fuel and oxidant mixture alone, corrections are readily obtained from absorbance measurements while a blank is aspirated into the flame.

A much more troublesome problem is encountered when the source of absorption or scattering originates in the sample matrix; here, the power of the transmitted beam, P, is reduced by the nonanalyte components, but the incident beam power, P_0, is not; a positive error in absorbance and thus concentration results. An example of a potential matrix interference due to absorption occurs in the determination of barium in alkaline-earth mixtures. Here, an intense and useful absorption line for barium atoms, occurring at 553.6 nm, lies in the center of a broad absorption band for CaOH, which extends from 540 to 560 nm (p. 576); interference by calcium in a barium analysis is to be expected. In this particular situation, the effect is readily eliminated by substituting nitrous oxide for air as the oxidant for the acetylene; the higher temperature decomposes the CaOH and eliminates the absorption band.

Fortunately, spectral interferences by matrix products are not widely encountered and often can be avoided by alteration of analytical parameters such as the temperature and fuel-to-oxidant ratio. Alternatively, if the source of interference is known, an excess of the interfering substance can be added to both sample and standards; provided the excess is large with respect to the concentration from the sample matrix, the contribution of the latter will become insignificant. Such an added substance is sometimes called a *radiation buffer*.

22C-6 CHEMICAL INTERFERENCES

Chemical interferences are more common than spectral interferences. Their effects can frequently be minimized by suitable choice of conditions.

Both theoretical and experimental evidence suggests that many of the processes occurring in the mantle of a flame are in approximate equilibrium. As a consequence, it becomes possible to regard the burned gases of the flame as a solvent medium to which the equilibrium law applies. The equilibria of principal interest include formation of compounds of low volatility, dissociation reactions, and ionizations.

Formation of compounds of low volatility. Perhaps the most common type of chemical interference is by anions that form compounds of low volatility with the analyte and thus reduce the rate at which it is atomized. Low results are the consequence. An example is the decrease in calcium absorbance that is observed with increasing concentrations of sulfate or phosphate. For a fixed calcium concentration, the absorbance is found to fall off nearly linearly with increasing sulfate or phosphate concentration until the anion-to-calcium ratio is about 0.5; the absorbance then levels off at about 30 to 50% of its original value and becomes independent of anion concentration.

Interferences due to formation of species of low volatility can often be eliminated or moderated by use of higher-temperature flames. Alternatively, *releasing agents,* which are cations that react preferentially with the interference and prevent its interaction with the analyte, can be employed. For example, addition of an excess of strontium or lanthanum ion minimizes the interference of phosphate in the determination of calcium by replacing the analyte in the compound formed with the interfering species.

Protective agents prevent interferences by forming stable but volatile species with the analyte. Three common reagents for this purpose are EDTA, 8-hydroxyquinoline, and APDC (the ammonium salt of 1-pyrrolidinecarbodithioic acid). The presence of EDTA has been shown to eliminate the interference of aluminium, silicon, phosphate, and sulfate in the determination of calcium.

Dissociation equilibria. In the hot gaseous environment of a flame, dissociation and recombination reactions are sources of potential chemical interference. For example, the decrease in the line intensity of sodium that is caused by the presence of HCl can be accounted for in terms of the equilibrium

$$NaCl(g) \rightleftharpoons Na(g) + Cl(g)$$

Chlorine atoms from the HCl tend to force this equilibrium to the left and thereby decrease the population of sodium atoms.

Ionization in flames. Ionization of atoms and molecules is small in combustion mixtures that involve air as the oxidant and generally can be neglected. Substitution of oxygen or nitrous oxide, however, results in temperatures that are sufficiently high to cause appreciable ionization. A significant concentration of electrons exists as a consequence of the equilibrium

$$M \rightleftharpoons M^+ + e^-$$

where M represents a neutral atom or molecule and M^+ is its ion.

It is important to appreciate that treatment of the ionization process as an equilibrium—with free electrons as one of the products—immediately implies that the degree of ionization of a metal will be strongly influenced by the presence of other ionizable metals in the flame. Thus, if the medium contains not only species M, but species B as well, and if B ionizes according to the equation

$$B \rightleftharpoons B^+ + e^-$$

then the degree of ionization of M will be decreased by the mass-action effect of the electrons formed from B.

The presence of atom-ion equilibria in flames has a number of important consequences in flame spectroscopy. For example, the intensity of atomic emission or absorption lines for the alkali metals, particularly potassium, rubidium, and cesium, is affected by temperature in a complex way. Increased temperatures cause an increase in the population of excited atoms; counteracting this effect, however, is a decrease in concentration of atoms as a result of ionization. Thus, under some circumstances a decrease in emission or absorption may be observed in hotter flames. It is for this reason that lower excitation temperatures are usually specified for the analysis of alkali metals.

The effects of shifts in ionization equilibria can frequently be eliminated by addition of an *ionization suppressor*, which provides a relatively high concentration of electrons to the flame; suppression of ionization by the analyte results. The effect of a suppressor is demonstrated by the calibration curves for strontium shown in Figure 22-8. Note the marked steepening of these curves as strontium ionization is repressed by the increasing concentration of potassium ions and electrons. Note also the enhanced sensitivity that results from the use of nitrous oxide instead of air as the oxidant; the higher temperature achieved with nitrous oxide undoubtedly

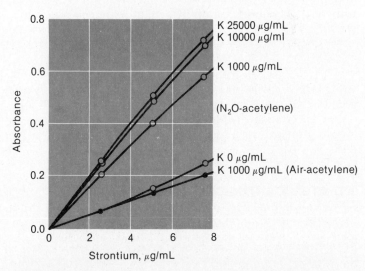

FIGURE 22-8 Effect of a potassium concentration on the calibration curve for strontium. [J. A. Bowman and J. B. Willis, *Anal. Chem.*, **39**, 1210 (1967). With permission.]

enhances the rate of decomposition and volatilization of the strontium compounds in the plasma.

22C-7 ANALYTICAL TECHNIQUES
Both calibration curves and the standard addition method are suitable for atomic absorption spectroscopy.

Calibration curves. While, in theory, absorbance should be proportional to concentration, deviations from linearity often occur. Thus, empirical calibration curves must be prepared. In addition, there are sufficient uncontrollable variables in the production of an atomic vapor to warrant measuring the absorbance of at least one standard solution each time an analysis is performed. Any deviation of the standard from the original calibration curve can then be employed to correct the analytical results.

Standard addition method. The standard addition method is extensively used in atomic absorption spectroscopy. Here, two or more aliquots of the sample are transferred to volumetric flasks. One is diluted to volume, and the absorbance of the solution is obtained. A known amount of analyte is added to the second, and its absorbance is measured after dilution to the same volume. Data for other standard additions may also be obtained. If a linear relationship between absorbance and concentration exists (and this should be established by several standard additions), the following relationships apply:

$$A_x = kC_x$$
$$A_T = k(C_s + C_x)$$

where C_x is the analyte concentration in the diluted sample and C_s is the contribution of the added standard to the concentration; A_x and A_T are the two measured absorbances. Combination of the two equations yields

$$C_x = C_s \frac{A_x}{(A_T - A_x)}$$

If several additions are made, A_T can be plotted against C_s. The resulting straight line can be extrapolated to $A_T = 0$. Substituting this value into the foregoing equation reveals that, at the intercept, $C_x = -C_s$.

The standard addition method has the advantage that it tends to compensate for variations caused by physical and chemical interferences in the sample solution.

22D ATOMIC EMISSION SPECTROSCOPY
Atomic emission spectroscopy has found widespread application to elemental analysis. The most important uses of flame emission spectroscopy have been in the analysis of sodium, potassium, lithium, and calcium, particularly in biological fluids and tissues. For reasons of convenience, speed, and relative freedom from interferences, flame emission spectroscopy has become the method of choice for these otherwise difficult-to-determine elements. The method has also been applied, with varying degrees of success, to the determination of perhaps half the ele-

ments in the periodic table. Thus, flame emission spectroscopy must be considered to be one of the important tools for analysis.[5]

22D-1 SOURCES

Much of the early work in flame emission analyses was accomplished with turbulent flow burners similar to the one described earlier in this chapter. Laminar flow burners, however, are becoming more and more widely used.

The recent introduction of the inductively coupled plasma source has significantly increased the attractiveness of emission methods, particularly for the heavier elements; thus, several analyses that were formerly carried out by atomic absorption are now being done by ICP methods.

22D-2 INSTRUMENTATION

Instruments for emission work are similar in construction to the flame absorption instruments except that a flame or plasma now acts as the radiation source; the hollow cathode lamp and chopper are, therefore, unnecessary. Some modern instruments are adaptable to flame or plasma emission as well as absorption analyses.

Spectrophotometers.   For nonroutine analysis, a recording, ultraviolet-visible spectrophotometer with a resolution of perhaps 0.5 Å is desirable.

Photometers.   Simple filter photometers often suffice for routine analyses of the alkali and alkaline-earth metals. A low-temperature flame is employed to eliminate excitation of most other metals. The spectra are simple as a consequence, and glass or interference filters can be used to isolate the desired emission line.

Several instrument manufacturers supply flame photometers designed specifically for the analysis of sodium, potassium, and lithium in blood serum and other biological samples. In these instruments, the radiation from the flame is split into three beams of approximately equal power. Each then passes into a separate photometric system consisting of an interference filter (which transmits an emission line of one of the elements while absorbing those of the other two), a phototube, and an amplifier. The outputs can be measured separately if desired. Ordinarily, however, lithium serves as an *internal standard* for the analysis. For this purpose, a fixed amount of lithium is introduced into each standard and sample. The ratios of outputs of the sodium and lithium transducer and the potassium and lithium transducer then serve as analytical parameters. This system provides improved accuracy because the intensities of the three lines are affected in the same way by most analytical variables such as flame temperature, fuel flow rates, and background radiation. Clearly, lithium cannot be a constituent of the sample.

Automated flame photometers.   Fully automated photometers now exist for the determination of sodium and potassium in clinical samples. In one of these, the

---

[5] For a more complete discussion of the theory and applications of flame emission spectroscopy, see: *Flame Emission and Atomic Absorption Spectroscopy,* vol. 1: *Theory;* vol. 2: *Components and Techniques;* vol. 3: *Elements and Matrices,* J. A. Dean and T. C. Rains, Eds. New York: Marcel Dekker, 1974; J. A. Dean, *Flame Photometry.* New York: McGraw-Hill, 1960; and B. L. Vallee and R. E. Thiers, in *Treatise on Analytical Chemistry,* I. M. Kolthoff and P. J. Elving, Eds., Part I, vol. 6, Chapter 65. New York: Interscience, 1965.

samples are withdrawn sequentially from a sample turntable, dialyzed to remove protein and particulates, diluted with the lithium internal standard, and aspirated into a flame. Sample and reagent transport is accomplished with a roller-type pump. Air bubbles serve to separate samples. Results are printed out on a paper tape. Calibration is performed automatically after every nine samples.

22D-3 INTERFERENCES

Atomic emission methods are affected by the same interferences as are encountered in atomic absorption spectroscopy (Section 22C-5). The influence of any particular interference will not be the same, however.

Spectral line interference. Interference between two overlapping atomic absorption peaks occurs only in the occasional situation where the peaks are within about 0.1 Å of one another. That is, the high degree of spectral specificity is more the result of the narrow line properties of the source than the high resolution of the monochromator. Atomic emission spectroscopy, in contrast, depends entirely upon the monochromator for selectivity; the probability of spectral interference due to line overlap is consequently greater.

Chemical interferences. Chemical interferences in flame emission studies are essentially the same as those encountered in flame absorption methods. They are dealt with by judicious choice of flame temperature and the use of protective agents, releasing agents, and ionization suppressors.

A major advantage of the inductively coupled plasma source is that chemical interferences are rare, relative to either flame emission or absorption methods. This freedom from interference results from the high temperature of the plasma, the relatively long residence time for the analyte, the inert environment provided, and the constant concentration of electrons due to the ionization of the argon.

Self-absorption. The center of a flame is hotter than the exterior; thus, atoms that emit in the center are surrounded by a cooler region which contains a higher concentration of unexcited atoms; *self-absorption* of the resonance wavelengths by the atoms in the cooler layer will occur. Doppler broadening of the emission line is greater than the corresponding broadening of the resonance absorption line, however, because the particles are moving more rapidly in the hotter emission zone. Thus, self-absorption tends to alter the center of a line more than its edges. In the extreme, the center may become less intense than the edges, or it may even disappear.

The inductively coupled plasma source is free of self-absorption and reversal effects because the sample channel is surrounded by a sheath of hot argon. Thus, no significant inhomogeneity of temperature exists among the sample particles. This property enhances the ranges of analyte concentration over which linearity is observed for the signal power versus concentration relationship. Thus, concentration ranges as great as four orders of magnitude can be accommodated.

Self-absorption and ionization sometimes result in S-shaped emission calibration curves with three distinct segments. At intermediate concentrations of potassium, for example, a linear relationship between intensity and concentration is observed (Figure 22-9). At low concentrations, curvature is due to the increased

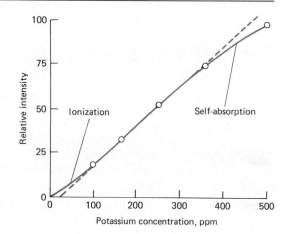

FIGURE 22-9 Effects of ionization and self-absorption on a calibration curve for potassium.

degree of ionization in the flame. Self-absorption, on the other hand, causes negative departures from a straight line at higher concentrations.

22D-4 ANALYTICAL TECHNIQUES
The analytical techniques for flame and plasma emission spectroscopy are similar to those described earlier for atomic absorption spectroscopy (Section 22C-7). Both calibration curves and the standard addition method are employed. In addition, internal standards may be used to compensate for flame variables.

22D-5 COMPARISON OF ATOMIC EMISSION AND ATOMIC ABSORPTION METHODS
For purposes of comparison, the main advantages and disadvantages of the two widely used flame methods, the electrothermal method, and the ICP method are listed in the paragraphs that follow.[6] The comparisons apply to analyses with versatile spectrophotometers that are readily adapted to the determination of numerous elements.

1. *Instruments.* A major advantage of emission methods is that the sample serves as the source. In contrast, absorption methods require an individual lamp for each element (or sometimes, for a limited group of elements). On the other hand, the quality of the monochromator for an absorption instrument does not have to be as great to achieve the same degree of selectivity because the lines emitted by the hollow cathode lamp are so narrow.
2. *Operator skill.* Emission methods generally require a higher degree of operator skill because of the critical nature of such adjustments as wavelength, the region of the source to be sampled, and fuel-to-oxidant ratio.
3. *Background correction.* Correction for band spectra arising from sample constituents is more easily, and often more exactly, carried out for emission methods.
4. *Precision and accuracy.* With electrothermal methods, uncertainties are

[6] For an excellent comparison of the two flame methods, see: E. E. Pickett and S. R. Koirtyohann, *Anal. Chem.*, **41** (14), 28A (1969).

in the range from 5 to 10% relative. In the hands of a skilled operator, uncertainties are about the same for the other three procedures (±0.5 to 1% relative). With less skilled personnel, atomic absorption methods appear to have an advantage.

5. *Interferences.* The two flame methods suffer from similar chemical interferences. Atomic absorption procedures are less subject to spectral line interferences; such interferences, however, are usually easily recognized and avoided in emission methods. Spectral band interferences were considered under background correction. As mentioned earlier, the plasma source is relatively free from chemical interferences.

6. *Concentration range.* Here, inductively coupled plasma spectroscopy appears to offer real advantage. As mentioned earlier, a concentration range of as much as four orders of magnitude has been reported.

7. *Detection limits.* The data in Table 22-2 (p. 586) provides a comparison of detection limits for the four methods under discussion. For a majority of elements, electrothermal atomic absorption provides the highest sensitivity; the inductively coupled plasma procedure is generally more sensitive than either of the two flame methods.

the analysis of real substances

23

Thus far we have stressed the problems associated with the final step in an analysis, specifically the measurement of some parameter that can be related to the concentration of the analyte. More often than not, this final step is relatively simple because it is performed on a solution that has been freed of interferences; consequently, the number of variables influencing the measurement are few, and present theory is often sufficient to account for the effects of these variables. Moreover, convenient and highly refined instruments are available for performing the measurement. Indeed, if every chemical analysis consisted of determining the concentration of a single element or compound in a simple and readily soluble homogeneous mixture, analytical chemistry could profitably be entrusted to the hands of a skilled technician; assuredly, a well-trained chemist could find more useful and challenging work.

Regardless of whether a chemist is employed in academic research or in the laboratories of industry, the materials that are encountered are *not*, as a rule, chemically simple. To the contrary, most substances that require analysis are complex, consisting of several or perhaps several tens of elements or compounds. Frequently these materials are far from ideal in matters of solubility, volatility, stability, and homogeneity. With such substances, several steps must precede the final measurement. As a matter of fact, this final step is often anticlimatic in a sense, being by far the easiest to perform.

To illustrate, consider the analysis for calcium, an element that occurs widely in nature and is important in many manufacturing processes. Several excellent methods exist for determining the calcium ion concentration of a simple aqueous solution, including precipitation as the oxalate, a compound that can be either titrated with a standard solution of permanganate or ignited to the carbonate for a gravimetric measurement. Calcium ion can also be determined by direct titration with ethylenediaminetetraacetic acid, by flame absorption, or by direct potentiometry with a calcium ion electrode. Any of these methods provides an accurate measure of the calcium content of a simple salt such as the carbonate. The chemist, however, is seldom interested in the calcium content of calcium carbonate. More likely, what is needed is the percentage of this element in a sample of animal tissue, a silicate rock, or a piece of glass. The analysis thereby acquires new and formidable complexities. For example, none of these materials is soluble in water or dilute aqueous reagents. Before the amount of calcium can be measured, therefore, the sample must be decomposed by high-temperature treatment with concentrated reagents. Unless care is taken, this step may cause losses of calcium; alternatively, the element may be introduced as a contaminant in the relatively large quantities of reagent usually required to complete the decomposition.

Even after the sample has been decomposed to give a solution of calcium ion, the excellent procedures mentioned previously cannot ordinarily be applied immediately to complete the analysis, for they are all based upon reactions or properties shared by several elements in addition to calcium. Thus, a sample of animal tissue, a silicate rock, or a glass would almost surely contain one or more components that would also precipitate with oxalate, react with ethylenediaminetetraacetic acid, or act as a chemical interference in a flame-absorption measurement; furthermore, the high ionic strength resulting from the reagents used for sample decomposition would complicate a direct potentiometric measurement. As a consequence, steps to free the calcium from potential interferences must usually precede the final measurement; these could well involve several additional operations.

We have chosen the term *real substances* to describe materials such as those in the preceding illustration. In this context, most of the samples encountered in an elementary quantitative analysis laboratory course definitely are not real, for they are generally homogeneous, usually stable even with rough handling, readily soluble, and, above all, chemically simple. Moreover, well-established and thoroughly tested directions exist for their analysis. From the pedagogical viewpoint, there is value in introducing analytical techniques with such substances, for they do allow the student to concentrate on the mechanical aspects of an analysis. Once these mechanics have been mastered, however, there is little point in the continued analysis of unreal substances; to do so creates the impression that any chemical analysis involves nothing more than the slavish adherence to a well-defined and narrow path, the end result of which is a number that is accurate to one or two parts in a thousand. All too many chemists retain this view far into their professional lives.

In truth, the pathway leading to knowledge of the composition of real substances is frequently more demanding of intellectual skills and chemical intuition than of mechanical aptitude. Furthermore, a compromise must often be struck between the time that can be afforded and the accuracy that is believed to be necessary. The chemist is frequently happy to settle for a part or two in one hundred instead of a part or two in a thousand, knowing that the latter may require several hours or even days in additional effort. In fact, with complex materials, even the former accuracy may be unrealistic.

The difficulties encountered in the analysis of real substances arise, of course, from their complexity as well as from differences in their composition. The chemist is frequently unable to find a clearly defined and well-tested analytical route to follow in the literature; thus, an existing procedure must be modified and tested or an entirely new method of analysis must be developed. In either case, each new component creates several new variables. Returning again to the analysis for calcium in calcium carbonate as an example, it is evident that the number of components is small and the variables likely to affect the results are reasonably few. Principal among the latter are the solubility of the sample in acid, the solubility of calcium oxalate as a function of pH, and the effect of the precipitation rate upon the purity and filtering characteristics of calcium oxalate. In contrast, the analysis for calcium in a real sample, such as a silicate rock that contains a dozen or more other elements, is far more complex. Here, the analyst has to consider the solubility not only of the calcium oxalate but also of the oxalates of the other cations that may be present; coprecipitation of each with the calcium oxalate also becomes a concern. Furthermore, a more drastic treatment is required to dissolve the sample, and steps must be taken to eliminate the ions that can interfere with the analysis. Each new step creates additional variables that make a theoretical treatment of the problem difficult if not impossible.

The analysis of a real substance is thus a challenging problem requiring knowledge, intuition, and experience. The development of a procedure for such materials is not to be taken lightly even by the experienced chemist.

23A CHOICE OF METHOD FOR ANALYSIS OF COMPLEX SUBSTANCES

The choice of method for the analysis of a complex substance requires good judgment and a sound knowledge of the advantages and limitations of the various available analytical tools; a familiarity with the literature on the subject is also essential. We cannot be too explicit concerning the selection of a method because there is no single best way that will apply under all circumstances. We can, however, suggest a somewhat systematic approach to the problem and present some generalities that will aid in making an intelligent decision.

23A-1 DEFINITION OF THE PROBLEM

A first step, which must precede any choice of method, involves a clear definition of the analytical problem at hand. The method of approach selected by the chemist will be largely governed by the answers to the following questions:

What is the concentration range of the species to be determined?
What degree of accuracy is demanded by the use to which data will be put?
What other components are present in the sample?
What are the physical and chemical properties of the gross sample?
How many samples are to be analyzed?

The concentration range of the element or compound of interest may well limit the choice of feasible methods. If, for example, the analyst is interested in an element present to the extent of a few parts per million, gravimetric or volumetric methods can generally be eliminated; spectrographic, potentiometric, and other sensitive methods become likely candidates. For components in the parts per million range, the chemist must guard against even small losses as a result of coprecipitation or volatility; contamination from reagents and apparatus also becomes a major concern. In contrast, if the analyte is a major component of the sample, these considerations become less important; furthermore, the classic analytical methods may well be preferable.

The answer to the question regarding the accuracy required is of vital importance in the choice of an analytical method and its performance. It is the height of folly to produce physical or chemical data with an accuracy significantly greater than that demanded by the use to which the data are to be put. We have pointed out that the relationship between time expended and accuracy achieved in an analysis is not ordinarily linear; a 100-fold increase in time (and often more) may be needed to improve the reliability of an analysis from, say, 2% to 0.2% relative. As a consequence, a few minutes spent at the outset of an analysis in careful consideration of what degree of accuracy is really needed represents an investment that a chemist can ill afford to neglect.

The demands of accuracy will frequently dictate the procedure chosen for an analysis. For example, if the allowable error in an aluminum analysis is only a few parts in a thousand, a gravimetric procedure will probably be required. On the other hand if an error of, say, 50 ppt can be tolerated, spectroscopic or electroanalytical procedures should be considered as well. The experimental details of the method are also affected by accuracy requirements. Thus, if precipitation with ammonia were chosen for the analysis of a sample containing 20% aluminum, the presence of 0.2% iron would be of serious concern where accuracy in the parts-per-thousand range was demanded; here, a preliminary separation of the two elements would be necessary. On the other hand, if an error of 50 ppt is tolerable, the separation of iron would not be necessary. Furthermore, this tolerance would govern other aspects of the method as well. For example, 1-g samples could be weighed to perhaps 10 mg and certainly no closer than 1 mg. In addition, less care would be needed in transferring and washing the precipitate and in other time-consuming operations of the gravimetric method. By choosing shortcuts intelligently, the chemist is not being careless but instead realistic in terms of economy of time. The question of accuracy, then, must be settled in clear terms at the very outset.

The third question to be resolved early in the planning state of an analysis is concerned with the chemical composition of the sample. An answer frequently can be reached by considering the origin of the material; in other situations, a partial or complete qualitative analysis must be undertaken. Regardless of its source, however, this information must be available before an intelligent selection of method can be made, since the various steps in any analysis are based on group reactions or group properties; that is, the analysis involves reactions or properties shared by several elements or compounds. Thus, measurement of the concentration of a given element by a method that is simple and straightforward in the presence of one group of elements or compounds may require many tedious and time-consuming separations before it can be used in the presence of others. A solvent that is suitable for one combination of compounds may be totally unsatisfactory when applied to another. Clearly, a knowledge of the qualitative chemical composition of the sample is a prerequisite for its quantitative analysis.

The chemist must consider the physical and chemical properties of the sample closely before attempting to derive a method for its analysis. The way it will be manipulated clearly depends upon its physical state under laboratory conditions; closely related are questions concerning its volatility. It must be established whether the sample is homogeneous and, if it is not, what steps are appropriate to achieve homogeneity. It is also important to know whether or not the sample is hygroscopic or efflorescent. It is essential to know what sort of treatment is sufficient to decompose or dissolve the sample without loss of analyte. Preliminary tests of one sort or another may be needed to provide this information.

Finally, the number of samples to be analyzed is an important aspect in the choice of method. If there are many, considerable time can be expended in calibrating instruments, preparing reagents, assembling equipment, and investigating shortcuts, since the cost of these operations can be spread over the large number of analyses. On the other hand, if at most a few samples are to be analyzed, a longer and more tedious procedure involving a minimum of these preparatory operations may actually prove to be the wiser choice from the economic standpoint.

Having answered these preliminary questions, the chemist is now in a position to consider possible approaches to the problem. At this point, based on past experience, the route to follow may be obvious. Also, it is prudent to speculate on the problems likely to be encountered in the analysis and how they can be solved. By now, some methods will probably have been eliminated from consideration and others put on the doubtful list. Ordinarily, however, the chemist will wish to turn to the analytical literature in order to profit from the experience of others. This, then, is the next logical step in choosing an analytical method.

23A-2 INVESTIGATION OF THE LITERATURE

The literature dealing with chemical analysis is extensive. For the chemist who will take advantage of them, published reports contain much of value. A list of reference books and journals concerned with various aspects of analytical chemistry appears in Appendix 1. This list is not intended to be an exhaustive catalog, but rather one that is adequate for most work. The list is divided into several categories. In many instances, the division is arbitrary, since some of the works could be logically placed in more than one category.

The chemist will probably begin a search of the literature by referring to one or more of the treatises on analytical chemistry or to those devoted to the analysis of specific types of materials. In addition, it is often helpful to consult a general reference work relating to the compound or element of interest. From this survey, a clearer picture of the problem at hand may develop—what steps are likely to be difficult, what separations must be made, what pitfalls must be avoided. Occasionally, all the answers needed or even a set of specific instructions for the analysis may be found. Alternatively, journal references that will lead directly to this information may be discovered. On other occasions, the chemist will acquire only a general notion of how to proceed. Several possible methods may appear suitable; others may have been eliminated. At this point, it is often helpful to consider reference works concerned with specific substances or specific techniques. Alternatively, the various analytical journals may be consulted. Monographs written on methods for completing the analysis are valuable in deciding among several possible techniques.

A major problem in using the analytical journals is that of locating articles which are pertinent to the problem at hand. The various reference books are useful since most are liberally annotated with references to the original journals. The key to a thorough search of the literature, however, is *Chemical Abstracts*. This journal contains short abstracts of all papers appearing in the major chemical publications of the world. Both yearly and cumulative indexes are provided to aid in the search; by looking under the element or compound to be determined and the type of substance to be analyzed, the chemist can make a thorough survey of the methods available. Completion of such a survey involves the expenditure of a great deal of time, however, and is often made unnecessary by consulting reliable reference works.

23A-3 CHOOSING OR DERIVING A METHOD

Having defined the problem and investigated the literature for possible approaches, the chemist must next decide upon the route to be followed in the laboratory. If the choice is simple and obvious, analysis can be undertaken directly. Frequently, however, the decision requires the exercise of considerable judgment and ingenuity; here experience, an understanding of chemical principles, and perhaps intuition all come into play.

If the substance to be analyzed occurs widely, the literature survey will probably have yielded several alternative methods for the analysis. Economic considerations may dictate the method that will yield the desired reliability with least expenditure of time and effort. As mentioned earlier, the number of samples to be analyzed will often be a determining factor in this choice.

Investigation of the literature will not invariably reveal a method designed specifically for the type of sample in question. Ordinarily, however, the chemist will have encountered procedures for materials that are at least analogous in composition to the one in question; the decision will then have to be made as to whether the variables introduced by differences in composition are likely to have any influence on results. This judgment is often difficult and fraught with uncertainty; recourse to the laboratory may be the only way of obtaining an unequivocal answer.

If it is decided that existing procedures are not applicable, consideration must be given to modifications that may overcome the problems imposed by the variation in composition. Again it may be possible to propose only tentative alterations, owing to the complexity of the system; whether these modifications will accomplish their purpose without introducing new difficulties can only be determined in the laboratory.

After giving due consideration to existing methods and their modifications, the chemist may decide that none will fit the problem and new procedures must be developed. Here, all of the facts with respect to the chemical and physical properties of the analyte must be marshalled and given consideration. Several possible ways of performing the desired measurement may be derived from this information. Each of the possibilities must then be examined critically, with consideration given to the influence of the other components in the sample as well as the reagents that must be used for solution or decomposition. At this point, the chemist must try to anticipate sources of error and possible interferences arising from interactions among the components and reagents; methods may have to be derived by which problems of this sort can be circumvented. In the end, it is to be hoped that one or more tentative methods that are worth testing will have been located. It is probable that the feasibility of some of the steps in the procedure cannot be determined on the basis of theoretical considerations alone; recourse must be made to preliminary laboratory testing of such steps. Certainly, critical evaluation of the entire procedure can come only from careful laboratory work.

23A-4 TESTING THE PROCEDURE

Once a procedure for an analysis has been selected, the problem usually arises as to whether the method can be employed directly, without testing, to the problem at hand. The answer to this question is not simple and depends upon a number of considerations. If the method chosen has been the subject of a single, or at most a few, literature references, there may be a real point to preliminary laboratory evaluation. With experience, the chemist becomes more and more cautious about accepting claims regarding the accuracy and applicability of a new method. All too often, statements found in the literature are overly optimistic; a few hours spent in testing the procedure in the laboratory may be enlightening.

Whenever a major modification of a standard procedure is undertaken or an attempt is made to apply it to a type of sample different from that for which it was designed, a preliminary laboratory test is advisable. The effects of such alterations simply cannot be predicted with certainty, and the chemist who dispenses with such precautions is sanguine indeed.

Finally, of course, a newly devised procedure must be extensively tested before it is adapted for general use. We must now consider the means by which a new method or a modification of an existing method can be tested for reliability.

Analysis of standard samples. Unquestionably, the best technique for evaluating an analytical method involves the analysis of one or more standard samples whose composition with respect to the element or compound of interest is reliably known. For this technique to be of value, however, it is essential that the standards closely resemble the samples to be analyzed with respect to both the concentration range of the analyte and the overall composition. Occasionally, standards of this sort can be readily synthesized from weighed quantities of pure compounds. Others may be purchased from sources such as the National Bureau of Standards;[1] these latter, however, are confined largely to common materials of commerce or widely distributed natural products.

As often as not, the chemist will be unable to acquire a standard sample that matches closely the substance to be analyzed. This situation is particularly true of complex materials in which the form of the analyte is unknown or variable and quite impossible to reproduce. In these circumstances, the best the chemist can do is to prepare a solution of known concentration whose composition approximates that of the sample after it has been decomposed and dissolved. Obviously, such a standard gives no information at all concerning the fate of the substance being determined during the important decomposition and solution steps.

Analysis by other methods. The results of an analytical method can sometimes be evaluated by comparison with some entirely different method. Clearly, a second method must exist; in addition, it should be based on chemical principles that differ considerably from the one under examination. Comparable results from the two serve as presumptive evidence that both are yielding satisfactory results, inas-

---

[1] United States Department of Commerce, *NBS Standard Reference Materials*, 1979–80 ed. NBS Special Publ. 260. Washington, D.C.: Government Printing Office, 1979.

much as it is unlikely that the same determinate errors would affect each. Such a conclusion will not apply to those aspects of the two methods that are similar.

Standard addition to the sample. When the foregoing approaches are inapplicable, the standard addition method may prove useful. Here, in addition to being used to analyze the sample itself, the proposed procedure is tested against portions of the sample to which known amounts of the analyte have been added. The effectiveness of the method can then be established by evaluating the extent of recovery of the added quantity. The standard addition method may reveal errors arising from the method of treating the sample or from the presence of the other elements or compounds.

23B ACCURACY OBTAINABLE IN ANALYSIS OF COMPLEX MATERIALS

To provide a clear idea of the accuracy that can be expected when the analysis of a complex material is carried out with a reasonable amount of effort and care, data on the determination of four elements in a variety of materials are presented in the tables that follow. These data were taken from a much larger set of results collected by W. F. Hillebrand and G. E. F. Lundell of the National Bureau of Standards and published in the first edition of their excellent book on inorganic analysis.[2]

The materials analyzed were naturally occurring substances and items of commerce; they had been especially prepared to give uniform and homogeneous samples and were then distributed among chemists who were, for the most part, actively engaged in the analysis of similar materials. The analysts were allowed to use the methods they considered most reliable and best suited for the problem at hand. In most instances, special precautions were taken so that the results are better than can be expected from the average routine analysis; on the other hand, they probably do not represent the ultimate in analytical perfection.

The number in the second column of each table is a best value obtained by the most painstaking analysis for the measured quantity. It is considered to be the "true value" for calculations of the absolute and relative errors shown in the fourth and fifth columns. The fourth column was obtained by discarding results that were extremely divergent, determining the deviation of the remaining individual data from the best value (second column), and averaging these deviations. The fifth column, the percent relative error, was obtained by dividing the data in the fourth column by the best value found in the second column and multiplying by 100.

The results for the 4 elements shown in these tables are typical of the data for 26 elements reported in the original publication. It is to be concluded that analyses reliable to a few tenths of a percent relative are the exception rather than the rule for the analysis of complex mixtures by ordinary methods and that, unless the chemist is willing to invest an inordinate amount of time in the analysis, errors on the order of 1 or 2% must be accepted. If the sample contains less than 1% of the analyte, even larger relative errors are to be expected.

Finally, it is clear from these data that the accuracy obtainable in the deter-

---

[2] W. F. Hillebrand and G. E. F. Lundell, *Applied Inorganic Analysis*, pp. 874–887. New York: Wiley, 1929.

mination of an element is greatly dependent upon the nature and complexity of the substrate. Thus, the relative error for the determination of phosphorus in two phosphate rocks was 1.1%; in a synthetic mixture, it was only 0.27%. The relative error in an iron determination in a refractory was 7.8%; in a manganese bronze having about the same iron content, it was only 1.8%. Here, the limiting factor in the accuracy is not in the completion step but rather in solution of the samples and the separation of interferences.

From these data, it is clear that the chemist is well advised to adopt a pessimistic viewpoint regarding the accuracy of an analysis, be it his or her own or one performed by someone else.[3]

[3] For additional data that corroborate this view, see: S. Abbey, *Anal. Chem.*, **53**, 529A (1981).

Table 23-1   Analysis of Iron in Various Materials*

Material	Iron Present (percent)	Number of Analysts	Average Error (absolute)	Average Error (percent relative)
Soda-lime glass	0.064 (Fe_2O_3)	13	0.01	15.6
Cast bronze	0.12	14	0.02	16.7
Chromel	0.45	6	0.03	6.7
Refractory	0.90 (Fe_2O_3)	7	0.07	7.8
Manganese bronze	1.13	12	0.02	1.8
Refractory	2.38 (Fe_2O_3)	7	0.07	2.9
Bauxite	5.66	5	0.06	1.1
Chromel	22.8	5	0.17	0.75
Iron ore	68.57	19	0.05	0.07

* From W. F. Hillebrand and G. E. F. Lundell, *Applied Inorganic Analysis*, p. 878. New York: John Wiley & Sons, Inc., 1929. With permission.

Table 23-2   Analysis for Manganese in Various Materials*

Material	Manganese Present (percent)	Number of Analysts	Average Error (absolute)	Average Error (percent relative)
Ferro-chromium	0.225	4	0.013	5.8
Cast iron	0.478	8	0.006	1.3
	0.897	10	0.005	0.56
Manganese bronze	1.59	12	0.02	1.3
Ferro-vanadium	3.57	12	0.06	1.7
Spiegeleisen	19.93	11	0.06	0.30
Manganese ore	58.35	3	0.06	0.10
Ferro-manganese	80.67	11	0.11	0.14

* From W. F. Hillebrand and G. E. F. Lundell, *Applied Inorganic Analysis*, p. 880. New York: John Wiley & Sons, Inc., 1929. With permission.

Table 23-3   Analysis for Phosphorus in Various Materials*

Material	Phosphorus Present (percent)	Number of Analysts	Average Error (absolute)	Average Error (percent relative)
Ferro-tungsten	0.015	9	0.003	20.
Iron ore	0.040	31	0.001	2.5
Refractory	0.069 (P_2O_5)	5	0.011	16
Ferro-vanadium	0.243	11	0.013	5.4
Refractory	0.45	4	0.10	22.
Cast iron	0.88	7	0.01	1.1
Phosphate rock	43.77 (P_2O_5)	11	0.5	1.1
Synthetic mixtures	52.18 (P_2O_5)	11	0.14	0.27
Phosphate rock	77.56 ($Ca_3(PO_4)_2$)	30	0.85	1.1

* From W. F. Hillebrand and G. E. F. Lundell, *Applied Inorganic Analysis*, p. 882. New York: John Wiley & Sons, Inc., 1929. With permission.

Table 23-4   Analysis for Potassium in Various Materials*

Material	Potassium Oxide Present (percent)	Number of Analysts	Average Error (absolute)	Average Error (percent relative)
Soda-lime glass	0.04	8	0.02	50.
Limestone	1.15	15	0.11	9.6
Refractory	1.37	6	0.09	6.6
	2.11	6	0.04	1.9
	2.83	6	0.10	3.5
Lead-barium glass	8.38	6	0.16	1.9

* From W. F. Hillebrand and G. E. F. Lundell, *Applied Inorganic Analysis*, p. 883. New York: John Wiley & Sons, Inc., 1929. With permission.

preliminary steps to an analysis

24

A chemical analysis is ordinarily preceded by steps that are necessary if the analytical data are to have significance. These steps include (1) sampling, (2) production of a homogeneous mixture for analysis, and (3) drying the sample or, alternatively, determining its moisture content. In rare circumstances, none of these steps is important or necessary; more commonly, one or more is vital in determining the accuracy and significance of the analytical result. This chapter deals briefly with each of these steps.

24A SAMPLING

Generally, a chemical analysis is performed on a fraction of the material whose composition is of interest. It is evident that the composition of this fraction must reflect as closely as possible the average composition of the bulk of the material if the analysis is to be of any value. The process by which a representative fraction is acquired is termed *sampling*. Often, sampling is the most difficult step in the entire analytical process. This statement is particularly applicable when the material to be sampled is an item of commerce weighing several tons or several hundreds of tons.

The end product of the sampling operation will be a quantity of material weighing a few grams or, at most, a few hundred grams that may be as little as one part in 10^7 or 10^8 of the entire material whose composition is sought. Yet the composition of this tiny fraction must, as closely as possible, be identical to the average composition of the total mass. Where, as with an ore or other items of commerce, the material is inherently a nonhomogeneous solid, the task of producing a representative sample is indeed formidable. Clearly, the reliability of the analysis cannot exceed that of the process by which the sample was acquired; the most painstaking work upon a poor sample is a waste of effort.

The literature on sampling of nonhomogeneous material is extensive.[1] We can only provide a brief outline of sampling methods. Basically, two steps are involved: (1) collection of a gross sample and (2) reduction of the gross sample to a size convenient for laboratory work.

24A-1 THE GROSS SAMPLE

Ideally, the gross sample is a miniature replica of the bulk of the material to be analyzed. It corresponds to the whole, not only in chemical composition but also in particle-size distribution.

Size of the gross sample. From the standpoint of convenience and economy, it is desirable that the gross sample be no larger than absolutely necessary. Basically, sample size is determined by (1) the uncertainty that can be tolerated between the composition of the sample and the whole, (2) the degree of heterogeneity of the material being sampled, and (3) the level of particle size at which heterogeneity begins. This last point warrants amplification. In a well-mixed, homogeneous solution of a gas or a liquid, heterogeneity exists only on a molecular scale, and the size of the molecules themselves will govern the minimum size of the gross sample. A particulate solid such as an ore or a soil represents the opposite situation. In such materials, the individual pieces of solid can be seen to differ in composition. Here, heterogeneity develops in particles that may have dimensions on the order of a centimeter or more. Intermediate between these extremes are colloidal materials and solidified metals. With the former, heterogeneity is first encountered in the particles of the dispersed phase; these typically have diameters in the range of 10^{-5} cm or less. In an alloy, heterogeneity first occurs among the crystal grains.

In order to obtain a truly representative gross sample, a certain number **n** of the particles referred to in (3) must be taken. The magnitude of this number is dependent upon (1) and (2) and may involve only a few particles, several millions, or even several millions of millions. Large numbers are of no great concern for homogeneous gases or liquids since heterogeneity among particles first occurs at the molecular level; thus, even a very small weight of sample will contain the requisite number. On the other hand, the individual particles of a particulate solid may weigh a gram or more; the gross sample may necessarily comprise several tons of material. Here, sampling is a costly, time-consuming procedure at best; determination of the smallest quantity required to provide the needed information will minimize the expense.

The composition of a sample removed randomly from a bulk of material is governed by the law of chance; thus, by suitable statistical manipulations, it is possible to predict the probability of a given fraction being similar to the whole. A

[1] See, for example, F. J. Welcher, Ed., *Standard Methods of Chemical Analysis*, 6th ed., vol. 2, part A, pp. 21–52. Princeton, NJ: Van Nostrand, 1963; *Book of Standards*. Philadelphia: American Society for Testing Materials (sampling of substances such as paints, fuels, petroleum products, and constructional materials); *Official Methods of Analysis*, 12th ed. Washington, DC: Association of Official Agricultural Chemists, 1975 (for soils, fertilizers, and foods); *Methods for Chemical Analysis of Metals*, 2d ed., pp. 57–72. Philadelphia: American Society for Testing Materials, 1956 (for metals and alloys). An extensive bibliography of specific sampling information has been compiled by C. A. Bicking, in *Treatise on Analytical Chemistry*, 2d ed., I. M. Kolthoff and P. J. Elving, Eds., part I, vol. 1, p. 299ff. New York: Interscience, 1978; B. Kratochvil and J. K. Taylor, *Anal. Chem.*, **53**, 924A (1981).

simple, idealized case will serve as an example. A carload of lead ore is made up of just two kinds of particles: galena (lead sulfide) and a gangue containing no lead. All particles have the same size. The car contains, let us say, 100 million particles, and we wish to know the fraction of these that comprises galena. The composition of the carload could, of course, be obtained exactly by counting all the galena particles, since the two components differ in appearance; this approach however, would probably involve several lifetimes of work. We must thus settle for the lesser accuracy involved in counting some reasonable fraction of the components. The number contained in this fraction will, of course, depend upon the error we are willing to tolerate in the measurement.

The relationship between the allowable error and the number of particles **n** to be counted can be stated as follows[2]:

$$\mathbf{n} = \frac{(1 - p)}{p\sigma^2} \tag{24-1}$$

where p is the fraction of galena particles; $(1 - p)$ is the fraction of gangue particles; and σ is the allowable relative standard deviation in the count of the galena particles. Thus, for example, if 80% of the particles were present as galena ($p = 0.8$) and the tolerable standard deviation was 1% ($\sigma = 0.01$), a random sampling of 2500 particles should be made. A standard deviation of 0.1% would require a sample containing 250,000 particles.

Let us now make the problem somewhat more realistic and assume that one of the components in the car contains a higher percentage P_1 of lead and the other component a lesser amount P_2. Furthermore, the average density d of the shipment differs from the densities d_1 and d_2 of these components. We are now interested in deciding the number of particles, and thus the weight to be taken, to assure a sample possessing the average percent of lead in the bulk P with a relative standard deviation due to sampling of σ. Equation 24-1 can be extended to include these stipulations[3]:

$$\mathbf{n} = p(1 - p) \left(\frac{d_1 d_2}{d^2}\right)^2 \left(\frac{P_1 - P_2}{\sigma P}\right)^2 \tag{24-2}$$

From this equation, we see that the demands of accuracy are costly, in terms of the sample size required, because of the inverse square relationship between the allowable standard deviation and the number of particles taken. Furthermore, a greater number of particles must be taken as the average percentage P of the element of interest becomes smaller. Finally, the degree of heterogeneity as measured by $(P_1 - P_2)$ has a profound effect, with the number of particles increasing as the square of the difference in composition of the two components of the mixture.

The problem of deciding upon a gross sample size is ordinarily more difficult than this example because most samples not only contain more than two components but also consist of a range of particle sizes. In most instances the first of

[2] For a discussion of the derivation and significance of this equation and the one that follows, see: A. .A. Benedetti-Pichler, in *Physical Methods in Chemical Analysis*, W. G. Berl, Ed., Vol. 3, pp. 183–194. New York: Academic Press, 1956; A. A. Benedetti-Pichler, *Essentials of Quantitative Analysis*, Chapter 19. New York: Ronald Press, 1956.
[3] Benedetti-Pichler, *Loc. cit.*

these problems can be met by dividing the sample into an imaginary two-component system. Thus, with an actual lead ore, one component might be selected to include all the various lead-bearing minerals of the ore, and the other to include all the residual components containing little or no lead. After assigning average densities and percentages of lead to each of these parts, the system would then be treated as if it had only two components.

The problem of variable particle size can be handled by calculation of the number of particles that would be needed if the sample consists of particles of a single size. Then, the actual sample weight is determined by taking into account the particle-size distribution. One approach would be to calculate the needed weight by assuming that all particles are the size of the largest. This procedure is not very efficient, however, for it usually calls for removal of a larger weight of material than necessary. Benedetti-Pichler gives alternative methods for computing the weight of sample to be chosen.[4]

An interesting conclusion from Equation 24-2 is that the number of particles comprising the gross sample is independent of particle size. The weight of the sample, of course, increases directly as the volume (or the cube of the particle diameter) so that reduction in the particle size of a given material has a large effect on the mass required in the gross sample.

Clearly, a great deal of information must be known about a substance in order to make use of Equation 24-2. Fortunately, reasonable estimates of the various parameters in the equation can often be made. These estimates can be based upon a qualitative analysis of the substance, visual inspection of the material, and information from the literature for substances of similar origin. Crude measurements of densities of the various sample components may also be necessary.

EXAMPLE 24-1. Suppose that the average particle in the carload of lead ore just considered was judged to be approximately spherical with a radius of about 3 mm. Roughly 4% of the particles appeared to be galena (~70% Pb), which has a density of 7.6; the remaining particles had a density of about 3.5 and contained little or no lead. How many pounds of ore should be removed if the sampling uncertainty is to be kept below 1% relative?

To apply Equation 24-2, we must first obtain values for the average density and percent lead.

$$d = 0.04 \times 7.6 + 0.96 \times 3.5 = 3.7$$

$$P = \frac{0.04 \times 7.6 \times 0.70 \text{ g Pb/cm}^3}{3.7 \text{ g sample/cm}^3} \times 100 = 5.8$$

Then

$$n = 0.04 \,(1 - 0.04) \left(\frac{7.6 \times 3.5}{3.7^2}\right)^2 \left(\frac{70 - 0}{0.01 \times 5.8}\right)^2$$

$$= 2.1 \times 10^5 \text{ particles required}$$

$$\text{wt sample} = 2.1 \times 10^5 \text{ particles} \times \frac{4}{3}\pi\,(0.3)^3 \,\frac{\text{cm}^3}{\text{particle}} \times \frac{3.7 \text{ g}}{\text{cm}^3} \times \frac{1}{454 \text{ g/lb}}$$

$$= 194 \text{ or about } 200 \text{ lb}$$

---

[4] A. A. Benedetti-Pichler, in *Physical Methods in Chemical Analysis*, W. G. Berl, Ed., Vol. 3, p. 192. New York: Academic Press, 1956.

Sampling homogeneous solutions of liquids and gases. For solutions of liquids or gases, the gross sample can be relatively small, since ordinarily nonhomogeneity first occurs at the molecular level, and even small volumes of sample will contain a tremendous number of particles. Whenever possible, the material to be analyzed should be well stirred prior to removal of the sample to make sure that homogeneity does indeed exist. With large volumes of solutions, mixing may be impossible; it is then best to sample several portions of the container with a "sample thief," a bottle that can be opened and filled at any desired location in the solution. This type of sampling, for example, is important in determining the constituents of liquids exposed to the atmosphere. Thus, the oxygen content of lake water may vary by a factor as large as 1000 over a depth difference of a few feet.

Industrial gases or liquids are often sampled continuously as they flow through pipes, with care being taken to ensure that the sample collected represents a constant fraction of the total flow and all portions of the stream are sampled.

Sampling particulate solids. The process of obtaining a random sample from a bulky particulate material is often difficult. It can be best accomplished while the material is being transferred. For example, every tenth shovelful or wheelbarrow load may be consigned to a sample pile, or portions of the material may be intermittently removed from a conveyor belt. Alternatively, the material may be forced through a riffle or a series of riffles that continuously isolate a fraction of the stream. Mechanical devices of this sort have been highly developed for handling coals and ores.

Sampling a large pile of a solid is difficult, inasmuch as there is seldom assurance that the material has a random distribution throughout. This problem is particularly acute where there is a considerable variation in composition with particle size. Inevitably, small particles tend to collect at the bottom and center of the pile, and the larger particles at the outside. Here, division of the material must involve some system that will maintain the same distribution of particle sizes in the sample as in the whole. Several stepwise partitions of the total are required.

A procedure called *coning and quartering* is often employed with amounts of material of the order of about 100 lb and containing particles smaller than about 4 mm in diameter. It is also applicable to smaller quantities in the laboratory. In this process, the solid is formed into a cone, with each new shovelful being deposited on the apex of the pile. Mixing is accomplished by repeatedly forming cones, and the solid from each old cone is removed systematically from around its base. The cone is then flattened by pressing down on its apex with a board or shovel to give a circular layer of material. This circle is divided into four equal quarters by drawing two perpendicular lines through its center. Alternate quarters are then discarded; the coning process is repeated, if necessary, on the residual two quarters.

Another effective way of mixing and dividing a solid weighing up to about 100 lb involves *rolling and quartering*. Here, a conical pile of the solid, consisting of particles that have been reduced to a diameter of 1 mm or less, is placed upon a tarpaulin, a rubberized sheet, or (for small samples) a piece of glazed paper. The cone is flattened and the solid is mixed by pulling first one corner of the sheet over to the opposite corner and then another corner over to its opposite corner. This rolling process should be repeated 100 times or more. Finally, the material is collected in the center of the sheet by simultaneously raising all four corners. The resultant pile is flattened and quartered, and a pair of opposite quarters is rejected. Rolling and quartering is also useful for reducing the size of a laboratory sample.

Sampling metals and alloys.   Samples of metals and alloys are obtained by sawing, milling, or drilling. In general, it is not safe to assume that chips of the metal removed from the surface will be representative of the entire bulk; sampling must include solid from the interior of the piece as well as from the surface. With billets or ingots of metal, a representative sample can be obtained by sawing across the piece at regularly spaced intervals and collecting the "sawdust" as the sample. Alternatively, the specimen may be drilled, again at various regularly spaced intervals, and the drillings collected as the sample; the drill should pass entirely through the block or halfway through from opposite sides. The drillings can then be broken up and mixed or melted together in a graphite crucible. A granular sample can often then be produced by pouring the melt into distilled water.

24A-2 PRODUCTION OF A LABORATORY SAMPLE

For nonhomogeneous materials, the gross sample may weigh several hundred pounds or more. Here, a considerable decrease in size is desirable before the sample is brought into the laboratory, where a few pounds at most are all that can be conveniently handled. The process of reducing the sample volume by a factor of 100 or more is ordinarily multistaged, involving repeated grinding, mixing, and dividing. Diminution in particle size is essential as the weight of sample is decreased to assure that the sample composition continues to be representative of the original material (Equation 24-2).

24A-3 TREATMENT OF THE LABORATORY SAMPLE[5]

The sample often requires further treatment in the laboratory before it can be analyzed, particularly if it is a solid. One of the objects of this pretreatment is to produce a material so homogeneous that any small portion removed for the analysis will be identical to any other fraction. Attainment of this condition usually involves decreasing the size of particles to a few hundredths of a millimeter and thorough mechanical mixing. Another object of the pretreatment is to convert the substance to a form which is readily attacked by the reagents employed in the analysis; with refractory materials particularly, grinding to a very fine powder is required. Finally, the sample may have to be dried or its moisture content determined.

Crushing and grinding of laboratory samples.   A certain amount of crushing or grinding is ordinarily required to decrease the particle size of solid samples. Unfortunately, these operations also tend to alter the composition of the sample; for this reason, the particle size should be reduced no more than is required for homogeneity and ready attack by reagents.

Several factors may cause appreciable changes in the composition of the sample as a result of grinding. Among these is the heat that is inevitably generated, which can cause losses of volatile components in the sample. In addition, grinding

---

[5] For further discussion, see: W. F. Hillebrand, G. E. F. Lundell, H. A. Bright, and J. I. Hoffman, *Applied Inorganic Analysis*, 2d ed., pp. 809–814. New York: Wiley, 1953; A. A. Benedetti-Pichler, *Essentials of Quantitative Analysis*, Chapters 18 and 19. New York: Ronald Press, 1956.

increases the surface area of the solid and thus increases its susceptibility to reactions with the atmosphere. For example, it has been observed that the iron(II) content of a rock may be altered by as much as 40% during grinding—apparently a direct result of the iron being oxidized to the $+3$ state.

The effect of grinding on the gain or loss of water from solids is considered in a later section.

Differences in the hardness of the components in a sample represent yet another potential source of error associated with crushing and grinding. Softer materials are converted to smaller particles more rapidly than are the hard ones; any loss of sample in the form of dust will thus cause an alteration in composition. Similarly, flying fragments tend to contain a higher fraction of the harder components.

Intermittent screening often increases the efficiency of grinding. In this operation, the ground sample is placed upon a wire or cloth sieve that will pass particles of the desired size. The residual particles are then returned for further grinding; the operation is repeated until the entire sample passes through the screen. This process will certainly result in segregation of the components with the hardest materials being last through the screen; it is obvious that grinding must be continued until the last particle has been passed. The need for further mixing after screening is also apparent.

A serious error can arise during grinding and crushing as a consequence of mechanical wear and abrasion of the grinding surfaces. Even though these surfaces are fabricated from hardened steel, agate, or boron carbide, contamination of the sample nevertheless is occasionally encountered. The problem is particularly acute in the analysis for minor constituents.

The so-called *Plattner diamond mortar*, shown in Figure 24-1, is used for crushing hard, brittle materials. It is constructed of hardened tool steel and consists of a base plate, a removable collar, and a pestle. The sample to be crushed is placed on the base plate inside the collar. The pestle is then fitted into place and struck several blows with a hammer. The sample is reduced to a fine powder; it is collected on a glazed paper after the apparatus has been disassembled.

A useful device for grinding solids that are not too hard is the *ball mill*. It consists of a porcelain crock of perhaps 2-L capacity, which can be sealed and ro-

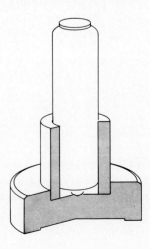

FIGURE 24-1 A Plattner diamond mortar.

tated mechanically. The container is charged with approximately equal volumes of the sample and flint or porcelain balls having a diameter of 20 to 50 mm. Grinding and crushing occur as the balls tumble within the rotating container. A finely ground and well-mixed powder can be produced in this way.

The mortar and pestle, the most ancient of grinding tools, still find wide use in the analytical laboratory. They now come in a variety of sizes and shapes and are fabricated from glass, porcelain, agate, mullite, and other hard materials.

Mixing solid laboratory samples. It is essential that solid materials be thoroughly mixed to assure random distribution of the components in the analytical sample. Several methods are commonly employed in the laboratory. One of these involves rolling the sample on a sheet of glazed paper. A pile of the substance is placed in the center and mixed by lifting one corner of the paper enough to roll the particles of the sample to the opposite corner. This operation is repeated many times, with the four corners of the sheet being lifted alternately.

Effective mixing of solids is also accomplished by rotating the sample for some time in a ball mill or a twin-shell dry-blender.[6]

It is worthwhile noting that with long standing, segregation of finely ground homogeneous materials upon the basis of particle size or density may occur. For example, analyses of layers of a set of student unknowns, which had not been used for several years, revealed a regular variation in the analyte concentration from top to bottom of the container. Apparently, segregation had occurred in this time as a consequence of vibrations in the storage area.

24B MOISTURE IN SAMPLES

The presence of water in the sample is a common and vexing problem that faces the chemist more often than not. Water may be present as a contaminant from the atmosphere or it may be chemically bound within the sample. Regardless of its origin, however, water plays a part in determining the composition of the sample. Unfortunately, and with solids particularly, the water content will vary with humidity, temperature, and state of subdivision. Thus, the constitution of a sample may change significantly with environment and method of handling.

To cope with this source of variability in composition, the chemist may attempt to remove moisture from the sample prior to the weighing step or, if this is not possible, to bring the water content to some reproducible level that can be duplicated at a later date if necessary. A third alternative involves determination of the water content at the time the samples are weighed for analysis; in this way the results can be corrected to a dry basis. In any event, most analyses are preceded by some sort of preliminary treatment designed to take into account the presence of water.

24B-1 FORMS OF WATER IN SOLIDS

It is convenient to distinguish among the several ways in which water is retained by a solid. Although developed primarily with respect to minerals, the classifica-

---

[6] Patterson Kelley Co. Inc., Stroudsburg, PA.

tion of Hillebrand[7] and his collaborators may be applied to other solids as well and forms the basis for the discussion that follows.

Essential water. That water which forms an integral part of the molecular or crystal structure of a component of the solid is classed as essential water. It exists in stoichiometric quantities. Thus, the *water of crystallization* in a stable solid hydrate (for example, $CaC_2O_4 \cdot 2H_2O$, $BaCl_2 \cdot 2H_2O$) qualifies as a type of essential water.

A second form is called *water of constitution*. Here, the water is not present as such in the solid but rather is formed as a product when the solid undergoes decomposition, usually as a result of heating. This type of water is typified by the processes

$$2KHSO_4 \longrightarrow K_2S_2O_7 + H_2O$$
$$Ca(OH)_2 \longrightarrow CaO + H_2O$$

Nonessential water. Nonessential water is not necessary for characterization of the chemical constitution of the sample and therefore does not occur in any sort of stoichiometric proportion. It is retained by the solid as a consequence of physical forces.

Adsorbed water is retained on the surface of solids in contact with a moist environment. The amount adsorbed is dependent upon humidity, temperature, and the specific surface area of the solid. Adsorption of water occurs to some degree upon all solids.

A second type of nonessential water is called *sorbed water* and is encountered with many colloidal substances such as starch, protein, charcoal, zeolite minerals, and silica gel. In contrast to adsorption, the quantity of sorbed water is often large, amounting to as much as 20% or more of the total weight of the solid. Interestingly enough, solids containing even this much water may appear as perfectly dry powders. Sorbed water is held as a condensed phase in the interstices or capillaries of the colloidal solid. The quantity contained in the solid is greatly dependent upon temperature and humidity.

A third type of nonessential moisture is *occluded water*. Here, liquid water is entrapped in microscopic pockets spaced irregularly throughout solid crystals. Such cavities often occur in minerals and rocks (also in gravimetric precipitates).

Water may also be dispersed in a solid as a *solid solution*. Here, the water molecules are distributed homogeneously throughout the solid. Natural glasses may contain as much as several percent of moisture in this form.

24B-2 EFFECT OF TEMPERATURE AND HUMIDITY ON WATER CONTENT OF SOLIDS
In general, the concentration of water contained in a solid tends to decrease with increasing temperature and decreasing humidity. The magnitude of these effects and the rate at which they manifest themselves differ considerably according to the manner in which water is retained.

Water of crystallization. The relationship between humidity and the water con-

---

[7] W. F. Hillebrand, G. E. F. Lundell, H. A. Bright, and J. I. Hoffman, *Applied Inorganic Analysis*, 2d ed., p. 815. New York: Wiley, 1952.

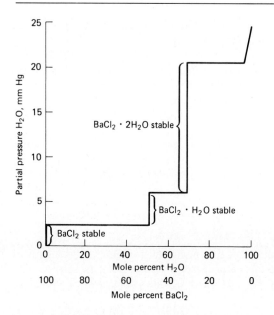

FIGURE 24-2 Vapor pressure-composition diagram for barium chloride and its hydrates, 25°C.

tent of a crystalline hydrate is shown by a vapor pressure–composition diagram; Figure 24-2 illustrates such a plot for barium chloride. It was obtained by measuring the pressure of water vapor over a mixture of barium chloride and its hydrates in a closed system. The mole percentage of water is shown along the abscissa; the equilibrium vapor pressure in torr (mm of Hg) is plotted as the ordinate. When anhydrous barium chloride is brought into equilibrium with a dry atmosphere, the partial pressure of water is zero. Addition of water to the system results in formation of an amount of the monohydrate and establishment of the equilibrium

$$BaCl_2 \cdot H_2O(s) \rightleftharpoons BaCl_2(s) + H_2O(g)$$

The vapor pressure of water in the system will be determined by this equilibrium, for which the equilibrium constant, K', is given by

$$K' = p'_{H_2O}$$

where p'_{H_2O} is the equilibrium pressure of water. As long as both the monohydrate and anhydrous salt are present, their activities are constant, and the partial pressure of water is *independent of the amounts* of these two compounds. This condition is shown by the horizontal line extending from just above 0 mole percent water to 50 mole percent. At 50 mole percent water, anhydrous barium chloride ceases to exist, and with its disappearance, the preceding equilibrium expression is no longer applicable. Increases in the amount of water result in formation of a new species, the dihydrate; the vapor pressure over the mixture is now governed by the equilibrium between this compound and the monohydrate

$$BaCl_2 \cdot 2H_2O(s) \rightleftharpoons BaCl_2 \cdot H_2O(s) + H_2O(g)$$

for which

$$K'' = p''_{H_2O}$$

As shown in Figure 24-2, p''_{H_2O} is larger than p'_{H_2O}. Again, the equilibrium pressure is constant as long as both the monohydrate and dihydrate are present.

When the mole percent of water in the system exceeds the molar ratio of water in the dihydrate (that is, 66.7%), the monohydrate disappears completely. Higher hydrates are not formed, however; instead, the dihydrate begins to dissolve. The result is a saturated solution that is in equilibrium with the solid dihydrate; that is,

$$BaCl_2(\text{sat'd soln}) \rightleftharpoons BaCl_2 \cdot 2H_2O(s) + H_2O(g)$$

As long as some dihydrate remains, we may write

$$K''' = p'''_{H_2O}$$

Again the equilibrium pressure of water will acquire a new value. This condition will be maintained until the solution is no longer saturated (at about 97 mole percent water). The solid dihydrate then disappears, and the vapor pressure of water increases continuously, approaching that of pure water (100 mole percent) at high dilutions.

A diagram such as Figure 24-2 is useful, for it shows clearly the stable forms of a substance at a given temperature as well as the conditions necessary to produce a given form. The behavior of hydrates under various atmospheric conditions can also be predicted. For example, Figure 24-2 indicates that barium chloride dihydrate is the stable form at 25°C when the partial pressure of water in its surroundings ranges between about 6 and 21 torr. These pressures correspond to a relatively humidity range of 25 to 88%.[8] The relative humidity in most laboratories will be well within this range except on very dry or very damp days; thus, the dihydrate will be stable when exposed to typical laboratory conditions. Moreover, if anhydrous barium chloride were in contact with an atmosphere that had a relative humidity within this range, absorption of moisture would occur until equilibrium had been achieved; that is, until all of the anhydrous salt had been completely converted to the dihydrate. Similarly, an aqueous solution of barium chloride would lose water to the atmosphere under these conditions until finally only crystals of the equilibrium species, the dihydrate, remained. The dihydrate would, of course, lose water under some circumstances. For example, if the relative humidity dropped below 25%, as might happen during a dry winter day, equilibrium would favor formation of the monohydrate. If the dihydrate were placed in a desiccator with a reagent that kept the partial pressure of water below 2 torr, quantitative conversion to the anhydrous salt would be the ultimate result. Thus, the composition of a sample containing a hydrate or a compound capable of forming a hydrate is greatly dependent upon the relative humidity of its environment.

As we have pointed out, temperature has a marked effect on equilibrium constants. In general, the equilibrium vapor pressure of water over a hydrate increases with temperature; thus, the horizontal lines in Figure 24-2 will be dis-

---

[8] Relative humidity is the ratio of the vapor pressure of water in the atmosphere compared with its vapor pressure in air that is saturated with moisture. At 25°C the partial pressure of water in saturated air is 23.76 torr. Thus, when air contains water at a partial pressure of 6 torr, the relative humidity is

$$\frac{6.00}{23.76} = 0.253 \quad \text{or} \quad 25.3\%$$

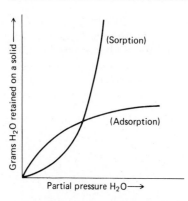

FIGURE 24-3   Typical adsorption and sorption isotherms.

placed to higher pressures when the temperature rises. Clearly, a temperature rise favors dehydration.

Adsorbed water.   The amount of water adsorbed on the surface of a solid also increases with the moisture content of the environment. The adsorption isotherm in Figure 24-3 illustrates this effect; here, the weight of water adsorbed on a typical solid is plotted against the partial pressure of water in the surrounding atmosphere. It is apparent from the diagram that the extent of adsorption is particularly sensitive to changes in water-vapor pressure at low partial pressures.

Quite generally, the amount of adsorbed water decreases with temperature increases and frequently approaches zero if the solid is dried at temperatures above 100°C.

A solid may lose or gain adsorbed moisture relatively rapidly; equilibrium is often reached after 5 or 10 min. The speed of the process is frequently apparent during the weighing of finely divided solids that have been dehydrated by heating; a continuous increase in weight will occur unless the solid is contained in a tightly stoppered vessel.

Sorbed water.   The quantity of moisture sorbed by a colloidal solid varies tremendously with atmospheric conditions, as may be seen in Figure 24-3. In contrast to the behavior of adsorbed water, however, equilibrium may require days or even weeks for attainment, particularly at room temperatures. Furthermore, the amount of water retained by the two processes is often quite different; typically, adsorption will involve quantities of water amounting to a few tenths of a percent of the solid while sorption may entail 10 or 20%.

The amount of water sorbed in a solid also decreases with temperature increases. However, complete removal of this type of moisture at 100°C is by no means a certainty, as is indicated by the drying curves for an organic compound shown in Figure 24-4. After drying this material for about 70 min at 105°C, constant weight was apparently reached. It is also clear, however, that additional moisture was removed by elevating the temperature. Even at 230°C, dehydration was probably not entirely complete.

Occluded water.   Occluded water is not in equilibrium with the atmosphere and is

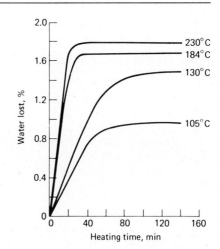

FIGURE 24-4 Removal of water from an organic compound at constant temperature. [From data of C. O. Willits. *Anal. Chem.*, **23**, 1058 (1951). With permission of the American Chemical Society.]

therefore insensitive to changes in humidity. Heating a solid containing occluded water may cause a gradual diffusion of the moisture to the surface, where it can evaporate; temperatures substantially higher than 100°C are often required for this process to occur at an appreciable rate. Frequently, heating is accompanied by *decrepitation*: the crystals of the solid are suddenly shattered by pressure of the steam created from moisture contained within the internal cavities.

24B-3 EFFECT OF GRINDING ON MOISTURE CONTENT
The moisture content, and thus the chemical composition, of a solid is frequently altered considerably during grinding and crushing. Both increases and decreases in percentage of water can occur.

Decreases in water content are sometimes observed during the grinding of solids containing essential water in the form of hydrates; for example, it has been reported that the water content of gypsum, $CaSO_4 \cdot 2H_2O$, is reduced from 20 to 5% by this treatment.[9] Undoubtedly, the change is a result of localized heating during the grinding and crushing of the particles.

Losses also occur when samples containing occluded water are ground; here, the grinding process ruptures some of the cavities and exposes the water so that it can evaporate. More commonly, perhaps, the grinding process is accompanied by an increase in moisture content, due primarily to the increase in surface exposed to the atmosphere. A corresponding increase in adsorbed water results. The magnitude of this effect is sufficient to alter the composition of a solid appreciably. For example, the water content of a piece of porcelain in the form of coarse particles was zero; after a period of grinding, it was 0.62%. Grinding a basaltic greenstone for 120 min changed its water content from 0.22 to 1.70%.[9]

From these remarks we may conclude that water determinations should be made upon solids before grinding whenever possible.

[9] W. F. Hillebrand, *J. Amer. Chem. Soc.*, **30**, 1120 (1908).

24B-4 DRYING THE ANALYTICAL SAMPLE

The methods for establishing the moisture content of a sample will depend upon its physical state and the information desired. Often the analytical chemist is called upon to determine the composition of a material as it is received. Here, the principal concern is that the moisture content of the material remain unchanged during preliminary treatment and storage. Where such changes are unavoidable or probable, it may be advantageous to determine the weight loss upon heating at some suitable temperature (say, 105°C) immediately upon receipt of the sample. Then, when the analysis is to be performed, the sample can be dried again at this same temperature so that the data can be corrected back to an "as received" basis.

Analyses are also performed, and results are reported, on an air-dry basis. Here, the material is allowed to acquire a constant weight while in contact with the atmosphere before samples are taken for analysis. Use of an air-dry sample weight is completely satisfactory for such nonhygroscopic substances as alloys. Other particulate materials, which do not tend to adsorb moisture strongly, can also be handled conveniently in this way.

We have already noted that the moisture content of some substances is markedly changed by variations in humidity and temperature. Colloidal substances containing large amounts of sorbed moisture are particularly susceptible to the effects of these variables. For example, the moisture content of a potato starch has been found to vary from 10 to 21% as a consequence of an increase in relative humidity from 20 to 70%.[10] With substances of this sort, comparable analytical data between laboratories or even within the same laboratory can be achieved only by carefully specifying a procedure for taking the moisture content into consideration; for example, the samples are frequently dried to constant weight at 105°C or at some other specified temperature. Analyses are then performed and results reported on this "dry basis." While such a procedure may not render the solid completely free of water, it will usually lower the moisture content to a reproducible level.

Often the only satisfactory procedure for obtaining an analysis on a "dry basis" will require a separate determination for moisture in a set of samples taken concurrently with the samples that are to be used for the analysis.

24C DETERMINATION OF WATER[11]

24C-1 DRYING PROCEDURES

Without question, oven drying is the most common method for determining the water content of samples. The amount evolved from a known weight of sample is established either from the loss in weight of the sample or by the gain in weight of an absorbent for water. The great virtue of the procedure is its simplicity; unfortunately, this simplicity does not necessarily extend to the interpretation of the data that the method provides, for several processes in addition to the evolution of water may also occur during the heating. Thus, one may also encounter volatilization of other components, decomposition of one or more of the constituents to give gase-

---

[10] I. M. Kolthoff and E. B. Sandell, *Textbook of Quantitative Inorganic Analysis*, 3d ed., p. 144. New York: Macmillan, 1952.

[11] For a more detailed discussion of this subject, see: C. O. Willits, *Anal. Chem.*, **23**, 1058–1080 (1951).

ous products, or perhaps air oxidation of a component in the sample. The first two of these effects will cause a decrease in sample weight; oxidation will cause an increase if the products of the reaction are nonvolatile and a decrease if they are volatile. Superimposed on these difficulties is the uncertainty with respect to the temperature required to cause complete evolution of water. Heating at 105°C will accomplish removal of adsorbed moisture and, in some instances, essential water as well. On the other hand, removal of sorbed and occluded water is often quite incomplete at this temperature. Many minerals, as well as such substances as alumina and silica, require temperatures of 1000°C or more.

Indirect determination.   In this method, the loss in weight of a solid during oven drying is measured; the assumption is then made that this loss equals the weight of water in the sample. The limitations to this procedure are apparent from the preceding paragraph.

In general, oven drying is carried out at as low a temperature as possible in order to minimize decomposition of the sample. The drying process is continued until the weight of the sample becomes constant at the chosen temperature; attainment of this state cannot be used as an unambiguous criterion of complete dehydration, however, as is clearly shown by the data in Figure 24-4.

The rate at which drying is completed can often be accelerated by sweeping a stream of dry air over the sample during the heating. This process can be carried out conveniently in a vacuum oven by reducing the internal pressure to a few torr; then, while pumping is continued, air that has been predried over a suitable desiccant is allowed to flow slowly and continuously through the drying chamber.

Direct determination.   Here, the water evolved from the sample is collected on an absorbent that is specific for water; the increase in weight of the absorbent is a direct measure of the amount of water present. The direct method circumvents many of the limitations inherent in indirect drying methods; accurate results can be ordinarily expected, provided the sample is heated at a sufficiently high temperature to remove all water. Errors, however, will result if the sample undergoes an air oxidation that yields water as a product; this problem is often encountered with substances containing organic components.

Figure 24-5 shows a typical arrangement for direct-moisture determination. The sample is weighed into a small porcelain boat, which is then placed in the Pyrex or Vycor combustion tube. Air, which has been dried by passage through concentrated sulfuric acid and then over a desiccant such as magnesium perchlorate, is forced over the sample. Heating is accomplished by a burner or a tube furnace. The exit gases are led through a U-tube containing magnesium perchlorate or other desiccant; this tube is weighed before and after the analysis. A second guard tube containing desiccant protects the absorbent tube from becoming contaminated by diffusion of water vapor from the atmosphere.

A very simple direct method exists for the determination of moisture in minerals, rocks, and other inorganic materials; the procedure is sometimes called *Penfield's method.*[12] The sample is placed in the end of a hard glass tube such as that shown in Figure 24-6. The water, which is driven off by ignition in a Bunsen flame,

---

[12] S. L. Penfield, *Amer. J. Sci.* (3), **48**, 31 (1894).

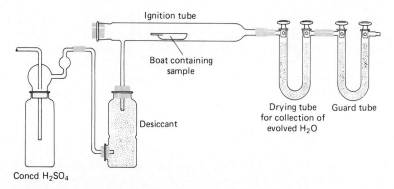

FIGURE 24-5   Apparatus for the determination of water.

collects in the center bulb of the tube, which is kept cool. After dehydration of the sample is judged to be complete, the sample end of the tube is softened, drawn off, and discarded, leaving the water in the bulb, which is weighed; the water is then removed by aspiration, and the bulb is again weighed.

24C-2 WATER BY DISTILLATION
Distillation, which is useful for the determination of water in materials that are readily air-oxidized, is widely employed for substances containing organic components such as fats, oils, waxes, cereals, plant materials, and foodstuffs. The sample to be analyzed is dissolved or suspended in an organic solvent, such as toluene or xylene, that is immiscible with and has a higher boiling point than water. Heating causes the water in the sample to be volatilized and distilled with the organic vapors. The distillate is condensed, and the volume of the aqueous phase is measured to give the water content.

   A typical distillation apparatus is illustrated in Figure 24-7. The condensed liquid is caught in a trap so constructed that the transferred water collects in the bottom while the organic liquid flows back into the distillation vessel. The trap is calibrated so that the volume of water can be determined directly.

24C-3 CHROMATOGRAPHIC METHODS
Gas-liquid chromatography, which is discussed in Chapter 29, has proved useful for the determination of the water content of various liquids.[13] Typically, water concentrations varying from a few parts per million to 1% or greater can be determined with relative errors in the 3 to 6% range.

[13] For example, see: J. M. Hogan, R. A. Engel, and H. F. Stevenson, *Anal. Chem.*, **42**, 249 (1970), and references therein.

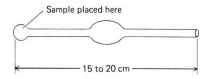

FIGURE 24-6   Penfield tube for water determination.

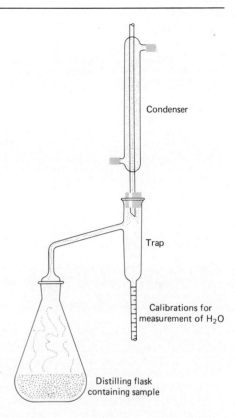

Condenser

Trap

Calibrations for
measurement of H_2O

Distilling flask
containing sample

FIGURE 24-7 Apparatus for the determination
of water by azeotropic distillation.

24C-4 CHEMICAL METHODS FOR WATER; THE KARL FISCHER REAGENT
A number of chemical methods for the determination of water have been devised.
Unquestionably the most important of these involves the use of Karl Fischer
reagent, a relatively specific reagent for water (see Section 15C-3).

24C-5 INFRARED METHODS
A near infrared absorption peak for water (1.94 μm) may be used for the determination of moisture in a variety of materials. A typical application is to the determination of water in various food products such as instant coffee, honey, potato chips,
and flour.[14] The samples are dispersed in dimethyl sulfoxide which extracts the
moisture nearly completely after 2 to 4 hr. Following extraction, the absorbance of
the liquid is measured in a 1-cm quartz cell. A linear relationship between absorbance and water concentration exists over the range of 0.00 to 0.70 mL of water per
100 mL of solution. The accuracy of the procedure appears to be equivalent to that
of the Karl Fischer method.

[14] D. M. Vomhof and J. T. Thomas, *Anal. Chem.*, **42**, 1230 (1970).

PROBLEMS*

* 24-1. The preparation of a heterogeneous catalyst involves coating spherical support particles with a layer of active material. A satisfactory product is entirely coated with a layer of catalyst; the existence of breaks in the layer is unacceptable. Complete the accompanying tabulation.

Number of Particles Counted	Number of Particles Counted That Were Satisfactory	Unsatisfactory	Percent Relative Standard Deviation of the Count	Absolute Standard Deviation of the Count
(a) 325	291			
(b) 325		34		
(c) 800		82		
(d) 675		68		
(e) 1200	1025			

24-2. A coating that weighs at least 3.00 mg is needed to impart adequate shelf life to a pharmaceutical pill. A random sampling of 250 pills revealed that 14 failed to meet this requirement.
  (a) Use this information to estimate the relative standard deviation for the measurement.
  (b) What is the 90% confidence interval for the number of unsatisfactory pills?
  (c) Assuming that the fraction of rejects remains unchanged, how many pills should be taken to assure a relative standard deviation of 10% in this measurement?

24-3. Changes in the method used to coat the pills lowered the percentage of rejects from 5.6% (Problem 24-2) to 2.0%. How many pills should be taken for inspection if the permissible relative standard deviation in the measurement is to be
  * (a) 25%?   * (c) 5%?
  (b) 10%?   (d) 1%?

* 24-4. Mishandling of a shipping container loaded with 750 cases of wine caused breakage of some bottles. An insurance adjuster proposed to settle the claim at 20.8% of the value of the shipment, based upon a random 250-bottle sample in which 52 were cracked or broken. Calculate
  (a) the relative standard deviation of the adjuster's evaluation.
  (b) the absolute standard deviation for the entire 750 cases (12 bottles/case).
  (c) the 90% confidence interval for the total number of bottles.
  (d) the size of a random sampling needed for a relative standard deviation of 5.0%, assuming a breakage rate of about 21%.

* 24-5. Approximately 15% of the particles in a shipment of silver-bearing ore are judged to be argenite, Ag_2S ($d = 7.3$, 87% Ag); the remainder are siliceous ($d = 2.6$) and contain essentially no silver.
  (a) Calculate the number of particles that should be taken for the gross sample if the relative standard deviation due to sampling is to be 1% or less.
  (b) Estimate the weight of the gross sample, assuming that the particles are spherical and have an average diameter of 4.0 mm.
  (c) The sample taken for analysis is to weigh 0.600 g and contain the same number of particles as the gross sample. To what diameter must the particles be ground to satisfy these criteria?

---

* Answers to problems or parts of problems marked with an asterisk are to be found at the end of the book.

24-6. The average diameter of the particles in a shipment of copper appears to be 5.0 mm. Approximately 5% of the particles are cuprite (80% Cu) with a density of 6; the remainder is estimated to have a density of 4 and contain 3% Cu.
   (a) How many particles of the ore should be sampled if the relative standard deviation due to sampling is to be 4% or less?
   (b) What should be the weight of the gross sample?
   (c) To what diameter must the particles be reduced in order to yield a sample for analysis that weighs 0.500 g and has the same number of particles as the gross sample?

* 24-7. The average particle diameter of an ore sample is 2.0 mm. It is estimated that the stibnite content ($d_{Sb_2S_3}$ = 4.5, 71.7% Sb) is approximately 2.0%; the remainder has a density of 3.0 and contains about 1% Sb.
   (a) How many particles of the ore should be taken if the relative standard deviation due to sampling is to be 1% or less?
   (b) What will be the weight of the gross sample?
   (c) To what diameter must the particles be reduced in order to yield a sample for analysis that weighs 0.750 g and has the same number of particles as the gross sample?

24-8. The seller of a mining claim took a random sample of an ore that weighed approximately 5 lb and had an average diameter of 5.0 mm. Inspection revealed that about 1% of the sample was argenite (see Problem 24-5) while the remainder had a density of about 2.6 and contained no silver. The prospective buyer insisted upon knowing the silver content of the claim with a relative error no greater than 5%. Establish whether the seller provided a sufficiently large sample to permit such an evaluation.

decomposing and dissolving the sample

25

Most analyses are completed by performing measurements on a solution (usually aqueous) of the analyte. Often, converting an analyte to a soluble form requires powerful reagents and strenuous treatment. For example, the determination of halogens or nitrogen in an organic compound requires vigorous treatment of the sample to rupture the strong bonds between these elements and carbon. Similarly, drastic conditions are usually required to destroy the silicate structure of a siliceous mineral, thus rendering its cations free for analysis.

The proper choice among the various reagents and techniques for decomposing and dissolving analytical samples can be critical to the success of an analysis, particularly where refractory substances are involved. This chapter describes some of the more common methods for obtaining aqueous solutions of samples that are difficult to decompose or dissolve.[1]

25A SOME GENERAL CONSIDERATIONS

Ideally, the reagent chosen should cause complete dissolution of the sample; attempts to leach one or more components from a mixture usually result in an incomplete separation from the unattacked residue.

Consideration must also be given to possible interferences introduced during the dissolution or decomposition processes. Contamination is a factor of particular importance in the determination of components that are present in low concentrations; in trace analysis, the most important consideration in choosing among possible solvents is frequently their purity and the amounts of reagent that must be used.

Volatilization of important constituents in a sample may occur during the so-

---

[1] For an extensive discussion of this topic, see: R. Bock, *A Handbook of Decomposition Methods in Analytical Chemistry.* New York: Wiley, 1979.

624

lution step unless proper precautions are taken. For example, treatment with acids can result in the loss of carbon dioxide, sulfur dioxide, hydrogen sulfide, hydrogen selenide, and hydrogen telluride; loss of ammonia is common in the presence of basic reagents. Treatment of a sample with hydrofluoric acid will result in vaporization of silicon and boron as their fluorides, while exposure of halogen-containing substances to strong oxidizing reagents may result in the evolution of chlorine, bromine, or iodine. Reducing conditions during the preliminary treatment of a sample can cause the volatilization of such compounds as arsine, phosphine, or stibine.

A number of elements form volatile chlorides that are partially or completely lost from hot hydrochloric acid solutions. Among these are arsenic and antimony trichloride, tin(IV) and germanium tetrachloride, and mercury(II) chloride. The oxychlorides of selenium and tellurium also volatilize to some extent from hot hydrochloric acid. The presence of chloride ion in hot, concentrated sulfuric or perchloric acid solutions can cause volatilization losses of bismuth, manganese, molybdenum, thallium, vanadium, and chromium.

Boric acid, nitric acid, and the halogen acids are lost from boiling aqueous solutions, while phosphoric acid distills from hot concentrated sulfuric or perchloric acids. Volatile oxides can also be lost from hot acidic solutions; these include the tetroxides of osmium and ruthenium as well as the heptoxide of rhenium.

25B LIQUID REAGENTS FOR DISSOLVING OR DECOMPOSING SAMPLES

The most common reagents for attacking analytical samples are the mineral acids or their aqueous solutions. Solutions of sodium or potassium hydroxide also find occasional use.

25B-1 HYDROCHLORIC ACID

Concentrated hydrochloric acid is an excellent solvent for many metal oxides as well as those metals that are more easily oxidized than hydrogen; often, it is a better solvent for oxides than the oxidizing acids. Concentrated hydrochloric acid is about 12 M, but upon heating, hydrogen chloride is lost until a constant-boiling 6 M solution remains (boiling point about 110°C).

25B-2 NITRIC ACID

Hot concentrated nitric acid will dissolve all common metals with the exception of aluminum and chromium, which become passive to the reagent as a consequence of surface oxide formation. When treated with concentrated nitric acid, tin, tungsten, and antimony form slightly soluble acids, which permits the separation of these elements from alloys by filtration immediately after the other components have dissolved.

25B-3 SULFURIC ACID

Hot concentrated sulfuric acid owes part of its effectiveness to its high boiling point (about 340°C); many samples decompose and dissolve rapidly at this elevated temperature. Organic compounds are dehydrated and oxidized by hot concentrated sulfuric acid. The reagent thus serves to eliminate such components from a sample. Most metals and many alloys are attacked by the hot acid.

25B-4 PERCHLORIC ACID

Hot concentrated perchloric acid, a potent oxidizing agent, attacks a number of iron alloys and stainless steels that are intractable to other mineral acids. Care must be taken in the use of the reagent, however, because of *its potentially explosive nature*. The cold concentrated acid is not hazardous nor are heated dilute solutions; on the other hand, violent explosions occur when hot concentrated perchloric acid comes in contact with organic material or easily oxidized inorganic substances. Because of this property, the concentrated reagent should be heated only in special hoods. Perchloric acid hoods are lined with glass or stainless steel, are seamless, and have a fog system for washing down the walls with water; their fan systems should be independent of other hoods. If proper precautions are taken,[2] perchloric acid is a safe and useful reagent.

Perchloric acid is marketed as the 60 or 72% acid. A constant-boiling mixture (72.4% $HClO_4$) is obtained at a temperature of 203°C.

25B-5 OXIDIZING MIXTURES

More rapid solvent action can sometimes be obtained by the use of mixtures of acids or by the addition of oxidizing agents to a mineral acid. *Aqua regia*, a mixture containing three volumes of concentrated hydrochloric acid and one of nitric acid, is well known. Addition of bromine or hydrogen peroxide to mineral acids often increases their solvent action and hastens the oxidation of organic materials in the sample. Mixtures of nitric and perchloric acid are also useful for this purpose and less dangerous than perchloric acid alone.

25B-6 HYDROFLUORIC ACID

The primary use of hydrofluoric acid is for the decomposition of silicate rocks and minerals where silica is not to be determined; the silicon is evolved as the tetrafluoride. After decomposition is complete, the excess hydrofluoric acid is driven off by evaporation with sulfuric acid or perchloric acid. Complete removal is often essential to the success of an analysis because the fluoride complexes of several cations are extraordinarily stable; the properties of these complexes may differ markedly from those of the parent cations. Thus, for example, precipitation of aluminum (as $Al_2O_3 \cdot xH_2O$) with ammonia is quite incomplete if fluoride is present even in a small amount. Frequently, removal of the last traces of fluoride ion from a sample is so difficult and time consuming as to negate the attractive features of the parent acid as a solvent for silicates.[3]

Hydrofluoric acid finds occasional use in conjunction with other acids in the attack on some of the more difficultly soluble steels.

Hydrofluoric acid can cause serious damage and painful injury when brought in contact with the skin; it must be handled with great respect. A burn may not become evident until hours after exposure.

[2] See: H. H. Willard and H. Diehl, *Advanced Quantitative Analysis*, p. 8. Princeton, N.J.: Van Nostrand, 1942; A. A. Schilt, *Perchloric Acid and Perchlorates*. Columbus, Ohio: G. Frederick Smith Chemical Company, 1979.

[3] For methods of removal of fluoride ion, see: H. H. Willard, L. M. Liggett, and H. Diehl, *Ind. Eng. Chem. Anal. Ed.*, **14**, 234 (1942).

25C DECOMPOSITION OF SAMPLES BY FLUXES

Many common substances—notably, silicates, some mineral oxides, and a few iron alloys—are attacked slowly, if at all, by the usual liquid reagents. Recourse to more potent fused-salt media, or *fluxes,* is then indicated. Fluxes will decompose most substances by virtue of the high temperature required for their use (300 to 1000°C) and the high concentration of reagent brought in contact with the sample.

Where possible, the employment of a flux is avoided, for several dangers and disadvantages attend its use. Among these is the possibility that significant contamination will be introduced in the rather large amount of flux (typically 10 times the sample weight) required for a successful fusion. Moreover, the aqueous solution from the fusion will have a high salt content, which may cause difficulties in the subsequent steps of the analysis. The high temperatures required for a fusion increase the danger of volatilization losses. Finally, the container in which the fusion is performed is almost inevitably attacked to some extent by the flux; again, contamination of the sample is the result.

For a sample containing only a small fraction of material that dissolves with difficulty, it is common practice to employ a liquid reagent first; the undecomposed residue is then isolated by filtration and fused with a relatively small quantity of a suitable flux. After cooling, the melt is dissolved and combined with the major portion of the sample.

25C-1 METHOD OF CARRYING OUT A FUSION

The sample in the form of a very fine powder is mixed intimately with perhaps a tenfold excess of the flux. Mixing is usually carried out in the crucible in which the fusion is to be performed. The time required for fusion may range from a few minutes to a matter of hours. The production of a clear melt signals completion of the decomposition, although often this condition is not totally obvious.

When the fusion is complete, the mass is allowed to cool slowly; then just before solidification, the crucible is rotated to distribute the solid around the walls to produce a thin layer that is easy to dislodge.

25C-2 TYPES OF FLUXES

With few exceptions the common fluxes used in analysis are compounds of the alkali metals. Basic fluxes employed for attack of acidic materials include carbonates, hydroxides, peroxides, and borates. The acidic fluxes are pyrosulfates, acid fluorides, as well as boric oxide. If an oxidizing flux is required, sodium peroxide can be used. As an alternative, small quantities of the alkali nitrates or chlorates can be mixed with sodium carbonate. The properties of the common fluxes are summarized in Table 25-1.

Sodium carbonate. Silicates and certain other refractory materials can be decomposed by heating to 1000 to 1200°C with sodium carbonate. This treatment generally converts the cationic constituents of a sample to acid-soluble carbonates or oxides; the nonmetallic constituents are converted to soluble sodium salts.

Carbonate fusions are normally carried out in platinum crucibles.

Potassium pyrosulfate. Potassium pyrosulfate provides a potent acidic flux that is

Table 25-1 Common Fluxes

Flux	Melting Point, °C	Type of Crucible for Fusion	Type of Substance Decomposed
Na_2CO_3	851	Pt	For silicates and silica-containing samples; alumina-containing samples; sparingly soluble phosphates and sulfates
Na_2CO_3 + an oxidizing agent such as KNO_3, $KClO_3$, or Na_2O_2	—	Pt (not with Na_2O_2), Ni	For samples requiring an oxidizing environment; that is, samples containing S, As, Sb, Cr, etc.
NaOH	318	Au, Ag, Ni	Powerful basic fluxes for silicates, silicon carbide, and certain minerals; main limitation: purity of reagents
KOH	380		
Na_2O_2	Decomposes	Fe, Ni	Powerful basic oxidizing flux for sulfides; acid-insoluble alloys of Fe, Ni, Cr, Mo, W, and La; platinum alloys; Cr, Sn, Zr minerals
$K_2S_2O_7$	300	Pt, porcelain	Acid flux for slightly soluble oxides and oxide-containing samples
B_2O_3	577	Pt	Acid flux for decomposition of silicates and oxides where alkali metals are to be determined
$CaCO_3$ + NH_4Cl	—	Ni	Upon heating the flux, a mixture of CaO and $CaCl_2$ is produced; used to decompose silicates for the determination of the alkali metals

particularly useful for the attack of the more intractable metal oxides. Fusions with this reagent are performed at about 400°C; at this temperature, the slow evolution of the highly acidic sulfur trioxide takes place:

$$K_2S_2O_7 \longrightarrow K_2SO_4 + SO_3$$

Potassium pyrosulfate can be prepared by heating potassium hydrogen sulfate:

$$2KHSO_4 \longrightarrow K_2S_2O_7 + H_2O$$

Other fluxes. Table 25-1 contains data for several other common fluxes. Noteworthy are boric oxide and the mixture of calcium carbonate and ammonium chloride. Both are employed to decompose silicates for the analysis of alkali metals. Boric oxide is removed after solution of the melt by evaporation to dryness with methyl alcohol; methyl borate, $B(OCH_3)_3$, distills.

25D DECOMPOSITION OF ORGANIC COMPOUNDS FOR ELEMENTAL ANALYSIS[4]

Analysis for the elemental composition of an organic sample generally requires drastic treatment to convert the elements of interest into a form susceptible to the common analytical techniques. These treatments are usually oxidative and involve conversion of carbon to carbon dioxide and hydrogen to water; occasionally, however, heating the sample with a potent reducing agent is sufficient to rupture the covalent bonds in the compound and free the analyte element from the carbonaceous residue.

Oxidation procedures can be grouped into two categories. *Wet-ashing* (or oxidation) makes use of liquid oxidizing agents such as sulfuric, nitric, or perchloric acids. *Dry-ashing* usually implies ignition of the organic compound in air or in a stream of oxygen. In addition, oxidations can be carried out in certain fused-salt media; sodium peroxide is the most common flux for this purpose.

In the sections that follow, we shall consider briefly some of the methods for decomposing organic substances prior to elemental analysis.

25D-1 WET-ASHING PROCEDURES

Solutions of strong oxidizing agents will decompose organic samples. The main problem associated with the use of these reagents is the potential for losses of the elements of interest by volatilization.

We have already encountered an example of wet-ashing in the Kjeldahl method for the determination of nitrogen in organic compounds (p. 249) where concentrated sulfuric acid is the oxidizing agent. This reagent is also frequently employed for decomposition of organic materials in which metallic constituents are to be determined. Nitric acid may be added periodically to the solution to hasten the rate at which oxidation occurs.[5] A number of elements are volatilized (at least partially) by this procedure, particularly if the sample contains chlorine; included are arsenic, boron, germanium, mercury, antimony, selenium, tin, the halogens, sulfur, and phosphorus.

An even more effective reagent than sulfuric/nitric acid mixtures is perchloric acid mixed with nitric acid. *Great care must be exercised in the use of this reagent,* however, because of the tendency of hot anhydrous perchloric acid to react explosively with organic material. It is essential to start with a mixture in which nitric acid predominates; this reagent attacks the easily oxidized components in the early stages. With continued heating, water and nitric acid are lost by decomposition and evaporation, and the solution becomes progressively a stronger oxidant. If the solution becomes too concentrated in perchloric acid before most of the oxidation is complete, it will darken in color or blacken. *Should darkening occur, the mixture should be immediately removed from the heat and diluted with water and nitric acid.* The heating can then be continued. As we mentioned earlier (p. 626), perchloric acid oxidations should be carried out only in a special hood. If properly performed, oxidations with a mixture of nitric and perchloric acids are

---

[4] For a thorough treatment of this topic, see: T. S. Ma and R. C. Rittner, *Modern Organic Elemental Analysis*. New York: Marcel Dekker, 1979.
[5] *Official Methods of Analysis of the AOAC*, 11th ed., p. 400. Washington, D.C.: Association of Official Analytical Chemists, 1970.

rapid, and losses of metallic ions are negligible.[6] It cannot be too strongly emphasized that proper precautions must be taken in the use of hot concentrated perchloric acid to prevent violent explosions.

25D-2 DRY-ASHING PROCEDURES

The simplest method for decomposing an organic sample is by heating with a flame in an open dish or crucible until all carbonaceous material has been oxidized to carbon dioxide. Red heat is often required to complete the oxidation. Analysis of the nonvolatile components follows solution of the residual solid. Unfortunately, a great deal of uncertainty always exists with respect to the completeness of recovery of supposedly nonvolatile elements from a dry-ashed sample. Some losses probably result from the mechanical entrainment of finely divided particulate matter in the convection currents around the crucible. In addition, volatile metallic compounds may be formed during the ignition. For example, copper, iron, and vanadium are appreciably volatilized when samples containing porphyrin compounds are heated.[7]

In summary, although dry-ashing is the simplest of all methods for decomposing organic compounds, it is often the least reliable; it should not be employed unless tests have demonstrated its applicability to a given type of sample.

25D-3 COMBUSTION-TUBE METHODS

Several common and important elemental components of organic compounds are converted to gaseous products as the sample is pyrolyzed. With suitable apparatus, it is possible to trap these volatile compounds quantitatively, thus making them available for the analysis of the element of interest. The heating is commonly performed in a glass or quartz combustion tube through which a stream of carrier gas is passed. The stream serves to transport the volatile products to parts of the apparatus where they can be separated and retained for measurement; the gas may also serve as the oxidizing agent. Elements susceptible to this type of treatment are carbon, hydrogen, oxygen, nitrogen, the halogens, sulfur, and oxygen. Table 25-2 gives details for some of these methods.

Automated combustion-tube analyzers are now available for the determination of carbon, hydrogen, and nitrogen or carbon, hydrogen, and oxygen in a single sample. The apparatus requires essentially no attention by the operator; the analysis is complete in less than 15 min. One such analyzer uses a mixture of helium and oxygen; the oxidation is catalyzed by a mixture of silver vanadate and silver tungstate. Halogens and sulfur are removed with a packing of silver salts. A packing of hot copper is located at the end of the combustion train to remove oxygen and convert nitrogen oxides to nitrogen. The exit gas, consisting of a mixture of water, carbon dioxide, nitrogen, and helium, is finally collected in a glass bulb. The analysis of this mixture is accomplished with three thermal conductivity measurements (see Section 29A-4). The first is made on the intact mixture, the second is made on the mixture after water has been removed by passage of the gases through a dehydrating agent, and the third is made on the mixture after removal of carbon dioxide with an absorbent. The relationship between thermal conductivity readings and

[6] T. T. Gorsuch, *Analyst*, **84**, 135 (1959); G. F. Smith, *Anal. Chim. Acta*, **8**, 397 (1953); G. F. Smith, *The Wet Chemical Oxidation of Organic Compositions Employing Perchloric Acid*. Columbus, Ohio: G. F. Smith Chemical Co., 1965.

[7] See: T. T. Gorsuch, *Analyst*, **84**, 135 (1959); R. E. Thiers in, *Methods of Biochemical Analysis*, vol. 5. New York: Interscience, 1957.

Table 25-2   Combustion-Tube Methods for the Elemental Analysis of Organic Substances

Element	Name of Method	Method of Oxidation	Method of Completion of Analysis
Halogens	Pregl	Sample burned in a stream of oxygen over a red-hot platinum catalyst; halogens converted primarily to HX and X_2	Gas stream passed through a carbonate solution containing SO_3^{2-} (to reduce halogens and oxyhalogens to halides); halide ion, X^-, then determined by usual procedures
	Grote	Sample burned in a stream of air over a hot silica catalyst; products are HX and X_2	Same as above
Sulfur	Pregl	Similar to halogen determination; combustion products are SO_2 and SO_3	Gas stream passed through aqueous H_2O_2 to convert sulfur oxides to H_2SO_4, which can then be titrated with standard base
	Grote	Similar to halogen determination; products are SO_2 and SO_3	Similar to above
Nitrogen	Dumas	Sample oxidized by hot CuO to give CO_2, H_2O, and N_2	Gas stream passed through concentrated KOH solution leaving only N_2, which is measured volumetrically
Carbon and hydrogen	Pregl	Similar to halogen analysis; products are CO_2 and H_2O	H_2O absorbed on a desiccant and CO_2 on Ascarite; determined gravimetrically or by thermal conductivity before absorption
Oxygen	Unter-zaucher	Sample pyrolyzed over carbon; oxygen converted to CO; H_2 used as carrier gas	Gas stream passed over I_2O_5 [$5CO + I_2O_5(s) \rightarrow 5CO_2 + I_2(g)$]; liberated I_2 titrated or CO_2 absorbed and weighed

concentration is linear, and the slope for each constituent is established by calibration with a pure compound such as acetanilide.

25D-4 COMBUSTION WITH OXYGEN IN SEALED CONTAINERS

A relatively straightforward method for the decomposition of many organic substances involves combustion with oxygen in a sealed container. The reaction products are absorbed in a suitable solvent before the reaction vessel is opened and are subsequently analyzed by ordinary methods.

A remarkably simple apparatus for performing such oxidations has been suggested by Schöniger (see Figure 25-1).[8] It consists of a heavy-walled flask of 300- to 1000-mL capacity fitted with a ground-glass stopper. Attached to the stopper

[8] W. Schöniger, *Mikrochim. Acta*, **1955**, 123; **1956**; 869. See also the review article by A. M. G. MacDonald, in *Advances in Analytical Chemistry and Instrumentation*, C. E. Reilley, Ed., vol. 4, p. 75. New York: Interscience, 1965.

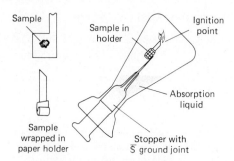

Sample

Sample in holder

Ignition point

Absorption liquid

Sample wrapped in paper holder

Stopper with $\bar{S}$ ground joint

FIGURE 25-1   Schöniger combustion apparatus. (Courtesy Arthur H. Thomas Company, Philadelphia, Pennsylvania)

is a platinum-gauze basket that holds from 2 to 200 mg of sample. If the substance to be analyzed is a solid, it is wrapped in a piece of low-ash filter paper cut in the shape shown in Figure 25-1. Liquid samples can be weighed into gelatin capsules, which are then wrapped in a similar fashion. A tail is left on the paper and serves as an ignition point.

A small volume of an absorbing solution (often sodium carbonate) is placed in the flask, and the air in the container is then displaced by oxygen. The tail of the paper is ignited, and the stopper is quickly fitted into the flask; the container is then inverted, as shown in Figure 25-1, to prevent the escape of the volatile oxidation products. The reaction ordinarily proceeds rapidly, being catalyzed by the platinum gauze surrounding the sample. During the combustion the flask is shielded to minimize the damage in case of explosion.

After cooling, the flask is shaken thoroughly and disassembled, and the inner surfaces are carefully rinsed. The analysis is then performed on the resulting solution. This procedure has been applied to the determination of halogens,[8] sulfur,[8] phosphorus,[9] fluorine,[10] and various metals[11] in organic compounds.

---

[9] R. Belcher and A. M. G. MacDonald, *Talanta*, **1**, 185 (1958).
[10] B. Z. Senkowski, E. G. Wollish, and E. G. E. Shafer, *Anal. Chem.*, **31**, 1574 (1959).
[11] R. Belcher, A. M. G. MacDonald, and T. S. West, *Talanta*, **1**, 408 (1958).

analytical separations

26

The physical and chemical properties upon which analytical methods are based are seldom, if ever, entirely specific. Instead, these properties are shared by numerous species; as a consequence, the elimination of interferences is more often the rule than the exception in a quantitative analysis.

Two general methods are available for coping with substances that interfere in an analytical measurement. The first involves alteration of the system to immobilize the potential interference and thereby prevent its participation in the measurement step; clearly, the alteration must not affect the species being determined. Immobilization is frequently accomplished by introducing a complexing agent that reacts selectively with the interfering substance. For example, in the iodometric determination of copper, iron(III) can be rendered unreactive toward iodide by complexation with fluoride or phosphate ion; neither anion inhibits the oxidation of iodide by copper(II). The introduction of a reagent to eliminate an interference is called *masking*. Numerous masking reagents have been discussed in earlier chapters.

The second method involves the physical separation of an interference from the analyte. Various methods for performing such separations are considered in this chapter and the next three.

26A NATURE OF THE SEPARATION PROCESS

All separation procedures have in common the distribution of the components in a mixture between two phases which subsequently can be separated mechanically. If the ratio between the amount of a particular component in each phase (the *distribution ratio*) differs significantly from that of another, a separation of the two is potentially feasible. To be sure, the complexity of the separation process depends upon the difference between the distribution ratios for the two components. Where the difference is extreme, a single-stage process suffices. For example, a single precipitation with silver ion is adequate for the isolation of chloride from many other anions. Here, the ratio of the chloride ion in the solid phase to that in equilibrium in the aqueous phase is immense, while comparable ratios for, say, nitrate or perchlorate ions approach zero.

A somewhat more complex situation is encountered when the distribution ratio for one component is essentially zero, as in the foregoing example, but the ratio for the other is not very large. Here, a multistage process is required. For example, uranium(VI) can be extracted into ether from an aqueous nitric acid solution. Although the distribution ratio in the two phases approaches unity for a single extraction, uranium(VI) can nevertheless be isolated by repeated or *exhaustive* extraction of the aqueous solution with fresh portions of ether.

The most complex procedures are required when the distribution ratios of the species to be separated are greater than zero and approach one another in magnitude; here, multistage *fractionation* techniques are necessary. These techniques do not differ in principle from their simpler counterparts; both are based upon differences in the distribution ratios of solutes between two phases. However, two factors account for the gain in separation efficiency associated with fractionation. First, the number of times that partitioning occurs between phases is increased enormously; second, distribution occurs between fresh portions of both phase. An exhaustive extraction differs from a fractionation in this latter respect; although many contacts between phases are provided during the former, fresh portions of only one phase are involved.

26B SEPARATION BY PRECIPITATION

The fundamental basis for all precipitation separations is the solubility difference between the analyte and the undesired components. Solubility-product considerations will generally provide guidance as to whether or not a given separation is theoretically feasible and will define the conditions required to achieve the separation. Unfortunately, other variables are influential in determining the success or failure of a separation based upon precipitation, and some of these are not susceptible to theoretical treatment within our present state of knowledge. We have seen, for example, that various coprecipitation phenomena may cause extensive contamination of a precipitate by an unwanted component, even though the solubility product of the contaminant has not been exceeded (Section 5B). Likewise, the rate of an otherwise feasible precipitation process may be so slow that it becomes useless as the basis for a separation. Finally, when the precipitate forms as a colloidal suspension, coagulation may be a difficult and slow process. These latter problems are particularly formidable when the isolation of a small quantity of a solid phase is attempted.

Many precipitating agents are employed for quantitative inorganic separations; we shall limit this discussion to those that have the most general applicability.

26B-1 SEPARATIONS BASED ON CONTROL OF ACIDITY

Enormous differences exist among the solubilities of the hydroxides, hydrous oxides, and acids of various elements. Moreover, the concentration of hydrogen or hydroxide ions in a solution can be varied by a factor of 10^{15} or more and can be read-

Table 26-1 Separations Based Upon Control of Acidity

Reagent	Species Forming Precipitates	Species Not Precipitated
Hot concd HNO_3	Oxides of W(VI), Ta(V), Nb(V), Si(IV), Sn(IV), Sb(V)	Most other metal ions
NH_3/NH_4Cl buffer	Fe(III), Cr(III), Al(III)	Alkali and alkaline earths, Mn(II), Cu(II), Zn(II), Ni(II), Co(II)
$HOAc/NH_4OAc$ buffer	Fe(III), Cr(III), Al(III)	Common dipositive ions
$NaOH/Na_2O_2$	Fe(III), most dipositive ions, rare earths	Zn(II), Al(III), Cr(VI), V(V), U(VI)

ily controlled by the use of buffers. As a consequence, many separation procedures based on pH control are, in theory, available to the chemist. In practice, these separations can be grouped in three categories: (1) those made in relatively concentrated solutions of strong acids, (2) those made in buffered solutions at intermediate pH values, and (3) those made in concentrated solutions of sodium or potassium hydroxide. Table 26-1 lists common separations that can be achieved by control of acidity.

26B-2 SULFIDE SEPARATIONS

With the exception of the alkali and alkaline-earth metals, most cations form sparingly soluble sulfides. Their solubilities differ greatly; since it is a relatively easy matter to control the sulfide ion concentration of an aqueous solution by adjustment of pH, separations based on formation of sulfides have found extensive use. Sulfides can be conveniently precipitated from homogeneous solution, with the anion being generated by hydrolysis of thioacetamide (see Table 5-2, p. 141).[1]

A theoretical treatment of the ionic equilibria influencing the solubility of

[1] See: E. H. Swift and F. C. Anson, in *Advances in Analytical Chemistry and Instrumentation*, C. E. Reilley, Ed., vol. 1, pp. 293–345. New York: Interscience, 1960.

Table 26-2 Precipitation of Sulfides

Elements	Conditions for Precipitation*	Conditions for No Precipitation*
Hg(II), Cu(II), Ag(I)	1, 2, 3, 4	
As(V), As(III), Sb(V), Sb(III)	1, 2, 3	4
Bi(III), Cd(II), Pb(II), Sn(II)	2, 3, 4	1
Sn(IV)	2, 3	1, 4
Zn(II), Co(II), Ni(II)	3, 4	1, 2
Fe(II), Mn(II)	4	1, 2, 3

* Conditions include:
1. 3-M HCl.
2. 0.3-M HCl.
3. Buffered to pH 6 with acetate.
4. Buffered to pH 9 with $NH_3/(NH_4)_2S$.

sulfide precipitates was considered in Section 4B-2. Such treatment may fail to provide realistic conclusions regarding the feasibility of separations, however, because of coprecipitation and the slow rate at which some sulfides form. As a consequence, resort must be made to empirical observations.

Table 26-2 shows some common separations that can be accomplished with hydrogen sulfide through control of pH.

26B-3 OTHER INORGANIC PRECIPITANTS

No other inorganic ions are as generally useful for separations as those discussed in the previous sections. Phosphate, carbonate, and oxalate ions are often employed as precipitants for cations; their behavior is nonselective, so preliminary separations must generally precede their use.

Chloride and sulfate ions are useful because of their relatively specific behavior. The former can be used to separate silver from most other metals, while the latter are frequently employed to separate a group of metals that includes lead, barium, and strontium.

26B-4 ORGANIC PRECIPITANTS

Selected organic reagents used for the isolation of various inorganic ions were discussed in Section 5D-3. Some organic precipitants such as dimethylglyoxime are useful because of their remarkable selectivity in forming precipitates with very few ions. Others, such as 8-hydroxyquinoline, form slightly soluble compounds with a host of cations. The solubilities of the hydroxyquinolates differ greatly; by control of reagent concentration, useful separations can be achieved. As with sulfide ion, the concentration of the precipitating reagent is readily controlled by adjustment of pH.

26B-5 SEPARATION OF CONSTITUENTS PRESENT IN TRACE AMOUNTS

A problem often encountered in trace analysis is that of isolating the minor constituent, which may be present in microgram quantities, from the major components of the sample. Although such a separation is sometimes based on a precipitation process, the techniques required differ from those used when the analyte is present in generous amounts.

Several problems attend the quantitative separation of a trace element by precipitation even when solubility losses are not important. Supersaturation may delay formation of the precipitate, and coagulation of small amounts of a colloidally dispersed substance is often difficult. In addition, it is likely that an appreciable fraction of the solid will be lost during transfer and filtration. To minimize these difficulties, a quantity of some other ion that also forms a precipitate with the reagent can be added to the solution. The precipitate formed by the added ion is called a *collector* and serves to carry the desired minor species out of solution. For example, in isolating manganese as the sparingly soluble manganese dioxide, a small amount of iron(III) is frequently added. The basic iron(III) oxide carries down even the smallest traces of the dioxide. A few micrograms of titanium can be removed from a large volume of solution by addition of aluminum ion and ammonia. Here, hydrous aluminum oxide serves as the collector. Copper sulfide is often employed to collect traces of zinc and lead ions. Many other uses of collectors are described by Sandell and Onishi.[2]

Sometimes the collector simply carries down the trace precipitate by physical entrainment. At other times the process must involve coprecipitation in which the minor component is adsorbed or incorporated in the collector precipitate as the result of mixed crystal formation.

Clearly, the collector must not interfere with the method selected for the subsequent analysis of the trace component.

26B-6 SEPARATION BY ELECTROLYTIC PRECIPITATION

Electrolytic precipitation constitutes a highly useful method for accomplishing separations. In this process, the more easily reduced species, be it the wanted or the unwanted component of the mixture, is isolated as a second phase. The method becomes particularly effective when the potential of the working electrode is controlled at a predetermined level (p. 438).

The mercury cathode (p. 440) has found wide application for the removal of many metal ions prior to the analysis of the residual solution. In general, metals more easily reduced than zinc are conveniently deposited in the mercury, leaving such ions as aluminum, beryllium, the alkaline earths, and the alkali metals in solution. The potential required to decrease the concentration of a metal ion to any desired level is readily calculated from polarographic data.

26C EXTRACTION METHODS

The distribution of a solute between two immiscible phases is an equilibrium process that can be treated by the law of mass action. Equilibrium constants for this process vary enormously among solutes, thus making possible many useful separations based on extraction. The extraction technique has been widely used to separate the components of organic systems. For example, carboxylic acids are readily separated from phenolic compounds by extracting a nonaqueous solution of the sample with dilute aqueous sodium bicarbonate. The carboxylic acids are almost completely transferred to the aqueous phase, while the phenolic constituents remain in the organic phase. As will be shown in Section 26D-1 a number of useful separations of inorganic species are also based upon extraction.

26C-1 THEORY

The partition of a solute between two immiscible solvents is governed by the *distribution law*. If we assume that the solute species A distributes itself between an aqueous and an organic phase, the resulting equilibrium may be written as

$$A_{aq} \rightleftharpoons A_{org}$$

where the subscripts aq and org refer to the aqueous and organic phases, respectively. Ideally the ratio of the activities of A in the two phases will be constant and independent of the total quantity of A. That is, at any given temperature

$$K = \frac{[A_{org}]}{[A_{aq}]} \tag{26-1}$$

[2] E. B. Sandell and H. Onishi, *Colorimetric Determination of Traces of Metals*, 4th ed. New York: Interscience, 1978.

where the equilibrium constant K is the *partition coefficient* or *distribution coefficient*. The terms in brackets are strictly the activities of A in the two solvents, but molar concentrations can frequently be substituted without serious error. Often, K is approximately equal to the ratio of the solubility of A in each solvent.

The solute may exist in different states of aggregation in the two solvents. Then, the equilibrium becomes

$$x(A_y)_{aq} \rightleftharpoons y(A_x)_{org}$$

and the partition coefficient takes the form

$$K = \frac{[(A_x)_{org}]^y}{[(A_y)_{aq}]^x}$$

Partition coefficients make it possible to establish the experimental conditions required to transfer a solute from one solvent to another. For example, consider a simple system that is adequately described by Equation 26-1. Suppose, further, that we have V_{aq} mL of an aqueous solution containing a_0 mmol of A and that we propose to extract this with V_{org} mL of an immiscible organic solvent. At equilibrium, a_1 mmol of A will remain in the aqueous layer, and we may write

$$[A_{aq}]_1 = \frac{a_1}{V_{aq}}$$

It follows, then, that

$$[A_{org}] = \frac{(a_0 - a_1)}{V_{org}}$$

Substitution of these quantities into Equation 26-1 gives upon rearrangement

$$a_1 = \left(\frac{V_{aq}}{V_{org}K + V_{aq}}\right) a_0 \tag{26-2}$$

The number of millimoles, a_2, remaining after a second extraction of the water with an identical volume of solvent will, by the same reasoning, be

$$a_2 = \left(\frac{V_{aq}}{V_{org}K + V_{aq}}\right) a_1$$

When this expression is substituted into Equation 26-2, we obtain

$$a_2 = \left(\frac{V_{aq}}{V_{org}K + V_{aq}}\right)^2 a_0$$

After n extractions, the number of millimoles remaining is given by the expression

$$a_n = \left(\frac{V_{aq}}{V_{org}K + V_{aq}}\right)^n a_0 \tag{26-3}$$

Equation 26-3 can be rewritten in terms of the initial and final aqueous concentration of A by substituting the relationships

$$a_n = [A_{aq}]_n V_{aq} \quad \text{and} \quad a_0 = [A_{aq}]_0 V_{aq}$$

where $[A_{aq}]_n$ is the concentration in the aqueous phase after n extractions. Substitution of these relationships into Equation 26-3 gives

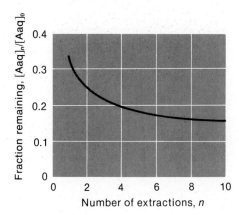

FIGURE 26-1 Plot of Equation 26-4, assuming $K = 2$ and $V_{aq} = 100$. The total volume of the organic solvent was also assumed to be 100, so that $V_{org} = 100/n$.

$$[A_{aq}]_n = \left(\frac{V_{aq}}{V_{org}K + V_{aq}}\right)^n [A_{aq}]_0 \tag{26-4}$$

As shown in the example that follows, the exponential nature of Equation 26-4 indicates that a more efficient extraction is achieved with several small volumes of solvent than a single large one.

EXAMPLE 26-1. The distribution coefficient of I_2 between CCl_4 and H_2O is 85. Calculate the concentration of I_2 remaining after extracting 50.0 mL of an aqueous 1.00×10^{-3} M solution of I_2 with (a) 50.0 mL CCl_4, (b) two 25.0-mL portions of CCl_4, and (c) five 10.0-mL portions.

    a. Substituting into Equation 26-4 gives

$$[I_{2aq}]_1 = \left(\frac{50.0}{50 \times 85 + 50}\right)^1 \times 1.00 \times 10^{-3} = 1.16 \times 10^{-5}$$

    b. $[I_{2aq}]_2 = \left(\dfrac{50.0}{25 \times 85 + 50}\right)^2 \times 1.00 \times 10^{-3} = 5.28 \times 10^{-7}$

    c. $[I_{2aq}]_5 = \left(\dfrac{50.0}{10 \times 85 + 50}\right)^5 \times 1.00 \times 10^{-3} = 5.29 \times 10^{-10}$

Figure 26-1 demonstrates that the improved efficiency brought about by multiple extractions falls off rapidly as the number of subdivisions increases. Clearly, little is gained by dividing the extracting solvent into more than five or six portions.

26C-2 TYPES OF EXTRACTION PROCEDURES

Simple extraction. A separation by extraction can be a simple process, provided the partition coefficient for one species is reasonably favorable (of the order of 10 or greater) while that for the other is unfavorable. Typically, a useful separation is possible if the second coefficient is in the range of 0.1 to 0.001 (or smaller). Under these circumstances, the extraction can be carried out in a separatory funnel, with the original solution being extracted successively with up to five to six portions of fresh solvent.

Exhaustive or continuous extraction. A relatively simple apparatus permits continuous extraction of a solution automatically. Such an apparatus is useful for the removal of extractable components from those with partition ratios that approach zero. Even components with relatively unfavorable partition ratios will be separated after an extraction period of several hours.

Countercurrent fractionation.[3] Automated devices have been developed that permit 200 or more successive extractions to be performed mechanically. Here, a *countercurrent* scheme causes distribution to occur at each step between fresh portions of both phases. An exhaustive extraction differs from the countercurrent technique in that fresh portions of only one phase are introduced in the former.

The countercurrent method permits separation of components with nearly identical partition ratios. For example, Craig[4] has demonstrated that 10 amino acids can be separated by countercurrent extraction even though their partition coefficients differ by less than 0.1.

26D APPLICATIONS OF EXTRACTION PROCEDURES

An extraction is often more attractive than a classic precipitation for separating inorganic species. The processes of equilibration and separation of phases in a separatory funnel are inherently less tedious and time-consuming than precipitation, filtration, and washing. In addition, problems of coprecipitation and postprecipitation are avoided. Finally, and in contrast to the precipitation process, extraction procedures are ideally suited for the isolation of trace quantities of a species.

26D-1 SOME EXAMPLES OF INORGANIC SEPARATIONS

Ether extractions of metal chlorides. The data in Table 26-3 indicate that a substantial number of metal chlorides can be extracted into ether from 6 *M* hydrochloric acid solution; equally important, a large number of metal ions are either unaffected or extracted only slightly under these conditions. Thus, many useful separations are possible. One of the most important of these is the separation of iron(III) (99% extracted) from a host of other cations. The greater part of iron from steel or iron ore samples can be removed by extraction prior to analysis for such trace elements as chromium, aluminum, titanium, or nickel. The species extracted has been shown to be the ion pair $H_3O^+FeCl_4^-$. It has also been found that the percentage of iron transferred to the organic phase is dependent upon the hydrochloric acid content of the aqueous phase (little is removed from solutions that are below 3 *M* and above 9 *M* HCl) and, to some extent, upon the iron content. Unless special precautions are taken, extraction of the last traces of iron is incomplete.[5]

Extraction of nitrates. Certain nitrate salts are selectively extracted by ether as well as other organic solvents. For example, uranium is conveniently separated from such elements as lead and thorium by ether extraction of an aqueous solution that is saturated with ammonium nitrate and has a nitric acid concentration of about 1.5 *M*; the uranium must be in the +6 oxidation state. Bismuth and iron(III) nitrates are also extracted to some extent under these conditions.

[3] For an extended discussion of this topic, see: L. C. Craig and D. Craig, in *Techniques of Organic Chemistry*, 2d ed., A. Weissberger, Ed., vol. III, part I, pp. 149–332. New York: Interscience, 1956.
[4] L. C. Craig, *Anal. Chem.*, **22**, 1346 (1950).
[5] See: S. E. Q. Ashley and W. M. Murray, *Ind. Eng. Chem., Anal. Ed.*, **10**, 367 (1938).

Table 26-3 Ethyl Ether Extractions of Various Chlorides from 6 M Hydrochloric Acid*

Percent Extracted	Elements and Oxidation State
90–100	Fe(III), 99%; Sb(V), 99%**; Ga(III), 97%; Ti(III), 95%**; Au(III), 95%
50–90	Mo(VI), 80–90%; As(III), 80%**†; Ge(IV), 40–60%
1–50	Te(IV), 34%; Sn(II), 15–30%; Sn(IV), 17%; Ir(IV), 5%; Sb(III), 2.5%*
< 1 > 0	As(V),* Cu(II), In(III), Hg(II), Pt(IV), Se(IV), V(V), V(IV), Zn(II)
0	Al(III), Bi(III), Cd(II), Cr(III), Co(II), Be(II), Fe(II), Pb(II), Mn(II), Ni(II), Os(VIII), Pd(II), Rh(III), Ag(I), Th(IV), Ti(IV), W(VI), Zr(IV)

* Data from Ernest H. Swift, *Introductory Quantitative Analysis*, p. 431. Englewood Cliffs, N.J.: Prentice-Hall, Inc., 1950. With permission.
** Isopropyl ether employed rather than ethyl ether.
† 8 M HCl rather than 6 M.

Extraction of chelate compounds. Many of the organic reagents mentioned in Section 5D-3, as well as others, form chelates with various metal ions; these chelates are frequently soluble in such solvents as chloroform, carbon tetrachloride, benzene, and ether. Thus, quantitative transfer of the metallic ions to the organic phase is possible.

A reagent that has widespread application for extractive separations is 8-hydroxyquinoline (p. 146). Most of its metal chelates are soluble in several organic solvents. The reaction which occurs when an aqueous solution of a divalent metal ion M^{2+} is extracted with an organic solvent containing 8-hydroxyquinoline (symbolized as HQ) can be formulated as

$$2(HQ)_{org} + (M^{2+})_{aq} \rightleftharpoons (MQ_2)_{org} + 2(H^+)_{aq}$$

where the subscript indicates the phase. The equilibrium is clearly pH dependent; thus, separations among metals having different formation constants with the ligand are possible through control of the pH of the aqueous phase. The method has proved particularly useful for separation of traces of metals.[6]

Another useful reagent for separating minute quantities of metal ions is dithizone[7] (diphenylthiocarbazone). Its reaction with a divalent metallic ion can be written as

dithizone

[6] See: E. B. Sandell and H. Onishi, *Colorimetric Determination of Traces of Metals*, 4th ed., Part I, pp. 420–431. New York: Interscience, 1978.
[7] See: E. B. Sandell and H. Onishi, *Colorimetric Determination of Traces of Metals*, 4th ed., Part I, pp. 604–611. New York: Interscience, 1978.

Both dithizone and its metal chelates are soluble in a variety of organic solvents. As with 8-hydroxyquinoline, the equilibrium between the metal ion and the reagent is pH dependent; thus, by controlling the pH of the aqueous phase, various separations of metallic ions are possible.

The dithizone complexes of many metal ions are intensely colored. Spectrophotometric measurement of the organic extract often serves to complete the analysis after the separation has been performed.

Information concerning the use of other organic chelating agents for separations by extraction can be found in several reference works.[8]

26E ION-EXCHANGE SEPARATION

Ion exchange is a process involving an interchange among ions of like sign between a solution and an essentially insoluble solid in contact with the solution. Many substances, both natural and synthetic, act as ion exchangers. The ion-exchange properties of clays and zeolites have been recognized and studied for more than a century. Synthetic ion-exchange resins were first produced in 1935 and have since found widespread laboratory and industrial application for water softening, water deionization, solution purification, and ion separation.

26E-1 ION-EXCHANGE RESINS

Synthetic ion-exchange resins are high-molecular-weight polymeric materials containing large numbers of an ionic functional group per molecule. Cation-exchange resins can be either a strong acid type containing sulfonic acid groups $(RSO_3^- H^+)$ or a weak acid type containing carboxylic acid groups (RCOOH); the former have wider application. Anion-exchange resins contain basic functional groups, generally amines, attached to the polymer molecule. Strong base exchangers are quaternary amines $(RN(CH_3)_3^+ OH^-)$; weak base types contain secondary or tertiary amines.

A cation-exchange process is illustrated by the equilibrium

$$xRSO_3^- H^+ + M^{x+} \rightleftharpoons (RSO_3^-)_x M^{x+} + xH^+$$
$$\text{solid} \qquad \text{solution} \qquad \text{solid} \qquad \text{solution}$$

where M^{x+} represents a cation and R represents *a part* of a resin molecule containing one sulfonic acid group. The analogous process involving a typical anion-exchange resin can be written

$$xRN(CH_3)_3^+ OH^- + A^{x-} \rightleftharpoons [RN(CH_3)_3^+]_x A^{x-} + xOH^-$$
$$\text{solid} \qquad \text{solution} \qquad \text{solid} \qquad \text{solution}$$

where A^{x-} is an anion.

26E-2 ION-EXCHANGE EQUILIBRIA

Ion-exchange equilibria can be treated by the law of mass action. For example,

---

[8] G. H. Morrison and H. Freiser, *Solvent Extraction in Analytical Chemistry*. New York: Wiley, 1957; A. K. De, S. M. Khophar, and R. A. Chalmers, *Solvent Extraction of Metals*. New York: D. Van Nostrand, 1970; E. B. Sandell and H. Onishi, *Colorimetric Determination of Traces of Metals*, 4th ed. New York: Interscience, 1978.

when a dilute solution of calcium ions is passed through a column packed with a sulfonic acid resin, the following equilibrium develops:

$$Ca_{aq}^{2+} + 2H_{res}^{+} \rightleftharpoons Ca_{res}^{2+} + 2H_{aq}^{+}$$

Application of the mass law leads to

$$K = \frac{[Ca_{res}^{2+}][H_{aq}^{+}]^2}{[Ca_{aq}^{2+}][H_{res}^{+}]^2}$$

Here, the bracketed terms are molar concentrations (for a more rigorous treatment, activities). Note that $[Ca_{res}^{2+}]$ and $[H_{res}^{+}]$ are concentrations of the two ions in the solid phase. In contrast to most solids, however, these concentrations can vary from zero to some maximum value when all the negative sites in the resin are occupied by one species only.

Many applications of ion exchange resins involve the situation in which one of the ions predominates in *both* phases. For example, in the removal of calcium ions from a dilute and somewhat acidic solution, the calcium ion concentration will be much smaller than that of hydrogen ions in both the aqueous and the resin phase; the concentration of hydrogen ions thus does not change significantly as a result of the exchange process. Under these circumstances, $[Ca_{res}^{2+}] \ll [H_{res}^{+}]$ and $[Ca_{aq}^{2+}] \ll [H_{aq}^{+}]$ and the foregoing equilibrium constant expression can be rearranged to give

$$\frac{[Ca_{res}^{2+}]}{[Ca_{aq}^{2+}]} = K \frac{[H_{res}^{+}]^2}{[H_{aq}^{+}]^2} = K_d \tag{26-5}$$

where K_d is a distribution coefficient analogous to the constant describing an extraction equilibrium in Equation 26-1.

26E-3 APPLICATION OF ION EXCHANGERS TO ANALYTICAL SEPARATIONS

Ion-exchange resins find several important applications in analytical chemistry. The most important of these is in chromatography, which will be considered in Section 28A-4. A brief description of other analytical applications of these useful materials follows.

Separation of interfering ions of opposite charge. Ion-exchange resins are useful for removal of interfering ions, particularly where these ions have a charge opposite to that of the analyte. For example, iron(III), aluminum(III), and other cations cause difficulty in the gravimetric determination of sulfate by virtue of their tendency to coprecipitate with barium sulfate. Passage of a solution to be analyzed through a column containing a cation-exchange resin results in retention of all cations and the liberation of a corresponding number of protons. The sulfate ion, on the other hand, passes freely through the column, and the analysis can be performed on the effluent. In a similar manner, phosphate ion, which interferes in the analysis of barium or calcium ions, can be removed by passing the sample through an anion-exchange resin.

Concentration of traces of an electrolyte. A useful application of ion exchangers involves the concentration of traces of an ion from a very dilute solution. Cation-

Table 26-4   Separation of Some Inorganic Species by Distillation

Analyte	Sample Treatment	Volatile Species	Method of Collection
CO_3^{2-}	Acidification	CO_2	$Ba(OH)_2(aq) + CO_2(g) \rightarrow BaCO_3(s) + H_2O$ or on Ascarite
SO_3^{2-}	Acidification	SO_2	$SO_2(g) + H_2O_2(aq) \rightarrow H_2SO_4(aq)$
S^{2-}	Acidification	H_2S	$Cd^{2+}(aq) + H_2S(g) \rightarrow CdS(s) + 2H^+$
F^-	Addition of SiO_2 and acidification	H_2SiF_6	Basic solution
Si	Addition of HF	SiF_4	Basic solution
H_3BO_3	Addition of H_2SO_4 and methanol	$B(OCH_3)_3$	Basic solution
$Cr_2O_7^{2-}$	Addition of concd HCl	CrO_2Cl_2	Basic solution
NH_4^+	Addition of NaOH	NH_3	Acidic solution
As, Sb	Addition of concd HCl and H_2SO_4	$AsCl_3$ $SbCl_3$	Water
Sn	Addition of HBr	$SnBr_4$	Water

exchange resins, for example, have been employed to collect traces of metallic elements from large volumes of natural waters. The ions are then liberated from the resin by treatment with acid; the result is a considerably more concentrated solution for analysis.

Conversion of salts to acids or bases. An interesting application of ion-exchange resins is the determination of the total salt content of a sample. This analysis can be accomplished by passing the sample through the acid form of a cation-exchange resin; absorption of the cations causes the release of an equivalent quantity of hydrogen ion, which can be collected in the washings from the column and then titrated. Similarly, a standard acid solution can be prepared from a salt; for example, a cation-exchange column in the acid form can be treated with a weighed quantity of sodium chloride. The salt liberates an equivalent quantity of hydrochloric acid from the resin, which is then collected in the washings and diluted to a known volume.

    In an analogous way, standard sodium hydroxide solutions can be prepared by treatment of an anion-exchange resin with a known quantity of sodium chloride.

26F SEPARATIONS OF INORGANIC SPECIES BY DISTILLATION

Distillation permits the separation of components in a mixture whose partition coefficients between solution and vapor phases differ significantly. If one species has a partition coefficient that is large compared with the other components of the mixture, the separation process is simple. Table 26-4 lists some inorganic species that can be conveniently separated by simple distillation.

PROBLEMS*

* 26-1. After 10.0 mL of an aqueous solution containing 1.235 mg of a drug was extracted with 5.0 mL of toluene, the aqueous layer was found to contain 0.346 mg of the compound.

  (a) Calculate the distribution coefficient for the drug between the two solvents.

  (b) How many 10.0-mL extractions would be required to recover 99.0% of the drug from 20.0 mL of water? What volume of toluene would be used?

  (c) What volume of toluene would be required if 5.00-mL portions of toluene were used for the same extraction as in (b)?

  (d) Repeat the calculations in (b) for an extraction with 2.00-mL portions of toluene.

  (e) How many extractions and what volume of toluene would be required if 10.0-mL portions of toluene were used to extract the same aqueous solution and it was desired to remove 99.9% of the drug?

26-2. The trihalomethanes ($CHCl_3$, $CHCl_2Br$, $CHClBr_2$, and $CHBr_3$) occur as contaminants in drinking water. The maximum concentration of these materials allowed by EPA is 0.10 mg/L. A standard method for determining these impurities involves extracting the compounds with a hydrocarbon solvent such as cyclohexane followed by gas chromatographic analysis of the extract. Assume that studies showed that extraction of 100 mL of water with a single 5.00-mL portion of cyclohexane resulted in a 32.8% transfer of $CHCl_3$ to the organic medium.

  (a) Calculate the distribution coefficient for the process.

  (b) How many 5.00-mL extractions would be required to increase the recovery to 95%? What volume of cyclohexane would be used?

  (c) Repeat the calculations in (b) for 3.00-mL extractions.

  (d) Repeat the calculations in (b) for 1.00-mL extractions.

  (e) Repeat the calculations in (b), (c), and (d) for a 99% recovery of $CHCl_3$.

  (f) What should be the magnitude of the distribution coefficient in order to obtain a 99% recovery of $CHCl_3$ using no more than a total of five 2.00-mL portions of cyclohexane?

* 26-3. The distribution coefficient for a metal halide between water and ether was determined to be 12.3. Calculate the concentration of the cation M^{2+} remaining after 50.0 mL of 0.125 M M^{2+} were extracted with the following quantities of ether: (a) one 40.0-mL portion, (b) two 20.0-mL portions, (c) four 10.0-mL portions, and (d) eight 5.00-mL portions.

26-4. The distribution coefficient for the 8-hydroxyquinoline chelate of the cation M^{3+} between water and methyl isobutyl ketone is 7.76. Calculate the percent M^{3+} remaining in 50.0 mL of the aqueous layer, which was originally 0.715 M in M^{3+}, after extraction with: (a) one 25.0-mL portion of the organic solvent, (b) two 12.5-mL portions, (c) five 5.00-mL portions, and (d) ten 2.5-mL portions.

* 26-5. What volume of ether would be required to decrease the concentration of M^{2+} in Problem 26-3 to 1.00×10^{-4} M if 25.0 mL of 0.0500 M M^{2+} were extracted with: (a) 25.0-mL portions of ether, (b) 10.0-mL portions, and (c) 3.00-mL portions?

26-6. What volume of methyl isobutyl ketone would be required to decrease the concentration of M^{3+} (Problem 26-4) to 1.00×10^{-5} M if 40.0 mL of 0.0200 M M^{3+} were treated with an excess of 8-hydroxyquinoline and extracted with: (a) 50.0-mL portions of the solvent, (b) 25.0-mL portions of the solvent, and (c) 10.0-mL portions of the solvent?

* 26-7. A 0.2000 M aqueous solution of the weak organic acid HA was prepared from the

* Answers to problems or parts of problems marked with an asterisk are to be found at the end of the book.

pure compound, and three 50.0-mL aliquots were transferred to 100-mL volumetric flasks. Solution 1 was diluted to 100 mL with 2.0 N $HClO_4$ while solution 2 was diluted to the mark with 2.0 N NaOH. Solution 3 was diluted to the mark with water. Aliquots of 25.0 mL each were extracted with 25.0 mL of n-hexane. The extract from the basic solution contained no detectable trace of A-containing species, which indicated that A^- is not soluble in the organic solvent. The extract from solution 1 contained no ClO_4^- or $HClO_4$ but was found to be 0.0846 M in HA (by extraction with standard NaOH and back-titration with standard HCl). The extract of solution 3 was found to be 0.0372 M in HA. Assume HA does not associate or dissociate in the organic solvent, and calculate

(a) the distribution coefficient for HA between the two solvents.

(b) the concentration of the *species* HA and A^- in the aqueous solution 3 after extraction.

(c) the dissociation constant for HA in water.

26-8. To determine the equilibrium constant for the reaction

$$I_2 + X^- \rightleftharpoons I_2X^-$$

a 0.00825 M solution of aqueous I_2 was prepared, and 50.0 mL of this solution were extracted with 20.0 mL of hexane. After extraction, spectrophotometric measurements showed that the I_2 concentration of the *aqueous phase* was 1.04×10^{-4} M. An aqueous solution was then prepared that was 0.00793 M in I_2 and 0.0750 M in NaX. After extraction of 50.0 mL of this solution with 20.0 mL of hexane, the concentration of I_2 *in the hexane* was found by spectrophotometric measurement to be 9.47×10^{-4} M.

(a) What is the partition coefficient for I_2 between hexane and water?

(b) What is the formation constant for I_2X^-?

* 26-9. The total cation content of natural water is often determined by exchanging the cations for hydrogen ions on a strong acid ion exchange resin. A 25.0-mL sample of a natural water was diluted to 100 mL with distilled water and 2.0 g of a cation exchange resin were added. After stirring, the mixture was filtered and washed with three 15.0-mL portions of water. The filtrate and washings required 16.1 mL of 0.0213 N NaOH to give a methyl orange end point.

(a) Calculate the number of milliequivalents of cation present in exactly 1 L of sample. (Here, the equivalent weight of a cation is its formula weight divided by its charge.)

(b) Report the results in terms of mg $CaCO_3$/L.

26-10. An organic acid was isolated and purified by recrystallization of its barium salt. To determine the equivalent weight of the acid, a 0.346-g sample of the salt was dissolved in about 100 mL of water. The solution was passed through a strong acid ion-exchange resin and was washed with water; the eluate and washings were titrated with 20.2 mL of 0.0996 N NaOH to a phenolphthalein end point.

(a) Calculate the equivalent weight of the organic acid.

(b) A potentiometric titration curve of the solution resulting when a second sample was treated in the same way revealed two end points, one appearing at pH 5 and the other at pH 9.0. What is the molecular weight of the acid?

* 26-11. Describe the preparation of exactly 1 L of 0.1000 N HCl from primary standard grade NaCl using a cation-exchange resin.

26-12. An aqueous solution containing $MgCl_2$ and HCl was analyzed by first titrating a 25.00-mL aliquot of the sample to a bromocresol green end point with 22.76 mL of 0.02376 N NaOH. A 10.00-mL aliquot was then diluted to 50 mL with distilled water and passed through a strong acid ion-exchange resin. The eluate and washings required 33.20 mL of NaOH to reach the same end point. Report the molar concentrations of HCl and $MgCl_2$ in the sample.

an introduction to chromatographic

separations

27

Without question, the most widely used means of performing analytical separations is *chromatography*, a procedure that finds application to all branches of science. Chromatography was invented and named by the Russian botanist Mikhail Tswett shortly after the turn of the century. He employed the technique to separate various plant pigments such as chlorophylls and xanthophylls by passing a solution of these compounds through a glass column packed with finely divided calcium carbonate. The separated species appeared as colored bands on the column, which accounts for the name he chose for the method.

The applications of chromatography have grown explosively in the last four decades, owing not only to the development of several new types of chromatographic techniques but also to the growing need by scientists for better methods for separating complex mixtures. The tremendous impact of these methods on science is attested by the 1952 Nobel prize that was awarded to A. J. P. Martin and R. L. M. Synge for their discoveries in the field.

27A A GENERAL DESCRIPTION OF CHROMATOGRAPHY
Chromatography encompasses a diverse and important group of separation methods that permit the scientist to separate, isolate, and identify closely related

components of complex mixtures; many of these separations are impossible by other means.[1]

The term "chromatography" is difficult to define rigorously, owing to the variety of systems and techniques to which it has been applied. All of these methods, however, make use of a *stationary phase* and a *mobile phase*. Components of a mixture are carried through the stationary phase by the flow of the mobile one; separations are based on differences in migration rates among the sample components.

27A-1 TYPES OF STATIONARY PHASES

For successful chromatography, the components to be separated must be soluble in the mobile phase. They must also be capable of interacting with the stationary phase either by dissolving in it, being adsorbed by it, or reacting chemically with it. As a consequence, during the separations, the components become distributed between the two phases.

Column chromatography refers to methods in which the stationary phase is contained in a narrow glass or metal tube. The mobile phase, which may be a liquid or a gas, is then forced through the solid under pressure or allowed to percolate through it by gravity. In *planar chromatography*, the stationary phase is supported on a flat glass or plastic plate; here, the mobile phase moves through the solid either by capillary action or under the influence of gravity. In either type of chromatography the stationary phase may be a finely divided solid or may consist of an immobilized liquid that is immiscible with the mobile phase. Several procedures are employed to fix the stationary liquid in place. For example, a finely divided solid coated with a thin layer of liquid may be held in a glass or metal tube through which the mobile phase flows or percolates. Ordinarily, the solid plays no direct part in the separation, functioning only to hold the stationary liquid phase in place by adsorption. Alternatively, the inner walls of a capillary tube can be coated with a thin layer of liquid; a gaseous mobile phase is then caused to flow through the tube. Finally, the stationary liquid phase can be held in place on the fibers of paper or on the surface of finely ground particles held on a glass plate.

Table 27-1 classifies common chromatographic methods according to the nature of the stationary and mobile phases. Most of the theoretical discussion in this chapter will be concerned with partition, adsorption, and gas-liquid chromatography. With suitable modification, these concepts can be adapted to the other types as well.

27A-2 LINEAR CHROMATOGRAPHY

All chromatographic separations are based upon differences in the extent to which solutes are partitioned between the mobile and the stationary phase. The equilibrium involved can be described quantitatively by means of a partition coefficient K,

---

[1] General references on chromatography include: E. Heftmann, *Chromatography*, 3d ed. New York: Van Nostrand-Reinhold, 1975; B. L. Karger, L. R. Snyder, and C. Horvath, *An Introduction to Separation Science.* New York: Wiley, 1973; J. M. Miller, *Separation Methods in Chemical Analysis.* New York: Wiley, 1975; R. Stock and C. B. F. Rice, *Chromatographic Methods*, 3d ed. London: Chapman & Hall, 1974; and *Chromatographic and Allied Methods*, O. Mikeš, Ed. New York: Wiley, 1979.

Table 27-1 Classification of Chromatographic Separations

Name	Type Mobile Phase	Type Stationary Phase	Method of Fixing the Stationary Phase
Gas-liquid	Gas	Liquid	Adsorbed on a porous solid held in a tube or adsorbed on the inner surface of a capillary tube
Gas-solid	Gas	Solid	Held in a tubular column
Partition or liquid	Liquid	Liquid	Adsorbed on a porous solid held in a tubular column
Adsorption	Liquid	Solid	Held in a tubular column
Paper	Liquid	Liquid	Held in the pores of a thick paper
Thin layer	Liquid	Liquid or solid	Finely divided solid held on a glass plate; liquid may be adsorbed on particles
Gel	Liquid	Liquid	Held in the interstices of a polymeric solid
Ion exchange	Liquid	Solid	Finely divided ion-exchange resin held in a tubular column

which for chromatography is defined as

$$K = \frac{C_S}{C_M} \tag{27-1}$$

Here, C_S is the analytical concentration of a solute in the stationary phase and C_M is its concentration in the mobile phase. At the low concentrations encountered in chromatography, K is found to be at least approximately constant so that a linear relationship exists between C_S and C_M. Chromatography performed under these circumstances is termed *linear chromatography;* we shall treat only this type.

27A-3 LINEAR ELUTION CHROMATOGRAPHY

In elution chromatography, a single portion of the sample dissolved in the mobile phase is introduced at the head of a column (see Figure 27-1), whereupon the components of the sample distribute themselves between the two phases. Introduction of additional mobile phase (the *eluent*) forces the solvent containing a part of the sample down the column, where further partition between the mobile phase and fresh portions of the stationary phase occurs. Simultaneously, partitioning between the fresh solvent and the stationary phase takes place at the site of the original sample. Continued additions of solvent carry solute molecules down the column in a continuous series of transitions between the mobile and the stationary phases. Because solute movement can only occur in the mobile phase, however, the average *rate* at which a solute migrates *depends upon the fraction of time it spends in that phase.* This fraction is small for solutes with partition ratios that favor retention in the stationary phase and is large where retention in the mobile phase is more likely. Ideally, the resulting differences in rates cause the components in a mixture to separate into bands located along the length of the column (see Figures 27-1 and 27-2). Isolation can then be accomplished by passing a sufficient quantity of mobile phase through the column to cause these various bands to pass out the

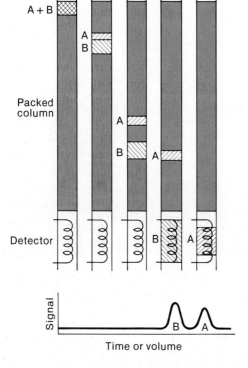

FIGURE 27-1 Schematic diagram of an elution chromatographic separation of components B and A of a mixture.

end, where they can be collected. Alternatively, the column packing can be removed and divided into portions containing the various components of the mixture.

The process whereby the solute is washed through the column by addition of fresh solvent is called *elution*. If a detector that responds to solute concentration is placed at the end of the column and its signal is plotted as a function of time (or of volume of the added mobile phase), a series of symmetric peaks is obtained, as shown in the lower part of Figure 27-1. Such a plot, called a *chromatogram*, is useful for both qualitative and quantitative analysis. The positions of peaks may serve to identify the components of the sample; the areas under the peaks can be related to their amounts.

27A-4 THEORIES OF ELUTION CHROMATOGRAPHY

Figure 27-2 shows concentration profiles for solutes A and B on a chromatographic column at an early and a late state of elution. The partition coefficient for A is the larger of the two; thus, A lags behind during the migration process. It is apparent that movement down the column increases the distance between the two peaks. At the same time, however, broadening of both bands takes place, which lowers the efficiency of the column as a separating device. Zone broadening is unavoidable; fortunately, however, it occurs more slowly than zone separation. Thus, as shown in Figure 27-2, a clean resolution of species is possible provided the column is sufficiently long.

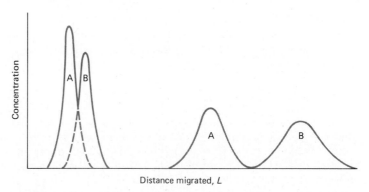

FIGURE 27-2 Concentration profiles for solutes A and B at different points in their migration down a column.

It is important to appreciate that a useful theory of chromatography must be able to account not only for the rate at which solutes migrate but also for the rate at which the zones broaden during migration, since the effectiveness of separation is equally dependent upon the two phenomena. The original theory of chromatography, called the *plate theory*, was able to describe migration rates in quantitative terms. Unfortunately, however, its utility was limited because it failed to describe the effects of the numerous variables responsible for zone broadening. As a consequence, the plate theory has now been supplanted by the *kinetic* or *rate theory*, which is capable of accounting for these variables as well.

It is useful to consider the plate theory briefly in order to indicate the genesis of two terms employed in the kinetic theory. This theory, which was originally developed by Martin and Synge,[2] envisages a chromatographic column as being composed of a series of discrete but contiguous, narrow, horizontal layers called *theoretical plates*. At each plate, equilibration of the solute between the mobile and the stationary phase is assumed to take place. Movement of the solute and solvent is then viewed as a series of stepwise transfers from one plate to the next.

The efficiency of a chromatographic column as a separation device improves as the number of equilibrations increases, that is, as the number of theoretical plates increases. Thus, the *number of theoretical plates, N,* is used as a measure of column efficiency. A second term, the *height equivalent of a theoretical plate, H,* also serves this purpose. The relationship between these two parameters is

$$N = \frac{L}{H} \tag{27-2}$$

where L is the length of the column packing. Note that H decreases as the efficiency of a column becomes greater. That is, as H becomes smaller, the number of equilibrations that occur in a given length of column becomes larger. It is important to note that H and N are retained as efficiency parameters in the rate theory and that Equation 27-2 continues to apply. It should be appreciated, however, that *a plate as a physical entity does not exist* in a column. Thus, the plate and the plate height should be viewed as criteria for column efficiency only.

[2] A. J. P. Martin and R. L. M. Synge, *Biochem. J.*, **35**, 1358 (1941).

27B THE RATE THEORY OF CHROMATOGRAPHY

The rate theory of chromatography successfully describes the effects of variables which affect the width of an elution band as well as the time of its appearance at the end of a column.[3]

27B-1 ZONE SHAPES

Examination of a typical chromatogram (Figures 27-1 and 27-3) reveals a similarity to normal error or Gaussian curves (Figure 3-3, p. 52) that are obtained when replicate values of a measurement are plotted as a function of the frequency of their occurrence. As we have shown in Section 3C-1, such plots can be rationalized by assuming that the uncertainty associated with any single measurement is the summation of a much larger number of small, individually undetectable and random uncertainties, each of which may be positive or negative in sign. The most common occurrence is for these uncertainties to cancel one another, thus leading to the mean value. With less likelihood, the summation may cause results that are greater or smaller than the mean. The consequence is a symmetric distribution of data around the mean value. In a similar way, the typical Gaussian shape of a chromatogram can be attributed to the additive combination of the random motions of the myriad solute particles in the chromatographic band or zone.

Let us first consider the behavior of an individual solute particle, which during migration undergoes many thousands of transfers between the stationary and the mobile phase. The time it spends in either phase after a transfer is highly irregular and depends upon its accidental gain of sufficient thermal energy from its environment to accomplish a reverse transfer. Thus, in some instances, the residence time in a given phase may be transitory; in others, the period may be relatively long. Recall that the particle can move *only during residence in the mobile phase;* as a result, its migration down the column is also highly irregular. Because of variability in the residence time, the average rate at which individual particles move relative to the mobile phase varies considerably. Certain individual particles travel rapidly by virtue of their accidental inclusion in the mobile phase for a majority of the time. Others, in contrast, may lag because they happen to have been incorporated in the stationary phase for a greater-than-average time. The consequence of these random individual processes is a symmetric spread of velocities around the mean value, which represents the behavior of the average and most common particles.

The breadth increases as the zone moves down the column because more time is allowed for spreading to occur. Thus, the zone breadth is directly related to residence time in the column and inversely related to the velocity at which the mobile phase flows.

Standard deviation as a measure of zone breadth. The breadth of a Gaussian curve is, of course, related to the standard deviation σ (p. 53); approximately 96% of the area under such a curve is included within plus or minus two standard deviations ($\pm 2\sigma$) of its maximum. Thus, a standard deviation derived from a chromatogram serves as a convenient quantitative measure of zone broadening.

---

[3] For a detailed presentation of the rate theory, see: J. C. Giddings, *J. Chem. Educ.*, **44**, 704 (1967); J. C. Giddings, *Dynamics of Chromatography*, Part I. New York: Marcel Dekker, 1965; and J. C. Giddings, in *Chromatography*, 3d ed., E. Heftmann, Ed., Chapter 3. New York: Van Nostrand-Reinhold, 1975.

It is important to understand that the standard deviation for a chromatographic peak can be expressed in units of time or, alternatively, in distance along the length of the column. Thus, a standard deviation derived from the zone profiles shown in Figure 27-2 would be in terms of length (usually in centimeters). More commonly, the abscissa of a chromatogram is in units of seconds or minutes (see Figure 27-3), and a standard deviation derived from such a curve would carry units of time. It is useful to distinguish between these units; thus, we shall use σ for a standard deviation in units of length and τ when the units are time.

Figure 27-3 illustrates a simple means for approximating τ (or σ) from an experimentally derived chromatogram. Tangents to the two sides of the Gaussian curve are extended to form a triangle with the abscissa. The area of this triangle corresponds to approximately 96% of the area under the curve. Thus, the intercepts occur at approximately $\pm 2\tau$ from the maximum; that is, $W = 4\tau$.

Calculation of H and N from zone breadth. The broadening of a chromatographic zone is conveniently expressed as the square of the standard deviation (the *variance*) per unit length of column. By definition,

$$H = \frac{\sigma^2}{L} \tag{27-3}$$

where H, the plate height in centimeters, is the measure of efficiency and L is the length of column, also in centimeters; note that σ must carry units of length (centimeters) to be consistent with H and L. Substitution of Equation 27-2 and rearrangement provide an alternative way of describing the separation characteristics of a column. That is,

$$N = \frac{L^2}{\sigma^2} \tag{27-4}$$

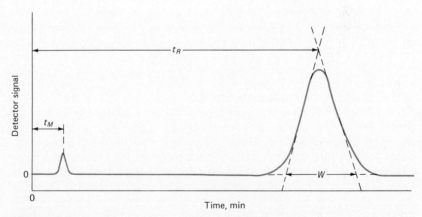

FIGURE 27-3 Determination of the standard deviation τ from a chromatographic peak. Here, $W = 4\tau$; t_R is the retention for a solute that is retained by the column packing, and t_M is the time for one that is not. Thus t_M is equal approximately to the time required for a molecule of the mobile phase to pass through the column.

where N is the number of plates contained in a column of length L. Clearly, column efficiency increases as the number of plates increases and the plate height decreases.

Experimental evaluation of N and H. Most experimental chromatograms, such as that in Figure 27-3, are obtained with time as the abscissa. From Figure 27-3, it is seen that t_R, the *retention time,* is the time required after sample injection for the solute peak to appear at the end of the chromatographic column. The average rate at which the solute particles travel is then L/t_R. Thus, the relationship between the standard deviation in units of length and units of time is given by the quotient of the standard deviation in centimeters and the rate of migration in centimeters per second, or

$$\tau = \frac{\sigma}{L/t_R} \tag{27-5}$$

Here, τ is the standard deviation in units of time and t_R is the time required for 1σ of the zone to emerge from the column. As noted earlier, however, for a Gaussian curve such as that in Figure 27-3, $W = 4\tau$. Substitution of this relationship into Equation 27-5 yields, upon rearrangement,

$$\sigma = \frac{LW}{4t_R}$$

Substitution into Equation 27-3 gives

$$H = \frac{LW^2}{16t_R^2}$$

To obtain N, we substitute into Equation 27-2 and rearrange, giving

$$N = 16 \left(\frac{t_R}{W}\right)^2 \tag{27-6}$$

Thus, N can be calculated from the two time measurements t_R and W; to obtain H, the length of the column must also be known.

27B-2 SOURCES OF ZONE BROADENING
Chromatographic peaks are generally broadened by three kinetically controlled processes, *eddy diffusion, longitudinal diffusion,* and *nonequilibrium mass transfer.* The magnitudes of these effects are determined by such controllable variables as flow rate, particle size of packing, diffusion rates, and thickness of the stationary phase. A number of equations have been developed that relate the efficiency of chromatographic columns to the extent to which these three processes occur.[4] The earliest and simplest of these, which is known as the *van Deemter* equation, was derived for gas-liquid chromatography; it provides an approximate

[4] For a summary of these various equations, see: J. M. Miller, *Separation Methods in Chemical Analysis,* pp. 120–126. New York: Wiley, 1975.

relationship between the flow rate u of the mobile phase and the plate height:

$$H = A + \frac{B}{u} + Cu \qquad (27\text{-}7)$$

Here, the quantity A is associated with eddy diffusion, B with longitudinal diffusion, and C with nonequilibrium mass transfer.

Eddy diffusion. Zone broadening from eddy diffusion is the result of the multitude of pathways by which a molecule can find its way through a packed column. As shown in Figure 27-4, the lengths of these pathways differ; thus, the residence times in the column for molecules of the same species are also variable. Solute molecules thus do not emerge simultaneously from the column; a broadening of the elution band results.

The quantity A in Equation 27-7 describes the effect of eddy diffusion and can be related to particle size, geometry, and tightness of packing of the stationary phase. As a first approximation, A is independent of flow rate.

Longitudinal diffusion. Longitudinal diffusion results from the tendency of molecules to migrate from the concentrated center part of a band toward more dilute regions on either side. This type of diffusion, which can occur in both the mobile and the stationary phase, causes further band broadening. Longitudinal diffusion is most important where the mobile phase is a gas, because diffusion rates in gases are several orders of magnitude greater than those in liquids. The amount of diffusion increases with time; thus, the extent of broadening increases as the flow rate decreases.

The second term (the longitudinal diffusion term) in Equation 27-7 is inversely proportional to flow rate.

Nonequilibrium mass transfer. Chromatographic bands are also broadened because the flow of the mobile phase is ordinarily so rapid that true equilibrium between phases cannot be realized. For example, at the front of the zone, where the mobile phase encounters fresh stationary phase, equilibrium is not instantly achieved, and solute is therefore carried somewhat farther down the column than

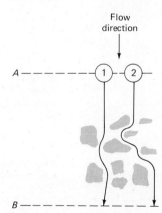

FIGURE 27-4 Typical pathways of two solute molecules during elution. Note that distance traveled by molecule 2 is greater than that traveled by molecule 1. Thus, molecule 2 would arrive at B later than molecule 1.

would be expected under true equilibrium conditions. Similarly, at the end of the zone, solutes in the stationary phase encounter fresh mobile phases. Again, the rate of transfer of solute molecules is not instantaneous; thus, the tail of the zone is more drawn out than it would be if time existed for equilibration. The net effect is a broadening at both ends of the solute band.

The effects of nonequilibrium mass transfer become smaller as the flow rate is decreased (Equation 27-7) because more time is available for equilibrium to be approached. Furthermore, a closer approach to true equilibrium is to be expected if the channels through which the mobile phase flows are narrow so that solute molecules do not have far to diffuse in order to reach the stationary phase. For the same reason, the layers of immobilized liquid on a stationary phase should be as thin as possible.

The last term in Equation 27-7, which is of considerable importance at the high flow rates of a gaseous mobile phase, describes the effect of nonequilibrium on band broadening.

Net effect of the zone broadening processes. Figure 27-5 shows the contribution of each term in the van Deemter equation as a function of mobile-phase velocity (broken lines) as well as their net effect (solid line) on H. Clearly, the optimum efficiency is realized at a flow rate corresponding to the minimum in the solid curve. Experimental curves of this type permit evaluation of A, B, and C for any column. Such data provide hints for improving the performance of a given type of packing.

The van Deemter equation provides only a first approximation of plate height, and several modifications have been developed that give a more precise description of the variables affecting column efficiency (see the references in footnote 3, p. 652).

27C SEPARATIONS ON COLUMNS

The discussion thus far has focused upon the efficiency of a chromatographic column, that is, how many plates it contains. We must now give consideration to the relationship between the number of plates in a column and the time required for any given separation and the efficiency of the separation.

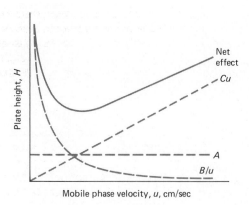

FIGURE 27-5 Effect of variables in Equation 27-7 on plate height.

27C-1 RATES OF SOLUTE MIGRATION

For the chromatogram shown in Figure 27-3, zero on the time axis corresponds to the instant the sample was injected onto the column and elution was started. The peak at t_M is for a species that is *not* retained by the column packing; its rate of motion will be the same as the average rate for the molecules of the mobile phase. The *retention time*, t_R, for the solute responsible for the second peak is the time required for that peak to reach the detector at the end of a column.

The average linear rate of migration $\bar{v}$ of the solute is given by

$$\bar{v} = \frac{L}{t_R} \tag{27-8}$$

Similarly, the average linear rate of movement u of molecules of the mobile phase will be

$$u = \frac{L}{t_M} \tag{27-9}$$

Relationship between retention time and partition coefficient. It is of interest to relate the retention time of a solute to its partition coefficient between the stationary and mobile phases (Equation 27-1). As we have mentioned earlier, a solute migrates only when it is in the mobile phase. Thus, the rate of migration can be expressed as a fraction of the velocity of the mobile phase. That is,

$$\bar{v} = u \left(\begin{matrix} \text{fraction of time the solute} \\ \text{spends in the mobile phase} \end{matrix}\right)$$

This fraction, however, equals the average number of moles of the solute in the mobile phase at any instant compared with the total number of moles in the column. That is,

$$\bar{v} = u \times \frac{\text{no. moles solute in mobile phase}}{\text{total number moles solute}}$$

or

$$\bar{v} = u \frac{C_M V_M}{C_M V_M + C_S V_S} = u \left(\frac{1}{1 + C_S V_S / C_M V_M}\right)$$

where C_M and C_S are the concentrations of the solute in the mobile and stationary phases, respectively; similarly, V_M and V_S are the total volumes of the two phases in the column.

Substitution of Equation 27-1 into this expression gives

$$\bar{v} = u \left(\frac{1}{1 + K V_S / V_M}\right) \tag{27-10}$$

The capacity factor. The *capacity factor* is an important constant that is related to the migration rate of the solute. It is defined as

$$k' = \frac{K V_S}{V_M} \tag{27-11}$$

Substitution of Equation 27-11 into 27-10 yields

$$\bar{v} = u \left(\frac{1}{1 + k'} \right) \tag{27-12}$$

In order to show how k' can be derived from a chromatogram, we substitute Equations 27-8 and 27-9 into expression 27-12, which gives

$$\frac{L}{t_R} = \frac{L}{t_M} \left(\frac{1}{1 + k'} \right) \tag{27-13}$$

This equation rearranges to

$$k' = \frac{t_R - t_M}{t_M} \tag{27-14}$$

As shown in Figure 27-3, t_R and t_M are readily obtained from a chromatogram.

27C-2 COLUMN RESOLUTION

The ability of a column to resolve two solutes is of prime interest in chromatography. Clearly, this property is related to the relative magnitude of the partition coefficients for the two species. The *selectivity factor*, α, of a column for two solutes is defined as

$$\alpha = \frac{K_Y}{K_X} \tag{27-15}$$

where K_Y is the partition coefficient for the more strongly retained solute Y and K_X is the constant for the less strongly held or more rapidly moving species X. By this definition, α must always be greater than unity.

Substitution of Equation 27-11 into 27-15 and rearrangement provides a relationship between the selectivity factor and the capacity factor. That is,

$$\alpha = \frac{k_Y'}{k_X'} \tag{27-16}$$

where k_Y' and k_X' are the capacity factors for Y and X, respectively. Substitution of Equation 27-14 gives an expression that permits the determination of α from an experimental chromatogram. That is,

$$\alpha = \frac{(t_R)_Y - t_M}{(t_R)_X - t_M} \tag{27-17}$$

Figure 27-6 shows chromatograms for species X and Y on three columns having different resolving powers. The *resolution*, R_s, for each of these columns is defined as

$$R_s = \frac{2\Delta Z}{W_X + W_Y} = \frac{2[(t_R)_Y - (t_R)_X]}{W_X + W_Y} \tag{27-18}$$

All of the terms on the right side of this equation are defined in Figure 27-6.

It is evident from Figure 27-6 that a resolution of 1.5 gives an essentially complete separation of X and Y, whereas a resolution of 0.75 does not. At a resolu-

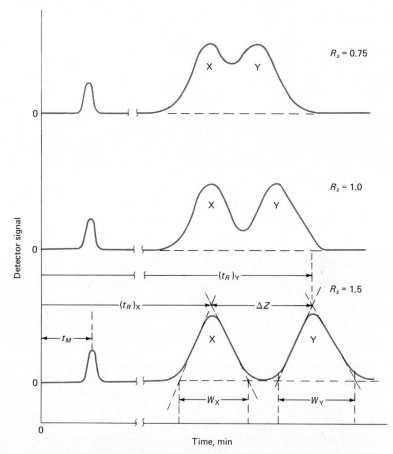

FIGURE 27-6 Separations at three resolutions. Here, $R_s = 2\Delta Z/(W_X + W_Y)$.

tion of 1.0, zone X contains about 4% Y and, conversely, at a resolution of 1.5, the overlap is about 0.3%. The resolution for a given stationary phase can be improved by lengthening the column, thus increasing the number of plates. An adverse consequence of the added plates, however, is an increase in the time required for the resolution.

The equations thus far presented permit derivation of two useful relationships, one of which predicts the resolution to be expected with a given number of plates; the second gives the time required for this separation to be completed.[5]

The resolution is given by

$$R_s = \frac{\sqrt{N}}{4}\left(\frac{\alpha - 1}{\alpha}\right)\left(\frac{k'_Y}{1 + k'_Y}\right) \tag{27-19}$$

[5] See: D. A. Skoog and D. M. West, *Principles of Instrumental Analysis*, 2d ed., pp. 677–679. Philadelphia: Saunders, 1980.

The time required to elute the more strongly held species Y is given by

$$(t_R)_Y = \frac{16R_s^2 H}{u} \left(\frac{\alpha}{\alpha - 1}\right)^2 \frac{(1 + k_Y')^3}{(k_Y')^2} \tag{27-20}$$

where u is the linear rate of movement of the mobile phase.

27C-3 OPTIMIZATION OF COLUMN PERFORMANCE

Equations 27-19 and 27-20 are of considerable importance in column chromatography because they serve as guides to the choice of conditions that will lead to a desired degree of resolution with a minimum expenditure of time. An examination of these equations reveals that each is made up of three parts. The first describes the efficiency of the column in terms of $\sqrt{N}$ or H. The second, which is the quotient containing α, is a selectivity term and depends solely on the properties of the two solutes. The third component is the capacity term, which is the quotient containing k_Y'; this term depends on the properties of both the solute and the column.

Column efficiency. It is evident from Equation 27-19 that resolution is proportional to the square root of the number of plates in the column. Increasing the number of plates to achieve a separation can be expensive in terms of time, however, unless the increase is accomplished by a reduction in H (Equation 27-20) and not by lengthening the column. The earlier section on rate theory described the variables that can be controlled to maximize efficiency or minimize H.

Variation in the capacity factor. The capacity factor k_Y' for a solute can frequently be varied over a considerable range by altering the composition of the mobile or the stationary phase. The former is usually easier and is the one that is most often varied to optimize a given separation. From Equation 27-11, it is evident that k_Y' can also be increased or decreased by a change in the ratio of the volume of the stationary phase to the mobile. An increase in k_Y' accompanies a reduction in particle size of the stationary support because the consequent increase in surface area results in an increase in V_S.

   Increases in k_Y' enhance resolution but at the expense of elution time. To understand the two effects, it is convenient to write Equations 27-19 and 27-20 in the form

$$R_s = Q \frac{k_Y'}{1 + k_Y'}$$

and

$$(t_R)_Y = Q' \frac{(1 + k_Y')^3}{(k_Y')^2}$$

where Q and Q' contain the rest of the terms in the equations. Figure 27-7 is a plot of R_s/Q and $(t_R)_Y/Q'$ as a function of k_Y', assuming that Q and Q' remain approximately constant. It is clear that values of k_Y' greater than about 10 are to be avoided because they provide little increase in resolution but markedly increase the time required for separations. The minimum in the elution time curve occurs at values of k_Y' between 2 and 3. Often, then, the optimal value of k_Y' lies in the range of 2 to 5.

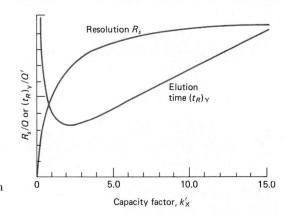

FIGURE 27-7 Effect of capacity factor k'_X on resolution R_s and elution time $(t_R)_Y$. It is assumed that Q and Q' remain constant with variation in k'_Y.

Variation in the selectivity factor. As would be expected, resolution is improved as the selectivity factor becomes larger. A clean separation is generally possible in a minimal time when the parameter is greater than 2. On the other hand, resolution is lost as α approaches 1, unless a very long column is employed. As shown by Equation 27-20, however, α values that approach 1 increase enormously the time required for separation. For example, an 84-fold increase in time is required for the same resolution when α is equal to 1.01 as compared with 1.1.

EXAMPLE 27-1. Substances A and B were found to have retention times of 16.40 and 17.63 min, respectively, on a 30.0-cm column. An unretained species passed through the column in 1.30 min. The peak widths (at base) for A and B were 1.11 and 1.21 min, respectively. Calculate: (a) the column resolution; (b) the average number of plates in the column; (c) the plate height; (d) the length of column required to achieve a resolution of 1.5; and (e) the time required to elute substance B on the longer column.

Employing Equation 27-18, we find
(a) $R_s = 2(17.63 - 16.40)/(1.11 + 1.21) = 1.06$
(b) Equation 27-6 permits computation of N. Thus,

$$N = 16 \left(\frac{16.40}{1.11}\right)^2 = 3493 \quad \text{and} \quad N = 16 \left(\frac{17.63}{1.21}\right)^2 = 3397$$

$$N_{av} = (3493 + 3397)/2 = 3445 = 3.4 \times 10^3$$

(c) $H = L/N = 30.0/3445 = 8.7 \times 10^{-3}$ cm
(d) k' and α do not change with increasing N and L. Thus, substituting N_1 and N_2 into Equation 27-19 and dividing one of the resulting equations by the other yields

$$\frac{(R_s)_1}{(R_s)_2} = \frac{\sqrt{N_1}}{\sqrt{N_2}}$$

where subscripts 1 and 2 refer to the original and the longer columns, respectively. Substituting the appropriate values for N_1, $(R_s)_1$, and $(R_s)_2$ gives

$$\frac{1.06}{1.5} = \frac{\sqrt{3445}}{\sqrt{N_2}}$$

$$N_2 = 3445 \left(\frac{1.5}{1.06}\right)^2 = 6.9 \times 10^3$$

But

$$L = N \times H = 6.9 \times 10^3 \times 8.7 \times 10^{-3} = 60 \text{ cm}$$

(e) Substituting $(R_s)_1$ and $(R_s)_2$ into Equation 27-20 and dividing yields

$$\frac{(t_R)_1}{(t_R)_2} = \frac{(R_s)_1^2}{(R_s)_2^2} = \frac{17.63}{(t_R)_2} = \frac{(1.06)^2}{(1.5)^2}$$

and

$$(t_R)_2 = 35 \text{ min}$$

Thus, to obtain the improved resolution requires that the time of separation be doubled.

27D SUMMARY OF IMPORTANT RELATIONSHIPS FOR CHROMATOGRAPHY

The number of quantities, terms, and relationships employed in chromatography is large and often confusing. Table 27-2 serves to summarize the most important definitions and equations used in this text.

27E QUALITATIVE AND QUANTITATIVE ANALYSIS BY CHROMATOGRAPHY

Chromatography has grown to be the premiere method for separating closely related chemical species. In addition, it can be employed for qualitative identification and quantitative determination of separated species. This section considers some of the general characteristics of chromatography as a tool for completion of an analysis.

27E-1 QUALITATIVE ANALYSIS

A chromatogram provides only a single piece of qualitative information about each species in a sample, namely, its retention time or its position on the stationary phase after a certain elution period. Additional data can, of course, be derived from chromatograms involving different mobile and stationary phases and various elution temperatures. Still, the number of data points for a species obtainable by chromatography is small compared with the the number provided by a single IR, NMR, or mass spectrum. Furthermore, spectral abscissa data can be determined with a much higher precision than can their chromatographic counterpart (t_R).

The foregoing should not be interpreted to mean that chromatography lacks important qualitative applications. Indeed, it is a widely used tool for recognizing the presence of components of mixtures containing a limited number of possible species whose identities are known. For example, the presence or absence of 30 or more amino acids in a protein hydrolysate can be ascertained with a relatively high degree of certainty from their positions after development on a thin layer chromatographic plate. Even here, however, confirmation of identity would require spectral or chemical investigation of the isolated components.

It is important to add that positive spectroscopic identification would ordinarily be impossible on as complex a sample as the foregoing without a preliminary chromatographic separation. Thus, chromatography is often a vital precursor to qualitative spectroscopic analyses.

27E-2 QUANTITATIVE ANALYSIS

Chromatography owes its precipitous growth during the past two decades in part to

Table 27-2 Summary of the Most Important Chromatographic Quantities and Relationships

Experimental Quantities

Name	Symbol	Determined from
Migration time, nonretained species	t_M	Chromatogram (see Figure 27-6)
Retention times, species X and Y	$(t_R)_X$, $(t_R)_Y$	Chromatogram (see Figure 27-6)
Peak widths, species X and Y	W_X, W_Y	Chromatogram (see Figure 27-6)
Length of column packing	L	Direct measurement
Flow rate	F	Direct measurement
Volume of stationary phase	V_S	Packing preparation data
Concentration of solute in mobile and stationary phases	C_M, C_S	Analysis and preparation data

Derived Quantities

Name	Calculation of Derived Quantities	Relationship to Other Quantities
Linear mobile phase velocity	$u = L/t_M$	
Volume of mobile phase	$V_M = t_M F$	
Capacity factor	$k' = (t_R - t_M)/t_M$	$k' = \dfrac{K V_S}{V_M}$
Partition coefficient	$K = \dfrac{k' V_M}{V_S}$	$K = \dfrac{C_S}{C_M}$
Selectivity factor	$\alpha = \dfrac{(t_R)_Y - t_M}{(t_R)_X - t_M}$	$\alpha = \dfrac{k'_Y}{k'_X} = \dfrac{K_Y}{K_X}$
Resolution	$R_s = \dfrac{2[(t_R)_Y - (t_R)_X]}{W_X + W_Y}$	$R_s = \dfrac{\sqrt{N}}{4}\left(\dfrac{\alpha - 1}{\alpha}\right)\left(\dfrac{k'_Y}{1 + k'_Y}\right)$
Number of plates	$N = 16\left(\dfrac{t_R}{W}\right)^2$	$N = \dfrac{L}{H} = 16 R_s^2\left(\dfrac{\alpha}{\alpha - 1}\right)^2\left(\dfrac{1 + k'_Y}{k'_Y}\right)^2$
Retention time	$(t_R)_Y = \dfrac{16 R_s^2 H}{u}\left(\dfrac{\alpha}{\alpha - 1}\right)^2 \dfrac{(1 + k'_Y)^3}{(k'_Y)^2}$	

its speed, simplicity, relatively low cost, and wide applicability as a separating tool. It is doubtful, however, if its use would have become as widespread had it not been for the fact that it can also provide useful quantitative information about the separated species. It is important, therefore, to discuss some of the general aspects of quantitative chromatography.

Quantitative chromatography is based upon a comparison of either the height or the area of the analyte peak with that of one or more standards. If conditions are properly controlled, both of these parameters vary linearly with concentration.

Analyses based on peak height. The height of a chromatographic peak is obtained by connecting the baselines on either side of the peak by a straight line and measuring the perpendicular distance from this line to the peak. This measurement can ordinarily be made with reasonably high precision and yields accurate results, provided variations in column conditions do not alter the peak widths during the period required to obtain chromatograms for sample and standards. The variables that must be controlled closely are column temperature, eluent flow rate, and rate of sample injection. In addition, care must be taken to avoid overloading the column. The effect of sample injection rate is particularly critical for the early peaks of a chromatogram. Relative errors of 5 to 10% due to this cause are not unusual with syringe injection.

Analyses based on peak areas.   Peak areas are independent of broadening effects due to the variables mentioned in the previous paragraph. From this standpoint, therefore, areas are a more satisfactory analytical parameter than peak heights. On the other hand, peak heights are more easily measured and, for narrow peaks, more accurately determined.

Many modern chromatographic instruments are equipped with ball and disc or electronic integrators which permit precise estimation of peak areas. If such equipment is not available, a manual estimate must be made. A simple method, which works well for symmetric peaks of reasonable widths, is to multiply the height of the peak by its width at one-half the peak height. Other methods involve the use of a planimeter or cutting out the peak and determining its weight relative to the weight of a known area of recorder paper. McNair and Bonelli measured the precision of these various techniques on chromatograms for ten replicate samples.[6] They reported the following relative standard deviations: electronic integration, 0.44%; ball and disc integration, 1.3%; weight of paper, 1.7%; height times width at one-half height, 2.6%; and planimeter, 4.1%.

Calibration with standards.   The most straightforward method for quantitative chromatographic analyses involves the preparation of a series of standard solutions that approximate the composition of the unknown. Chromatograms for the standards are then obtained and peak heights or areas are plotted as a function of concentration. A plot of the data should yield a straight line passing through the origin; analyses are based upon this plot. Frequent restandardization is necessary for highest accuracy.

The most important source of error in analyses by the method just described is usually the uncertainty in the volume of sample; occasionally the rate of injection is also a factor. Ordinarily, samples are small (~ 1 μL), and the uncertainties associated with injection of a reproducible volume of this size with a microsyringe may amount to several percent relative. The situation is exacerbated in gas-liquid chromatography, where the sample must be injected into a heated sample port; here, evaporation from the needle tip may lead to large variations in the volume injected.

Errors in sample volume can be reduced to perhaps 1 to 2% relative by means of a rotary sample valve such as that shown in Figure 27-8. Here, the sample loop ACB in (a) is filled with sample; rotation of the valve by 45 deg then introduces a reproducible volume of sample (the volume originally contained in ACB) into the mobile-phase stream.

---

[6] H. M. McNair and E. J. Bonelli, *Basic Gas Chromatography*, p. 158. Walnut Creek, CA: Varian Aerograph, 1968.

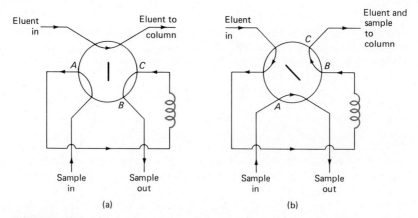

FIGURE 27-8 A rotary sample valve. (a) Valve position for filling sample loop *ACB* and (b) for introduction of sample into column.

The internal standard method. The highest precision for quantitative chromatography is obtained by use of internal standards because the uncertainties introduced by sample injection are avoided. In this procedure, a carefully measured quantity of an internal standard substance is introduced into each standard and sample, and the ratio of analyte to internal standard peak areas (or heights) serves as the analytical parameter. For this method to be successful, it is necessary that the internal standard peak be well separated from the peaks of all other components of the sample ($R_s > 1.25$); the standard peak should, on the other hand, appear close to the analyte peak. With a suitable internal standard, precisions of 0.5 to 1% relative are reported.

The area normalization method. Another approach that avoids the uncertainties associated with sample injection is the area normalization method. Complete elution of all components of the sample is required, a restriction that has limited the procedure to gas-liquid chromatography. In the normalization method, the areas of all eluted peaks are computed; after correcting these areas for differences in the detector response to different compound types, the concentration of the analyte is found from the ratio of its area to the total area of all peaks. The following example illustrates the procedure.

> **EXAMPLE 27-2.** The following area data were obtained from a chromatogram of a mixture of butyl alcohols (the detector sensitivity corrections were obtained in separate experiments with known amounts of pure alcohols).

Alcohol	Peak Area, cm²	Detector Response Factor	Corrected Areas, cm²
n-butyl	2.74	0.603	1.652
i-butyl	7.61	0.530	4.033
s-butyl	3.19	0.667	2.128
t-butyl	1.66	0.681	1.130
			8.943

Each entry in column 4 is the product of the data in columns 2 and 3. To normalize,

% n-butyl alcohol = 1.652 × 100/8.943 = 18.5

% i-butyl alcohol = 4.033 × 100/8.943 = 45.1

% s-butyl alcohol = 2.128 × 100/8.943 = 23.8

% t-butyl alcohol = 1.130 × 100/8.943 = 12.6 / 100.0

PROBLEMS*

* 27-1. The following data apply to a column for partition chromatography:

length of packing	22.6 cm
flow rate	0.287 mL/min
V_M	1.26 mL
V_S	0.148 mL

A chromatogram of a mixture of species A, B, C, and D provided the following data:

	Retention Time, min	Width of Peak Base (W), min
nonretained	4.2	—
A	6.4	0.45
B	14.4	1.07
C	15.4	1.16
D	20.7	1.45

Calculate
(a) the number of plates from each peak.
(b) the mean and the standard deviation for N.
(c) the plate height for the column.
* 27-2. From the data in Problem 27-1, calculate for A, B, C, and D
(a) the capacity factor.
(b) the partition coefficient.
* 27-3. From the data in Problem 27-1 for species B and C, calculate
(a) the resolution.
(b) the selectivity factor, α.
(c) the length of column necessary to give a resolution of 1.5.
(d) the time required to separate B and C with a resolution of 1.5.
* 27-4. From the data in Problem 27-1 for species C and D, calculate
(a) the resolution.
(b) the length of column required to give a resolution of 1.5.
(c) the time required to separate C and D with a resolution of 1.5.
27-5. The following data were obtained by gas-liquid chromatography on a 40-cm packed column:

* Answers to problems or parts of problems marked with an asterisk are to be found at the end of the book.

Compound	t_R, min	W, min
air	2.5	—
methylcyclohexane	10.7	1.3
methylcyclohexene	11.6	1.4
toluene	14.0	1.8

Calculate
(a) an average number of plates from the data.
(b) the standard deviation for the average in (a).
(c) an average plate height for the column.

27-6. Referring to Problem 27-5, calculate the resolution for
(a) methylcyclohexene and methylcyclohexane.
(b) methylcyclohexene and toluene.
(c) methylcyclohexane and toluene.

27-7. If a resolution of 1.5 was required to resolve methylcyclohexane and methylcyclo-hexene in Problem 27-5,
(a) how many plates would be required?
(b) how long would the column have to be if the same packing were employed?
(c) what would be the retention times for the three compounds on the column in Problem 27-7b?
(d) what would be the resolution for the last two compounds on the column in Problem 27-7b?

27-8. If V_S and V_M for the column in Problem 27-5 were 20.6 and 64.2 mL, respectively, and a nonretained air peak appeared after 2.5 min, calculate the
(a) capacity factor for each of the three compounds.
(b) partition coefficient for each of the three compounds.
(c) selectivity factor for methylcyclohexane and methylcyclohexene.
(d) selectivity factor for methylcyclohexene and toluene.

* 27-9. A column was found to have the following constants for Equation 27-7: $A = 0.013$ cm, $B = 0.30$ cm^2/s, and $C = 0.015$ s. What are the optimum flow rate and the corresponding minimum plate height?

27-10. What would be the effect on a chromatographic peak of introducing the sample at too slow a rate?

* 27-11. From distribution studies species M and N are known to have partition coefficients between water and hexane of 6.50 and 6.31 ($K = [M]_{H_2O}/[M]_{hex}$). The two species are to be separated by elution with hexane in a column packed with silica gel with water on its surface. The ratio V_S/V_M for the packing is known to be 0.422.
(a) Calculate the capacity factor for each of the solutes.
(b) Calculate the selectivity factor.
(c) How many plates will be needed to provide a resolution of 1.5?
(d) How long a column is needed if the plate height of the packing is 1.02×10^{-2} cm?
(e) If a flow rate of 7.10 cm/s is employed, what time will be required to elute the two species?

27-12. Repeat the calculations in Problem 27-11 assuming the two constants have values of 6.50 and 6.11.

* 27-13. The relative peak areas obtained from a gas chromatogram of a mixture of methyl acetate, methyl propionate, and methyl *n*-butyrate were 18.1, 43.6, and 29.9, respectively. Calculate the percentage of each compound if the respective relative detection responses were 0.60, 0.78, and 0.88.

27-14. The relative areas for the five gas chromatographic peaks shown in Figure 29-5c are given below. Also shown are the relative responses of the detector to the five compounds. Calculate the percentage of each component in the mixture.

Compound	Peak Area, Relative	Detection Response, Relative
A	28.2	0.70
B	32.1	0.72
C	46.9	0.75
D	44.3	0.73
E	29.6	0.78

27-15. For a gas-chromatographic column, the values for A, B, and C in the van Deemter equation were 0.15 cm, 0.36 cm^2 s^{-1}, and 4.3×10^{-2} s. Calculate the minimum plate height and the best flow rate.

27-16. For a gas-chromatographic column, the values of A, B, and C in the van Deemter equation were 0.060 cm, 0.57 cm^2 s^{-1}, and 0.13 s. Calculate the minimum plate height and the optimum flow rate.

liquid chromatography

28

This chapter deals with five of the chromatographic methods listed in Table 27-1. Included are partition (or liquid-liquid), adsorption (or liquid-solid), ion-exchange, thin-layer, and paper chromatography. All use a liquid mobile phase and are thus types of liquid chromatography. The first three are column methods; the latter pair are examples of planar chromatography.

28A COLUMN CHROMATOGRAPHY

Early liquid chromatographic columns were glass tubes with diameters of perhaps 10 to 50 mm that held 50- to 500-cm columns of solid particles of the stationary phase. To assure reasonable flow rates, the particles of the solid were kept larger than 150 to 200 μm in diameter. Ordinarily, the head of liquid above the packing sufficed to force the mobile phase down the column. Flow rates were, at best, a few tenths of a milliliter per minute; thus, separations tended to be time consuming.

Attempts to speed up the classic procedure by application of vacuum or by pumping were not effective; increases in flow rates acted to increase plate heights beyond the minimum in Figure 27-5 (p. 656) and thus to decrease efficiencies.

Early in the development of liquid chromatography, it was realized that large increases in column efficiency could be expected to accompany decreases in the particle size of packings. It was not until the late 1960s, however, that the technology of producing and using packings with particle diameters as small as 10 μm

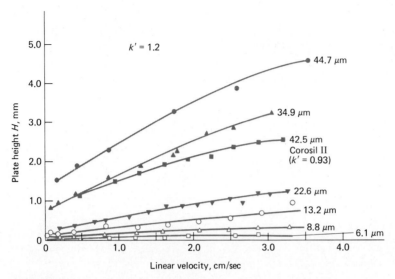

FIGURE 28-1   Effect of particle size of packing and flow rate upon
plate height H. Column dimensions: 30 cm × 2.4 mm. Solute:
N,N-diethyl-p-aminoazobenzene. Mobile phase: mixture of hexane,
methylene chloride, and isopropyl alcohol. The packing labeled
Corosil®II is a pellicular silica gel (p. 674). [R. E. Majors, *J.
Chromatogr. Sci.* **11**, 88 (1973). With permission.]

was developed. This technology required sophisticated instruments that con-
trasted markedly with the simple devices employed in classic liquid chromatography. The name *high-performance liquid chromatography* (HPLC) is often employed to distinguish these newer procedures from the classic methods, which still
find considerable use for preparative purposes.[1]

28A-1 HIGH-PERFORMANCE LIQUID CHROMATOGRAPHY (HPLC)

Figure 28-1 shows plots of experimental liquid chromatographic data relating plate
height to flow rate and particle diameter of packing materials. In none of the plots is
the minimum shown by Figure 27-5 reached; generally, in liquid chromatography,
this minimum is observed only at prohibitively low flow rates. Furthermore, the
van Deemter equation (Section 27B-2) does not adequately describe the relationship between efficiency and velocity of the mobile phase; here, a considerably
more complex equation, developed by Giddings, must be employed.[2]

It is apparent from Figure 28-1 that separation efficiency is greatly enhanced
with small particle diameters. Reasonable flow rates with packings consisting of
such particles can only be realized, however, by pumping at high pressures. As a

---

[1] A large number of books on liquid chromatography are available. Among these are: L. R. Snyder and
J. J. Kirkland, *Introduction to Modern Liquid Chromatography*, 2d ed. New York: Wiley, 1979; *Practical
High-Performance Liquid Chromatography*, C. F. Simpson, Ed. London: Heyden, 1977; and R. P. W.
Scott, *Contemporary Liquid Chromatography*, in *Techniques of Chemistry*, A. Weissberger, Ed., vol.
XI. New York: Wiley-Interscience, 1976.
[2] J. C. Giddings, *Anal. Chem.*, **35**, 1338 (1963).

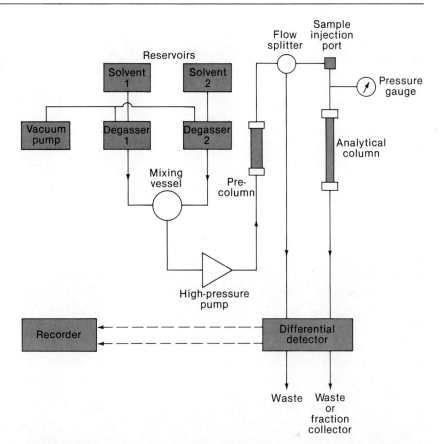

FIGURE 28-2 Schematic diagram showing liquid flow in a typical liquid chromatographic instrument. Many instruments do not employ a flow splitter and produce only a single output stream.

consequence, the equipment required for HPLC is considerably more elaborate than the simple gravity-fed column used in classic liquid chromatography. Figure 28-2 is a block diagram showing the components of such equipment. Each of these components is discussed in the paragraphs that follow.[3]

Solvent reservoir and degassing system. A modern HPLC apparatus is equipped with one or more glass or stainless steel reservoirs, each of which contains 1 to 2 L of a solvent. The reservoirs are often equipped with a means of removing dissolved gases—usually oxygen and nitrogen—that interfere by forming bubbles in the column and the detector systems. These bubbles cause band spreading; in addition, they often interfere with the performance of the detector.

---

[3] For a detailed discussion of HPLC systems, see: *Instrumentation for High-Performance Liquid Chromatography*, J. F. K. Huber, Ed. New York: Elsevier, 1978; H. Engelhardt, *High Performance Liquid Chromatography. Chemical Laboratory Practice*. New York: Springer-Verlag, 1979; H. Veening, *J. Chem. Educ.*, **50**, A429, A481, A529 (1973); and H. Kern and K. Imhof, *Amer. Lab.*, **10** (2), 131 (1978).

Degassers may consist of a vacuum pumping system, a distillation system, or a device for heating and stirring the solvent. It is not necessary that the degasser be an integral part of the HPLC system as shown in Figure 28-2. A convenient way of treating solvents before introduction into the reservoir is to filter them through a millipore filter under vacuum. This treatment removes gases as well as suspended matter in the solvent.

A separation that employs but a single solvent system is termed an *isocratic elution*. Frequently, separation efficiency is greatly enhanced by *gradient elution*. Here two (and sometimes more) solvent systems that differ significantly in polarity are employed. After elution is begun, the ratio of the two solvents is varied in a programmed way, sometimes continuously and sometimes in a series of steps. Modern HPLC equipment is often equipped with devices that introduce solvents from two or more reservoirs into a mixing chamber at rates that vary continuously; the volume ratio of the solvents may then be altered linearly or exponentially with time.

Figure 28-3 illustrates chromatographic separations for a mixture of chlorobenzenes. Isocratic elution with a 50:50 (v/v) methanol/water solution yielded curve (b). Curve (a) is for gradient elution which was initiated with a 40:60 mixture of the two solvents; the methanol concentration was then increased at the rate of 8%/min. Note that gradient elution shortened the time of separation significantly without sacrifice in resolution of the early peaks.

Pumps. Most HPLC pumps have outputs of at least 1000 psi (lbs/in²), and preferably 4000 to 6000 psi, with a flow delivery rate of at least 3 mL/min. The flow rate should be constant to ±2%. Two types of mechanical pumps are employed: one is a screw-driven syringe and the other is a reciprocating pump. The former produces a pulse-free delivery that is readily controlled; it suffers from lack of capacity and inconvenience in changing solvents. Reciprocating pumps, which are much more widely used, produce a pulsed flow which must be subsequently damped.

Pneumatic pumps are also employed. In the simplest of these, the mobile phase is contained in a collapsible container housed in a vessel that can be pressurized by a compressed gas. Pumps of this type are simple, inexpensive, and pulse free; they suffer from limited capacity and pressure output as well as a dependence of flow rate on solvent viscosity. In addition, they are not amenable to gradient elution.

It should be noted that the high pressures generated by the pumping devices do not constitute an explosion hazard because liquids are not very compressible. Thus, rupture of a component of the system results only in solvent leakage.

Precolumns. Some HPLC instruments are equipped with a so-called precolumn, which contains a packing chemically identical to that in the analytical column. The particle size is much larger, however, so that the pressure drop across the precolumn is negligible with respect to the rest of the system. The purpose of the precolumn is to remove impurities from the solvent and thus prevent contamination of the analytical column. In addition, the precolumn saturates the mobile phase with the liquid making up the stationary phase; in this way, stripping of the stationary phase from the packing of the analytical column is eliminated.

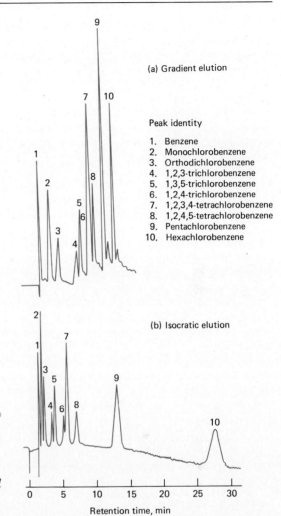

(a) Gradient elution

Peak identity

1. Benzene
2. Monochlorobenzene
3. Orthodichlorobenzene
4. 1,2,3-trichlorobenzene
5. 1,3,5-trichlorobenzene
6. 1,2,4-trichlorobenzene
7. 1,2,3,4-tetrachlorobenzene
8. 1,2,4,5-tetrachlorobenzene
9. Pentachlorobenzene
10. Hexachlorobenzene

(b) Isocratic elution

Retention time, min

FIGURE 28-3 Improvement in separation efficiency by gradient elution. Column: 1 m × 2.1 mm id, precision-bore stainless; packing: 1% Permaphase® ODS. Sample: 5 μL of chlorinated benzenes in isopropanol. Detector: UV photometer (254 nm). Conditions: temperature, 60°C, pressure, 1200 psi. [J. J. Kirkland, *Modern Practice of Liquid Chromatography*, p. 88. New York: Interscience, 1971. With permission.]

Sample injection systems. Rotary sampling valves, similar in construction to that shown in Figure 27-8, are often employed in HPLC.

Sample injection can also be accomplished by means of a syringe and a self-sealing septum of silicone, neoprene, or Teflon. Alternatively, flow through the column can be stopped momentarily, a cap on the column removed, and the sample injected directly onto the head of the column with a syringe.

Columns. Columns for HPLC are manufactured from heavy-walled glass tubing or precision-bore stainless steel; the former are limited to pressures below about 600 psi. The typical column ranges from 15 to 150 cm in length with an inside diameter of about 2 to 3 mm. Longer columns (up to a meter or more) are usually prepared by connecting sections of shorter columns. Coiled columns are also encountered, although some loss in efficiency results from this configuration.

The most common type of packing for HPLC columns is finely divided silica gel. Occasionally, alumina and Celite (a diatomaceous earth) are employed. Uniform particle sizes in the 3- to 10-μm range are desirable.

A second type of packing material, called *pellicular* particles, consists of small beads (often of glass and having diameters of about 40 μm), coated with a 1- to 3-μm layer of a porous material such as silica gel, alumina, or an ion-exchange resin. The rate at which equilibrium between phases is approached is high in these thin layers (p. 656); thus, improved efficiency results. Pellicular packings have the disadvantage of limited sample capacity (roughly one-tenth that of porous packings).

Temperature control. Most liquid chromatography is performed at room temperature and without thermostating. Water jacketed columns are available commercially when precise temperature control is desired.

Detectors. No highly sensitive, universal detector system, such as those found in gas chromatography (Section 29A-4), is available for HPLC. Thus, the system used will depend upon the nature of the sample. Table 28-1 lists the common detectors and some of their properties.

Perhaps the most common detectors for HPLC are based upon absorption of ultraviolet or visible radiation. Both photometers and spectrophotometers are available commercially. The former usually makes use of the 254- and 280-nm lines from a mercury source; at these wavelengths many organic functional groups absorb (see Section 21B-1). Spectrophotometric detectors are considerably more versatile than photometers because the radiation can be tailored to the anticipated absorption peaks of the sample components. Obviously, photometric detection requires that the sample components absorb and the solvent transmits radiation within the wavelength range of the instrument.

Figure 28-4 is a schematic diagram of an ultraviolet photometric detector.

Infrared spectrophotometric detectors are also available from commercial sources.

Table 28-1 Characteristics of Liquid Chromatography Detectors*

Detector Basis	Type**	Maximum Sensitivity†	Flow Rate Sensitive?	Temperature Sensitivity	Useful with Gradient?	Available Commercially?
UV absorption	S	5×10^{-10}	No	Low	Yes	Yes
IR absorption	S	10^{-6}	No	Low	Yes	Yes
Fluorometry	S	10^{-10}	No	Low	Yes	Yes
Refractive index	G	5×10^{-7}	No	$\pm 10^{-4}$ °C	No	Yes
Conductometric	S	10^{-8}	Yes	± 1 °C	No	Yes
Moving wire	G	10^{-8}	Yes	None	Yes	Yes
Mass spectrometry	S	10^{-10}	No	None	—	Yes
Polarography	S	10^{-10}	Yes	± 1 °C	—	No
Radioactivity	S	—	No	None	Yes	No

* Most of these data were taken from: L. R. Snyder and J. J. Kirkland, *Introduction to Modern Liquid Chromatography*, 2d ed., New York: Wiley-Interscience, 1979, p. 162. With permission.
** G = general; S = selective.
† Sensitivity for a favorable sample in grams per milliliter.

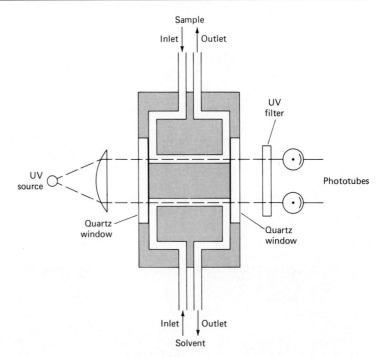

FIGURE 28-4   Ultraviolet detector for HPLC.

Another widely used detector is based upon the changes in refractive index of the solvent in the presence of solute molecules. In contrast to absorbance detectors, the refractive index indicator is general rather than selective and responds to the presence of all solutes.

28A-2 HPLC ADSORPTION CHROMATOGRAPHY

All of the pioneering work in chromatography was based upon liquid-solid adsorption in which the stationary phase was the surface of a finely divided solid. In such a packing, the solute competes with the eluting solvent for sites on the surface of the solid; retention is the result of adsorption forces. Currently, liquid-solid HPLC is used extensively for the separation of neutral organic compounds.

Stationary and mobile phases.   Silica gel is by far the most common adsorbent for liquid-solid chromatography. Alumina also finds considerable use. Packing materials of specified particle diameter ranges are available from several commercial sources; typical for HPLC is a 10-μm silica gel packing in which 80% of the particles range between 8 and 12 μm. One company offers columns packed with specially prepared spherical silica gel particles, with 95% of these particles having diameters within ± 1 μm of the nominal particle size of 3 μm.

The choice of mobile phase is all-important for successful liquid-solid chromatography; by varying the solvent, the capacity factor (k') for solutes can be varied until they fall in the ideal range of 1 to 10 (p. 660).

FIGURE 28-5 A liquid-solid chromatogram for adrenal steroid-cortisones. Peaks: (A) cortisol; (B) cortisone; (C) corticosterone. Column: 25 cm × 4.6 mm packed with 10 μm silica gel. Pressure: 225 psi. Flow rate: 59.8 mL/hr. Mobile phase: 75% heptane/25% ethanol. (Courtesy of Whatman, Inc., Clifton, New Jersey)

Applications. Large differences exist in the tendencies of compounds to be adsorbed, and these differences serve as the basis for adsorption chromatography. For example, a positive correlation can be discerned between adsorption properties and the number of hydroxyl groups in an organic molecule. A similar correlation exists with double bonds. Compounds containing certain functional groups are more strongly held than others. The tendency to be adsorbed decreases in the order: acid > alcohol > carbonyl > ester > hydrocarbon. The nature of the adsorbent is also influential in determining the order of adsorption. Much of the available knowledge in this field is empirical; the choice of adsorbent and solvent for a given separation frequently must be made on a trial-and-error basis.

Figure 28-5 shows a typical liquid-solid separation by HPLC.

28A-3 HPLC PARTITION CHROMATOGRAPHY
Partition or liquid-liquid chromatography was originated in 1941 in the Nobel prize-winning work of Martin and Synge.[4] In their studies, they were able to demonstrate that a properly prepared column can have a plate height as small as 0.002 cm. Thus, a 10-cm column of this type may contain as many as 5000 theoretical plates. High separation efficiencies are to be expected even with relatively short columns.

Solid supports. The most widely used solid support for partition chromatography has been silicic acid or silica gel. This material adsorbs water strongly; the stationary phase is often aqueous as a consequence. For some separations, the inclusion of a buffer or a strong acid (or base) in the water film has proved helpful. Polar solvents such as aliphatic alcohols, glycols, or nitromethane, alone or mixed with water, have also served as the stationary phase on silica gel. Other support media

---

[4] A. J. P. Martin and R. L. M. Synge, *Biochem. J.*, **35**, 1358 (1941).

include alumina, diatomaceous earth, starch, cellulose, and powdered glass; water and a variety of organic liquids have been used to coat these solids.

The mobile phase may be a pure solvent or a mixture of solvents; its polarity must be markedly different from the stationary liquid so that the two are immiscible. Ordinarily, the more polar of the two solvents is incorporated on the solid and serves as the stationary phase; in *reverse-phase* chromatography, the stationary phase is nonpolar and the mobile phase is polar. The choice of liquid pairs is largely empirical. As mentioned earlier, gradient elution is frequently employed to enhance separation efficiency.

Bonded phase packings. A type of packing that is becoming increasingly popular for reverse-phase HPLC consists of pure silica gel particles onto which an organic group has been chemically attached. As an example, a hydrocarbon surface can be formed by the reaction of chlorooctadecyl silane with the OH groups on the surface of silica gel. That is, where R is the octadecyl group and the Si in the circle attached

$$
\text{SiOH} + \text{Cl} - \overset{\overset{\displaystyle R}{|}}{\underset{\underset{\displaystyle R}{|}}{\text{Si}}} - R \longrightarrow \text{Si} - O - \overset{\overset{\displaystyle R}{|}}{\underset{\underset{\displaystyle R}{|}}{\text{Si}}} - R
$$

to an OH group represents one of many SiOH groups on the surface of the gel particle. Other groups that have been bonded to silica gel include aliphatic amines, ethers, and nitrates as well as aromatic hydrocarbons.

The behavior of the chemically bonded silica surface appears to be intermediate between a solid surface at which adsorption occurs and an immobilized liquid at which a liquid-liquid equilibrium exists. Chemically bonded surfaces offer a considerable advantage over ordinary solid-supported liquids in that the stationary phase cannot be stripped of its liquid by the mobile phase. On the other hand, chemically bonded surfaces suffer from limited loading capacities.

Applications. Partition chromatography has become a powerful tool for the separation of closely related substances. Typical examples include the resolution of the numerous amino acids formed in the hydrolysis of a protein, the separation and analysis of closely related aliphatic alcohols, and the separation of sugar derivatives.

28A-4 ION-EXCHANGE CHROMATOGRAPHY

We have already considered some of the applications of ion-exchange resins to analytical chemistry (Section 26E-3). In addition, ion-exchange resins have been successfully employed as the stationary phase in elution chromatography. An early example of such an application was provided by Beukenkamp and Reiman, who described the separation of sodium and potassium ions on a column packed with a sulfonic acid resin in its acidic form.[5] When the sample was introduced at the top of the column, equilibria were established that can be described by the equation

---

[5] J. Beukenkamp and W. Reiman III, *Anal. Chem.*, **22**, 582 (1950).

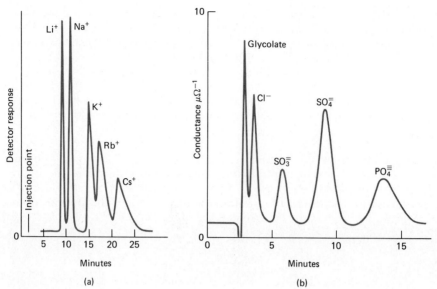

FIGURE 26-6 Ion-exchange chromatographs. (a) Sulfonic acid packing. Eluent: 0.01 N HCl. [H. Small, T. S. Stevens, and W. C. Bauman, *Anal. Chem.*, **47**, 1803 (1975)]. (b) Anion exchanger packing. Eluent: NaHCO$_3$/Na$_2$CO$_3$ solutions. [T. M. Stevens, V. T. Turkelson, and W. A. Albe, *Anal. Chem.*, **49**, 1177 (1977).] With permission.

$$B_{aq}^+ + H_{res}^+ \rightleftharpoons B_{res}^+ + H_{aq}^+ \tag{28-1}$$

where B$^+$ represents either Na$^+$ or K$^+$.

The two cations can then be eluted with hydrochloric acid. In this process, the hydrogen ion concentration far exceeds the sodium or potassium ion concentration in either phase and is essentially constant. Thus, the equilibrium constant expression for the foregoing reaction can be written in the form

$$\frac{[B_{res}^+]}{[B_{aq}^+]} = K\frac{[H_{res}^+]}{[H_{aq}^+]} = K_D \tag{28-2}$$

Equation 28-2 can then be employed in the same way as Equation 27-1 (Section 27A-2), and the general theory of chromatography, which we have already described, can be applied to a stationary phase consisting of an ion-exchange resin.

Note that K_D in Equation 28-2 represents the affinity of the resin for the ion B$^+$ *relative* to another ion (here, H$^+$). Where K_D is large, a strong tendency exists for the solid phase to retain ion B; where K_D is small, the reverse obtains. Selection of a common reference ion such as H$^+$ permits comparison of distribution ratios for different ions on a given type of resin. Such experiments reveal that polyvalent ions are much more strongly held than singly charged species. Within a given charge group, however, differences appear that are related to the size of the hydrated ion as well as other properties. Thus, for a typical sulfonated cation-exchange resin, values for K_D decrease in the order Cs$^+$ > Rb$^+$ > K$^+$ > NH$_4^+$ > Na$^+$ > H$^+$ > Li$^+$. For divalent cations, the order is Ba^{2+} > Pb^{2+} > Sr^{2+} > Ca^{2+} > Cd^{2+} > Cu^{2+} > Zn^{2+} > Mg^{2+}.

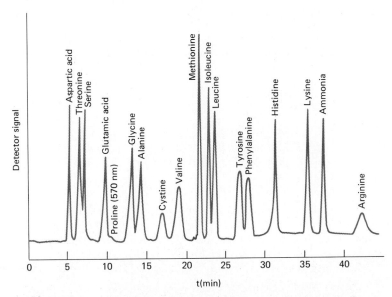

FIGURE 28-7 Separation of amino acids on an ion-exchange column. Packing: cation exchange with particle size of 8 μm. Pressure: 2700 psi. [J. R. Benson, *Amer. Lab.*, 4, (10) 53, 1972. Copyright 1972 by International Scientific Communications, Inc., Fairfield, Connecticut.]

The techniques for fractionation of ions with K_D values that are relatively close to one another are analogous to those described for adsorption and partition chromatography. For example, Figure 28-6a shows an HPLC separation of the alkali-metal ions on a pellicular sulfonic acid cation-exchange resin; Figure 28-6b illustrates an ion-exchange separation of anions.

Ion-exchange chromatography finds widespread use in amino acid analyzers. For example, Figure 28-7 shows a chromatogram of a mixture containing 1.0×10^{-8} mol of 17 amino acids. Separation time was only 42 min. Gradient elution with buffers of different pH was employed.

28A-5 COMPARISON OF HIGH-PERFORMANCE LIQUID CHROMATOGRAPHY WITH GAS-LIQUID CHROMATOGRAPHY

Table 28-2 provides a comparison between high-performance liquid chromatography (HPLC) and gas-liquid chromatography (GLC), which is discussed in the next chapter. Where either is applicable, gas-liquid chromatography offers the advantages of speed and simplicity of equipment. On the other hand, HPLC is applicable to nonvolatile substances (including inorganic ions) and thermally unstable materials whereas gas-liquid chromatography is not. The two methods tend to be complementary.

28B PLANAR CHROMATOGRAPHY

Two types of planar chromatography exist. In *thin-layer chromatography*, separations take place in a layer of finely divided solid that has been fixed upon a flat sup-

Table 28-2 Comparison of High-Performance Liquid and Gas-Liquid Chromatography

Characteristics possessed by both methods:
  Efficient, highly selective, and widely applicable
  Only small sample required
  Ordinarily nondestructive of sample
  Readily adapted to quantitative analysis
Particular advantages of high-performance liquid chromatography:
  Can accommodate nonvolatile and thermally unstable samples
  Generally applicable to inorganic ions
Particular advantages of gas-liquid chromatography:
  Simple and inexpensive equipment
  Rapid

porting surface.[6] In *paper chromatography*, a sheet or strip of heavy filter paper serves as the medium.[7] Paper chromatography was first used for separations in the middle of the nineteenth century. It was not until the late 1940s, however, that the usefulness of the technique became fully appreciated by scientists. Thin-layer chromatography was developed in the late 1950s and by now has become the more important of the two planar methods.

    Both paper and thin-layer chromatography provide remarkably simple and inexpensive means for separating and identifying the components of small samples of complex inorganic, organic, and biochemical substances. Furthermore, the methods, particularly thin-layer chromatography, permit reasonably accurate quantitative determination of the concentrations of the components of such mixtures.

28B-1 THIN-LAYER CHROMATOGRAPHY

Stationary phase. Solid absorbents (or sometimes adsorbents) for thin-layer chromatography are similar in chemical composition and particle size to those already described in the discussion of the various types of column chromatography. The most widely used substance is silica gel, which often functions as a support for water or other polar solvents for liquid-liquid separations. If the silica layer is oven-dried after preparation, however, much of the moisture is lost and the surface becomes a predominately solid adsorbent for liquid-solid separations. In this latter mode, care must be taken to avoid exposing the surface to the atmosphere, for it has been demonstrated that water adsorption from air occurs in a few minutes; the surface then reverts to a support for a liquid.

    Ion-exchange resins are also employed as the stationary phase in thin-layer chromatography.

Preparation of thin-layer plates. A thin-layer plate can be prepared by spreading an aqueous slurry of the finely ground solid onto the clean surface of a glass or mylar

---

[6] References: J. G. Kirchner, in *Thin Layer Chromatography*, A. Weissberger, Ed. New York: Wiley-Interscience, 1978; and J. T. Touchstone and M. R. Dobbins, *Practice of Thin Layer Chromatography*. New York: Wiley, 1978.
[7] References: R. Stock and C. B. F. Rice, *Chromatographic Methods*, 3d ed., Chapter 3. London: Chapman and Hall, 1974; and J. Sherma and G. Zweig, *Paper Chromatography*. New York: Academic Press, 1971.

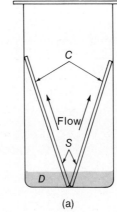
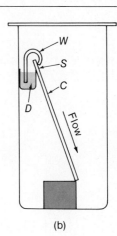

FIGURE 28-8 Two types of apparatus for thin-layer chromatography: (a) ascending flow; (b) descending flow. S: Initial position of sample: D: developer; C: chromatographic surface; W: cotton wick.

(a)

(b)

plate or microscope slide. A binder often is incorporated into the slurry to enhance adhesion of the solid particles to the glass and to one another. The plate is then allowed to stand until the layer has set and adheres tightly to the surface; for some purposes, it may be heated in an oven for several hours.

Plate development. A drop of the sample is placed near one edge of the plate (most plates have dimensions of 5×20 or 20×20 cm), and its position is marked with a pencil. After the sample solvent has evaporated, the plate is placed in a closed container saturated with vapors of the developing solvent. One end of the plate is then wetted with the developer by means of one of the arrangements shown in Figure 28-8. After the developer has traversed one-half to two-thirds the length of the plate, the latter is removed and dried and the positions of the components are determined in any of several ways.

Figure 28-9 illustrates the separation of amino acids in a mixture by development in two directions (*two-dimensional planar chromatography*). The sample was placed in one corner of a square plate and development was performed in the ascending direction with solvent A. This solvent was then removed by evaporation, and the plate was rotated 90 deg, following which ascending development with solvent B was performed. After solvent removal, the positions of the amino acids were determined by spraying with ninhydrin, a reagent that forms a pink to purple product with amino acids. The spots were identified by comparison of their positions with those of standards.

Identification of species. Several methods are employed to locate the various sample components after separation. Two common methods, which can be applied to most organic mixtures, involve spraying with solutions of iodine or sulfuric acid; both react with organic compounds to yield dark reaction products. Several specific reagents (such as ninhydrin) are also useful for locating separated species. Fluorescent substances can be detected with an ultraviolet lamp. Alternatively, a fluorescent material is incorporated into the stationary phase. After development, the plate can be examined with ultraviolet light. Here, all of the plate fluoresces except where the nonfluorescing sample components are located.

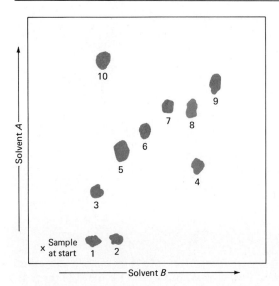

FIGURE 28-9 Two-dimensional thin-layer chromatogram (silica gel) of some amino acids. Solvent A: toluene-2-chloroethanol-pyridine. Solvent B: chloroform-benzyl alcohol-acetic acid. Amino acids: (1) aspartic acid; (2) glutamic acid; (3) serine; (4) β-alanine; (5) glycine; (6) alanine; (7) methionine; (8) valine; (9) isoleucine; and (10) cysteine.

Quantitative analysis. A semiquantitative estimate of the amount of a component present can be obtained by comparing the area of a spot with that of a standard. Better data can be obtained by scraping the spot from the plate, extracting the analyte from the resulting solid, and measuring the analyte by a suitable physical or chemical method. Finally, a scanning densitometer can be employed to measure radiation emitted from the spot by fluorescence or reflection.[8]

28B-2 PAPER CHROMATOGRAPHY

Separations by paper chromatography are performed in the same way as those on thin-layer plates. Usually, special papers, which are highly purified and reproducible as to porosity and thickness, are employed. Such papers contain sufficient adsorbed water that paper chromatography can be classified as a liquid-liquid type. Other liquids can be made to displace the water, however, thus providing a different type of stationary phase. For example, paper treated with silicone or paraffin oil permits reverse-phase paper chromatography, in which the mobile phase is a polar solvent. Also available commercially are special papers that contain an adsorbent or an ion-exchange resin, thus permitting adsorption and ion-exchange paper chromatography.

---

[8] For a description of commercially available densitometers as well as other thin-layer equipment, see: P. F. Lott and R. J. Hurtubise, *J. Chem. Educ.*, **48**, A437 (1971); and D. Rogers, *Amer. Lab.* **11** (5), 77 (1979).

gas-liquid chromatography

29

In *gas-liquid chromatography* (GLC), the components of a vaporized sample are fractionated as consequence of partition between a mobile *gaseous* phase and a stationary phase held in a column. The concept of gas-liquid chromatography was first described in 1941 by Martin and Synge, who were also responsible for liquid-partition chromatography. More than a decade was to elapse, however, before the value of GLC was demonstrated experimentally.[1] Since that time, the growth in applications of the procedure has been phenomenal.[2]

29A APPARATUS

Numerous instruments, varying in sophistication and ranging in price from several hundreds to several thousands of dollars, are available for gas chromatography. The basic components of these instruments are shown in Figure 29-1. A description of each component follows.

---

[1] A. J. James and A. J. P. Martin, *Analyst,* **77,** 915 (1952); *Biochem. J.,* **50,** 679 (1952).

[2] For detailed discussions of GLC, see: *Modern Practice of Gas Chromatography,* R. L. Grob, Ed. New York: Wiley-Interscience, 1977; A. B. Littlewood, *Gas Chromatography,* 2d ed. New York: Academic Press, 1970; J. Q. Walker, M. T. Jackson, Jr., and J. B. Maynard, *Chromatographic Systems,* 2d ed., Part II. New York: Academic Press, 1977; and H. M. McNair and E. J. Bonelli, *Basic Gas Chromatography.* Walnut Creek, CA: Varian Aerograph, 1971.

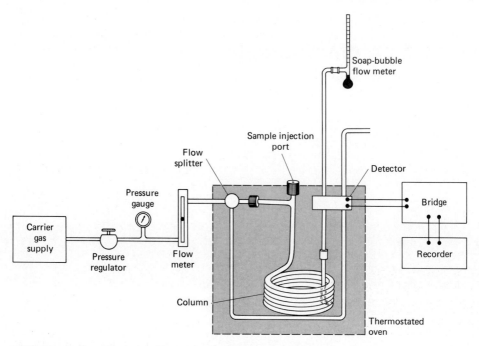

FIGURE 29-1 Schematic diagram of a gas chromatograph.

29A-1 CARRIER GAS SUPPLY

Carrier gases, which must be chemically inert, include helium, argon, nitrogen, and hydrogen; helium is the most widely used. All of these gases are available in pressurized tanks. Associated with the gas supply are pressure regulators, gauges, and flowmeters. In addition, the carrier gas system often contains a molecular sieve to remove water or other impurities.

Flow rates are controlled by a pressure regulator. Inlet pressures usually range from 10 to 50 psi (above room pressure), which lead to flow rates of 25 to 50 mL/min. Generally, it is assumed that the flow rate will be constant if the inlet pressure remains constant. Flow rates can be established by a rotameter at the column head; this device, however, is not as accurate as a simple soap-bubble meter, which is located at the end of the column. A soap film is formed in the path of the gas when a rubber bulb containing an aqueous solution of soap or detergent is squeezed; the time required for this film to move between two graduations on the buret is measured and converted to a flow rate.

29A-2 SAMPLE INJECTION SYSTEM

Column efficiency requires that the sample be of suitable size and be introduced as a "plug" of vapor; slow injection of oversized samples causes band spreading and poor resolution. A microsyringe is used to inject liquid samples through a rubber or silicone diaphragm into a heated sample port located at the head of the column (the sample port is ordinarily about 50°C above the boiling point of the least volatile component of the sample). For ordinary analytical columns, sample sizes vary from a few tenths of a microliter to 10 μL. Capillary columns require much smaller samples ($\sim 10^{-3}$ μL); here, a sample splitter system is employed to deliver only a small fraction of the injected sample to the column head, with the remainder going to waste.

Gas samples are best introduced by means of a sample valve such as that shown in Figure 27-8 (p. 665). Solid samples are introduced as solutions or, alternatively, are sealed into thin-walled vials that can be inserted at the head of the column and crushed from the outside.

29A-3 COLUMNS

Two types of columns are employed in GLC. One type is fabricated from capillary tubing (0.3 to 0.5 mm i.d.), the bore of which is coated with a very thin film (~ 1 μm) of the liquid phase. Capillary columns have a negligible pressure drop and can thus be of great length (10 to 100 m or more); the ratio of V_S/V_M (Equation 27-11) for these columns ranges from 100 to 300, which in part accounts for their high efficiencies. Columns of several hundred thousand theoretical plates have been described. These columns, however, have very low sample capacities (<0.01 μL). The capacity of capillary columns can be increased by coating the inside of the tube with a porous material such as graphite, a metal oxide, or a silicate. The added surface area increases the amount of liquid retained by the tube and, thus, the capacity of the column.

Packed columns are fabricated from glass or metal tubes 1 to 8 mm in inside diameter; they are designed to hold solid packings that range from 2 to 20 m in length. The tubes ordinarily are folded or coiled so that they can be conveniently fitted into a thermostat. Typically, columns have V_S/V_M ratios of 15 to 20 and contain 100 to 1000 theoretical plates per foot. The best packed columns have theoretical plates in excess of 20,000.

Solid support for packed columns. The ideal support would consist of small, uniform, spherical particles with good mechanical strength and a specific surface area of at least 1 m^2/g. In addition, the material should be inert at elevated temperatures and be uniformly wetted by the liquid phase. No substance that meets all of these criteria perfectly is yet available.

The most common supports are derived from diatomaceous earths. Two types are available, namely, firebrick and kieselguhr. Firebrick, which is sold under trade names such as Chromosorb P, C 22, and Sterchamol, has the better strength and the larger specific surface area (~ 4 m^2/g); its disadvantage lies in the fact that it is more active and, therefore, cannot be employed with polar compounds. Kieselguhr is more fragile than firebrick and has a smaller specific surface area (~ 1 m^2/g) but is less reactive; it is sold under such trade names as Chromosorb W, Celite, Embacel, and Celatom.

Liquid phase. Desirable properties for the immobilized, liquid phase in a gas-liquid chromatographic column include: (1) *low volatility*, ideally, the boiling point of the liquid should be at least 200° higher than the maximum operating temperature for the column; (2) *thermal stability;* (3) *chemical inertness;* and (4) *solvent characteristics* such that α and k' (pp. 658 and 657) values for the solutes to be resolved fall within a suitable range.

No single liquid meets all of these requirements, the last in particular. As a consequence, it is common practice to have available several interchangeable columns, each with a different stationary phase. The choice among these is often a matter of trial and error, although some qualitative guidelines exist; factors affecting this choice are considered in the section devoted to applications.

Column preparation.   The support material is first screened to limit the particle size range. It is then made into a slurry with a volatile solvent that contains an amount of the stationary liquid calculated to produce a thin coating (5 to 10 μm) on all of the particles. After the solvent has been evaporated, the particles appear dry and are free flowing.

Columns are fabricated from glass, stainless steel, copper, or aluminum. They are filled by slowly pouring the coated support into the straight tube with gentle tapping or shaking to provide a uniform packing. Care must be taken to avoid channeling. After the column has been packed, it is bent or coiled into an appropriate shape to fit the oven (see Figure 29-1).

A properly prepared column may be employed for several hundred analyses. Numerous types of packed columns are available from commercial sources.

Column thermostating.   Column temperature is an important variable that must be controlled to a few tenths of a degree for precise work. Thus, the column is ordinarily housed in a thermostated oven. The optimum column temperature depends upon the boiling point of the sample and the degree of separation required. Roughly, a temperature equal to or slightly above the average boiling point of a sample results in an elution period of reasonable time (2 to 30 min). For samples with a broad boiling range, it is often desirable to employ temperature programming whereby the column temperature is increased either continuously or in steps as the separation proceeds.

In general, optimum resolution is associated with minimal temperature; the cost of lowered temperature, however, is an increase in elution time and therefore the time required to complete an analysis. Figure 29-2 illustrates this principle.

29A-4 DETECTION SYSTEMS

Detection devices for a gas-liquid chromatograph must respond rapidly and reproducibly to the low concentrations of solutes emitted from the column. The solute concentration in a carrier gas at any instant is only a few parts per thousand at most; frequently, the detector is called upon to respond to concentrations that are smaller than this by one or more orders of magnitude. In addition, the interval during which a peak passes the detector is usually one second or less; therefore, the detector must be capable of exhibiting its full response during this brief period.

Other desirable properties of the detector include linear response, good stability over extended periods, and uniform response for a wide variety of compounds. No single detector meets all of these requirements, although more than a dozen different types have been proposed. We shall describe the three most widely used of these.

Thermal conductivity detectors.   A relatively simple and broadly applicable detection system is based upon changes in the thermal conductivity of the gas stream;

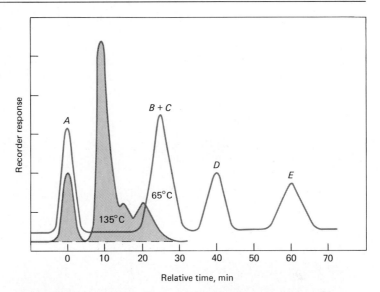

FIGURE 29-2 Effect of temperature on the separation of hexane isomers: *A*, 2,2-dimethylbutane; *B*, 2,3-dimethylbutane; *C*, 2-methylpentane; *D*, 3-methylpentane; *E*, n-hexane. [C. E. Bennett, S. Dal Nogare, and L. W. Safranski, in *Treatise on Analytical Chemistry*, I. M. Kolthoff and P. J. Elving, Eds., Part I, vol. 3, p. 1690. New York: Interscience, 1961. With permission.]

such a device is sometimes called a *katharometer*. The sensing element of a katharometer is an electrically heated source whose temperature at constant electrical power depends upon the thermal conductivity of the surrounding gas. The heated element may be a fine platinum or tungsten wire or, alternatively, a semiconducting thermistor. The resistance of the wire or thermistor gives a measure of the thermal conductivity of the gas; in contrast to the wire detector, the thermistor has a negative temperature coefficient.

In chromatographic applications, a double detector system is always employed, with one element being placed in the gas stream *ahead* of the sample injection chamber and the other immediately beyond the column. Alternatively, the gas stream may be split as shown in Figure 29-1. In either case, the thermal conductivity of the carrier gas is canceled, and the effects of variation in flow rate, pressure, and electrical power are minimized. The resistances of the twin detectors are usually compared by incorporating them into two arms of a simple Wheatstone bridge circuit such as that shown in Figure 29-3.

The thermal conductivities of hydrogen and helium are roughly six to ten times greater than those of most organic compounds. Thus, the presence of even small amounts of organic materials causes a relatively large decrease in the thermal conductivity of the column effluent; the detector undergoes a marked rise in temperature as a result. The conductivities of nitrogen and carbon dioxide more closely resemble those of organic constituents; detection by thermal conductivity is less sensitive with these substances as carrier gases.

Thermal conductivity detectors are simple, rugged, inexpensive, nonselec-

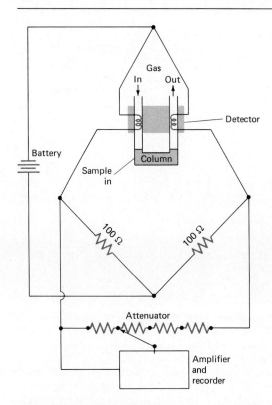

FIGURE 29-3 Schematic diagram of a thermal conductivity detector-recorder system.

tive, accurate, and nondestructive of the sample. Other devices, however, are considerably more sensitive.

Flame ionization detectors. Most organic compounds, when pyrolyzed at the temperature of a hydrogen/air flame, produce ionic intermediates that provide a mechanism by which electricity can be carried through the flame. By employing an apparatus such as that shown in Figure 29-4, these ions can be collected and the resulting ion current measured. The electrical resistance of a flame is very high (perhaps $10^{12} \, \Omega$), and the resulting currents are therefore minuscule; an electrometer must be employed for their measurement.

The ionization of carbon compounds in a flame is a poorly understood process, although it is known that the number of ions produced is roughly proportional to the number of reduced carbon atoms in the flame. Functional groups such as carbonyl, alcohol, and amine produce fewer ions than fully reduced carbon or none at all; thus, the response diminishes in the presence of these groups.

The hydrogen flame detector currently is one of the most popular and most sensitive detectors. It is more complicated and more expensive than the thermal conductivity detector but has the advantage of higher sensitivity. In addition, it has a wide range of linear response. It is, of course, destructive of the sample.

Electron-capture detectors. With electron-capture detectors, the effluent from the column is passed over a β-emitter (a radioactive compound that emits electrons)

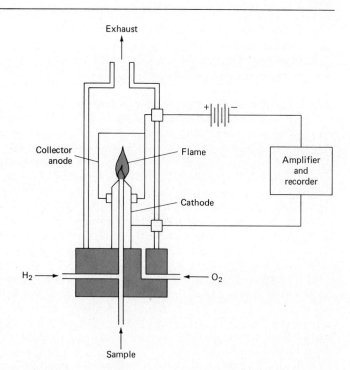

FIGURE 29-4 Hydrogen flame-ionization detector.

such as nickel-63 or tritium (adsorbed on platinum or titanium foil). An electron from the emitter causes ionization of the carrier gas (often nitrogen) and the production of a burst of electrons. In the absence of organic species, a constant standing current between a pair of electrodes results from this ionization process. The current decreases, however, in the presence of organic molecules that tend to capture electrons. The response is nonlinear unless the potential across the detector is pulsed.

The electron-capture detector is selective in its response and is highly sensitive toward electronegative functional groups such as halogens, peroxides, quinones, and nitro groups. It is insensitive to compounds such as amines, alcohols, and hydrocarbons. An important application of the electron-capture detector has been for the detection and determination of chlorinated insecticides.

Electron-capture detectors are highly sensitive and possess the advantage of not altering the sample significantly (in contrast to flame detectors).

29B APPLICATIONS OF GLC

In evaluating the importance of GLC, it is necessary to distinguish between the two roles the method plays. The first is as a tool for performing separations; in this capacity, it is unsurpassed when applied to complex organic, metal-organic, and biochemical systems. The second, and distinctly different function, is that of providing the means for completion of an analysis. Here, retention times or volumes are employed for qualitative identification, while peak heights or areas provide quantitative information. For these purposes, GLC is considerably more limited than some of the other methods considered in earlier chapters. As a consequence, an important trend in the field has been in the direction of combining the remarkable fractionation qualities of GLC with the superior analytical properties of such instruments as mass, ultraviolet, infrared, and NMR spectrometers.

29B-1 CHOICE OF THE STATIONARY LIQUID PHASE

Several hundred liquids have been suggested for use as the stationary phase in GLC. The successful separation of closely related compounds is often critically dependent upon the proper choice among these.

The retention time for a solute depends directly upon its partition coefficient (Equations 27-10 and 27-20), which, in turn, is related to the nature of the stationary phase. Clearly, to be useful in GLC the immobilized liquid must generate different partition coefficients among solutes; additionally, however, these coefficients must be neither extremely small nor extremely large. Solutes with small coefficients pass through the column so rapidly that no significant separation occurs. On the other hand, the time required to remove solutes with large coefficients from the columns becomes prohibitive.

To have a reasonable residence time in the column, a solute must show at least some degree of compatability (solubility) with the stationary phase. Thus, the polarities of the two substances should be at least somewhat alike. For example, a stationary liquid such as squalene (a high-molecular-weight saturated hydrocarbon) might be chosen for separation of members of a nonpolar homologous series such as hydrocarbons or ethers. On the other hand, a more polar liquid such as polyethylene glycol would be more effective for separating alcohols or amines. For aromatic hydrocarbons, benzyldiphenyl might prove appropriate.

Table 29-1   Some Common Stationary Phases

Trade Name	Chemical Composition	Maximum Temperature, °C	Polarity*	Type of Separation
Squalene	$C_{30}H_{62}$	150	NP	Hydrocarbons
OV-1	Polymethyl siloxane	350	NP	General purpose nonpolar
DC 710	Polymethylphenyl siloxane	300	NP	Aromatics
QF-1	Polytrifluoropropyl methyl siloxane	250	P	Amino acids, steroids, nitrogen compounds
XE-30	Polycyanomethyl siloxane	275	P	Alkaloids, halogenated compounds
Carbowax 20M	Polyethylene glycol	250	P	Alcohols, esters, essential oils
DEG adipate	Diethylene glycol adipate	200	SP	Fatty acids, esters
	Dinonyl phthalate	150	SP	Ketones, ethers, sulfur compounds

* NP = nonpolar; SP = semipolar; P = polar.

Among solutes of similar polarity, the elution order usually follows the order of boiling points; where these differ sufficiently, clean separations are feasible. Solutes with nearly identical boiling points but different polarities frequently require a liquid phase that will selectively retain one (or more) of the components by dipole interaction or by adduct formation. Another important interaction that often enhances selectivity is hydrogen bond formation. For this effect to operate, the solute must have a polar hydrogen atom and the solvent must have an electronegative group (oxygen, fluorine, or nitrogen) or the converse must be true.

Table 29-1 lists a few of the most widely used stationary liquid phases.

29B-2 QUALITATIVE ANALYSIS

Gas chromatograms are widely used as criteria of purity for organic compounds. Contaminants, if present, are revealed by the appearance of additional peaks; the areas under these peaks provide rough estimates of the extent of contamination. The technique is also useful for evaluating the effectiveness of purification procedures.

In theory, retention-time data should be useful for the identification of components in mixtures. In fact, however, the applicability of the data is limited by the number of variables that must be controlled in order to obtain reproducible results. Nevertheless, GLC provides an excellent means of confirming the presence of a suspected compound in a mixture, provided an authentic sample of the substance is available. No new peaks in the chromatogram of the mixture should appear upon addition of the known compound, and enhancement of an existing peak should be observed. The evidence is particularly convincing if the effect can be duplicated on different columns and at different temperatures.

We have seen (p. 658) that the selectivity factor α for compounds X and Y is given by the relationship

$$\alpha = \frac{K_Y}{K_X} = \frac{(t_R)_Y - t_M}{(t_R)_X - t_M} \cong \frac{(t_R)_Y}{(t_R)_X}$$

If a standard substance is chosen as compound Y, then α can provide an index for identification of compound X that is largely independent of column variables other than temperature; that is, numerical tabulations of separation factors for pure compounds relative to a common standard can be prepared and then used for the characterization of samples on any column. The amount of such data available in the literature is presently limited.

29B-3 QUANTITATIVE ANALYSIS

The detector signal from a gas-liquid chromatographic column has had wide use for quantitative and semiquantitative analyses. An accuracy of 1 to 3% relative is attainable under carefully controlled conditions. As with most analytical tools, reliability is directly related to the control of variables; the nature of the sample also plays a part in determining the potential accuracy.

The general discussion of quantitative chromatographic analysis given in Section 27E-2 applies to gas chromatography as well as to other types; therefore, no further consideration of this topic is given here.

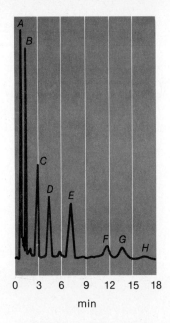

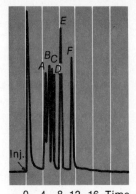

0   4   8   12 16 Time
180° 212° 230° Holding temp.
      196° 228°

(b) Sedative mixture: *A*, butalbital;
    B, amobarbital; *C*, pentobarbital;
    D, secobarbital; *E*, glutethimide;
    F, phenobarbital. All at 500 ng.

(a) Fatty acid methyl esters:
    A, ethyl benzene; *B*, caprylate;
    C, laurate; *D*, myristate;
    E, palmitate; *F*, stearate;
    G, oleate; *H*, linoleate.

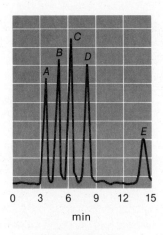

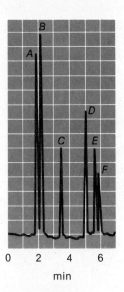

(c) Steroids: *A*, DHA;
    B, estraionol E$_2$;
    C, estrone E$_1$;
    D, EPI-testosterone;
    E, estriol; E$_3$.

(d) Carbohydrates:
    A, fructose;
    B, dextrose;
    C, phenyl beta
        D-glucopyranoside;
    D, sucrose;
    E, lactose;
    F, maltose;

FIGURE 29-5   Some examples of gas chromatographic separations. (See Table 29-2 for details about the separations.)

Table 29-2 Conditions for Chromatograms Shown in Figure 29-5

Chromato-gram	Column*	Packing	Detector**	Temper-ature, °C	Carrier	Flow, mL/min
(a)	10′ × ⅛″ S 5′ × ⅛″ S	Chromosorb 102 Molecular sieve 5A	TCD	65–200	He	30
(b)	6′ × ¼″ G	1.5% OV-17 on Chromosorb G	ECD	22		
(c)	6′ × ⅛″ S	Chromosorb W	TCD	190	He	30
(d)	6′ × ¼″ G	1.5% OV-17 on HP Chromosorb G	FID	180–230		
(e)	6′ × 3.4 mm G	5% OV-210 + 2.5% OV-17 on Supelcoport	ECD	260	A	
(f)	6′ × 3.4 mm	2% OV-17 on Chromosorb W	FID	105–325	N₂	40

* S = stainless steel; G = glass.
** TCD = thermal conductivity; ECD = electron capture; FID = flame ionization.

29C EXAMPLES OF APPLICATIONS OF GAS CHROMATOGRAPHY

Figure 29-5 illustrates some typical applications of gas chromatography to analytical problems. Data on the conditions employed for obtaining the various chromatograms in Figure 29-5 are found in Table 29-2.

chemicals, apparatus, and unit operations

for analytical chemistry

30

This chapter is concerned with practical aspects of the unit operations that are used in most analyses as well as the description of the apparatus and chemicals required for these operations.

30A CHOOSING AND HANDLING CHEMICALS AND REAGENTS

Of constant concern in analytical chemistry is the quality of reagents, inasmuch as the accuracy of an analysis is often affected by this factor.

30A-1 CLASSIFICATION OF COMMERCIAL CHEMICALS

Technical or commercial grade. Chemicals labeled technical or commercial grade are of indeterminate quality and should be used only where high purity is not of paramount importance. Thus, the potassium dichromate and the sulfuric acid used in the preparation of cleaning solution can be of this grade. In general, however, technical- or commercial-grade chemicals are not used in analytical work.

USP grade. USP chemicals have been found to conform to the tolerances set forth in the *United States Pharmacopoeia*.[1] The specifications are designed to limit con-

---

[1] U.S. Pharmacopoeial Convention, *Pharmacopoeia of the United States of America*, 20th rev. Easton, PA: Mack, 1980.

taminants that are dangerous to health; thus, chemicals passing the USP tests still may be quite heavily contaminated with impurities that are not physiological hazards.

Reagent grade. Reagent-grade chemicals conform to the minimum specifications of the Reagent Chemical Committee of the American Chemical Society;[2] these are used, wherever possible, in analytical work. Some suppliers label their products with the maximum limits of impurity allowed by these specifications; others print the actual results of analyses for the various impurities.

Primary standard grade. *Primary standards* are substances that are obtainable in extraordinarily pure form (see Section 6B-1). Primary standard-grade reagents, which are available commercially, have been carefully analyzed, and the assay value is printed on the label. An excellent source for primary standard chemicals is the National Bureau of Standards. This agency also supplies *reference standards* —complex mixtures that have been exhaustively analyzed.[3]

Special purpose reagents. A number of manufacturers produce chemicals designed for specific uses. Some examples include spectrophotometric solvents, HPLC solvents, products for liquid scintillation counting, reagents for nonaqueous atomic spectroscopy, and reagents for electron microscopy. Detailed information is provided for these materials that is pertinent to their applications. Thus, for spectrophotometric solvents, data are supplied as to absorbance at various wavelengths, UV cutoff wavelengths, infrared windows, and assay value.

30A-2 HANDLING REAGENTS AND SOLUTIONS
The availability of reagents and solutions with established purity is of prime importance for successful analytical work. A freshly opened bottle of a reagent-grade chemical can be used with confidence in most applications; whether the same confidence is justified when this bottle is half full depends entirely upon the way it has been handled after being opened. Only conscientious adherence to the rules given here will prevent contamination of reagents.

1. Select the best available grade of chemical for analytical work. If there is a choice, pick the smallest bottle that will supply the necessary quantity.
2. Replace the top of every container immediately after removal of reagent. Do not rely on someone else to do this.
3. Hold stoppers between the fingers; stoppers should never be set on the desk top.
4. Unless specifically directed to the contrary, *never return any excess reagent or solution to a bottle.* The minor saving represented by the return of an excess is likely to be a false economy in view of the risk of contaminating the entire bottle.

[2] Committee on Analytical Reagents, *Reagent Chemicals. American Chemical Society Specifications*, 5th ed. Washington, D.C.: American Chemical Society, 1974.
[3] United States Department of Commerce, *Catalog and Price List of Standard Materials Issued by the National Bureau of Standards*. NBS Misc. Publ. 260 and its semiannual supplements. Washington, D.C.: Government Printing Office, 1968–

5. Again, unless specifically instructed otherwise, do not insert spoons, spatulas, or knives into a bottle containing a reagent chemical. Instead, shake the capped bottle vigorously or tap it sharply on a wooden table to dislodge the contents. Then pour out the desired quantity. In some instances, these measures will not loosen the contents and a clean porcelain spoon must be used.
6. Keep the reagent shelf and the laboratory balances clean. Be sure to clean up any spilled chemicals immediately.

30B CLEANING AND MARKING LABORATORY WARE

Beakers and crucibles should be marked for identification purposes. The etched area on the sides of beakers, filtering crucibles, and flasks can be marked semipermanently with a pencil. Special marking inks are available for porcelain surfaces; the marking is baked permanently into the glaze. A saturated solution of iron(III) chloride can also be used, although it is not as satisfactory as the commercial preparations.

Care must be taken to ensure that glass and porcelain ware is thoroughly clean before use. The apparatus should be washed with hot detergent and then rinsed first with copious amounts of tap water and finally with several small portions of distilled water. A properly cleaned object will be coated with a uniform and unbroken film of water. It is seldom necessary to dry glassware before use; in fact, this practice should be discouraged because it wastes time and can be a cause of contamination.

If a grease film persists after thorough cleaning with detergent, a cleaning solution consisting of sodium dichromate in concentrated sulfuric acid may be used. Extensive rinsing is required after using this solution to remove the last traces of dichromate ions, which adhere strongly to glass or porcelain surfaces. Cleaning solution is most effective when warmed to about 70°C; at this temperature it rapidly attacks plant and animal matter and is thus a *potentially dangerous preparation*. Any spillages should be diluted promptly with copious volumes of water.

PREPARATION OF CLEANING SOLUTION. In a 500-mL, heat-resistant conical flask mix 10 to 15 g of sodium or potassium dichromate with about 15 mL of water. Add concentrated sulfuric acid *slowly;* swirl the flask thoroughly between increments. The contents of the flask will become a semisolid red mass; add just enough sulfuric acid to dissolve this mass. Allow the solution to cool somewhat before attempting to transfer it to a storage bottle. The solution may be reused until it acquires the green color of chromium(III) ion, at which time it should be discarded. *CAUTION: Cleaning solution is highly corrosive and must be used with extreme care.*

30C EVAPORATION OF LIQUIDS

In the course of an analysis the chemist often finds it necessary to decrease the volume of a solution without loss of a nonvolatile solute. An arrangement such as illustrated in Figure 30-1 is generally satisfactory for most evaporations. A ribbed cover glass permits vapors to escape and protects the solution from accidental contamination. Less satisfactory is the use of glass hooks to provide space between the lip of the container and a conventional watch glass.

FIGURE 30-1 Arrangement for the evaporation of liquids.

The evaporation process is occasionally difficult to control, owing to the tendency of some solutions to superheat locally. The bumping that results, if sufficiently violent, can cause a partial loss of the sample. This danger is minimized by careful and gentle heating. The introduction of glass beads, where permissible, is also helpful.

Unwanted constituents of a solution can frequently be eliminated by evaporation. For example, chloride and nitrate ions can be removed by adding sulfuric acid and evaporating until copious white fumes of sulfur trioxide are observed. Nitrate ion and nitrogen oxides can be eliminated by adding urea to an acidic solution, evaporating the solution to dryness, and gently igniting the residue. If large quantities of ammonium chloride must be removed, it is better to add concentrated nitric acid after evaporating the solution to a small volume. Rapid oxidation of the ammonium ion occurs upon heating; the solution is then evaporated to dryness.

Unwanted organic substances are frequently eliminated by adding sulfuric acid and evaporating the solution until sulfur trioxide fumes are observed. Nitric acid may be added at this point to hasten oxidation of the last traces of organic matter.

30D THE MEASUREMENT OF MASS

The measurement of mass is one of the commonest operations carried out by the chemist. Performance of many analyses will require the acquisition of highly reliable weighing data at one stage or another. For such measurements an *analytical balance*, which provides highly accurate information, is employed. Approximate weighing data are completely satisfactory for other purposes; these measurements are ordinarily obtained with a less precise but more rugged auxiliary *laboratory balance*.

30D-1 THE DISTINCTION BETWEEN MASS AND WEIGHT

The reader should clearly recognize the difference between the concepts of mass and weight. The more fundamental of these is *mass* — an invariant measure of the quantity of matter in an object. The *weight* of an object, on the other hand, is the force of attraction exerted between the object and its surroundings, principally the earth. Because gravitational attraction is subject to slight geographical variation with altitude as well as latitude, the weight of an object is likewise a somewhat variable quantity. For example, the weight of a crucible would be less in Denver than in Atlantic City because the attractive force between it and the earth is less at the higher altitude. Similarly, it would weigh more in Seattle than in Panama because the earth is somewhat flattened at the poles, and the force of attraction increases appreciably with latitude. The mass of this crucible, on the other hand, remains constant regardless of the location in which it is measured.

Weight and mass are simply related to each other through the familiar expression

$$W = Mg$$

where the weight, W, is given by the product of the mass, M, of the object and the acceleration due to gravity, g.

Chemical analyses are always based on mass in order to free the results from dependence on locality. The mass of an object is ordinarily obtained by means of a *balance* that permits comparison of the mass of the object with that of a known mass; because g affects both the known and unknown to the same extent, an equality in weight also indicates an equality in mass.

The distinction between weight and mass is not always observed; in common usage, the operation of comparing masses is called *weighing* and the objects of known mass as well as the results of the process are called *weights*. Hereafter, we shall use the two terms synonymously to indicate *mass*.

30D-2 THE ANALYTICAL BALANCE[4]

An analytical balance is capable of detecting a weight difference that is about one one-millionth of its maximum load or better. For example, one widely used analytical balance can tolerate a load up to 160 g; the standard deviation of measurements with this balance is about ± 0.1 mg ($\pm 10^{-4}$ g).

Analytical balances are of two types, namely, *single-pan* and *equal-arm*. All modern balances are of the former type.

Analytical balances are also classified according to their sensitivities and load limits. The typical *macro analytical* balance has a maximum loading of 160 to 200 g, although oversize balances that can accommodate loads up to 2000 g are occasionally encountered. *Semimicro balances* have a maximum loading of 50 to 160 g with a standard deviation of about ± 0.01 mg. *Micro analytical* balances will tolerate loads of 10 to 20 g; a typical standard deviation with such a balance is ± 0.001 mg.

The discussion that follows will deal exclusively with macro analytical balances, the type that is found in nearly every laboratory.

30D-3 COMPONENTS OF AN ANALYTICAL BALANCE

Analytical balances differ considerably in appearance, design, and performance characteristics. Nevertheless, all contain common components; these will be considered in this section.

Beams. The heart of a balance is the beam, which is a lever that pivots about a central support. Placement of the object to be weighed on one end creates a mo-

---

[4] For a description of commercially available balances, see: R. F. Hirch, *J. Chem. Educ.*, **44**, A1023 (1967); **45**, A7 (1968); and G. W. Ewing, *J. Chem. Educ.*, **53**, A252, A292 (1976).

ment of force that tends to rotate the beam about its pivot. This motion is offset by a counterforce, the magnitude of which is a measure of the mass of the object. Depending upon balance design, this counterforce is either applied at the opposite end of the beam or it is removed from the side upon which the object has been placed. The counterforce is ordinarily gravitational and is derived from the mass of standard weights. However, a measured electromagnetic force may also be employed; more and more balances based on this design are appearing on the market.

Figure 30-2 depicts examples of single-pan and equal-arm balances. Common to both is the light-weight beam (1), which oscillates about a central prism-shaped knife-edge (2) as a result of the forces described in the preceding paragraph. The knife-edge and the plane surface upon which it rests form a bearing having minimal friction. The performance of a balance is ultimately limited by the mechanical perfection of this bearing; thus, both parts are precision formed and manufactured from unusually hard materials (agate or synthetic sapphire). Care must be taken in using any type of balance to avoid damage to the knife-edge and its bearing surface.

One analytical balance manufacturer uses taut metal bands in place of knife-edges or bearings. The beam of the balance is firmly attached to the bands. Movement of the beam results in a slight torsional twist of the metal bands. The advantages claimed for this arrangement are ruggedness, freedom from wear, and imperviousness to dust and dirt.

Stirrups and pans. The single-pan balance shown in Figure 30-2a contains a second bearing around which motion can occur. Here, the flat bearing surface is contained in the *stirrup* (3), which rests on an outer knife-edge (4). The stirrup serves to couple the pan (5), which holds the object to be weighed to the beam.

The equal-arm balance shown in Figure 30-2b requires two stirrups rather than one; thus, the single-pan balance has one less moving part than its double-pan counterpart. Other things being equal, this smaller number of bearings is an advantage because it reduces the friction associated with beam motion.

Beam deflection detectors. Another necessary component of all balances is a device that detects and measures the displacement of the beam from its equilibrium position. The simplest detector is the mechanical pointer (6) and scale (7) of an equal-arm balance (Figure 30-2b). For small angles of deflection, the horizontal displacement of the pointer tip along the scale is directly proportional to the difference in loading of the two pans. The *sensitivity* of the balance, which can be defined in terms of scale division displacement per unit of mass,[5] is directly proportional to the length of the pointer.

One method of increasing the sensitivity of a balance is to use a so-called "optical lever." Here, a small mirror is mounted on the top of the beam so that it is turned through an angle by the deflection process. A light is reflected off the mirror and focused on a scale some distance away. In this way, the sensitivity may be enhanced by as much as 100.

---

[5] A more general definition of sensitivity is the ratio of the change in response to the quantity measured. See: L. B. Macurdy, *et al.*, *Anal. Chem.*, **26**, 1190 (1954).

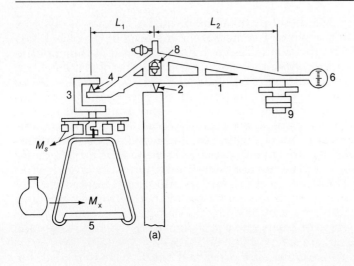

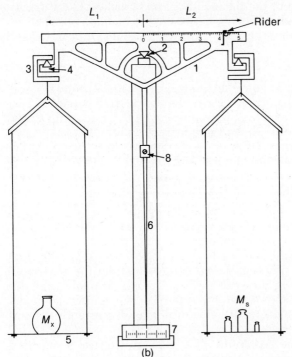

FIGURE 30-2 Schematic diagram of (a) a single-pan balance and (b) an equal-arm balance.

The typical single-pan balance employs an optical lever to detect beam deflections. A glass plate or reticle scribed with a vertical scale is attached to one end of the beam (see 6 in Figure 30-2a). Light from a small lamp is then passed through the screen; the resulting image is enlarged and reflected onto a frosted glass surface.

Weights. An analytical balance is supplied with a combination of weights such that any desired loading up to the design maximum can be realized. As shown in Figure 30-2a, these weights (M_s) are attached to the beam of most single-pan balances; a mechanical means is provided for their individual removal. With equal-arm balances, weights are usually added manually to the pan opposite to the object being weighed.

Dampers. *Dampers* are devices that shorten the time required for the beam of a balance to come to rest after it has been set in motion. A single-pan balance is ordinarily equipped with an *air damper,* consisting of a piston attached to the beam that moves within a concentric cylinder mounted to the balance case (see 9 and 10 in Figure 30-3). When the beam is set in motion, the enclosed air undergoes slight expansions and contractions because of the close spacing between piston and cylinder; the beam comes rapidly to rest as a result of this opposition to its motion. An equal-arm balance is more likely to be equipped with a *magnetic damper,* which consists of a metal plate (generally aluminum) secured to the end of the beam and positioned between the poles of a permanent magnet. The forces induced by the motion of the beam act to oppose further motion; the beam rapidly acquires its rest position as a result.

Other components. Most analytical balances are equipped with *beam arrests* and *pan arrests*. The beam arrest is a mechanical device that lifts the beam so that the central knife-edge is freed from contact with its bearing surface; in addition, the stirrups are simultaneously freed from contact with the outer knife-edge or edges. When engaged, pan arrests support most of the weight of the pans and thus prevent them from swinging. Their purpose is to prevent damage to the bearings of the bal-

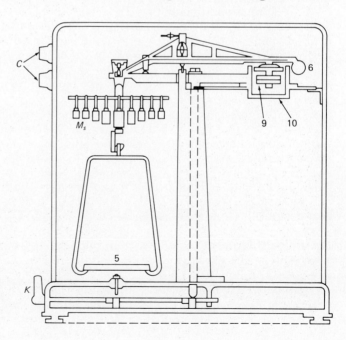

FIGURE 30-3 Schematic diagram of a single-pan balance. (Courtesy of Mettler Instrument Corporation.)

ance when objects are added to or removed from the pans. Both arrests should be engaged whenever a balance is not in use.

To discriminate between small weight differences (<1 mg), it is necessary to protect the balance from air currents in the laboratory. Thus, an analytical balance is enclosed in a case equipped with doors to permit the introduction and removal of objects.

30D-4 STABILITY AND SENSITIVITY OF A BALANCE

An important property of a balance is *stability;* that is, when the beam is set in motion around the central knife-edge by the momentary application of a slight force, it must ultimately return to its original position. For a balance to be stable, the center of gravity of the beam and its accouterments must lie *below the central knife-edge;* only then will there exist a restoring force (the weight of the beam) that will act to offset the displacement.

The sensitivity of a balance is inversely proportional to the distance between the central knife-edge and the center of gravity of the beam and pans (for a sensitive balance this distance is no more than a few thousandths of a centimeter). Balances are often equipped with a small weight that can be moved to adjust the sensitivity. These weights are labeled (8) in Figure 30-2.

30D-5 METHODS OF WEIGHING

The classic method of weighing with an equal-arm balance involves determining the position of the deflection indicator in the absence of a load (this point is called the *zero point* of the balance). The object to be weighed is then placed on one pan and weights are added incrementally to the other side of the beam until the indicator is restored to the zero point.

It is important to note that the sensitivity of an equal-arm balance decreases with increased loading. In contrast, variations of sensitivity with load do not occur in single-pan balances because they are designed for *substitution weighing.* This technique is illustrated for the single-pan balance shown in Figure 30-2a. At the outset, the side of the beam holding the pan is equipped with a full set of weights, M_s; a fixed counter weight (9) provides the counterforce to level the beam. With an object M_x on the pan, the beam is restored to its zero point or a rest point by the *removal of weights.* Note that the loading of this instrument, and thus its sensitivity, is constant and independent of the mass of M_x. Constant sensitivity offers the considerable advantage that the scale divisions for beam deflection can be made to read directly in weight. Single-pan analytical balances are calibrated in this way. The weight of an object can be obtained in a matter of seconds with a single-pan balance; the same operation may require minutes with an equal-arm instrument.

30D-6 SINGLE-PAN ANALYTICAL BALANCE

Operational features. The essential parts of a single-pan balance are shown in Figure 30-3. The pan (5) and a set of nonmagnetic, stainless steel weights (M_s) for loads that are larger than 100 mg are located on one side of the fulcrum; a fixed weight (9), which also serves as the piston of an air damper (10), is located on the other side. Restoration of the beam to its original position after an object has been placed on the pan is accomplished by removing weights (M_s) by means of knobs (C)

in an amount equal to the mass of the object. In actual operation, sufficient weights are removed to leave the system with a residual imbalance of 100 mg or less with the object still heavy. The optical device (6) translates the resulting deflection of the beam into units of mass and displays the fractional weight to the operator as an image on a frosted-glass surface.

Weighing. The weight of an object is very simply obtained with a modern single-pan balance (Note).

To zero the empty balance, rotate the arrest knob (K in Figure 30-3) to the full release position. Then manipulate the zero-adjusting control until the illuminated scale indicates a reading of zero. Next, arrest the balance, and place the object to be weighed on the pan. Turn the arrest knob to its partial release position. Rotate the dial controlling the heaviest likely weight for the object until the illuminated scale changes position or the notation "remove weight" appears; then, turn the knob back one stop. Repeat this procedure with the other dials, working successively through the lighter weights. Then, turn the arrest knob to its full release position, and allow the balance to achieve equilibrium. The weight of the object is found by taking the sum of the weights indicated on the dials and that which appears on the illuminated scale. A vernier is helpful in reading this scale to the nearest tenth of a milligram.

> NOTE
> Directions for operation of single-pan balances vary with make and model. Consult with the instructor for any modifications to this procedure that may be required.

30D-7 SUMMARY OF RULES GOVERNING THE USE OF AN ANALYTICAL BALANCE

Continued good performance from an analytical balance depends entirely upon the treatment accorded it. Similarly, reliable data are obtained only when careful attention is paid to the details of the weighing operation. Because weighing data are required for virtually every quantitative analysis, it is desirable to summarize the rules and precautions that relate to their acquisition.

To avoid damage or minimize wear to the balance and weights.

1. Be certain that arresting mechanisms for the beam are engaged whenever the loading on the balance is being changed and when the balance is not in use.
2. Center the load on the pan as much as possible.
3. Protect the balance from corrosion. Only vitreous materials and nonreactive metal or plastic objects should be placed directly on the pans.
4. The weighing of volatile materials requires special precautions (p. 709).
5. Do not attempt to adjust the balance without the prior consent of the instructor.
6. The balance and case should be kept scrupulously clean. A camel's-hair brush is useful for the removal of spilled material or dust.
7. Free the beam of an equal-arm balance by releasing the beam support slowly, followed by the pan arrests. Reverse this order when arresting

the beam; engage the pan arrests as the pointer passes through the center of the deflection scale.

8. Handle weights for an equal-arm balance gently and with special forceps. Weights should never be touched because moisture from the hands can initiate corrosion. Keep weights enclosed in their box when not in use.

9. If possible, avoid bringing weights for an equal-arm balance into the laboratory.

To obtain reliable weighing data.

10. Do not attempt to weigh an object until it has returned to room temperature.

11. Do not touch a dried object with bare hands; handle it with tongs or use finger pads to prevent the uptake of moisture.

30D-8 BOUYANCY CORRECTION

When the density of the object weighed differs considerably from that of the weights, a *buoyancy* error is introduced. This discrepancy arises from the difference in buoyant force of the medium (air) as it acts on the object and the weights. A correction for buoyancy is seldom needed when solid objects are weighed because the density of most solids approaches that of the weights. With liquids, gases, or low-density solids, however, a correction must be applied. The corrected weight is calculated by means of the equation

$$W_1 = W_2 + W_2 \left(\frac{d_{air}}{d_1} - \frac{d_{air}}{d_2} \right) \qquad (30\text{-}1)$$

where W_1 is the corrected weight of the object and W_2 is the mass of the weights. The terms d_1 and d_2 represent the densities of the object and the weights, respectively, and d_{air} is the density of the air displaced by them; its value is 0.0012 g/mL. Table 30-1 gives the density d_2 of the various metals used in the construction of weights. The weights of a single-pan balance are ordinarily made of stainless steel.

EXAMPLE 30-1. A sample of an organic liquid having a density of 0.92 g/mL was weighed into a glass bottle. The weight of the empty bottle was 8.6500 g against stainless steel weights. After addition of the sample, it was 9.8600 g. Correct the weight of the sample for buoyancy.

The apparent weight of the liquid is (9.8600 − 8.6500) or 1.2100 g. Inasmuch

Table 30-1 Density of Metals Used in the Manufacture of Weights

Metal	Density (g/mL)
Aluminum	2.7
Stainless steel	7.8
Brass	8.4
Tantalum	16.6
Gold	19.3
Platinum-iridium	21.5

as the same buoyant force acted on the glass container during both weighings, we need to consider only the buoyant force acting on the 1.2100 g of liquid. Using the density of stainless steel from Table 30-1 and 0.0012 g/mL for the density of air, we obtain the corrected weight with Equation 30-1

$$W_1 = 1.2100 + 1.2100 \left(\frac{0.0012}{0.92} - \frac{0.0012}{7.8} \right)$$

$$= 1.2100 + 1.2100 \,(0.0013 - 0.00015)$$

$$= 1.2114 \text{ g}$$

30D-9 OTHER SOURCES OF ERROR IN THE WEIGHING OPERATION

Attempts to weigh an object that differs in temperature from its surroundings will result in significant error. Failure to permit sufficient time for a heated object to return to room temperature is the commonest source of this error. Two effects are responsible: First, convection currents within the balance case exert a buoyant effect on the pan and object. Second, warm air entrapped within a closed container (such as a weighing bottle) weighs less than the same volume at a lower temperature. Both effects cause the apparent weight of the object to be low. For a typical weighing bottle or a glass crucible, the error may be as large as 10 or 15 mg. Heated objects should always be cooled for at least 30 min before being weighed to avoid this error.

Occasionally, a porcelain or glass object will acquire a static charge that is sufficient to cause the balance to perform erratically; this problem is particularly serious when the humidity is low. Spontaneous discharge frequently occurs after a short period. The use of a faintly damp chamois to wipe the object is a recommended preventative.

The optical scale of a single-pan balance should be checked regularly for accuracy, particularly under loading conditions that require essentially the full range of the scale; a standard 100-mg weight is useful in performing this check.

30D-10 ELECTRONIC BALANCES

Most of the balances now being offered by manufacturers are electronic or partially electronic, and it seems likely that the purely mechanical single-pan balances, which we have just described, will disappear from the laboratory as they wear out and are replaced.

Null detectors. Electronic balances are generally equipped with a null detector, a device that senses when the beam is displaced from its null or balanced position. Some of these devices are optical and consist of a vane attached to the beam, a small lamp, and a photodetector. Movement of the vane increases or decreases the radiation reaching the detector; the resulting electrical changes are displayed on a current meter. Inductive null detectors are also used. Here, a soft iron bob attached to the beam hangs within the windings of two secondary coils of a transformer. Movement of the bob changes the ratio of current in the two secondaries; this change is then sensed with a suitable meter.

Restoring force. Electronic balances generally employ an electromagnetic restoring force to bring the beam to the null point as indicated by the meter. In some instruments, coarse balance is achieved by addition of weights as in an ordinary mechanical single-pan balance. Fine adjustment is then carried out electromagnetically. Other balances are entirely electromagnetic. In either case, the currents required are directly proportional to the weight on the pan. These currents are readily displayed digitally in units of grams.

Types of electronic balances. Less expensive electronic instruments are manual in the sense that the operator turns a series of knobs until the detector indicates the null condition. The readout may be displayed digitally or may be obtained from the position of the knobs. The most sophisticated instruments are self balancing. Here, the operator places the object on the pan, operates an on/off control, and reads the weight from the digital display or from a paper printout. In such instruments, the out-of-balance signal activates a servo mechanism which varies the electromagnetic restoring force until balance is realized.

Some of these modern balances have been designed to provide remarkably accurate information under adverse conditions involving vibrations and rough handling.

30D-11 AUXILIARY BALANCES

A variety of balances with lesser sensitivities than the analytical balance are employed in the laboratory. These offer the advantage of ruggedness, speed, large capacity, and convenience; such balances should be used in lieu of the analytical balance whenever maximum sensitivity is not required.

Auxiliary balances designed for top loading are particularly convenient. A sensitive top-loading balance permits weighing of loads as large as 150 to 200 g with a precision of about 1 mg—an order of magnitude less than the macro analytical balance. Balances of this type with maximum loads ranging up to as high as 25,000 (± 0.05) g are available commercially. Most are equipped with *taring* devices, which will bring the balance reading to zero when a sample container is placed on the pan. Some are fully automated and require no manual dialing or weight handling; often the restoring force is electromagnetic. A digital readout is typical.

Triple-beam and double-pan balances with sensitivities less than most top-loading balances are also useful. The former is a single-pan balance in which three decades of weights can be added by moving sliding objects along three calibrated scales. Balances of this type ordinarily show precisions that are an order or two of magnitude less than that of a top-loading instrument. Their advantages include simplicity, ruggedness, and low cost.

30E EQUIPMENT AND MANIPULATIONS ASSOCIATED WITH WEIGHING

Most solids absorb atmospheric moisture and, as a consequence, change in composition. This effect assumes appreciable proportions when a large surface area is exposed, as with a sample or a reagent chemical that has been ground to a fine powder. It is ordinarily necessary to dry such solids before weighing to free the results from dependence upon the humidity of the surrounding atmosphere.

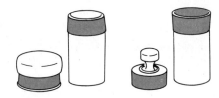

FIGURE 30-4 Typical weighing bottles. Some common sizes include 25 × 40, 40 × 80, and 70 × 33 mm.

30E-1 EQUIPMENT FOR DRYING AND WEIGHING

Weighing bottles. Solids are conveniently dried and stored in weighing bottles, two common varieties of which are illustrated in Figure 30-4. Ground-glass contacting surfaces ensure a snug fit between container and lid. In the newer design the lid acts as a cap; with this style of weighing bottle, there is less possibility of the sample being entrained on, and subsequently lost from, the ground-glass surface. Weighing bottles usually have numbers from 1 to 100 etched on their sides for identification purposes. Plastic weighing bottles are available; ruggedness is the principal advantage possessed by these bottles over their glass counterparts.

Desiccator, desiccants. Oven drying is the most convenient method for removing absorbed moisture from a solid. The technique, of course, is not appropriate for samples that undergo decomposition at the temperature of the oven. Furthermore, with some solids the temperatures attainable in ordinary drying ovens are insufficient to effect complete removal of the bound water.

While they cool, dried materials are stored in desiccators, which help prevent the uptake of moisture. As illustrated in Figure 30-5, the base section of a desiccator contains a quantity of a chemical drying agent. Samples are placed on a perforated plate that is supported by a constriction in the wall. Lightly greased ground-glass surfaces provide a tight seal between lid and base.

Several substances find use as drying agents in desiccators. Among these are anhydrous calcium chloride, calcium sulfate (Drierite[6]), anhydrous magnesium perchlorate (Anhydrone[7] or Dehydrite[8]), and phosphorous pentoxide.

30E-2 MANIPULATIONS ASSOCIATED WITH DRYING AND WEIGHING

Use of the desiccator. Whether it is being replaced or removed, the lid of the desiccator is properly moved by a sliding, rather than a lifting, motion. An airtight seal is achieved by slight rotation and direct downward pressure upon the positioned lid.

When a heated object is placed in a desiccator, the increased pressure of the enclosed air is often sufficient to break the seal between the lid and base. If heating has caused the grease on the ground-glass surfaces to soften, there is the further danger that the lid may slide off and break. Upon cooling, the opposite effect is likely to occur, the interior of the desiccator now being under a partial vacuum. Both of these conditions can cause the contents of the desiccator to be physically

[6] ® W. A. Hammond Drierite Co.
[7] ® J. T. Baker Chemical Company.
[8] ® Arthur H. Thomas Company.

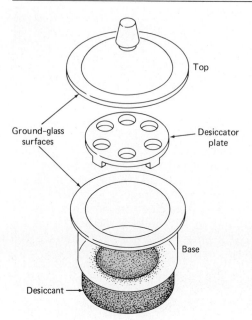

Ground-glass surfaces

Top

Desiccator plate

Base

Desiccant

FIGURE 30-5 Components of a typical desiccator.

lost or contaminated. Although it defeats the purpose of the desiccator somewhat, it is wise to allow some cooling to occur before finally sealing the lid. It also helps to break the seal several times during cooling in order to relieve any vacuum that may be developing. Finally, it is prudent to lock the lid in place with one's thumbs while moving the desiccator to ensure against accidental breakage.

Very hygroscopic materials should be stored in containers equipped with snugly fitting covers, and the covers should remain in place during storage in the desiccator. Other substances may be stored with container covers removed.

Manipulation of weighing bottles.   The moisture bound on the surface of many solid materials can be removed by heating at 105 to 110°C for about an hour. Figure 30-6 depicts the arrangement recommended for drying a sample in a weighing bottle. The bottle is contained in a beaker, which in turn is covered by a ribbed watch glass (or a watch glass supported on glass hooks). The sample is thus protected from accidental contamination while free access of air is maintained. This arrangement also satisfactorily accommodates crucibles containing precipitates that can be freed of moisture by simple drying. The beaker should be marked to permit identification.

Weighing data can be significantly affected by moisture picked up as a consequence of handling a dried weighing bottle with one's fingers. For this reason, the bottle should be manipulated with tongs, chamois finger cots, clean cotton gloves, or strips of clean paper. The latter technique is illustrated in Figure 30-7, which shows the weighing of a sample by difference. The weight of the bottle and its contents is first obtained. Then the sample is transferred from the bottle to the container, with the utmost care being taken to avoid losses during transfer. Gentle tapping of the weighing bottle with its top provides adequate control over the

FIGURE 30-6 Arrangement for the drying of samples.

process; slight rotation of the bottle is also helpful. Finally, the bottle is again weighed.

Drying (or igniting) to constant weight. Drying a sample, a precipitate, or a container to constant weight is a process in which the object whose weight is to be determined is first heated at a specified temperature for an hour or more, following which it is cooled in a desiccator and finally weighed. The heating, cooling, and weighing cycle is then repeated as many times as is required to achieve successive weights that do not differ from one another by more than 0.2 to 0.3 milligram. Weighing to constant weight provides some assurance that the chemical or physical processes brought about by the heat treatment are complete.

Weighing of hygroscopic substances. Hygroscopic substances often equilibrate rapidly, taking up moisture almost to capacity in a short time. Where this effect is pronounced, the approximate amount of each sample should be introduced to individual weighing bottles. After drying and cooling, the exact weight is determined by difference, care being taken to transfer the sample and replace the top of the weighing bottle as rapidly as possible.

Weighing of liquids. The weight of a liquid is always obtained by difference. Samples that are noncorrosive and relatively nonvolatile can be weighed into containers fitted with snugly fitted covers, such as weighing bottles; the mass of the container is subtracted from the total weight. If the sample is volatile or corrosive, it should be sealed in a weighed glass ampoule. The bulb of the ampoule is first

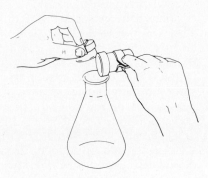

FIGURE 30-7 Method for transferring the sample.

heated. Then the neck is immersed in the sample. As cooling occurs, the liquid is drawn into the bulb. The neck is then sealed off with a small flame. After cooling, the bulb and contents are weighed, with any glass removed during the sealing also included in this weighing. The ampoule is then broken in the vessel where the sample is desired. A small volume correction for the glass from the ampoule is required for the most precise work.

30F EQUIPMENT AND MANIPULATIONS FOR FILTRATION AND IGNITION

30F-1 EQUIPMENT

Simple crucibles.   Simple crucibles serve as containers only and are of two general types. The weight of the more common variety remains constant within the limits of experimental error while in use. These crucibles are fabricated from porcelain, aluminum oxide, silica, or platinum and are employed to convert precipitates into suitable weighing forms. Nickel, iron, silver, or gold crucibles serve as containers for high-temperature fusions of difficultly soluble samples. Crucibles of this type may be appreciably attacked by the atmosphere and by their contents. Both effects will cause a change in weight; the latter will also contaminate the contents with species derived from the crucible. The analyst selects the crucible whose components will offer the least interference in subsequent steps of the analysis.

Filtering crucibles.   Filtering crucibles are also of two types. In one type, the filtering medium is an integral part of the crucible; in the other, a perforated bottom supports a glass or asbestos filter mat that is not a part of the crucible. The latter is called a *Gooch crucible.*

In *sintered* or *fritted glass crucibles,* a porous glass disk is ealed permanently to the bottom of a glass cylinder. The crucibles are normally available in three porosities and are labeled *F, M,* or *C* for fine, medium, or coarse. Under most circumstances, glass crucibles should not be heated above 150 to 200°C although with extreme care, their range can be extended to 500°C.

Filtering crucibles made entirely of fused quartz may be taken to high temperatures and can be cooled rapidly without damage.

Unglazed porcelain is also useful as a filtering medium. Crucibles of this type have the versatility of temperature range but are not as costly as fused quartz. They require no preliminary preparation comparable to that for the Gooch crucible and are far less hygroscopic. Filtering crucibles made of aluminum oxide offer similar advantages.

Gooch crucibles employing a mat of asbestos fiber were historically of considerable importance because they could be ignited at high temperatures. Federal regulations on the use of asbestos may prevent future applications of this useful material in the laboratory.

Small circles of glass matting are available commercially and can be used instead of asbestos in a Gooch crucible; they are used in pairs to protect against accidental disintegration while liquid is added. Glass mats can tolerate temperatures in excess of 500°C and are substantially less hygroscopic than asbestos.

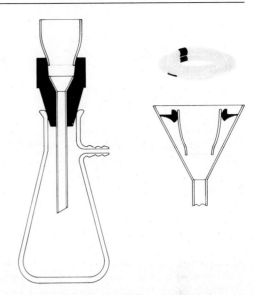

FIGURE 30-8 Adapters for filtering crucibles.

A partial vacuum is used to draw the supernatant liquid through a filtering crucible; this procedure frequently shortens the time needed for filtration. Connection is made between the crucibles and a heavy-walled filtering flask with a rubber adapter (see Figure 30-8). A diagram for the complete filtration train is shown in Figure 30-13.

Filter paper. Paper is an important filtering medium. Because it is appreciably hygroscopic, a paper filter is always destroyed by ignition if its contents are to be weighed. Ashless paper is manufactured from fibers that have been washed with hydrochloric and hydrofluoric acids and neutralized with ammonia. The residual ammonium salts in many papers are sufficient to affect analyses for amine nitrogen. Typically, 9- or 11-cm circles of such paper will leave an ash weighing less than 0.1 mg, an amount that is ordinarily negligible. Ashless paper is manufactured in various grades of porosity (see appendix 8).

Gelatinous precipitates such as hydrous iron(III) oxide present special problems, owing to their tendency to clog the pores of the paper upon which they are being retained. This problem can be minimized by mixing a dispersion of ashless filter paper pulp with the precipitate prior to filtration. The pulp may be prepared by briefly treating a piece of ashless paper with concentrated hydrochloric acid and washing the disintegrated mass free of acid. Tablets of pulp are also commercially available.

Table 30-2 summarizes the characteristics of common filtering media.

Heating equipment. Many precipitates may be weighed directly after the low-temperature removal of moisture; drying ovens, which are heated electrically and are capable of maintaining uniform temperatures to within a degree or less, are convenient for this purpose. The maximum attainable temperature will range from 140 to 260°C, depending upon the make and model. For many precipitates 110°C is a satisfactory drying temperature.

The efficiency of a drying oven is increased significantly by the forced circu-

Table 30-2   Comparison of Filtering Media for Gravimetric Analysis

Characteristic	Paper	Asbestos (Gooch)	Glass Crucible	Porcelain Crucible	Aluminum Oxide Crucible
Speed of filtration	Slow	Rapid	Rapid	Rapid	Rapid
Convenience and ease of preparation	Somewhat troublesome and inconvenient	Somewhat troublesome and inconvenient	Convenient	Convenient	Convenient
Maximum ignition temperature	None	1200°C	200–500°C	1100°C	1450°C
Chemical reactivity	Carbon from paper has reducing properties	Inert	Inert	Inert	Inert
Control of porosity	Many porosities available	Little control possible	Several porosities available	Several porosities available	Several porosities available
Convenience with gelatinous precipitates	Satisfactory	Unsuitable; filter tends to clog	Unsuitable; filter tends to clog	Unsuitable; filter tends to clog	Unsuitable; filter tends to clog
Cost	Low	Low	High	High	High

lation of air. Further refinement is achieved by predrying the air to be circulated and by using vacuum ovens through which a small flow of predried air is maintained.

Ordinary heat lamps are useful for laboratory drying and provide temperatures capable of charring filter paper. A convenient method of treating precipitates collected on paper involves the use of heat lamps for initial drying and charring, followed by ignition at elevated temperatures in a muffle furnace.

Burners are convenient sources of intense heat. The maximum temperature attainable from a burner depends upon its design as well as the combustion properties of the gas used. Of the three common laboratory burners, the Meker provides the greatest heat, followed by the Tirrill and Bunsen types in that order.

Heavy-duty electric furnaces are capable of maintaining temperatures of 1100°C or higher with a control superior to that obtainable with a burner. Special long-handled tongs and heat-resistant gloves are required for protection while transferring objects to and from such furnaces.

A rough judgment of the temperature of an object can be gained from observation of its color; Table 30-3 will serve as a guide.

30F-2 MANIPULATIONS ASSOCIATED WITH THE FILTRATION AND IGNITION PROCESSES

Preparation of crucibles.   A crucible used to convert a precipitate into a form suitable for weighing must maintain a substantially constant weight throughout the drying or ignition process. To demonstrate this property, the crucible first is cleaned thoroughly (filtering crucibles are conveniently cleaned by backwashing with suction) and then brought to constant weight, using the same heating and cooling cycle as will be required for the precipitate. Agreement within 0.2 mg between consecutive cycles can be considered as constant weight for most purposes.

Filtration and washing of precipitates. The actual filtration process consists of three operations: decantation, washing, and transfer. Decantation involves gently pouring off the bulk of liquid phase while leaving the precipitated solid essentially undisturbed. The pores of any filtering medium become clogged with precipitate; hence, the longer the transfer of solid can be delayed, the more rapid will be the overall filtration process. A stirring rod is employed to direct the flow of the decantate; see Figure 30-9b. When the flow is discontinued, any liquid at the tip of the pouring spout is removed by touching the tip with the stirring rod and returning the rod into the beaker. Wash liquid is then added to the beaker and thoroughly mixed with the precipitate; after the solid has again settled, this liquid is also decanted through the filter. It can be seen that the principal washing of the precipitate is carried out *before* the solid is transferred; a more thoroughly washed precipitate and a more rapid filtration are the result.

The transfer process is illustrated in Figures 30-9c and 30-9d. The bulk of the precipitate is moved from beaker to filter by suitably directed streams of liquid. As before, a stirring rod is used to provide direction for the flow of liquid to the filtering medium.

The last traces of precipitate that cling to the walls of the beaker are dislodged with a *rubber policeman,* a small section of rubber tubing that has been crimped shut at one end; this device is fitted on the end of a stirring rod. A rubber policeman should be wetted with wash liquid before use. Any solid collected is combined with the main portion on the filter. Small pieces of ashless paper can be used to wipe the last traces of hydrous oxide precipitates from the wall of the beaker; these papers are ignited along with the cone containing the bulk of the precipitate.

Many precipitates have the exasperating property of *creeping* or spreading over wetted surfaces against the force of gravity. Filters are never filled to more than three-quarters of their capacity, owing to the possibility of losses of solid as a result of creeping. Sometimes creeping can be prevented by addition of a small amount of a nonionic detergent to the solution before filtration.

A gelatinous precipitate should not be allowed to dry before the washing cycle is complete because the dried mass shrinks and develops cracks. Any liquid that is subsequently added merely passes through these channels and accomplishes little or no washing.

Table 30-3 Estimation of Temperature by Color*

Temperature (°C)	Approximate Color of Object at This Temperature
700	Dull red
900	Cherry red
1100	Orange
1300	White

* Adapted from T. B. Smith, *Analytical Processes*, 2d ed., p. 431. London: Edward Arnold, 1940.

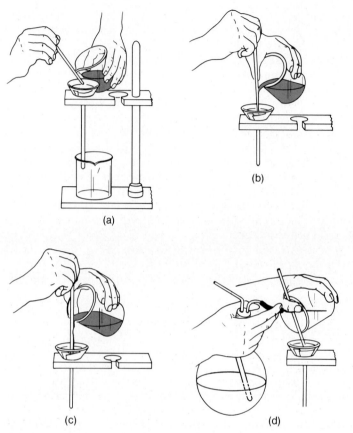

(a)

(b)

(c)

(d)

FIGURE 30-9   The filtering operation. Techniques for the decantation and transfer of precipitates are shown here.

30F-3 DIRECTIONS FOR FILTRATION AND IGNITION WITH ASHLESS FILTER PAPER

Preparation of a filter paper.   Figure 30-10 illustrates the sequence followed in folding a filter paper and seating it in a 58-deg funnel (or 60-deg fluted funnel). The paper is first folded exactly in half (a), firmly creased, and next loosely folded into quarters (b). A triangular portion is torn from one of the two single corners (c) parallel to the second fold. The paper is then opened to form a cone (d); after fitting in the funnel, the second fold is creased. Seating is completed by dampening the cone with water from a wash bottle and *gently patting* with a finger. When the cone is properly seated, there will be no leakage of air between the paper and funnel, and the stem of the funnel will be filled with an unbroken column of liquid.

Transfer of paper and precipitate to crucible.   Upon completion of the filtration and washing steps, the filter cone and its contents must be transferred from the funnel to a weighed crucible. Ashless paper has very low wet strength; thus, considerable care must be exercised in performing this operation. The danger of

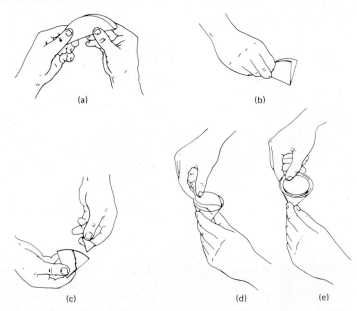

FIGURE 30-10 Technique for folding and seating a filter paper.

tearing can be reduced considerably if the paper is allowed to dry partially prior to its removal from the funnel.

Figure 30-11 illustrates the preferred method of transfer. First, the triple-thick portion is drawn across the filter (a) to flatten the cone along its upper edge; the corners then are folded inward (b). Next, the top edge is folded over (c). Finally, the paper and contents are eased into the crucible (d) so that the bulk of the precipitate is near the bottom.

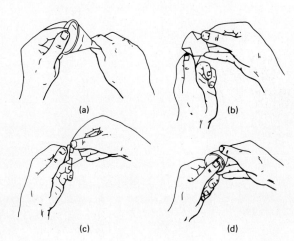

FIGURE 30-11 Method for transferring a filter paper and precipitate to a crucible.

Ashing of a filter paper. If a heat lamp is to be used, the crucible is placed on a clean, nonreactive surface; a wire screen covered with a layer of aluminum foil is satisfactory. The lamp is then positioned about one-half inch from the top of the crucible and is turned on. Charring of the paper will take place without further intervention; the process is considerably accelerated if the paper can be moistened with no more than one drop of concentrated ammonium nitrate solution. Removal of the remaining carbon is accomplished with a burner, as described in the following paragraphs.

Considerably more attention is required when a burner is employed to ash a filter paper. Because the burner can produce much higher temperatures, there exists the danger of expelling moisture so rapidly in the initial stages of heating that mechanical loss of the precipitate occurs. A similar possibility arises if the paper is allowed to flame. Finally, so long as carbon is present, the possibility exists that the precipitate will be partially reduced; reduction is a serious problem where reoxidation following ashing of the paper is not convenient.

To minimize these difficulties, the crucible is placed as illustrated in Figure 30-12; the tilted position allows for the ready access of air. A clean crucible cover should be located nearby, ready for use if necessary. Heating is then commenced with a small burner flame. The temperature is gradually increased as moisture is evolved and the paper begins to char. The smoke that is given off serves as a guide with respect to the intensity of heating that can be safely tolerated. Normally, smoke will appear in thin wisps. An increase in its volume indicates that the paper is about to flash; heating should be temporarily discontinued if this condition is observed. If, despite precautions, a flame does appear, it should be immediately snuffed out with the crucible cover. (The cover may become discolored due to the condensation of carbonaceous products; these must ultimately be removed by ignition to confirm the absence of entrained particles of precipitate.) Finally, when no further smoking can be detected, the residual carbon is removed by gradually increasing the flame. Strong heating, as necessary, can then be undertaken. Care must be exercised to avoid heating the crucible in the reducing portion of the flame.

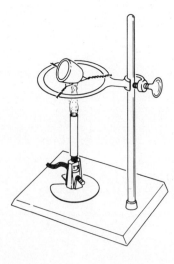

FIGURE 30-12 Ignition of the precipitate. Arrangement of the crucible for the preliminary charring of the paper is illustrated.

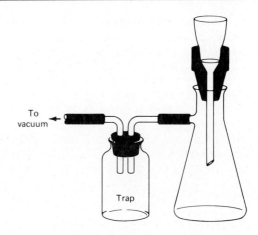

To vacuum

Trap

FIGURE 30-13   Train for vacuum filtration.

The foregoing sequence will ordinarily precede the final ignition of the sample in a muffle furnace, in which a reducing atmosphere is equally undesirable.

30F-4 DIRECTIONS FOR THE USE OF FILTERING CRUCIBLES
When paper does not serve as the filtering medium, a suction filtration train such as that in Figure 30-13 is used. The trap serves to isolate the filter flask from the vacuum source and thus prevent contamination by the filtrate.

30F-5 RULES FOR THE MANIPULATION OF HEATED OBJECTS
1. A crucible that has been subjected to the full flame of a burner or to a muffle furnace should be allowed to cool momentarily on a heat-resistant ceramic plate or a wire gauze before being moved to the desiccator.
2. Hot objects should not be placed directly on the desk top but should be set on a ceramic plate or clean wire gauze.
3. Manipulations should be practiced first to assure that adequate control can be maintained with the implements to be used.
4. The tongs and forceps employed in handling heated objects should be kept scrupulously clean. The tips should not be allowed to come in contact with the desk top.

30G THE MEASUREMENT OF VOLUME

30G-1 UNITS OF VOLUME
The fundamental unit of volume is the *liter* (L), defined as one cubic decimeter. The *milliliter* (mL) is one one-thousandth of a liter and is used when the liter represents an inconveniently large volume unit.

30G-2 EFFECT OF TEMPERATURE UPON VOLUME MEASUREMENTS
The volume occupied by a given mass of liquid varies with temperature. So also does the volume of the container that holds the liquid. The accurate measurement of volume may require taking both of these temperature effects into account.

Most volumetric measuring devices are constructed of glass, which fortunately has a small temperature coefficient. Thus, for example, a soft glass vessel will change in volume by about 0.003% per degree; with heat-resistant glass the change is about one-third of this value. Clearly, variations in the volume of a container due to changes in temperature need be considered only for the most exacting work.

The coefficient of expansion for dilute aqueous solutions is approximately 0.025% per degree. The magnitude of this figure is such that a temperature variation of about 5°C will measurably affect the reliability of ordinary volumetric measurements.

EXAMPLE 30-2. A 40.00-mL sample is taken from a liquid refrigerated at 5°C. Calculate the volume this sample will occupy at 20°C.

$$V_{20°} = V_{5°} + 0.00025(20 - 5)(40.00) \qquad (30\text{-}2)$$
$$= 40.00 + 0.15$$
$$= 40.15 \text{ mL}$$

Volumetric measurements must be referred to some standard temperature; to minimize the need for calculations such as these, 20.0°C has been chosen for this reference point. The ambient temperature in most laboratories is close to 20°C; thus, the need seldom arises for a temperature correction in ordinary analytical work with aqueous solutions. The coefficient of cubic expansion for many organic liquids, however, is considerably greater than that for water or dilute aqueous solutions. Good precision in the volumetric measurement of these liquids may require corrections for temperature variations of a degree or less.

EXAMPLE 30-3. A calibrated 50.00-mL pipet was used to deliver aliquots of a standard solution of KOH in ethanol. The temperature of the reagent was 25.2°C. The coefficient of expansion for ethanol is 0.11%/°C. What is the relative error in the analysis if no temperature correction is performed?

$$V_{20°} = 50.00 - 0.0011 \times 50 \times 5.2 = 49.71$$
$$\text{rel error} = (49.71 - 50.00) \times 1000/50.00 = -5.7 \text{ ppt}$$

30G-3 APPARATUS FOR THE PRECISION MEASUREMENT OF VOLUME

The reliable measurement of volume is performed with the *pipet*, the *buret*, and the *volumetric flask*. Volumetric equipment is marked by the manufacturer to indicate not only the manner of calibration (usually with a TD for "to deliver" or a TC for "to contain") but also the temperature for which the calibration strictly refers. Pipets and burets are ordinarily designed and calibrated to deliver specified volumes, whereas volumetric flasks are calibrated on a "to contain" basis.

Pipets. Pipets are devices that permit the transfer of accurately known volumes from one container to another. Some of the common types are pictured in Figure 30-14; Table 30-4 provides further details on their use.

Volumetric or *transfer* pipets (Figure 30-14a) deliver a single fixed volume in the range of 0.5 to 200 mL; many are color coded by volume to facilitate sorting and identification. *Measuring* pipets are of two types, Mohr and serological (Figure 30-14b and c). Measuring pipets are calibrated in convenient units to allow delivery of any volume up to the maximum capacity; sizes from 0.1 to 25 mL are avail-

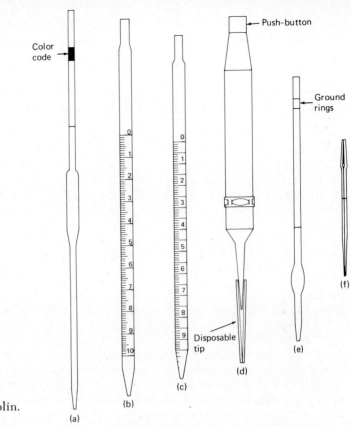

FIGURE 30-14 Typical pipets. (a) Volumetric. (b) Mohr. (c) Serological. (d) Syringe. (e) Ostwald-Folin. (f) Lambda.

Table 30-4 Pipets

Name	Type of Calibration	Function	Available Capacities (mL)	Type of Drainage
Volumetric	TD	Delivery of a fixed volume	1–200	Free drainage
Mohr	TD	Delivery of a variable volume	1–25	Drain to lower calibration line
Serological	TD	Same	0.1–10	Blow out last drop*
Serological	TD	Same	0.1–10	Drain to lower calibration line
Ostwald-Folin	TD	Delivery of a fixed volume	0.5–10	Blow out last drop*
Lambda	TC	To contain a fixed volume	0.001–2	Wash out with suitable solvent
Lambda	TD	Delivery of a fixed volume	0.001–2	Blow out last drop*
Syringe	TD	Delivery of a variable or a fixed volume	0.001–1	Tip emptied by syringe

* A frosted ring near the top of recently manufactured pipets indicates that the last drop is to be blown out.

able commercially. For situations that call for repeated delivery of a particular volume, a variety of *automatic pipets* that deliver volumes of 0.1 to 50 mL can be purchased. Hand-held syringe pipets (Figure 30-14d) that deliver volumes in the range between 1 and 1000 μL (1 mL) are available; a useful feature of this type of pipet is that the liquid is contained in a disposable plastic tip. A measured volume of liquid is drawn into the tip by means of a spring-operated piston activated by a push button at the top of the syringe. The liquid is then transferred by reversing the action of the piston by means of the push button. Remarkable precision is claimed for these devices (± 0.02 μL for a 1-μL measurement and ± 0.3 μL at 1000 μL).

Volumetric and measuring pipets are filled to an initial calibration mark at the outset; the manner in which the transfer is completed depends upon the particular type. Because of the attraction between most liquids and glass, a drop tends to remain in the tip of a drained transfer pipet. This drop is blown from some pipets but not from others. Table 30-4 and Figure 30-14 summarize the several varieties most likely to be encountered in an analytical laboratory.

Burets. Burets, like measuring pipets, enable the analyst to deliver any volume up to the maximum capacity. The precision attainable with a buret is appreciably better than that with a measuring pipet.

A buret consists of a calibrated tube containing the liquid and a valve arrangement by which the flow from a tip can be controlled. Principal differences among burets are to be found in the type of valve employed. The simplest consists of a closely fitting glass bead within a short length of rubber tubing. Only when the tubing is deformed can liquid flow past the bead. Burets equipped with glass stopcocks rely upon a lubricant between the ground-glass surfaces of stopcock and barrel for a liquid-tight seal. Some solutions, notably bases, will cause a stopcock to freeze upon long contact; thorough cleaning is necessary after each use.

Valves made of Teflon are commonly encountered; these are inert to attack by most common reagents and require no lubricant.

More elaborate burets are also available, which are connected directly to a reagent reservoir. The simplest of these devices employs an ordinary buret equipped with a three-way stopcock; one setting of the stopcock fills the buret from the reservoir while the other is used for titrations. Some have an added feature which zeros the buret automatically. The most sophisticated burets are electronic and usually involve the use of an electrically driven syringe. Push buttons permit automatic refilling of the buret and controlling the flow of reagent during the titration. Usually, the volume delivered is read out on a digital meter.

Volumetric flasks. Volumetric flasks are manufactured with capacities ranging from 5 mL to 5 L and are usually calibrated to contain a specified volume when filled to the line etched on the neck. They are used in the preparation of standard solutions and the dilution of samples to known volumes prior to taking aliquot portions with a pipet. Some are also calibrated on a "to deliver" basis. These are readily distinguishable by two reference lines; if delivery of the stated volume is desired, the flask is filled to the upper of the two lines.

30G-4 CLEANING OF VOLUMETRIC EQUIPMENT
Only clean glass surfaces will support a uniform film of liquid; the presence of dirt or oil will tend to cause breaks in this film. The appearance of water-breaks is a cer-

tain indication of an unclean surface. If the volume calibrations are to have meaning, water-breaks must be absent; thus, volumetric equipment must be kept scrupulously clean when in use.

Ordinarily, a brief soaking in warm detergent solution suffices to remove dirt and grease that cause water-breaks. Long soaking is to be avoided, however, because a rough area or ring often develops at the detergent/air interface. This ring is impossible to remove and causes a film break that destroys the usefulness of the equipment.

When detergent solution fails, soaking for a few minutes in warm cleaning solution will usually provide a satisfactory interior surface to volumetric ware.

After cleaning, the apparatus must be thoroughly rinsed (several times) with tap water and then with three or four portions of distilled water. It is seldom necessary to dry volumetric ware. As a general rule, calibrated glass equipment should never be heated; permanent distortion of the glass, which leads to a volume change, may result.

30G-5 AVOIDANCE OF PARALLAX

When a liquid is confined in a narrow tube such as a buret or a pipet, the surface exhibits a marked curvature called a *meniscus*. It is common practice to use the bottom of the meniscus as the point of reference in calibrating and using volumetric ware. This minimum can often be established more exactly if an opaque card or piece of paper is held behind the graduations (see Figure 30-15).

In judging volumes, the eye must be level with the liquid; otherwise the reading will be in error due to *parallax*. Thus, if one's eye level is above that of the liquid, it will appear that a smaller volume has been taken than is actually the case. An error in the opposite direction can be expected if the point of observation is too low (see Figure 30-15).

30G-6 DIRECTIONS FOR THE USE OF A PIPET

The following instructions pertain specifically to the manipulation of transfer pipets, but with minor modifications they may be used for other types as well. Liquids are drawn into pipets through the application of a slight vacuum. The mouth should *never* be used for suction because of the danger of accidentally ingesting liquids. Instead, a rubber suction bulb or a rubber tube connected to an aspirator pump should be employed for filling the pipet (see Figure 30-16a).

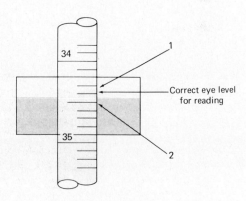

FIGURE 30-15 Method for reading a buret. The eye should be level with the meniscus. The black-white line on the card should be slightly below the meniscus. The reading shown is 34.38 mL. If viewed from 1, the reading will appear smaller than 34.38: from 2, it will appear larger.

Cleaning.   Holding the pipet vertically, suck detergent or cleaning solution to a level that is an inch or two above the mark. Allow to stand for a brief period (it may be desirable to repeat this operation several times). Rinse with several portions of tap water in the same way. Inspect for film breaks, and repeat the cleaning cycle if necessary. Finally, fill the pipet to perhaps one-third its capacity with distilled water and, while holding it nearly horizontal, carefully rotate the pipet so that the interior surfaces are wetted. Repeat this rinsing at *least* twice more.

Measurement of an aliquot.   As in rinsing, draw in a small quantity of the liquid to be sampled, and thoroughly rinse the interior surfaces. Repeat with at *least* two more portions. Then carefully fill the pipet somewhat past the graduation mark, employing a suction bulb as in Figure 30-16a. Quickly place a *forefinger* over the upper end of the pipet to arrest the outflow of liquid (Figure 30-16b). Make certain

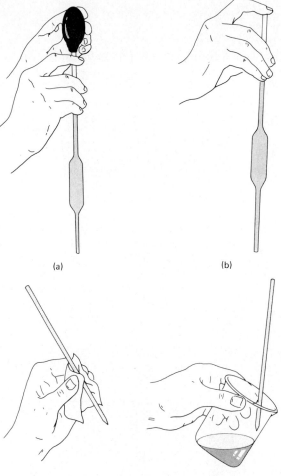

(a)   (b)

(c)   (d)

FIGURE 30-16   Technique for using a volumetric pipet.

that there are no bubbles in the bulk of the liquid or foam at the surface. Tilt the pipet slightly from the vertical, and wipe the exterior free of adhering liquid (Figure 30-16c). Touch the tip of the pipet to the wall of a glass vessel (*not the actual receiving vessel*), and slowly allow the liquid level to drop by partially releasing the forefinger (Note 1). Halt further flow as the bottom of the meniscus coincides exactly with the graduation mark. Then place the tip of the pipet well into the receiving vessel, and allow the sample to drain. When free flow ceases, rest the tip against an inner wall for a full 10 s (Figure 30-16d). Finally, withdraw the pipet with a rotating motion to remove any droplet still adhering to the tip. *The small volume remaining inside the tip is not to be blown or rinsed into the receiving vessel.*

> NOTES
> 1. The liquid can best be held at a constant level in the pipet if one's forefinger is slightly moist; too much moisture, however, makes control difficult.
> 2. It is good practice to avoid handling a pipet by the bulb to eliminate the possibility of changing the temperature of its contents.
> 3. Pipets should be thoroughly rinsed with distilled water after use.

30G-7 DIRECTIONS FOR THE USE OF A BURET

Before being placed in service, a buret must be scrupulously clean. In addition, it must be established that the valve is liquid tight.

Cleaning. Thoroughly clean the tube with detergent and a long brush. If water-breaks persist after rinsing, clamp the buret in an inverted position with the end dipped in a beaker of warm cleaning solution. Connect a hose from the buret tip to a vacuum line. Gently pull the cleaning solution into the buret, stopping well short of the stopcock (Note 1). Allow to stand for at least 15 min and then drain. Rinse thoroughly with tap water followed by distilled water, and again inspect for water-breaks. Repeat the treatment if necessary.

Lubrication of a glass stopcock buret. Carefully remove all old grease from the stopcock and barrel with a paper towel, and dry both parts completely. Lightly grease the stopcock, with care being taken to avoid the area near the hole. Insert the stopcock into the barrel, and rotate it vigorously with a slight inward pressure. When the proper amount of lubricant has been used, the area of contact between stopcock and barrel appears nearly transparent, the seal is liquid tight, and no grease has worked its way in to the tip.

> NOTES
> 1. Cleaning solution often disperses more stopcock lubricant than it oxidizes and leaves a buret with a heavier grease film than before treatment. For this reason, cleaning solution should *not* be allowed to come in contact with lubricated stopcock assemblies.
> 2. Grease films unaffected by cleaning solution may yield to treatment with such organic solvents as acetone or benzene. Thorough washing with detergent should follow such treatment. The use of silicone lubricants is not recommended; contamination by such preparations is difficult or impossible to remove.
> 3. So long as the flow of liquid is not impeded, fouling of the buret tip with lubricant is not a serious matter. Removal is best accomplished with organic solvents. A stoppage in the middle of a titration can be freed by *gently* warming the tip with a lighted match.
> 4. Before returning a buret to service after reassembly, it is advisable to test for leakage. Simply fill the buret with water, and establish that the volume reading does not change with time.

Filling. Make certain that the stopcock is closed. Add 5 to 10 mL of solution, and carefully rotate the buret to wet the interior completely. Allow the liquid to drain through the tip. Repeat this procedure at *least* two more times. Then fill the buret well above the zero mark. Free the tip of air bubbles by rapidly rotating the stopcock and allowing small quantities of solution to pass. Finally, lower the level of the solution to or somewhat below the zero marking (Note). After allowing about a minute for drainage, take an initial volume reading.

> NOTE
> A convenient way of proceeding here is to withdraw the liquid to about 1 mm above the zero graduation and allow about 1 min for drainage. Then bring the level to exactly 0.00 mL.

Titration. Figure 30-17 illustrates the preferred method for the manipulation of a stopcock. Any tendency for lateral movement of the stopcock will be in the direction of firmer seatings.

With the tip well within the titration vessel, introduce the solution from the buret in increments of 1 mL or so. Swirl (or stir) the sample constantly to ensure efficient mixing. Decrease the volume of the additions as the titration progresses. In the immediate vicinity of the end point, add reagent a drop at a time. When it is judged that only a few more drops are needed, rinse the walls of the titration vessel before completing the titration. Allow at least 30 s to elapse between the last addition of reagent and the reading of the buret.

> NOTES
> 1.   When unfamiliar with a particular titration, many analysts prepare an extra sample. No care is lavished on its titration since its functions are to reveal the nature of the end point and to provide a rough estimate of titrant requirements. This deliberate sacrifice of one sample often results in an overall saving of time.
> 2.   Instead of being rinsed near the end of the titration, a flask can be carefully tipped and rotated so that the bulk of the liquid picks up any droplets adhering to the walls.
> 3.   Volume increments smaller than a normal drop may be taken by allowing a small volume of liquid to form on the tip of the buret and then touching the tip to the wall of the flask. This droplet is then combined with the bulk of the solution as in Note 2.

FIGURE 30-17 Recommended techniques for the manipulation of a buret stopcock.

30G-8 DIRECTIONS FOR THE USE OF A VOLUMETRIC FLASK

Before use, volumetric flasks should be washed with detergent and, if necessary, cleaning solution, followed by thorough rinsing. Only rarely need they be dried. If required, however, drying is best accomplished by clamping the flasks in an inverted position. The insertion of a glass tube connected to a vacuum line will hasten the process.

Weighing directly into a volumetric flask. The direct preparation of a standard solution requires that a known weight of solute be introduced into a volumetric flask. To minimize the possibility of loss during transfer, insert a powder funnel into the neck of the flask. The funnel must subsequently be washed free of solid.

When heating is required to dissolve the solid, the foregoing procedure is not applicable. Instead, the solid is weighed into a beaker or flask and is dissolved; the solution is cooled to room temperature. Quantitative transfer of this solution is then accomplished by the method described in the next section.

Quantitative transfer of solutions to a volumetric flask. Employing a small funnel, transfer the liquid to the volumetric flask following the procedure shown in Figure 30-9b and the directions for decantation given in Section 30F-2. After transfer is complete, touch the tip of the stirring rod to the pouring spout, return the rod into the beaker, and rinse the walls of the beaker and the rod with distilled water. Transfer the rinsing to the flask, again pouring the liquid down the rod. Repeat this operation at *least* twice more.

Dilution to the mark. After transferring the solute, fill the flask about half full, and swirl the contents to achieve solution. Add more solvent and again mix well. Bring the liquid level almost to the mark, and allow time for drainage; then, use a medicine dropper to make such final additions of solvent as are necessary. Firmly stopper the flask, and invert repeatedly to assure uniform mixing. Finally, transfer the contents to a storage bottle that is either initially dry or has been thoroughly rinsed with several small portions of the solution from the flask.

> NOTE
> If, as sometimes happens, the liquid level accidentally exceeds the calibration mark, the solution can be saved by correcting for the excess volume. Use a gummed label to mark the actual position of the meniscus. After the flask has been emptied, carefully refill to the etched graduation mark with water. Then, with a buret, determine the additional volume needed to duplicate the actual volume of the solution. This volume, of course, should be added to the nominal value for the flask when calculating the concentration of the solution.

30G-9 CALIBRATION OF VOLUMETRIC WARE

The reliability of a volumetric analysis depends upon agreement between the volumes actually and purportedly contained (or delivered) by the apparatus. Calibration simply verifies this agreement or provides a correction if it is lacking. The latter involves either the assignment of corrections to the existing volume markings or the striking of new markings that agree more closely with the nominal values.

In general, a calibration consists of determining the mass of a liquid of known density contained (or delivered) by the volumetric vessel. Although calibration appears to be a straightforward process, a number of important variables must be controlled. Principal among these is the temperature, which influences a calibration in two ways. First, and most important, the volume occupied by a given mass of liquid varies with temperature. Second, the volume of the apparatus itself is variable, owing to the tendency of the glass to expand or contract with changes in temperature.

We noted earlier (p. 704) that the effect of buoyancy upon weighing data is most pronounced when the density of the object is significantly less than that of the weights. As a general rule, a buoyancy correction must be applied to data where water is the calibration fluid.

Finally, the liquid employed for calibration requires consideration. Water is the liquid of choice for most work. Mercury is also useful, particularly where small volumes are involved. Because mercury does not wet glass surfaces, the volume contained by the apparatus will be identical with that which is delivered. In addition, the convex meniscus of mercury gives rise to a small correction that must be applied to give the corresponding volume for a liquid forming a concave meniscus. The magnitude of this correction is dependent upon the diameter of the apparatus at the graduation mark.

The calculations associated with calibrations, while not difficult, are somewhat involved. First, the raw weighing data are corrected for buoyancy, using Equation 30-1 (p. 704). Next, the volume of the apparatus at the temperature (t) of calibration is obtained by dividing the density of the liquid at that temperature into the corrected weight. Finally, this volume is corrected to the standard temperature of 20°C by means of Equation 30-2 (p. 704).

Table 30-5 is provided to ease the computational burden of calibration. Corrections for buoyancy with respect to stainless steel or brass weights (the difference in density between the two is small enough to be neglected) and for the volume change of the water, as well as its glass container, have been incorporated into these data. Multiplication by the appropriate factor from the table converts the mass of water measured at some other temperature to the volume it would occupy at 20°C.

EXAMPLE 30-4. A 25-mL pipet was found to deliver 24.976 g of water when calibrated against stainless steel weights at 25°C. Use the data in Table 30-5 to calculate the volume delivered by the pipet at this temperature and at 20°C.

At 25°C $V = 24.976$ g $\times$ 1.0040 g/mL = 25.08 mL
At 20°C $V = 24.976$ g $\times$ 1.0037 g/mL = 25.07 mL

General directions for calibration work. All volumetric apparatus should be painstakingly freed of water-breaks before being tested. Burets and pipets need not be dried. Volumetric flasks should be thoroughly drained and dried (at room temperature).

The water used for calibration should be drawn well in advance of use to permit it to reach thermal equilibrium with its surroundings. This condition is best assured by noting the temperature of the water at frequent intervals and waiting until no further changes are observed.

An analytical balance can be used for calibrations involving 50 mL or less. Weighings need never be more reliable than the nearest milligram; this order of reproducibility is within the capabilities of many modern top-loading, single-pan lab-

Table 30-5   Volume Occupied by 1.000 g Water Weighed in Air Against Stainless Steel Weights*

Temperature, t (°C)	Volume (mL)	
	At Temperature, t	Corrected to 20°C
10	1.0013	1.0016
11	1.0014	1.0016
12	1.0015	1.0017
13	1.0016	1.0018
14	1.0018	1.0019
15	1.0019	1.0020
16	1.0021	1.0022
17	1.0022	1.0023
18	1.0024	1.0025
19	1.0026	1.0026
20	1.0028	1.0028
21	1.0030	1.0030
22	1.0033	1.0032
23	1.0035	1.0034
24	1.0037	1.0036
25	1.0040	1.0037
26	1.0043	1.0041
27	1.0045	1.0043
28	1.0048	1.0046
29	1.0051	1.0048
30	1.0054	1.0052

* Corrections for buoyancy (stainless steel weights) and the change in volume of the container have been applied.

oratory balances. Weighing bottles or small, well-stoppered conical flasks are convenient receivers for small volumes.

Calibration of a volumetric pipet.   Determine the empty weight of a stoppered receiver to the nearest milligram. Transfer a volume of water of known temperature to the receiver with the pipet (Section 30G-6), weigh the receiver and contents to the nearest milligram, and calculate the weight of water delivered from the difference in these weights. Calculate the volume with the aid of Table 30-5. Repeat the calibration several times.

Calibration of a buret.   Fill the buret with water of known temperature, and make certain that no bubbles are entrapped in the tip. Withdraw water until the level is just above the zero mark. After allowing 1 min for drainage, bring the meniscus to the 0.00 mL mark, and touch the tip to the wall of a beaker to remove any adhering drop. Allow the buret to stand for 10 min and recheck the reading; if the stopcock is tight, there should be no noticeable change. During this interval, weigh (to the nearest milligram) a 125-mL conical flask fitted with a rubber stopper.

Once the tightness of the stopcock has been established, slowly (about 10 mL/min) transfer approximately 10 mL into the flask. Touch the tip to the wall of the flask. Wait 1 min, record the volume, and refill the buret. Weigh the flask and its contents to the nearest milligram; the difference between this and the initial weight gives the mass of water actually delivered. Convert this mass into the volume delivered, using Table 30-5. Compute the correction in this interval by subtracting the apparent volume from the true volume. The difference is the correction that must be applied to the *apparent volume* to give the *true volume*. Repeat this calibration until agreement with ± 0.02 mL is obtained.

Starting again from the zero mark, repeat the calibration, using about 20 mL. Test the buret at 10-mL intervals over its entire volume. Prepare a plot of the correction to be applied as a function of the volume delivered. Corrections associated with any interval may be determined from the plot.

Calibration of a volumetric flask. If an equal-arm balance is to be used, weigh the clean, dry flask, placing it on the right pan (Note). Set a beaker on the left pan, and add lead shot until balance is achieved. Remove the flask, and in its place substitute known weights until the same point of balance is reached. Carefully fill the flask with water of known temperature until the meniscus coincides with the graduation mark. Return the flask to the right pan and the beaker to the left. Repeat the process of counter-weighing with lead shot, followed by substituting weights for the flask. The difference between the two weighings gives the mass of water contained by the apparatus. Calculate the corresponding volume with the aid of Table 30-5.

NOTES
1. The substitution method described here is employed to eliminate any error arising from unequal arm lengths of the balance. If a top-loading, single-pan balance is available, the substitution technique is unnecessary and the flask and its contents can be weighed directly.
2. A glass tube with one end drawn to a tip is useful in making final adjustments of the liquid level.

Calibration of a volumetric flask relative to a pipet. The calibration of a flask relative to a pipet provides an excellent method for partitioning a sample into aliquots. The following directions pertain specifically to a 50-mL pipet and a 500-mL flask; other combinations are equally convenient.

With a 50-mL pipet, carefully transfer 10 volumes to a 500-mL volumetric flask. Mark the location of the meniscus with a gummed label. Coat the label with paraffin to assure permanence. When a sample is diluted to this mark, the same 50-mL pipet will deliver a one-tenth aliquot of the total sample.

30H THE LABORATORY NOTEBOOK

30H-1 RULES FOR KEEPING THE LABORATORY NOTEBOOK

1. The notebook should be permanently bound with consecutively numbered pages (numbering of pages should be carried out on or before the first laboratory period).
2. Entries should be legible and well spaced from one another. Most notebooks have more than ample room, so crowding of data is unnecessary.

3. The first few pages of the notebook should be reserved for a table of contents, which should be conscientiously kept up to date.
4. *All data should be entered in ink directly into the notebook.*
   a. Entries should be liberally identified with labels. If a series of weights refers to a set of empty crucibles, it should be labeled "Empty Crucible Weight" or something similar. The significance of an entry is obvious when it is recorded but may become unclear with the passage of time.
   b. Each notebook page should be dated as it is used.
   c. An erroneous entry should *not* be erased, nor should it be obliterated. Instead, it should be crossed out with a *single* horizontal line with the corrected entry located as nearby as possible. Numbers should never be written over; in time it may become impossible to decide what the correct number is.
   d. Pages should not be removed from the notebook. It is sufficient to

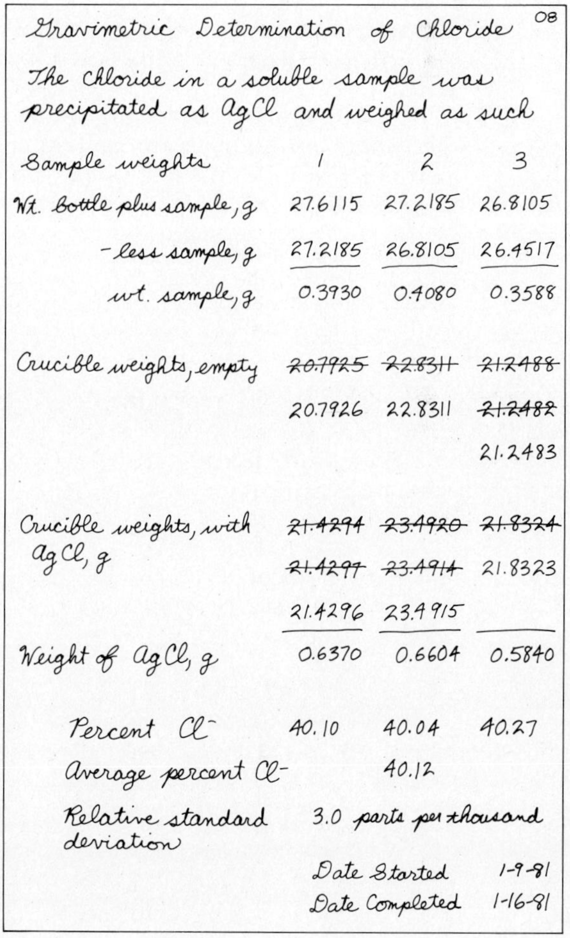

08

Gravimetric Determination of Chloride

The Chloride in a soluble sample was precipitated as AgCl and weighed as such

Sample weights	1	2	3
Wt. bottle plus sample, g	27.6115	27.2185	26.8105
– less sample, g	27.2185	26.8105	26.4517
wt. sample, g	0.3930	0.4080	0.3588
Crucible weights, empty	~~20.7925~~	~~22.8311~~	~~21.2488~~
	20.7926	22.8311	~~21.2482~~
			21.2483
Crucible weights, with AgCl, g	~~21.4294~~	~~23.4920~~	~~21.8324~~
	~~21.4297~~	~~23.4914~~	21.8323
	21.4296	23.4915	
Weight of AgCl, g	0.6370	0.6604	0.5840
Percent Cl⁻	40.10	40.04	40.27
Average percent Cl⁻		40.12	
Relative standard deviation	3.0 parts per thousand		

Date Started 1-9-81
Date Completed 1-16-81

FIGURE 30-18 Data page.

draw a single line diagonally across a page that is to be disregarded. A brief notation of the reason for striking out the page is useful.

30H-2 SUGGESTED FORM

A satisfactory format involves the consecutive use of all pages in the notebook. One of the pages should contain the following:

1. The title of the experiment—for example, "The Gravimetric Determination of Chloride."
2. A brief statement of the principles upon which the analysis is based.
3. The original data entries and the result calculated for each sample in the set.
4. A report of the best value for the set and a statement of the precision attained in the analysis.

A sample data page is shown in Figure 30-18. The second page should show the following:

1. Equations for the principal reactions in the analysis.
2. An equation that shows the calculation employed in computing the results.
3. The calculations of the results and their precision.
4. The summary of observations that appear to bear upon the validity of a particular result or the analysis as a whole. These observations should be recorded *at the time of observation.*

selected methods of analysis

31

This chapter provides specific instructions for carrying out selected chemical analyses. Before starting an analysis, the student should read and understand any discussion that precedes the procedure as well as any notes that follow; sections of earlier chapters that have been cited as sources for additional information should also be consulted.

Before undertaking an analysis, it is excellent practice to distinguish between those quantities that must be measured with great care and those that do not. Failure to use sufficient care in making critical measurements will ruin an analysis; conversely, it is a waste of time to make a measurement with precision where an approximation will suffice.

1 GRAVIMETRIC METHODS

METHOD 1-1. DETERMINATION OF WATER IN BARIUM CHLORIDE DIHYDRATE

Discussion. Read pages 612 and 619. The water in a crystalline hydrate such as $BaCl_2 \cdot 2H_2O$ is readily determined gravimetrically. A known weight of sample is heated at a suitable temperature; its water content is taken as the difference in its weight before and after heating. Less frequently, the evolved water is collected and weighed.

Procedure. Throughout this determination, perform all weighings to the nearest 0.1 mg. Follow the procedure on page 707 for handling and weighing the samples.

Carefully clean two or three weighing bottles. Dry them for at least 1 hr at 105 to 110°C; use a covered beaker to prevent accidental contamination (see Figure 30-6). Determine the weight of each bottle after it has cooled to room temperature in a desiccator. Repeat this cycle of heating, cooling, and weighing until successive weighings agree within 0.2 mg (this process is called drying to constant weight, see p. 709. Next, introduce a quantity of unknown to each bottle (Note 1) and reweigh. Heat the samples for about 2 hr at 105 to 110°C; then cool and weigh as before. Repeat the heating cycle until constant weights for the bottles and their contents have been attained. Report the percentage of water in the sample.

NOTES
1. If the unknown consists of the pure dihydrate $BaCl_2 \cdot 2H_2O$, take samples weighing 1 to 1.5 g. If the unknown is a mixture of the dihydrate and some anhydrous diluent, obtain the proper sample size from the instructor.
2. Barium chloride can be heated to elevated temperatures without danger of decomposition. If desired, the analysis can be performed in crucibles with a Bunsen flame as the source of heat.
3. Magnesium sulfate hepthahydrate can be used instead of barium chloride; heating to 140°C is needed to eliminate the last traces of moisture.

METHOD 1-2. GRAVIMETRIC DETERMINATION OF CHLORIDE IN A SOLUBLE SAMPLE

Discussion. The chloride content of a soluble salt can be determined by precipitation as silver chloride:

$$Ag^+ + Cl^- \longrightarrow AgCl(s)$$

The precipitate is collected in a filtering crucible, washed, and brought to constant weight by drying at 105 to 110°C. Precipitation is carried out in acid solution to eliminate potential interference from anions of weak acids (for example, CO_3^{2-}) that form precipitates with silver in neutral media. A moderate excess of silver ion is required to diminish the solubility of the precipitate; a large excess will cause serious coprecipitation and should be avoided.

Silver chloride forms first as a colloid; it is coagulated with heat in the presence of a relatively high electrolyte concentration. A small quantity of nitric acid is added to the wash liquid to maintain the electrolyte concentration and prevent peptization during washing; the acid is volatilized during the subsequent heat treatment. See pages 134 to 138 for further information concerning the characteristics of colloidal precipitates.

Like other silver halides, silver chloride in suspension is susceptible to photodecomposition, the reaction being

$$AgCl(s) \longrightarrow Ag(s) + \frac{1}{2} Cl_2(aq)$$

The precipitate acquires a violet color due to the accumulation of finely divided silver. If photochemical decomposition occurs in the presence of excess silver ion, the additional reaction

$$3Cl_2(aq) + 3H_2O + 5Ag^+ \longrightarrow 5AgCl(s) + ClO_3^- + 6H^+$$

will cause the analytical results to be high. In the absence of silver ion, the results will be low. Dry silver chloride is virtually unaffected by exposure to light.

Some photodecomposition of silver chloride is inevitable as the analysis is ordinarily performed. The effect is negligible, provided the solid is not exposed to direct sunlight. Storage in a dark space is also a worthwhile precaution.

Iodide, bromide, and thiocyanate, if present in the sample, will be precipitated along with silver chloride and cause high results. In addition, the chlorides of tin and antimony are likely to precipitate as hydrous oxides under the conditions of the analysis.

With the current high price of silver, it is imperative that every effort be made to minimize losses of silver nitrate during the experiment and to recover the silver chloride residue after the experiment is complete. Thus, any unused silver nitrate solution as well as the filtrate and washings from the filtration should be collected and placed in a storage container. Instructions for recovering the silver chloride from crucibles are also included.

Reagents needed.

$Concentrated\ HNO_3$
$6\ M\ HNO_3$
$6\ M\ NH_3$
$0.20\ M\ AgNO_3\ (15\ to\ 20\ mL/sample)$

Procedure. Clean three fritted glass filtering crucibles (medium [M] or fine [F] but not coarse [C]) by filling each with about 5 mL of concentrated HNO_3 and letting them stand for a few minutes. Attach each to the filtering apparatus (p. 717) and draw the acid through the crucible. Rinse with three portions of tap water and then turn off the suction. Add 5 mL of 6 M NH_3 and let stand for a few minutes. Draw the NH_3 through the crucible and rinse six to eight times with small portions of distilled (or deionized) water. Provide each crucible with an identifying mark, and bring it to constant weight (p. 709) by heating at 110°C for at least 1 hr during periods of waiting in the analysis (Note 1).

Dry the unknown at 105 to 110°C for 1 to 2 hr in a weighing bottle (see Figure 30-6, p. 709). Store in a desiccator while cooling. Weigh individual 0.15- to 0.2-g samples (to the nearest 0.1 mg by difference) into 400-mL beakers (p. 708). To each add about 100 mL of distilled (or deionized) water and 2 to 3 mL of 6 M HNO_3. Slowly and with good stirring add 0.2 M $AgNO_3$ to the cold solution until the precipitate is observed to coagulate (Notes 2 and 3); then add an additional 3 to 5 mL. Heat almost to boiling, and digest (p. 138) the precipitate at this temperature for about 10 min. Check for completeness of precipitation by adding a few drops of $AgNO_3$ to the supernatant liquid; should additional AgCl appear, add 3 mL of $AgNO_3$, digest, and again check for completeness of precipitation. Return any unused $AgNO_3$ solution to the waste container for silver residues. Cover the beaker and store in a dark place for at least 1 to 2 hr, preferably until the next laboratory period.

Read pages 713 to 719 on filtering and washing precipitates. Then, decant the supernatant through a weighed filtering crucible. Wash the precipitate several times (while it is still in the beaker) with a cold solution consisting of 2 to 5 mL of 6 M HNO_3 per liter of distilled water; decant these washings through the filter also. Finally, transfer the bulk of the precipitate to the crucible, using a rubber policeman to dislodge any particles that adhere to the walls of the beaker. Continue washing until the filtrate is found to be substantially free of $AgNO_3$ (Note 4). Retain the filtrate and washings and transfer to the silver residue waste container. Dry the precipitates at 105 to 110°C for at least 1 hr. Store crucibles in a desiccator until they have cooled to room temperature. Determine the weight of the crucibles and their contents. Repeat the cycle of heating, cooling, and weighing until constant weight has been achieved (to the nearest 0.2 mg). Report the %Cl in the sample.

After the experiment is complete, remove the precipitates from the crucibles by tapping them over a piece of glazed paper. Transfer the collected AgCl to the waste container for silver residues. Clean the crucibles as instructed at the beginning of the procedure. Collect the washings and transfer to the silver waste container.

NOTES
1. Gooch crucibles with two glass fiber circles may be substituted (see p. 710).
2. To determine an approximate amount of $AgNO_3$ needed, calculate the volume that would be required if the sample were pure NaCl.
3. Use a separate stirring rod for each sample and leave it in the solution throughout the determination.
4. Washings are readily tested for the presence of Ag^+ by collecting a few milliliters in a test tube and adding a few drops of HCl. Washing is judged complete when little or no turbidity is observed with this test.

METHOD 1-3. GRAVIMETRIC DETERMINATION OF SULFATE IN A SOLUBLE SAMPLE
The analysis of a soluble sulfate is based upon precipitation with barium ion

$$Ba^{2+} + SO_4^{2-} \longrightarrow BaSO_4(s)$$

The barium sulfate, which forms as a crystalline precipitate, is collected on a suit-

Table 31-1   Interferences Attending the Gravimetric Determination of Sulfate as BaSO₄

Effect upon Analysis		Nature of Interference
Low results	1.	Excessive amounts of mineral acid present. (Solubility of $BaSO_4$ is appreciably greater in strong acid media.)
	2.	Coprecipitation of sulfuric acid. (Note that this is a source of error in a gravimetric determination of sulfate but not of barium, since this H_2SO_4 is driven off during ignition.)
	3.	Coprecipitation of alkali metal and various divalent ions. (Sulfates of these ions usually weigh less than the equivalent amount of $BaSO_4$, which should have formed.)
	4.	Coprecipitation of ammonium ion. (Ammonium sulfate is volatilized upon ignition of the precipitate.)
	5.	Coprecipitation of iron as a basic iron(III) sulfate.
	6.	Partial reduction of $BaSO_4$ to BaS if filter paper is charred too rapidly.
	7.	Presence of trivalent chromium. [May not achieve complete precipitation of $BaSO_4$ owing to formation of soluble complex sulfates of chromium(III).]
High results	1.	Absence of mineral acid. (The slightly soluble carbonate or phosphate of barium can precipitate under these conditions.)
	2.	Coprecipitation of barium chloride.
	3.	Coprecipitation of anions, particularly nitrate and chlorate, as barium salts.

able filter, washed with water, and strongly ignited. See pages 138 to 139 for further discussion concerning the characteristics of crystalline precipitates.

Superficially, this method appears straightforward. In fact, however, it is subject to numerous interferences, due chiefly to the tendency of barium sulfate to occlude foreign anions and cations. Table 31-1 summarizes the more common interferences affecting this analysis. Purification by reprecipitation is not feasible because there is no practical solvent for barium sulfate. It is therefore necessary to eliminate the principal interferences by preliminary treatment of the sample and then to precipitate the barium sulfate from hot, dilute solutions. Even so, the excellent agreement often observed between theoretical and experimental results is due in considerable measure to cancellation of errors.

The procedure that follows is particularly useful for the determination of sulfate in samples that contain alkali metal chlorides. It has been reported that the slow addition of barium chloride to a hot solution of such a sample can cause negative errors as large as 1.0 to 1.5% owing to coprecipitation of the alkali metal sulfates.[1] It has been found empirically, however, that *rapid* addition of the reagent to a hot solution of sample largely eliminates this negative bias. As expected, the rapid addition of the reagent yields smaller crystals of barium sulfate; the particle size is still large enough, however, for convenient filtration.

Reagents needed.

> 6 M HCl
> 1.3% (w/v) BaCl$_2$·2H$_2$O (filter if not clear); 100 mL/sample

Procedure.   Clean and mark three porcelain crucibles and covers (p. 696). Ignite them for 15 min at the highest temperature obtainable with Tirrill or Meker burners. Cool in a desiccator for at least 1 hr, and then weigh. Handle the crucibles only with tongs. Repeat the ignition until a constant weight (± 0.2 mg) is achieved.

Dry the unknown (see Figure 30-6) for at least 1 hr at 105 to 110°C; allow it to cool to room temperature in a desiccator. Weigh (to the nearest 0.1 mg) individual 0.5- to 0.7-g samples into 400-mL beakers (p. 708). Dissolve each in 200 mL of distilled (or deionized) water to which 4 mL of 6 M HCl have been added.

For each sample, heat 100 mL of the BaCl$_2$ solution nearly to boiling, and add quickly with vigorous stirring to the hot solution of the sample (Note 1).

Digest the precipitated BaSO$_4$ at just below the boiling point for 1 to 2 hr (Notes 1 and 2). (See pages 713 to 717 regarding filtering and washing precipitates.) Decant the hot supernatant through a fine ashless paper (Note 3). Wash the precipitate three times with hot water, decanting the washings through the filter. Finally, quantitatively transfer the precipitate to the paper. Use a rubber policeman to remove the last traces of precipitate from the beaker. Place papers and the contents in porcelain crucibles that have been ignited to constant weight; gently char the papers (p. 716). Ignite to constant weight (p. 709). Report the percentage of sulfate in the sample.

---

[1] W. Rieman III and G. Hagen, *Anal. Chem.*, **14**, 150 (1942).

METHOD 1-4. DETERMINATION OF NICKEL IN STEEL

The nickel in a steel sample can be precipitated from a slightly alkaline medium with an alcoholic solution of dimethylglyoxime (p. 147). Tartaric acid is introduced to complex iron(III) and prevent its interference. After drying at 110°C, the organic nickel compound [$Ni(C_4H_7O_2N_2)_2$] serves as a convenient weighing form.

Owing to the bulky character of the precipitate, only a small quantity of nickel can be conveniently handled. The sample weight taken is governed by this consideration. The excess of precipitating agent must be controlled not only because its solubility in water is low but also because the nickel compound becomes appreciably more soluble as the alcohol content of the precipitating medium is increased.

Reagents needed.

> $6 M HCl$
> $6 M HNO_3$
> $6 M NH_3$
> $Concentrated NH_3$
> $15\% (w/v) tartaric acid; 30 mL/sample$
> $1\% (w/v) dimethylglyoxime in ethanol; 20 mL/sample$

Procedure. Dry three marked glass filtering crucibles to constant weight (p. 709) for at least 1 hr at 110°C (Note 1).

Weigh individual samples containing between 30 and 35 mg of Ni into 400-mL beakers, and dissolve by warming with about 50 mL of $6 M$ HCl. Carefully introduce about 15 mL of $6 M$ HNO_3, and boil gently to expel the oxides of nitrogen. Dilute the resulting solution to 200 mL, and heat nearly to boiling. Introduce about 30 mL of the 25% tartaric acid (Note 2), and neutralize with concentrated NH_3 until a faint odor of NH_3 can be detected in the vapors over the solution; add 1 to 2 mL in excess. If the solution is not clear at this stage, proceed as directed in Note 3. Make the solution slightly acidic with HCl (no odor of NH_3), heat to 60 to 80°C, and add 20 mL of a 1% (w/v) alcoholic solution of dimethylglyoxime. Then, with good stirring (Note 4), introduce sufficient $6 M$ NH_3 until a slight excess is present, as indicated by the odor, plus an additional 1 to 2 mL. Digest for 30 to 60 min at about 60°C, cool for at least 1 hr, and filter through medium porosity (M) filtering crucibles that have been previously brought to constant weight (p. 709). Wash with water until free of Cl^- (Note 5). Finally, bring crucibles and contents to constant weight by drying at 110 to 120°C. Report the %Ni in the sample. The dried precipitate has the composition $NiC_8H_{14}O_4N_4$.

2. If the tartaric acid solution is not clear, it should be filtered prior to use.
3. If a precipitate is formed upon the addition of base, the solution should be acidified, treated with additional tartaric acid, and again made alkaline. Alternatively, the precipitate can be removed by filtration. Thorough washing of the entire filter paper with a hot, dilute NH_3/NH_4Cl solution is required; the washings should be combined with the remainder of the sample.
4. Use a separate stirring rod for each sample and leave it in the solution throughout the determination.
5. Washings can be tested for Cl^- by collecting a few milliliters in a test tube, acidifying with HNO_3, and adding a few drops of 0.1 M $AgNO_3$. Washing is complete when little or no turbidity is observed.

2 PRECIPITATION TITRATIONS WITH SILVER NITRATE

Chapter 7 should be consulted for a discussion of volumetric precipitation methods in general and argentometric titrations in particular. Specific directions follow for the determination of chloride ion by the Fajans and the Mohr methods. The Fajans titration can be satisfactorily performed with silver nitrate solutions ranging from 0.1 to 0.01 M. The Mohr method gives best results with 0.1 M silver nitrate. The high cost of silver nitrate makes the use of more dilute solutions attractive.

PREPARATION 2-1. INDICATORS FOR ARGENTOMETRIC TITRATIONS

Procedure. *2',7'-dichlorofluorescein (Fajans Method).* Dissolve 0.1 g of dichlorofluorescein in 100 mL of 75% (v/v) ethanol/water solution.

   Potassium chromate (Mohr Method). Dissolve 5 g of potassium chromate in 100 mL of water.

PREPARATION 2-2. PREPARATION OF A STANDARD SILVER NITRATE SOLUTION

Discussion. Silver nitrate is obtainable in primary standard purity. It has a high equivalent weight and dissolves readily in water. The solid as well as its solutions must be scrupulously protected from organic matter and from sunlight; elemental silver is the product of reduction by the former and photodecomposition by the latter. *The reagent is expensive;* every effort should be made to avoid waste. Unused solutions should be collected rather than discarded; similarly, appreciable amounts of silver chloride should also be collected.

Procedure. Use a top-loading balance to transfer the approximate weight (Note 1) of $AgNO_3$ to a clean, dry weighing bottle. Dry at 110°C for 1 hr but not much longer (Note 2); cool to room temperature in a desiccator. Weigh (to the nearest 0.1 mg) the bottle and its contents. Quickly transfer the contents to a volumetric flask; reweigh the bottle and any residual solid. Dissolve the $AgNO_3$, dilute to the mark, and mix well (Note 3).

   NOTES
   1. Consult with the instructor concerning the volume and concentration of $AgNO_3$ to be used.

Concentration	Weight (g) of AgNO₃ to be Taken for		
Required, M	1000 mL	500 mL	250 mL
0.1	16.9	8.5	4.2
0.05	8.5	4.2	2.1
0.01	1.7	0.9	0.5

2. Silver nitrate is perceptibly decomposed by prolonged heating. Some discoloration may occur, even after 1 hr at 110°C; the effect on the purity is ordinarily negligible.
3. Silver nitrate solutions should be stored in a dark place when not actually in use.

METHOD 2-1. DETERMINATION OF CHLORIDE BY THE FAJANS METHOD

Discussion.   The Fajans method is described on page 188.

Reagents needed.

> Standard $AgNO_3$ solution
> Dichlorofluorescein solution (see Preparation 2-1)
> Dextrin

Procedure.   Dry the unknown at 110°C for 1 hr. Weigh samples into conical flasks and dissolve in an appropriate volume of distilled (or deionized) water (Note 1). Add about 0.1 g of dextrin and 5 drops of indicator. Titrate (Note 2) with $AgNO_3$ to the first permanent appearance of the pink color of the indicator.

> NOTES
> 1. Weigh (to the nearest 0.1 mg) 0.25-g samples if 0.1 M AgNO₃ is to be used and about half as much for 0.05 M reagent. Use about 200 mL of water for the former and 100 mL for the latter. For 0.01 M AgNO₃, it is recommended that a 0.4-g sample be weighed (to the nearest 0.1 mg) into a liter volumetric flask and that 50.0-mL aliquots be taken for titration.
> 2. Silver chloride is particularly sensitive to photodecomposition in the presence of the indicator; the titration will fail if attempted in direct sunlight. Where this problem exists, the approximate equivalence point should first be ascertained by a trial titration, this value being used to calculate the volume of AgNO₃ required for the other samples. The addition of indicator and dextrin should be delayed until the bulk of the silver nitrate has been added to subsequent samples, after which the titration should be completed without delay.

METHOD 2-2. DETERMINATION OF CHLORIDE BY THE MOHR METHOD

Discussion.   The Mohr method is described on page 185.

Reagents needed.

> Standard $AgNO_3$ solution
> 5% K_2CrO_4 indicator solution (Preparation 2-1)
> Sodium hydrogen carbonate
> Calcium carbonate

Procedure.   Dry the unknown for 1 hr at 110°C. Weigh (to the nearest 0.1 mg) 0.25-g samples into conical flasks and dissolve in about 100 mL of distilled (or deionized) water (Note 1). Add $NaHCO_3$, a pinch at a time, until effervescence ceases. Then introduce 1 to 2 mL of 5% K_2CrO_4 and titrate (Note 2) to the first permanent appearance of a buff color due to Ag_2CrO_4. Determine an indicator blank by suspending a small amount of $CaCO_3$ in about 100 mL of water containing the same volume of indicator as was used in the titration. Correct the titration data for the volume needed to develop the same buff color in the blank (Note 1).

NOTES
1. The Mohr method gives best results with 0.1 M $AgNO_3$ solutions. If 0.05 M reagent is used, the sample size should be scaled down accordingly. In addition, particular care is needed to establish a reliable value for the blank.
2. Mohr titrations should be performed at room temperature. Elevated temperatures significantly increase the solubility of Ag_2CrO_4; its sensitivity as an indicator for this titration undergoes a corresponding decrease.

3   NEUTRALIZATION TITRATIONS

Neutralization titrations are performed with standard solutions of strong acids or strong bases. In principle, a single solution (acid or base) should suffice; in practice, however, it is convenient to have both a standard acid and a standard base available to locate end points more exactly. One solution is standardized against a primary standard; the normality of the other is then found by determining the acid-base ratio (that is, the volume of acid required to neutralize 1.000 mL of base).

    The instructions that follow include methods for the standardization of both an acid or a base against appropriate primary standards. Directions for determining the acid-base ratio are also provided.

WATER FOR NEUTRALIZATION TITRATIONS

Atmospheric carbon dioxide is in equilibrium with aqueous carbonic acid; the concentration of the latter is about 1.5×10^{-5} M at ordinary temperatures. The quantities of acids and bases used in the procedures described in this section are generally sufficiently great that this small concentration of carbonic acid leads to no significant error. On the other hand, distilled water is sometimes supersaturated with the gas and thus contains sufficient carbonic acid to cause detectable errors. To test the water to be used in neutralization titrations, draw about 500 mL from the source, add 5 drops of phenolphthalein, and titrate with 0.1 N NaOH. Less than 0.2 to 0.3 mL of base should be needed to form the first faint pink color. If a larger volume is required, standard solutions should be prepared from water that has been boiled briefly to remove carbon dioxide and cooled to room temperature. Furthermore, the water used to dissolve and dilute the samples should also be boiled and cooled; as an alternative, corrections can be made by titrating blanks that are identical in composition except for the sample. The instructions that follow assume that dissolved CO_2 can be neglected.

PREPARATION 3-1. PREPARATION OF INDICATOR SOLUTIONS

Discussion. Acid-base indicators exist for virtually any pH range between 1 and 13.[2] The theory of indicator behavior is discussed in Section 8A-1. Directions follow

---

[2] See, for example, L. Meites, *Handbook of Analytical Chemistry*, p. 3-35. New York: McGraw-Hill, 1963.

for the preparation of indicator solutions that will permit the performance of most common analyses.

Procedure. Stock solutions of indicators generally contain 0.5 to 1.0 g of indicator per liter.

Methyl orange and methyl red. Dissolve the sodium salt directly in distilled water.

Phenolphthalein and thymolphthalein. Dissolve the solid indicator in a solution that is 80% ethyl alcohol (v/v).

Sulfonphthaleins. Dissolve the sulfonphthaleins in water by adding sufficient NaOH to react with the sulfonic acid group of the indicator. To prepare stock solutions, triturate 100 mg of the solid indicator with the specified volume of 0.1 N NaOH; dilute to 100 mL with distilled water. The volumes of 0.100 N base in milliliters required are as follows: *bromocresol green*, 1.45; *bromothymol blue*, 1.6; *bromophenol blue*, 1.5; *thymol blue*, 2.15; *cresol red*, 2.65; *phenol red*, 2.85.

The sodium salts of several sulfonphthaleins are available commercially. These substances can be dissolved directly in water.

PREPARATION 3-2. PREPARATION OF 0.1 *N* HYDROCHLORIC ACID

Procedure. Add about 8 mL of concentrated HCl to approximately 1 L of distilled or deionized water. Mix thoroughly, and store in a glass-stoppered bottle.

> NOTE
> For very dilute HCl (<0.05 N) solutions, it is advisable to eliminate CO_2 from the water by a preliminary boiling.

PREPARATION 3-3. PREPARATION OF CARBONATE-FREE 0.1 *N* SODIUM HYDROXIDE

Discussion. See Section 10A-3.

Procedure. If directed by the instructor, prepare a bottle for protected storage as in Figure 10-1, page 248 (Note 1). Transfer 1 L of distilled or deionized water (if the water contains significant amounts of H_2CO_3, it should be boiled and cooled; see p. 739) to the storage bottle. Decant 4 to 5 mL of commercial 50% NaOH (Note 2), add to the water, and *mix thoroughly*. Protect the solution from unnecessary contact with the atmosphere.

> NOTES
> 1. If the base is to be used for a few titrations that are to be completed in a 1- to 2-week period, the solution can be prepared and stored in a tightly capped polyethylene bottle. After each removal of base, squeeze the bottle while tightening the cap to minimize the air space above the base.
> 2. Be sure that the solid sodium carbonate in the 50% NaOH has settled to the bottom of the container so none is transferred during the decantation step. If necessary, the concentrated base can be filtered through a glass mat held in a Gooch crucible. The NaOH is collected in a small test tube in the filtering flask.

STANDARDIZATION 3-1. DETERMINATION OF THE ACID-BASE RATIO

Discussion. If both acid and base solutions have been prepared, it is convenient to determine their volume ratio before standardizing one of them.

Reagents needed.

> *0.1 N HCl*
> *0.1 N NaOH*
> *Phenolphthalein indicator solution*

Procedure. Add to each buret three or four small portions (3 to 5 mL) of the solution it is to contain; after each addition, rotate the buret so that the entire inner surface is contacted by the solution. Drain each portion of the rinse solution through the tip. Then, fill each buret, and remove the air bubble in the tip by opening the stopcock briefly. Cover the top of the buret that contains the base with a test tube. After allowing 30 s for drainage, record the initial buret readings. Slowly, deliver a 35- to 40-mL portion of the acid into a 250-mL conical flask. Touch the tip of the buret to the inside wall of the flask, and rinse down with a little distilled water. Add 2 drops of phenolphthalein (Note 1), and then introduce NaOH until the solution is definitely pink. Now, add HCl dropwise until the solution is again colorless, rinse the inside surface of the flask with water, and again add NaOH until the solution acquires a light pink hue that persists for 30 s (Note 2). Fractional drops can be delivered by forming them on the buret tip and touching the tip to the wall of the flask; the wall of the flask is then rinsed down with water. Record the final buret readings. Repeat the experiment, and calculate the volume ratio between the acid and the base (Note 3). Duplicate titrations should yield values that lie within 1 to 2 ppt of the mean.

> NOTES
> 1. The volume ratio can also be obtained with an indicator that has an acidic transition range, such as bromocresol green. With this particular indicator, the end point color has the faintest tinge of green. If appreciable carbonate is present, the ratios obtained with phenolphthalein will differ significantly from those with bromocresol green.
> 2. The end point slowly fades as CO_2 is absorbed from the atmosphere.
> 3. Use the acid-base ratio and the normality of the solution that is standardized to calculate the concentration of the other.

STANDARDIZATION 3-2 STANDARDIZATION OF HYDROCHLORIC ACID AGAINST SODIUM CARBONATE

Discussion. See Section 10A-2.

Reagents needed.

> *0.1 N HCl*
> *0.1 N NaOH (optional)*
> *Primary standard grade Na_2CO_3*
> *Bromocresol green indicator solution*
> *0.05 M NaCl, 100 mL*

Procedure. Dry a quantity of primary standard Na_2CO_3 for 2 hr at 110°C (Figure 30-6, p. 709), and cool in a desiccator. Weigh 0.2- to 0.25-g samples (to the nearest 0.1 mg) into 250-mL conical flasks, and dissolve in about 50 mL of distilled water. Introduce 3 drops of bromocresol green, and titrate until the solution just changes from blue to green. Boil the solution for 2 to 3 min, cool to room temperature (Note), and complete the titration. If a dilute base solution has also been prepared, the end point can be established with more certainty by adding small increments of the base and acid until a minimal volume of acid causes the color to change from blue to green. The volume of base used must, of course, be recorded; this volume is multiplied by the acid-base ratio and then subtracted from the total amount of acid to establish the volume needed for the titration of the Na_2CO_3.

Determine an indicator correction by titrating approximately 100 mL of 0.05 M NaCl and 3 drops of indicator. Subtract the volume of the blank from the titration data.

NOTE
The indicator should change from green to blue as a result of loss of CO_2 in the heating step. If it does not, an excess of acid was added originally. This excess can be back-titrated with base, provided its combining ratio with the acid has been established; otherwise the sample must be discarded.

STANDARDIZATION 3-3. STANDARDIZATION OF SODIUM HYDROXIDE AGAINST POTASSIUM HYDROGEN PHTHALATE

Discussion. See Section 10A-4.

Reagents needed.

> *0.1 N NaOH*
> *0.1 N HCl (optional)*
> *Primary standard grade potassium hydrogen phthalate*
> *Phenolphthalein indicator solution*

Procedure. Dry a quantity of primary standard potassium hydrogen phthalate for 2 hr at 110°C (Figure 30-6), and cool in a desiccator.

Weigh 0.7- to 0.9-g samples (to the nearest 0.2 mg) into 250-mL conical flasks, and dissolve in 50 to 75 mL of distilled (or deionized) water. Add 2 drops of phenolphthalein, and titrate with base until the pink color of the indicator persists for 30 s. If an HCl solution has also been prepared, the end point can be established with more certainty by adding small increments of acid and base until a minimal volume of NaOH causes development of the pink color. The volume of acid must, of course, be recorded; a correction to the volume of base can then be calculated from the acid-base ratio.

METHOD 3-1. DETERMINATION OF POTASSIUM HYDROGEN PHTHALATE

Discussion. The sample consists of potassium hydrogen phthalate, $KHC_8H_4O_4$, mixed with neutral, soluble salts.

Reagents needed.

Standardized 0.1 N NaOH
Standardized 0.1 N HCl (optional)
Phenolphthalein indicator solution

Procedure. Dry the unknown for 2 hr at 110°C, and cool in a desiccator. Weigh samples that will require 30- to 40-mL titrations (see instructor) into 250-mL conical flasks, and dissolve in 50 to 75 mL of distilled water. Introduce 2 drops of phenolphthalein, and titrate with standard 0.1 N NaOH. Consider the first pink that persists for 30 s as the end point. A standard acid solution, if available, may be employed to establish the end point more exactly. Calculate the % $KHC_8H_4O_4$ in the sample.

METHOD 3-2. DETERMINATION OF THE EQUIVALENT WEIGHT OF A WEAK ACID

Discussion. The equivalent weight of an acid is an aid in establishing its identity (p. 254); its value is readily determined by titrating a weighed quantity of the purified acid with standard sodium hydroxide.

Reagents needed.

Standardized 0.1 N NaOH
Standardized 0.1 N HCl (optional)
Phenolphthalein indicator solution

Procedure. Weigh 0.3-g samples (to the nearest 0.1 mg) of the purified acid into 250-mL conical flasks, and dissolve in 50 to 75 mL of distilled water (Note). Add 2 drops of phenolphthalein, and titrate with standard base to the first persistent pink color. Calculate the equivalent weight of the acid, assuming 100.0% purity for the sample.

> NOTE
> Acids that have limited solubility in water can be dissolved in ethanol or an ethanol/water mixture. As received, the alcohol may be measurably acidic and should be rendered faintly alkaline to phenolphthalein before being used as a solvent. As an alternative, a sparingly soluble acid can be dissolved in a known volume of standard base, with the excess being determined by back-titration with standard acid.

METHOD 3-3. DETERMINATION OF THE ACID CONTENT OF VINEGAR

Discussion. The total acid content of vinegar is conveniently determined by titration with standard base. Even though other acids are also present, the results of the analysis are customarily reported in terms of acetic acid, the principal acidic constituent. Vinegars contain approximately 5% acid (w/v), expressed as acetic acid.

Reagents needed.

Standardized 0.1 N NaOH
Standardized 0.1 N HCl (optional)
Phenolphthalein indicator solution

Procedure. Pipet 25 mL of vinegar into a 250-mL volumetric flask, and dilute to the mark with distilled water. Mix thoroughly, and pipet 50-mL aliquots into 250-mL flasks. Add 50 mL of water and 2 drops of phenolphthalein, and titrate with standard 0.1 N NaOH to the first permanent pink color.

Calculate the total acidity as g CH_3COOH per 100 mL of sample.

METHOD 3-4. DETERMINATION OF SODIUM CARBONATE IN AN IMPURE SAMPLE

Discussion. The titration of sodium carbonate is discussed in connection with its use as a primary standard (Section 10A-2). This analysis is conveniently performed concurrently with the acid standardization.

Reagents needed.

Standardized 0.1 N HCl
Standardized 0.1 N NaOH (optional)
Bromocresol green indicator solution

Procedure. Dry the unknown for 2 hr at 110°C, and then cool in a desiccator. Weigh samples that will require 30- to 40-mL titrations (see instructor) into 250-mL flasks. Dissolve in 50 to 75 mL of water, and add 2 drops of bromocresol green. Titrate with standard acid until the indicator just begins to turn green. Boil the solution for 2 to 3 min, cool, and complete the titration. If additional acid is not required after boiling, either discard the sample or back-titrate the excess acid with standard base.

Calculate the %Na_2CO_3 in the sample.

METHOD 3-5. DETERMINATION OF SODIUM CARBONATE AND SODIUM HYDROGEN CARBONATE IN A MIXTURE

Discussion. See pages 252 to 254.

Reagents needed.

Standardized 0.1 N HCl
Standardized 0.1 N NaOH
Bromocresol green indicator solution
Phenolphthalein indicator solution
10% (w/v) BaCl$_2$; 10 mL/sample or blank

Procedure. If the sample is a solid, weigh dried portions into 250-mL volumetric flasks, dissolve in 50 to 75 mL of water, and dilute to the mark. Mix well.

If the sample is a solution, pipet suitable aliquots into a 250-mL volumetric flask, and dilute to volume. Mix thoroughly.

To determine the total number of milliequivalents of the two components, transfer 25-mL aliquots to 250-mL conical flasks, and titrate with standard 0.1 N HCl, following the directions given in Method 3-4.

To determine the bicarbonate content, pipet additional 25-mL portions into flasks; *treat each aliquot individually from here.* Pipet 50 mL of standard 0.1 N NaOH into an aliquot. Immediately, add 10 mL of the $BaCl_2$ and 2 drops of phenolphthalein. Titrate the excess NaOH at once with standard 0.1 N HCl to the disappearance of the pink color.

Titrate a blank consisting of 25 mL of water, 10 mL of the $BaCl_2$ solution, and *exactly* the same volume of NaOH as used with the samples. The difference between the volume of HCl needed for the blank and the sample corresponds to the $NaHCO_3$ present.

Calculate the $\%NaHCO_3$ and Na_2CO_3 in the sample.

METHOD 3-6. DETERMINATION OF AMMONIUM ION

Discussion. A method for the isolation of ammonium ion as ammonia by distillation is discussed on page 249 in connection with the Kjeldahl method.

Reagents needed.

> *Standardized 0.1 N HCl*
> *Standardized 0.1 N NaOH or 4% (w/v) H_3BO_3*
> *Bromocresol green indicator solution*
> *Concentrated NaOH; 45 g NaOH in 75 mL H_2O for each sample*
>   Cool to room temperature before use. (*Caution!* this solution
>   is highly corrosive; if it comes in contact with the skin, wash
>   the area immediately with copious amounts of water.)
> *Granulated zinc*
> *Litmus paper*

Procedure. Transfer the sample, which should contain 2 to 4 meq of ammonium salt, to a 500-mL Kjeldahl flask, and add enough water to give a total volume of about 200 mL.

Arrange a distillation apparatus similar to those shown in Figure 10-2. Use a buret or pipet to measure precisely 50.00 mL of standard 0.1 N HCl into the receiver flask (Note 1). Clamp the flask so that the tip of the adapter extends just below the surface of the standard acid. Circulate water through the jacket of the condenser.

With the Kjeldahl flask tilted, slowly pour about 85 mL of the concentrated NaOH down the side of the container to minimize mixing with the solution in the flask (Note 2). Add several pieces of granulated Zn (Note 3) and a small piece of litmus paper. *Immediately* connect the flask to the spray trap. Very *cautiously* mix the solution by gentle swirling. After mixing is complete, the litmus paper should indicate that the solution is basic.

Immediately bring the solution to a boil, and distill at a steady rate until one-half to one-third of the original solution remains. Control the rate of heating during this period to prevent the receiver acid from being drawn back into the distillation flask. After the distillation is judged complete, lower the receiver flask until the tip of the adapter is well clear of the standard acid. Then discontinue heating, disconnect the apparatus, and rinse the inside of the condenser with small portions of water. Disconnect the adapter, and rinse it thoroughly. Add 2 drops of bromocresol green and titrate the residual HCl with standard 0.1 N NaOH to the color change of the indicator.

Calculate the %$(NH_4)_2SO_4$ in the sample.

NOTES
1. A modification of this procedure makes use of about 50 mL of 4% boric acid solution in place of the standard HCl in the receiver flask (p. 251). The distillation is then carried out as before. When complete, the ammonium borate produced is titrated with a standard 0.1 N HCl solution, and 2 to 3 drops of bromocresol green are used as indicator.
2. The dense caustic solution should form a second layer on the bottom of the flask. Mixing must be avoided at this point in order to prevent loss of the volatile NH_3.
3. Granulated Zn (10 to 20 mesh) is added to minimize bumping during the distillation. It reacts slowly with the alkali to give small bubbles of H_2 that prevent superheating of the liquid.

METHOD 3-7. DETERMINATION OF NITROGEN IN AN ORGANIC MIXTURE; THE KJELDAHL METHOD

Discussion. These directions are suitable for the analysis of protein in samples such as blood meal, wheat flour, macaroni, dry cereals, and pet foods. Prereduction is not required for such materials. A simple modification will permit analysis of samples that contain more highly oxidized forms of nitrogen.[3] The Kjeldahl analysis is discussed on page 249.

Reagents needed.

> *Standardized 0.1 N HCl*
> *Standardized 0.1 N NaOH or 4% (w/v) H_3BO_3*
> *Bromocresol green indicator solution*
> *Concentrated H_2SO_4*
> *Concentrated NaOH; 45 g NaOH in 75 mL H_2O for each sample*
> > *Cool before use; see Method 3-6.*
> *K_2SO_4*
> *Catalyst (Note 2)*

Procedure. Weigh three samples (see instructor for size) onto individual 9-cm filter papers (Note 1); after folding the paper around the samples, drop each into a 500-mL Kjeldahl flask (the paper wrapping will prevent the sample from clinging to the neck of the flask). Add 25 mL of concentrated H_2SO_4, 10 g of powdered K_2SO_4, and the catalyst (Note 2); clamp the flask in an inclined position in a hood. Heat the mixture carefully until the H_2SO_4 is boiling; discontinue heating briefly if

[3] See: *Official Methods of Analysis*, 12th ed., p. 16. Washington, D.C.: Association of Official Analytical Chemists, 1975.

foaming becomes excessive; never allow the foam to reach the neck of the flask. After the foaming ceases and the acid is boiling vigorously, the samples can be left unattended while the distillation apparatus is prepared. Continue heating until the solution becomes colorless or light yellow; as long as 2 to 3 hr may be required. If necessary, *cautiously* replace the acid lost by evaporation.

Remove the flame, and allow the flask to cool to room temperature; swirl the liquid if it begins to solidify. Cautiously dilute with 250 mL of water, and cool to room temperature under a water tap. If Hg was used as the catalyst, introduce 25 mL of a 4% Na_2S solution. Then, complete the analysis as described beginning in the second paragraph of the procedure for Method 3-6.

Calculate the percent protein in the sample (Note 3).

NOTES
1. For the most accurate results, a blank, which includes the filter paper, should be carried through all steps of the analysis. Acid-washed filter paper should not be used because it often is contaminated with ammonium salts.
2. As a catalyst, one may use any of the following: a drop of mercury, 0.5 g of HgO, a crystal of $CuSO_4$, 0.1 g of Se, or 0.2 g $CuSeO_3$. Alternatively, the catalyst may be omitted.
3. For samples of proteinaceous materials, the percent protein is obtained by multiplying the percent nitrogen by a factor that depends upon the nature of the sample. The factor is 5.7 for cereals, 6.25 for meats, and 6.38 for dairy products.

4 NEUTRALIZATION TITRATIONS IN GLACIAL ACETIC ACID

It was pointed out in Chapter 11 that the substitution of a nonaqueous solvent system for water often permits the determination of species that are insufficiently strong as acids or bases for titration in an aqueous environment. The experiments that follow illustrate the use of standard perchloric acid solutions in glacial acetic acid. Further information about these titrations is found on page 270.

Dilute solutions of perchloric acid in glacial acetic acid have been used safely for many years. *Under no circumstance, however, should these solutions be heated or allowed to evaporate for extended periods; they should be diluted with water and disposed of promptly after use.*

It should also be noted that glacial acetic acid is a skin irritant; should any of the reagent come in contact with skin, the affected area should be washed with copious amounts of water immediately.

As noted on page 270, the coefficient of expansion of glacial acetic acid is 0.11% per °C. Thus, the temperature of the reagents should be determined at the time of standardization; Equation 11-12 (p. 270) can then be employed to correct the titrant volumes for the analysis to the temperature at the time of standardization.

PREPARATION 4-1. PREPARATION OF 0.1 *N* HClO$_4$ IN GLACIAL ACETIC ACID

Procedure. Mix 4.3 mL of 72% perchloric acid with about 150 mL of glacial acetic acid, and add 10 mL of acetic anhydride. Dilute to approximately 500 mL with acetic acid. Allow the solution to stand overnight or longer before use to permit the reaction between acetic anhydride and water (introduced with the perchloric acid) to become complete.

PREPARATION 4-2. PREPARATION OF 0.1 N SODIUM ACETATE IN GLACIAL ACETIC ACID

Procedure. Dissolve 4.1 g anhydrous $NaC_2H_3O_2$ in glacial acetic acid, and dilute to about 500 mL with the acid.

PREPARATION 4-3. INDICATOR

Procedure. Dissolve 0.2 g of methyl violet in 100 mL of glacial acetic acid or chlorobenzene.

STANDARDIZATION 4-1. DETERMINATION OF THE ACID-BASE RATIO

Reagents needed.

> *0.1 N $HClO_4$ in acetic acid*
> *0.1 N $NaC_2H_3O_2$ in acetic acid*
> *Methyl violet indicator solution*

Procedure. Fill burets with the $HClO_4$ and the $NaC_2H_3O_2$ solutions, and measure about 30 mL of the former into a flask. Add 2 drops of the methyl violet indicator, and titrate with the $NaC_2H_3O_2$ solution to a faint violet color. Practice the end point determination by making further small additions of acid and base to the solution.

Repeat the determination with a fresh portion of the $HClO_4$ solution. Calculate the volume ratio of acid to base.

STANDARDIZATION 4-2. STANDARDIZATION OF THE PERCHLORIC ACID REAGENT

Reagents needed.

> *Primary standard grade potassium hydrogen phthalate*
> *0.1 N $HClO_4$ in acetic acid*
> *0.1 N $NaC_2H_3O_2$ in acetic acid*
> *Methyl violet indicator solution*
> *Glacial acetic acid*

Procedure. Dry potassium hydrogen phthalate, $KHC_8H_4O_4$, for 1 to 2 hr at 110°C and cool. Weigh 0.5- to 0.6-samples (to the nearest 0.2 mg) into flasks, and add 60 mL of glacial acetic acid to each. Cautiously heat the mixtures until all solid is dissolved; then cool to room temperature.

Measure and record the temperature of the perchloric acid reagent. Then fill burets with the $HClO_4$ and the $NaC_2H_3O_2$ solutions. Add 2 drops of methyl violet indicator to each of the samples, and titrate to the same end point as in the determination of the acid-base ratio; make use of both the acid and base solutions to establish the end point exactly.

Calculate the normality of the acid and the base.

METHOD 4-1. ANALYSIS OF AMINES BY TITRATION IN AN ANHYDROUS MEDIUM

Reagents needed.

> *Standardized HClO$_4$ in acetic acid*
> *Standardized NaC$_2$H$_3$O$_2$ in acetic acid*
> *Methyl violet indicator solution*

Procedure. Dissolve a sample containing 2 to 4 meq of the amine in 50 mL of glacial acetic acid. Measure and record the temperature of the standard acid and base. Add 2 drops of methyl violet to the sample, and titrate to the disappearance of the violet color. Back-titrate with standard NaC$_2$H$_3$O$_2$ until the first faint violet color reappears.

Correct the volumes of the two standard solutions to the standardization temperature with the aid of Equation 11-12. Then, calculate the number of milliequivalents of amine in the sample.

METHOD 4-2. DETERMINATION OF AMINO ACIDS BY TITRATION IN AN ANHYDROUS MEDIUM

The behavior of amino acids in aqueous solution was considered in Section 9C-3. When acetic acid is substituted as solvent, the acid function of these solutes is completely repressed; they behave simply as weak bases. Owing to their limited solubility in acetic acid, amino acid samples are ordinarily dissolved in a known excess of standard HClO$_4$, and the excess is determined by back-titration with standard NaC$_2$H$_3$O$_2$.

Reagents needed.

> *Standardized HClO$_4$ in acetic acid*
> *Standardized NaC$_2$H$_3$O$_2$ in acetic acid*
> *Methyl violet indicator solution*

Procedure. Weigh samples containing 2 to 4 meq of the amino acid into flasks. Measure and record the temperature of the standard HClO$_4$ and NaC$_2$H$_3$O$_2$ solutions; then, pipet exactly 50.00 mL of the HClO$_4$ into each sample. After solution is complete, add 2 drops of the methyl violet indicator, and titrate with the standard NaC$_2$H$_3$O$_2$ to the first violet color. Correct the data to the standardization temperature, and calculate the number of equivalents of amino acid present per gram of sample.

5A COMPLEX FORMATION TITRATIONS WITH EDTA

PREPARATION 5A-1. INDICATORS

Procedure. *Eriochrome black T (Erio T)*. Dissolve 100 mg of the solid in a solution consisting of 15 mL of triethanolamine and 5 mL of absolute ethanol. The solution should be freshly prepared every 2 weeks; refrigeration slows its deterioration.

  Calmagite.   Prepare a 0.1% (w/v) aqueous solution of the indicator.

PREPARATION 5A-2. DIRECT PREPARATION OF STANDARD 0.01 *M* EDTA

Procedure.   Dry the purified dihydrate ($Na_2H_2Y\cdot2H_2O$) at 80°C to remove superficial moisture (Notes 1 to 4). After cooling, weigh about 3.8 g (to the nearest milligram) into a 1-L volumetric flask, using a powder funnel. Rinse the funnel well before removing from the flask. Add 600 to 800 mL of water and swirl periodically until the EDTA has dissolved. The rate of dissolution is slow and 15 min or longer may be required to complete the process. Examine the bottom of the flask for undissolved salt; if any remains, continue the periodic shaking until solution is complete. Dilute to the mark and mix well.

> NOTES
> 1. W. J. Blaedel and H. T. Knight (*Anal. Chem.*, **26**, 741 [1954]) give specific instructions for the purification of commercial preparations of the disodium salt.
> 2. The direct preparation of standard solutions requires the total exclusion of polyvalent cations. If any doubt exists regarding the quality of the distilled or deionized water, pretreatment by passing it through a cation-exchange resin is recommended.
> 3. If desired, the anhydrous salt can be employed instead of the dihydrate. The weight taken should be adjusted accordingly.
> 4. If desired, an EDTA solution that is approximately 0.01 *M* can be prepared and standardized against a Mg^{2+} solution of known strength or against primary standard $CaCO_3$. For the latter, the procedure described in Method 5A-2 is followed.

PREPARATION 5A-3. PREPARATION OF 0.10 *M* MAGNESIUM COMPLEX OF EDTA

Discussion.   A solution of the magnesium complex of EDTA is useful for the determination of cations which form complexes that are more stable than the magnesium complex but for which no indicator is available. Here, the magnesium is displaced by part of the analyte cations. The remaining uncomplexed analyte and the liberated magnesium are then titrated with EDTA, Eriochrome black T, or Calmagite being used as the indicator. Note that the concentration of the magnesium solution is not important; it is only necessary that the magnesium and EDTA be present in an exact 1 to 1 ratio.

  Method 5A-2 requires a solution of the magnesium/EDTA complex.

Procedure.   To 3.722 g of $Na_2H_2Y\cdot2H_2O$ in 50 mL of distilled water add an equivalent quantity (2.465 g) of $MgSO_4\cdot7H_2O$. Introduce a few drops of phenolphthalein, followed by sufficient sodium hydroxide to turn the solution faintly pink. Dilute the solution to about 100 mL. When properly prepared, portions of this solution should assume a dull violet color when treated with pH-10 buffer and a few drops of Eriochrome black T (Erio T) or Calmagite indicator. Furthermore, a single drop of 0.01 *M* Na_2H_2Y should cause a color change to blue, whereas an equal quantity of 0.01 *M* Mg^{2+} should cause a change to red. If necessary, the composition of the solution can be adjusted by adding Mg^{2+} or Na_2H_2Y until these criteria are met.

PREPARATION 5A-4. PREPARATION OF A pH-10 BUFFER

Procedure. Dilute 570 mL of concentrated NH_3 and 70 g NH_4Cl to about 1 L.

METHOD 5A-1. DETERMINATION OF MAGNESIUM BY DIRECT TITRATION WITH EDTA

Reagents needed.

> *Standard 0.01 M EDTA*
> *pH-10 buffer*
> *Eriochrome black T indicator solution*

Procedure. The sample will be issued as an aqueous solution; transfer it to a clean 500-mL volumetric flask, and dilute to the mark. Mix thoroughly. Take 50.00-mL aliquots, and add 1 to 2 mL of pH-10 buffer and 4 to 5 drops of Erio T or Calmagite indicator solution to each. Titrate with 0.01 M Na_2H_2Y to a color change from red to pure blue. Express the results of the analysis in terms of milligrams of Mg^{2+} per liter of the diluted solution.

> NOTES
> 1. The color change of the indicator is slow in the vicinity of the end point. Care must be taken to avoid overtitration.
> 2. Other alkaline earths, if present, will also be titrated along with magnesium and should be removed prior to the analysis; $(NH_4)_2CO_3$ is a suitable reagent. Most polyvalent cations also interfere and should be precipitated as hydroxides.

METHOD 5A-2. COMPLEXOMETRIC DETERMINATION OF CALCIUM BY DISPLACEMENT TITRATION

Discussion. See the discussion under Preparation 5A-3.

Reagents needed.

> *Standard 0.01 M EDTA*
> *0.1 M Mg/EDTA solution*
> *6 N HCl*
> *6 N NaOH*
> *Methyl red indicator solution*
> *Eriochrome black T indicator solution*
> *pH-10 buffer*

Procedure. Weigh the sample (Note 1) into a 500-mL beaker. Cover with a watch glass, and carefully introduce 5 to 10 mL of 6 N HCl. After the sample has dissolved, add 50 mL of distilled or deionized water and boil gently for a few minutes to remove CO_2. Cool the solution, and neutralize with 6 N NaOH (Note 2). Transfer quantitatively to a 500-mL volumetric flask, and dilute to the mark. Take 50.00-mL aliquots for titration, treating each as follows: add approximately 2 mL of pH-10 buffer, 1 mL of the Mg/EDTA solution, and 4 to 5 drops of Erio T or Calmagite indicator. Titrate with standard 0.01 M Na_2H_2Y until a color change from red to blue occurs. Report the percentage of CaO in the sample.

NOTES
1. The sample taken should contain about 150 to 160 mg of Ca^{2+}.
2. To neutralize the solution, introduce a few drops of methyl red, and add base until the red color is discharged.
3. Interferences with this method are substantially the same as with the direct determination of magnesium and are eliminated in the same way.

METHOD 5A-3. COMPLEXOMETRIC DETERMINATION OF CALCIUM BY BACK-TITRATION

Discussion. This method is sometimes employed in lieu of Method 5A-2. Here an excess of standard EDTA is introduced. The excess is then back-titrated with a $MgSO_4$ solution that has been standardized against the EDTA solution. Again, this method takes advantage of the Erio T end point with Mg^{2+}.

Reagents needed.

> Standard 0.01 M EDTA
> 0.01 M $MgSO_4$; 2.5 g $MgSO_4 \cdot 7H_2O$ in 1 L of H_2O
> 6 N HCl
> 6 N NaOH
> Methyl red indicator solution
> Eriochrome black T indicator solution
> pH-10 buffer

Procedure. Standardize the $MgSO_4$ solution by titration of 25.0-mL aliquots with 0.01 M EDTA following Method 5A-1.

Prepare the sample as directed for the displacement titration of calcium. To each 50.00-mL aliquot add about 2 mL of pH-10 buffer and 4 to 5 drops of Erio T or Calmagite indicator. Run in an excess of 0.01 M Na_2H_2Y solution from a buret, and record the volume taken. Titrate the excess chelating agent with standard 0.01 M $MgSO_4$ solution until a color change from blue to red occurs.

METHOD 5A-4. DETERMINATION OF WATER HARDNESS BY TITRATION WITH EDTA

Discussion. See page 300.

Reagents needed.

> Standard 0.01 M EDTA
> 6 N HCl
> 6 N NaOH
> Methyl red indicator solution
> Eriochrome black T indicator solution
> pH-10 buffer solution

Procedure. Acidify 100-mL aliquots with a few drops of HCl, and boil gently for a few minutes to remove CO_2. Cool, add a few drops of methyl red, and neutralize the solution with NaOH. Introduce 2 mL of pH-10 buffer and 4 to 5 drops of Erio T or Calmagite indicator, and titrate with standard 0.01 M Na_2H_2Y until the color changes from red to pure blue (Note). Report the results of the analysis in terms of milligrams of $CaCO_3$/L of water.

NOTE

If the color change of the indicator is sluggish, the absence of magnesium is indicated. In this event add 1 to 2 mL of the 0.1 M MgY^{2-} solution before the titration.

METHOD 5A-5. ANION-EXCHANGE SEPARATION OF NICKEL, COBALT, AND ZINC

Discussion. In this experiment, nickel, cobalt, and zinc are separated on an anion-exchange column. The cobalt and zinc ions are subsequently titrated with standard EDTA. The nickel is not determined.

In 9 M hydrochloric acid, the principal forms of the three metal ions are $CoCl_4^{2-}$, $ZnCl_4^{2-}$, and $NiCl^+$. The first two are retained on an anion-exchange resin, whereas the positively charged nickel species is not. After the nickel passes through the column, cobalt can be eluted with 4 M HCl; at this chloride ion concentration, a positively charged cobalt complex forms, which is then free to pass through the column. Finally, elutions with water convert the $ZnCl_4^{2-}$ into a positively charged species.

The cobalt solution is freed of most of the hydrochloric acid by evaporation; it is then buffered to a pH of 5.5 to 6.0 and titrated with EDTA, employing the commercially available indicator Napthyl Azoxine S (NAS).

The zinc-containing eluate is buffered to a pH of 10 and titrated with EDTA, using Eriochrome black T for end point detection.

Reagents needed.

> *Standard 0.01 M EDTA* (see Preparation 5A-2)
> *Eriochrome black T indicator* (see Preparation 5A-1)
> *NAS indicator.* Suspend 0.5 g of Napthyl Azoxine S (Eastman Kodak Company Catalogue No. 8643) in 50 mL of water. Add enough NH_3 to dissolve the indicator.
> *9 M HCl*
> *4 M HCl*
> *pH-10.0 buffer* (see Preparation 5A-4)
> *Hexamethylenetetramine*
> *Dowex 1-X8 anion-exchange resin* (50-100 mesh)

Procedure. Prepare a column by inserting a plug of glass wool into the bottom of a 50-mL buret. Add a slurry of Dowex 1-X8 resin in 9 M HCl to the column until the depth of the resin is 8 to 10 cm. *Do not allow the liquid level to drop below the resin level;* keep about 2 cm of 9 M HCl above the resin.

Wash the column with two 5-mL portions of 9 M HCl using a flow rate of 2 to 3 mL/min. Leave 1 cm of HCl above the resin level. Pipet exactly 2 mL of the sample onto the column and elute the Ni with about 75 mL of 9 M HCl; use 15-mL portions and a flow rate of 2 to 3 mL/min. The yellow-green $NiCl^+$ complex will flow through the column, leaving a blue Co band. Discard the Ni solution.

Replace the collection container with a clean 250-mL beaker and elute the Co with five 10-mL portions of 4 M HCl using a flow rate of 2 to 3 mL/min. Evaporate the eluate to near dryness (<5 mL) in the hood. Place a stirring rod in the beaker to avoid bumping.

During the evaporation step, elute the Zn by passing 50 to 100 mL of distilled water through the column in 10- to 15-mL portions at a flow rate of 3 to 4 mL/min.

To determine the Co, dilute the cooled sample with 100 mL of water and add sufficient solid hexamethylenetetramine to give a pH of 5.5 to 6.0. Add three drops of NAS indicator and titrate with the standard EDTA solution to a color change from orange-yellow to red. Calculate the milligrams of Co per milliliter of sample.

To determine the Zn, add 10 to 20 mL of the pH-10 buffer and 1 to 2 drops of Eriochrome black T indicator. Titrate with the standard EDTA to a color change from red to blue. Report the milligrams of Zn per milliliter of sample.

5B COMPLEX FORMATION TITRATIONS WITH MERCURY(II)

Chloride ion can be titrated with mercury(II):

$$Hg^{2+} + 2Cl^- \rightleftharpoons HgCl_2$$

Mercury(II) chloride, the product, is soluble and only slightly dissociated. Diphenylcarbazide serves as indicator for the titration, with the first permanent color of the complex that forms between its conjugate base and mercury(II) being taken as the end point.

Considerable attention must be paid to the acidity of the solution. At very low pH values, the concentration of the indicator species decreases to the point where sensitivity is lost. The opposite effect occurs in solutions with high pH values; here, the concentration of In^- is sufficiently large to generate color with mercury(II) derived from $HgCl_2$, the product of the titration; low results are thus obtained. The optimum nitric acid concentration appears to be about 0.02 M.

PREPARATION 5B-1. PREPARATION OF AN APPROXIMATELY 0.02 M NITRIC ACID SOLUTION

Discussion. It is desirable that the standard mercury(II) solution be 0.02 M in nitric acid. Furthermore, all standardizations and determinations are performed in this medium.

Procedure. Dissolve 2.5 mL of concentrated HNO_3 in 2.0 L of distilled or deionized water.

PREPARATION 5B-2. PREPARATION OF 0.1 M MERCURY(II) NITRATE SOLUTION

Reagents needed.

$Hg(NO_3)_2 \cdot H_2O$
0.02 M HNO_3

Procedure. Dissolve about 16 g of $Hg(NO_3)_2 \cdot H_2O$ in 200 mL of 0.02 M HNO_3. Filter the solution if it is turbid; then dilute to about 1 L with 0.02 M HNO_3.

PREPARATION 5B-3. PREPARATION OF DIPHENYLCARBAZIDE INDICATOR

Procedure. Dilute 35 mL of ethanol to about 50 mL with water and saturate with diphenylcarbazide. This solution will acquire a red color on standing, which does not decrease its effectiveness as an indicator.

STANDARDIZATION 5B-1. STANDARDIZATION OF MERCURY(II) NITRATE SOLUTION

Reagents needed.

> *Approximately 0.1 M Hg(NO₃)₂ solution in 0.02 M HNO₃*
> *Primary standard NaCl*
> *0.02 M HNO₃, about 100 mL/standardization*
> *Diphenylcarbazide indicator solution*

Procedure. Dry a quantity of standard NaCl for about 1 hr at 110°C, and cool in a desiccator. Weigh 0.15- to 0.20-g samples (to the nearest 0.1 mg) into 250-mL conical flasks and dissolve in 100 mL of 0.02 M HNO_3. Add 5 drops of diphenylcarbazide indicator and titrate with $Hg(NO_3)_2$ to the first permanent pink color of the solution (Note). Calculate the molarity of the solution.

> NOTE
> Users of this titration are divided over the matter of the color change to be used for the end point. The first appearance of color, which is recommended here, is reproducible and readily observed; it occurs a few drops short of the equivalence point, at which the change from violet to blue-violet is difficult to judge. No error is incurred as long as the same color change is used for both standardization and titration.

METHOD 5B-1. DETERMINATION OF CHLORIDE

Reagents needed.

> *Standardized 0.1 M Hg(NO₃)₂ solution in 0.02 M HNO₃*
> *0.02 M HNO₃, about 100 mL/sample*
> *Diphenylcarbazide indicator solution*

Procedure. Dry the unknown for 1 hr at 110°C, and permit it to return to room temperature in a desiccator. Weigh (to the nearest 0.1 mg) 0.4- to 0.5-g samples into 250-mL conical flasks. Dissolve in 100 mL of 0.02 M HNO_3. Add 5 drops of indicator, and titrate to the first permanent pink color of the solution. Calculate the %Cl in the sample.

6 TITRATIONS WITH POTASSIUM PERMANGANATE

This section contains methods for the analysis of iron and calcium by titration with a standard solution of potassium permanganate.

PREPARATION 6-1. PREPARATION OF APPROXIMATELY 0.1 N POTASSIUM PERMANGANATE

Discussion. The precautions that must be observed in the preparation and storage of permanganate solutions are discussed on pages 366 to 370.

Procedure. Dissolve 3.2 g of KMnO$_4$ in about 1 L of distilled water. Heat to boiling, and keep hot for about 1 hr. Cover, and let stand overnight. To remove the MnO$_2$, filter the solution through a fine-porosity sintered glass funnel or crucible (Note 1) or through a Gooch crucible with a glass mat (p. 710). Store the solution in a clean, glass-stoppered bottle, and keep in the dark when not in use.

> NOTES
> 1. After filtration, the MnO$_2$ collected on the fritted plate can be removed with dilute H$_2$SO$_4$ containing a few milliliters of 3% H$_2$O$_2$ followed by rinsing with copious amounts of water.
> 2. If the standardization and the analysis of the unknown are performed on the same day, the heating and filtering steps can be omitted.

STANDARDIZATION 6-1. STANDARDIZATION OF POTASSIUM PERMANGANATE AGAINST SODIUM OXALATE

Discussion. Directions are provided for standardization by the McBride method and the method of Fowler and Bright; see page 369 for a discussion of these methods as well as other primary standards for permanganate solutions.

Reagents needed.

> *0.1 N KMnO$_4$*
> *Primary standard Na$_2$C$_2$O$_4$*
> *0.9 M H$_2$SO$_4$; 250 mL/standardization or blank*

Procedure. Dry primary standard grade Na$_2$C$_2$O$_4$ for at least 1 hr at 110 to 120°C. Cool in a desiccator, and weigh 0.2- to 0.3-g samples (to the nearest 0.1 mg) of the dried Na$_2$C$_2$O$_4$ into 400-mL beakers; dissolve in approximately 250 mL of 0.90 M H$_2$SO$_4$ by stirring with a thermometer.

Method of McBride. Heat to 80 to 90°C, and titrate with the KMnO$_4$, stirring vigorously with a thermometer. The reagent should be introduced slowly so that the pink color is discharged before further additions are made (Notes 1 and 2). If the solution temperature drops below 60°C, reheat. The end point is the first persistent pink color (Notes 3 and 4). Determine an end point blank by titrating an equal volume of the water and sulfuric acid.

Method of Fowler and Bright. Introduce from a buret sufficient permanganate to consume 90 to 95% of the oxalate (about 40 mL of 0.1 N KMnO$_4$ for a 0.3-g sample; a preliminary titration by the McBride method will provide the approximate volume required). Let stand until the solution is decolorized. Warm to 55 to 60°C, and complete the titration, taking the first pale pink color that persists for 30 s as the end point. Determine an end point correction by titrating 250 mL of 0.90 M sulfuric acid at this same temperature. Correct for the blank, and calculate the normality.

NOTES
1. Any $KMnO_4$ spattered on the sides of the titration vessel should be washed down immediately with a stream of water.
2. If the addition of $KMnO_4$ is too rapid, some MnO_2 will be produced in addition to Mn^{2+}; evidence for MnO_2 formation is a faint brown discoloration of the solution. The presence of the precipitate is not a serious problem so long as sufficient oxalate remains to reduce the MnO_2 to Mn^{2+}; the titration is temporarily discontinued until the solution clears. The solution must be free of MnO_2 at the equivalence point.
3. To measure the volume of $KMnO_4$ take the surface of the liquid as a point of reference. Alternatively, provide sufficient backlighting with a flashlight or match to permit reading of the meniscus in the conventional manner.
4. Permanganate solutions should not be allowed to stand in burets any longer than necessary because decomposition to MnO_2 may occur. Freshly formed MnO_2 can be removed from burets and glassware with a solution of $1\ M\ H_2SO_4$ containing a small amount of 3% H_2O_2.

METHOD 6-1. DETERMINATION OF IRON IN AN ORE BY TITRATION WITH POTASSIUM PERMANGANATE

Discussion. The common iron ores are hematite (Fe_2O_3), magnetite (Fe_3O_4), and limonite ($3Fe_2O_3\cdot3H_2O$). Volumetric methods for the analysis of iron samples containing these substances consist of three steps: (1) solution of the sample, (2) reduction of the iron to the divalent state, and (3) titration with a standard oxidant.

Iron ores are often completely decomposed in concentrated hydrochloric acid. The rate of attack by this reagent is increased by the presence of a small amount of tin(II) chloride, which probably acts by reducing sparingly soluble iron(III) oxides on the surface of the particles to more soluble iron(II) species. Because iron(III) tends to form stable chloride complexes, hydrochloric acid is a much more efficient solvent than either sulfuric or nitric acid.

Many iron ores contain silicates that may not be entirely decomposed by treatment with hydrochloric acid. Where decomposition is complete, the white residue of hydrated silica that remains behind in no way interferes with the analysis. Incomplete decomposition is indicated by a dark residue that remains after prolonged treatment with the acid.

Part or all of the iron will exist in the trivalent state in the dissolved sample; prereduction must therefore precede titration with the oxidant. Any of the methods described earlier may be used (Section 15A-1), for example, the Jones or the Walden reductors.

Perhaps the most satisfactory of all prereductants for iron is tin(II) chloride. The only other common elements reduced by this reagent are vanadium, copper, molybdenum, tungsten, and arsenic. The excess reducing agent is removed from the solution by the addition of mercury(II) chloride:

$$Sn^{2+} + 2HgCl_2 \rightleftharpoons Hg_2Cl_2(s) + Sn^{4+} + 2Cl^-$$

The slightly soluble mercury(I) chloride produced will not reduce permanganate, nor will the excess mercury(II) chloride reoxidize the iron(II). Care must be exerted, however, to prevent occurrence of the alternative reaction,

$$Sn^{2+} + HgCl_2 \rightleftharpoons Hg(l) + Sn^{4+} + 2Cl^-$$

Metallic mercury reacts with permanganate and causes a high result in an iron analysis. Formation of mercury is favored by an appreciable excess of tin(II); it is prevented by careful control of this excess and by the rapid addition of excess mercury(II) chloride. A proper reduction is indicated by the appearance of a slight, white precipitate after addition of the latter reagent. A gray precipitate indicates the presence of mercury; the total absence of a precipitate indicates that an insufficient amount of tin(II) chloride was added. In either of these events, the sample must be discarded.

The reaction of iron(II) with permanganate proceeds smoothly and rapidly to completion. In the presence of hydrochloric acid, however, oxidation of chloride ion by permanganate causes high results. This reaction, which normally does not proceed rapidly enough to cause serious errors, is *induced* by the presence of divalent iron. Its effects are avoided by preliminary removal of chloride by evaporation with sulfuric acid or by use of the *Zimmermann-Reinhardt reagent*. The latter is a solution of manganese(II) in fairly concentrated sulfuric and phosphoric acids. Chloride ion is apparently oxidized by an intermediate manganese(III) ion formed during the reduction with iron(II). Both components of the Zimmermann-Reinhardt reagent inhibit the formation of chlorine by decreasing the potential of the manganese(III)/manganese(II) couple. The manganous ion acts through the common ion effect. Phosphate forms a stable complex with manganese(III). In addition, it complexes with the iron(III) produced in the titration and prevents the intense yellow color of iron(III) chloride complexes from interfering with the end point.[4]

Reagents needed.

Standardized 0.1 N $KMnO_4$

Concentrated HCl

Concentrated H_2SO_4

85% H_3PO_4

0.25 M $SnCl_2$ (for 100 titrations). Dissolve 60 g of iron-free $SnCl_2 \cdot 2H_2O$ in 100 mL of concentrated HCl; warm if necessary. After solution is complete, dilute to about 1 L, and store in a well-stoppered bottle. A few pieces of mossy Sn in the bottle will prevent air oxidation of Sn(II).

5% (w/v) $HgCl_2$ (for 100 titrations). Dissolve 50 g of $HgCl_2$ in about 1 L of water.

Zimmermann-Reinhardt reagent (for 100 titrations). Dissolve 300 g of $MnSO_4 \cdot 4H_2O$ in about 1 L of water. Cautiously add 400 mL of concentrated H_2SO_4 and 400 mL of 85% phosphoric acid. Dilute to about 3 L.

0.2 M $KMnO_4$. Dissolve about 3 g of $KMnO_4$ in 100 mL of water.

Procedure. *Sample preparation.* Dry the ore for at least 3 hr at 105 to 110°C, cool in a desiccator, and weigh individual samples into 500-mL conical flasks. A sample of optimum size will require 25 to 40 mL of the standard $KMnO_4$. Add 10 mL of concentrated HCl and 3 mL of the 0.25 M $SnCl_2$ solution (Note 1). Cover the flask

---

[4] The mechanism of the action of Zimmermann-Reinhardt reagent has been the subject of much study. For a discussion of this work, see: H. A. Laitinen, *Chemical Analysis.* pp. 369–372. New York: McGraw-Hill, 1960.

with a small watch glass, and heat at just below boiling until the sample is decomposed, as indicated by the disappearance of all of the dark particles (Note 2). A pure white residue may remain. A blank consisting of 10 mL of HCl and 3 mL of $SnCl_2$ should be heated for the same length of time. If any of the solutions becomes yellow during the heating, add another milliliter or two of $SnCl_2$. After the decomposition is complete, remove the excess $SnCl_2$ by adding approximately 0.2 M $KMnO_4$ dropwise until the solution is just yellow. Dilute to about 15 mL. Add $KMnO_4$ to the blank until the solution just turns pink. Then just decolorize with 1 drop of the $SnCl_2$. *Carry samples individually through subsequent steps to minimize air oxidation of iron(II).*

Reduction of iron. Heat the solution containing the sample nearly to boiling, and add 0.25 M $SnCl_2$ drop by drop until the yellow color disappears. Add 2 drops in excess (Note 3). Cool to room temperature, and *rapidly* add 10 mL of the $HgCl_2$ solution. A small quantity of a silky white precipitate of Hg_2Cl_2 should appear. If no precipitate forms or if the precipitate is gray due to the presence of Hg, the sample should be discarded (Note 4). The blank solution should also be treated with 10 mL of the $HgCl_2$ solution.

Titration. After 2 to 3 min add 25 mL of Zimmermann-Reinhardt reagent and 300 mL of water. Titrate *immediately* with the $KMnO_4$ to the first faint pink that persists for 15 to 20 s. Do not titrate rapidly at any time. Correct the volume of $KMnO_4$ for the blank titration.

Calculate the $\%Fe_2O_3$ in the sample.

NOTES

1. The $SnCl_2$ hastens the dissolution of the ore by reducing iron(III) oxides to iron(II). Absence of sufficient $SnCl_2$ is indicated by the appearance of the yellow iron(III) chloride complexes in the solution.
2. If dark particles persist after treatment with acid for several hours, filter the solution through ashless paper, wash with 5 to 10 mL of 6 M HCl, and retain the filtrate and washings. Place the paper in a small platinum crucible and ignite. Mix 0.5 to 0.7 g of finely ground anhydrous Na_2CO_3 with the residue, and heat until a clear liquid melt is obtained. Cool, add 5 mL of water, followed by the cautious addition of an equal volume of 6 M HCl. Warm the crucible until the flux is dissolved, and combine the contents with the original filtrate. Evaporate the solution to about 15 mL, and proceed with the analysis.
3. The solution may not become entirely colorless but may instead acquire a pale yellow-green hue. Further additions of $SnCl_2$ will not alter this color. If too much $SnCl_2$ is inadvertently introduced, add 0.2 M $KMnO_4$ until the yellow color is restored, and repeat the reduction.
4. The absence of precipitate means that an excess of $SnCl_2$ was not present and, thus, the reduction was incomplete. Formation of Hg occurs when too much $SnCl_2$ is added; the Hg consumes $KMnO_4$.

METHOD 6-2. DETERMINATION OF CALCIUM IN A LIMESTONE BY TITRATION WITH POTASSIUM PERMANGANATE

Discussion. In common with a number of other cations, calcium is conveniently precipitated with oxalate ion. The solid is filtered, washed free of excess precipitating reagent, and dissolved in dilute acid; the liberated oxalic acid is then titrated with a standard solution of permanganate or some other oxidizing agent. This method is applicable to samples that contain magnesium and the alkali metals. Most other cations must be absent, however, since they either precipitate or coprecipitate as oxalates and cause positive errors in the analysis.

To obtain satisfactory results from this procedure, the mole ratio of calcium to oxalate must be exactly 1 in the precipitate and thus in the solution at the time of the titration. Several precautions must be observed to assure this condition. For example, calcium oxalate formed in a neutral or ammoniacal solution is likely to be contaminated with calcium hydroxide or a basic calcium oxalate; the presence of either causes low results. This problem can be eliminated by adding the oxalate to an acidic solution of the sample and slowly forming the precipitate by the dropwise addition of ammonia. The coarsely crystalline solid that is produced is readily filtered. Losses due to the solubility of calcium oxalate are negligible at a pH of about 4, provided washing is restricted to freeing the precipitate of excess oxalic acid.

A potential source of positive error in the analysis is the coprecipitation of sodium oxalate, which occurs when the sodium ion concentration exceeds that of calcium. The error from this source can be eliminated by double precipitation.

Magnesium, if present in high concentrations, may also contaminate the calcium oxalate precipitate. This interference is minimized if the excess of oxalate is sufficient to allow the formation of a soluble magnesium complex and if filtration is performed promptly after the completion of the precipitation. When the magnesium content exceeds that of calcium in the sample, a double precipitation may be required.

Limestones are composed principally of calcium carbonate; dolomitic limestones contain large concentrations of magnesium carbonate in addition. Also present in smaller amounts are calcium and magnesium silicates as well as carbonates and silicates of iron, aluminum, manganese, titanium, the alkalies, and other metals.

Hydrochloric acid will often decompose limestones completely; only silica, which does not interfere with the analysis, remains undissolved. Some limestones are more readily decomposed if first ignited; a few will yield only to a carbonate fusion.

The following method is remarkably effective for the analysis of calcium in most limestones. Iron and aluminum, in amounts equivalent to the calcium, do not interfere. Small amounts of titanium and manganese can be tolerated.[5]

Reagents needed.

> *Standardized 0.1 N KMnO$_4$*
> *Concentrated HCl*
> *3 M H$_2$SO$_4$*
> *Saturated Br$_2$ water*
> *6% (w/v) (NH$_4$)$_2$C$_2$O$_4$·H$_2$O (filter if not clear); 100 mL/sample*
> *Methyl red indicator solution*

Procedure. Dry the unknown for 1 to 2 hr at 110°C, and cool in a desiccator. If the material can be readily decomposed with acid, weigh (to the nearest 0.1 mg) 0.25- to 0.3-g samples into 250-mL beakers, and cover each beaker with a watch glass. Add 10 mL of water and 10 mL of concentrated HCl dropwise, with care being taken to avoid loss by spattering. Proceed with the analysis as described in the next

---

[5] For further details of the method, see: J. J. Lingane, *Ind. Eng. Chem., Anal. Ed.*, **17**, 39 (1945).

paragraph. If the limestone is not completely decomposed by acid, weigh the sample into a small porcelain crucible and ignite. Raise the temperature slowly to 800 to 900°C, and maintain this temperature for 30 min. Allow the crucible and contents to cool, place in a 250-mL beaker, add 5 mL of water, and cover with a watch glass. Carefully add 10 mL of concentrated HCl, and heat to boiling. Remove the crucible with a stirring rod and rinse thoroughly with water.

Add 5 drops of saturated bromine water to oxidize any iron present, and boil for 5 min to remove the excess Br_2. Dilute to 50 mL, heat to boiling, and add 100 mL of hot, 6% (w/v) $(NH_4)_2C_2O_4$ solution. Add 3 to 4 drops of methyl red, and precipitate the CaC_2O_4 by the dropwise addition of 6 M NH_3. The rate of addition should be 1 drop every 3 or 4 s until the solution turns to the intermediate orange-yellow color of the indicator (pH 4.5 to 5.5). Allow the solution to stand for 30 min but no longer and filter (see Note); a Gooch crucible with a glass filtering mat or a medium-porosity filtering crucible can be used. Wash the precipitate with several 10-mL portions of cold water. Rinse the outside of the crucible, and return it to the beaker in which the CaC_2O_4 was originally formed. Add 150 mL of water and 50 mL of 3 M H_2SO_4.

Heat the solution to 80 to 90°C, and titrate with 0.1 N $KMnO_4$. The solution should be kept above 60°C throughout the titration.

Report the %CaO in the sample.

NOTE
If Mg^{2+} is absent, the period of standing need not be limited to 30 min.

7 TITRATIONS WITH CERIUM(IV)

PREPARATION 7-1. PREPARATION OF ORTHOPHENANTHROLINE INDICATOR

Discussion. The orthophenanthroline complex of iron(II) is a useful indicator for titrations with cerium(IV). Its properties are listed in Table 14-2.

Procedure. To obtain a 0.025 M solution of the indicator, dissolve 1.485 g of orthophenanthroline in 100 mL of water that contains 0.695 g of $FeSO_4 \cdot 7H_2O$. Indicator solutions are available commercially.

PREPARATION 7-2. PREPARATION OF APPROXIMATELY 0.1 N CERIUM(IV)

Discussion. For a discussion of the preparation and properties of standard solutions of cerium(IV), read pages 370 to 373.

Procedure. Carefully add 50 mL of concentrated H_2SO_4 to 500 mL of water, and then add 63 g of $Ce(SO_4)_2 \cdot 2(NH_4)_2SO_4 \cdot 2H_2O$ with continual stirring. Cool, filter if the solution is not clear, and dilute to about 1 L.

If $Ce(NO_3)_4 \cdot 2NH_4NO_3$ is used, weigh about 55 g into a 1-L beaker. Add about 60 mL of 95% H_2SO_4, and stir for 2 min. *Cautiously* introduce 100 mL of water, and again stir for 2 min. Repeat the operations of adding water and stirring until all of the salt has dissolved. Then dilute to about 1 L.

STANDARDIZATION **7-1.** STANDARDIZATION OF CERIUM(IV) AGAINST SODIUM
OXALATE

Reagents needed.

> *0.1 N cerium(IV) solution*
> *Primary standard Na$_2$C$_2$O$_4$*
> *Iron(II) orthophenanthroline indicator solution*
> *0.017 M ICl catalyst.* Mix 25 mL of 0.04 *M* KI, 40 mL of concen-
>   trated HCl, and 20 mL of 0.025 *M* KIO$_3$. Add 5 to 10 mL of
>   toluene or chloroform and shake thoroughly in a separatory
>   funnel. Titrate with either KI or KIO$_3$ until the organic
>   layer is barely pink after shaking. The former should be
>   added if the organic layer is colorless and the latter if
>   it is too pink. Discard the organic layer.
> *Concentrated HCl*

Procedure.   Dry primary standard grade Na$_2$C$_2$O$_4$ at 110 to 120°C for about 1 hr,
and cool in a desiccator. Weigh 0.25- to 0.3-g portions (to the nearest 0.1 mg) into
250-mL beakers, and dissolve in 75 mL of water. Add 20 mL of concentrated HCl
and 1.5 mL of 0.017 *M* ICl catalyst. Heat to 50°C, and add 2 to 3 drops of iron(II)
orthophenanthroline indicator. Titrate with cerium(IV) until the solution turns colorless or pale blue and the pink does not return within 1 min (Note). The temperature should be between 45 and 50°C throughout.

> NOTE
> Do not permit the temperature to exceed 50°C because the indicator may be destroyed.

METHOD **7-1.** DETERMINATION OF IRON IN AN ORE BY TITRATION WITH CERIUM(IV)

Discussion.   Tetravalent cerium oxidizes iron(II) smoothly and rapidly at room
temperature. Orthophenanthroline is an excellent indicator for the titration. In contrast to the analysis based upon oxidation with permanganate, consumption of the
reagent by chloride ion is of no concern.

   The problems associated with the solution of the sample and the prereduction of the iron are the same as in the permanganate method, the only major difference being that here there is no need for the Zimmermann-Reinhardt reagent.
The material in the discussion section of Method 6-1 should, therefore, be read before undertaking the determination with cerium(IV).

Reagents needed.

> *Standardized cerium(IV) solution*
> *Iron(II) orthophenanthroline indicator solution*
> *For remaining reagents see Method 6-1*

Procedure.   Follow the directions in the paragraphs labeled *Sample preparation*
and *Reduction of iron* given in Method 6-1 (pp. 758 and 759).

   About 2 to 3 min after adding the HgCl$_2$, introduce 50 mL of 6 *M* HCl,

15 mL of 85% H_3PO_4, 200 mL of water, and a drop or two of orthophenanthroline; titrate to the color change of the indicator. For accurate work, a blank should be carried through the entire procedure.

Report the %Fe_2O_3 in the sample.

8 TITRATION WITH POTASSIUM DICHROMATE

PREPARATION 8-1. PREPARATION OF SODIUM DIPHENYLAMINE SULFONATE SOLUTION

Procedure. Dissolve 0.2 g of sodium diphenylamine sulfonate in 100 mL of water.

PREPARATION 8-2. PREPARATION OF 0.1 N POTASSIUM DICHROMATE

Discussion. For a discussion of the preparation and properties of standard solutions of potassium dichromate, read Section 15B-3.

Procedure. Dry primary standard $K_2Cr_2O_7$ for 2 hr at 150 to 200°C. After it cools, weigh 4.9 g of the solid (to the nearest milligram) into a 1-L volumetric flask, dissolve, and dilute to the mark with distilled water; mix well.

If the purity of the salt is suspect, recrystallize three times from water before drying.

METHOD 8-1. DETERMINATION OF IRON IN AN ORE

Discussion. Potassium dichromate reacts rapidly with iron(II) to yield iron(III) and chromium(III) ions. The sodium salt of diphenylamine sulfonic acid (p. 359) serves as an excellent indicator for the titration; the color change is from the green of chromium(III) to the violet of the oxidized indicator.

The problems associated with the solution of the sample and the prereduction of the iron are the same as in the permanganate method, the only major difference being that here there is no need for the Zimmermann-Reinhardt reagent. The material in the discussion section of Method 6-1 should, therefore, be read before undertaking the determination with potassium dichromate.

Reagents needed.

> Standard 0.1 N $K_2Cr_2O_7$
> Sodium diphenylamine sulfonate indicator solution
> 3 M H_2SO_4
> 85% H_3PO_4
> For remaining reagents see Method 6-1

Procedure. Follow the directions in the paragraphs labeled *Sample preparation* and *Reduction of iron* given in the procedure for Method 6-1 (pp. 758 and 759).

About 2 to 3 min after adding $HgCl_2$, introduce 60 mL of 3 M H_2SO_4, 15 mL of 85% H_3PO_4, and 100 mL of water. Cool, add 8 drops of diphenylamine sulfonate indicator, and titrate with dichromate to the violet-blue end point.

Report the % Fe_2O_3 in the sample.

9 IODIMETRIC TITRATIONS

The oxidizing properties of iodine, the composition and stability of triiodide solutions, and the application of this reagent to volumetric methods are discussed in Section 15B-4.

PREPARATION 9-1. PREPARATION OF STARCH INDICATOR

Procedure. Make a paste by rubbing about 2 g of soluble starch in about 30 mL of water. Pour this mixture into 1 L of boiling water, and heat until a clear solution results. Cool, and store in stoppered bottles. A fresh solution should be prepared every few days. For most titrations, 3 to 5 mL of the indicator should be sufficient.

PREPARATION 9-2. PREPARATION OF APPROXIMATELY 0.1 N TRIIODIDE SOLUTION

Procedure. Weigh about 40 g KI into a 100-mL beaker. Add 12.7 g of I_2 and 10 mL of water. Stir for several minutes (Note 1). Then add an additional 20 mL of water and again stir for several minutes. Carefully decant most of the liquid into a storage bottle containing about 1 L of water. Be sure to avoid transferring any undissolved iodine that may be present in the last milliliter or two of the concentrated solution (Note 2).

> NOTES
> 1. Iodine dissolves only slowly in the concentrated KI solution. Thus, the KI/I_2 mixture should be stirred several minutes before being transferred to the storage bottle.
> 2. Any solid I_2 that is inadvertently transferred to the storage bottle will cause the normality of the solution to increase gradually. Filtration through a sintered glass crucible will eliminate this source of difficulty.

STANDARDIZATION 9-1. STANDARDIZATION OF A TRIIODIDE SOLUTION AGAINST PRIMARY STANDARD BARIUM THIOSULFATE MONOHYDRATE

Discussion. See page 378.

Reagents needed.

>*Approximately 0.1 N I_2 solution*
>*Primary standard grade $Ba_2S_2O_3 \cdot H_2O$ (Note 1)*
>*Starch indicator solution (Preparation 9-1)*

Procedure. Weigh 0.8- to 1.0-g portions of air-dried $BaS_2O_3 \cdot H_2O$ into 250-mL conical flasks (Note 1). Add 100 mL of water and 5 mL of starch, and titrate to the first faint blue color that persists for at least 30 s (Note 2).

NOTES

1. For a procedure for preparation of pure $BaS_2O_3 \cdot H_2O$, see: W. M. MacNevin and O. H. Kriege, *Anal. Chem.*, **25**, 767 (1953).
2. The barium thiosulfate will not dissolve completely, and the approach to the end point is signaled by the disappearance of the solid phase.

STANDARDIZATION 9-2. STANDARDIZATION OF A TRIIODIDE SOLUTION AGAINST PRIMARY STANDARD ARSENIC(III) OXIDE

Discussion. Arsenic(III) oxide is a carcinogen. Thus, OSHA regulations must be carefully followed if it is to be used. The *compound is highly toxic* and should accordingly be handled with great care.

The use of arsenic(III) for the standardization of iodine solution is discussed on page 377.

Reagents needed.

> *Approximately 0.1 N I_2 solution*
> *Primary standard grade As_2O_3*
> *1 M NaOH*
> *6 M HCl*
> *$NaHCO_3$*
> *Phenolphthalein indicator solution* (Preparation 3-1)
> *Starch indicator solution* (Preparation 9-1)

Procedure. Dry a quantity of primary standard As_2O_3 (CAUTION; see Discussion) for about 1 hr at 110°C, and cool in a desiccator. For standardization of 0.1 N solutions, weigh 0.15- to 0.2-g samples (to the nearest 0.1 mg), and dissolve in 10 mL of 1 M NaOH. When solution is complete, dilute with about 75 mL of water, and add 2 drops of phenolphthalein. Promptly introduce 6 M HCl until the red color just disappears (Note). Then, add about 1 mL of acid in excess. Carefully add 3 to 4 g of solid $NaHCO_3$, in small portions at first to avoid losses of solution due to effervescence of the CO_2. Add 5 mL of starch indicator, and titrate the solution to the first faint purple or blue color that persists for at least 30 s.

NOTE

Alkaline solutions of arsenic(III) are readily air oxidized. Thus, the neutralization should be completed promptly.

METHOD 9-1. DETERMINATION OF ANTIMONY IN STIBNITE

Discussion. The analysis of stibnite, a common antimony ore, illustrates the application of a direct iodimetric method. Stibnite is primarily antimony sulfide that contains silica and other contaminants. Provided the material is free of iron and arsenic, the determination of its antimony content is a straightforward process. The sample is decomposed in hot, concentrated hydrochloric acid to eliminate the sulfide as H_2S. Some care is required in this step to prevent losses of the volatile antimony trichloride; the addition of potassium chloride increases the tendency for the formation of nonvolatile chloro complexes such as $SbCl_4^-$ and $SbCl_6^{3-}$.

The reaction of trivalent antimony with iodine is quite analogous to that of trivalent arsenic. Here, however, an additional step is required to prevent precipitation of such basic salts as antimony oxychloride, SbOCl, as the solution is neutralized. These species react incompletely with iodine and cause erroneously low results. The problem is readily overcome by the addition of tartaric acid prior to dilution. The tartrate complex $(SbOC_4H_4O_6^-)$ that forms is rapidly and completely oxidized by iodine.

Reagents needed.

> Standardized 0.1 N I_2
> Concentrated HCl
> 6 M HCl
> 6 M NaOH
> KCl
> NaHCO$_3$
> Tartaric acid
> Phenolphthalein indicator solution (Preparation 3-1)
> Starch indicator solution (Preparation 9-1)

Procedure. Dry the unknown for 1 hr at 110°C. After it cools, weigh into 500-mL conical flasks sufficient quantities of the ore to consume 30 to 40 mL of 0.1 N I_3^-. Add about 0.3 g of KCl and 10 mL of concentrated HCl. Heat the mixture to just below boiling (in a hood), and maintain this temperature until only a white or slightly gray residue of SiO_2 remains.

Add 3 g of solid tartaric acid to the solution; heat for another 10 to 15 min. While swirling the solution, slowly add water from a pipet until the volume is about 100 mL. The addition of water should be slow enough to prevent formation of white SbOCl. If reddish Sb_2S_3 forms, stop the addition of water, and heat further to remove H_2S; add more acid if necessary.

Add 3 drops of phenolphthalein to the solution and 6 M NaOH until the first pink color is observed. Add 6 M HCl dropwise until the solution is decolorized and then add 1 mL in excess. Add 4 to 5 g of NaHCO$_3$, with care being taken to avoid losses of solution during the addition. Add 5 mL of starch indicator, and titrate to the first blue color that persists for 30 s or longer.

Report the %Sb_2S_3.

10 IODOMETRIC METHODS OF ANALYSIS

Numerous methods are based upon the reducing properties of iodide ion; the reaction product, iodine, is ordinarily titrated with a standard thiosulfate solution. Iodometric methods are discussed in Section 15C-2.

PREPARATION 10-1. PREPARATION OF APPROXIMATELY 0.1 N SODIUM THIOSULFATE

Discussion. Read page 387.

Procedure. Boil about 1 L of distilled water for at least 5 min. Cool, and add about 25 g of $Na_2S_2O_3 \cdot 5H_2O$ and 0.1 g of Na_2CO_3. Stir until solution is complete; then, transfer to a clean stoppered bottle (glass or plastic), and store in the dark.

STANDARDIZATION 10-1. STANDARDIZATION OF SODIUM THIOSULFATE AGAINST POTASSIUM IODATE

Discussion. See page 388.

Reagents needed.

> *Primary standard grade KIO₃*
> *KI*
> *6 M HCl*
> *Starch indicator solution* (Preparation 9-1)

Procedure. Dry primary standard grade KIO_3 for at least 1 hr at 110°C, and cool in a desiccator. Weigh (to the nearest 0.1 mg) 0.12- to 0.15-g samples into 250-mL conical flasks. Dissolve in 75 mL of water, and add about 2 g of iodate-free KI. *Treat each sample individually from this point in order to minimize errors resulting from air oxidation of iodide in the acidic solution.* After the iodide has dissolved, add 2 mL of 6.0 M HCl, and titrate immediately with $Na_2S_2O_3$ until the color of the solution becomes pale yellow. Add 5 mL of starch indicator, and titrate to the disappearance of the blue color.

To minimize the weighing error, this procedure can be modified by weighing a 0.6-g sample of KIO_3 into a 250-mL volumetric flask, dissolving in water, and diluting to the mark. A 50-mL aliquot of this solution can then be used for each standardization.

STANDARDIZATION 10-2. STANDARDIZATION OF SODIUM THIOSULFATE AGAINST COPPER

Discussion. Copper wire or foil meeting ACS specifications and having a purity of 99.9% is commercially available. It can be used advantageously for the standardization of thiosulfate when the reagent is to be used subsequently for copper determination, because any determinate method errors in the analysis tend to be canceled.

A discussion of the reaction of copper(II) with iodide ion is found under Method 10-1.

Reagents needed.

> *Primary standard grade copper wire*
> *6 M HNO₃*
> *Concentrated NH₃*
> *3 M H₂SO₄*
> *85% H₃PO₄*
> *5% (w/v) urea; 25 mL/standardization*
> *KI*
> *KSCN*
> *Starch indicator solution* (Preparation 9-1)

Procedure. Wipe the copper free of dust and grease with filter paper. Subsequently, handle the wire with filter paper strips. The wire can be cut into appropriate length samples with scissors. The wire should not be dried.

Weigh 0.20- to 0.25-g samples into previously weighed weighing bottles or watch glasses, and obtain the weight. Transfer each sample into a 250-mL flask, add 5 mL of 6 M HNO_3, and cover with a small watch glass. Warm the mixture in a hood until solution is complete. Add 25 mL of water and 5 mL of 5% urea (w/v). Boil for about 1 min. Cool, and rinse the watch glass, collecting the washings in the flask.

Add concentrated NH_3 dropwise and with thorough mixing until the first dark-blue color of the tetraammine copper(II) complex appears. The solution should smell faintly of NH_3. Add 3 M H_2SO_4 dropwise until the color of the complex just disappears. Then add 2.0 mL of concentrated H_3PO_4. Cool to room temperature.

From this point on treat each sample individually to minimize the effect of air oxidation of iodide ion. Add 4.0 g of KI to the sample, and titrate immediately with the $Na_2S_2O_3$ until the I_2 color is no longer distinct. Add 5 mL of starch solution, and titrate until the blue begins to fade. Add 2 g of KSCN, and swirl vigorously for several seconds. Complete the titration with vigorous mixing. The end point is signaled by the disappearance of the blue color of the starch/iodine complex.

METHOD 10-1. DETERMINATION OF COPPER IN AN ORE

Discussion. The analysis for copper is a typical iodometric method. The directions that follow will permit the determination of copper in samples of ore.

Copper(II) is quantitatively reduced to copper(I) by iodide ion. The reaction can be expressed as

$$2Cu^{2+} + 4I^- \rightleftharpoons 2CuI(s) + I_2$$

Consideration of the following standard potentials makes it evident that the formation of sparingly soluble CuI plays an important part in driving the reaction to completion.

$$Cu^{2+} + e \rightleftharpoons Cu^+ \qquad E^0 = 0.15 \text{ V}$$
$$I_2 + 2e \rightleftharpoons 2I^- \qquad E^0 = 0.54 \text{ V}$$
$$Cu^{2+} + I^- + e \rightleftharpoons CuI(s) \qquad E^0 = 0.86 \text{ V}$$

The first two potentials suggest that iodide should have no tendency to reduce Cu^{2+}; the formation of CuI, however, forces the reduction reaction.

Much systematic experimentation has been devoted to establishing ideal conditions for a copper analysis.[6] These studies have revealed that the solution should be at least 4% with respect to potassium iodide. Furthermore, a pH less than 4 is expedient; at a higher pH, formation of basic copper(II) species causes a slower and less complete oxidation of iodide ion. In the presence of copper(I) ion, hydrogen ion concentrations greater than about 0.3 M must be avoided to prevent air oxidation of iodide ion.

[6] See: E. W. Hammock and E. H. Swift, *Anal. Chem.,* **21**, 975 (1949).

It has been found experimentally that the titration of iodine by thiosulfate in the presence of copper(I) iodide tends to yield slightly low results because small but appreciable quantities of iodine are physically adsorbed upon the solid. The adsorbed iodine is released only slowly, even in the presence of thiosulfate ion; transient and premature end points result. This difficulty is largely overcome by the addition of thiocyanate ion, which also forms a sparingly soluble copper(I) salt. Part of the copper(I) iodide is converted to the corresponding thiocyanate at the surface of the solid:

$$CuI(s) + SCN^- \rightleftharpoons CuSCN(s) + I^-$$

Accompanying this reaction is the release of the adsorbed iodine, thus making it available for titration. Early addition of thiocyanate must be avoided, however, because of the tendency for that ion to reduce iodine slowly.

The iodometric method is convenient for the assay of copper in an ore. Ordinarily, samples dissolve readily in hot, concentrated nitric acid. Care must be taken to remove any nitrogen oxides formed in the process because these species catalyze the air oxidation of iodide. A convenient way to remove these oxides involves the addition of urea. The reaction is

$$(NH_2)_2CO + 2HNO_2 \longrightarrow 2N_2 + CO_2 + 3H_2O$$

Some samples require the addition of hydrochloric acid to complete the solution step. The chloride ion must then be removed by evaporation with sulfuric acid because iodide ion will not reduce copper(II) quantitatively from its chloro complexes.

Of the elements ordinarily associated with copper in nature, only iron, arsenic, and antimony interfere with the iodometric method. Fortunately, difficulties caused by these elements are readily eliminated. Iron is rendered nonreactive by the addition of such complexing agents as fluoride or pyrophosphate; because these ions form more stable complexes with iron(III) than with iron(II), the potential for this system is altered to the point where appreciable oxidation of iodide cannot occur. Interference by arsenic and antimony is prevented by converting these elements to the +5 state during the solution step. Ordinarily, the hot nitric acid used to dissolve the sample will cause conversion to the desired oxidation state, although a small amount of bromine water can be added in case of doubt; the excess bromine is then expelled by boiling. As has been pointed out, arsenic in the +5 state does not oxidize iodide ion, provided the solution is not too acidic (p. 376). Antimony behaves similarly. Thus, by maintaining the pH of the solution at 3 or greater, interference by these elements is avoided. We have seen, however, that oxidation of iodide by copper is incomplete at pH values greater than 4. Thus, when copper is to be determined in the presence of arsenic or antimony, control of the pH between 3 and 4 is essential. Ammonium hydrogen fluoride, NH_4HF_2, is a convenient buffer for this purpose. The anion of the salt dissociates as follows:

$$HF_2^- \rightleftharpoons HF + F^- \qquad K = 0.26$$
$$HF \rightleftharpoons H^+ + F^- \qquad K = 7.2 \times 10^{-4}$$

The first dissociation provides equal quantities of hydrogen fluoride and fluoride ions which then buffer the solution to a pH somewhat greater than 3. In addition to acting as a buffer, the salt also serves as a source of fluoride ions to complex any iron(III) that may be present.

Reagents needed.

> *Standardized 0.1 N* $Na_2S_2O_3$
> *Concentrated* HNO_3
> *3 M* H_2SO_4
> *Concentrated* NH_3
> *5% (w/v) urea*
> *KI*
> NH_4HF_2
> *KSCN*
> *Starch indicator solution* (Preparation 9-1)

Procedure. Weigh appropriately sized samples (150 to 250 mg Cu) of the finely ground and dried ore into 150-mL beakers, and add 20 mL of concentrated HNO_3. Cover with a watch glass, and heat in a hood until all of the Cu is in solution. If the volume becomes less than 5 mL, add more HNO_3. Continue the heating until only a white or slightly gray siliceous residue remains (Note 1). Evaporate to about 5 mL.

Add 25 mL of distilled water and 5 mL of 5% urea, and boil to dissolve soluble salts and to expel oxides of nitrogen. If the residue is small and nearly colorless, no filtration is necessary. Otherwise, filter the suspension, and collect the filtrate in a 250-mL conical flask. Wash the paper with several small portions of hot 1:100 HNO_3; then, discard. Evaporate the filtrate and washings to about 25 mL, cool, and slowly add concentrated NH_3 to the first appearance of the deep blue tetraammine copper(II) complex. A faint odor of NH_3 should be detectable over the solution. If it is not, add another drop of NH_3 and repeat the test. Avoid an excess (Note 2).

From this point, treat each sample individually. Add 2.0 ± 0.1 g NH_4HF_2 (CAUTION! Note 3), and swirl until completely dissolved. Then add 4 g of KI, and titrate immediately with 0.1 N $Na_2S_2O_3$. When the color of the iodine is nearly discharged, add 2 g of KSCN and 5 mL of starch. Swirl vigorously for several seconds. Continue the titration with vigorous mixing until the blue starch/iodine color is decolorized and does not return for several minutes.

NOTES
1. If the ore is not readily decomposed by the HNO_3, add 5 mL of concentrated HCl and heat (Hood) until only a small white or gray residue remains. Do not evaporate to dryness. Cool, add 10 mL of concentrated H_2SO_4, and evaporate in a hood until copious white fumes of SO_3 are observed. Cool and carefully add 15 mL of water and 10 mL of saturated bromine water. Boil the solution vigorously in a hood until all of the bromine has been removed. Cool and proceed with the filtration step in the second paragraph.
2. If too much NH_3 is added, neutralize the excess with 3 M H_2SO_4.
3. Ammonium hydrogen fluoride is a highly toxic and corrosive chemical. Avoid contact with the skin. If exposure does occur, *immediately* rinse the affected area with copious amounts of water.

METHOD **10-2.** DETERMINATION OF COPPER IN BRASS

Discussion. The iodometric method discussed under Method 10-1 can be adapted to the determination of copper in brass, an alloy consisting principally of copper, zinc, lead, and tin. Several other elements, iron and nickel, for example, may be tolerated in minor amounts.

The method is relatively simple and applicable to the analysis of brasses containing less than 2% of iron. It involves solution in nitric acid, removal of nitrate by fuming with sulfuric acid, adjustment of the pH by neutralization with ammonia, acidification with a measured quantity of phosphoric acid, and, finally, iodometric determination of the copper.

It is instructive to consider the fate of each major constituent during the course of this treatment. Tin is oxidized to the $+4$ state by the nitric acid and precipitates slowly as the slightly soluble hydrous tin(IV) oxide, $SnO_2 \cdot 4H_2O$. This precipitate, which frequently forms as a colloid, is sometimes called *metastannic acid*. It has a tendency to adsorb copper(II) and other cations from the solution. Lead, zinc, and copper are oxidized to soluble divalent salts by the nitric acid, and iron is converted to the trivalent state. Evaporation and fuming with sulfuric acid redissolves the metastannic acid but may cause part of the lead to precipitate as the sulfate; the copper, zinc, and iron are unaffected. Upon dilution with water, lead is nearly completely precipitated as lead sulfate while the other elements remain in solution. None, except copper and iron, is reduced by iodide. Interference from the iron is eliminated by complexing with phosphate ion.

Reagents needed.

> Standardized 0.1 N $Na_2S_2O_3$
> 6 M HNO_3
> Concentrated H_2SO_4
> Concentrated NH_3
> 3 M H_2SO_4
> 85% H_3PO_4
> KI
> KSCN
> Starch indicator solution (Preparation 9-1)

Procedure. Weigh (to the nearest 0.1 mg) about 0.3-g samples of the clean, dry metal into 250-mL conical flasks, and add 5 mL of 6 M HNO_3. Warm the solution (Hood) until decomposition is complete. Then add 10 mL of concentrated H_2SO_4, and evaporate in a hood to copious white fumes of SO_3. Allow the mixture to cool. Carefully add 30 mL of water, boil for 1 to 2 min, and cool.

With good mixing, add concentrated NH_3 dropwise until the first dark blue color of the tetraammine copper(II) complex appears. The solution should smell faintly of NH_3. Add 3 M H_2SO_4 dropwise until the color of the complex just disappears. Then add 2.0 mL of 85% phosphoric acid. Cool to room temperature.

From this point, treat each sample individually. Add 4.0 g of KI to the sample, and titrate immediately with standard $Na_2S_2O_3$ until the iodine color is no longer distinct. Add 5 mL of starch solution, and titrate until the blue begins to fade. Add 2 g of KSCN, swirl vigorously for several seconds, and complete the titration with good mixing.

METHOD 10-3. DETERMINATION OF DISSOLVED OXYGEN BY THE WINKLER METHOD

Discussion. The Winkler method for the determination of dissolved oxygen in natural water is an interesting example of iodometry. In this procedure, the sample is first treated with an excess of manganese(II), potassium iodide, and sodium hydroxide. The white manganese(II) hydroxide that forms reacts rapidly with oxygen to form brown manganese(III) hydroxide. That is,

$$4Mn(OH)_2(s) + O_2 + 2H_2O \longrightarrow 4Mn(OH)_3(s)$$

When acidified, the manganese(III) oxidizes iodide to iodine. Thus,

$$2Mn(OH)_3(s) + 2I^- + 6H^+ \longrightarrow I_2 + 3H_2O + 2Mn^{2+}$$

The liberated iodine is titrated in the usual way.

Success of the method is critically dependent upon the manner in which the sample is manipulated; at all stages, every effort must be made to assure that oxygen is neither introduced to nor lost from the sample. Biological Oxygen Demand (BOD) bottles are designed to minimize the entrapment of air. Ordinary 250-mL glass-stoppered bottles can be used.

The sample should be free of any solutes that will oxidize iodide or reduce iodine. Numerous modifications have been developed to permit the use of the Winkler method in the presence of such species.

Reagents needed.

Manganese(II) sulfate reagent. Dissolve 48 g of $MnSO_4·4H_2O$ in sufficient water to give 100 mL of solution.

Potassium iodide/sodium hydroxide. Dissolve 15 g of KI in about 25 mL of water, add 66 mL of 50% NaOH, and dilute to 100 mL.

Standard 0.025 N sodium thiosulfate. See Preparation and Standardization 10-1 (p. 766); approximately 6.2 g of $Na_2S_2O_3·5H_2O$ are needed to prepare 1 L of solution.

Concentrated H_2SO_4

Starch indicator solution (Preparation 9-1)

Procedure. Transfer the sample to a BOD bottle, with care being taken to minimize exposure to air. Use a tube to introduce the water to the bottom of the bottle; remove the tube slowly while the bottle is overflowing.

Add 1 mL of $MnSO_4$ reagent with a dropper; discharge the reagent well below the surface (some overflow will occur). Similarly, introduce 1 mL of the KI/NaOH solution. Place the stopper in the bottle; be sure that no air becomes entrapped. Invert the bottle to distribute the precipitate uniformly.

When the precipitate has settled at least 3 cm below the stopper, introduce 1 mL of concentrated (18 M) H_2SO_4 well below the surface (Note 1). Replace the stopper and mix until the precipitate dissolves (Note 2). With a graduated cylinder, measure exactly 200 mL of the acidified sample into a 500-mL conical flask (Note 3). Titrate with 0.025 N $Na_2S_2O_3$ until the iodine color becomes faint. Then introduce 5 mL of starch indicator and complete the titration.

Report the milliliters of O_2 (STP) in each liter of sample (Note 4).

NOTES
1. Care should be taken to avoid exposure to the overflow, as the solution is quite alkaline.
2. A magnetic stirrer is effective in bringing about solution of the precipitate.
3. For most purposes, the accuracy required is such that careful measurement of the sample volume with a graduated cylinder suffices. Furthermore, correction of the volume for the two additions of reagents is unnecessary.
4. Interesting results can often be obtained if samples are taken from different depths in a lake (particularly in summer) or above and below the rapids in a stream. Analysis of water samples drawn sequentially from a little used faucet is also of interest.

11 TITRATIONS WITH POTASSIUM BROMATE

The uses of potassium bromate for the determination of organic compounds are described in Section 15B-5. Directions follow for two typical applications, including the determination of phenol in water and the determination of ascorbic acid in vitamin C tablets.

PREPARATION 11-1. DIRECT PREPARATION OF STANDARD 0.05 N POTASSIUM BROMATE

Procedure. Dry reagent-grade $KBrO_3$ for 1 hr at 100 to 110°C, and cool. Weigh approximately 0.7 g (to the nearest 0.2 mg) into a 500-mL volumetric flask, dilute to the mark, and mix thoroughly. Calculate the normality on the basis of a six-electron change for BrO_3^-.

STANDARDIZATION 11-1. STANDARDIZATION OF 0.05 N SODIUM THIOSULFATE AGAINST POTASSIUM BROMATE

Discussion. The standard bromate solution is employed to generate a known amount of I_2.

$$BrO_3^- + 6I^- + 6H^+ \longrightarrow Br^- + 3I_2 + 3H_2O$$

The iodine is then titrated with the sodium thiosulfate solution.

Reagents needed.

Standard 0.05 N KBrO₃
0.05 N Na₂S₂O₃. See Preparation 10-1 (p. 766); employ 12.5 g
    $Na_2S_2O_3 \cdot 5H_2O$ for 1 L of solution.
3 M H₂SO₄
KI
Starch indicator solution (Preparation 9-1)

Procedure. Pipet 25-mL aliquots of the $KBrO_3$ solution into conical flasks. *From this point, treat each aliquot individually.* Add about 2 to 3 g of KI and 5 mL of 3 M H_2SO_4. Titrate the liberated I_2 until the solution is just faintly yellow. Add 5 mL of starch solution, and continue the titration until the blue color disappears. Calculate the normality of the thiosulfate solution.

METHOD 11-1. DETERMINATION OF PHENOL BY BROMINATION

Discussion. The phenol content of waste water from manufacturing processes is readily determined by adding a measured excess of standard bromate solution to the sample followed by an excess of bromide. Upon acidification, the following reactions occur (see also p. 379).

$$BrO_3^- + 5Br^- + 6H^+ \longrightarrow 3Br_2 + 3H_2O$$
$$C_6H_5OH + 3Br_2 \longrightarrow C_6H_2Br_3OH + 3H^+ + 3Br^-$$

After allowing time for the bromination to occur completely, the excess bromate is determined iodometrically as in Standardization 11-1.

Reagents needed.

> *Standard 0.05 N KBrO₃*
> *Standardized 0.05 N Na₂S₂O₃*
> *3 M H₂SO₄*
> *KBr*
> *KI*
> *Starch indicator solution* (Preparation 9-1)

Procedure. Transfer a sample containing between 1 and 1.5 mmol of phenol to a 250-mL volumetric flask, and dilute to the mark. Pipet 25-mL aliquots of this solution into 250-mL glass-stoppered conical flasks, and add exactly 25 mL of the standard $KBrO_3$ reagent. Add about 1 g of KBr to each flask and about 5 mL of 3 M H_2SO_4. Stopper each flask *immediately* after addition of the acid to prevent loss of Br_2. Mix, and let stand for at least 10 min. Weigh 2 to 3 g of KI (not accurately) for each sample. Add the KI rapidly to each flask, and restopper it immediately thereafter. Swirl the solution until the KI is dissolved. Then, titrate with standard $Na_2S_2O_3$ until the solution is faintly yellow. Add 5 mL of starch, and complete the titration.

Report the number of milligrams phenol contained in the original sample.

METHOD 11-2. DETERMINATION OF ASCORBIC ACID IN VITAMIN C TABLETS BY TITRATION WITH POTASSIUM BROMATE

Discussion. Ascorbic acid is cleanly oxidized to dehydroascorbic acid by bromine.

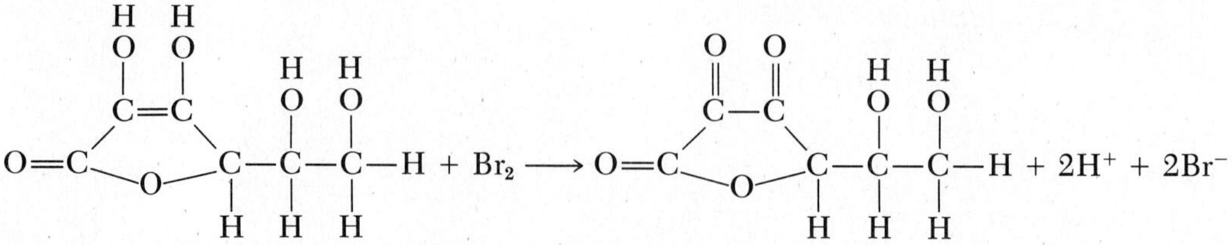

An excess of bromate is employed and the excess is determined iodometrically by titration with standard thiosulfate solution. The procedure must be carried out promptly once the sample is dissolved to avoid air oxidation of the ascorbic acid.

Reagents needed.

> Standard 0.05 N KBrO$_3$
> Standardized 0.05 N Na$_2$S$_2$O$_3$
> Starch indicator solution (Preparation 9-1)
> 3 M H$_2$SO$_4$
> KBr
> KI

Procedure. Weigh (to the nearest milligram) three vitamin C tablets. Pulverize them in a mortar, and transfer the powder to a dry weighing bottle. Weigh 0.2- to 0.25-g samples (to the nearest 0.1 mg) into 250-mL conical flasks. *From this point, treat each sample individually* to avoid air oxidation of the ascorbic acid. Dissolve the sample in a solution made up of 10 mL of 3 M H$_2$SO$_4$, 25 mL of water, and 5 g KBr (Note). Titrate immediately with KBrO$_3$ to the first faint yellow due to excess Br$_2$. Record the volume of oxidant, add 5 g of KI, 5 mL of starch, and back-titrate with the Na$_2$S$_2$O$_3$ solution. Report the average milligrams of ascorbic acid, C$_6$H$_8$O$_6$, in each tablet.

> NOTE
> For most samples, some undissolved binder will remain as a suspension throughout the titration.

12 TITRATIONS WITH POTASSIUM IODATE

The oxidizing properties and applications of iodate ion are described in Section 15B-6.

PREPARATION 12-1. DIRECT PREPARATION OF STANDARD 0.025 *M* POTASSIUM IODATE

Discussion. The potassium iodate solution will be employed for two purposes: (1) the standardization of a sodium thiosulfate solution and (2) the oxidation of I$^-$ and I$_2$ in the sample to ICl$_2^-$.

Procedure. Dry reagent-grade KIO$_3$ for 1 hr at 100 to 110°C, and cool. Weigh about 2.68 g (to the nearest milligram) into a 500-mL volumetric flask, dilute to the mark, and mix thoroughly.

STANDARDIZATION 12-1. STANDARDIZATION OF SODIUM THIOSULFATE AGAINST 0.025 *M* POTASSIUM IODATE

Discussion. For standardization of thiosulfate, a measured quantity of the reagent reacts with an excess of iodide to form a known quantity of iodine, which is then titrated with the thiosulfate solution. Here,

$$IO_3^- + 5I^- + 6H^+ \rightleftharpoons 3I_2 + 3H_2O$$

In its reaction with thiosulfate, each iodine molecule undergoes a two-electron reduction so that the equivalent weight of potassium iodate is one-sixth the gram formula weight; the normality of the solution is six times its molarity.

Reagents needed.

> *Standard 0.025 M KIO$_3$*
> *0.1 N Na$_2$S$_2$O$_3$.* See Preparation 10-1 (p. 766)
> *KI*
> *6 M HCl*
> *Starch indicator solution* (Preparation 9-1)

Procedure. Pipet 25-mL aliquots of the 0.025 M KIO$_3$ into 250-mL conical flasks;
add about 50 mL of water and 2 g of iodate-free KI to each. *From here, treat
each aliquot individually.* Add about 2 mL of 6 M HCl and titrate with the Na$_2$S$_2$O$_3$
reagent until the color becomes pale yellow. Add 5 mL of starch indicator (p. 764),
and titrate until the blue color disappears.

METHOD 12-1. DETERMINATION OF IODINE AND IODIDE ION IN AN AQUEOUS MIXTURE

Discussion. The iodine in the mixture is determined by direct titration of an aliquot of the sample with standard thiosulfate solution. The concentration of iodine
plus iodide is then determined by the direct titration of another sample with the
standard iodate solution. Here, the solution contains a high concentration of chloride ion, and the reaction product is exclusively ICl$_2^-$:

$$2I^- + IO_3^- + 6Cl^- + 6H^+ \longrightarrow 3ICl_2^- + 3H_2O$$
$$2I_2 + IO_3^- + 10Cl^- + 6H^+ \longrightarrow 5ICl_2^- + 3H_2O$$

Starch indicator does not function in the presence of high concentrations of hydrochloric acid. Thus, the end point is observed as the disappearance of the iodine
color in a small amount of toluene or chloroform.

Reagents needed.

> *Standard 0.025 M KIO$_3$*
> *Standardized 0.1 N Na$_2$S$_2$O$_3$*
> *Concentrated HCl*
> *Starch indicator solution* (Preparation 9-1)
> *Toluene or chloroform*

Procedure. Obtain the unknown in a clean, 500-mL volumetric flask. Dilute to
the mark and mix thoroughly.

To determine the I$_2$ in the unknown, titrate 50.0-mL aliquots of the sample
with standard Na$_2$S$_2$O$_3$ until the solution is faint yellow in color. Add 5 mL of starch
indicator and complete the titration.

To determine the combined concentration of I$_2$ and I$^-$ in the sample, fill a
buret with the unknown solution and introduce about 25.0 mL into a 250-mL con-

ical flask. Add about 40 mL of concentrated HCl and 10 mL of toluene or chloroform. Titrate with KIO_3 until the purple color disappears from the organic layer. As the end point is approached, make dropwise additions of reagent, and swirl vigorously between additions. To eliminate the likelihood of overtitration, introduce enough of the sample solution to give the iodine color in the nonaqueous phase again. Titrate as before with KIO_3. Repeat, as necessary, to establish the end point.

Report the grams I_2 and KI in the sample.

13 APPLICATION OF PERIODATE SOLUTIONS

METHOD 13-1. DETERMINATION OF GLYCEROL

Discussion.   General aspects of periodate oxidations are found in Section 15B-7. Instructions follow for the determination of glycerol; this substance is cleanly oxidized to formic acid and formaldehyde by a measured excess of periodate in dilute acid solution. Reaction:

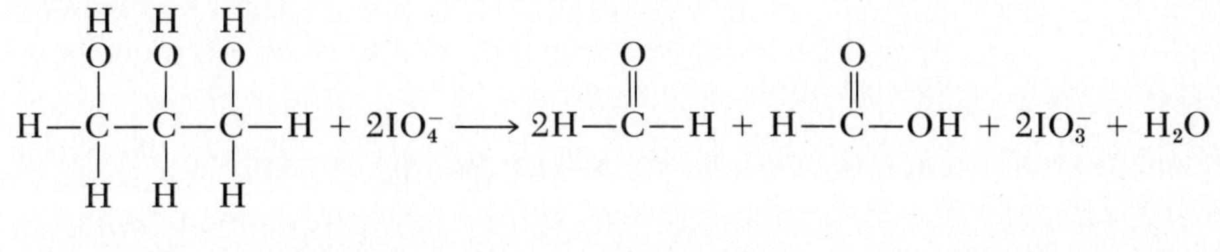

$$\begin{array}{ccc} H & H & H \\ O & O & O \\ | & | & | \\ H-C-C-C-H \\ | & | & | \\ H & H & H \end{array} + 2IO_4^- \longrightarrow 2H-\overset{O}{\overset{\|}{C}}-H + H-\overset{O}{\overset{\|}{C}}-OH + 2IO_3^- + H_2O$$

The oxidation is complete in 1 hr at room temperature. The reaction can be completed in either of two ways:

(1) The reaction mixture is acidified, an excess of potassium iodide is introduced, and the iodine liberated by the excess periodate as well as the iodate formed in the reaction is titrated with standard sodium thiosulfate:

$$IO_4^- + 7I^- + 8H^+ \longrightarrow 4I_2 + 4H_2O$$
$$IO_3^- + 5I^- + 6H^+ \longrightarrow 3I_2 + 3H_2O$$

(2) The reaction mixture is rendered slightly alkaline, a measured excess of sodium arsenite is introduced, and the excess As(III) is titrated with a standard iodine solution. In the alkaline medium, periodate is reduced by the As(III) while the iodate is not.

The first method suffers from the disadvantage of requiring a large back-titration with thiosulfate but has the virtue of eliminating the need for arsenic(III), the use of which is now subject to close regulation. Only the first procedure is described here.

Reagents needed.

0.03 M KIO_4. Dissolve about 3.5 g of KIO_4 in 500 mL of 0.05 *M* H_2SO_4; warm, as necessary, to bring about solution.

Standardized 0.15 N $Na_2S_2O_3$. See Preparation and Standardization
10-1. For this solution, use about 37 g of $Na_2S_2O_3 \cdot 5H_2O$/L.
Starch indicator solution (Preparation 9-1)
Potassium iodide, about 5 g/sample
3 M HCl

Procedure. Transfer 25.0-mL aliquots of periodate solution to individual 250-mL
conical flasks. Introduce 25.0-mL aliquots of a solution that is between 0.01 and
0.02 *M* in glycerol. Permit at least 1 hr for completion of the oxidation (Note). *Treat
each sample individually from this point.* Introduce about 15 mL of 3 *M* HCl and
about 5 g of KI. Titrate immediately with standard thiosulfate until the solution becomes a pale straw yellow. Then introduce 3 to 5 mL of starch solution and complete the titration. Report the milligrams of glycerol in each milliliter of the sample.

> NOTE
> The periodate solution is conveniently standardized during the time required for the glycerol
> oxidation. Use 25 mL of water in place of the glycerol.

14 POTENTIOMETRIC METHODS

The uses of potential measurements in analytical chemistry are discussed in
Chapter 16. A general procedure for carrying out potentiometric titrations is given
here; directions for typical applications follow.

GENERAL INSTRUCTIONS FOR PERFORMING POTENTIOMETRIC TITRATIONS

1. Dissolve the sample in 50 to 250 mL of water. Rinse the electrodes with
   distilled water, and immerse them in the sample solution. Provide magnetic (or mechanical) stirring. Position the buret so that the reagent can be
   delivered without splashing.
2. Connect the electrodes to the pH meter, commence stirring, measure, and
   record the initial potential as well as the initial volume.
3. Measure and record the potential and the volume after each addition of
   reagent. Introduce fairly large volumes (say, 5 mL) at the outset. Withhold
   each succeeding addition until the potential remains constant within 1 to
   2 mV (or 0.05 pH unit) for 30 s. A stirring motor will occasionally cause
   erratic potential readings; it may be advisable to turn off the motor during
   the actual measuring process. Judge the volume of reagent to be added by
   calculating an approximate value of $\Delta E/\Delta V$ after each addition. In the
   immediate vicinity of the equivalence point, introduce the reagent in
   exact 0.1-mL increments. Continue the titration 2 to 3 mL past the equivalence point. Increase the volumes added as $\Delta E/\Delta V$ once again acquires
   small values.
4. Locate the end point by one of the methods described on pages 421 to
   424.

METHOD 14-1. POTENTIOMETRIC DETERMINATION OF CHLORIDE AND IODIDE IN A MIXTURE

Discussion. The potentiometric titration of halide mixtures is discussed in Section 7A-3. The indicator electrode can be simply a polished silver wire or a com-

mercial billet-type variety. A calomel electrode can be used as reference, although diffusion of Cl^- from the salt bridge may cause the results to be slightly high. To avoid this error, the calomel electrode can be placed in a saturated KNO_3 solution; contact is made with the solution to be titrated by means of a KNO_3 bridge. Alternatively, the solution can be made acidic with several drops of nitric acid; a glass electrode can then be employed as the reference because the pH of the solution will remain constant throughout the titration.

Multiple end points are often revealed by a potentiometric titration when the sample contains more than one reacting species. The titration of halide mixtures with silver nitrate is an important example. Figure 7-3 (p. 183) is a theoretically derived titration curve for a mixture of chloride and iodide with silver nitrate. Because the measured potential is directly proportional to the negative logarithm of the silver ion concentration, the experimental potential curve has the same shape, although the ordinate units are different.

Experimental curves for the titration of halide mixtures do not show the sharp discontinuity at the first equivalence point that appears in Figure 7-3. More important, the volume of silver nitrate required to reach the first end point is generally somewhat greater than theoretical; the total volume, however, approaches the correct amount. These observations apply to the titration of chloride/bromide, chloride/iodide, and bromide/iodide mixtures and can be explained by assuming that the more soluble silver halide is coprecipitated during forming of the less soluble compound. An overconsumption of reagent in the first part of the titration thus occurs.

Despite the coprecipitation error, the potentiometric method is useful for the analysis of chloride/iodide mixtures. When approximately equal quantities are present, relative errors can be kept to about 1 to 2%.[7] Less satisfactory results are obtained with mixtures containing bromide ion.

Reagents needed.

Standard 0.01 M $AgNO_3$, 250 mL. See Preparation 2-2 (p. 737).
2.5 M KNO_3
KNO_3
Agar

Procedure. To prepare a KNO_3 salt bridge, bend an 8-mm glass tube into a U-shape with long enough arms to reach near the bottoms of two 100-mL beakers. Prepare an agar suspension for filling the tube by heating 50 mL of water to boiling and stirring in 1.8 g of powdered agar. Continue the heating and stirring until a uniform suspension is formed. Dissolve 12 g of KNO_3 in the hot suspension. Cool until gel formation just begins and then warm slightly until the gel disappears. Clamp the tube in an upright position, and with a medicine dropper, fill it with the warm mixture. Cool the tube under the tap to form the gel. When not in use, the two ends of the bridge should be immersed in a 2.5 M KNO_3 solution.

Transfer a 25-mL aliquot of the sample to a clean 100-mL beaker and add a drop or two of concentrated HNO_3. Place about 25 mL of 2.5 M KNO_3 in a second

---

[7] For further details, see: I. M. Kolthoff and N. H. Furman, *Potentiometric Titrations*, 2d ed., pp. 154–158. New York: Wiley, 1931.

beaker and join the two by means of the agar salt bridge. Immerse the silver electrode in the analyte solution and the calomel reference electrode in the second beaker. Titrate with the $AgNO_3$ as described in *General Instructions*. Use small increments of titrant near the two end points. Plot the data and establish the end point for each ion (p. 421). Plot a theoretical titration curve, assuming the measured concentrations of the two constituents to be correct. Report the number of milligrams of I^- and Cl^- in the sample or as instructed.

Rinse the electrodes and transfer the solution containing the precipitate to the container for silver wastes. Any unused $AgNO_3$ should also be placed in this container.

METHOD 14-2. POTENTIOMETRIC TITRATION OF A WEAK ACID

Discussion. A glass/calomel electrode system is convenient for locating end points in acid-base titrations and for establishing the approximate dissociation constants of weak acids and bases (see Section 16D-4). Ordinarily, prior to a series of titrations the electrode system is standardized by means of a buffer of known pH.

Reagents needed.

> *Standardized 0.1 N carbonate-free NaOH, 500 mL.* See Preparation
>   3-3 (p. 740) and Standardization 3-3 (p. 742).
> *Standard buffer solution(s)*
> *Phenolphthalein indicator solution* (p. 740)

Procedure. Dissolve between 2 and 4 meq of the acid in about 100 mL of water. Some of the less soluble organic acids may require a larger volume of water. Alternatively, the acid may be dissolved in 5 to 10 mL of ethyl alcohol (Note) and then diluted to 50 to 100 mL.

Calibrate the electrode system by rinsing the electrodes with water and immersing them in one or more buffer solutions of known pH. Follow the instrument instructions for adjusting the meter to read this pH.

Rinse the electrodes thoroughly and immerse them in the sample solution. Add 1 to 2 drops of phenolphthalein. Titrate the solution as directed in *General Instructions* (p. 778). Some samples will contain more than one replaceable hydrogen; be alert for more than one break in the titration curve. Note the volume at which the indicator changes color.

Plot the titration data, and determine the end point or points. Compare these with the phenolphthalein end point. The derivative method for end point determination may also be used.

Calculate the number of milliequivalents of H^+ present in the sample. Evaluate the approximate dissociation constant(s) for the species titrated.

> NOTE
> Ethanol is often contaminated with small amounts of acid. If ethanol is used, blank titrations are recommended.

METHOD 14-3. DETERMINATION OF THE SPECIES PRESENT IN A PHOSPHATE MIXTURE

Discussion. The sample will contain one or two of the following components: HCl, H_3PO_4, NaH_2PO_4, Na_2HPO_4, Na_3PO_4, and NaOH. The object is to determine which compatible species are present as well as their amounts.

The analysis of some mixtures can be completed by titrations with either standard HCl or with standard NaOH. Others will require titration of separate aliquots, one with acid and the other with base. The appropriate titrant(s) can be determined from the initial pH of the sample and the titration curve for H_3PO_4 (p. 234). A study of pages 221 to 235 may assist in interpreting the data.

Reagents needed.

> *Standardized 0.1 N HCl and/or standardized 0.1 N NaOH, 500 mL.*
> See Preparations 3-2 and 3-3 and Standardizations 3-2 and 3-3.
> *Bromocresol green indicator solution* (p. 740)
> *Phenolphthalein indicator solution* (p. 740)

Procedure. The sample will be issued as a solution, which should be diluted to volume in a 250-mL volumetric flask. Mix well.

Pipet exactly 50 mL of the sample into a beaker, and determine its pH. Titrate with either standard HCl or NaOH until one or two end points have been passed; follow *General Instructions* (p. 778). If necessary, titrate a second aliquot with the reagent not employed in the first titration. On the basis of the titration curve(s), choose acid-base indicators that would be suitable for the titrations, and perform the duplicate analyses with these.

Identify and report the total number of millimoles of the component(s) in the unknown solution. Also, calculate and report approximate values for any dissociation constant for H_3PO_4 that can be obtained from the titration data.

METHOD 14-4. DIRECT POTENTIOMETRIC DETERMINATION OF FLUORIDE ION

Discussion. The solid state fluoride electrode (p. 409) has been widely used for the determination of fluoride ion in a variety of materials. In this section, methods for the determination of this ion in drinking water and toothpaste are given. For both, a total ionic strength adjustment buffer is added to the samples and standards to bring them to approximately the same ionic strength; in this way, the concentration rather than the activity of fluoride ion is determined. The pH of the buffer is about 5, a level at which fluoride ion is the predominant fluorine-containing species. The buffer also contains an aminocarboxylic acid complexing agent, cyclohexylenedinitrilotetraacetic acid, which forms strong complexes with heavy metal ions such as iron(III) and aluminum(III), thus freeing the fluoride ion from its complexes with these cations.

Before undertaking this experiment, it is suggested that Section 16C on direct potentiometry be studied.

Apparatus. The apparatus for this experiment will consist of a solid-state fluoride-ion electrode, a saturated calomel electrode, and a pH meter. For the toothpaste analysis, a sleeve-type calomel electrode is required because the measurement is made upon a suspension that tends to clog the liquid junction. After each series of samples has been analyzed, the interface must be renewed by loosening the sleeve momentarily.

Reagents needed.

Total ionic strength adjustment buffering solution (TISAB).
This solution can be purchased from a commercial source under the label of TISAB.[8] It can be prepared by dissolving with stirring approximately 57 mL of glacial acetic acid, 58 g of NaCl, and 4 g of cyclohexylenedinitrilotetraacetic acid in 500 mL of distilled water in a 1-L beaker. Cool the beaker in a water or ice bath and carefully add 5 M NaOH until the solution reaches a pH of 5.0 to 5.5 (glass electrode). Dilute to about 1 L, mix, and store in a stoppered plastic bottle.
100 ppm fluoride standard solution. Prepare a 100 ppm standard solution of F^- by weighing into a 1-L volumetric flask 0.22 g (to the nearest milligram) of NaF (CAUTION! NaF is highly toxic) that has been dried for 2 hr at 110°C and cooled in a desiccator. Dissolve, dilute to the mark, mix well, and store in a plastic bottle. Calculate its exact concentration in ppm F^-. This solution is available commercially.

Procedure.

Determination of fluoride in drinking water. Transfer 50.0-mL samples of the water to 100-mL volumetric flasks and dilute to volume with the TISAB solution. Prepare a 5 ppm F^- solution by diluting 25.0 mL of the 100 ppm solution to 500 mL in a volumetric flask. Transfer 5.00-, 10.0-, 25.0-, and 50.0-mL aliquots of the 5 ppm solution to 100-mL volumetric flasks, add 50 mL of the TISAB solution, and dilute to 100 mL (these solutions correspond to 0.5, 1.0, 2.5, and 5.0 ppm F^- in the sample). After thorough rinsing and drying with tissue, immerse the electrodes in the 0.5 ppm standard solution and stir mechanically for 3 min. Measure and record the potential. Repeat for each of the remaining standards and the samples. Construct a plot of the potential versus the log of the concentration of the standards. Calculate and report the slope of this plot. Determine the parts per million of F^- in the unknown samples from the measured potentials.

Determination of fluoride in a sample of toothpaste.[9] Weigh approximately 0.2-g (to the nearest 1 mg) samples of the toothpaste into 250-mL beakers. Add 50 mL of the TISAB solution, and boil for 2 min with good mixing. After cooling, transfer the suspension quantitatively to a 100-mL volumetric flask, and dilute to the mark. Follow the directions beginning with the second sentence in the preceding paragraph. Report the %F^- in the sample.

---

[8] Orion Research Inc., Cambridge, MA 02139.
[9] This procedure was taken from T. S. Light and C. C. Cappuccino, *J. Chem. Educ.*, **52**, 247 (1975).

METHOD 14-5. DETERMINATION OF ORGANIC SALTS BY POTENTIOMETRIC TITRATION IN GLACIAL ACETIC ACID

Discussion. The most common way of establishing end points in nonaqueous neutralization titrations involves potentiometric measurements with a glass/calomel electrode system.

Many organic salts can be dissolved in glacial acetic acid and titrated directly with standard perchloric acid prepared in the same solvent. Salts that are not readily soluble will often dissolve in an excess of standard perchloric acid reagent; this excess is then back-titrated with a standard solution of sodium acetate in glacial acetic acid.

Reagents needed.

> *0.1 N HClO₄ in acetic acid.* See Preparation 4-1 (p. 747).
> *0.1 N NaC₂H₃O₂ in acetic acid.* This reagent will not be needed
> if the sample is soluble in glacial acetic acid. If it is required,
> see Preparation 4-2.

Procedure. Standardize the $0.1 N$ $HClO_4$ solution, following the instructions in Standardization 4-2 (p. 748). Determine the end point potentiometrically as follows. Rinse the glass and calomel electrodes with water and dry with a tissue immediately before the titration. Then follow the *General Instructions* for performing a potentiometric titration. Use the millivolt scale of the meter.

If the $NaC_2H_3O_2$ solution is required, determine the acid-base ratio following the instructions in Standardization 4-1. Again, determine the end point potentiometrically.

The sample will be the sodium, potassium, or ammonium salt of an organic acid. Weigh 0.1- to 0.2-g portions (to the nearest 0.1 mg) into beakers, and dissolve in 25 mL of glacial acetic acid. Heat, if necessary, until solution is complete; then cool to room temperature. Titrate immediately with standard $HClO_4$, using the glass/calomel electrode system and the millivolt scale of the pH meter.

If the sample is not readily soluble, add an accurately measured volume of the standard $HClO_4$ solution (about 20 mL) to the suspension of the salt in acetic acid, and stir for about 10 min. The presence of the $HClO_4$ should bring the salt into solution. Back-titrate the excess $HClO_4$ with standard $NaC_2H_3O_2$, and determine the end point potentiometrically, as in the acid-base ratio determination.

Repeat the experiment, and report the average equivalent weight of the sample.

METHOD 14-6. POTENTIOMETRIC TITRATION OF AMINO ACIDS

Reagents needed.

> *0.1 N HClO₄ in glacial acetic acid.* See Preparation 4-1
> (p. 747).

Procedure. Follow the procedure described in Method 14-5.

METHOD 14-7. POTENTIOMETRIC TITRATION OF IRON(II) WITH CERIUM(IV)

Discussion. The iron in a sample of matter can be determined conveniently by potentiometric titration with a standard solution of tetravalent cerium. Prereduction of the iron is accomplished with tin(II) chloride (see p. 757). The potentiometric measurements can be made with a platinum/calomel electrode system. Alternatively, differential measurements are possible with a pair of platinum electrodes arranged as shown in Figure 16-15 (p. 429).

Reagents needed.

> *Standardized 0.1 N cerium(IV).* See pp. 761 to 762.
> *0.25 M SnCl₂.* See p. 758.
> *5% (w/v) HgCl₂*

Procedure.
Platinum/calomel electrode. Rinse the electrodes and connect to the meter. Use the millivolt scale. Samples should be carried individually and expeditiously through the prereduction and titration steps to avoid air oxidation of the iron(II). Blanketing the surface of solution with nitrogen throughout the titration may lead to more reproducible results.

Reduce the Fe to the plus two state following the procedure in the paragraph labeled *Reduction of iron* on page 759. Then immerse the electrode in the solution and titrate immediately following the *General Instructions* on page 778. Plot the potential data versus the reagent volume and calculate the Fe content of the sample.

Differential electrode system. In this procedure, a pair of platinum electrodes is used; one is housed in a small glass tube, as shown on page 429.

Prepare the reagent and sample as directed in the preceding section. Add about 1 mL of the reagent, stir, and measure the potential. Then homogenize the solution around the shielded electrode by squeezing the rubber bulb, and add more reagent. Again record the potential and volume of reagent before mixing the solution around the shielded electrode with the bulk of the solution. Continue this process, decreasing the volume increments to 0.1 mL in the vicinity of the end point. Carry the titration 2 to 5 mL beyond the end point.

Plot the data, and calculate the %Fe.

NOTE
A small hole in the rubber bulb will minimize the danger of drawing solutions into the bulb. The hole is covered with a finger to push liquid from the tube and is uncovered at all other times.

15 ELECTROGRAVIMETRIC METHODS

METHOD 15-1. DETERMINATION OF COPPER IN AN AQUEOUS SOLUTION

Discussion. Read Section 17B.

Reagents needed.

> *Concentrated H₂SO₄*
> *6 M HNO₃*
> *Ethyl alcohol or acetone*

Procedure. Transfer the sample quantitatively to a 250-mL volumetric flask, dilute to the mark, and mix well (Note 1). Aliquots containing 0.2 to 0.3 g Cu should be taken. Transfer the solution to a 180-mL electrolytic beaker. Add 3 mL of concentrated H_2SO_4 and 2 mL of freshly boiled and cooled 6 M HNO_3 (Note 2). If necessary, add enough water to give a volume of about 100 mL.

Prepare the platinum electrodes by immersing them in hot 6 M HNO_3 for 5 min (Note 3). Wash thoroughly with distilled water, rinse several times with small portions of ethyl alcohol or acetone, and dry in an oven at 110°C for 2 to 3 min. Cool and weigh the cathode to the nearest 0.1 mg on an analytical balance.

Attach the cathode to the negative terminal of the electrolytic apparatus and the anode to the positive terminal. Elevate the beaker containing the solution so that all but a few millimeters of the cathode are covered. Start the stirring motor, and adjust the potential so that a current of about 2 A passes through the cell (Note 4). When the blue color has entirely disappeared from the solution, add sufficient water to raise the liquid level by a detectable amount, and continue the electrolysis with a current of about 0.5 A. If no further Cu deposit appears on the newly covered portion of the cathode within 15 min, the electrolysis is complete. If additional Cu does deposit, continue testing for completeness of deposition from time to time as directed above.

When no more Cu is deposited after a 15-min period, stop the stirring and slowly lower the beaker while continuously playing a stream of wash water on the electrodes. Rinse the electrodes thoroughly with a fine stream of water. *Maintain the applied potential until rinsing is complete* (Note 5). Disconnect the cathode, and immerse it in a beaker of distilled water. Then rinse it with several portions of alcohol or acetone. Dry the cathode in an oven for 2 to 3 min at 110°C, cool, and weigh.

Report the milligrams of Cu in the sample.

NOTES

1. The sample should not contain Cl^- because its presence results in attack on the Pt anode. This reaction not only is destructive but also will lead to errors because the dissolved Pt will codeposit with the Cu at the cathode.
2. The HNO_3 is boiled to remove oxides of nitrogen which tend to delay the deposition of Cu and may also give rise to formation of copper oxide on the deposit. The nitrate ion is added to serve as a depolarizer (p. 439), which is reduced in lieu of hydrogen ions. The formation of hydrogen leads to less adherent deposits.
3. Grease and organic material, if present, can be removed by bringing the electrode to red heat in a flame. The cathode surface should not be touched with the fingers because grease and oil cause nonadherent deposits.
4. Alternatively, the electrolysis may be carried out without stirring. Here, the current should be kept below 0.5 A. Several hours will be required to complete the analysis.
5. It is important to maintain the application of a potential until the electrodes have been removed from the solution and washed free of acid. If this precaution is not observed, some Cu may redissolve.

METHOD 15-2. DETERMINATION OF COPPER IN AN ORE

Reagents needed.

Concentrated HNO₃
9 M H₂SO₄

Procedure. Weigh appropriately sized samples of the finely ground and dried ore (about 1.5 g for a sample containing 10 to 30% Cu) into 150-mL beakers, and add 15 mL of concentrated HNO_3. Cover with ribbed watch glasses and heat in a hood until all of the copper is dissolved. If the volume becomes less than 5 mL, add more HNO_3. Continue heating until a white or slightly gray siliceous residue remains (Note). Cool the solution, add 10 mL of 9 M H_2SO_4, and heat until dense white fumes of SO_3 appear. (The heating removes oxides of nitrogen; see Note 2, Method 15-1).

   Dilute the solution to about 80 mL, heat to boiling, and filter through a medium-porosity filter paper into a 180-mL electrolytic beaker. Wash the paper with three 5-mL portions of water. Add 2 mL of freshly boiled and cooled 6 M HNO_3. Proceed with the electrolysis as described beginning in the second paragraph of the procedure for Method 15-1.

> NOTE
> If the ore is not readily decomposed, follow the instructions given in Note 1 of Method 10-1. Then follow the instructions in the second paragraph of this procedure.

16 COULOMETRIC TITRATIONS

METHOD 16-1. COULOMETRIC DETERMINATION OF CYCLOHEXENE

Discussion.[10]   In a largely nonaqueous environment and in the presence of mercury(II) as a catalyst, most olefins react rapidly enough with bromine to make their direct titration feasible. A convenient way of performing this titration is to introduce an excess of bromide ions and generate the bromine at an anode connected to a constant current source. The generator electrode processes are

$$2Br^- \rightleftharpoons 2Br_2 + 2e$$
$$2H^+ + 2e \rightleftharpoons H_2$$

The hydrogen does not react with bromine rapidly enough to constitute an interference. The bromine reacts with an olefin such as cyclohexene to give the addition product; that is,

The amperometric method (see Section 18C) provides a convenient way to deter-

[10] This procedure was described by D. H. Evans, *J. Chem. Educ.*, **45**, 88 (1968).

mine the end point for this titration. Here, two similar platinum electrodes are maintained at a potential difference of 0.2 to 0.3 V. This potential is not sufficient to cause the generation of bromine and hydrogen. Thus, at the outset the indicator electrodes are polarized and no current is observed. The first excess of bromine depolarizes the system, giving a current that is proportional to the bromine concentration. The indicator electrode reactions are then

$$2Br^- \rightleftharpoons Br_2 + 2e$$
$$Br_2 + 2e \rightleftharpoons 2Br^-$$

The current is readily measured with a microammeter.

A convenient way to perform replicate analyses is to generate sufficient bromine in a blank to give a readily measured current, say, 20.0 μA. Upon introduction of the sample, this current immediately decreases and approaches zero as the bromine is consumed. Generation is again commenced, and the time needed to regain a current of 20.0 μA is measured. A second portion of the sample is then added to the same solution and the process is repeated. Several samples can thus be analyzed without changing the solvent system.

The procedure that follows has been designed for the determination of the cyclohexene content of a methanol solution. It can be applied to numerous other olefins.

Reagents needed.

Solvent. Mix 300 mL of glacial acetic acid, 130 mL of methanol, and 65 mL of water; dissolve 9 g of KBr and 0.5 g of mercury(II) acetate in this mixture. CAUTION! Mercury compounds are highly toxic; additionally, the solvent mixture is a skin irritant. If inadvertent contact with the solvent occurs, wash the affected area with copious amounts of water.

Methanol

Procedure. Obtain the sample in a 100-mL volumetric flask and dilute to a few milliliters below the mark with methanol. Before diluting and taking aliquots of this sample, be sure the temperature of the mixture lies between 18 and 22°C. (The temperature coefficient of expansion is 0.11%/°C; thus, significant volumetric errors can result if temperature is not controlled.) Dilute to the mark and mix well.

Add sufficient acetic acid/methanol solvent to the electrolysis vessel to cover the indicator and generator electrodes. Apply about 0.2 V to the indicator electrodes, and generate bromine until a current of about 20 μA is indicated on the microammeter. Stop the generator, record the indicator electrode current to the nearest 0.1 μA, and set the timer to zero. Introduce a 10-mL aliquot of the sample with a transfer pipet; the indicator current should decrease and approach zero. Titrate the sample by turning on the generator. As the indicator current rises and approaches the previously recorded current, add the reagent in smaller and smaller increments by turning the generator on for shorter and shorter periods. When the original indicator current is reached, read and record the time. Then, reset the timer, add another 10-mL aliquot (or larger or smaller aliquots if the time for the first titration was too short or too long for convenient and accurate measurement). Titrate additional aliquots if desired.

Report the number of milligrams of cyclohexene in the original sample.

17 VOLTAMMETRY

Various aspects of polarography and amperometric titrations were considered in Chapter 18. Typical applications of the procedures are found in this section. Details concerning operation of the instruments for these determinations are not given because the wide variety of equipment available makes it impossible to present meaningful general instructions.

METHOD 17-1. POLAROGRAPHIC DETERMINATION OF COPPER AND ZINC IN BRASS

Discussion. The polarographic procedure makes possible the rapid determination of Cu and Zn in brass. The accuracy, however, is substantially less than that attainable by classic methods of analysis for these elements.

The sample is dissolved in a minimum amount of nitric acid, which converts at least part of the tin in the sample to a precipitate of metastannic acid $(SnO_2 \cdot 4H_2O)$; this precipitate is not removed. The solution is then made basic with an ammonia/ammonium chloride buffer, which results in precipitation of lead as the basic oxide. The polarogram of the supernatant liquid consists of two waves for copper at -0.2 and -0.5 V versus the saturated calomel electrode (SCE); one corresponds to the reduction of Cu^{2+} to Cu^+ and the second to the formation of Cu. The analysis is based upon the total diffusion current for the two waves. The zinc concentration is obtained from its wave at -1.3 V. For instruments equipped with a current compensation device, the copper wave is first measured at the highest feasible sensitivity. Its wave is then suppressed by the compensator and the Zn wave is obtained at the highest possible sensitivity setting.

Reagents needed.

Standard 2.5×10^{-2} M Cu^{2+}. Cut pure copper wire into short pieces with scissors and weigh 0.4 g (to the nearest milligram) into a weighing bottle. Transfer to a small flask and dissolve in 5 mL of concentrated HNO_3. Boil to remove oxides of nitrogen, cool, dilute with water, and transfer quantitatively into a 250-mL volumetric flask. Dilute to the mark.

Standard 2.5×10^{-2} M Zn^{2+}. Dry reagent-grade ZnO at 110°C, cool in a desiccator, and weigh about 0.5 g (to the nearest milligram) into a small beaker. Dissolve in 25 mL of water containing 5 mL of boiled, concentrated HNO_3. Transfer to a 250-mL volumetric flask and dilute to the mark.

0.1% gelatin. Dissolve about 0.1 g of gelatin in 100 mL of boiling water.

NH_3/NH_4Cl buffer. Prepare 500 mL of a solution that is approximately 1 M in NH_3 and 2 M in NH_4Cl.

Concentrated HNO_3

Procedure. With a buret, transfer 0-, 1-, 8-, and 15-mL aliquots of the standard Cu^{2+} solution to 50-mL volumetric flasks; add 5 mL of the gelatin solution and 30 mL of the buffer. Dilute to the mark. For each standard, rinse the polarographic cell with three small portions of the solution and then fill. Bubble nitrogen through the solution for 10 to 15 min to remove oxygen. Apply a potential of about -1.6 V (Note), and adjust the sensitivity of the instrument so that the detector is at nearly full scale. Then, obtain a polarogram by varying the potential between 0 and -1.5 V (versus the SCE). Obtain a diffusion current by measuring the limiting current at a potential just beyond the second wave; measure the current for the blank at this same potential and calculate i_d/C.

Repeat the foregoing measurements with the standard Zn solution.

Analyze the brass sample by dissolving 0.10- to 0.15-g samples (weighed to the nearest 0.5 mg) in 2 mL of concentrated HNO_3 and boiling briefly to remove oxides of nitrogen. Cool, add 10 mL of water, and transfer quantitatively to a 50-mL volumetric flask. Dilute to the mark and mix well. Transfer a 10.0-mL aliquot to a 50.0-mL volumetric flask, add 5 mL of gelatin, 30 mL of buffer, and dilute to the mark. Obtain the polarogram in the same way as for the standards. Determine diffusion currents from the Cu and Zn waves and calculate the percent of each element in the sample.

NOTE

The blank should not be treated in this way. Obtain its polarogram at the sensitivity setting used for the standard of lowest concentration.

METHOD 17-2. DETERMINATION OF LEAD BY AN AMPEROMETRIC TITRATION

Discussion. Amperometric titrations are discussed in Section 18B. In the procedure that follows, the lead concentration of an aqueous solution is determined by titration with a standard potassium dichromate solution. The reaction is

$$2Pb^{2+} + Cr_2O_7^{2-} + H_2O \longrightarrow 2PbCrO_4(s) + 2H^+$$

The titration can be performed with a dropping electrode that is maintained at either 0 or -1.0 V (versus SCE). In the former case, the current remains near zero until equivalence; then it rises rapidly as a consequence of reduction of the excess dichromate ion. At -1.0 V both dichromate and lead ions are reduced. Thus, the current decreases to a minimum and then rises as the equivalence point is passed. In principle, the titration error should be less with the V-shaped curve. The advantage of the titration at 0 V is that oxygen does not have to be removed.

Reagents needed.

Buffered supporting electrolyte. Dissolve 10 g of KNO_3 and 8.2 g of sodium acetate in about 500 mL of water. Add glacial acetic acid until a pH of 4.2 is reached (pH meter). About 10 mL of the acid will be required.

0.1% (w/v) gelatin. Dissolve 0.1 g gelatin in 100 mL of boiling water.

Standard 0.01 M $K_2Cr_2O_7$. Weigh 1.47 g of primary standard-grade $K_2Cr_2O_7$ (to the nearest milligram) into a 500-mL volumetric flask. Dissolve, dilute to the mark, and mix well.

Procedure. The titration can be carried out in a 100-mL beaker. A saturated KNO_3 salt bridge similar to the one described in Method 14-1 (p. 778) should be used to provide contact between the saturated calomel electrode and the analyte solution.

Obtain the sample in a 100-mL volumetric flask, dilute to the mark, and mix well (the resulting solution will be 0.01 to 0.02 M in Pb^{2+}). Transfer a 10.00-mL aliquot to the titration vessel, add 25 mL of the buffer, and 5 mL of the gelatin. Determine the current at zero applied potential. Then add $K_2Cr_2O_7$ in 1-mL increments, measuring the current after each addition. Continue the additions to 5 mL beyond the equivalence point. Correct the currents for the volume change, and plot the data. Determine the end point and calculate the number of milligrams of Pb in the sample.

Repeat the titrations at -1.0 V. Here, however, it will be necessary to bubble nitrogen through the solution for 10 to 15 min before the titration and while additions of reagents are made. The flow of nitrogen must, of course, be stopped during the current measurement. Again, correct the currents for dilution, plot the data, establish the end point, and report the number of milligrams of Pb in the sample.

18 METHODS BASED UPON ABSORPTION OF RADIATION

Typical analytical applications of absorption measurements have been described in Chapter 21. Directions follow for (1) a determination of iron that requires the generation of a calibration curve, (2) a method for the analysis of manganese in steel that makes use of a standard addition, and (3) an experiment involving the resolution of a mixture that contains two species with mutually overlapping absorption spectra.

METHOD 18-1. DETERMINATION OF IRON IN WATER

Discussion. A standard, sensitive method for the determination of iron in household or industrial water supplies is based upon the formation of the orange-red iron(II) orthophenanthroline complex. Orthophenanthroline is a weak base; in acidic solution the principal species is the phenanthrolium ion PhH^+. Thus, the complex formation reaction is best described by the equation

$$Fe^{2+} + 3PhH^+ \rightleftharpoons Fe(Ph)_3^{2+} + 3H^+$$

(For the structure of $Fe(Ph)_3^{2+}$, see page 358). The equilibrium constant for this reaction is 2.5×10^6 at 25°C. Quantitative formation of the complex is observed in the pH region between 2 and 9. Ordinarily, a pH of about 3.5 is recommended to prevent the precipitation of various iron salts such as phosphates. Careful control of pH is not, however, required.

When orthophenanthroline is used for the analysis of iron, an excess of reducing agent is added to keep the iron in the $+2$ state; either hydroxylamine hydrochloride or hydroquinone is convenient for this purpose. Once formed, the color of the complex is stable for long periods of time.

The experiment can be performed with a spectrophotometer at 508 nm or with a photometer equipped with a green filter. Consult the manufacturer's instructions for operating details of the instrument to be used.

Reagents needed.

Standard iron solution, 0.01 mg Fe/mL. Weigh 0.0702 g of reagent-grade $FeSO_4 \cdot (NH_4)_2SO_4 \cdot 6H_2O$ into a 1-L volumetric flask. Dissolve in 50 mL of water containing 1 to 2 mL of concentrated H_2SO_4. Dilute to the mark, and mix well.

Hydroxylamine hydrochloride reductant. Dissolve 10 g of $H_2NOH \cdot HCl$ in about 100 mL of water.

Orthophenanthroline in water. Dissolve 0.1 g orthophenanthroline monohydrate in 100 mL of water. If necessary, warm gently and stir well. If the solution darkens, it must be discarded. Store in a dark place. (One milliliter of this solution is sufficient for no more than about 0.09 mg of iron.)

Sodium acetate buffering reagent. Dissolve about 10 g NaOAc in 100 mL of water.

Procedure. To prepare a calibration curve, measure a 25.0-mL aliquot of the standard iron solution into a 100-mL volumetric flask and about 25 mL of distilled water into a second. Add 1 mL of the hydroxylamine solution, 10 mL of the sodium acetate, and 10 mL of the orthophenanthroline solution to each. Allow the mixtures to stand 5 min, and then dilute each to the mark.

Clean the cells for the instrument. Rinse one with at least three small portions of the blank and the second with the standard. Determine the absorbance of the standard with reference to the blank.

Prepare at least three other solutions employing volumes of the standard iron solution such that an absorbance range of about 0.1 to 1.0 is covered. Plot a calibration curve.

To analyze the sample, pipet a 10-mL aliquot into a 100-mL volumetric flask; treat in exactly the same way as the standards, measuring the absorbance with reference to the blank. Obtain replicate measurements after adjusting the sample size such that the absorbance falls between 0.1 and 1.0.

Calculate the milligrams of Fe per liter (ppm).

METHOD 18-2. DETERMINATION OF MANGANESE IN STEEL

Discussion. Small quantities of manganese are readily determined colorimetrically by oxidation of Mn(II) to the highly colored permanganate ion. Potassium periodate is an effective oxidizing reagent for this purpose:

$$5IO_4^- + 2Mn^{2+} + 3H_2O \longrightarrow 2MnO_4^- + 5IO_3^- + 6H^+$$

Permanganate solutions containing an excess of periodate are relatively stable.

Interferences to this procedure are few. The presence of colored ions can be compensated for by employing as a blank an aliquot of the sample that has not been oxidized. This method of correction is not effective in the presence of appreciable quantities of cerium(III) or chromium(III) ions; both are oxidized by the periodate to a greater or lesser extent, and their reaction products absorb in the region commonly employed for measuring the permanganate.

The accompanying method is applicable to most steels except those containing large amounts of chromium. The sample is dissolved in nitric acid; any carbon present is removed by oxidation with peroxodisulfate. Phosphoric acid is added to complex the iron(III) and prevent the color of this species from interfering with the analysis. The standard addition method (p. 537) is used to establish the relationship between absorbance and concentration.

A spectrophotometer set at 525 nm or a photometer with a green filter may be used for absorbance measurements. Consult the instruction manual of the instrument to be used for specific operating instructions.

Reagents needed.

> *Standard manganese(II) solution.* Dissolve approximately 0.100 g
> of Mn (weighed to the nearest 0.2 mg) in about 10 mL of 6 M
> HNO_3. Boil the solution gently to eliminate oxides of nitrogen;
> cool, then transfer the solution quantitatively to a 1-L volu-
> metric flask, and dilute to the mark. Alternatively, the
> standard solution may be prepared from $MnSO_4 \cdot H_2O$, which can
> be dried at 110°C without loss of water.
>
> 6 M HNO_3
> 85% H_3PO_4
> $(NH_4)_2S_2O_8$
> $NaHSO_3$
> KIO_4

Procedure. Weigh replicate 1.0-g samples of the steel (to the nearest milligram), and dissolve in 50 mL of 6 M HNO_3 with gentle boiling; heating for 5 min should suffice. Cautiously add about 1 g of ammonium peroxodisulfate, and boil gently for 10 to 15 min. If the solution is pink or contains brown MnO_2, add approximately 0.1 g of $NaHSO_3$ or NH_4HSO_3 and heat for another 5 min. Cool and dilute the solution to exactly 100 mL in a volumetric flask.

Pipet three 20.0-mL aliquots of each sample into small beakers. Treat as follows:

	H_3PO_4, mL	Std Mn Soln, mL	KIO_4, g
aliquot 1	5	0.00	0.4
aliquot 2	5	5.00	0.4
aliquot 3	5	0.00	0.0

Boil each solution gently for 5 min; cool and dilute to 50.0 mL in volumetric flasks. Determine the absorbance of aliquots 1 and 2, with aliquot 3 serving as the reference for setting zero absorbance.

Report the average percentage of manganese in the sample.

METHOD 18-3. SPECTROPHOTOMETRIC DETERMINATION OF THE pH OF A BUFFER MIXTURE

Discussion. In this method, the absorptivities of the acid and base forms of an indicator are determined at two wavelengths. The concentration of each form in an unknown buffer is then determined by spectral measurement at the two wavelengths (p. 537); the hydronium ion concentration can then be calculated from these data.

The accompanying directions make use of bromocresol green as the indicator; several other indicators are equally satisfactory.

The behavior of bromocresol green in aqueous solution can be described by the reaction

$$HIn + H_2O \rightleftharpoons In^- + H_3O^+$$

yellow   blue

and

$$K_a = 1.6 \times 10^{-5} = \frac{[H_3O^+][In^-]}{[HIn]}$$

Spectrophotometric determination of $[In^-]$ and $[HIn]$ permits the calculation of $[H_3O^+]$.

Reagents needed.

> *Bromocresol green indicator.* Dissolve approximately 40 mg (weighed to the nearest 0.1 mg) of bromocresol green in a minimum amount of dilute NaOH, and dilute to 500 mL in a volumetric flask.
>
> *0.4 M HCl*
> *0.4 M NaOH*

Procedure.

Determination of individual absorption spectra. Transfer 25.0-mL aliquots of stock bromocresol green indicator to each of two 100-mL volumetric flasks. To one add 25.0 mL of 0.4 M HCl; to the other add 25.0 mL of 0.4 M NaOH. Dilute each to the mark; mix thoroughly. Obtain the absorption spectra for the acid and conjugate-base forms of the indicator between 400 and 600 nm, using water as a blank; record absorbance values at 10-nm intervals routinely and at smaller intervals as necessary to define maxima or minima. Evaluate the absorptivity for each species at wavelengths that correspond to their respective absorption maxima.

Determination of the pH of an unknown buffer. Transfer a 25.0-mL aliquot of the stock bromocresol green indicator to a 100-mL volumetric flask. Add 50.0 mL of the unknown buffer, dilute to the mark, and measure the absorbance of the diluted solution at the wavelengths for which absorptivity data were calculated. Report the pH of the buffer.

19 MOLECULAR FLUORESCENCE

METHOD 19-1. DETERMINATION OF QUININE IN TONIC WATER OR BITTER LEMON[11]

Discussion. Solutions of quinine fluoresce strongly when excited by radiation having a wavelength of 350 nm. The relative intensity of the fluorescence peak, which occurs at 450 nm, provides a sensitive method for the determination of quinine in beverages. The concentration of this compound in beverage mixes ordinarily ranges from 25 to 60 ppm.

For a discussion of molecular fluorescence see pages 556 to 562.

Reagents needed.

> 0.05 M H_2SO_4
>
> *Standard quinine sulfate solution, 1 ppm.* Weigh 0.100 g of quinine sulfate (weighed to the nearest 0.5 mg) into a 1-L volumetric flask, and dilute to the mark with 0.05 M H_2SO_4. Pipet exactly 10 mL of this solution into a 1-L volumetric flask and dilute to the mark with 0.050 M H_2SO_4. This solution must be prepared daily and should be kept in the dark when not in use.

Procedure. Measure the relative fluorescence intensity at 450 nm or with a suitable filter that transmits in this range. To find a suitable working range of concentrations, measure the relative fluorescent intensity of the 1 ppm standard. Then dilute 10 mL of this reagent with 10 mL of 0.05 M H_2SO_4, and again measure the relative fluorescent intensity. For these preliminary measurements, the dilutions can be done with graduated cylinders. Repeat the dilution and measurement process until the relative intensity approaches that for a blank of 0.05 M H_2SO_4. Plot the data and choose a suitable concentration range for the analysis (where the plot is linear). Using a pipet or a buret and volumetric flasks, prepare three or four standards within the chosen range; obtain relative fluorescence intensities for each. Plot the data.

Obtain a tonic sample from the instructor and make dilutions of it with 0.05 M H_2SO_4 until the fluorescence intensity falls within the range of the calibration curve. Calculate the parts per million quinine sulfate in the sample.

20 ATOMIC SPECTROSCOPY

METHOD 20-1. DETERMINATION OF LEAD IN BRASS BY ATOMIC ABSORPTION SPECTROSCOPY

Discussion. Brass and other copper-based alloys contain from zero to ten percent lead as well as zinc, tin, and copper. Atomic absorption spectroscopy permits the

---

[11] This procedure was adapted from J. E. O'Reilly, *J. Chem. Educ.*, **52**, 610 (1975).

rapid quantitative estimation of these elements. The weighed sample is dissolved in nitric acid which contains enough hydrochloric acid to prevent precipitation of the tin as metastannic acid. After suitable dilution, the sample is aspirated into flame and the various elements are determined with a hollow cathode lamp for each constitutent.

In this method, lead is determined by the standard addition method described on page 590. Here, a standard solution is added in varying quantities to a series of fixed aliquots of the sample and the absorbance is determined for each. The data are then plotted and extrapolated to zero absorbance. The concentration C_x is then given by $-C_s$, where C_s is the intercept on the abscissa scale.

Reagents needed.

> *Standard solution of Pb^{2+}, 100 $\mu g/mL$.* Dry a quantity of reagent-grade $Pb(NO_3)_2$ for 1 hr at 110°C. Cool; then weigh 0.17 g (to the nearest 0.1 mg) into a 1-L volumetric flask. Dissolve in water containing 1 to 3 mL of concentrated HNO_3; dilute to the mark with distilled or deionized water.

Procedure. Weigh samples of brass (Note 1) into 100-mL beakers. Dissolve in a mixture consisting of about 4 mL of concentrated HNO_3, 4 mL of water, and 4 mL of concentrated HCl (Note 2). Boil gently to remove oxides of nitrogen. Cool; transfer the solutions to 250-mL volumetric flasks, and dilute to the mark.

Use a buret to measure 0-, 5-, 10-, 15-, and 20-mL portions of the standard Pb solution into individual 50-mL volumetric flasks. Transfer 10-mL aliquots of the sample into each flask with a pipet; dilute to the mark.

Set the monochromator to 283.3 nm. Adjust the instrument to read zero absorption with a water blank and maximum absorption with the sample containing the 20-mL standard addition. Measure the absorption of the remaining solutions (Note 3).

Plot the instrument readings as the ordinate versus the concentration of lead in each of the standard additions and extrapolate the resulting straight line to zero absorbance. Determine the micrograms of Pb^{2+} in the solutions and calculate the %Pb in the original sample.

NOTES
1. The amount of sample that will be convenient will depend upon the lead content of the brass as well as the sensitivity of the instrument to be used; 6 to 10 mg of Pb per sample is a reasonable amount. Consult with instructor.
2. Brasses that contain large amounts of tin will require additional HCl to prevent the formation of metastannic acid (p. 771). Upon prolonged standing the diluted samples may develop some turbidity; small amounts have no effect on the analysis for lead.
3. Take at least three readings for each measurement.

METHOD 20-2. DETERMINATION OF LEAD IN A POTTERY GLAZE

Discussion. Glazes used in the decoration of pottery may contain metals—notably lead, cadmium, and barium—that are harmful to humans. Atomic absorption provides an excellent means of determining whether such species can be leached from a glaze and become a health hazard. The procedure recommended by the Association of Official Analytical Chemists calls for analysis of a dilute acetic acid solution that has been in contact with the glaze for 24 hr at room temperature.

Reagents needed.

4% acetic acid. Dilute 40 mL of glacial acetic acid to 1.0 L with distilled or deionized water.

Standard lead solution. See Method 20-1; use 4% acetic acid instead of water as solvent.

Working solutions. Dilute 0, 5.00, 10.00, 15.00, and 20.00 mL of the standard lead solution to 100 mL with 4% acetic acid.

Procedure.   Using a graduated cylinder, fill the sample with 4% acetic acid to within one-fourth inch of overflowing (measure distance along surface of the utensil, not vertical distance). Record the volume of acid. Cover each unit to prevent evaporation. Let stand for 24 hr at room temperature.

Determine the absorption of both the standards and the test solution. Report the micrograms of Pb per milliliter.

METHOD 20-3. DETERMINATION OF SODIUM, POTASSIUM, AND CALCIUM IN MINERAL WATER BY FLAME EMISSION SPECTROSCOPY

Discussion.   A common method for determining the alkali metals and calcium in natural water and blood serum is based upon the emission spectrum for these elements generated in a natural gas/air flame. The accompanying directions are suitable for the analysis of the three elements in typical water samples. Radiation buffers (p. 587) are used to minimize the effect of each ion upon the emission intensity of the others.

Reagents needed.

Standard calcium solution, approximately 500 ppm. Weigh about 1.25 g (to the nearest mg) of $CaCO_3$, which has been dried at 110°C, into a 500-mL beaker. Add about 200 mL of water and 10 mL of concentrated HCl. Cover the beaker with a watch glass during addition of acid to prevent loss of solution as CO_2 is evolved. After the solution is complete, transfer quantitatively to a 1-L volumetric flask and dilute to the mark.

Standard potassium and sodium solutions, approximately 500 ppm. Weigh about 0.95 g of dried KCl and 1.25 g of dried NaCl (to the nearest milligram) into 1-L volumetric flasks. Dissolve in water and dilute to the mark.

Radiation buffer for sodium determination. Prepare a saturated solution with reagent-grade $CaCl_2$, KCl, and $MgCl_2$, in that order.

Radiation buffer for potassium determination. Prepare a saturated solution with reagent-grade NaCl, $CaCl_2$, and $MgCl_2$, in that order.

Radiation buffer for calcium determination. Prepare a saturated solution with reagent-grade NaCl, KCl, and $MgCl_2$, in that order.

Procedure. *Preparation of working curves.* Introduce 5.00 mL of the appropriate radiation buffer to each of a series of 100-mL volumetric flasks. Then add volumes of the standard solution that will cover a concentration range from 0 to 100 ppm. Dilute to 100 mL with distilled water and mix well. Measure the emission intensity for these samples; take at least three readings for each. Between each set of measurements, aspirate distilled water through the burner. Correct the average values for background luminosity, and prepare a working curve from these data.

Analysis of a water sample. Prepare aliquot portions of the sample as directed in the previous paragraph. If necessary, use a standard to calibrate the response of the spectrophotometer to the working curve. Then measure the emission intensity for the unknown. After correcting the data for background, determine the concentration by comparison with the working curve.

21 GAS CHROMATOGRAPHY

METHOD 21-1. QUANTITATIVE DETERMINATION OF THE COMPONENTS IN A MIXTURE OF ISOMERIC KETONES

Discussion. In this experiment, a gas chromatographic column is used for the separation and quantitative determination of the components in a mixture of the three isomeric ketones

2-hexanone	3-methyl-2-pentanone	3,3-dimethyl-2-butanone

The separation can be conveniently carried out on a 5-foot column packed with SE-30 on Chromosorb. The quantitative measurements are based on peak areas employing the area normalization procedure described on page 665. The detector response to the three compounds is found by calibration with a standard mixture.

Reagents needed.

2-hexanone
3-methyl-2-pentanone
3,3-dimethyl-2-butanone

Procedure. *Determination of optimum column temperature.* Prepare a mixture of the three components in a volume ratio of $1:2:3$ and store in a sealed vial. Set the column temperature to about 80°C, inject 2 μL of the mixture, and mark the point of injection on the recorder chart. Allow the three components to elute.

Repeat the experiment using temperatures of about 100°C and 120°C.

Choose the optimum temperature for an analysis. Identify each peak on the basis of its relative size.

Quantitative determination. To calibrate the detector, prepare a $1:1:1$ mixture of the components and obtain chromatograms for three replicate aliquots under the conditions chosen in the previous section. Then, repeat the experiment with three aliquots of the unknown mixture.

Data treatment. Determine the areas of each of the peaks obtained with the standard and the unknown. One method that can be used is to treat the peaks as isosceles triangles and multiply the height in millimeters by the width at one-half the peak height. Alternatively, the peaks can be carefully cut out with scissors and their weights compared with the weight of a rectangular piece of the paper of known area.

Calculate a pooled standard deviation for the area measurements. Calculate the relative sensitivity of the detector for the three components. Then, calculate the volume percent of each compound in the unknown using the method illustrated on page 665.

answers to problems

appendixes

index

answers to problems

CHAPTER 2

2-1. (a) acid; NH_3

(b) acid; Cl^-
(c) acid; HSO_4^-

2-2. (a) acid; CN^-

(b) acid; $H_2PO_4^-$
(c) base; HPO_4^{2-}

2-3. (a) $2H_2SO_4 \rightleftharpoons H_3SO_4^+ + HSO_4^-$
(c) $2C_2H_5OH \rightleftharpoons C_2H_5OH_2^+ + C_2H_5O^-$

2-4. 1.50×10^{24} electrons

2-6. (a) 66.2 mmol
(b) 1.63 mmol
(c) 66.0 mmol

2-8. (a) 1.91×10^4 mg
(b) 7.21×10^3 mg
(c) 7.52×10^5 mg

2-10. (a) $3.41 \times 10^{-4}\ M$
(b) $6.82 \times 10^{-4}\ M$
(c) $1.02 \times 10^{-3}\ M$
(d) $1.36 \times 10^{-2}\ \%$

(d) acid or base;
conjugate base, SO_4^{2-}
conjugate acid, H_2SO_4
(e) base; HCN
(f) base; $CH_3NH_3^+$

(d) acid or base;
conjugate acid, H_3PO_4
conjugate base, HPO_4^{2-}
(e) acid; $C_6H_5NH_2$
(f) acid; OCl^-

(d) 0.500 mmol
(e) 1.02×10^{-2} mmol

(d) 9.4×10^2 mg
(e) 5.1×10^6 mg

(e) 1.70×10^{-2} mmol
(f) 38.1 ppm
(g) 3.166
(h) 2.991

2-12. (a) $[Mg^{2+}] = 5.33 \times 10^{-2}$
    $[Ca^{2+}] = 1.02 \times 10^{-2}$
 (b) $pMg = 1.273$
    $pCa = 1.991$
2-14. (a) $[Na^+] = 0.440$, $pNa = 0.357$
    $[SO_4^{2-}] = 0.115$, $pSO_4 = 0.94$
    $[OH^-] = 0.210$, $pOH = 0.678$
 (c) $[H^+] = 2.00$, $pH = -0.301$
    $[Cl^-] = 2.70$, $pCl = -0.431$
    $[Zn^{2+}] = 0.350$, $pZn = 0.456$
 (e) $[H^+] = 1.76 \times 10^{-5}$, $pH = 4.754$
    $[Ba^{2+}] = 6.04 \times 10^{-4}$, $pBa = 3.219$
    $[ClO_4^-] = 1.23 \times 10^{-3}$, $pClO_4 = 2.912$
2-15. (a) $1.1 \times 10^{-12}\ M$; (c) $2.4 \times 10^{-8}\ M$; (e) $3.6 \times 10^{-5}\ M$; (g) $9.1\ M$
2-16. $M_{K_2SO_4} = 0.552$; $M_{K^+} = 1.10$
2-18. (a) Dissolve 30 g glycol in water and dilute to 500 mL
 (b) Dissolve 30 g glycol in 470 g water
 (c) Dissolve 30 mL of glycol in water and dilute to 500 mL
2-20. (a) Dissolve 32.7 g I_2 in methanol and dilute to 6.00 L
 (b) Dissolve 4.44 g K_2SO_4 in water and dilute to 600 mL
 (c) Dilute 96.0 mL of 1.25 M reagent to 600 mL
 (d) Dilute 141 mL of 5% $AgNO_3$ to 600 mL
 (e) Dissolve 0.103 g of $BaCl_2 \cdot 2H_2O$ in water and dilute to 6.00 L
2-22. (a) Dissolve 72.8 g $K_2Cr_2O_7$ and dilute to 1.50 L
 (b) Dissolve 373 g KCl and dilute to 50.0 L
 (c) Dilute 151 mL of 0.175 M $Zn(NO_3)_2$ to 500 mL
 (d) Dilute 238 mL of 0.317 M $AlBr_3$ to 20.0 L
2-24. (a) 1.37×10^3 g HNO_3
 (b) Dilute 58 mL of concentrated HNO_3 to 6.0 L
2-26. Dilute 205 mL concentrated H_3PO_4 to 500 mL
2-29. (a) 8.77 g KIO_3 (d) 1.52 g $La(NO_3)_3$
 (b) 995 g KIO_3 (e) 1.49 g $La(IO_3)_3$
 (c) 3.10 g $La(IO_3)_3$
2-32. (a) 2.06 g $Ba(OH)_2$ (d) 49.6 mL $AgNO_3$
 (b) 29.3 mL H_2SO_4 (e) 168 mL $Ba(OH)_2$
 (c) 32.7 mL HCl
2-34. 14.9 g $BaCl_2 \cdot 2H_2O$
2-36. (a) 1.80 g $SrCO_3$ (b) 0.439 g CO
2-38. (a) $[Ba^{2+}][SO_4^{2-}] = K_{sp}$ (g) $[Tl^+][N_3^-] = K_{sp}$
 (c) $[Bi^{3+}][I^-]^3 = K_{sp}$ (i) $[Hg_2^{2+}][Cl^-]^2 = K_{sp}$
 (e) $[Zn^{2+}][OH^-]^2 = K_{sp}$
2-39. (a) $S = [Ba^{2+}] = \sqrt{K_{sp}}$ (g) $S = [Tl^+] = \sqrt{K_{sp}}$
 (c) $S = [Bi^{3+}] = (K_{sp}/27)^{1/4}$ (i) $S = [Hg_2^{2+}] = (K_{sp}/4)^{1/3}$
 (e) $S = [Zn^{2+}] = (K_{sp}/4)^{1/3}$
2-40. (a) $S = [SO_4^{2-}]$ (g) $S = [N_3^-]$
 (c) $S = [I^-]/3$ (i) $S = [Cl^-]/2$
 (e) $S = \frac{1}{2}[OH^-]$
2-41. (a) $7.2 \times 10^{-7}\ M$ (c) $2.1 \times 10^{-11}\ M$
 (b) $4.3 \times 10^{-11}\ M$
2-43. (a) $1.6 \times 10^{-4}\ M$ (c) $2.9 \times 10^{-8}\ M$
 (b) $1.3 \times 10^{-5}\ M$

2-45. (a) $OCl^- + H_2O \rightleftharpoons HOCl + OH^-$ $\qquad \dfrac{[HOCl][OH^-]}{[OCl^-]} = K_b$

(c) $C_6H_5O^- + H_2O \rightleftharpoons C_6H_5OH + OH^-$ $\qquad \dfrac{[C_6H_5OH][OH^-]}{[C_6H_5O^-]} = K_b$

(e) $HCO_3^- + H_2O \rightleftharpoons H_3O^+ + CO_3^{2-}$ $\qquad \dfrac{[CO_3^{2-}][H_3O^+]}{[HCO_3^-]} = K_2$

2-46. (a) $HOCl + H_2O \rightleftharpoons H_3O^+ + OCl^-$ $\qquad \dfrac{[H_3O^+][OCl^-]}{[HOCl]} = K_a$

(c) $HPO_4^{2-} + H_2O \rightleftharpoons H_3O^+ + PO_4^{3-}$ $\qquad \dfrac{[H_3O^+][PO_4^{3-}]}{[HPO_4^{2-}]} = K_3$

(e) $HCO_3^- + H_2O \rightleftharpoons H_2CO_3 + OH^-$ $\qquad \dfrac{[H_2CO_3][OH^-]}{[HCO_3^-]} = K_b$

2-47. (a) $K_b = 3.3 \times 10^{-7}$ $\qquad\qquad$ (e) $K_2 = 4.7 \times 10^{-11}$
$\quad$ (c) $K_b = 1.00 \times 10^{-4}$
2-48. (a) $K_a = 3.0 \times 10^{-8}$ $\qquad\qquad$ (e) $K_{b2} = 2.25 \times 10^{-8}$
$\quad$ (c) $K_3 = 4.2 \times 10^{-13}$
2-49. (a) $0.0800\ M$ $\qquad\qquad\qquad$ (e) $5.9 \times 10^{-2}\ M$ (quadratic method)
$\quad$ (c) $1.3 \times 10^{-5}\ M$
2-50. (a) $0.0500\ M$ $\qquad\qquad\qquad$ (e) $2.24 \times 10^{-3}\ M$ (approximate method)
$\quad$ (c) $9.38 \times 10^{-4}\ M$ $\qquad\qquad\qquad\quad 2.19 \times 10^{-3}\ M$ (quadratic method)
2-51. (a) $CuI(s) + I^- \rightleftharpoons CuI_2^-$ $\quad [CuI_2^-]/[I^-] = K_f$
$\quad$ (c) $I_2(s) + I^- \rightleftharpoons I_3^-$ $\qquad [I_3^-]/[I^-] = K_f$

2-52. (a) $\dfrac{[Cu(NH_3)_4^{2+}]}{[Cu^{2+}][NH_3]^4} = \beta_4$ $\qquad\qquad$ (c) $K = [Zn^{2+}]p_{H_2}/[H^+]^2$

2-53. (a) $H_2SO_3 + H_2O \rightleftharpoons H_3O^+ + HSO_3^-$ $\qquad \dfrac{[H_3O^+][HSO_3^-]}{[H_2SO_3]} = K_1$

$\qquad HSO_3^- + H_2O \rightleftharpoons H_3O^+ + SO_3^{2-}$ $\qquad \dfrac{[H_3O^+][SO_3^{2-}]}{[HSO_3^-]} = K_2$

(c) $Hg^{2+} + I^- \rightleftharpoons HgI^+$ $\quad [HgI^+]/[Hg^{2+}][I^-] = K_1$
$\quad\ Hgl^+ + I^- \rightleftharpoons HgI_2$ $\quad [HgI_2]/[HgI^+][I^-] = K_2$
$\quad\ HgI_2 + I^- \rightleftharpoons HgI_3^-$ $\quad [HgI_3^-]/[HgI_2][I^-] = K_3$
$\quad\ HgI_3^- + I^- \rightleftharpoons HgI_4^{2-}$ $\quad [HgI_4^{2-}]/[HgI_3^-][I^-] = K_4$

(e) $Ni^{2+} + CN^- \rightleftharpoons NiCN^+$ $\qquad\qquad [NiCN^+]/[Ni^{2+}][CN^-] = K_1$

$\qquad NiCN^+ + CN^- \rightleftharpoons Ni(CN)_2(s)$ $\qquad \dfrac{1}{[NiCN^+][CN^-]} = \dfrac{1}{K_{sp}} = K_2$

$\qquad Ni(CN)_2(s) + CN^- \rightleftharpoons Ni(CN)_3^-$ $\qquad [Ni(CN)_3^-]/[CN^-] = K_3$
$\qquad Ni(CN)_3^- + CN^- \rightleftharpoons Ni(CN)_4^{2-}$ $\qquad [Ni(CN)_4^{2-}]/[Ni(CN)_3^-][CN^-] = K_4$

CHAPTER 3

3-1. For A, $\quad E_{abs} = 0.085 \qquad E_{rel} = 2.1\%$
$\qquad$ C, $\quad E_{abs} = -4.7 \qquad\ E_{rel} = -12\%$
$\qquad$ E, $\quad E_{abs} = -0.16 \qquad E_{rel} = -0.32\%$

3-2.

	A	C	E
(a) mean, $\bar{x}$	61.37	12.20	9.996
(b) median	61.42	12.22	9.995
(c) range, w	0.39	0.23	0.073
(d) ave rel dev from $\bar{x}$, ppt	2.0	5.7	2.7
(e) abs dev from $\bar{x}$	0.12	0.07	0.027
(f) rel range, %	0.64	1.9	0.73
(g) std dev, s	0.17	0.099	0.033

3-3.

	Larger rel range	Larger abs dev	Larger rel dev	Larger std dev
(a)	E	A	E	A
(c)	C	C	C	C
(e)	C	C	C	C

3-4.

	Abs error	Rel error, %
Set A	−0.08	−0.13
Set C	−0.06	−0.49
Set E	−0.014	−0.14

3-5.

	s, Equation 3-7	s, Equation 3-6
Set A	0.19	0.17
Set C	0.11	0.099
Set E	0.035	0.033

3-6. rel error = (a) −2.0%; (c) −0.2%

3-7. wt sample = (a) 10.4 g; (c) 1.3 g

3-8. −25 ppt

3-9. (a) $s_1 = 0.10$, $s_2 = 0.099$, $s_3 = 0.11$, $s_4 = 0.13$, and $s_5 = 0.10$ mg Ca^{2+}/100 mL
    (b) pooled $s = 0.11$

3-11. $s_1 = 0.15$, $s_2 = 0.14$, $s_3 = 0.17$, and $s_4 = 0.18$ μg Pb/m^3; pooled $s = 0.16$ μg Pb/m^3

3-12. (a) ±0.064 ppb Au; (b) ±0.037 ppb Au; (c) ±0.029 ppb Au

3-15. (a) ±0.76 ppm CO; (b) 7 measurements

3-16. ±0.82 ppm

3-18. (a) 0.087% Hg; (b) 12.72 (±0.11)% Hg; (c) (1) yes, (2) no

3-19. (a)

	90% Confidence Interval with	
Subject Number	Pooled $s \rightarrow \sigma$	s from Individual Samples
1	3.16 ± 0.081	3.16 ± 0.10
2	4.08 ± 0.090	4.08 ± 0.12
3	3.75 ± 0.081	3.75 ± 0.10
4	3.49 ± 0.10	3.49 ± 0.22
5	3.32 ± 0.074	3.32 ± 0.082

3-20. $Q_{exp} = 0.59$; therefore, retain

3-22. Sample A, retain; Sample C, reject; Sample E, retain

3-23. No error demonstrated

3-24. No error demonstrated in parts (a), (b), (c), (d), and (e)
      Error suggested in (f)

3-26. No difference demonstrated in (a), (b), and (f)
      A difference suggested in (c), (d), and (e)

3-29. $\Delta x_{\min} =$ (a) 0.42 mg; (b) 0.30 mg; (c) 0.26 mg
3-30. (a) 0.34; (c) 0.57
3-31. (a) No improvement demonstrated
(c) Improvement suggested
3-32. (a) No precision difference demonstrated
(c) No precision difference demonstrated
3-33. (a) 3; (b) 4; (c) 6; (d) 3; (e) 2; (f) 5; (g) 2; (h) 6
3-35.

	s_y	$(s_y)_r$, %	y
(a)	0.00014	4.1×10^{-4}	$33.8083(\pm 0.0001)$
(b)	0.00014	0.29	$0.0489(\pm 0.0001)$
(c)	0.0036	0.083	$4.321(\pm 0.004)$
(d)	3.7	0.11	$3315(\pm 4)$
(e)	0.20	0.84	$23.8(\pm 0.2)$
(f)	4.0×10^{-5}	0.89	$4.48(\pm 0.04) \times 10^{-3}$

3-37.

	s_y	$(s_y)_r$, %	y
(a)	0.46	1.9	$24.5(\pm 0.5)$
(b)	0.22	0.25	$86.6(\pm 0.2)$
(c)	4.6	0.86	$543(\pm 5)$
(d)	2.2×10^{-5}	0.11	$0.01991(\pm 0.00002)$
(e)	0.0014	0.065	$2.143(\pm 0.001)$
(f)	4.9×10^{-11}	0.94	$5.15(\pm 0.05) \times 10^{-9}$

3-38.

	s_y	$(s_y)_r$, %	y
(a)	0.0055	0.29	$1.891(\pm 0.006)$
(c)	4.5×10^{-11}	1.3	$3.38(\pm 0.04) \times 10^{-9}$
(e)	0.0031	0.47	$0.651(\pm 0.003)$

3-39.

	s_y	y
(a)	1.5×10^{-3}	$2.943(\pm 0.001)$
(b)	3.5×10^{-4}	$-0.3048(\pm 0.0004)$
(c)	7.9×10^{-3}	$-4.785(\pm 0.008)$ or $-4.78(\pm 0.01)$
(d)	100	$4.4(\pm 0.1) \times 10^3$
(e)	3.0×10^{-10}	$6.44(\pm 0.03) \times 10^{-8}$
(f)	2.3×10^{-4}	$2.0000(\pm 0.0002)$

3-41. (b) $y = 0.16 + 0.232x$
(d) $s_b = 0.017$ and $s_r = 0.28$
(e) 15.1 mg SO_4^{2-}/L, $s_c = 1.4$ mg SO_4^{2-}/L, and $(s_c)_r = 9.3\%$
(f) 15.1 mg SO_4^{2-}/L, $s_c = 0.81$ mg SO_4^{2-}/L, and $(s_c)_r = 5.4\%$
3-42. (b) $E = 92.86 - 29.74$ pCa
(c) $s_r = 2.1$
(d) $s_b = 0.68$
(e) pCa $= 2.44$
(f) $s_c = 0.061$ and $(s_c)_r = 2.5\%$; $s_c = 0.043$ and $(s_c)_r = 1.8\%$
(g) $[Ca^{2+}] = 3.63 \times 10^{-3}$
(h) For $[Ca^{2+}]$, $(s_c)_r = 14.1\%$ and $s_c = 0.51 \times 10^{-3} M$
$(s_c)_r = 9.9\%$ and $s_c = 0.36 \times 10^{-3} M$

CHAPTER 4

4-1. (1) 4.29×10^{-11}; (b) 3.20×10^{-6}; (c) 6.2×10^{-23}; (d) 3.3×10^{-7}
4-3. (a) $9.1 \times 10^{-9} M$; (b) $4.2 \times 10^{-15} M$; (c) $2.1 \times 10^{-15} M$

4-5. (a) 4.0×10^{-4} M; (b) 6.5×10^{-10} M; (c) 1.7×10^{-3} M

4-7. (a) 1.15 mg/150 mL; (b) 14.5 mg/150 mL

4-9. (b) no precipitate

(c) no precipitate

4-11. (a) 3.6×10^{-9} M; (b) 2.1×10^{-2} M; (c) 8.1×10^{-13} M

4-12. (a) 1.3×10^{-4} M; (b) 1.6×10^{7} M; (c) 9.3×10^{-5} M

4-15. (a) 4.0×10^{-12}; (d) 8.0×10^{-30}

4-16. (a) 1.4×10^{-6} M; (b) 9.7×10^{-7} M; (c) 9.3×10^{-6} M; (d) 4×10^{-15} M

4-17. (a) 6.2×10^{-15} M; (b) 3.8×10^{-19} M; (c) 3.1×10^{-15} M; (d) 2×10^{-37} M

4-20. (a) $BiOOH$; (b) $Hf(OH)_4$; (c) $BiOOH$; (d) BiO^+; (e) $Hf(OH)_4$ and $Fe(OH)_2$

4-22. (a) 0.0250 M; (b) 0.0115 M; (c) 1.1×10^{-6} M

4-24. (b) $[HNO_2] + [NO_2^-] = 0.100$

$[H_3O^+] = [NO_2^-] + [OH^-]$

(d) $[H_3PO_4] + [H_2PO_4^-] + [HPO_4^{2-}] + [PO_4^{3-}] = 0.050$

$[H_3O^+] = [H_2PO_4^-] + 2[HPO_4^{2-}] + 3[PO_4^{3-}] + [OH^-]$

(g) $[SO_3^{2-}] + [HSO_3^-] + [H_2SO_3] = 0.080 + 0.030 = 0.110$

$[Na^+] + [H_3O^+] = [OH^-] + [HSO_3^-] + 2[SO_3^{2-}]$

Note that $[Na^+] = 0.030$

(i) $[H_3PO_4] + [H_2PO_4^-] + [HPO_4^{2-}] + [PO_4^{3-}] = 0.060$

$[Na^+] = 0.040 + 2 \times 0.020 = 0.080$

$[H_3O^+] + [Na^+] = [OH^-] + [H_2PO_4^-] + 2[HPO_4^{2-}] + 3[PO_4^{3-}]$

4-25. (b) $[I^-] = 2 \times 0.040 + [Ag^+] = 0.080 + S$

$[Mg^{2+}] = 0.040$

$2[Mg^{2+}] + [Ag^+] + [\cancel{H_3O^+}] = [I^-] + [\cancel{OH^-}]$

4-26. $[HOAc] + [OAc^-] = 0.100 + [Ag^+]$

$[H_3O^+] + [Ag^+] = [OAc^-] + [OH^-]$

4-28. (a) 3.6×10^{-2} M; (b) 5.0×10^{-4} M; (c) 1.3×10^{-4} M; (d) 7.9×10^{-5} M

4-29. (a) $1.55 \times 10^{-5} = 1.6 \times 10^{-5}$; (b) $1.60 \times 10^{-5} = 1.6 \times 10^{-5}$; (c) % error $= -3.1$ (note that the error is smaller than the uncertainty in K_{sp})

4-31. 1.2×10^{-5} M

4-33. (b) $1.4 \times 10^{-4} = 1 \times 10^{-4}$ M; (d) $5.6 \times 10^{-14} = 6 \times 10^{-14}$ M

4-34. (d) 2.1×10^{-5} M

4-35. (a) 6.4×10^{-1} M; (b) 9.0×10^{-2} M; (c) 2.4×10^{-3} M; (d) 2.7×10^{-5} M; (e) 1.4×10^{-7} M

4-37. 14.3 mL

4-39. 3.1×10^{-3} M

4-40. (a) $Cr(OH)_3$ forms first; (b) $[Cr^{3+}] = 1.5 \times 10^{-7}$ M

4-42. (a) feasible; (d) not feasible; (f) not feasible

4-43. (a) feasible; $[H_3O^+] = 0.18$ to 3.4×10^{-4}

(c) feasible; $[H_3O^+] = 4.8 \times 10^{-9}$ to 1.6×10^{-10}

(e) not feasible

4-44. (a) 5.01×10^{-3}; (c) 1.00×10^{-2}; (e) 0.100

4-45. (a) 2.8×10^{-11}; (c) 2.88×10^{-9}; (e) 3.8×10^{-20}

4-46. (a) 0.81; (c) 0.22; (e) 0.54

4-47. 0.0100

4-48. 3.7×10^{-8}

4-51. $S^{4/3} + \dfrac{K_w S^{2/3}}{3(K_{sp})^{1/3}} - \dfrac{(K_{sp})^{1/3}}{3} = 0$

CHAPTER 5

5-1. (a) $\dfrac{\text{fw } MgCl_2}{2 \text{ fw } AgCl}$ (g) $\dfrac{\text{fw } Fe_6S_{17}}{3 \text{ fw } Fe_2O_3}$

(c) $\dfrac{2 \text{ fw Ag}}{\text{fw Ag}_2\text{CrO}_4}$

(i) $\dfrac{\text{fw K}_2\text{SO}_4}{2 \text{ fw }(\text{C}_6\text{H}_5)_4\text{BK}}$

(e) $\dfrac{\text{fw Pb}_3\text{O}_4}{3 \text{ fw PbO}_2}$

5-2. (a) $\dfrac{\text{fw Cu}_2\text{HgI}_4}{\text{fw Hg}}$

(d) $\dfrac{\text{fw Cu}_2\text{HgI}_4}{2 \text{ fw Cu}}$

(b) $\dfrac{2 \text{ fw Cu}_2\text{HgI}_4}{\text{fw Hg}_2\text{S}}$

(e) $\dfrac{\text{fw Cu}_2\text{HgI}_4}{4 \text{ fw AgI}}$

(c) $\dfrac{\text{fw Cu}_2\text{HgI}_4}{\text{fw Cu}_2(\text{SCN})_2}$

(f) $\dfrac{3 \text{ fw Cu}_2\text{HgI}_4}{4 \text{ fw In}(\text{IO}_3)_3}$

5-4. AgI

5-6. (a) $\dfrac{\text{fw CaC}_2\text{O}_4 \cdot 2\text{H}_2\text{O}}{\text{fw CaC}_2\text{O}_4}$

(e) $\dfrac{\text{fw CaC}_2\text{O}_4 \cdot 2\text{H}_2\text{O}}{\text{fw CaO}}$

(c) $\dfrac{\text{fw CaC}_2\text{O}_4 \cdot 2\text{H}_2\text{O}}{\text{fw CaCO}_3}$

5-7. 5%
5-9. 1.28 g
5-11. 2.43 g
5-13. 1385 g
5-15. (a) 56.7%; (b) 92.9%; (c) 76.9%; (d) 89.4%; (e) 46.3%; (f) 41.1%
5-17. 16.6%
5-19. 61.61%
5-21. 0.0325 g/tablet
5-23. 3.21%
5-25. 32.8%
5-26. 0.0648 M
5-28. 0.653%
5-30. 18.3%
5-32. 3.7%
5-34. 31.5%
5-36. 17.2% KCl and 31.7% $(\text{NH}_4)_2\text{SO}_4$
5-38. 57.25%
5-40. 7.87% NH_4Cl and 73.6% NaIO_3
5-41. 0.14 g
5-43. (a) 0.44 g; (b) 0.50 g; (c) 29 mL; (d) 76.5 g; (e) 0.083 M; (f) 0.830 g
5-44. 30.05% Mg and 69.95% Al
5-46. 0.16 g
5-48. 0.88 g
5-50. (a) 85.51% Cu with an error of -1.0 ppt
     (b) 1.543% Sn with an error of 79 ppt
5-52. 12.9%

CHAPTER 6

6-1. (a) neutralization; equivalent weights equal to:
         fw HClO_4; fw $\text{MgO}/2$; fw $\text{H}_2\text{SO}_4/2$; fw $\text{Mg}_3(\text{PO}_4)_2/6$
     (b) oxidation-reduction; equivalent weights equal to:
         fw $\text{V}_2\text{O}_5/2$; fw Fe; fw $\text{Tl}_4\text{V}_2\text{O}_7/2$; fw $\text{Fe}_3\text{O}_4/3$

(c) neutralization; equivalent weights equal to:
fw $Na_2CO_3/2$; fw $CO_2/2$; fw $Al_2(CO_3)_3/6$; fw $KHIO_3$

(d) oxidation-reduction; equivalent weights equal to:
fw $Cr/3$; fw $Cr_3O_4/9$; fw $U_3O_8/6$; fw $K_2Cr_2O_7/6$

6-3. If y is the formula weight of the compound, its equivalent weight is given by: (a) y; (b) y/2; (c) y; (d) y/8; (e) y/4; (f) y/2; (g) y; (h) y/4; (i) y/13; (j) y/3

6-5. (a) 6.000 meq; (b) 4.499 meq; (c) 2.950 meq

6-7. (a) 0.0500 M; (b) 0.200 M; (c) 0.0500 N; (d) 4.90 mg Zn^{2+}/mL

6-9. (a) 0.211 g; (b) 0.199 g; (c) 0.0996 g; (d) 0.243 g; (e) 0.0933 g; (f) 0.0194 g

6-11. (a) 0.0400 N; (b) 0.0800 N; (c) 0.0400 N

6-12. (a) 2.12 mg/mL; (b) 3.90 mg/mL; (c) 0.320 mg/mL

6-15. (a) 0.350 mmol; (b) 0.350 mmol; (c) 0.117 mmol; (d) 0.175 mmol

6-17. The following amount of reagent is diluted to 500 mL: (a) 20.4 mL; (b) 13.5 mL; (c) 105 g

6-19. (a) 6.64 g

6-20. (b) 0.208 g; (e) 0.0480 g

6-21. (a) 2.51 g; (b) 2.62 g; (c) 3.30 g; (d) 4.01 g

6-23. (a) 0.0375 M; (b) 0.0300 M; (c) 0.0134 M; (d) 0.0120 M

6-25. (a) 18.8 mL; (b) 18.8 mL; (c) 22.40 mL; (d) 51.64 mL; (e) 28.78 mL

6-27. (a) 0.1076 N; (b) 0.1013 N; (c) 0.1124 N; (d) 0.1742 N

6-29. 0.0847 N

6-31. 0.04933 N

6-33. 44.09%

6-35. 75.84%

6-37. 43.1%

6-39. 87.8%

6-41. 5.48 g/L

6-43. 0.408 N

6-45. (a) 16.1%; (b) 33.4%

CHAPTER 7

7-1. (a) 0.0206 M; (b) 0.0217 M; (c) 0.0366 M

7-2. 30.54%

7-4. 24.7 mg/mL; (b) 91.03%

7-6. 29.57%

7-8. 100.5 mg

7-9. 41.52%

7-11. 14.9 mg/tablet

7-13. Only one chloride in $C_{10}H_5Cl_7$ is titrated

7-16. 18.1 mg/L

7-18. 0.500%

7-19. 10.68% Cl and 51.43% ClO_4

7-21. 63.6% KCl and 36.4% K_2SO_4

7-22. (a)

Vol, mL	$[Ag^+]$	pAg	$[SCN^-]$	pSCN
20.0	1.00×10^{-2}	2.00	1.1×10^{-10}	9.96
30.0	4.00×10^{-3}	2.40	2.8×10^{-10}	9.56
39.0	3.39×10^{-4}	3.47	3.2×10^{-9}	8.49
40.0	1.05×10^{-6}	5.98	1.05×10^{-6}	5.98
41.0	3.4×10^{-9}	8.47	3.28×10^{-4}	3.48
50.0	3.8×10^{-10}	9.42	2.86×10^{-3}	2.54
60.0	2.2×10^{-10}	9.66	5.00×10^{-3}	2.30

(c)

Vol, mL	[Ag$^+$]	pAg	[Cl$^-$]	pCl
10.0	4.00×10^{-2}	1.40	4.55×10^{-9}	8.34
20.0	1.60×10^{-2}	1.80	1.14×10^{-8}	7.94
29.0	1.36×10^{-3}	2.87	1.34×10^{-7}	6.87
30.0	1.35×10^{-5}	4.87	1.35×10^{-5}	4.87
31.0	1.39×10^{-7}	6.86	1.31×10^{-3}	2.88
40.0	1.59×10^{-8}	7.80	1.14×10^{-2}	1.94
50.0	9.10×10^{-9}	8.04	2.00×10^{-2}	1.70

(e)

Vol, mL	[Ba^{2+}]	pBa	[SO$_4^{2-}$]	pSO$_4$
10.0	1.71×10^{-2}	1.77	7.6×10^{-9}	8.12
20.0	7.50×10^{-3}	2.12	1.7×10^{-8}	7.76
29.0	$6.7 \ \times 10^{-4}$	3.17	1.9×10^{-7}	6.72
30.0	$1.1 \ \times 10^{-5}$	4.94	1.1×10^{-5}	4.94
31.0	1.97×10^{-7}	6.71	6.6×10^{-4}	3.18
40.0	2.17×10^{-8}	7.66	6.0×10^{-3}	2.22
50.0	1.19×10^{-8}	7.92	1.1×10^{-2}	1.96

7-23. 5.0 mL, [Ag$^+$] $= 8.2 \times 10^{-12}$ and pAg $= 11.09$
     40.0 mL, [Ag$^+$] $= 7.2 \times 10^{-7}$ and pAg $= 6.14$
     41.0 mL, [Ag$^+$] $= 1.1 \times 10^{-3}$ and pAg $= 2.96$
7-24. 30.0 mL, [Hg$_2^{2+}$] $= 4.55 \times 10^{-3}$ and pHg$_2$ $= 2.34$
     40.0 mL, [Hg$_2^{2+}$] $= 6.9 \ \times 10^{-7}$ and pHg$_2$ $= 6.16$
     50.0 mL, [Hg$_2^{2+}$] $= 2.2 \ \times 10^{-14}$ and pHg$_2$ $= 13.66$

CHAPTER 8

Many of the answers in this chapter are labeled (A) or (Q). The former refers to solutions in which the approximation has been made that [H$_3$O$^+$] and/or [OH$^-$] are small with respect to the analytical concentration of the acid and/or base. Those answers labeled (Q) were obtained by means of a more exact quadratic solution.

8-1. (a) -0.072; (b) 13.80; (c) -0.52
8-3. (a) 1.44; (b) 2.10; (c) 12.30; (d) 1.44; (e) 1.44; (f) 1.59
8-5. (a) 12.50; (b) 2.40; (c) 11.90; (d) 12.50; (e) 9.98; (f) 9.48
8-7. (Q) (a) 1.15; (b) 2.02; (c) 4.00
      (A) (a) 0.88; (b) 1.38; (c) 2.38
8-9. (A) or (Q) (a) 4.26; (b) 4.76; (c) 5.76
8-11. (A) or (Q) (a) 5.12; (b) 5.62; (c) 6.62
8-13. (Q) (a) 12.03; (b) 11.48; (c) 9.97
       (A) (a) 12.06; (b) 11.56; (c) 10.56
8-15. (a) 9.52 (A); (b) 9.02 (A); (c) 8.02 (A), 8.01 (Q)
8-17. (A) or (Q) (a) 8.86; (b) 8.36; (c) 7.40
       In part (c), the contribution of H$_2$O to [H$_3$O$^+$] should be taken into account (see p. 207).
8-19. (Q) (a) 1.74; (b) 2.35; (c) 1.79; (e) 2.31
       (A) (a) 1.23; (b) 1.25; (c) 0.821; (d) 1.70; (e) 1.54; (f) 12.00
8-21. (A) (a) 4.52; (b) 7.82; (c) 7.22; (d) 7.52; (e) 12.00; (f) 1.70
8-23. (A) (a) 10.84; (b) 8.77; (c) 9.37; (d) 5.41; (e) 9.25; (f) 12.43
8-25. (A) (a) 6.92; (b) 5.36; (c) 3.15; 3.19 (Q); (d) 3.11; (e) 5.43; (f) 3.56; 3.57 (Q);
       (g) 3.47; 3.48 (Q); (h) 4.70
8-27. (Q) (a) 2.61; (b) 3.25; (c) 2.72; (d) 2.59
       (A) (a) 2.32; (b) 3.20; (c) 2.70; (d) 2.56
8-29. (A) (a) 2.99; (c) 9.20; (e) 3.98

8-30. (A) (a) 2.89; (c) 9.33; (e) 3.88

8-31. M_{HA}/M_{A^-} (a) 0.029; (b) 0.065; (c) 1.3×10^2; (d) 0.022; (e) 4.3

8-33. M_B/M_{BH^+} (a) 3.0; (b) 1.6×10^2; (c) 3.2×10^4; (d) 0.54; (e) 1.7

8-34. (a) 0.89 (Q); (b) -0.85 (A); (c) 0.017 (Q); (d) 0.134 (Q)

8-36. (Q) (a) -0.34; (b) -3.38; (c) -0.046; (d) -0.432

8-38. (Q) (a) 0.46; (b) 3.26; (c) 0.047; (d) 0.648

8-40. 31.2 g

8-42. 9.3 g

8-44. 5.1×10^2 mL

8-46. 66 mL

8-48.

Vol, mL	pH	Vol, mL	pH
0.0	13.00	49.0	11.00
10.0	12.82	50.0	7.00
25.0	12.52	51.0	3.00
40.0	12.05	55.0	2.32
45.0	11.72	60.0	2.04

8-49.

	(a)		(b)		(c)
Vol, mL	pH (A)	pH (Q)	pH (A)	pH (Q)	pH (A)
0.0	2.15	2.16	2.43	2.44	3.12
5.0	2.34	2.49	2.91	2.96	4.28
15.0	2.92	2.95	3.50	3.50	4.86
25.0	3.29	3.31	3.86	3.87	5.23
40.0	3.89	3.90	4.47	4.47	5.83
45.0	4.25	4.25	4.82	4.82	6.18
49.0	4.98	4.99	5.55	5.56	6.92
50.0	8.00	8.00	8.28	8.28	8.96
51.0	11.00	11.00	11.00	11.00	11.00
55.0	11.68	11.68	11.68	11.68	11.68
60.0	11.96	11.96	11.96	11.96	11.96

8-51.

	(a)	(c)
Vol, mL	pH (A)	pH (A)
0.0	2.80	4.26
5.0	3.64	6.57
15.0	4.23	7.16
25.0	4.60	7.52
40.0	5.20	8.12
49.0	6.29	9.21
50.0	8.65	10.11
51.0	11.00	11.00
55.0	11.68	11.68
60.0	11.96	11.96

CHAPTER 9

Many of the answers in this chapter are labeled (A) or (Q). The former refers to solutions in which the approximation has been made that $[H_3O^+]$ and/or $[OH^-]$ are small with

respect to the analytical concentration of the acid and/or base. Those answers labeled (Q)
were obtained by means of a more exact quadratic solution.
9-1. (Q) (a) 1.66; (b) 3.83; (c) 1.75; (d) 2.11; (e) 1.81; (f) 1.50
   (A) (a) 1.53; (b) 3.83; (c) 1.65; (d) 2.08; (e) 1.72; (f) 1.29
9-3. The following answers were obtained with Equation 9-16:
   (a) 4.54; (b) 8.34; (c) 4.70; (d) 9.70; (e) 4.33
9-5. (Q) (a) 12.80; (b) 11.50; (c) 10.05; (d) 4.14
   (A) (a) 13.86; (b) 11.51; (c) 10.05; (d) 4.14
9-7. (a) 1.23 (Q); (b) 1.72 (Q); (c) 7.36 (7.31 if dissociation of water not taken into account)
9-9. (a) 1.86 (Q); (b) 4.71; (c) 9.69; (d) 12.33 (Q)
9-11. (a) 1.53; (b) 1.51; (c) 1.71; (d) 2.00
9-13. (a) 12.57; (b) 12.19; (c) 12.02; (d) 12.00
9-15. (A) (a) 10.21; (b) 10.10; (c) 1.91; (d) 12.19
   (Q) (a) 10.21; (b) 10.10; (c) 2.04; (d) 12.09
9-17.   Principal Species   Acid/Base Concn Ratio

   (a) T^{2-}, HT^- $[HT^-]/[T^{2-}] = 1.0$
   (b) HM^-, H_2M $[H_2M]/[HM^-] = 0.25$
   (c) Ox^{2-}, HOx^- $[HOx^-]/[Ox^{2-}] = 1.8$
   (d) HSu^-, H_2Su $[H_2Su]/[HSu^-] = 1.6$
   (e) T^{2-}, HT^- $[HT^-]/[T^{2-}] = 2.3$
9-19. (A) (a) 7.20; (b) 12.38; (c) 4.69; (d) 1.66; (f) 2.15
   (Q) (a) 7.20; (b) 12.29; (c) 4.69; (d) 1.73; (f) 2.27; (e) 9.76 (by Equation 9-16)
9-21. (A) (a) 7.50; (b) 12.08; (c) 1.85; (d) 1.85; (e) 7.50; (f) 12.08
   (Q) (a) 7.50; (b) 11.78; (c) 2.08; (d) 2.02; (e) 7.50; (f) 11.91
9-23. (a) Dissolve 22 g Na_2CO_3 in 1.0 L of the $NaHCO_3$ solution
   (b) Dissolve 18 g $NaHCO_3$ in 1.0 L of the Na_2CO_3 solution
   (c) Dissolve 26 g Na_3PO_4 in the H_3PO_4 solution
9-25. (a) Mix 724 mL of the HCl solution with 1276 mL of the Na_2CO_3 solution
   (b) Mix 937 mL of the HCl solution with 1063 mL of the Na_2SO_3 solution
9-26. (a)

pH	D	α_0	α_1	α_2
1.00	1.01×10^{-2}	9.91×10^{-1}	9.12×10^{-3}	3.93×10^{-6}
2.00	1.09×10^{-4}	9.15×10^{-1}	8.42×10^{-2}	3.63×10^{-4}
3.00	1.96×10^{-6}	5.10×10^{-1}	4.69×10^{-1}	2.02×10^{-2}
4.00	1.42×10^{-7}	7.06×10^{-2}	6.50×10^{-1}	2.80×10^{-1}
5.00	4.90×10^{-8}	2.04×10^{-3}	1.88×10^{-1}	8.10×10^{-1}
6.00	4.06×10^{-8}	2.46×10^{-5}	2.27×10^{-2}	9.77×10^{-1}
7.00	3.97×10^{-8}	2.52×10^{-7}	2.32×10^{-3}	9.98×10^{-1}
8.00	3.97×10^{-8}	2.52×10^{-9}	2.32×10^{-4}	1.00×10^{0}

(c)

pH	D	α_0	α_1	α_2
1.00	1.54×10^{-2}	6.51×10^{-1}	3.49×10^{-1}	1.89×10^{-4}
2.00	6.39×10^{-4}	1.56×10^{-1}	8.39×10^{-1}	4.55×10^{-3}
3.00	5.75×10^{-5}	1.74×10^{-2}	9.32×10^{-1}	5.05×10^{-2}
4.00	8.28×10^{-6}	1.21×10^{-3}	6.48×10^{-1}	3.51×10^{-1}
5.00	3.44×10^{-6}	2.91×10^{-5}	1.56×10^{-1}	8.44×10^{-1}
6.00	2.96×10^{-6}	3.38×10^{-7}	1.81×10^{-2}	9.82×10^{-1}
7.00	2.91×10^{-6}	3.44×10^{-9}	1.84×10^{-3}	9.98×10^{-1}
8.00	2.91×10^{-6}	3.44×10^{-11}	1.84×10^{-4}	1.00×10^{0}

(e) pH	D	α_0	α_1	α_2	α_3
1.00	1.01×10^{-3}	9.93×10^{-1}	7.40×10^{-3}	1.28×10^{-6}	5.14×10^{-12}
2.00	1.08×10^{-6}	9.31×10^{-1}	6.93×10^{-2}	1.20×10^{-4}	4.82×10^{-9}
3.00	1.76×10^{-9}	5.69×10^{-1}	4.24×10^{-1}	7.33×10^{-3}	2.95×10^{-6}
4.00	9.74×10^{-12}	1.03×10^{-1}	7.65×10^{-1}	1.32×10^{-1}	5.32×10^{-4}
5.00	2.10×10^{-13}	4.77×10^{-3}	3.56×10^{-1}	6.15×10^{-1}	2.47×10^{-2}
6.00	1.88×10^{-14}	5.31×10^{-5}	3.96×10^{-2}	6.85×10^{-1}	2.75×10^{-1}
7.00	6.48×10^{-15}	1.54×10^{-7}	1.15×10^{-3}	1.99×10^{-1}	8.00×10^{-1}
8.00	5.31×10^{-15}	1.88×10^{-10}	1.40×10^{-5}	2.43×10^{-2}	9.76×10^{-1}

9-27. (a)

pH	D	α_0	α_1	α_2
3.00	1.82×10^{-5}	5.49×10^{-2}	9.45×10^{-1}	6.08×10^{-5}
4.00	1.73×10^{-6}	5.78×10^{-3}	9.94×10^{-1}	6.39×10^{-4}
5.00	1.73×10^{-7}	5.77×10^{-4}	9.93×10^{-1}	6.38×10^{-3}
6.00	1.83×10^{-8}	5.46×10^{-5}	9.39×10^{-1}	6.04×10^{-2}
7.00	2.83×10^{-9}	3.54×10^{-6}	6.09×10^{-1}	3.91×10^{-1}
8.00	1.28×10^{-9}	7.82×10^{-8}	1.35×10^{-1}	8.65×10^{-1}
9.00	1.12×10^{-9}	8.90×10^{-10}	1.53×10^{-2}	9.85×10^{-1}
10.00	1.11×10^{-9}	9.03×10^{-12}	1.55×10^{-3}	9.98×10^{-1}
11.00	1.11×10^{-9}	9.04×10^{-14}	1.56×10^{-4}	1.00×10^{0}
12.00	1.11×10^{-9}	9.04×10^{-16}	1.56×10^{-5}	1.00×10^{0}

(c)

pH	D	α_0	α_1	α_2	α_3
3.00	7.00×10^{-9}	1.4×10^{-1}	8.6×10^{-1}	9.0×10^{-5}	2.7×10^{-13}
4.00	6.11×10^{-11}	1.6×10^{-2}	9.8×10^{-1}	1.0×10^{-3}	3.1×10^{-11}
5.00	6.07×10^{-13}	1.6×10^{-3}	9.9×10^{-1}	1.0×10^{-2}	3.1×10^{-9}
6.00	6.63×10^{-15}	1.5×10^{-4}	9.0×10^{-1}	9.5×10^{-2}	2.8×10^{-7}
7.00	1.23×10^{-16}	8.1×10^{-6}	4.9×10^{-1}	5.1×10^{-1}	1.5×10^{-5}
8.00	6.90×10^{-18}	1.4×10^{-7}	8.7×10^{-2}	9.1×10^{-1}	2.7×10^{-4}
9.00	6.38×10^{-19}	1.6×10^{-9}	9.4×10^{-3}	9.9×10^{-1}	3.0×10^{-3}
10.00	6.50×10^{-20}	1.5×10^{-11}	9.2×10^{-4}	9.7×10^{-1}	2.9×10^{-2}
11.00	8.19×10^{-21}	1.2×10^{-13}	7.3×10^{-5}	7.7×10^{-1}	2.3×10^{-1}
12.00	2.52×10^{-21}	4.0×10^{-16}	2.4×10^{-6}	2.5×10^{-1}	7.5×10^{-1}

(e)

pH	D	α_0	α_1	α_2	α_3
3.00	8.11×10^{-9}	1.23×10^{-1}	8.77×10^{-1}	5.56×10^{-5}	2.33×10^{-14}
4.00	7.22×10^{-11}	1.39×10^{-2}	9.86×10^{-1}	6.25×10^{-4}	2.62×10^{-12}
5.00	7.16×10^{-13}	1.40×10^{-3}	9.92×10^{-1}	6.29×10^{-3}	2.64×10^{-10}
6.00	7.56×10^{-15}	1.32×10^{-4}	9.40×10^{-1}	5.96×10^{-2}	2.50×10^{-8}
7.00	1.16×10^{-16}	8.61×10^{-6}	6.12×10^{-1}	3.88×10^{-1}	1.63×10^{-6}
8.00	5.22×10^{-18}	1.92×10^{-7}	1.36×10^{-1}	8.64×10^{-1}	3.63×10^{-5}
9.00	4.58×10^{-19}	2.18×10^{-9}	1.55×10^{-2}	9.84×10^{-1}	4.13×10^{-4}
10.00	4.53×10^{-20}	2.21×10^{-11}	1.57×10^{-3}	9.94×10^{-1}	4.18×10^{-3}
11.00	4.70×10^{-21}	2.13×10^{-13}	1.51×10^{-4}	9.60×10^{-1}	4.03×10^{-2}
12.00	6.40×10^{-22}	1.56×10^{-15}	1.11×10^{-5}	7.04×10^{-1}	2.96×10^{-1}

9-28. (a) $D = 1.33 \times 10^{-16}$, $\alpha_0 = 4.75 \times 10^{-4}$, $\alpha_1 = 0.842$, $\alpha_2 = 0.158$
   (d) $D = 3.07 \times 10^{-5}$, $\alpha_0 = 0.817$, $\alpha_1 = 0.183$, $\alpha_2 = 1.42 \times 10^{-4}$

9-29.

Vol, mL	pH	Vol, mL	pH	Vol, mL	pH
10.0	1.48	22.0	3.70	45.0	8.74
18.0	2.23	35.0	4.93	46.0	11.32
20.0	2.95	44.0	6.14	50.0	12.00

9-31.

Vol, mL	(a) H_2SO_3		(c) H_3PO_3	
	pH (A)	pH (Q)	pH (A)	pH (Q)
0.0	1.43	1.53	1.55	1.62
5.0	1.29	1.75	1.52	1.88
10.0	1.77	2.02	2.00	2.18
19.0	3.05	3.17	3.28	3.35
20.0	4.54		4.33	
21.0	5.91		5.31	
30.0	7.19		6.58	
39.0	8.48		7.86	
40.0	9.92		9.62	
41.0	11.34		11.34	
50.0	12.30		12.30	

CHAPTER 10

10-1. (a) fw $H_2SO_3/2$; (c) fw HIO_3; (e) fw $H_2CO_3/1$

10-2. (a) fw KOH; (c) fw $Ba(OH)_2/2$; (e) fw $Ca(OH)_2/2$

10-3. (a) 9.16 meq; (c) 0.808 meq; (e) 23.1 meq

10-4. (a) Dilute 20 mL concentrated HCl to 2.0 L

    (b) Dilute 21.57 g of the reagent to 1.000 L

    (c) Dilute 27.65 mL of the 2.170 N HCl to 500.0 mL

    (d) Dilute 24.90 of the 20.0% HCl to 1.00 L

10-6. (a) 0.0798 N; (b) 0.1025 N; (c) 0.1086 N; (d) 0.1125 N

10-8. (a) 0.1026 N; (b) 0.29%; (c) 3.9×10^{-4} N

10-10. (a) 0.19 to 0.24 g; (b) 0.86 to 1.1 g; (c) 0.19 to 0.24 g; (d) 0.41 to 0.53 g; (e) 0.53 to 0.69 g; (f) 0.64 to 0.82 g; (g) 0.42 to 0.54 g; (h) 0.47 to 0.60 g

10-11. (a) pH = 7.0, bromothymol blue

    (b) pH = 6.0, bromocresol purple

    (c) pH = 9.3, phenolphthalein

    (d) pH = 3.1, methyl yellow

    (e) pH = 4.6, methyl red

    (f) pH = 9.4, phenolphthalein

    (g) pH = 7.0, bromothymol blue

10-13. 108 g/eq

10-14. 4.33%

10-16. (a) 90.00%; (b) 47.48%; (c) 32.86%; (d) 10.20%

10-18. 44.90%

10-21. 77.5%

10-24. 3.885%

10-25. 24.3% protein; 44.7 g protein/can

10-27. 2.18 ppm

10-29. 6.90% $(NH_4)_2SO_4$ and 22.52% KNO_3

10-31. 32.91% $H_2C_2O_4 \cdot 2H_2O$ and 24.76% $Na_2C_2O_4 \cdot H_2O$

10-33. 20.3% Na_2CO_3 and 15.4% $NaHCO_3$

10-35. (a) 36.5 mL; (b) 36.5 mL; (c) 13.4 mL; (d) 48.8 mL; (e) 27.4 mL

10-37. (a) 164.3 mg Na_2CO_3 and 152.5 mg $NaHCO_3$

    (b) 177.3 mg $NaHCO_3$

    (c) 180.2 mg NaOH

    (d) 282.1 mg Na_2CO_3

    (e) 85.4 mg Na_2CO_3 and 137.5 mg NaOH

CHAPTER 11

11-1. (a) $2H_2O \rightleftharpoons H_3O^+ + OH^-$
(c) $2HCOOH \rightleftharpoons HCOOH_2^+ + HCOO^-$
(e) $2CH_3COOH \rightleftharpoons CH_3COOH_2^+ + CH_3COO^-$

11-2. (a) $2HCN \rightleftharpoons H_2CN^+ + CN^-$
(c) $2HF \rightleftharpoons H_2F^+ + F^-$

11-3. 2.0×10^{-12}

11-5. (a) $pH = pOH = 7.00$
(c) $pH = pHCOO = 3.1$
(e) $pH = pC_2H_3O_2 = 7.2$

11-6. (a) NaOH
(c) NaOOCH
(e) $NaC_2H_3O_2$

11-7. (a) $HC_8H_4O_4^- + H_2O \rightleftharpoons C_8H_4O_4^{2-} + H_3O^+$
(b) $HC_8H_4O_4^- + CH_3COOH \rightleftharpoons H_2C_8H_4O_4 + CH_3COO^-$
(c) $H_2NCONH_2 + HCOOH \rightleftharpoons H_2NCONH_3^+ + HCOO^-$

11-8. (a) All strong bases react completely with the solvent to form the conjugate base of the solvent.

$$B \;+\; HS \;\rightleftharpoons\; BH^+ \;+\; S^-$$
base 1   acid 2   acid 1   base 2

Thus, the strong base B cannot exist as such in the solvent and S^- is the strongest base.

(c) $HA + H_2O \rightleftharpoons H_3O^+ + A^-$ (1)
$HA + H_2NCH_2CH_2NH_2 \rightleftharpoons H_3NCH_2CH_2NH_2^+ + A^-$ (2)
Water is a weaker base than $H_2NCH_2CH_2NH_2$; equilibrium (1) lies farther to the left than (2). Therefore, more compounds will behave as weak bases of differing strengths in water than in $H_2NCH_2CH_2NH_2$.

(e) $H_2O + HBz \rightleftharpoons H_3O^+ + Bz^-$ (1)
$C_2H_5OH + HBz \rightleftharpoons C_2H_5OH_2^+ + Bz^-$ (2)
Although water and ethanol have roughly the same strengths as bases, equilibrium (2) lies considerably farther to the left than (1) because the lower dielectric constant for ethanol reduces the tendency for charge separation to occur.

11-9. (a) 5.27; (b) 10.89; (c) 8.68

11-10. (a) 1.98; (b) 7.00; (c) 0.82

11-13. (a) $\Delta pH = 10.0 - 4.0 = 6.0$
(b) $\Delta pH = 10.5 - 4.0 = 6.5$
(c) $\Delta pH = 15.1 - 4.0 = 11.1$
(d) $\Delta pH = 29.0 - 4.0 = 25.0$
(e) $\Delta pH = 3.14 - 3.08 = 0.06$
(f) $\Delta pH = 11.3 - 4.0 = 7.3$

Note that in (e) the dissociation of the solvent contributes significantly to the pH and must be taken into account. Thus, in the presence of 10^{-4} M strong acid

$$[HCOOH_2^+] = 10^{-4} + [HCOO^-] = 10^{-4} + 6 \times 10^{-7}/[HCOOH_2^+]$$

11-15.

Vol, mL	pH	Vol, mL	pH
0.0	11.85	40.0	3.60
10.0	6.18	41.0	3.12
25.0	5.48	45.0	2.45
39.0	4.11	50.0	2.18

11-16.

Vol, mL	pH	Vol, mL	pH
0.0	5.03	49.9	11.85
25.0	9.15	50.0	13.5
40.0	9.75	50.1	15.1
49.0	10.84	51.0	16.1
		60.0	17.1

CHAPTER 12

12-1. (a) Dissolve 2.800 g $Na_2H_2Y\cdot2H_2O$ in water and dilute to 500 mL
(b) 0.841 mg CaO/mL
(c) 2.285 mg $Zn_2P_2O_7$/mL

12-3. (a) 0.02568 M; (b) 23.18%

12-5. (a) 332 mg $CaCO_3$/L
(b) 204 mg $CaCO_3$/L and 108 mg $MgCO_3$/L

12-7. 15.66%

12-8. 84.69% Pb and 15.31% Cd

12-9. 27.48% Bi, 3.413% Cd, and 19.98% Pb

12-11. 7.61% Pb, 19.66% Zn, 48.72% Cu, and 24.01% Sn

12-13. (a) 4.6×10^9; (b) 1.1×10^{12}; (c) 7.4×10^{13}

12-15.

	α_4	δ	K''_{CoY}
(a)	5.4×10^{-3}	3.5×10^{-2}	3.8×10^{12}
(b)	3.5×10^{-1}	3.5×10^{-2}	2.5×10^{14}
(c)	5.4×10^{-3}	2.2×10^{-5}	2.4×10^9
(d)	3.5×10^{-1}	2.2×10^{-5}	1.6×10^{11}

12-17. $[Ag^+] = 7.9 \times 10^{-10}$, $[AgNH_3^+] = 1.6 \times 10^{-7}$, and $[Ag(NH_3)_2^+] = 1.0 \times 10^{-4}$

12-19. (a) ~ 10; (c) <2.0

12-20. (a) 3 to 5; (c) 4 to 5

12-21. (a) 5.00; (c) 4.99

12-22.

Vol, mL	pBa	Vol, mL	pBa
10.0	2.40	26.0	6.36
20.0	2.94	30.0	7.06
24.0	3.67	35.0	7.36
25.0	5.01		

12-24.

Vol, mL	pCu	Vol, mL	pCu
10.0	13.54	26.0	16.11
20.0	14.08	30.0	16.82
24.0	14.81	35.0	17.12
25.0	15.47		

CHAPTER 13

13-1.

	Oxidizing Agent	Reducing Agent
(a)	$Cl_2 + 2e \rightleftharpoons 2Cl^-$	$3I^- \rightleftharpoons I_3^- + 2e$
(b)	$Ag^+ + e \rightleftharpoons Ag$	$Zn \rightleftharpoons Zn^{2+} + 2e$
(c)	$Mn^{3+} + e \rightleftharpoons Mn^{2+}$	$H_2 \rightleftharpoons 2H^+ + 2e$

	Oxidizing Agent	Reducing Agent

(d) $Cr_2O_7^{2-} + 14H^+ + 6e \rightleftharpoons 2Cr^{3+} + 7H_2O$ $U^{4+} + 2H_2O \rightleftharpoons UO_2^{2+} + 4H^+ + 2e$
(e) $V(OH)_4^+ + 2H^+ + e \rightleftharpoons VO^{2+} + 3H_2O$ $V^{3+} + H_2O \rightleftharpoons VO^{2+} + 2H^+ + e$
(f) $MnO_4^- + 8H^+ + 5e \rightleftharpoons Mn^{2+} + 4H_2O$ $HNO_2 + H_2O \rightleftharpoons NO_3^- + 3H^+ + 2e$
(g) $H_2O_2 + 2H^+ + 2e \rightleftharpoons 2H_2O$ $3I^- \rightleftharpoons I_3^- + 2e$
(h) $Ce^{4+} + e \rightleftharpoons Ce^{3+}$ $H_2O_2 \rightleftharpoons O_2 + 2H^+ + 2e$
(i) $IO_3^- + 6H^+ + 5e \rightleftharpoons \frac{1}{2}I_2 + 3H_2O$ $I^- \rightleftharpoons \frac{1}{2}I_2 + e$
(j) $Sn^{4+} + 2e \rightleftharpoons Sn^{2+}$ $Ag + I^- \rightleftharpoons AgI + e$

13-3. (a) $Cl_2 + 3I^- \rightleftharpoons 2Cl^- + I_3^-$
    (b) $2Ag^+ + Zn \rightleftharpoons 2Ag + Zn^{2+}$
    (c) $2Mn^{3+} + H_2 \rightleftharpoons 2Mn^{2+} + 2H^+$
    (d) $Cr_2O_7^{2-} + 3U^{4+} + 2H^+ \rightleftharpoons 2Cr^{3+} + 3UO_2^{2+} + H_2O$
    (e) $V(OH)_4^+ + V^{3+} \rightleftharpoons 2VO^{2+} + 2H_2O$
    (f) $2MnO_4^- + 5HNO_2 + H^+ \rightleftharpoons 2Mn^{2+} + 5NO_3^- + 3H_2O$
    (g) $3I^- + H_2O_2 + 2H^+ \rightleftharpoons I_3^- + 2H_2O$
    (h) $2Ce^{4+} + H_2O_2 \rightleftharpoons O_2 + 2Ce^{3+} + 2H^+$
    (i) $IO_3^- + 5I^- + 6H^+ \rightleftharpoons 3I_2 + 3H_2O$
    (j) $Sn^{4+} + 2Ag + 2I^- \rightleftharpoons Sn^{2+} + 2AgI$

13-5. (a) fw $Cl_2/2$ (f) fw $MnO_4^-/5$
    (b) fw Ag^+ (g) fw $H_2O_2/2$
    (c) fw Mn^{3+} (h) fw Ce^{4+}
    (d) fw $Cr_2O_7^{2-}/6$ (i) fw $IO_3^-/5$
    (e) fw $V(OH)_4^+$ (j) fw $Sn^{4+}/2$

13-7. (a) 7×10^{27} (f) 10^{96}
    (b) 7×10^{52} (g) 9×10^{41}
    (c) 1×10^{51} (h) 4×10^{25}
    (d) 10^{101} (i) 8×10^{55}
    (e) 7×10^{10} (j) 2.1×10^{10}

13-9. (a) 0.691 V   (d) -0.19 V
    (b) 0.200 V   (e) 0.494 V
    (c) 0.500 V

13-11. (a) -0.216 V   (d) -0.107 V
     (b) 0.71 V   (e) 0.604 V
     (c) 0.129 V   (f) 0.876 V

13-13. (a) anode, 0.312 V   (d) anode, 0.318 V
     (b) anode, 0.098 V   (e) anode, 0.0591 V
     (c) cathode, 0.080 V

13-15. $F_2 > Ce^{4+} > O_2 > Fe^{3+} > Fe(CN)_6^{3-} > H^+ > Ag(CN)_2^- > Cd^{2+} > Cr^{3+}$

13-17. (a) left; (b) left; (c) left; (d) right; (e) right

13-19. (a) -0.336 V; Cd anode in a galvanic cell
     (b) -0.657 V; Ag anode in a galvanic cell
     (c) 0.571 V; Ag anode in a galvanic cell
     (d) -2.121 V; Zn anode in a galvanic cell
     (e) 0.177 V; Ag|AgCl(s), KCl (1.00 M) anode in a galvanic cell

13-21. All of the cells are galvanic.
     (a) 0.533 V; (b) 0.339 V; (c) 0.161 V; (d) 0.559 V; (e) 0.300 V

13-23. $[Ag^+] = 2.6 \times 10^{-16}$ and $K_{sp} = 3.3 \times 10^{-17}$

13-25. -0.175 V

13-27. 1.8×10^{-4}

13-29. -0.984 V

13-31. 8×10^{13}

13-33. 6.6×10^{-4}

13-35. -1.104 V
13-37. (a) -1.198 V; (b) -2.62 V

CHAPTER 14

14-1. (a) 0.360 V
    (b) 0.751 V
    (c) 1.28 V
    (d) 0.821 V
    (e) 0.336 V

14-3. (a) 8×10^{20}
    (b) 1×10^{31}
    (c) 10^{96}
    (d) 7×10^{17}
    (e) 5.1×10^{13}

14-5. (a) 3.2×10^{-9} M
    (b) 2×10^{-19} M
    (c) 4×10^{-16} M
    (d) 7.8×10^{-13} M
    (e) 9.9×10^{-9} M

14-7.

E(vs. SHE), V

Vol, mL	(a)	(b)	(c)	(d)	(e)
5.0	0.799	0.237	0.86	0.504	0.140
15.0	0.743	0.266	0.89	0.532	0.168
19.0	0.695	0.289	0.92	0.556	0.192
20.0	0.360	0.751	1.28	0.821	0.336
21.0	0.192	1.21	1.43	1.085	0.479
25.0	0.172	1.23	1.44	1.107	0.500
30.0	0.163	1.24	1.44	1.125	0.509

14-9.

Vol, mL	E, V	Vol, mL	E, V
5.0	0.331	35.0	1.03
15.0	0.387	39.0	1.08
19.0	0.435	40.0	1.42
20.0	0.68	41.0	1.49
21.0	0.92	45.0	1.50
26.0	0.98	50.0	1.51

CHAPTER 15

15-1. (a) $2V(OH)_4^+ + C_2O_4^{2-} + 4H^+ \rightarrow 2VO^{2+} + 2CO_2 + 6H_2O$
    (b) $2Fe^{3+} + H_2S \rightarrow 2Fe^{2+} + S + 2H^+$
    (c) $2Mn^{2+} + 5BiO_3^- + 4H^+ \rightarrow 2MnO_4^- + 5BiO^+ + 2H_2O$
    (d) $3U^{4+} + 2MnO_4^- + 2H_2O \rightarrow 3UO_2^{2+} + 2MnO_2 + 4H^+$
    (e) $4Fe^{2+} + O_2 + 4H^+ \rightarrow 4Fe^{3+} + 2H_2O$
    (f) $H_2O_2 + Cl_2 \rightarrow O_2 + 2Cl^- + 2H^+$
    (g) $H_2MoO_4 + Ag + Cl^- + 2H^+ \rightarrow MoO_2^+ + AgCl + 2H_2O$

15-3. (a) $CH_2OHCH_2OH + H_5IO_6 \rightarrow 2H_2CO + IO_3^- + 3H_2O + H^+$
    (c) $CH_2OH(CHOH)_4CH_2OH + 5H_5IO_6 \rightarrow$
$$5IO_3^- + 2CH_2O + 4HCOOH + 5H^+ + 11H_2O$$
    (e) $CH_2OH(CHOH)_4CHO + 5H_5IO_6 \rightarrow$
$$5IO_3^- + 5HCOOH + H_2CO + 5H^+ + 10 H_2O$$

15-4. 0.106 N or 0.0212 M
15-6. Dissolve 4.946 g As_2O_3 in dilute NaOH, acidify with HCl, and dilute to 2.000 L
15-8. 0.0663 N

15-10. 45.8%

15-12. 13.9%

15-14. 0.0457 N and 6.74% Mn

15-16. 36.3% Fe_2O_3 and 9.89% TiO_2

15-18. 16.7%

15-20. 9.38%

15-22. 22.7 mg

15-24. 2.8×10^2 ppm

15-26. 80.1 mg

15-28. 19.95%

15-30. 3.75 mg As/mL, 12.7 mg I_2/mL, 0.801 mg N_2H_4/mL, and 1.62 mg KCNS/mL

15-32. 3.07 mg nitroglycerine/tablet

15-34. 1.697%

15-36. 99.5%

15-38. 2×10^{47}

15-40. 2.2×10^{45}

15-41. $[Fe^{2+}] = 6 \times 10^{-5}$

15-42. 22.6% KI and 54.7% KBr

CHAPTER 16

16-1. (a) 0.428 V

(b) $SCE||VO_3^-(xM),AgVO_3(sat'd)|Ag$

(c) $pVO_3 = (E_{cell} - 0.187)/0.0591$

(d) 5.72

16-3. (a) $SCE||Fe(CN)_6^{4-}(xM),Cu_2Fe(CN)_6(sat'd)|Cu$
$pFe(CN)_6 = (E_{cell} + 0.241 - E^0_{Cu_2Fe(CN)_6}) \times 4/0.0591$

(b) $SCE||IO_3^-(xM),Hg_2(IO_3)_2(sat'd)|Hg$
$pIO_3 = (E_{cell} + 0.241 - E^0_{Hg_2(IO_3)_2})/0.0591$

(c) $SCE||Tl^{3+}(1.00 \times 10^{-4}\ M),Tl^+(xM)|Pt$
$pTl(I) = (E_{cell} - 0.89) \times 2/0.0591$

16-5. 9.17

16-7. (a) -0.453 V; (c) -0.039 V; (e) -0.229 V

16-8. $E^0 = 0.605$ V

16-10. (a) 0.145 V; (c) 0.027 V

16-11. (a) 3.89; (b) 8.37; (c) 11.43; (d) 14.26

16-13. 3.2×10^{-15}

16-15.

Vol, mL	E_{cell}, V	Vol, mL	E_{cell}, V
5.00	−0.053	49.00	+0.025
10.00	−0.043	50.00	+0.38
15.00	−0.036	51.00	+1.10
25.00	−0.025	55.00	+1.14
40.00	−0.007	60.00	+1.16

16-17.

Vol, mL	E_{cell}, V	Vol, mL	E_{cell}, V
0.00	0.405	25.00	0.496
10.00	0.414	25.10	0.525
20.00	0.430	26.00	0.584
24.00	0.452	30.00	0.624
24.90	0.481		

16-18. (a) 10.68 and 2.10×10^{-11} M; (b) 2.02×10^{-11} to 2.18×10^{-11} M; (c) 3.8%
16-20. 3.19×10^{-4} M and 3.50

CHAPTER 17

17-1. (a) -0.736 V; (b) 0.42 V; (c) -2.01 V; (d) -2.07 V
17-3. (a) 0.266 V; (c) 0.039 V
17-4. (a) 4.9×10^{-6} M; (b) -0.274 V
17-6. (a) Cd; (b) -3.2 V; (c) -0.82 to -1.04 V
17-9. (a) Separation not feasible
      (b) Separation is feasible
      (c) 0.037 to 0.058 V. Note that the cell is galvanic.
17-10. (a) 0.203 V; (b) -0.787 V; (c) -0.326 V; (d) -0.120 V
17-12. (a) 18.2 min; (b) 9.10 min
17-14. 23.4%
17-16. 74.0 ppm
17-18. 0.0400%
17-20. Sample 1, 5.40 ppm
17-23. 11.4% KCl and 8.17% KBr
17-25. 35.15%
17-27. 15.97% CCl_4 and 19.08% $CHCl_3$
17-29. 0.103 g
17-31. 29.4 μg

CHAPTER 18

18-1. (a) The decomposition potential is the potential on a current-voltage curve at which
      the current first becomes greater than the residual current that is associated with a
      blank. The half-wave potential is the potential on a current-voltage curve that cor-
      responds to a current that is exactly one-half the diffusion current.
      (b) A limiting current is a current that is essentially constant and independent of ap-
      plied potential. It arises as a consequence of a limitation in the rate at which react-
      ants can be brought to an electrode surface.
      A diffusion current is a current whose magnitude is limited by the rate at which
      reactants *diffuse* to the electrode surface. In order to observe a diffusion current, it
      is necessary to minimize the transport of reactant to the electrode by mechanical
      mixing and electrostatic attraction.
18-2. (a) Generally, a limiting current increases linearly with analyte concentration.
      (b) If the analyte bears a charge that is opposite to that of the electrode, decreases in
      electrolyte concentration result in increases in the limiting current. If the charges
      are the same, however, the effect is reversed. For an uncharged analyte, limiting
      currents are not effected by the concentration of the electrolyte.
18-3.

concn, C	i, μA	i_d, μA
3.00×10^{-4}	6.4	4.0
9.9×10^{-4}	15.6	13.2
4.00×10^{-3}	55.6	53.2

18-6. (a) 46 mg/100 mL; (c) 129 mg/100 mL
18-7. (a) 0.369 mg/mL; (c) 0.144 mg/mL

CHAPTER 19

19-1.

	λ, nm	λ, Å	λ, cm	λ, μm	ν, Hz	σ, cm^{-1}
(a)	280	2800	2.80×10^{-5}	0.280	1.07×10^{15}	3.57×10^{4}
(b)	913	9130	9.13×10^{-5}	0.913	3.29×10^{14}	1.09×10^{4}
(c)	0.03	0.3	3×10^{-9}	3×10^{-5}	1×10^{19}	3×10^{8}
(d)	2.75×10^{3}	2.75×10^{4}	2.75×10^{-4}	2.75	1.09×10^{14}	3.64×10^{3}
(e)	6×10^{7}	6×10^{8}	6	6×10^{4}	5×10^{9}	0.2
(f)	0.16	1.6	1.6×10^{-8}	1.6×10^{-4}	1.9×10^{18}	6.4×10^{7}

19-2. (a) 7.10×10^{14} Hz   (c) 8.02×10^{17} Hz
      (b) 1.00×10^{14} Hz   (d) 1.20×10^{8} Hz

19-6. (a) 67.9%; (b) 11.6%; (c) 94.6%
19-7. (a) 0.435; (b) 0.532; (c) 0.092
19-8. (a) 82.4%; (b) 34.0%; (c) 97.3%
19-9. (a) 0.736; (b) 0.833; (c) 0.393
19-10. (a) 60.6%; (b) 54.2%; (c) 90.0%
19-11. (a) $c = 3.60 \times 10^{-6}\ M$ (e) $A = 0.250$
       (c) $b = 1.50$ cm   (g) $C = 6.18 \times 10^{-3}$ g/L
19-13. (a) cm^{-1} mg^{-1} L   (c) cm^{-1} g^{-1} L
19-14. 6.99×10^{3}
19-15. 6.48×10^{3}
19-18. 2.52 ppm
19-19. (a) 1.80×10^{3} (d) 0.201 or 20.1%
       (b) 1.14×10^{-2} (e) 0.396 or 39.6%
       (c) $2.58 \times 10^{-4}\ M$ or 40.8 ppm

CHAPTER 20

20-1. (a) The two lamps are filled with different gases as their names imply. The radiation source in the deuterium lamp is somewhat larger and more intense than that for the hydrogen lamp.

(c) A grating monochromator contains a reflection or transmission grating for dispersion of radiation while a prism monochromator employs a prism. The advantages of a grating are its linear dispersion and its cheapness. The advantage of a prism is its greater freedom from scattered and higher-order radiation.

(e) A photovoltaic cell consists of a semiconducting layer sandwiched between two metallic electrodes. When radiation impinges on the photovoltaic surface, electrons in the semiconductor coating are promoted to a conduction band and produce a current that is proportional to radiant power. A phototube consists of a semicylindric cathode and a wire anode housed in a vacuum tube with a transparent window. When radiation strikes the cathode surface, electrons are emitted and are accelerated to the anode under the influence of a dc potential of 90 V or more. The resulting current is proportional to the radiant power.

The advantages of the photovoltaic cell are its ruggedness, simplicity, and low cost. Its disadvantages are low sensitivity and its tendency to exhibit fatigue. The advantages of the phototube are its greater sensitivity and freedom from fatigue. Its disadvantages include higher cost and the dark currents which exist even in the absence of radiation.

(g) Colorimeters depend upon the human eye to achieve a color match between the analyte and standards. Photometers use a photoelectric detector to measure absorbance, which is related to the concentration of the analyte. The advantages of colorimeters lie in their simplicity and low cost. The advantages of photometers are found in their greater sensitivity, precision, and freedom from interference by other colored species.

(i) In a single-beam instrument P and P_0 are measured successively to give A, whereas in a double-beam instrument the two quantities are measured simultaneously, or nearly so, thus freeing the measurement from the effects of all but the most short term fluctuations in the source, amplifier, and detector. The double-beam design is also advantageous when absorbance or transmittance is to be recorded as a function of wavelength. The disadvantages of double-beam instruments are their greater complexity and cost.

20-2. (a) 33.8%   (c) 69.7%
    (b) 0.471   (d) 11.4%

CHAPTER 21

21-1. (a) The color of a solution is determined by the color it transmits. Thus, a solution appears red because it transmits red light and absorbs the complementary green radiation. It is the absorbed radiation, however, that is related to the concentration of the analyte. Thus, radiation that is complementary to the color transmitted by the solution is required. Such radiation is transmitted by a filter that is complementary in color to that of the solution of the analyte.

(c) The slope ratio method requires the use of a large enough excess of one or the other of the reactants so that the equilibria between the two lie far towards completion. This requirement precludes the study of any intermediate species and permits evaluation of the overall formation stoichiometry only.

(e) The dissociation constant for picric acid is so large (5.1×10^{-1}) that little change in its concentration, and thus in absorbance of the solution, occurs during the course of the titration. As a consequence, no end point would be observed.

21-3. (a) violet; (c) orange
21-4. (a) blue; (c) green; (e) orange
21-5. (a) filter D; (c) filters B + A; (e) filters B + C or C + D
21-6. (a) filter D or filters C + D or C + E; (c) filters A + B
21-7. (a) 70 nm; (c) 50 nm
21-8. 15.5 ppm
21-10. 0.017 to 0.36 M
21-13. (a) A plot of the data reveals a linear relationship between absorbance and concentration.
    (b) $A = -5.0 \times 10^{-4} + 0.0395c$
    (c) 0.0017
    (d) 1.1×10^{-4}
21-14. (a) 3.63 ppm; $s_c = 0.050$ ppm and 0.036 ppm
    (c) 1.74 ppm; $s_c = 0.052$ ppm and 0.039 ppm
    (e) 18.29 ppm; $s_c = 0.051$ ppm and 0.037 ppm
21-15. (a) 0.520 ppm; (c) 2.08 ppm; (e) 4.16 ppm
21-16. (b) $A = -1.2 \times 10^{-3} + 0.160c$
    (c) $s_b = 5.9 \times 10^{-4}$
    (d) $s_r = 2.4 \times 10^{-3}$
21-17. (a) 1.34 ppm; $s_c = 0.017$ and 0.012 ppm
    (c) 2.98 ppm; $s_c = 0.016$ and 0.011 ppm
    (e) 0.695 ppm; $s_c = 0.018$ and 0.013 ppm
21-19. 8.50×10^{-4}%
21-20. (a) 1.70%; (c) 0.770%
21-21. (a) 6.0 ppm Co^{2+} and 3.18 ppm Ni^{2+}
    (c) 3.7 ppm Co^{2+} and 10.5 ppm Ni^{2+}

(e) 8.5 ppm Co^{2+} and 3.98 ppm Ni^{2+}

21-22.

	$[CrO_4^{2-}]$	$[Cr_2O_7^{2-}]$	A_{345}	A_{370}	A_{400}
(a)	2.10×10^{-4}	2.95×10^{-4}	0.390	1.227	0.451
(c)	1.40×10^{-4}	1.30×10^{-4}	0.259	0.767	0.288

21-24. Data for the spectrum can be obtained by adding the absorbances of A and B.

21-25.

	A_{475}	A_{700}
(a)	0.815	0.720
(c)	0.858	0.686
(e)	0.642	0.810

21-26. (a) 9.69×10^{-5} M in A and 2.62×10^{-5} M in B
     (c) 7.13×10^{-5} M in A and 4.94×10^{-5} M in B
     (e) 5.51×10^{-5} M in A and 3.17×10^{-5} M in B

21-27. (a) yellow; (b) blue; (c) 625 nm; (d) 0.344; (e) 555 nm

21-28. (a) 6.02; (b) 0.342

21-31. The data for construction of plots follow.

(a)

λ, nm	A	λ, nm	A	λ, nm	A
440	0.234	505	0.258	600	0.495
470	0.248	535	0.280	615	0.506
480	0.252	555	0.342	625	0.500
485	0.253	570	0.409	635	0.488
490	0.253	585	0.456	650	0.450
				680	0.342

(c)

λ, nm	A	λ, nm	A	λ, nm	A
440	0.119	505	0.131	600	0.680
470	0.112	535	0.204	615	0.719
480	0.113	555	0.342	625	0.722
485	0.115	570	0.482	635	0.714
490	0.116	585	0.588	650	0.665
				680	0.511

21-32. The data for construction of plots follow.

(a)

λ, nm	A	λ, nm	A	λ, nm	A
440	0.334	535	0.346	615	0.319
470	0.368	555	0.342	625	0.305
485	0.374	570	0.345	635	0.291
490	0.372	585	0.340	650	0.262
505	0.369	600	0.334	680	0.195

(c)

λ, nm	A	λ, nm	A	λ, nm	A
440	0.162	535	0.233	615	0.638
470	0.164	555	0.342	625	0.638
485	0.167	570	0.454	635	0.628
490	0.168	585	0.538	650	0.584
505	0.179	600	0.610	680	0.448

21-33. (a) 1.5%; (b) 1.4%; (c) 2.1%; (d) 2.2%; (e) 1.6%

21-34. (a) 1.4%; (b) 3.4%

21-35. See Figure 21A.

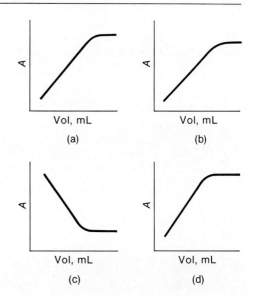

FIGURE 21A   Titration curves from Problem
21-35.

21-37. If a small amount of Cu^{2+} is added to the analyte solution, the end point will be indicated by a sudden increase in A.

21-39. $8.22 \times 10^{-4}\ M$

21-40. (a) $FeSCN^{2+}$; (b) 888; (c) 3.49×10^3

21-42. $Cu(Q)_2^{2+}$

21-44. $Fe(Ph)_3^{2+}$

21-46. (a) $Co(R)_3^{2+}$; (b) 5.6×10^{16}

21-48. 1.53×10^6

CHAPTER 24

24-1.

	$(\sigma)_{rel}$, %	$(\sigma)_{abs}$, particles
(a)	1.9	6
(b)	16	6
(c)	10	9
(d)	12	8
(e)	1.2	13

24-3. (a) 784 pills; (c) 2.0×10^4 pills

24-4. (a) $(\sigma)_{rel} = 12\%$; (b) $(\sigma)_{abs} = 1.1 \times 10^3$ bottles
(c) 9000 ± 1800 bottles; (d) 1.5×10^3 bottles

24-5. (a) 3.5×10^4 particles; (b) 3.9×10^3 g or 8.5 lb; (c) 0.22 mm

24-7. (a) 2.2×10^5 particles; (b) 2.8×10^3 g or 6.1 lb; (c) 0.13 mm

CHAPTER 26

26-1. (a) 5.14; (b) 4 extractions and 40.0 mL;
(c) 6 extractions and 30.0 mL

(d) 12 extractions and 24.0 mL
(e) 6 extractions and 60.0 mL

26-3. (a) $1.15 \times 10^{-2} M$ (c) $8.72 \times 10^{-4} M$
    (b) $3.57 \times 10^{-3} M$ (d) $2.04 \times 10^{-4} M$

26-5. (a) 75.0 mL; (b) 40.0 mL; (c) 21.0 mL

26-7. (a) 5.49; (b) $[HA]_{aq} = 6.78 \times 10^{-3} M$ and $[A^-]_{aq} = 0.0560 M$; (c) 4.62×10^{-1}

26-9. (a) 13.7 meq/L; (b) 686 mg $CaCO_3$/L

CHAPTER 27

27-1. (a) $N_A = 3236$, $N_B = 2898$, $N_C = 2820$, and $N_D = 3261$
    (b) $\bar{N} = 3054 = 3.0 \times 10^3$ $s_N = 227$ plates
    (c) 7.4×10^{-3} cm/plate

27-2. (a) $k'_A = 0.52$, $k'_B = 2.4$, $k'_C = 2.7$, and $k'_D = 3.9$
    (b) $K_A = 4.5$, $K_B = 21$, $K_C = 23$, and $K_D = 33$

27-3. (a) 0.897; (b) 1.10; (c) 63 cm; (d) 43.1 min

27-4. (a) 4.06; (b) 3.1 cm; (c) 2.8 min

27-9. 4.5 cm/s and 0.15 cm/plate

27-11. (a) $k'_M = 2.74$ and $k'_N = 2.66$
    (b) 1.03
    (c) 7.9×10^4 plates
    (d) 8.1×10^2 cm
    (e) $(t_R)_M = 7.1$ min and $(t_R)_N = 6.9$ min

appendixes

APPENDIX 1

SELECTED REFERENCES TO THE LITERATURE OF ANALYTICAL CHEMISTRY

TREATISES

As used here, a *treatise* is a comprehensive presentation of one or more broad areas of analytical chemistry.

N. H. Furman and F. J. Welcher, Eds., *Standard Methods of Chemical Analysis*, 6th ed. New York: Van Nostrand, 1962–66. In five parts; largely devoted to specific applications.

I. M. Kolthoff and P. J. Elving, Eds., *Treatise on Analytical Chemistry*. New York: Wiley, 1959–81. Part I (12 volumes) is devoted to theory; Part II (17 volumes) deals with analytical methods for the elements; and Part III (4 volumes) treats industrial analytical chemistry. Early volumes of the second edition of this monumental work began to appear in 1978.

G. Svehla, C. L. Wilson, and D. W. Wilson, Eds., *Comprehensive Analytical Chemistry*. New York: Elsevier, 1959–   . To 1981, 14 volumes of this work have appeared.

A. Weissberger, Ed., *Techniques of Chemistry*, Volume I, *Physical Methods of Chemistry*, 4th ed. New York: Interscience, 1971–1972. This work consists of 11 individually bound books dealing with various instruments employed for the measurement of chemical systems.

OFFICIAL METHODS OF ANALYSIS
These publications are often single volumes that provide a useful source of analytical methods for the determination of specific substances in articles of commerce. The methods have been developed by various scientific societies and serve as standards in arbitration as well as in the courts.

Standard Methods for the Examination of Water and Wastewater, 14th ed. New York: American Public Health Association, 1976.

ASTM Book of Standards. Philadelphia: American Society for Testing Materials. This 48-volume work is revised annually and contains methods for both physical testing and chemical analysis. Volume 12 is one of the more useful volumes for the chemist and is entitled *Chemical Analysis of Metals and Metal Bearing Ores.*

N. H. Hanson, *Official, Standardized and Recommended Methods of Analysis,* 2d ed. London: Society for Analytical Chemistry, 1973.

Official Methods of Analysis, 13th ed. Washington, D.C.: Association of Official Analytical Chemists, 1980. This is a very useful source of methods for the analysis of such materials as drugs, foods, pesticides, agricultural materials, cosmetics, vitamins, and nutrients.

REVIEW SERIALS
Analytical Chemistry, Fundamental Reviews. These reviews appear biennially as a supplement to the April issues of *Analytical Chemistry* in even-numbered years. Most of the significant developments occurring in the past 2 years in some 30 or more areas of analytical chemistry are covered.

Analytical Chemistry, Application Reviews. These reviews appear as part of *Analytical Chemistry* biennially in odd-numbered years. The articles are devoted to recent analytical work in 18 specific areas, such as water analysis, clinical chemistry, petroleum products, and air pollution.

Critical Reviews in Analytical Chemistry. This publication appears quarterly and provides in-depth reviews of various aspects of analytical chemistry.

D. Glick, Ed., *Methods of Biochemical Analysis.* This annual publication consists of a series of review articles covering the newest developments in the analysis of biochemical substances.

TABULAR COMPLICATIONS
L. Meites, Ed., *Handbook of Analytical Chemistry.* New York: McGraw-Hill, 1963.

G. Milazzo, S. Caroli, and V. K. Sharma, *Tables of Standard Electrode Potentials.* New York: Wiley, 1978.

E. Hoegfeldt and D. D. Perrin, *Stability Constants of Metal-Ion Complexes.* London: The Chemical Society, 1979 and 1981. Two volumes.

ADVANCED ANALYTICAL AND INSTRUMENTAL TEXTBOOKS
H. G. Bauer, G. D. Christian, and J. E. O'Reilly, Eds., *Instrumental Analysis.* Boston: Allyn and Bacon, 1978.

G. W. Ewing, *Instrumental Methods of Chemical Analysis,* 4th ed. New York: McGraw-Hill, 1975.

H. A. Laitinen and W. E. Harris, *Chemical Analysis*, 2d ed. New York: McGraw-Hill, 1975.

E. D. Olsen, *Modern Optical Methods of Analysis*. New York: McGraw-Hill, 1975.

D. A. Skoog and D. M. West, *Principles of Instrumental Analysis*, 2d ed. Philadelphia: Saunders, 1980.

H. Strobel, *Chemical Instrumentation*, 2d ed. Boston: Addison-Wesley, 1973.

H. H. Willard, L. L. Merritt, Jr., and J. A. Dean, *Instrumental Methods of Analysis*, 5th ed. Princeton, N.J.: Van Nostrand, 1974.

MONOGRAPHS

Hundreds of monographs devoted to limited areas of analytical chemistry are available. In general, these are authored by experts and are excellent sources of information. Representative monographs in various areas are listed below.

GRAVIMETRIC AND TITRIMETRIC METHODS

M. R. F. Ashworth, *Titrimetric Organic Analysis*. New York: Interscience, 1965. Two volumes.

L. Erdey, *Gravimetric Analysis*. Oxford: Pergamon, 1965.

J. S. Fritz, *Acid-Base Titrations in Nonaqueous Solvents*. Boston: Allyn and Bacon, 1973.

W. F. Hillebrand, G. E. F. Lundell, H. A. Bright, and J. I. Hoffman, *Applied Inorganic Analysis*, 2d ed. New York: Wiley, 1953.

I. M. Kolthoff, V. A. Stenger, and R. Belcher, *Volumetric Analysis*. New York: Interscience, 1942–57. Three volumes.

T. S. Ma and R. C. Rittner, *Modern Organic Elemental Analysis*. New York: Dekker, 1979.

W. Wagner and C. J. Hull, *Inorganic Titrimetric Analysis*. New York: Dekker, 1971.

ORGANIC ANALYSIS

S. Siggia and J. G. Hanna, *Quantitative Organic Analysis via Functional Groups*, 4th ed. New York: Wiley, 1979.

F. T. Weiss, *Determination of Organic Compounds: Methods and Procedures*. New York: Wiley-Interscience, 1970.

SPECTROMETRIC METHODS

D. F. Boltz and J. A. Howell, *Colorimetric Determination of Nonmetals*, 2d ed. New York: Wiley-Interscience, 1978.

W. J. Price, *Analytical Absorption Spectrometry*. London: Heyden, 1972.

E. B. Sandell and H. Onishi, *Colorimetric Determination of Traces of Metals*, 4th ed. New York: Interscience, 1978.

F. D. Snell, *Photometric and Fluorometric Methods of Analysis*. New York: Wiley, 1978. Two volumes.

J.A. Dean and T. C. Rains, Eds., *Flame Emission and Atomic Absorption Spectroscopy*. New York: Dekker, 1974. Three volumes.

ELECTROANALYTICAL METHODS

J. J. Lingane, *Electroanalytical Chemistry*, 2d ed. New York: Interscience, 1954.

D. T. Sawyer and J. L. Roberts, Jr., *Experimental Electrochemistry for Chemists*. New York: Wiley, 1974.

A. J. Bard and L. R. Faulkner, *Electrochemical Methods*. New York: Wiley, 1980.

ANALYTICAL SEPARATIONS

E. Heftmann, *Chromatography*, 3d ed. New York: Van Nostrand-Reinhold, 1975.

B. L. Karger, L. R. Snyder, and C. Horvath, *An Introduction to Separation Science*. New York: Wiley, 1973.

O. Mikeš, Ed., *Chromatographic and Allied Methods.* New York: Wiley, 1979.
J. M. Miller, *Separation Methods in Chemical Analysis.* New York: Wiley, 1975.

MISCELLANEOUS

R. G. Bates, *Determination of pH: Theory and Practice.* New York: Wiley, 1964.
R. Bock, *Decomposition Methods in Analytical Chemistry.* New York: Wiley 1979.
G. H. Morrison, Ed., *Trace Analysis.* New York: Interscience, 1965.
D. D. Perrin, *Masking and Demasking Chemical Reactions.* New York: Wiley, 1970.
M. Pinta, *Modern Methods for Trace Element Analysis.* Ann Arbor, Mich.: Ann Arbor Science, 1978.
W. Rieman and H. F. Walton, *Ion Exchange in Analytical Chemistry.* Oxford: Pergamon, 1970.
W. J. Williams, *Handbook of Anion Determination.* London: Butterworths, 1979.

PERIODICALS

Numerous journals are devoted to analytical chemistry; these are primary sources of information in the field. Some of the best-known titles are listed below.

American Laboratory*
Analyst, *The*
Analytical Biochemistry
Analytical Chemistry
Analytica Chimica Acta
Analytical Letters
Applied Spectroscopy
Chemical Instrumentation
Clinical Chemistry
Journal of the Association of Official Analytical Chemists
Journal of Chromatographic Science
Journal of Chromatography
Journal of Electroanalytical Chemistry and Interfacial Electrochemistry
Microchemical Journal
Separation Science
Spectrochimica Acta
Talanta
Zeitschrift für Analytische Chemie

* The boldface portion of the title is the *Chemical Abstracts* abbreviation for the journal.

APPENDIX 2
SOME STANDARD AND FORMAL ELECTRODE POTENTIALS

Half-reaction*	E^0, V	Formal potential, V
$Ag^+ + e \rightleftharpoons Ag(s)$	+0.799	0.228, 1 M HCl; 0.792, 1 M HClO$_4$; 0.77, 1 M H$_2$SO$_4$
$AgBr(s) + e \rightleftharpoons Ag(s) + Br^-$	+0.073	
$AgCl(s) + e \rightleftharpoons Ag(s) + Cl^-$	+0.222	0.228, 1 M KCl
$Ag(CN)_2^- + e \rightleftharpoons Ag(s) + 2CN^-$	−0.31	

APPENDIX 2   (CONTINUED)
SOME STANDARD AND FORMAL ELECTRODE POTENTIALS

Half-reaction*	E^0, V	Formal potential, V
$Ag_2CrO_4(s) + 2e \rightleftharpoons 2Ag(s) + CrO_4^{2-}$	+0.446	
$AgI(s) + e \rightleftharpoons Ag(s) + I^-$	−0.151	
$Ag(S_2O_3)_2^{3-} + e \rightleftharpoons Ag(s) + 2S_2O_3^{2-}$	+0.017	
$Al^{3+} + 3e \rightleftharpoons Al(s)$	−1.662	
$H_3AsO_4 + 2H^+ + 2e \rightleftharpoons H_3AsO_3 + H_2O$	+0.559	0.577, 1 M HCl, HClO$_4$
$Ba^{2+} + 2e \rightleftharpoons Ba(s)$	−2.906	
$BiO^+ + 2H^+ + 3e \rightleftharpoons Bi(s) + H_2O$	+0.320	
$BiCl_4^- + 3e \rightleftharpoons Bi(s) + 4Cl^-$	+0.16	
$Br_2(l) + 2e \rightleftharpoons 2Br^-$	+1.065	1.05, 4 M HCl
$Br_2(aq) + 2e \rightleftharpoons 2Br^-$	+1.087†	
$BrO_3^- + 6H^+ + 5e \rightleftharpoons \frac{1}{2}Br_2(l) + 3H_2O$	+1.52	
$BrO_3^- + 6H^+ + 6e \rightleftharpoons Br^- + 3H_2O$	+1.44	
$Ca^{2+} + 2e \rightleftharpoons Ca(s)$	−2.866	
$C_6H_4O_2 \text{ (quinone)} + 2H^+ + 2e \rightleftharpoons C_6H_4(OH)_2$	+0.699	0.696, 1 M HCl, HClO$_4$, H$_2$SO$_4$
$2CO_2(g) + 2H^+ + 2e \rightleftharpoons H_2C_2O_4$	−0.49	
$Cd^{2+} + 2e \rightleftharpoons Cd(s)$	−0.403	
$Ce^{4+} + e \rightleftharpoons Ce^{3+}$		1.70, 1 M HClO$_4$; 1.61, 1 M HNO$_3$; 1.44, 1 M H$_2$SO$_4$; 1.28, 1 M HCl
$Cl_2(g) + 2e \rightleftharpoons 2Cl^-$	+1.359	
$HClO + H^+ + e \rightleftharpoons \frac{1}{2}Cl_2(g) + H_2O$	+1.63	
$ClO_3^- + 6H^+ + 5e \rightleftharpoons \frac{1}{2}Cl_2(g) + 3H_2O$	+1.47	
$Co^{2+} + 2e \rightleftharpoons Co(s)$	−0.277	
$Co^{3+} + e \rightleftharpoons Co^{2+}$	+1.808	
$Cr^{3+} + e \rightleftharpoons Cr^{2+}$	−0.408	
$Cr^{3+} + 3e \rightleftharpoons Cr(s)$	−0.744	
$Cr_2O_7^{2-} + 14H^+ + 6e \rightleftharpoons 2Cr^{3+} + 7H_2O$	+1.33	
$Cu^{2+} + 2e \rightleftharpoons Cu(s)$	+0.337	
$Cu^{2+} + e \rightleftharpoons Cu^+$	+0.153	
$Cu^+ + e \rightleftharpoons Cu(s)$	+0.521	
$Cu^{2+} + I^- + e \rightleftharpoons CuI(s)$	+0.86	
$CuI(s) + e \rightleftharpoons Cu(s) + I^-$	−0.185	
$F_2(g) + 2H^+ + 2e \rightleftharpoons 2HF(aq)$	+3.06	
$Fe^{2+} + 2e \rightleftharpoons Fe(s)$	−0.440	
$Fe^{3+} + e \rightleftharpoons Fe^{2+}$	+0.771	0.700, 1 M HCl; 0.732, 1 M HClO$_4$; 0.68, 1 M H$_2$SO$_4$
$Fe(CN)_6^{3-} + e \rightleftharpoons Fe(CN)_6^{4-}$	+0.36	0.71, 1 M HCl; 0.72, 1 M HClO$_4$, H$_2$SO$_4$
$2H^+ + 2e \rightleftharpoons H_2(g)$	0.000	−0.005, 1 M HCl, HClO$_4$
$Hg_2^{2+} + 2e \rightleftharpoons 2Hg(l)$	+0.788	0.274, 1 M HCl; 0.776, 1 M HClO$_4$; 0.674, 1 M H$_2$SO$_4$
$2Hg^{2+} + 2e \rightleftharpoons Hg_2^{2+}$	+0.920	0.907, 1 M HClO$_4$
$Hg^{2+} + 2e \rightleftharpoons Hg(l)$	+0.854	
$Hg_2Cl_2(s) + 2e \rightleftharpoons 2Hg(l) + 2Cl^-$	+0.268	0.242, sat'd KCl; 0.282, 1 M KCl; 0.334, 0.1 M KCl
$Hg_2SO_4(s) + 2e \rightleftharpoons 2Hg(l) + SO_4^{2-}$	+0.615	
$HO_2^- + H_2O + 2e \rightleftharpoons 3OH^-$	+0.88	

APPENDIX 2 (CONTINUED)
SOME STANDARD AND FORMAL ELECTRODE POTENTIALS

Half-reaction*	E^0, V	Formal potential, V
$I_2(s) + 2e \rightleftharpoons 2I^-$	+0.5355	
$I_2(aq) + 2e \rightleftharpoons 2I^-$	+0.615†	
$I_3^- + 2e \rightleftharpoons 3I^-$	+0.536	
$ICl_2^- + e \rightleftharpoons \frac{1}{2}I_2(s) + 2Cl^-$	+1.056	
$IO_3^- + 6H^+ + 5e \rightleftharpoons \frac{1}{2}I_2(s) + 3H_2O$	+1.196	
$IO_3^- + 6H^+ + 5e \rightleftharpoons \frac{1}{2}I_2(aq) + 3H_2O$	+1.178†	
$IO_3^- + 2Cl^- + 6H^+ + 4e \rightleftharpoons ICl_2^- + 3H_2O$	+1.24	
$H_5IO_6 + H^+ + 2e \rightleftharpoons IO_3^- + 3H_2O$	+1.601	
$K^+ + e \rightleftharpoons K(s)$	-2.925	
$Li^+ + e \rightleftharpoons Li(s)$	-3.045	
$Mg^{2+} + 2e \rightleftharpoons Mg(s)$	-2.363	
$Mn^{2+} + 2e \rightleftharpoons Mn(s)$	-1.180	
$Mn^{3+} + e \rightleftharpoons Mn^{2+}$		1.51, 7.5 M H_2SO_4
$MnO_2(s) + 4H^+ + 2e \rightleftharpoons Mn^{2+} + 2H_2O$	+1.23	1.24, 1 M $HClO_4$
$MnO_4^- + 8H^+ + 5e \rightleftharpoons Mn^{2+} + 4H_2O$	+1.51	
$MnO_4^- + 4H^+ + 3e \rightleftharpoons MnO_2(s) + 2H_2O$	+1.695	
$MnO_4^- + e \rightleftharpoons MnO_4^{2-}$	+0.564	
$N_2(g) + 5H^+ + 4e \rightleftharpoons N_2H_5^+$	-0.23	
$HNO_2 + H^+ + e \rightleftharpoons NO(g) + H_2O$	+1.00	
$NO_3^- + 3H^+ + 2e \rightleftharpoons HNO_2 + H_2O$	+0.94	0.92, 1 M HNO_3
$Na^+ + e \rightleftharpoons Na(s)$	-2.714	
$Ni^{2+} + 2e \rightleftharpoons Ni(s)$	-0.250	
$H_2O_2 + 2H^+ + 2e \rightleftharpoons 2H_2O$	+1.776	
$O_2(g) + 4H^+ + 4e \rightleftharpoons 2H_2O$	+1.229	
$O_2(g) + 2H^+ + 2e \rightleftharpoons H_2O_2$	+0.682	
$O_3(g) + 2H^+ + 2e \rightleftharpoons O_2(g) + H_2O$	+2.07	
$Pb^{2+} + 2e \rightleftharpoons Pb(s)$	-0.126	-0.14, 1 M $HClO_4$; -0.29, 1 M H_2SO_4
$PbO_2(s) + 4H^+ + 2e \rightleftharpoons Pb^{2+} + 2H_2O$	+1.455	
$PbSO_4(s) + 2e \rightleftharpoons Pb(s) + SO_4^{2-}$	-0.350	
$PtCl_4^{2-} + 2e \rightleftharpoons Pt(s) + 4Cl^-$	+0.73	
$PtCl_6^{2-} + 2e \rightleftharpoons PtCl_4^{2-} + 2Cl^-$	+0.68	
$Pd^{2+} + 2e \rightleftharpoons Pd(s)$	+0.987	
$S(s) + 2H^+ + 2e \rightleftharpoons H_2S(g)$	+0.141	
$H_2SO_3 + 4H^+ + 4e \rightleftharpoons S(s) + 3H_2O$	+0.450	
$S_4O_6^{2-} + 2e \rightleftharpoons 2S_2O_3^{2-}$	+0.08	
$SO_4^{2-} + 4H^+ + 2e \rightleftharpoons H_2SO_3 + H_2O$	+0.172	
$S_2O_8^{2-} + 2e \rightleftharpoons 2SO_4^{2-}$	+2.01	
$Sb_2O_5(s) + 6H^+ + 4e \rightleftharpoons 2SbO^+ + 3H_2O$	+0.581	
$H_2SeO_3 + 4H^+ + 4e \rightleftharpoons Se(s) + 3H_2O$	+0.740	
$SeO_4^{2-} + 4H^+ + 2e \rightleftharpoons H_2SeO_3 + H_2O$	+1.15	
$Sn^{2+} + 2e \rightleftharpoons Sn(s)$	-0.136	-0.16, 1 M $HClO_4$
$Sn^{4+} + 2e \rightleftharpoons Sn^{2+}$	+0.154	0.14, 1 M HCl
$Ti^{3+} + e \rightleftharpoons Ti^{2+}$	-0.369	
$TiO^{2+} + 2H^+ + e \rightleftharpoons Ti^{3+} + H_2O$	+0.099	0.04, 1 M H_2SO_4
$Tl^+ + e \rightleftharpoons Tl(s)$	-0.336	-0.551, 1 M HCl; -0.33, 1 M $HClO_4$, H_2SO_4
$Tl^{3+} + 2e \rightleftharpoons Tl^+$	+1.25	0.77, 1 M HCl

APPENDIX 2   (CONTINUED)
SOME STANDARD AND FORMAL ELECTRODE POTENTIALS

Half-reaction*	E^0, V	Formal potential, V
$UO_2^{2+} + 4H^+ + 2e \rightleftharpoons U^{4+} + 2H_2O$	$+0.334$	
$V^{3+} + e \rightleftharpoons V^{2+}$	-0.256	-0.21, 1 M HClO$_4$
$VO^{2+} + 2H^+ + e \rightleftharpoons V^{3+} + H_2O$	$+0.359$	
$V(OH)_4^+ + 2H^+ + e \rightleftharpoons VO^{2+} + 3H_2O$	$+1.00$	1.02, 1 M HCl, HClO$_4$
$Zn^{2+} + 2e \rightleftharpoons Zn(s)$	-0.763	

* Sources for E^0 values: A. J. deBethune and N. A. S. Loud, *Standard Aqueous Electrode Potentials and Temperature Coefficients at 25°C*. Skokie, Ill.: Clifford A. Hampel, 1964, and G. Milazzo, S. Caroli, and V. K. Sharma, *Tables of Standard Electrode Potentials*. New York: Wiley, 1978. Source of formal potentials: E. H. Swift and E. A. Butler, *Quantitative Measurements and Chemical Equilibria*. San Francisco: W. H. Freeman and Company. Copyright © 1972.
† These potentials are hypothetical because they correspond to solutions that are 1.00 M in Br$_2$ or I$_2$. The solubilities of these two compounds at 25°C are 0.18 M and 0.0020 M, respectively. In saturated solutions containing an excess of Br$_2$(l) or I$_2$(s), the standard potentials for the half-reactions Br$_2$(l) + 2e $\rightleftharpoons$ 2Br$^-$ or I$_2$(s) + 2e $\rightleftharpoons$ 2I$^-$ should be used. On the other hand, at Br$_2$ and I$_2$ concentrations less than saturation, these hypothetical electrode potentials should be employed.

APPENDIX 3
SOLUBILITY PRODUCT CONSTANTS*

Substance	Formula	K_{sp}
Aluminum hydroxide	$Al(OH)_3$	2×10^{-32}
Barium carbonate	$BaCO_3$	5.1×10^{-9}
Barium chromate	$BaCrO_4$	1.2×10^{-10}
Barium iodate	$Ba(IO_3)_2$	1.57×10^{-9}
Barium manganate	$BaMnO_4$	2.5×10^{-10}
Barium oxalate	BaC_2O_4	2.3×10^{-8}
Barium sulfate	$BaSO_4$	1.3×10^{-10}
Bismuth oxide chloride	$BiOCl$	7×10^{-9}
Bismuth oxide hydroxide	$BiOOH$	4×10^{-10}
Cadmium carbonate	$CdCO_3$	2.5×10^{-14}
Cadmium hydroxide	$Cd(OH)_2$	5.9×10^{-15}
Cadmium oxalate	CdC_2O_4	9×10^{-8}
Cadmium sulfide	CdS	2×10^{-28}
Calcium carbonate	$CaCO_3$	4.8×10^{-9}
Calcium fluoride	CaF_2	4.9×10^{-11}
Calcium oxalate	CaC_2O_4	2.3×10^{-9}
Calcium sulfate	$CaSO_4$	2.6×10^{-5}
Copper(I) bromide	$CuBr$	5.2×10^{-9}
Copper(I) chloride	$CuCl$	1.2×10^{-6}
Copper(I) iodide	CuI	1.1×10^{-12}
Copper(I) thiocyanate	$CuSCN$	4.8×10^{-15}
Copper(II) hydroxide	$Cu(OH)_2$	1.6×10^{-19}
Copper(II) sulfide	CuS	6×10^{-36}
Iron(II) hydroxide	$Fe(OH)_2$	8×10^{-16}
Iron(II) sulfide	FeS	6×10^{-18}

APPENDIX 3   (CONTINUED)
SOLUBILITY PRODUCT CONSTANTS*

Substance	Formula	K_{sp}
Iron (III) hydroxide	$Fe(OH)_3$	4×10^{-38}
Lanthanum iodate	$La(IO_3)_3$	6.2×10^{-12}
Lead carbonate	$PbCO_3$	3.3×10^{-14}
Lead chloride	$PbCl_2$	1.6×10^{-5}
Lead chromate	$PbCrO_4$	1.8×10^{-14}
Lead hydroxide	$Pb(OH)_2$	2.5×10^{-16}
Lead iodide	PbI_2	7.1×10^{-9}
Lead oxalate	PbC_2O_4	4.8×10^{-10}
Lead sulfate	$PbSO_4$	1.6×10^{-8}
Lead sulfide	PbS	7×10^{-28}
Magnesium ammonium phosphate	$MgNH_4PO_4$	3×10^{-13}
Magnesium carbonate	$MgCO_3$	1×10^{-5}
Magnesium hydroxide	$Mg(OH)_2$	1.8×10^{-11}
Magnesium oxalate	MgC_2O_4	8.6×10^{-5}
Manganese(II) hydroxide	$Mn(OH)_2$	1.9×10^{-13}
Manganese(II) sulfide	MnS	3×10^{-13}
Mercury (I) bromide	Hg_2Br_2	5.8×10^{-23}
Mercury(I) chloride	Hg_2Cl_2	1.3×10^{-18}
Mercury(I) iodide	Hg_2I_2	4.5×10^{-29}
Silver arsenate	Ag_3AsO_4	1×10^{-22}
Silver bromide	$AgBr$	5.2×10^{-13}
Silver carbonate	Ag_2CO_3	8.1×10^{-12}
Silver chloride	$AgCl$	1.82×10^{-10}
Silver chromate	Ag_2CrO_4	1.1×10^{-12}
Silver cyanide	$AgCN$	7.2×10^{-11}
Silver iodate	$AgIO_3$	3.0×10^{-8}
Silver iodide	AgI	8.3×10^{-17}
Silver oxalate	$Ag_2C_2O_4$	3.5×10^{-11}
Silver sulfide	Ag_2S	6×10^{-50}
Silver thiocyanate	$AgSCN$	1.1×10^{-12}
Strontium oxalate	SrC_2O_4	5.6×10^{-8}
Strontium sulfate	$SrSO_4$	3.2×10^{-7}
Thallium(I) chloride	$TlCl$	1.7×10^{-4}
Thallium(I) sulfide	Tl_2S	1×10^{-22}
Zinc hydroxide	$Zn(OH)_2$	1.2×10^{-17}
Zinc oxalate	ZnC_2O_4	7.5×10^{-9}
Zinc sulfide	ZnS	4.5×10^{-24}

* Taken from L. Meites, *Handbook of Analytical Chemistry*, p. 1-13. New York: McGraw-Hill, 1963.

APPENDIX 4
DISSOCIATION CONSTANTS FOR ACIDS*

Name	Formula	Dissociation Constant, 25°C		
		K_1	K_2	K_3
Acetic	CH_3COOH	1.75×10^{-5}		
Arsenic	H_3AsO_4	6.0×10^{-3}	1.05×10^{-7}	3.0×10^{-12}
Arsenious	H_3AsO_3	6.0×10^{-10}	3.0×10^{-14}	
Benzoic	C_6H_5COOH	6.14×10^{-5}		
Boric	H_3BO_3	5.83×10^{-10}		
1-Butanoic	$CH_3CH_2CH_2COOH$	1.51×10^{-5}		
Carbonic	H_2CO_3	4.45×10^{-7}	4.7×10^{-11}	
Chloroacetic	$ClCH_2COOH$	1.36×10^{-3}		
Citric	$HOOC(OH)C(CH_2COOH)_2$	7.45×10^{-4}	1.73×10^{-5}	4.02×10^{-7}
Ethylene- diamine- tetraacetic	H_4Y	1.0×10^{-2}	2.1×10^{-3} $K_4 = 5.5 \times 10^{-11}$	6.9×10^{-7}
Formic	$HCOOH$	1.77×10^{-4}		
Fumaric	$trans\text{-}HOOCCH:CHCOOH$	9.6×10^{-4}	4.1×10^{-5}	
Glycolic	$HOCH_2COOH$	1.48×10^{-4}		
Hydrazoic	HN_3	1.9×10^{-5}		
Hydrogen cyanide	HCN	2.1×10^{-9}		
Hydrogen fluoride	H_2F_2	7.2×10^{-4}		
Hydrogen peroxide	H_2O_2	2.7×10^{-12}		
Hydrogen sulfide	H_2S	5.7×10^{-8}	1.2×10^{-15}	
Hypochlorous	$HOCl$	3.0×10^{-8}		
Iodic	HIO_3	1.7×10^{-1}		
Lactic	$CH_3CHOHCOOH$	1.37×10^{-4}		
Maleic	$cis\text{-}HOOCCH:CHCOOH$	1.20×10^{-2}	5.96×10^{-7}	
Malic	$HOOCCHOHCH_2COOH$	4.0×10^{-4}	8.9×10^{-6}	
Malonic	$HOOCCH_2COOH$	1.40×10^{-3}	2.01×10^{-6}	
Mandelic	$C_6H_5CHOHCOOH$	3.88×10^{-4}		
Nitrous	HNO_2	5.1×10^{-4}		
Oxalic	$HOOCCOOH$	5.36×10^{-2}	5.42×10^{-5}	
Periodic	H_5IO_6	2.4×10^{-2}	5.0×10^{-9}	
Phenol	C_6H_5OH	1.00×10^{-10}		
Phosphoric	H_3PO_4	7.11×10^{-3}	6.34×10^{-8}	4.2×10^{-13}
Phosphorous	H_3PO_3	1.00×10^{-2}	2.6×10^{-7}	
o-Phthalic	$C_6H_4(COOH)_2$	1.12×10^{-3}	3.91×10^{-6}	
Picric	$(NO_2)_3C_6H_2OH$	5.1×10^{-1}		
Propanoic	CH_3CH_2COOH	1.34×10^{-5}		
Pyruvic	$CH_3COCOOH$	3.24×10^{-3}		
Salicylic	$C_6H_4(OH)COOH$	1.05×10^{-3}		
Sulfamic	H_2NSO_3H	1.03×10^{-1}		
Sulfuric	H_2SO_4	strong	1.20×10^{-2}	
Sulfurous	H_2SO_3	1.72×10^{-2}	6.43×10^{-8}	
Succinic	$HOOCCH_2CH_2COOH$	6.21×10^{-5}	2.32×10^{-6}	

Name	Formula	Dissociation Constant, 25°C		
		K_1	K_2	K_3
Tartaric	HOOC(CHOH)$_2$COOH	9.20×10^{-4}	4.31×10^{-5}	
Trichloroacetic	Cl$_3$CCOOH	1.29×10^{-1}		

* Taken from L. Meites, *Handbook of Analytical Chemistry*, p. 1-21. New York: McGraw-Hill, 1963.

APPENDIX 5
DISSOCIATION CONSTANTS FOR BASES*

Name	Formula	Dissociation Constant, K, 25°C
Ammonia	NH$_3$	1.76×10^{-5}
Aniline	C$_6$H$_5$NH$_2$	3.94×10^{-10}
1-Butylamine	CH$_3$(CH$_2$)$_2$CH$_2$NH$_2$	4.0×10^{-4}
Dimethylamine	(CH$_3$)$_2$NH	5.9×10^{-4}
Ethanolamine	HOC$_2$H$_4$NH$_2$	3.18×10^{-5}
Ethylamine	CH$_3$CH$_2$NH$_2$	4.28×10^{-4}
Ethylenediamine	NH$_2$C$_2$H$_4$NH$_2$	$K_1 = 8.5 \times 10^{-5}$
		$K_2 = 7.1 \times 10^{-8}$
Hydrazine	H$_2$NNH$_2$	1.3×10^{-6}
Hydroxylamine	HONH$_2$	1.07×10^{-8}
Methylamine	CH$_3$NH$_2$	4.8×10^{-4}
Piperidine	C$_5$H$_{11}$N	1.3×10^{-3}
Pyridine	C$_5$H$_5$N	1.7×10^{-9}
Trimethylamine	(CH$_3$)$_3$N	6.25×10^{-5}

* Taken from L. Meites, *Handbook of Analytical Chemistry*, p. 1-21. New York: McGraw-Hill, 1963.

APPENDIX 6
STEPWISE FORMATION CONSTANTS*

Ligand	Cation	$\log K_1$	$\log K_2$	$\log K_3$	$\log K_4$	$\log K_5$	$\log K_6$
CH$_3$COO$^-$	Ag$^+$	0.4	−0.2				
	Cd^{2+}	1.3	1.0	0.1	−0.4		
	Cu^{2+}	2.2	1.1				
	Hg^{2+}	$\log K_1 K_2 = 8.4$					
	Pb^{2+}	2.7	1.5				
NH$_3$	Ag$^+$	3.3	3.8				
	Cd^{2+}	2.6	2.1	1.4	0.9	−0.3	−1.7
	Co^{2+}	2.1	1.6	1.0	0.8	0.2	−0.6

APPENDIX 6 (CONTINUED)
STEPWISE FORMATION CONSTANTS*

Ligand	Cation	$\log K_1$	$\log K_2$	$\log K_3$	$\log K_4$	$\log K_5$	$\log K_6$
	Cu^{2+}	4.3	3.7	3.0	2.3	−0.5	
	Ni^{2+}	2.8	2.2	1.7	1.2	0.8	0.0
	Zn^{2+}	2.4	2.4	2.5	2.1		
Br^-	Ag^+	$AgBr(s) + Br^- \rightleftharpoons AgBr_2^-$			$\log K_{s5} = -4.7$		
		$AgBr_2^- + Br^- \rightleftharpoons AgBr_3^-$			$\log K_3 = 0.7$		
	Hg^{2+}	9.0	8.3	1.4	1.3		
	Pb^{2+}	1.2					
Cl^-	Ag^+	$AgCl(s) + Cl^- \rightleftharpoons AgCl_2^-$			$\log K_{s2} = -4.7$		
		$AgCl_2^- + Cl^- \rightleftharpoons AgCl_3^-$			$\log K_3 = 0.0$		
	Bi^{3+}	2.4	2.0	1.4	0.4	0.5	
	Cd^{2+}	1.5	0.4	0.4			
	Cu^+	$Cu^+ + 2Cl^- \rightleftharpoons CuCl_2^-$			$\log K_1K_2 = 4.9$		
	Fe^{2+}	0.4	0.0				
	Fe^{3+}	1.5	0.6	−1.0			
	Hg^{2+}	6.7	6.5	0.9	1.0		
	Pb^{2+}	1.6	$Pb^{2+} + 3Cl^- \rightleftharpoons PbCl_3^-$		$\log K_1K_2K_3 = 1.7$		
	Sn^{2+}	1.1	0.6	0.0			
CN^-	Ag^+	$Ag^+ + 2CN^- \rightleftharpoons Ag(CN)_2^-$			$\log K_1K_2 = 21.1$		
	Cd^{2+}	5.5	5.1	4.6	3.6		
	Hg^{2+}	18.0	16.7	3.8	3.0		
	Ni^{2+}	$Ni^{2+} + 4CN^- \rightleftharpoons Ni(CN)_4^-$			$\log K_1K_2K_3K_4 = 22$		
EDTA	See Table 12-2						
F^-	Al^{3+}	6.1	5.0	3.8	2.7	1.6	0.5
	Fe^{3+}	5.3	4.0	2.8			
OH^-	Al^{3+}	8.9	$Al(OH)_3(s) + OH^- \rightleftharpoons Al(OH)_4^-$		$\log K_{s4} = 1.0$		
	Cd^{2+}	2.3					
	Cu^{2+}	6.5					
	Fe^{2+}	3.9					
	Fe^{3+}	11.1	10.7				
	Hg^{2+}	10.3					
	Ni^{2+}	4.6					
	Pb^{2+}	6.2	$Pb(OH)_2(s) + OH^- \rightleftharpoons Pb(OH)_3^-$		$\log K_{s4} = -1.3$		
	Zn^{2+}	4.4	$Zn(OH)_2(s) + 2OH^- \rightleftharpoons Zn(OH)_4^{2-}$		$\log K_{s4} = -0.9$		
I^-	Cd^{2+}	2.4	1.6	1.0	1.1		
	Cu^+	$CuI(s) + I^- \rightleftharpoons CuI_2^-$			$\log K_{s2} = -3.1$		
	Hg^{2+}	12.9	11.0	3.8	2.3		
	Pb^{2+}	1.3	$PbI_2(s) + I^- \rightleftharpoons PbI_3^-$		$\log K_{s3} = -4.7$		
			$PbI_3^- + I^- \rightleftharpoons PbI_4^-$		$\log K_4 = -3.8$		
$C_2O_4^{2-}$	Al^{3+}	$\log K_1K_2 = 13$		3.8			
	Fe^{3+}	9.4	6.8	4.0			
	Mg^{2+}	3.4	1.0				
	Mn^{2+}	3.9	1.9				
	Pb^{2+}	$Pb^{2+} + 2C_2O_4^{2-} \rightleftharpoons Pb(C_2O_4)_2^{2-}$			$\log K_1K_2 = 6.5$		
SO_4^{2-}	Al^{3+}	3.2	1.9				
	Cd^{2+}	2.3					
	Cu^{2+}	2.4					
	Fe^{3+}	3.0	1.0				
SCN^-	Ag^+		$AgSCN(s) + SCN^- \rightleftharpoons Ag(SCN)_2^-$		$\log K_{s2} = -7.2$		

APPENDIX 6   (CONTINUED)
STEPWISE FORMATION CONSTANTS*

Ligand	Cation	$\log K_1$	$\log K_2$	$\log K_3$	$\log K_4$	$\log K_5$	$\log K_6$
	Cd^{2+}	1.0	0.7	0.6	1.0		
	Co^{2+}	2.3	0.7	-0.7	0.0		
	Cu^{2+}		$CuSCN(s) + SCN^- \rightleftharpoons Cu(SCN)_2^-$			$\log K_{s2} = -3.4$	
	Fe^{3+}	2.1	1.3				
	Hg^{2+}	$\log K_1 K_2 = 17.3$		2.7	1.8		
	Ni^{2+}	1.2	0.5	0.2			

* Taken from L. Meites, *Handbook of Analytical Chemistry*, p. 1-39, New York: McGraw-Hill, 1963.

APPENDIX 7
DESIGNATIONS AND POROSITIES FOR FILTERING CRUCIBLES*

Type	Coarse		Medium	Fine	
Glass, Pyrex® †		C(60)	M(15)	F(5.5)	
Glass, Kimax® ‡	EC(170–220)	C(40–60)	M(10–15)	F(4–55)	VF(2–2.5)
Porcelain,					
Coors U.S.A.® §		Medium(15)		Fine(5)	Very fine(1.2)
Porcelain, Selas‖	XF(100)	XFF(40)	#10(8.8)	Extra fine #01(6)	
Aluminum oxide					
ALUNDUM®¶	Extra coarse(30)	Coarse(20)	Medium(5)	Fine(0.1)	

* Nominal maximum pore diameter in microns is given in parentheses.
† Corning Glass Works, Corning, N.Y.
‡ Owens-Illinois, Toledo, Ohio.
§ Coors Porcelain Company, Golden, Colo.
‖ Selas Corporation of America, Dresher, Pa.
‖ Selas Corporation of America, Dresher, Pa.

APPENDIX 8
DESIGNATIONS CARRIED BY ASHLESS FILTER PAPERS*

Manufacturer	Fine Crystals	Moderately Fine Crystals	Coarse Crystals		Gelatinous Precipitates	
Schleicher and						
Schuell†	507, 590	589 white	589	589	589 black	589-1H
	589 blue	ribbon	green	black	ribbon	
	ribbon		ribbon	ribbon		
Munktell‡	OOH	OK OO	OOR		OOR	
Whatman§	42	44, 40	41		41	41H

APPENDIX **8** (CONTINUED)
DESIGNATIONS CARRIED BY ASHLESS FILTER PAPERS*

Manufacturer	Fine Crystals	Moderately Fine Crystals	Coarse Crystals	Gelatinous Precipitates
Eaton-Dikeman[‖]	90	80	60	50

* Manufacturers' literature should be consulted for more complete specifications. Tabulated are manufacturer designations of papers suitable for filtration of the indicated type of precipitate.
† Schleicher and Schuell, Inc., Keene, N.H.
‡ E. H. Sargent and Company, Chicago, Ill., agents.
§ H. Reeve Angel and Company, Inc., Clifton, N.J., agents.
‖ Eaton-Dikeman Company, Mount Holly Springs, Pa.

APPENDIX 9
FOUR-PLACE LOGARITHMS OF NUMBERS

n	0	1	2	3	4	5	6	7	8	9
10	0000	0043	0086	0128	0170	0212	0253	0294	0334	0374
11	0414	0453	0492	0531	0569	0607	0645	0682	0719	0755
12	0792	0828	0864	0899	0934	0969	1004	1038	1072	1106
13	1139	1173	1206	1239	1271	1303	1335	1367	1399	1430
14	1461	1492	1523	1553	1584	1614	1644	1673	1703	1732
15	1761	1790	1818	1847	1875	1903	1931	1959	1987	2014
16	2041	2068	2095	2122	2148	2175	2201	2227	2253	2279
17	2304	2330	2355	2380	2405	2430	2455	2480	2504	2529
18	2553	2577	2601	2625	2648	2672	2695	2718	2742	2765
19	2788	2810	2833	2856	2878	2900	2923	2945	2967	2989
20	3010	3032	3054	3075	3096	3118	3139	3160	3181	3201
21	3222	3243	3263	3284	3304	3324	3345	3365	3385	3404
22	3424	3444	3464	3483	3502	3522	3541	3560	3579	3598
23	3617	3636	3655	3674	3692	3711	3729	3747	3766	3784
24	3802	3820	3838	3856	3874	3892	3909	3927	3945	3962
25	3979	3997	4014	4031	4048	4065	4082	4099	4116	4133
26	4150	4166	4183	4200	4216	4232	4249	4265	4281	4298
27	4314	4330	4346	4362	4378	4393	4409	4425	4440	4456
28	4472	4487	4502	4518	4533	4548	4564	4579	4594	4609
29	4624	4639	4654	4669	4683	4698	4713	4728	4742	4757
30	4771	4786	4800	4814	4829	4843	4857	4871	4886	4900
31	4914	4928	4942	4955	4969	4983	4997	5011	5024	5038
32	5051	5065	5079	5092	5105	5119	5132	5145	5159	5172

n	0	1	2	3	4	5	6	7	8	9
33	5185	5198	5211	5224	5237	5250	5263	5276	5289	5302
34	5315	5328	5340	5353	5366	5378	5391	5403	5416	5428
35	5441	5453	5465	5478	5490	5502	5514	5527	5539	5551
36	5563	5575	5587	5599	5611	5623	5635	5647	5658	5670
37	5682	5694	5705	5717	5729	5740	5752	5763	5775	5786
38	5798	5809	5821	5832	5843	5855	5866	5877	5888	5899
39	5911	5922	5933	5944	5955	5966	5977	5988	5999	6010
40	6021	6031	6042	6053	6064	6075	6085	6096	6107	6117
41	6128	6138	6149	6160	6170	6180	6191	6201	6212	6222
42	6232	6243	6253	6263	6274	6284	6294	6304	6314	6325
43	6335	6345	6355	6365	6375	6385	6395	6405	6415	6425
44	6435	6444	6454	6464	6474	6484	6493	6503	6513	6522
45	6532	6542	6551	6561	6571	6580	6590	6599	6609	6618
46	6628	6637	6646	6656	6665	6675	6684	6693	6702	6712
47	6721	6730	6739	6749	6758	6767	6776	6785	6794	6803
48	6812	6821	6830	6839	6848	6857	6866	6875	6884	6893
49	6902	6911	6920	6928	6937	6946	6955	6964	6972	6981
50	6990	6998	7007	7016	7024	7033	7042	7050	7059	7067
51	7076	7084	7093	7101	7110	7118	7126	7135	7143	7152
52	7160	7168	7177	7185	7193	7202	7210	7218	7226	7235
53	7243	7251	7259	7267	7275	7284	7292	7300	7308	7316
54	7324	7332	7340	7348	7356	7364	7372	7380	7388	7396
55	7404	7412	7419	7427	7435	7443	7451	7459	7466	7474
56	7482	7490	7497	7505	7513	7520	7528	7536	7543	7551
57	7559	7566	7574	7582	7589	7597	7604	7612	7619	7627
58	7634	7642	7649	7657	7664	7672	7679	7686	7694	7701
59	7709	7716	7723	7731	7738	7745	7752	7760	7767	7774
60	7782	7789	7796	7803	7810	7818	7825	7832	7839	7846
61	7853	7860	7868	7875	7882	7889	7896	7903	7910	7917
62	7924	7931	7938	7945	7952	7959	7966	7973	7980	7987
63	7993	8000	8007	8014	8021	8028	8035	8041	8048	8055
64	8062	8069	8075	8082	8089	8096	8102	8109	8116	8122
65	8129	8136	8142	8149	8156	8162	8169	8176	8182	8189
66	8195	8202	8209	8215	8222	8228	8235	8241	8248	8254
67	8261	8267	8274	8280	8287	8293	8299	8306	8312	8319
68	8325	8331	8338	8344	8351	8357	8363	8370	8376	8382

APPENDIX **9** (CONTINUED)
FOUR-PLACE LOGARITHMS OF NUMBERS

n	0	1	2	3	4	5	6	7	8	9
69	8388	8395	8401	8407	8414	8420	8426	8432	8439	8445
70	8451	8457	8463	8470	8476	8482	8488	8494	8500	8506
71	8513	8519	8525	8531	8537	8543	8549	8555	8561	8567
72	8573	8579	8585	8591	8597	8603	8609	8615	8621	8627
73	8633	8639	8645	8651	8657	8663	8669	8675	8681	8686
74	8692	8698	8704	8710	8716	8722	8727	8733	8739	8745
75	8751	8756	8762	8768	8774	8779	8785	8791	8797	8802
76	8808	8814	8820	8825	8831	8837	8842	8848	8854	8859
77	8865	8871	8876	8882	8887	8893	8899	8904	8910	8915
78	8921	8927	8932	8938	8943	8949	8954	8960	8965	8971
79	8976	8982	8987	8993	8998	9004	9009	9015	9020	9025
80	9031	9036	9042	9047	9053	9058	9063	9069	9074	9079
81	9085	9090	9096	9101	9106	9112	9117	9122	9128	9133
82	9138	9143	9149	9154	9159	9165	9170	9175	9180	9186
83	9191	9196	9201	9206	9212	9217	9222	9227	9232	9238
84	9243	9248	9253	9258	9263	9269	9274	9279	9284	9289
85	9294	9299	9304	9309	9315	9320	9325	9330	9335	9340
86	9345	9350	9355	9360	9365	9370	9375	9380	9385	9390
87	9395	9400	9405	9410	9415	9420	9425	9430	9435	9440
88	9445	9450	9455	9460	9465	9469	9474	9479	9484	9489
89	9494	9499	9504	9509	9513	9518	9523	9528	9533	9538
90	9542	9547	9552	9557	9562	9566	9571	9576	9581	9586
91	9590	9595	9600	9605	9609	9614	9619	9624	9628	9633
92	9638	9643	9647	9652	9657	9661	9666	9671	9675	9680
93	9685	9689	9694	9699	9703	9708	9713	9717	9722	9727
94	9731	9736	9741	9745	9750	9754	9759	9763	9768	9773
95	9777	9782	9786	9791	9795	9800	9805	9809	9814	9818
96	9823	9827	9832	9836	9841	9845	9850	9854	9859	9863
97	9868	9872	9877	9881	9886	9890.	9894	9899	9903	9908
98	9912	9917	9921	9926	9930	9934	9939	9943	9948	9952
99	9956	9961	9965	9969	9974	9978	9983	9987	9991	9996

index

Page numbers in **bold face** refer to specific laboratory directions

841

Compounds Recommended for the Preparation of Standard Solutions of Some Common Elements*

Element	Compound	FW	Solvent**	Notes
Aluminum	Al metal	26.98	Hot dil HCl	a
Antimony	$KSbOC_4H_4O_6 \cdot \frac{1}{2}H_2O$	333.93	H_2O	c
Arsenic	As_2O_3	197.84	dil HCl	i,b,d
Barium	$BaCO_3$	197.35	dil HCl	
Bismuth	Bi_2O_3	465.96	HNO_3	
Boron	H_3BO_3	61.83	H_2O	d,e
Bromine	KBr	119.01	H_2O	a
Cadmium	CdO	128.40	HNO_3	
Calcium	$CaCO_3$	100.09	dil HCl	i
Cerium	$(NH_4)_2Ce(NO_3)_6$	548.23	H_2SO_4	
Chromium	$K_2Cr_2O_7$	294.19	H_2O	i,d
Cobalt	Co metal	58.93	HNO_3	a
Copper	Cu metal	63.55	dil NHO_3	a
Fluorine	NaF	41.99	H_2O	b
Iodine	KIO_3	214.00	H_2O	i
Iron	Fe metal	55.85	HCl, hot	a
Lanthanum	La_2O_3	325.82	HCl, hot	f
Lead	$Pb(NO_3)_2$	331.20	H_2O	a
Lithium	Li_2CO_3	73.89	HCl	a
Magnesium	MgO	40.31	HCl	
Manganese	$MnSO_4 \cdot H_2O$	.169.01	H_2O	g
Mercury	$HgCl_2$	271.50	H_2O	b
Molybdenum	MoO_3	143.94	$1\,M$ NaOH	
Nickel	Ni metal	58.70	HNO_3, hot	a
Phosphorus	KH_2PO_4	136.09	H_2O	
Potassium	KCl	74.56	H_2O	a
	$KHC_8H_4O_4$	204.23	H_2O	i,d
	$K_2Cr_2O_7$	294.19	H_2O	i,d
Silicon	Si metal	28.09	NaOH, concd	
	SiO_2	60.08	HF	
Silver	$AgNO_3$	169.87	H_2O	a
Sodium	NaCl	58.44	H_2O	i
	$Na_2C_2O_4$	134.00	H_2O	i,d
Strontium	$SrCO_3$	147.63	HCl	a
Sulfur	K_2SO_4	174.27	H_2O	
Tin	Sn metal	118.69	HCl	
Titanium	Ti metal	47.90	H_2SO_4, 1:1	a
Tungsten	$Na_2WO_4 \cdot 2H_2O$	329.86	H_2O	h
Uranium	U_3O_8	842.09	HNO_3	d
Vanadium	V_2O_5	181.88	HCl, hot	
Zinc	ZnO	81.37	HCl	a

*The data in this table were taken from a more complete list assembled by B.W. Smith and M.L. Parsons, *J. Chem. Educ.*, **50**, 679 (1973). Unless otherwise specified, compounds should be dried to constant weight at 110°C.

**Unless otherwise specified, acids are concentrated analytical grade.

a Conforms well to the criteria listed in section 6B-1 and approaches primary standard quality.

b Highly toxic.

c Loses $\frac{1}{2}$ H_2O at 110°C. After drying, fw = 324.92. The dried compound should be weighed quickly after removal from the desiccator.

d Available as a primary standard from the National Bureau of Standards.

e H_3BO_3 should be weighed directly from the bottle. It loses 1 H_2O at 100°C and is difficult to dry to constant weight.

f Absorbs CO_2 and H_2O. Should be ignited just before use.

g May be dried at 110°C without loss of water.

h Loses both waters at 110°C. fw = 293.82. Keep in desiccator after drying.

i Primary standard.